CORROSION

Source Book

A collection of outstanding articles from the technical literature

Other Source Books in This Series:

Superalloys
Applications of the Laser in Metalworking
Brazing and Brazing Technology
Copper and Copper Alloys
Electron Beam and Laser Welding
The Ferritic Stainless Steels
Gear Design, Technology and Performance
Innovative Welding Processes
Materials Selection (Volumes I and II)
Materials for Elevated-Temperature Applications
Nitriding
Powder Metallurgy
Selection and Fabrication of Aluminum Alloys
Wear Control Technology
Foundry Technology
Production to Near Net Shape
Quality Control
Rapid Solidification Technology
Titanium and Titanium Alloys
Tool and Die Failures

CORROSION

Source Book

A collection of outstanding articles from the technical literature

Compiled by
Consulting Editor

SEYMOUR K. COBURN
Corrosion Engineer
Corrosion Consultants, Inc.

Senior Corrosion Specialist
Metallurgical Department
Pittsburgh Testing Laboratory

American Society for Metals
Metals Park, Ohio 44073

National Association of Corrosion Engineers
Houston, Texas 77218

(713) 492-0535

Library of Congress Catalog Card No.: 83-073371
ISBN: 0-87170-177-4
SAN 204-7586

PRINTED IN THE UNITED STATES OF AMERICA

DEDICATION

To Elsa and Karen, Kenneth, Monica, Valerie, and Marshall,

and

to C. P. Larrabee for initiating studies, and me, to the weathering steels, and to S. C. Lore for introducing them worldwide while supporting my studies.

Contributors to This Source Book

VINOD SHANKAR AGARWALA
Banaras Hindu University, India

WILLIAM H. AILOR, JR.
Reynolds Metals Company

D. B. ANDERSON
The International Nickel Co., Inc.

BEN BALALA
California Div. of Bay Toll Crossings

H. B. BAMBERGER
Reactive Metals

L. D. BARRETT
International Rustproof Co.

DEAN M. BERGER
Gilbert Associates, Inc.

C. E. BIRD
NCRL, CSIR

E. F. BLADHOLM
Southern California Edison

JACK BLASINGAME
F. W. Gartner Co.

E. G. BOHLMANN
Oak Ridge National Laboratory

W. K. BOYD
Battelle Memorial Institute

ANTON deS. BRASUNAS
University of Missouri

W. D. BRENTNALL
TRW, Inc.

BORIS BRESLER
University of California

JOHN A. BURGBACHER
Shell Oil Co.

JERRY D. BYRD
Carboline Company

DAVID A. CARTER
Betz Laboratories, Inc.

JOSEPH F. CHITTUM
Chevron Research Co.

SEYMOUR K. COBURN
U. S. Steel Corp.

KENNETH G. COMPTON
Consultant

W. J. COPENHAGEN
Council for Scientific and
Industrial Research
Capetown, South Africa

ISRAEL CORNET
University of California

JAMES R. COWLES
Agra Engineering Co.

JOHN M. DONOHUE
Betz Laboratories, Inc.

EDGAR W. DREYMAN
Petro-Chemical Associates

RICHARD W. DRISKO
Naval Civil Engineering Laboratory

J. O. EDWARDS
CANMET

K. D. EFIRD
International Nickel Co., Inc.

R. H. ESPY
Armco, Inc.

A. O. FISHER
Monsanto Co.

M. G. FONTANA
Ohio State University

W. PATRICK GALLAGHER
United States Steel Corp.

P. J. GEGNER
PPG Industries

D. H. GELFER
American Corrosion Control Div.

J. B. GILMOUR
CANMET

T. GINSBERG
Union Carbide Corp.

L. W. GLEEKMAN
Wyandotte Chemicals Corp.

R. I. HAMILTON
CANMET

MICHAEL D. HASSER
Carboline Company

D. A. HAUSMANN
American Pipe and Construction Co.

NOTE: Affiliations given were applicable at date of contribution.

F. H. HAYNIE
National Air Pollution Administration

F. P. HELMS
Union Carbide Corp.

A. L. HENDRICKS
Wisconsin Protective Coating Co.

C. MALCOLM HENDRY
Napko Corporation

E. R. HINDEN
Reliance Universal, Inc.

J. G. HINES
ICI, Ltd.

YOSHIHIRO HISAMATSU
University of Tokyo

E. W. HORVICK
Zinc Institute

BERNARD HUSOCK
Harco Corp.

R. T. JONES
U. S. Steel Corp.

J. D. KEANE
Steel Structures Painting Council

BILL R. KEENEY
Halliburton Co.

JOHN A. KNOX
Halliburton Co.

GREGORY KOBRIN
E. I. du Pont de Nemours
and Co., Inc.

E. V. KUNKEL
Celanese Corp. of America

W. LANDEGREN
Gränges Metallverken

W. J. LANTZ
Chairman, NACE Technical Unit
Committee T-6B

C. P. LARRABEE
Retired, U. S. Steel Corp.

REGINALD M. LASATER
Halliburton Co.

R. A. LEGAULT
Inland Steel Research Laboratories

JOSEPH A. LEHMANN
Electro Rust-Proofing Corporation

HENRY LEIDHEISER, JR.
Lehigh University

JORAM LICHTENSTEIN
Southern California Edison Co.

A. W. LOGINOW
U. S. Steel Corp.

A. K. LONG
Celanese Coatings Co.

GLENN A. MARSH
Union Oil Co. of California

MASCIALE
Mobil Chemical Co.

WILLIAM L. MATHAY
U. S. Steel Corp.

EINAR MATTSON
Swedish Corrosion Institute

H. B. McLAUGHLIN
Laser Technology, Inc.

G. D. MENKE
TRW, Inc.

I. METIL
Imco Laboratories, Inc.

J. H. MILO
Metal Improvement Co.

GEORGE E. MOLLER
International Nickel Company, Inc.

JOHN F. MONTLE
Carboline Company

E. M. MOORE
Arabian-American Oil Co.

P. J. MORELAND
ICI, Ltd.

C. G. MUNGER
Consultant

CHARLES C. NATHAN
Betz Laboratories, Inc.

T. R. NEWMAN
Nalco Chemical Company

HAJIME OKAMURA
University of Tokyo

E. H. PHELPS
U. S. Steel Corp.

FRED M. REINHART
Naval Civil Engineering Laboratory

OLEN L. RIGGS, JR.
Continental Oil Co.

JANE H. RIGO
U. S. Steel Corp.

HARRY C. ROGERS
General Electric Co.

R. R. ROSENTHAL, JR.
Carboline Co.

HERBERT C. RUTEMILLER
Alcoa Research Laboratories

M. H. SANDLER
U. S. Army Aberdeen Research
and Development Center

JOSEPH SANSONE
M. I. T.

T. F. SCHAFFER, JR.
U. S. Steel Corp.

EDWARD SCHASCHL
Union Oil Co. of California

GLENN SCHIEFELBEIN
Stainless Foundry and
Engineering, Inc.

ROBERT J. SCHMITT
U. S. Steel Corp.

HENRY L. SHULDENER
Water Service Laboratories

OLIVER W. SIEBERT
Monsanto Company

D. M. SMITH
Armco Steel Corp.

REBECCA H. SPARLING
General Dynamics

D. O. SPROWLS
Alcoa Research Laboratories

SIDNEY SUSSMAN
Water Service Laboratories

H. D. TARLAS
Carboline Co.

K. B. TATOR
Kenneth Tator Associates

PAUL D. THOMPSON
U. S. Department of Agriculture

I. J. TOTH
TRW, Inc.

H. E. TOWNSEND
Bethlehem Steel Corp.

KRISHNA CHANDRA TRIPATHI
Banaras Hindu University, India

HERBERT H. UHLIG
M. I. T.

J. B. UPHAM
National Air Pollution Administration

RAY VICKERS
Smith Industries, Inc.

FREDERICK H. VORHIS
The Pfaudler Co.

J. B. VRABALE
U. S. Steel Corp.

J. J. WARGA
Arabian-American Oil Co.

AARON WEISSTUCH
Betz Laboratories, Inc.

T. P. WILHELM
Glidden-Durkee

FOREWORD

During the presidency of Franklin D. Roosevelt and the slow ascent of the Defense Department in his administration as the war clouds of Europe were developing, and most significantly since the Pauly Report of the late 1940's, every new president is briefed upon assuming office concerning the state of our defenses and our ability to control any military threat. The key elements in that report deal with our requirements for chromium, nickel, cobalt, zirconium, titanium, uranium, copper, zinc, phosphate rock, etc., and of more recent concern, natural gas and low sulfur crude oil.

Interestingly, none of the foregoing has any significant impact on our ability to erect buildings or construct highways, raise cattle or grow farm crops. On the other hand, without the family of stainless steels or the chemical process industry, oil refinery operations, pharmaceutical plants, food processing companys, and fertilizer and pulp and paper plants would be severely handicapped. Given a little time our steel bridges, transmission towers, automobiles, highway signs and guard rails, bicycles, etc., would rust away slowly from atmospheric attack unless properly maintained.

To prevent this breakdown, a small cadre of chemists, chemical engineers, electrical engineers, metallurgists, bacteriologists, and materials engineers exists as a core group of "corrosion scientists and engineers" dedicated to the understanding of the deterioration of materials. This group, by virtue of its specialized knowledge, is now formulating the concepts, theories, and experiments necessary to shed light on this vexing problem which can seriously affect our life style. Such items as sophisticated aircraft and automobiles, television sets and personal computers, X-rays and CAT scanners along with the mundane steel bridges, transmission towers, and highway signs and guardrails constructed from a variety of processed raw materials could suffer for lack of parts and replacements.

In this kaleidoscope of life the laws of thermodynamics cannot be repealed. Carbon steel structures under appropriate conditions of exposure will deteriorate. The protective coatings have a finite service life. The metallic coatings applied to steel have a longer service life but a finite life nevertheless. Design engineers, while recognizing the physical limitations of a material in a structure or in a piece of equipment, must enhance their knowledge concerning the corrosion aspects of the materials they specify. As it is a near impossibility to "major" in two fields simultaneously while in college, and similarly impossible during the working years, it becomes the responsibility of thoughtful people such as the technical societies to expedite the spread of knowledge.

I consider it an honor and a privilege to have been invited to construct one small bridge to man's accumulated knowledge in the field of corrosion mitigation. This Source Book is an invitation to all its readers to assume the moral and intellectual responsibility to aid in the conservation of nature's resources for today and tomorrow.

SEYMOUR K. COBURN
Corrosion Engineer, Ret'd.
United States Steel Corporation

Corrosion Consultants, Inc.
President
Pittsburgh, Pennsylvania 15217

PREFACE

This Source Book is designed to present "case histories" of metal failures caused by the severity of the environment and/or the lack of understanding by the design engineer of the nature of the corrosion process. It has been stated by lubrication engineers that improper lubrication is the cause of a vast amount of damage to operating machinery and, therefore, the cause of much loss to the gross national product. Workers in the sophisticated fields of metal fatigue and fracture mechanics also point to the large monetary losses caused by a lack of awareness of the operating principles of these disciplines. But it remained for the National Bureau of Standards (NBS) together with aid from Battelle-Columbus to establish an approximate loss due to corrosion of $82 billion in 1975. This came to 4.9 percent of the gross national product. The 1971 Hoar Report in England indicated corrosion losses equivalent to 3.5 percent of their GNP. It is estimated that with our present knowledge, $33 billion, amounting to 2 percent of our GNP, can be saved.

To simplify the learning curve and spread some of this knowledge, the American Society for Metals and the National Association of Corrosion Engineers have cooperated in the preparation of this Source Book. The assumption has been made that the readership has a grasp of the fundamentals of general inorganic chemistry from which the electrochemical theory of corrosion derives its support. A literature survey was made to seek out case history type articles in the belief that such a presentation is the quickest means for informing engineers, metallurgists, designers, college engineering students and the general reader concerning the nature of corrosion and the means available for its prevention.

While research in the field of corrosion science is necessary to advance our knowledge at the fundamental level, fortunately the existing principles established by the early giants such as Luigi Galvani (1737-1798), Alessandro Volta (1745-1827), Sir Humphry Davy (1778-1829), and Michael Faraday (1791-1867) remain as the foundation for problem solving efforts today. Therefore, this Source Book is intended to serve in outline form as a textbook of sorts. While the principles of corrosion technology are being presented in a painless form, "shop floor" articles were selected not for their date of publication, but for their illustrative value and their continuing utility. Because of the types of articles sought and the attention to corrosion principles paid by the authors, the collection came principally from the publications of the National Association of Corrosion Engineers.

We begin with the statement concerning the eight visual forms of corrosion by a pragmatic teacher who indicates that virtually all corrosion failures result from carelessness by the user and poor choice of materials or configuration by the designer. This is followed by a description of these forms by a supplier of water treatment chemicals. There follows a discussion of the primary sources of corrosion, namely, the atmospheres: urban, rural, marine and those in between, including the soil. Attention is called to the ongoing testing by committees of the American Society for Testing and Materials who sponsored joint corrosion studies as early as 1906.

Back in the mid-1920's meetings were held at the NBS to consider the serious problems caused by the intense corrosion of underground piping by D. C. traction current. A solution was sought by encasing the pipes in a wire mesh. The idea was based on the recommendation made by Sir Humphry Davy in 1826 to protect the iron hulls of the British Navy by attaching bars of zinc. Thus was born the first commercial use of cathodic protection whose subsequent use has resulted in enormous savings not only in numerous underground applications but, also, in numerous seawater applications such as offshore oil drilling platforms. The successful application of this technique supports Uhlig's contention that "Cathodic protection is perhaps the most important of all approaches to corrosion control." Five papers are devoted to this subject.

The next most impressive and significant means for the protection of steel structures from corrosion deals with the interiors of pipes and chemical equipment exposed to fluid flow where corrosion is controlled with inhibitors. Imagine periodically adding no more than a few parts per million of a substance to prevent attack of a potable water system, or an automotive antifreeze, or a refinery operation to prevent premature replacement. Inorganic inhibitors range from the simple silicates, phosphates and molybdates to simple soaps and complex organic nitrogen compounds. Five papers have been included to illustrate the wide-ranging effectiveness of this remarkable approach to corrosion control.

So long as carbon steel continues as the workhorse for such structures as buildings, bridges, light standards, transmission towers and the like, it requires protection from the elements. One of the most durable approaches is to coat it with zinc by the galvanizing process. Sheet steel is coated on high-speed lines while large objects are coated by immersion in molten baths of zinc. The coating thickness varies from 1 to 5 mils; the service life ranges upwards of 50 years depending upon the environment. Five papers are devoted to the use of zinc as a protective coating.

Aluminum also can be used as a structural material where engineering design justifies. And, like steel, aluminum has a variety of compositions that contribute to its versatility in different atmospheres and under different loading conditions. Like other metals, aluminum can be misused owing to a lack of familiarity with its chemical resistance to different chemicals. And, like zinc, it can be applied to steel sheet on a high-speed line to make aluminized steel having refractory properties as well as weather resistance. The latest development is to combine aluminum with an almost equal quantity of zinc in a high-speed line to produce a coated steel product combining the qualities of both metals to make steel a more versatile product. It is licensed worldwide under the name Galvalume by Bethlehem Steel Corporation.

One of the more remarkable developments in post World War II metallurgy is the expanding use of the HSLA compositions, more familiarly known as the high-strength low-alloy steels under the trade mark USS COR-TEN steels produced by United States Steel Corporation and licensed worldwide. The writer has great familiarity with these steels having been involved from field testing and early use in railroad hopper cars to their presently expanding use in bridges, buildings, transmission towers and highway applications. Because of their rich earthy colors, they have become a favorite medium for sculptors. The subject is covered in four papers.

The stainless steels were discovered by German and English metallurgists at about the same time shortly before World War I. Because of their exceptional corrosion resistance and satisfying appearance, they have found wide application from building exteriors to restaurant sinks. But, more significantly, their resistance to corrosion has made possible the manufacture of numerous pharmaceuticals and sensitive organic compounds that cannot tolerate contamination by trace quantities of heavy metals from equipment corrosion. The most important characteristic of which users of some stainless steels must be aware, however, is their propensity to exhibit stress corrosion cracking under certain circumstances that only now are being more carefully defined. Much effort has gone into the study of this phenomenon which all metals experience under appropriate conditions. One of the newer techniques being used to study the condition with stainless steels is that of anodic polarization. Several papers addressing the problem have been included.

No survey of protective practices of metals would be complete without a thorough discussion of the role of paints. This term, in the minds of many, has a connotation of being decorative in function. But over the past 30 years through the efforts of the Steel Structures Painting Council and the action of many paint producers, the true function of paint has been that of a "protective coating." Few people stop to realize that a thin film—no more than two to five mils thick—of a soft organic resin that is easily scratched can provide resistance to the abuse of wind and rain, sun and airborne dust, and temperature fluctuations every 24 hours, and can continue to serve effectively from five to ten or more years. Research has shown the importance of proper surface preparation, the importance of primers and tie coats, and the use of synthetic top coats formulated from complex polymers. To make the systems more sophisticated, inhibitive pigments and reactive chemicals are incorporated so that chemical reactions occur at the point of application on the substrate. A collection of papers has been included to inform the reader of the factors that must be understood to obtain optimum performance from the entire system.

As a result of the presence of American soldiers in Australia during World War II, a little known protective coating development was brought to the United States in 1949 and modified for the American market. According to one of the early American investigators, the "inorganic zinc coatings have been one of the technological developments of our time which have made a positive impact on society." His reference is to the use of 90-95 volume percent zinc dust mixed into sodium silicate or silicate derivatives to form an inorganic zinc-rich coating that has great abrasion and atmospheric resistance, as well as resistance to a large number of organic compounds. To effect good adherence the steel substrate must present a "white metal" condition. Were an organic zinc-rich composition to be used, then a "commercial" blast condition would be satisfactory. Both zinc-rich systems serve in the capacity of primers; however, the inorganic system can be very effective without a topcoat. Because of a widespread interest in this subject, seven papers were selected.

In any metallurgical treatment of the subject, one cannot fail to pay heed to a failure mechanism alluded to earlier, namely, that of stress corrosion cracking. Unfortunately, there are few warning signs because corrosion is not immediately evident, just the resultant cracking. Stainless steel when stressed to 80 percent of its yield strength will stress crack in a chloride environment at high temperatures. Fracture will occur at locations where stress has been imposed, such as at welds or at the site of bending. Aluminum can stress crack under similar conditions. Copper is susceptible in the presence of ammonia and organic amines or any nitrogen compounds capable of being converted to an amine. Carbon steel is vulnerable in the presence of alkalies and nitrates. To relieve stress, heat treatment has proved helpful; shot peening with glass beads has proven valuable, too. Because of the importance of this subject, six papers were selected for review.

One of the lesser known, though important considerations, is that of hydrogen embrittlement. As one investigator put it, "Hydrogen is like dust in the house—it is extremely difficult to get rid of completely, and everything that is done seems to produce a little." Laboratory experiments have shown that the amount of hydrogen necessary to cause damage is often beyond the limits of detection. Two papers are devoted to this subtle problem.

Copper and its alloys are important metals; however, they suffer from two problems that are not well-known: cavitation and fluid velocity. A paper describing each of these characteristics is included.

Piling is very important in construction. On land it is used in the form of H-piles and pipe piles, whereas in the water it is commonly used in the form of interlocking sheet piles. Five papers that should be of great interest to architects and design engineers have been devoted to the corrosion aspects.

One of the most costly ongoing problems facing highway engineers is the early deterioration of concrete bridge decks caused by the corrosion of the reinforcing bars due to the use of deicing salts. Research has revealed the mechanism as well as several solutions. These are described in three papers.

A number of special topics that are important from the standpoint of corrosion have been covered by single papers. They involve erosion-corrosion, fasteners, valves, microbial attack, fatigue, solar energy, test procedures, monitoring of corrosion with field instruments, and use of exotic metals and titanium in the chemical industry.

As indicated earlier, blame for many failures was placed on the design engineer. Two papers are intended to show that corrosion can be reduced at the design stage with examples given to prove this contention. The final paper is a review of some classic blunders in which the rules of corrosion prevention obviously were ignored.

To complete this survey through the field of corrosion literature, a list of useful books, handbooks, and journals has been included. Valuable bibliographies also are listed following this preface. These are useful assets in making a literature search should the reader have a problem of the type discussed.

In conclusion, the reader can obtain additional information by attending local, regional, and national meetings of the American Society for Metals and the National Association of Corrosion Engineers.

References

For the newcomer wishing to become familiar with corrosion technology and interested in undertaking a systematic reading program of the review type, useful insights will be derived by first consulting the following references. From the titles it is apparent that the order of listing is from the textbook to the handbook to the specialized topics. Following the book list are monthly journals.

Corrosion: Causes and Prevention, Speller, F. N., McGraw-Hill, New York, 1951.

Protection Against Atmospheric Corrosion: Theories and Methods, Barton, K., John Wiley, New York, 1976.

Corrosion Resistance of Metals and Alloys, McKay, R. J., and Worthington, R., Reinhold, New York, 1936.

Metallic Corrosion Passivity and Protection, Evans, U. R., Edward Arnold, London, 1948.

The Corrosion and Oxidation of Metals, Evans, U. R., St. Martin's Press, New York, 1963.

The Corrosion and Oxidation of Metals, 1st Supplemental Volume, St. Martin's Press, 1968, 2d, 1976.

Corrosion and Corrosion Control, Uhlig, H. H., 2d ed., John Wiley, New York, 1971.

Corrosion, Volumes I and II, Shreir, L. L., ed., 2d ed., Newnes-Butterworths, London, 1976.

An Introduction to Corrosion and Protection of Metals, Wranglen, G., Stockholm, 1972.

Corrosion Engineering, Fontana, M. G., and Greene, N. D., McGraw-Hill, New York, 1967.

Theory of Corrosion: Protection of Metals, Tomashov, N. D., MacMillan, New York, 1966.

Fundamentals of Corrosion, Scully, J. C., Pergamon Press, New York, 1966.

Corrosion: A Compilation, Fontana, M. G., The Press of Hollenbeck, Columbus, Ohio, 1957.

Corrosion Testing Procedures, Champion, F. A., Chapman and Hall, London, 1964.

Corrosion Testing, La Que, F. L., Marburg Lecture, Proceedings of the American Society for Testing and Materials, 1951.

Design and Corrosion Control, Pludek, V. R., John Wiley, New York, 1977.

Marine Corrosion: Causes and Prevention, La Que, F. L., John Wiley, New York, 1975.

Metals Handbook, Properties and Selection: Irons and Steels, Vol. I, 9th ed., American Society for Metals, 1978.

Corrosion Handbook, Uhlig, H. H., ed., John Wiley, New York, 1948.

Handbook on Corrosion Testing and Evaluation, Ailor, W. H., ed., John Wiley, New York, 1971.

Atmospheric Factors Affecting the Corrosion of Engineering Metals, STP 646, Coburn, S.K., ed., American Society for Testing and Materials, 1978.

Metal Corrosion in the Atmosphere, STP 435, American Society for Testing and Materials, 1968.

Underground Corrosion, Romanoff, M., Circular 579, U.S. Govt., Printing Office, Washington, D.C., 1957.

Corrosion Inhibitors, Rozenfeld, I. L., McGraw-Hill, New York, 1981.

Corrosion Inhibitors, Nathan, C. C., ed., National Association of Corrosion Engineers, 1973.

Protective Coatings for Metals, Burns, R. M., and Bradley, W. W., Reinhold, 3d ed., New York, 1967.

Coatings for Corrosion Protection, Cochran, E. W. and Tonini, D., eds., American Society for Metals, 1979.

Corrosion Control by Coatings, Leidheiser, H., Jr., ed., Science Press, Princeton, N. J., 1979.

The Corrosion of Light Metals, Godard, H. P., ed., John Wiley, New York, 1967.

Corrosion of Stainless Steels, Sedriks, A. J., John Wiley, 1979.

Zinc: Its Corrosion Resistance, Slunder, C. J., and Boyd, W. K., Zinc Institute, New York, 1971.

The Corrosion of Copper, Tin and Their Alloys, Leidheiser, H., Jr., John Wiley, New York, 1971.

Corrosion Data Survey—Metals and Nonmetals, Hamner, N. E., compiler, National Association of Corrosion Engineers, 5th ed., 1974.

Stress Corrosion Cracking and Embrittlement, Robertson, W. D., John Wiley, New York, 1956.

The Theory of Stress Corrosion Cracking in Alloys, Scully, J. C., North Atlantic Treaty Organization (NATO), Brussels, 1971.

Handbook of Corrosion Protection for Steel Pile Structures in Marine Environments, Dismuke, T. D., Coburn, S. K. and Hirsch, C. M., eds., American Iron and Steel Institute, Washington, D.C., 1981.

Corrosion Chemistry, Brubaker, G. R. and Phipps, P. B., eds., American Chemical Society, Washington, D.C., 1979.

Corrosion-Erosion Behavior of Materials, Natesan, K., ed., The Metallurgical Society of AIME, Warrendale, Pa., 1980.

Proceedings of the American Society for Testing and Materials, 1950—See reports of Committees A-10 on stainless steels, A-5 on galvanized and aluminized steel sheets and wire, B-3 on nonferrous metals and alloys, B-6 on die-cast metals and alloys, B-7 on light metals, alloys and cast materials and B-8 on electrodeposited finishes.

Good Painting Practice, Volumes I and II, Steel Structures Painting Council, Bigos, J., ed., Pittsburgh, 1st ed., 1954 (2d ed. available in 1983).

Corrosion of Building Materials, Knofel, D., Van Nostrand Reinhold, New York, 1975.

The Chemistry of Building Materials, Diamont, R. M. E., Business Books, London, 1970.

Journals

Corrosion, National Association of Corrosion Engineers, Houston.

Materials Performance, National Association of Corrosion Engineers, Houston.

British Corrosion Journal, The Metals Society, London.

Corrosion Science, Pergamon Press, Oxford, United Kingdom.

Anti-Corrosion: Methods and Materials, Sawell Publications, Ltd., London.

CONTENTS

SECTION I: PROTECTION AGAINST CORROSION

SECTION II: SPECIFIC ALLOYS

ZINC

ALUMINUM

HIGH STRENGTH LOW-ALLOY STEELS

STAINLESS STEELS

SECTION III: COATINGS

PROTECTIVE COATINGS

CONCRETE

MISCELLANEOUS

SECTION I

Protection Against Corrosion

GENERAL CONSIDERATIONS

CORROSION IN DIFFERENT MEDIA

CATHODIC PROTECTION

INHIBITOR PROTECTION

PLANNING AGAINST CORROSION

Our Amazing Ignorance of the Causes of Corrosion

- **Erstaunlich, unsere Ignoranz der Korrosions-Ursachen**
- **Notre stupéfiante ignorance des causes de la corrosion**
- **Nuestra espantosa ignorancia de las causes de la corrosión**

Corrosion costs industry anywhere from $15 to $30 billion a year. The exact figure will never be known, but the sum is staggering.

In view of the critical shortages of so many things these days, the question arises: Is this loss necessary?

■ "Virtually all premature corrosion failures occur for reasons which were already well known and these failures can be prevented," contends Dr. Mars G. Fontana, professor and chairman of the Department of Metallurgical Engineering at Ohio State University, Columbus, Ohio.

"In my years of experience," he continues, "I have seen only a handful of failures due to causes which required materials that had not yet been developed."

Other authorities in the field of corrosion agree with Dr. Fontana that costs can be reduced drastically. But the losses stemming from corrosive destruction and the expensive replacement of complex engineering components are only part of the overall picture.

Prof. Dr. M. G. Fontana

$10 000 per day. Other losses, often difficult or impossible to measure, occur from contamination of materials being processed in corroded equipment loss of efficiency in operating equipment, and slowdowns or shutdowns of production lines. In one case of pitting corrosion, lost production was valued at $10 000 per day.

Still another area where the costs of corrosion are virtually impossible to assess with any degree of accuracy is with military and aerospace hardware that will not operate properly when called upon at the critical time.

Corrosion needs to be looked upon as something more than just an inescapable and necessary evil. It can be prevented or at least reduced substantially so that metals will have a more durable life cycle.

"Unfortunately, many design people and other engineers are woefully ignorant of corrosion and its ramifications and do not consider the corrosion aspects as they should," says Dr. Fontana. "Many violate the most simple fundamentals or axioms of corrosion."

Education Needed. "Probably the most important single perspective that needs emphasising is that there is an overwhelming need for education of materials people and of all engineering

Erosion-corrosion destroyed this pump impeller in just three weeks. • Korrosion durch Erosion zerstörte dieses Pumpenflügelrad in nur 3 Wochen. • La corrosion érosive a détruit ce rotor de pompe en trois semaines net. • Este rotor de bomba quedó destruído por la erosión en sólo tres semanas.

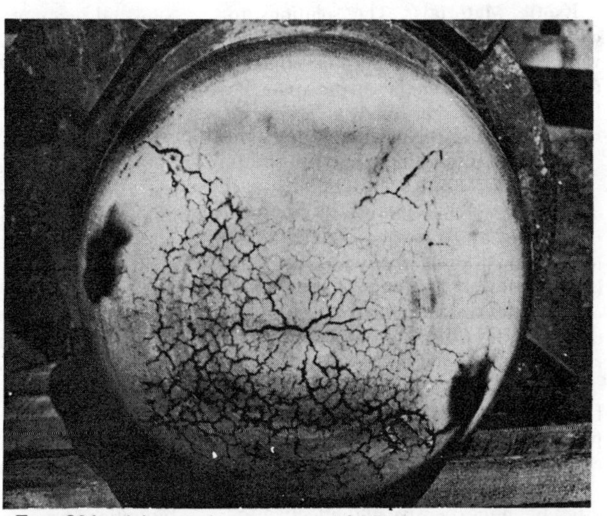

Type 304 stainless steel pressure vessel failed due to stress corrosion. • Spannungskorrosion verursachte die Zerstörung dieses Druckkessels aus rostfreiem Stahl des Typs 304. • Ce récipient sous pression en acier inoxydable 304 a succombé à la corrosion sous tension. • La tensocorrosion produjo a falla del acero inoxidable de tipo 304 en un autoclave.

disciplines regarding the pestilence of corrosion."

Fighting corrosion should be an interdisciplinary effort. The design engineer should work closely with the people who specify materials. The metallurgist, particularly, should become involved in the problems affecting corrosion. The logical place to combat corrosion is on the drawing board while it is in the design stage, or in the laboratory while a material is still under development.

"The metallurgist," notes Dr. Fontana, "develops alloys with tensile strengths of 300 000 or 400 000 psi (2 100 000 to 2 800 000 kN/m²) through the use of sophisticated physical metallurgy and then finds that they're susceptible to stress corrosion when exposed to sea water or even absolutely pure water."

In recent years, the number of corrosion problems has increased significantly. In addition to trying to find solutions to old problems, new programmes have fostered new problems.

Salt and Pollution. Water desalination programmes, for example have introduced a host of critical problems, many of which have yet to be solved.

In the area of pollution control, whether it be water or air, each installation has its own set of corrosion problems.

Ocean engineering has given impetus to the development of systems that will be able to withstand the severe marine environments.

The recent trends in consumerism have given rise to a greater need for corrosion resistance in many products where manufacturers may risk legal liability.

And in the field of medicine, surgical implants of various kinds have opened up still another challenging area where corrosion resistance is a must.

Exotic Materials? The solutions to corrosion problems need not lean toward complex systems and sophisticated materials. Many of the answers can be found in the use of everyday materials such as carbon steel and aluminium.

In this regard, Dr. Fontana suggests: "We need to spend more effort toward devising economical and reliable techniques for using inexpensive materials to slow the spiral of corrosion costs. I have never seen more than a handful of corrosion failures which resulted from a defect in the manufacture of the material."

Then he adds, "Virtually all corrosion failures result from carelessness on the part of the user or poor choice of material or configuration by the designer."

As for the consumer, he is more interested in an inexpensive and effective solution to a problem than he is in the materials used or how the product was made. For this reason, it behooves the corrosion engineer to be well versed in the properties of structural materials.

Not Metals Alone. Corrosion is most commonly associated with metals. But corrosion is a factor with other types of engineering materials, namely, glass, wood, plastics, rubber, aggregates such as concrete, and the many combinations of composites.

Another common belief is that corrosion is caused only be electrochemical activity. Actually, the definition is much broader. In Dr. Fontana's words, it's the "degradation of a material caused by an environment."

As such, the environment may be something other than water or air. It can be a gas, a molten salt, a liquid metal, an organic liquid or any one of a number of others. It may even be a corrosive environment such as ultraviolet light, gamma rays, neutron radiation or fission fragments.

It Can be Useful. And contrary to popular opinion, not all forms of corrosion are destructive or degradating. Under controlled conditions, "corrosion" can be made useful as in the pickling of metals, electrochemical machining, electropolishing of metals, "blueing" of steel, and metallographical etching.

It is interesting to note that in a survey over a two-year period by the Du Pont Co. in which 313 failures were recorded, 56,9% were attributed to corrosion and 43,1% were mechanical. The corrosion failures, broken down into the eight basic forms, are listed in the table in order of their frequency.

Of interest, too, is that at Du Pont where the personnel are well versed in the problems of corrosion, two-metal failures had an incidence of zero.

Industries Differ. The patterns of failures in other industries would be quite dissimilar from those in the chemical industry. The automotive industry's chief nemeses would be general corrosion, followed by pitting, stress corrosion and two-metal corrosion.

In the building industry, most of the problems could be covered by three forms of corrosion—general, pitting and two-metal.

The eight forms of corrosion classified in the table serve as a checklist for corrosion engineers. They provide the clues to the corrosion mechanisms involved which in turn suggest possible solutions.

Eight Ways to Fight It. Similarly, the means for the prevention of corrosion can be categorized into eight approaches. These include: More resistant alloys including optimum heat treatment; cathodic and anodic protection; metal purification; nonmetallics; altering the environment as with inhibitors; designing for optimum geometric configuration; organic coatings; metallic or other inorganic coatings.

"Corrosion processes are known also for their unusual, nonobvious and complex interactions," Dr. Fontana points out. "Many times it is in fact these 'secondary' interactions that are limiting considerations in corrosion."

He cites, as one example, the unwanted reaction products. Frequently, even though the corrosion rate of the materials is quite low, the reaction products may still produce limiting conditions in a system.

Copper Migrates. Copper dissolved from a copper alloy in a heat exchanger will migrate to other areas within the system and settle on a metal of lower electrochemical potential, such as aluminium, thus causing an accelerated condition of pitting.

The dissolution of lead and mercury serves as another example of unwanted reaction products of corrosion. The deadly effects of these two elements are well known.

Another type of secondary interaction is found in the misapplication of inhibitors. Used properly, inhibitors can curb the anodic and cathodic processes, thus reducing the corrosion rate. On the other hand, misuse of an inhibitor can, in fact, accelerate corrosion.

Aluminium Cracks Concrete. The secondary interaction is again brought into play by the forces which the corrosion products themselves exert. At times, these forces can be just as damaging, or more so, than the corrosion process itself. Cracks in concrete are often the result of corrosion products of imbedded pieces of aluminium.

Dr. Fontana cites the effects that the structure of the material can have on the corrosion process. One of these is the effect of an impurity or inclusion in an alloy. In certain chloride solutions, for example, nitrogen accelerates stress corrosion cracking of stainless steel.

Contamination of titanium by iron degrades the corrosion resistance of the titanium in some aqueous solutions.

Iron impurities in aluminium lead to pitting.

And pure zirconium contaminated by nitrogen will have an accelerated rate of corrosion in high hydrogen water and steam.

Rough Surface Hurts. Surface structures and treatments, likewise, can influence the behavior of a material. For example, chemical or mechanical roughening of a stainless steel surface accelerates stress corrosion.

The use of sharp or dull tools on zirconium alloys has a drastic effect on their corrosion behavior in hot water and steam.

Significant progress is being made in the understanding and ability of measuring the corrosion phenomena. The techniques are quite sophisticated and the equipment requirements are both complex and expensive. This is what Dr. Fontana labels as "the price of progress." At present, no single institution or company has the kind of funds required to wage an all-out concerted drive against corrosion.

Funds Needed. The losses from corrosion to industry, government and society in general are so great that a substantial investment in so vital a common cause would pay handsome dividends. What is needed is an infusion of funds from industry as well as government agencies.

A major stumbling block toward achieving real progress on corrosion problems has been the lack of a single source or centre which would gather, develop, classify and disseminate up-to-date corrosion information.

Some of the techniques of corrosion research include anodic and cathodic polarization, linear polarization, strain electrometry, diffraction, ellipsometry, electrochemistry, fracture mechanics and others. Soon, two other developments—electron spectrochemical analysis and Auger spectroscopy—may be used.

Not to be overshadowed in the fight against corrosion are such techniques as the control of the ferrite/austenite ratio in austenitic cast stainless alloys, which is already being practiced on a large scale.

A New Technique. Anodic protection, as opposed to cathodic protection, is quite new and the commercial applications have come into being only within the past several years. In this technique, the application of an external anodic current forms a protective film over the material.

The use of Teflon as the 'noble metal' among plastics is increasing for cases where corrosion is severe. Teflon tubing has found use in scores of installations for heat exchangers in pickling and plating tanks. And Teflon-lined piping and valves are used where the corrosive service was too tough for other materials.

The development of a cast stainless alloy, CD 4M Cu, under the sponsorship of the U.S. Alloy Casting Institute, shows excellent resistance to general corrosion.

The high-purity ferritic stainless steels, made possible by electron beam melting on a production scale, are no longer plagued by embrittling impurities. They are now far more resistant to chloride attack and pitting.

The New Steels. Also, the weathering low-alloy steels and the dispersion-hardened nickel alloys have both contributed significantly to the battle.

Much has been done—more, much more still needs to be done. Heading the list is the job of educating designers and materials people so that the effort could be one of interaction, with help from industry and government.

As mentioned earlier, there is a real need for a central group that gathers all available information, classifies it, and disseminates it.

Many Areas to Explore. Much more work is required in other vital areas such as on hydrogen in metals, electrochemical research in the higher temperature range, the corrosion mechanism of composite materials, the oxidation resistance of turbine materials for both aircraft and automotive applications, crevice effects of components with restricted geometries, and better understanding of the metals used in medical implants.

In addition to research, other areas require attention. One such area involves the inclusion of corrosion limitations in design manuals in much the same manner as is done with strength and hardness data.

It is a tall order for those in the corrosion field where the stakes in money and lives are enormous. ■

Advances in corrosion technology have prolonged the life of North Sea drilling rigs like this. • Fortschritte in der Korrosions-Technologie haben das Leben von Nordsee-Bohrinseln wie dieser bedeutend verlängert. • Les progrès en technologie de corrosion ont prolongé la vie d'îles de forage telles que celle-ci en Mer du Nord. • Hoy las torres de perforación, como esta en el Mar del Norte, tienen mayor duración gracias a los adelantos en la tecnología de la corrosión.

Corrosion caused this bridge to collapse with a heavy loss of life. • Für diesen Brücken-Einsturz, verbunden mit grossen Verlusten an Menschenleben war Korrosion die Ursache. • La corrosion a fait écrouler ce pont en causant de nombreuses victimes. • La corrosión produjo la caída de este puente con una gran pérdida de vidas.

Source: *Iron Age Metalworking International*, July 1974, 38-40

Classifications of Corrosion Failures

T. R. Newman, Chief Metallurgist
Nalco Chemical Company

Accurate laboratory diagnosis of corrosion problems depends on several factors. First, a representative sample of the corroded metal and of any corrosion products must be submitted for examination. Second, information regarding the orientation of the failed part, as well as any pattern of attack, must be available. Then, the corrosive environment must be examined to determine chemical composition, flow rates, and temperatures. Frequently, the history of the system and operational procedures provide useful clues. After these variables are studied, it is then necessary to classify the type of metal deterioration and reconstruct the failure mechanism.

The broadest classifications of corrosion are *general* (uniform) *corrosion, localized corrosion,* and *cracking.* In *general corrosion*, the most common type, the metal thins uniformly, and therefore the expected life of equipment can be estimated with reasonable accuracy. *Localized corrosion,* which includes pitting, is a more insidious form of metal deterioration since it can cause a failure with only slight metal loss in the equipment. *Cracking* failures can occur unexpectedly and cause costly unscheduled shutdowns. Both pitting and cracking conditions can go undetected by corrosion coupon studies.

GENERAL CORROSION

General corrosion is characterized by a chemical or electro-chemical reaction that occurs uniformly over the exposed surface. Anodic and cathodic sites shift constantly so that corrosion spreads over the entire metal surface. Identifying general corrosion is usually simple, but determining its cause is often difficult.

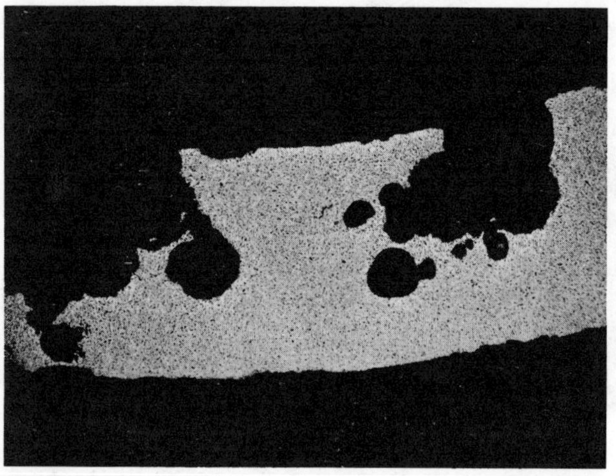

Figure 1—*Results of acid attack on low carbon steel [12X magnification].*

Figure 2 — *Magnetite needles formed as a result of attack by caustic on steel [500X magnification].*

Studies to find the cause of uniform corrosion should be initiated when a noticeable reduction in equipment life occurs. A thorough study of any changes in the environment, operating procedures, and metals must be made to determine the cause. Chemical dissolution by acids, bases, or chelants frequently results in general corrosion (Figures 1 and 2).

LOCALIZED CORROSION

Pitting

Localized corrosion is generally classified as pitting when the diameter of a cavity at the metal surface is the same or less than the depth (Figure 3). General corrosion becomes pitting when the anodic and cathodic sites stop shifting and become fixed on the metal surface. This can be caused by variations in the metal, such as surface defects, emerging dislocations, or incomplete surface films or coatings. Pits usually grow in the direction of gravity (downward on horizontal surfaces). Few pits start on vertical surfaces or upward from the bottom of horizontal surfaces. Often the initial cause of the pit is not the cause of its propagation.

Figure 3 — *Pitting of austenitic stainless steel caused by sulfate reducing bacteria [12X magnification].*

Frequently differential oxygen concentration cells are responsible for starting a pit. Once pitting begins, the environment within a pit starts to change. The positively charged anodes in the pit attract negative chloride ions, causing a buildup of acidic metal chlorides. For this reason pitting is sometimes called auto-catalytic. Conditions within a pit become more aggressive, and the rate of penetration increases with time.

Crevice Corrosion

Crevice corrosion is a special type of pitting. The anode of a corrosion cell is fixed by the geometry in a crevice or under a deposit. To function as a corrosion site, a crevice must be wide enough to permit entry of the liquid, but narrow enough to maintain a stagnant zone. Metals or alloys that depend on oxide films or passive layers for corrosion resistance are particularly susceptible to crevice corrosion.

Galvanic Corrosion

When two dissimilar metals are in contact with each other and exposed to a conductive environment, a potential exists between them, and a current flows. The less resistant metal becomes anodic, and the more resistant, cathodic. Attack on the less resistant metal increases, while on the more resistant one, it decreases. A table of electromotive forces will list the relative resistance of metals, though slight variations in relative positions may occur. Generally, the farther apart two metals are in this table, the greater the possibility of galvanic corrosion. The greatest metal loss of the less resistant metal occurs at the junction of two dissimilar metals. In highly conductive water, such as sea water, attack will be confined to the immediate vicinity of metal contact. In poorly conductive water, such as distilled water, the attack will taper off away from the juncture. The most commonly encountered instances of galvanic corrosion result from deposits of a heavy metal, such as copper, on ferrous or light metals.

Stray-Current Corrosion

Stray-current corrosion differs from other forms in that the source of the current causing the corrosion is external to the affected equipment. This cause of metal deterioration is frequently mis-diagnosed. Stray-current corrosion can cause local metal loss in buried or submerged metal structures, but it occurs much less frequently in underwater transporting equipment than in underground structures. Stray-current corrosion is almost always associated with direct current. At the anodic areas, metal goes into solution and the electrolyte tends to become acidic. It is most commonly encountered in soils containing water.

Selective Leaching

Selective leaching describes a corrosion process also called "parting" or de-alloying. More specifically, it can be called de-zincification in the case of brasses, de-nickelification in cupro-nickels, etc. (Figure 4). Selective leaching may occur in a plug form or in a more evenly distributed layer type. Stagnant conditions and regions under deposits are conducive to selective leaching. In brasses it can occur at pH extremes in water; high dissolved solids and high temperature also promote selective leaching. The overall dimensions of a part do not change drastically, but appreciable weakening can occur.

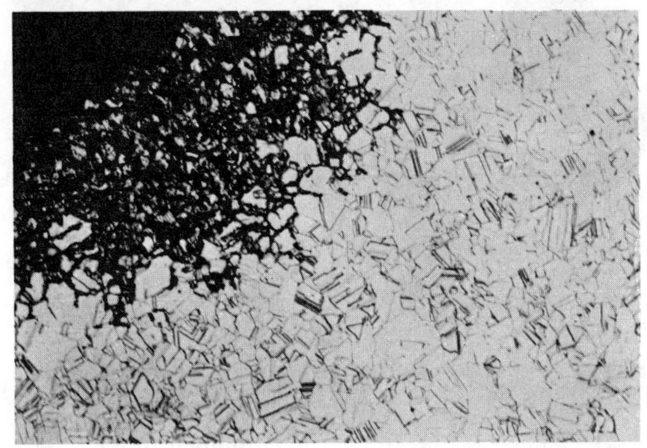

Figure 4—*Dezincification of brass [100X magnification].*

There are several reasons for de-alloying, but the end result is always the same. In some copper alloys, a porous copper structure is left behind in place of the original brass or cupro-nickel.

Gray cast iron is susceptible to a type of de-alloying commonly referred to as graphitic corrosion. In gray cast iron, the ferrite matrix is anodic to the graphite flakes. The matrix is converted to iron oxides, which are held in place by the network of graphite flakes. This usually happens in slightly acidic waters or waters high in dissolved solids.

Erosion and Erosion-Corrosion

Erosion is a strictly mechanical phenomenon, while erosion-corrosion is a combination of mechanical action and chemical or electro-chemical reaction. Pure erosion seldom occurs in aqueous systems. Erosion-corrosion is characterized by grooves, gullies, waves, rounded holes and valleys, and usually exhibits a directional pattern. In copper alloy heat exchanger tubes, the attack frequently results in the formation of horseshoe-shaped depressions (Figure 5). Erosion-corrosion is the acceleration of metal loss because of the relative movement between a fluid and a metal surface. Generally, the movement is rapid, and the effects of mechanical wear are involved. Metal is removed as dissolved ions or as solid corrosion products that are swept from the surfaces.

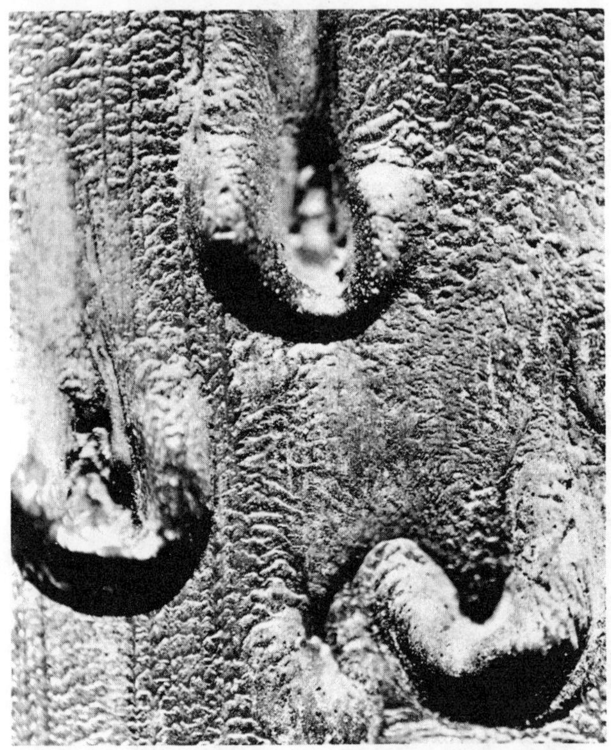

Figure 5—*Erosion-corrosion of brass [12X magnification].*

Most metals are susceptible to erosion-corrosion. Soft ones, such as copper, aluminum, and lead, are particularly susceptible because they are readily worn mechanically. Many metals depend upon a passive surface film for their corrosion resistance, and when erosive forces remove these films, erosion-corrosion proceeds. Particles or gas bubbles carried in suspension can increase metal loss due to erosion-corrosion. The presence of dissolved oxygen in waters can cause a great increase in the erosion-corrosion of copper.

Cavitation

Cavitation is a particular kind of erosion-corrosion caused by the formation and collapse of vapor bubbles in a liquid contacting a metal surface. The resultant shock forces reach high levels in local areas and can tear out jagged chunks of brittle materials or deform soft metals. Where the environment is corrosive, severity of cavitation damage increases.

Fretting Corrosion

Another special case of erosion-corrosion, fretting corrosion, occurs when two heavily loaded metals rub rapidly together, causing damage to one or both metals. Vibration is usually responsible for the damage, but corrosion is also a factor because the frictional heat increases oxidation. In addition, mechanical removal of protective corrosion products continually exposes fresh metal. Fretting corrosion occurs more frequently in air than in water.

Intergranular Corrosion

Metals are composed of grains or crystals which form as solidification occurs. A crystal grows until it meets another advancing crystal. The regions of disarray between crystals are called grain boundaries, which differ in composition from the crystal center. Intergranular corrosion is the selective attack of the grain boundary or an adjacent zone (Figure 6). The most common example of intergranular corrosion is that of sensitized austenitic stainless steels in heat affected zones at welds. Intergranular corrosion usually leaves the surface roughened, but definite diagnosis must be made by microscopic examination.

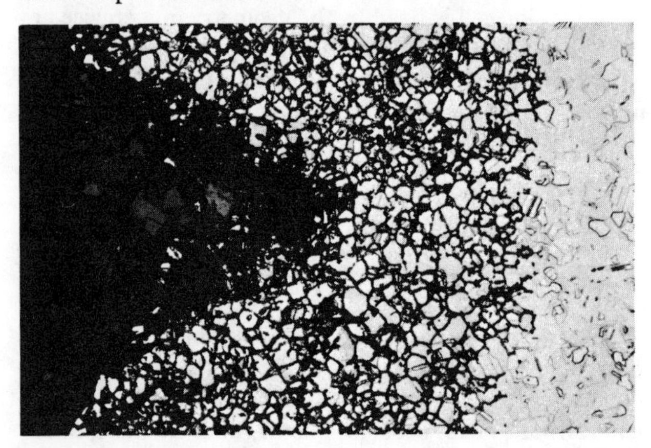

Figure 6—*Intergranular corrosion of brass [100X magnification].*

CRACKING

When a metal part fails by cracking, it is generally obvious that it cracked, but the exact type of cracking and the cause are less obvious. To determine the type of cracking, microscopic examination is necessary. In some instances the environment plays a minor role, while in others its role is major.

Overload

When a metal part has been subjected to a single stress beyond its tensile strength, it can fail by overload. The fracture can be either ductile or brittle, depending on factors such as the metal's hardness and operating temperature. In most cases, a single fracture results.

Fatigue

Subjecting a material to repeated stresses ultimately results in cracking. The environment may have an effect on the fatigue limit of a metal, though this is usually a minor factor. Generally, a fatigue failure is a single fracture, which is transgranular in most common metals. There is normally only a single fracture because stresses on other regions of the surface are relieved when the fracture occurs. Characteristic chevron patterns or beach marks can appear on the fracture face.

Corrosion Fatigue

The simultaneous action of corrosion and cyclic stresses can result in a failure known as corrosion fatigue (Figure 7). The combined effect of these two factors is much greater than the effect of either one alone. The cracking usually begins at surface defects, pits, or irregularities, develops at more than one point, and propagates transgranularly. They have a wedge-shaped profile. The width of the wedge can be related to the stress frequency. Fine cracks result from high-frequency stresses, while broader ones are caused by low-frequency stresses. No metal is immune to this type of cracking.

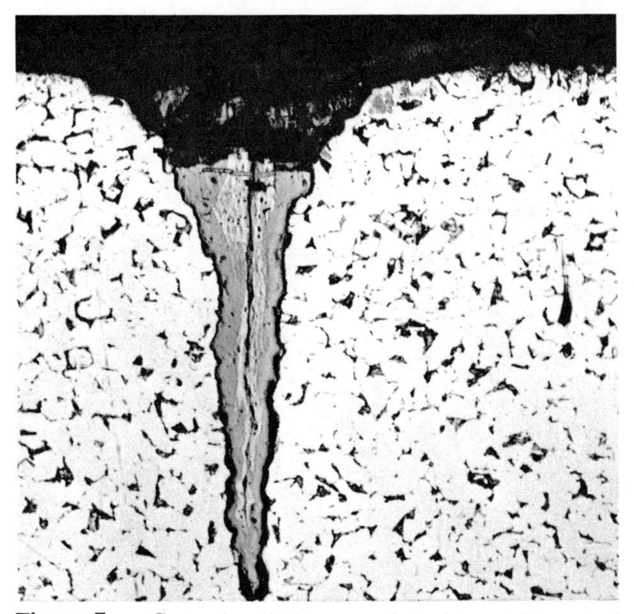

Figure 7 — *Corrosion fatigue crack in low carbon steel [150X magnification].*

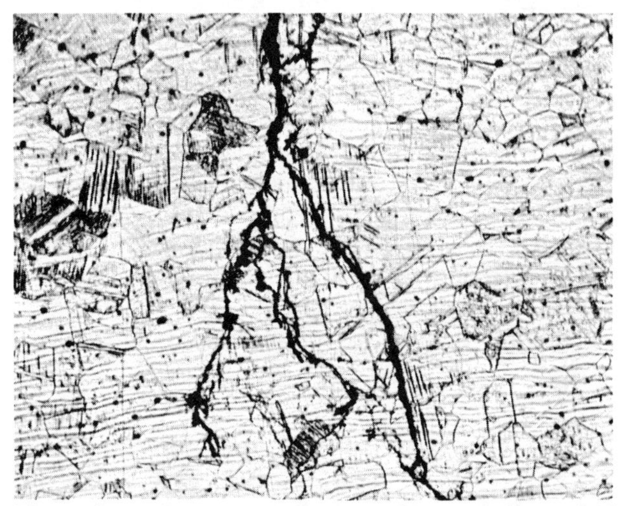

Figure 8 — *Stress corrosion cracking of Type 316 stainless steel [150X magnification].*

Stress Corrosion Cracking

Stress corrosion cracking is the result of the combined action of static stresses and corrosion. The static stresses may be residual or applied service stresses. The environment plays an important role in this type of cracking. The resulting cracks are branched, and can propagate either transgranularly or intergranularly, and sometimes both ways (Figure 8). Caustic cracking of steel is a case of stress corrosion cracking that is sometimes called caustic embrittlement (Figure 9). Generally speaking, it is not necessarily the concentration of the corrodent in the bulk environment that causes the cracking, but the increased concentration occurring in crevices or in alternately wetted and dried regions.

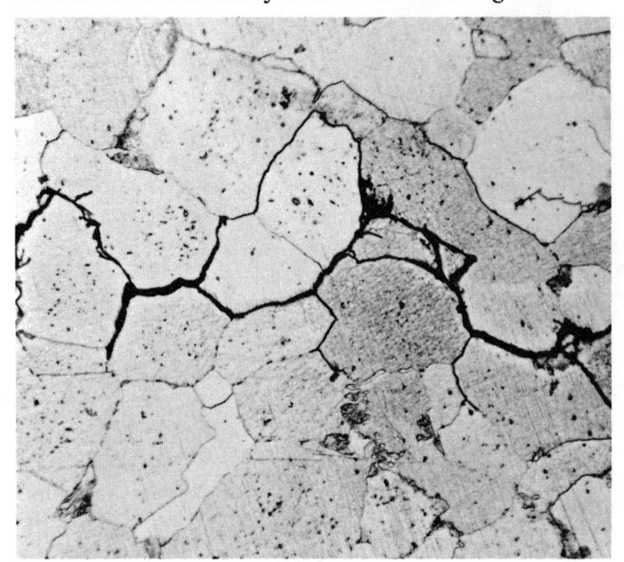

Figure 9— *Caustic cracking of steel [500X magnification].*

Hydrogen Damage

At moderate temperatures, hydrogen damage can occur as a result of a corrosion reaction on a surface or cathodic protection. Atomic hydrogen diffuses into the metal and collects at internal voids or laminations where it combines to form more voluminous molecular hydrogen. In steels, blisters sometimes occur. At higher temperatures and pressures, atomic hydrogen can diffuse into steel and collect at grain boundaries. Either molecular hydrogen is then formed, or the hydrogen reacts with iron carbides to form methane, resulting in cracking and decarburization. Hydrogen cracking is intergranular and highly branched, but not continuous.

EXFOLIATION

Exfoliation is a type of subsurface corrosion that occurs and propagates as cracks approximately parallel to the surface. It leaves the metal in a laminated, flaky, or blistered condition, and appears most frequently in aluminum alloys or cupro-nickels.

Electrochemical and galvanic corrosion of coated steel surfaces

Corrosion takes place when steel becomes the anode in an electrolytic cell. Here is how this process occurs.*

Dean M. Berger, *Gilbert/Commonwealth*

☐ All metals assume their most stable state in nature. Unfortunately, the metals used by engineers—such as steel, zinc and copper—assume their most stable form as oxides, sulfides and similar compounds. By introducing sufficient energy, these compounds can be converted to pure metals. Then, these metals may be processed further into an unending series of items, the most common of which are steel structures. Steel offers the engineer the strength required for many applications.

Corrosion of structural grades of iron and steel, however, proceeds rapidly unless the metal is amply protected. This susceptibility to corrosion of iron and steel is of great concern because annual U.S. losses have been estimated at nearly $70 billion.

Metal corrosion

Corrosion, whether in the atmosphere, underwater or underground, is caused by a flow of electricity from one metal to another or to a recipient (i.e., soil) of some kind; or from one part of the surface of a piece of metal to another part.

An electrolyte is needed for this flow to occur. Water, especially salt water, is, of course, an excellent electrolyte. Simply stated, energy (electricity) passes from a negative area to a positive one via the electrolyte.

So, to have corrosion take place in metals, there must be an electrolyte, plus a metallic area or region with a negative charge in relation to a second area, and a second area positive in opposition to the first [*1*].

The recipient may be soil. This may happen because of the various compositions within a given soil: Soil frequently contains dispersed metallic particles or bacteria pockets that provide a natural electrical pathway to buried metal. If an electrolyte is present and the soil is negative in relation to the metal, the electric path will occur from the metal to the soil. And corrosion will result.

Water readily dissolves a small amount of oxygen

*Coatings are only mentioned briefly here. For more information, see an article by the author on zinc-rich primers, which appeared here previously (Dec. 14, 1981, p. 101). An article by Gary N. Kirby on maintenance paints will be published July 26, also in Materials Engineering Forum.

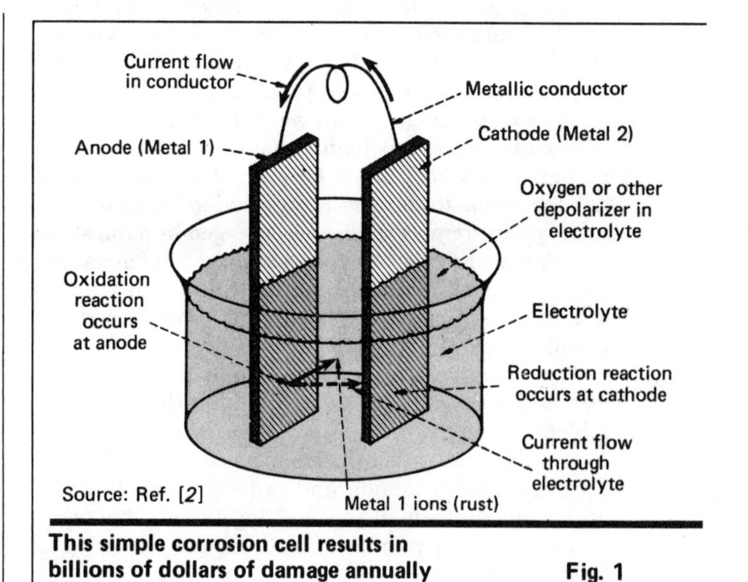

This simple corrosion cell results in billions of dollars of damage annually **Fig. 1**

from the atmosphere, and may become highly corrosive. When the free oxygen dissolved in water is removed, the water is practically noncorrosive, unless it becomes acidic or unless anaerobic bacteria incite corrosion. If oxygen-free water is kept neutral or slightly alkaline, it will be practically noncorrosive to steel. Thus, steam boilers and water-supply systems are effectively protected by deaeration of water.

Electrochemical corrosion

The cell shown in Fig. 1 illustrates the corrosion process in its simplest form. Oxygen is usually present as a depolarizing agent. Hydrogen gas is evolved when a metal corrodes in acid and when the corrosion rate is relatively rapid. A cathode having a layer of adsorbed gas bubbles as a consequence of the corrosion-cell reaction is said to be polarized. This reduces the consumption of metal by corrosion.

As can be seen in Fig. 1, the components form a closed electrical circuit. In the simplest case, the anode would be one metal, perhaps iron, the cathode another, say copper, and the electrolyte might or might not have the same composition at both electrodes. Alternatively, the electrodes could be of the same metal if the electrolyte composition varied.

For the cell shown, an electrical current would flow

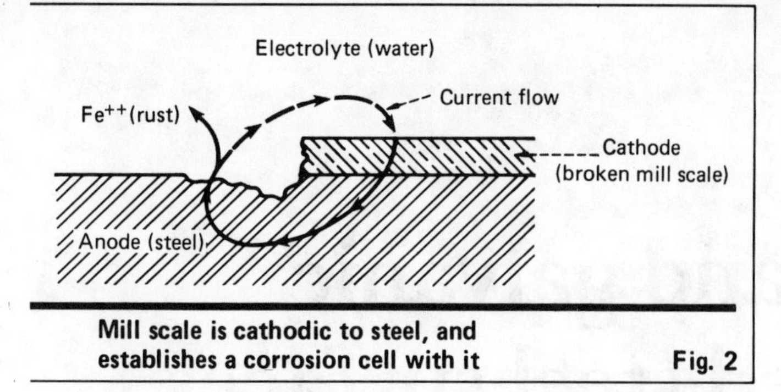

Mill scale is cathodic to steel, and establishes a corrosion cell with it **Fig. 2**

through the metallic conductor and the electrolyte. The anode would corrode (rust, if iron); this is an oxidation reaction. Simultaneously, a nondestructive chemical reaction (reduction) would proceed at the cathode, in most cases producing hydrogen gas on it.

The difference in potential that causes these currents is due mainly to contact between dissimilar metallic conductors or to differences in solution concentration, mainly with respect to dissolved oxygen in natural waters. Almost any lack of homogeneity of the metal surface or its environment may initiate attack, by causing a difference in potential. The result is corrosion that is usually localized.

Atmospheric corrosion differs from the action that occurs in water or underground, in that there is always a plentiful supply of oxygen. Such corrosion is mainly electrochemical, rather than being a chemical attack by the elements. The anodic and cathodic areas, however, are usually quite small and close together, so that corrosion is apparently uniform, rather than occurring as severe pitting, as is true for water or soil.

The larger the anodic area is in relation to the cathode, the faster the rate of corrosion. Anodes and cathodes exist on all iron and steel surfaces. They are caused by surface imperfections, grain orientation, lack of homogeneity of the metal, variation in the environment, localized shear and torque during manufacture, mill scale, or existing red iron oxide rust.

Rust equation

The formation of rust may be expressed as:

$$4Fe + 3O_2 + H_2O \rightarrow 2\,Fe_2O_3 \cdot H_2O.$$

The most stable form of rust is Fe_2O_3. At higher temperatures (900 to 1,300°F), Fe_2O_3 reverts to Fe_3O_4.

In an acidic environment, even without oxygen, the anodic metal is attacked rapidly. When acid corrosion results in salt formation, the reaction is slowed because of salt deposition on the surface being attacked.

Galvanic corrosion

Better known simply as dissimilar-metal corrosion, this occurs in the most unusual places and often causes the most considerable of professional problems.

The galvanic series of metals details how the galvanic current will flow between two metals, and which will corrode when they are in contact or near each other in the ground (see box). Metals near each other in the series do not have a strong effect on each other. The farther apart any two metals are, the stronger the corroding effect on the higher one in the galvanic series.

It is possible for certain metals to reverse their positions in some environments, but the galvanic series will generally hold in natural waters and in the atmosphere [*2*]. (The galvanic series should not be confused with the similar electromotive-force series, which shows exact potentials based on highly standardized conditions that rarely exist in nature.)

While the preceding galvanic series generally defines the available driving force to promote corrosion, the actual rate may vary. Electrolytes may be poor conductors, or long distances may introduce a large resistance into the corrosion-cell circuit. More frequently, scale forms a partially insulating layer over the anode.

The passivity of stainless steels or other metals or alloys is due to the presence of a corrosion-resistant oxide film on their surfaces. In most natural environments, such metals remain passive and thus tend to be cathodic to ordinary iron and steel. Change to an active state usually occurs only when chloride concentrations are

Galvanic series

Corroded end (anodic)

Magnesium
Magnesium alloys
Zinc
Aluminum 2S
Cadmium
Aluminum 17ST
Steel or iron
Cast iron
Chromium-iron (active)
Ni-resist
18-8 Chromium-nickel-iron (active)
18-8-3 Chromium-nickel-molybdenum-iron (active)
Lead-tin solder
Lead
Tin
Nickel (active)
Inconel (active)
Hastelloy C (active)
Brass
Copper
Bronzes
Copper-nickel alloys
Monel
Silver solder
Nickel (passive)
Inconel (passive)
Chromium-iron (passive)
18-8 Chromium-nickel-iron (passive)
18-8-3 Chromium-nickel-molybdenum-iron (passive)
Hastelloy C (passive)
Silver
Graphite
Gold
Platinum

Protected end (cathodic)

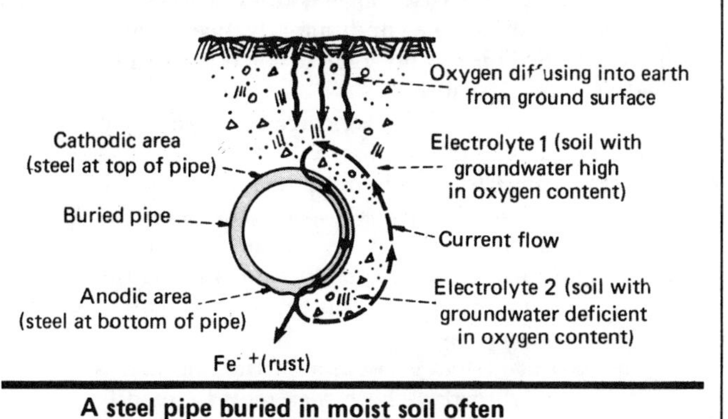

A steel pipe buried in moist soil often will corrode along its lower surface **Fig. 3**

high for stainless steel, as, for example, in seawater.

Accelerated corrosion of steel and iron can be produced by stray currents. Stray d.c. currents in the soil or water associated with nearby cathodic protection systems, industrial activities, or electric railways can be intercepted and carried for considerable distances by buried steel structures. Corrosion takes place when stray currents are discharged from the steel to the environment, and structural damage occurs rapidly [3].

Mill scale

If rust or mill scale is present on the surface of the steel, galvanic corrosion will occur. This is due to a dissimilarity with the base metal. The metal is the anode.

The difference in potential generated between steel and mill scale often amounts to 0.2 to 0.3 V; this couple is nearly as powerful a generator of corrosion currents as is the copper-steel couple. Fig. 2 shows how a pit forms where a break occurs in the scale. When contact between dissimilar materials is unavoidable and their surfaces are painted, it is important to paint both materials— especially the cathode. If only the anode is coated, any weak points, such as pinholes in the coating, will probably result in intense pitting [3].

In general, mill scale is magnetic and contains three layers of iron oxide, but the boundaries between the oxides are not sharp. The outer layer is essentially ferric oxide, Fe_2O_3, which is relatively stable and does not react easily. The layer closest to the steel surface and sometimes intermingled with the surface's crystalline structure is ferrous oxide, FeO. This substance is unstable, and the iron in it is easily oxidized to the ferric state. This process, accompanied by an increase in volume, may result in loosening of the scale.

The intermediate layer of magnetic oxide is best represented by Fe_3O_4. The actual thickness of mill scale on structural steel, which depends upon rolling conditions, varies from about 0.002 to about 0.020 in., and consists mainly of Fe_3O_4 and FeO. Much of the mill scale formed at high initial rolling temperature is knocked off in subsequent rolling [3].

Soil conditions

Differences in soil conditions, such as moisture content and resistivity, commonly are responsible for creating anodic and cathodic areas. Cathodes develop at points of relatively high oxygen concentration and anodes at points of low concentration.

Strained portions of metal tend to be anodic and unstrained portions cathodic. Thus, under all ordinary circumstances where iron and steel are exposed to natural environments, the basic conditions essential to corrosion are present to a greater or lesser degree.

A metal pipe buried in moist soil may corrode on the bottom (see Fig. 3). A variation in oxygen content at different levels in the electrolyte causes this. Thus, anodic and cathodic areas will develop, and a corrosion cell, called concentration cell, will form.

Concentration cells

Severe corrosion, leading to pitting, is often caused by concentration cells, particularly where differences in dissolved-oxygen concentration occur. When a part of the metal is in contact with water relatively low in dissolved oxygen, it is, of course, anodic to adjoining areas in contact with water higher in dissolved oxygen.

This lack of oxygen may be caused by exhaustion of dissolved oxygen in a crevice (see Fig. 4). The low-oxygen area is always anodic. Fig. 4 also illustrates another type of concentration cell; this cell, at the mouth of a crevice, is created by differences in concentration of the metal in solution. These two effects sometimes work together, as in a re-entrant angle in a riveted seam.

As a pit (perhaps caused by broken mill scale) becomes deeper, an oxygen concentration cell is started by the depletion of oxygen in the pit, and the rate of penetration is accelerated.

Coating systems

Using these systems is vital to protect steel. Coatings help prevent corrosion by providing:

1. Sacrificial or galvanic protection.
2. Passivation of the steel (inhibitive pigments).
3. A barrier against the environment.

Sacrificial coatings—Zinc-rich primers are applied at 3.0-mil dry-film thickness to provide galvanic protection. These primers are very effective, even in chemical environments, since the zinc will dissipate before the steel is attacked. Adequate high-performance topcoats are recommended to prolong the life of the coating.

Corrosion inhibitors—Most paint primers contain a partially soluble inhibitor pigment, e.g., zinc chromate, that reacts with the steel substrate to form the iron salt. Such salt slows down corrosion. Chromates, phosphates, molybdates, borates, silicates and plumbates are commonly used. Some pigments passivate by contributing alkalinity, thereby slowing down attack on steel.

Barrier coatings—Protective coatings are the most widely used and recognized forms of barrier material. These barriers may vary in thickness from thin paint films of only a few mils, to heavy mastic coatings applied at about $\frac{1}{4}$ to $\frac{1}{2}$ in., to acidproof brick linings several inches thick. Barrier coatings are effective because they keep moisture and oxygen away from the steel substrate.

Coating breakdown

Most coating films permit chemicals, moisture and oxygen to permeate them and attack the steel. This

Source: *Chemical Engineering*, June 28, 1982, 109-112

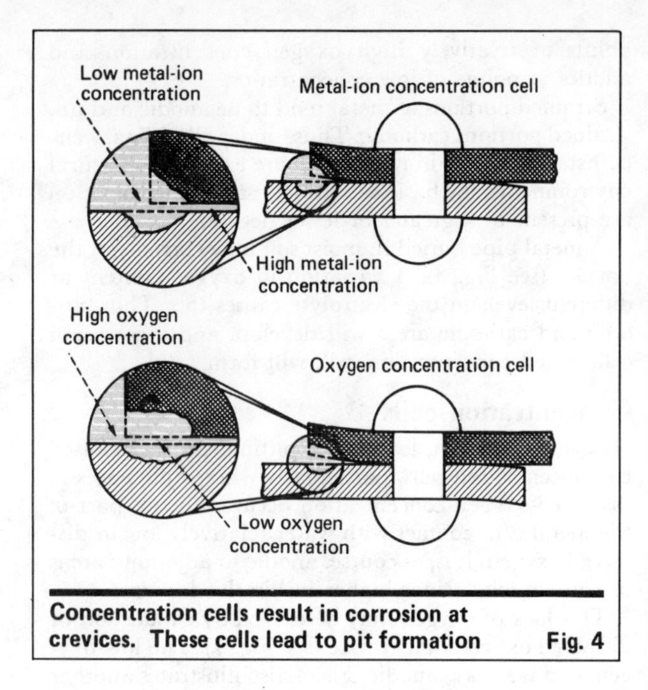

Concentration cells result in corrosion at crevices. These cells lead to pit formation **Fig. 4**

phenomenon is accentuated at temperatures between 150 and 210°F. Gaseous penetration occurs not only through pinholes and other micropores, but also through the coating film itself. The movement of penetrants through the film is fostered principally by osmotic and electroendosmotic pressures and the constant thermally induced movements and vibrations of the coating-film molecules [*4*].

The breakdown of protective coatings over steel substrates may be analyzed in a stepwise procedure. These steps usually are recognized visually and are a part of corrosion analysis:

1. Blistering—Early stages of corrosion, recognized as blistering, are often neglected. They have been described many times as "rust spotting." Standards for determining and evaluating the degree of rust spotting are found in the Steel Structures Painting Council standard Vis-2, or American Soc. for Testing and Materials (ASTM) D 610-68.

Frequently, blistering occurs without external evidence of rusting. The mechanism of blistering is attributed to osmotic attack or a dilation of the coating film at the interface with the steel, under the influence of moisture. Water and chemical gases pass through the film, dissolve ionic material from either the film or the substrate, and cause an osmotic pressure. This establishes a solute concentration gradient, with water building up at these sites until the film eventually blisters. Visual blistering standards are found in the standard, ASTM D 714-56.

Blistering is also dependent upon electrochemical reactions. Water diffuses through a coating also by an electroendosmotic gradient. After corrosion has started, moisture is pulled through the coating by an electrical potential gradient between the corroding area and the protected areas in electrical contact. Therefore, osmosis starts the blistering and, as soon as corrosion begins, electroendosmotic reactions accelerate the corrosion process greatly.

2. Rusting—After one rust spot has been observed or a few blisters are found, the condition advances to general rust spotting. This second stage of corrosion can be described as general rusting.

3. Coating disbonding—The next stage of corrosion is the total disbonding of the coating. This results in direct exposure of the steel to the environment. No longer is the coating protecting the steel substrate; corrosion can now occur at an uninhibited rate. Disbonding occurs because of chemical attack on the substrate, forcing the coating off. One might think of this process as creating one very large blister.

Once the coating system is lost as a barrier, the steel is left to be attacked directly. This attack most frequently is not uniform but rather localized, as many electrolytic cells.

4. Pitting and flaking—Pitting develops when the anodic (corroding) area is small in relation to the cathodic area (a corrosion cell). In a short time, the pitting undercuts mill scale, and flaking occurs. Pitting causes structural failure from localized weakening effects, even while there is still considerable sound metal remaining.

As the corrosion cell becomes more active, the rusting and pitting become more advanced. Deep pits in steel may eventually penetrate completely to create holes. Such penetrations generally result in structural loss. Within the corrosion cell, pitting occurs to such a degree that undercutting, flaking and delamination of the steel is seen. Most of these pits have a conical configuration. After a small hole develops, the electrolyte now can seek other fresh surfaces on the reverse side. This enables corrosion to occur on both surfaces, front and back.

5. Structural loss—As the chemicals attack the unprotected substrate, corrosion occurs at its most rapid and aggressive rate. Large gaping holes are found, causing considerable structural damage. These holes are rapidly enlarged because the electrolyte is ever present on both front and back surfaces of the steel. Catastrophic failure can result.

Richard Greene, Editor

References

1. Introduction to Corrosion, Carboline Co., St. Louis, Mo., 1968.
2. "Paint Manual," 3rd ed., U.S. Bureau of Reclamation, U.S. Dept. of Interior, Denver, Colo., Chap III. 1976, Stock No. 024-003-00104-0.
3. Speller, F. N., "Corrosion: Causes and Prevention," McGraw-Hill Book Co., New York, 1951.
4. Hare, C. H., Anti-Corrosion Barrier and Inhibitive Pigments, Unit 27, Federation of Societies for Coating Technology, Philadelphia, Pa., 1979 (pamphlet).

The author

Dean M. Berger is a coating specialist in the chemical engineering department of Gilbert/Commonwealth, P.O. Box 1498, Reading, PA 19603, telephone 215-775-2600. He received his B.S. in chemistry from North Central College, and has done graduate work at the University of Wisconsin. His experience includes 20 years in research and development with PPG Industries and work with Union Carbide Corp. He has devoted many years to the Steel Structures Painting Council, National Assn. of Corrosion Engineers, American Soc. for Testing and Materials, and Federation of Societies for Coating Technology.

Plenary Lecture—1982

This special lecture, sponsored by the Technical Practices Committee, was given by Einar Mattsson at Corrosion/82. Mattsson has been head of the Swedish Corrosion Institute since 1976. Mattsson has authored more than 100 publications, mainly on corrosion and surface treatment.

The Atmospheric Corrosion Properties of Some Common Structural Metals—A Comparative Study*

*EINAR MATTSSON**

The atmospheric corrosion under outdoor conditions takes place in electrochemical corrosion cells, which can operate only in the presence of an electrolyte, that is, when the surface is wet. The moisture film will contain various species deposited from the atmosphere such as oxygen, SO_X, CO_2, NO_X, and Cl^-, and species originating from the corroding metal. The thermodynamic possibilities for atmospheric corrosion reactions and the formation of different corrosion products are surveyed in potential-pH-diagrams and concentration-pH-diagrams. For the common structural metals—steel, zinc, aluminum, and copper—efforts are made to correlate fundamental corrosion properties with test results and with application aspects.

Introduction

THE ATMOSPHERIC CORROSION of metals has been studied all over the world for many decades, in several countries as the first subject chosen when starting up corrosion research. The reason is, of course, the extensive use of metals for structures out-of-doors, like buildings, bridges, pylons, fences, cars, ships, etc. Most of the studies have been field tests at sites in different types of climate, and much useful information has been obtained from them. Also, advanced research tools have lately been employed for laboratory investigations on atmospheric corrosion and its mechanism. Still, this phenomenon has not yet been fully clarified in all its complexity. In this lecture, the main features of atmspheric corrosion will be surveyed for the commonly used metal (steel, zinc, aluminum, and copper). Efforts will be made to correlate the fundamental corrosion properties of these metals with test results and with application aspects.

*Plenary lecture presented during Corrosion/82, March, 1982, Houston, Texas.

*Swedish Corrosion Institute, Stockholm, Sweden.

Conditions for Atmospheric Corrosion

Atmospheric corrosion outdoors is generally an electrochemical process. It takes place in corrosion cells with anodes and cathodes. The cells can operate only in the presence of an electrolyte, which means that atmospheric corrosion occurs only when the surface is wet. It has not been clarified, however, what the minimum thickness of the electrolyte film required for operation of the corrosion cell is.

The time of wetness varies with the climatic conditions at the site. The metal surface may be wetted if hygroscopic salts (deposited or formed by corrosion) absorb water form the atmosphere. Such absorption occurs above a certain relative humidity, called the critical relative humidity. Its value depends on the metal and on the surface contaminants. The surface may also be wetted by dew or rain. Barton, *et al*,[1] have estimated roughly the amount of water on the metal surface as follows

Conditions	Amount Water (g/m^2)
Critical Relative Humidity	0.01
100% Relative Humidity	1
Covered by Dew	10
Wet from Rain	100

The moisture film on the surface will contain various species deposited from the atmosphere and originating from the corroding metal.

Oxygen will be readily absorbed from the air so that at least the outer region of the thin water film on the metal surface may be considered as saturated with oxygen.

Sulfur oxides, SO_X, will be deposited in rain, by gas absorption, or included in particulates. They mainly originate from combustion of coal or oil. The adsorption of sulfur dioxide on metal surfaces depends on the relative humidity, the metal, and the presence of corrosion products. At a relative humidity of 80% or higher practically all sulfur dioxide molecules hitting a rusty surface are adsorbed.[2] In various types of atmospheres, the deposition rates are of the following orders of magnitude:[1,3]

Type of Atmosphere	Deposition Rate ($mg\ SO_2/m^2$, day)
Rural	10 to 30
Urban	up to 100
Industrial	up to 200

CO_2 occurs in the atmosphere in a concentration of 0.03 to 0.05% by volume, varying slightly with the hours of the day and the seasons of the year due to its cycle in nature.[4] At equilibrium, the percentages mentioned correspond to a concentration of the order of 10^{-5} mole/L in the water film.

Nitrogen oxides, NO_X, are also deposited from the atmosphere. They are formed by combustion processes or by electric discharge phenomena. Although the influence of NO_X on atmospheric corrosion has not yet been extensively studied, it seems to be of importance only on special occasions.

Chlorides are deposited mainly in marine atmospheres as droplets or as evaporated residues of sea water spray which has been carried by the wind from the sea. Chloride deposition decreases strongly with increasing distance from the beach as the droplets and crystals are filtered off when the wind passes through vegetation, or they may settle by gravitation. The deposition rate is generally in the range 0.3 to 300 mg Cl^-/m^2/day.[3]

The concentration of the various species in the electrolyte on the surface vary greatly with respect to parameters such as: deposition rates; corrosion rate; intervals between rain washing; presence of rain shelter; and drying conditions.

To determine which concentration of the corroding metal may occur in the electrolyte film, the supply of corrosion products has been calculated on the assumption of a corrosion rate of 1 μm/y and the amounts of water on the surface mentioned earlier. The following results have been obtained:

	Amount of Water (g/m^2)	Supply of Corrosion Products (mole/L, day)
Critical Relative Humidity	0.01	30
100% Relative Humidity	1	0.3
Covered by Dew	10	0.03
Wet from Rain	100	0.003

One may conclude that the concentration in the electrolyte film will be low during a rainy period, while highly concentrated solution may form after a long period without rain washing.

The pH value of the water film is also difficult to specify, but from what has been found in analyses of rain water and of water collected from metal surfaces exposed outdoors it seems likely to be in the range 2 to 7; near the lower value when the air is heavily polluted with SO_X and higher when it is clean.[5]

The thermodynamic possibilities for atmospheric corrosion reactions on the metal surface and the formation of different corrosion products will be surveyed in so-called potential-pH-diagrams. This type of diagram was developed by Pourbaix, who worked out an atlas of diagrams for various metals in contact with pure H_2O.[6] As an example, the diagram for the system Cu-H_2O is shown in Figure 1. The potential-pH-

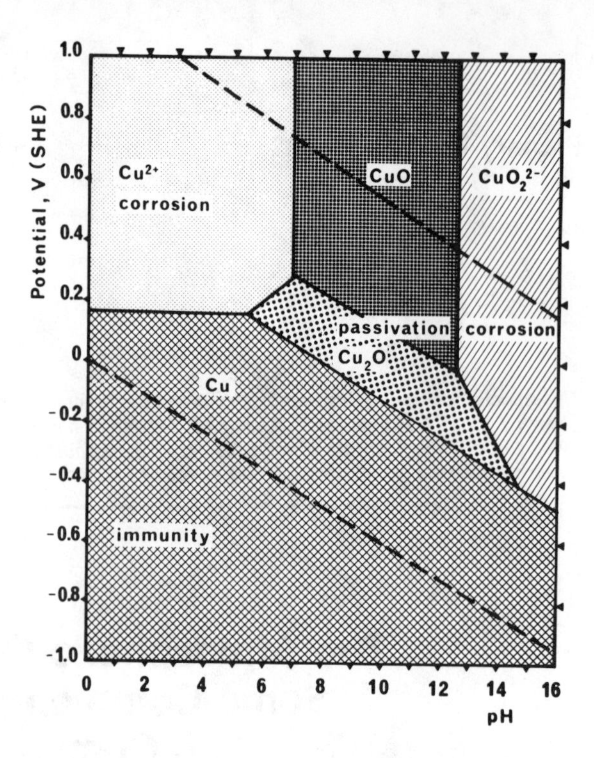

FIGURE 1 — Potential-pH-diagram, Cu-H_2O; 25 C, 10^{-6} M Cu.[6]

diagram shows the stability domains for various species with respect to the redox potential and the pH value of the corrosive agent. The stability domain for water is enclosed between the two dashed lines. The positions of the borderlines between solid species and species in solution depend on the concentrations (or rather activities) of the species in solution. On atmospheric corrosion, this concentration in the water film on the metal surface is fluctuating and generally rather high. So, in the diagrams used in this presentation, it has been set at 10^{-1} mole/L. Even considerable deviations from this value, however, will generally not radically change the stability domains in the diagram, because the potential and the pH value are proportional to the logarithm of the concentration (activity). When constructing diagrams representative of atmospheric corrosion, one also has to take into account species other than those in the metal-water system dealt with in the Pourbaix atlas. Thus, one has to consider H_2CO_3, SO_4^{2-}, Cl^-, and NO_3^-, which under certain conditions form solid phases or complexes with the corroding metal. The stability domains for these species can be calculated from stability constants available in the literature.[7] It should be pointed out, however, that the potential-pH-diagram only gives information on the thermodynamic possibilities for reactions, not on the kinetics.

As for the kinetics, the corrosion rate depends on the amount and composition of the electrolyte on the surface. The instantaneous corrosion rate varies strongly with time, its maximum value being several powers of ten greater than its minimum value (Figure 2).[8] Nevertheless, the cumulative attack over a lengthy period does not depend so strongly on the time of wetness. For when the surface becomes wet, the large amount of corrosive surface contaminants accumulated during a long dry period will generally cause a higher corrosion rate than the smaller amount accumulated during a shorter dry period. In fact, the cumulative damage averaged over the surface and a time period (t) covering one or several years generally follows a continuous curve (Figure 3). This damage will be called the penetration (depth) (p). As regards the corro-

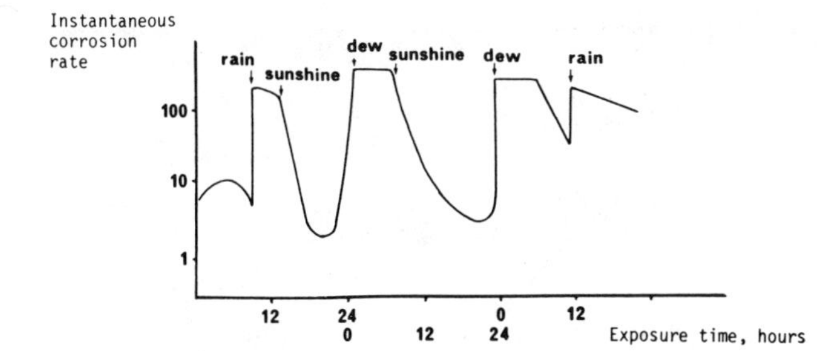

FIGURE 2 — The instantaneous corrosion rate during a few days of exposure.[8]

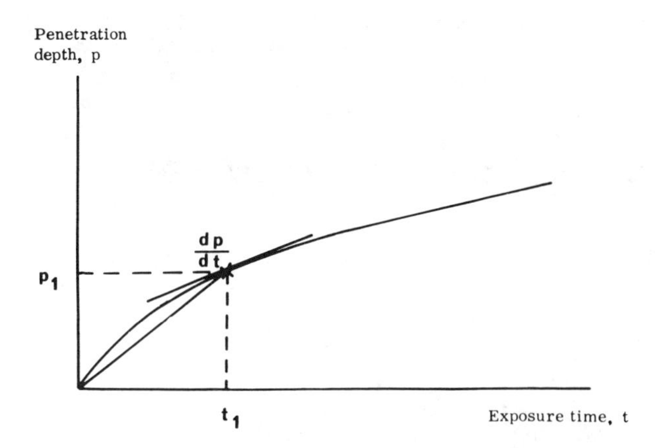

FIGURE 3 — Penetration depth vs exposure time curve: p = cumulative penetration depth averaged over the surface; p_1/t_1 = mean corrosion rate during the period t_1; and (dp/dt) t = t_1 = differential corrosion rate at the exposure time t_1.

sion rate, two different concepts may be distinguished in addition to the instantaneous corrosion rate mentioned: (1) the mean corrosion rate during a certain exposure (t_1), *i.e.*, p_1/t_1; and (2) the differential corrosion rate at a certain exposure time (t_1), *i.e.*, (dp/dt) t = t_1—generally, the differential corrosion rate approaches a constant value after some years of exposure.

Carbon Steel

Corrosion Behavior

The rust layer formed on unalloyed steel generally consists of two regions:[9] (1) an inner region, next to the interface steel/rust consisting primarily of dense, amorphous FeOOH with some crystalline Fe_3O_4; and (2) an outer region consisting of loose, crystalline α-FeOOH and γ-FeOOH.

The penetration depth increases with the time of exposure, as shown in Figure 4.[10] The curves there follow a power law:

$$p = kt^n \quad (1)$$

where: p = penetration depth; k = a constant; t = time of exposure; and n = a constant.

Such a time dependence has been reported by many authors.[11] It suggests that the transport of reactants through a

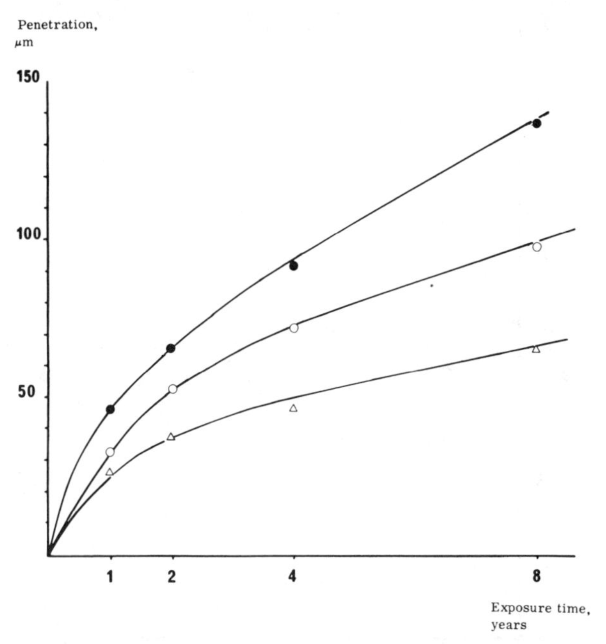

FIGURE 4 — Penetration depth vs exposure time for unalloyed carbon steel.[10] ● Muhlheim/Ruhr—Industrial atmosphere; ○ Cuxhaven—marine atmosphere; and △ Olpe—rural atmosphere.

growing protective rust layer determines the rate of the corrosion process. The power law Equation (1) can also be written:

$$\log p = \log k + n \log t \quad (2)$$

This equation is represented by a straight line in a bilogarithmic diagram (Figure 5).[10] Pourbaix has suggested that the penetration depth for up to 20 to 30 years can be estimated by extrapolation of lines determined for an initial 4 year period.[11] According to Bohnenkamp, Burgmann, Schwenk and Grimme,[10,12] however, such an extrapolation may lead to underestimations of the penetration, as after some years of exposure (1 to 8 years) the constant n will approach 1, which means that the time dependence changes into a linear relationship, that is, the corrosion rate becomes constant. This change will occur when the rust layer reaches a constant thickness, the rate of formation then being equal to the rate at which rust is dissolved, dusted away, or flaked off.

As can be seen in Figures 4 and 5 , the penetration is dependent on the type of atmosphere; the higher the SO_2 pollution, the higher the corrosion rate. On the basis of a survey of the literature, Krause reported the following corrosion rates in various types of atmosphere:[13]

Atmosphere	Mean Corrosion Rate (μm/y)
Rural	4 to 65
Urban	23 to 71
Industrial	26 to 175
Marine	26 to 104

Theoretical Background

The thermodynamic reaction possibilities for the rusting of steel can be surveyed in the potential-pH-diagram for the system Fe-H_2O (Figure 6). As can be seen, the metal is not stable in water solutions. In the domain marked "FeOOH" a more or less effective protection by a passivating coating of

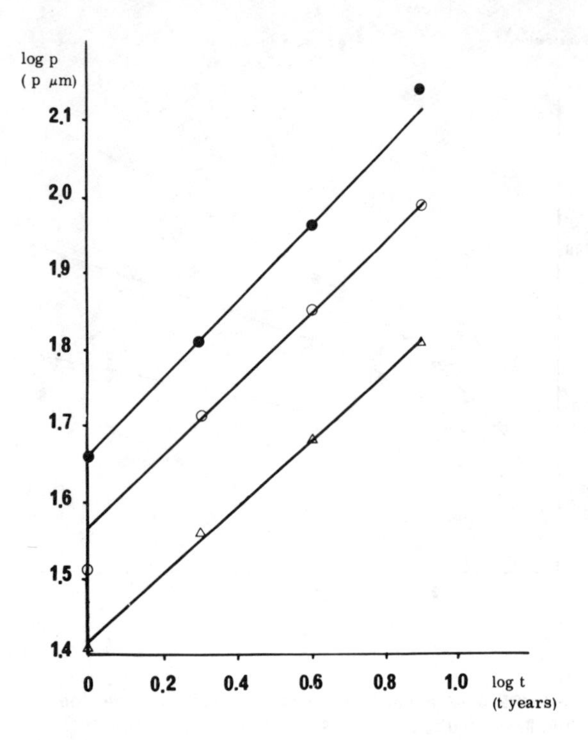

FIGURE 5 — Bilogarithmic diagram of penetration depth (p) and exposure time (t) for unalloyed carbon steel; the same results as in Figure 4.[10] ● Muhlheim/Ruhr—Industrial atmosphere; ○ Cuxhaven—marine atmosphere; and △ Olpe—rural atmosphere.

FeOOH can be expected, while corrosion is likely to occur under conditions corresponding to the domain marked "Fe^{2+}," because then soluble Fe^{2+} ions are stable.

Rusting may be initiated on the steel surface under hygroscopic deposits where the electrolyte is formed by absorption. Initiation may also occur at surface inclusions like MnS, which dissolve when the surface becomes wet.[14]

The rust thus formed will almost completely absorb the SO_2 reaching the surface,[2] and the rust will also catalyze the oxidation of the absorbed SO_2 to SO_4^{2-}.[14] This will lead to the formation of sulfate agglomerates, so-called sulfate nests in the rust.

When the surface becomes wet by rain, dew, or moisture absorption, the sulfate nests in combination with the surrounding area form corrosion cells (Figure 7). The electrolyte is mostly very concentrated and has a low water activity.[1,14] Anodes are located inside the sulfate nests where the pH value and the redox potential become low. The conditions here correspond to a position in the domain "Fe^{2+}" in the potential-pH-diagram and local attack will take place in the steel surface. The surrounding area acts as cathode. This is true even if the surface is covered with oxide containing crystalline magnetite (Fe_3O_4), because magnetite is a good electronic conductor. The following equations might in principle describe the reactions taking place in the corrosion cells:

At the cathode:

$$1/2\,O_2 + H_2O + 2e^- \rightarrow 2OH^- \quad (3)$$

$$Fe^{3+} + e^- \rightarrow Fe^{2+} \quad (4)$$

At the anode:

$$Fe \rightarrow Fe^{2+} + 2e^- \quad (5)$$

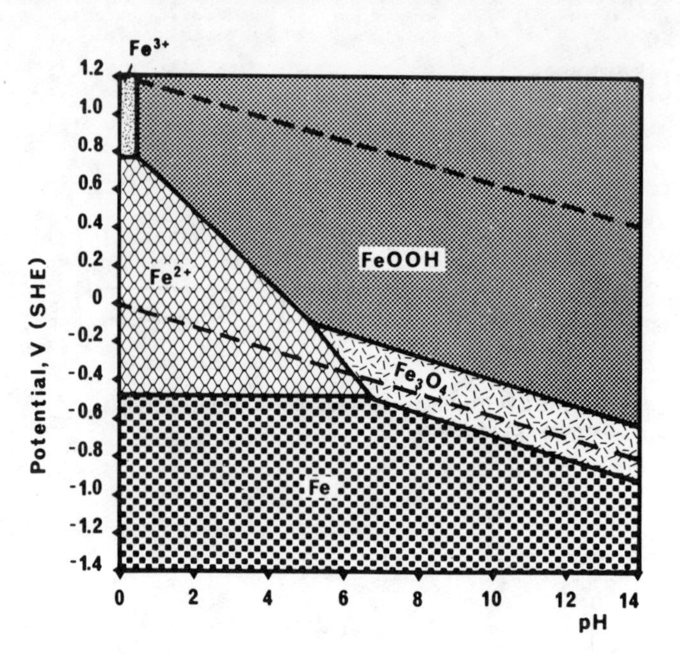

FIGURE 6 — Potential-pH-diagram, Fe-H_2O; 25 C, 10^{-1} M Fe.

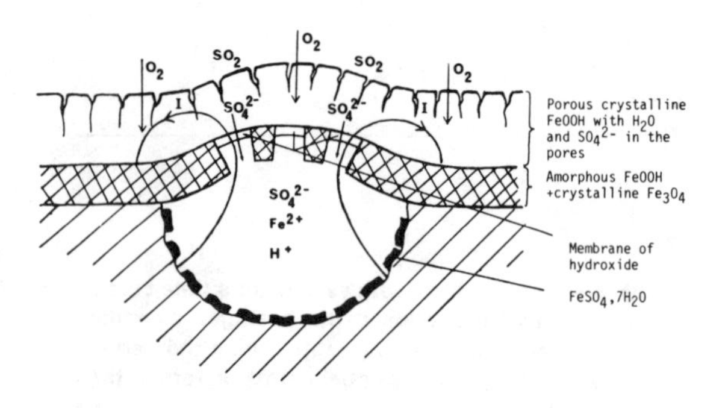

FIGURE 7 — Sketch of corrosion cell at sulfate nest on steel.

In the rust layer:

$$2Fe^{2+} + 3H_2O + 1/2\,O_2 \rightarrow 2FeOOH + 4H^+ \quad (6)$$

The sulfate nest becomes enclosed within a semipermeable membrane of hydroxide formed through hydrolysis of the iron ions. The electric current in the corrosion cell causes migration of SO_4^{2-} ions into the nest. This will stabilize the existence of the nest.

The protective ability of the rust is influenced by the conditions prevailing when it is formed.[8] If the rust formed becomes infected with sulfate nests, as may happen when the steel is first exposed during the winter season, the rust becomes little protective for some time. If, on the other hand, the first rust is formed during the summer season, it will generally become less infected by sulfate, and as a consequence more protective.

Chloride contamination in the rust may also influence the atmospheric corrosion of steel. One effect of the chloride is to decrease the critical relative humidity for absorption of moisture from the atmosphere.[15]

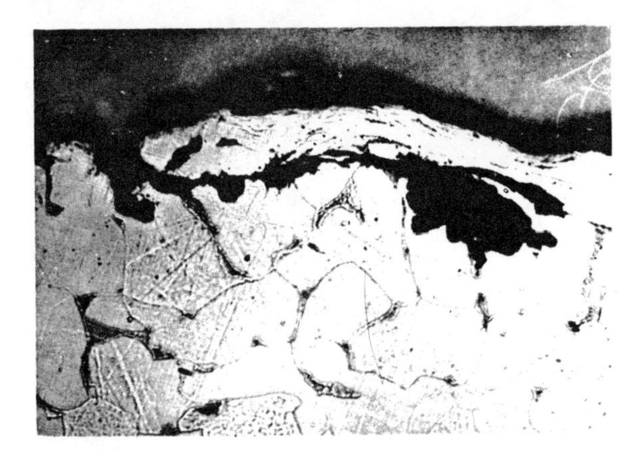

FIGURE 8 — Section through steel surface after blasting.[18]

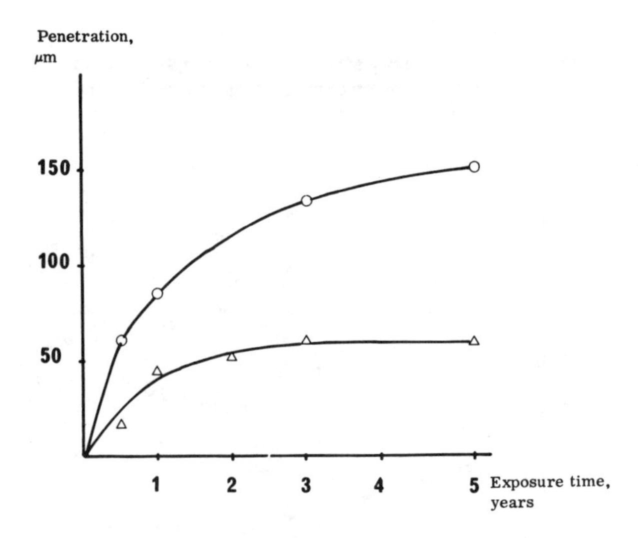

FIGURE 9 — Penetration depth vs exposure time in the urban atmosphere at Letnany, Prague.[21] ○ unalloyed carbon steel; and △ weathering steel.

Application Aspects

Carbon steel is not able to develop a protective coating by itself. So unalloyed steel to be used in outdoor atmospheres is generally given an additional surface protection; it may be a coating of antirust paint, zinc, or aluminum.

To ensure a satisfactory life of an antirust paint coating, one has to clean the steel surface carefully from rust and other contaminants before painting. Otherwise, corrosion cells may build up under the paint coating around sulfate nests or hygroscopic chloride inclusions while water and oxygen permeate through the paint; as a result, the paint coating delaminates and flakes off.

At present, cleaning before painting is generally carried out by blasting, wire brushing, or abrading.[16] It has been found, however, that not even thorough dry blasting will completely remove all SO_4^{2-} and Cl^- contamination from a rusty steel surface which has been exposed to a polluted atmosphere.[17] As shown by Gronvall, Thureson, and Victor and also by Calabrese and Allen,[19] salt residues may be entrapped by plastic deformation of surface irregularities into folds and pockets (Figure 8). The entrapped salt residues will initiate activity in corrosion cells, which may cause breakdown of a subsequently applied paint coating. One can remove the water soluble residues by water washing of the blasted surface. This complicated procedure is comparatively costly, so there is an interest in replacing the dry blasting by wet blasting, which also directly removes the entrapped salt residues if they are water soluble.

Weathering Steel

By alloying the steel with elements like Cu, P, Cr, Si, Ni, and Mn, one improves their corrosion resistance in outdoor atmospheres. Such low alloyed steels are called weathering steels.

Corrosion Behavior

The rust developing on weathering steels under urban or rural conditions is more protective than the rust on unalloyed carbon steel. The process seems to be effective, however, only if the steel is exposed to SO_2 and alternately washed by rain and dried up.[20]

The penetration depth for weathering steels also follows the power law, Equation (1), but with different values of the constants k and n. The constants vary with the composition of the steel as well as with the environment. In urban and rural atmospheres, the penetration depth approaches a nearly constant value after some years, which indicates that the rust layer then offers effective protection (Figure 9).[21]

Theoretical Background

In principle, the corrosion mechanism is similar for weathering and unalloyed carbon steels. On the former, however, the rust forms a more dense and compact layer which more effectively screens the steel surface from the corrosive components of the atmosphere. It seems as though the sulfate nests decrease and ultimately disappear if the exposure conditions are favorable, that is, when frequent dry to wet alternations occur.[22] Such climatic changes are believed to favor the bursting of the membranes around the sulfate nests and the dissolution and washing away of their sulfate. Many proposals have been made to explain the protective mechanism, but so far, no theory seems to have been generally accepted.

Application Aspects

The protective rust layer formed on weathering steels is generally considered a nice-looking, dark brown patina. So under urban and rural conditions, the weathering steels may be used without any additional protective coating of antirust paint, zinc, or aluminum. Power pylons and shells of self-supporting chimneys are examples of structures successfully made of weathering steel.

During the first years of exposure, before the protective rust layer has developed, the rusting proceeds at a comparatively high rate. Then rusty water is produced which may stain masonry, pavements, etc. So, precautions have to be taken against detrimental staining effects. One can, for instance, make the stained ground area exchangeable, or one can color the exposed masonry surfaces brown from the outset, so that staining will not be visible.

When the atmosphere is heavily polluted with SO_2, however, and under climatic conditions with long wet periods, the sulfate nests do not disappear with time. On the contrary, they will increase in size.[21] Then the weathering steels behave rather much like unalloyed carbon steels. The same is true for weathering steels in marine atmospheres, especially in sheltered positions.

Zinc

Zinc is a commonly used metal, mainly as a protective coating on carbon steel.

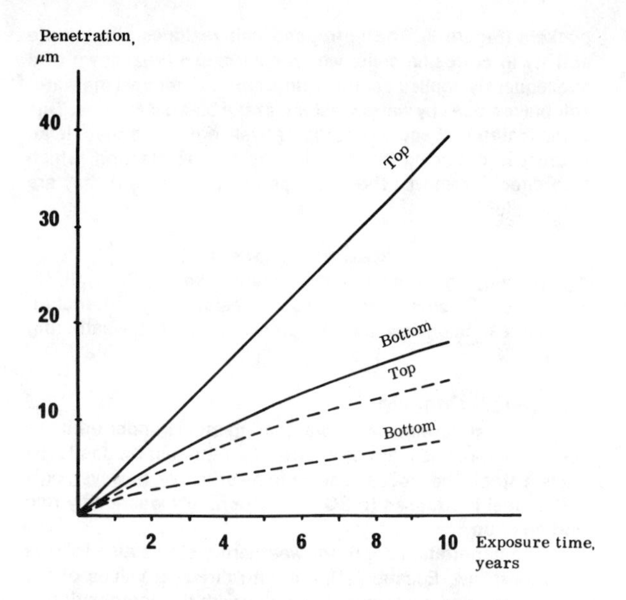

FIGURE 10 — Predictive curves for the atmospheric corrosion of galvanized steel based on data obtained during 4 to 5 years exposure.[24]  in East Chicago—urban atmosphere; and ----- at Kure Beach (245 m lot)—marine atmosphere.

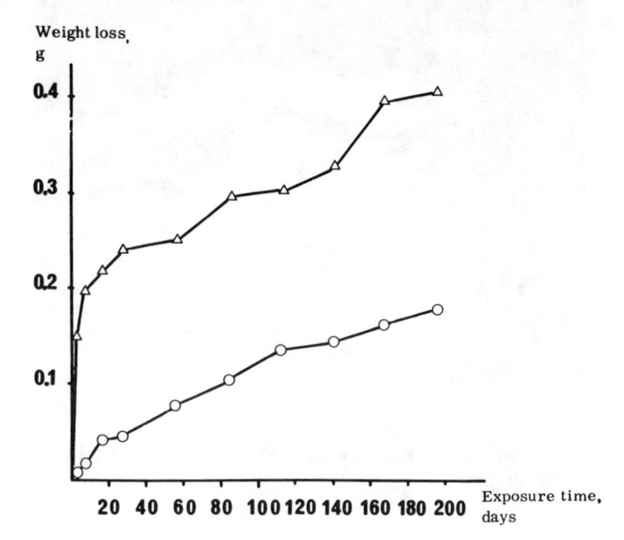

FIGURE 11 — Memory effect on the atmospheric corrosion of zinc.[25] △ samples exposed August 4, 1947; and ○ samples exposed February 17, 1948.

Corrosion Behavior

The penetration depth is generally reported to be an approximately linear function of the exposure time in rural and urban atmospheres.[23] Legault and Pearson,[24] however, have found that this is true only for skyward surfaces exposed to urban atmospheres (Figure 10). Groundward surfaces in urban, and the skyward as well as the groundward ones in marine atmospheres, show nonlinear relations with time in accordance with the power law Equation (1). It may be noted that the corrosion rate is higher on the skyward surfaces than on the groundward ones. Further, it was concluded that Equation (1) can reliably predict the long-term atmospheric corrosion behavior of zinc on the basis of two weight loss determinations during the initial two year period of exposure.

As reported by Ellis, the corrosion rate is very dependent on the weather conditions during the early part of the exposure.[25] Long-lasting rainfall or a relative humidity at or near 100% during the first few days leads to a high corrosion rate, while drier conditions lead to lower corrosion rates. The surface will "memorize" the initial conditions and corrosion rates for at least one year (Figure 11).

The following corrosion rates of zinc have been reported for various types of atmosphere:[23,26]

Atmosphere	Diffential Corrosion Rate (μm/y)
Rural	0.2 to 3
Urban and industrial	2 to 16
Marine	0.5 to 8.

Theoretical Background

Zinc is a relatively base metal. The stability domains of various zinc containing species in the system Zn-CO_2-H_2O at 25 C are shown in Figure 12. The diagram is valid for a total H_2CO_3 content of 10^{-5} mole/L in the moisture film, that is a solution in equilibrium with the CO_2 content (about 0.03%) in outdoor atmospheres. As shown by the diagram, there is a stability domain for $ZnCO_3$ in the pH range 6 to 7. In a supplementary diagram (Figure 13 A), how the width of the stability domain for $ZnCO_3$ varies with the H_2CO_3 content is shown. Basic zinc carbonate does not seem to be stable, but it has been identified on zinc surfaces[27] and may occur as a metastable species. From these diagrams, it is obvious that a coating containing $ZnCO_3$ can form on zinc in rural atmospheres provided the pH value is not too low, and this coating is apparently protective. From Figure 13 C, it is also clear that basic zinc chloride coating is also likely to be protective. As shown in Figure 13 B, however, stable basic zinc sulfate is not likely to form when the air is heavily polluted with

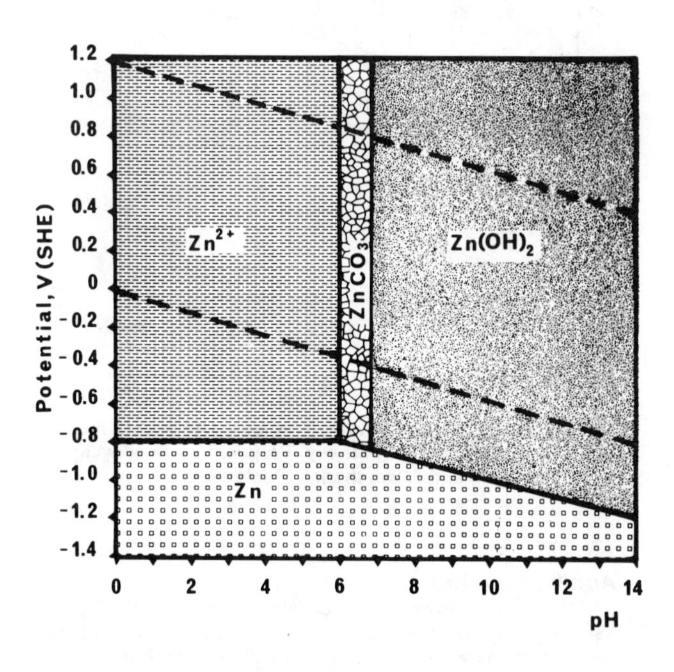

FIGURE 12 — Potential-pH-diagram, Zn-CO_2-H_2O; 25 C, 10^{-1} M Zn, 10^{-5}M H_2CO_3.

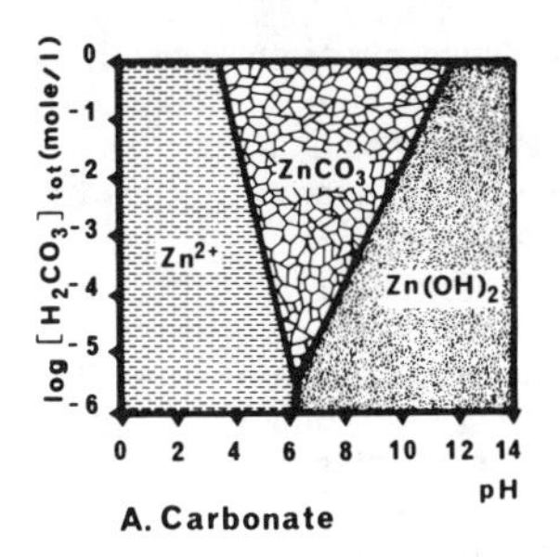

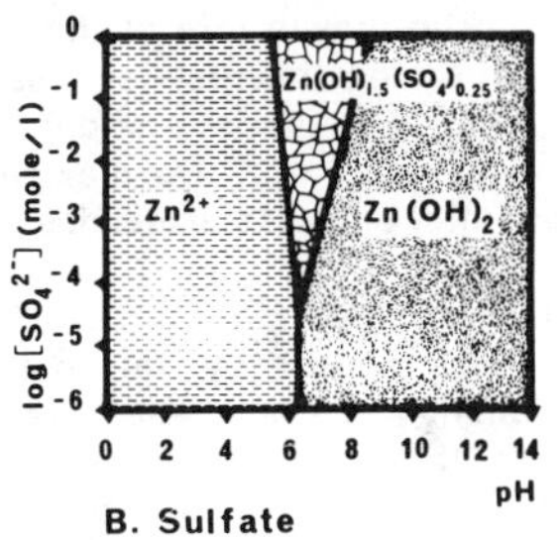

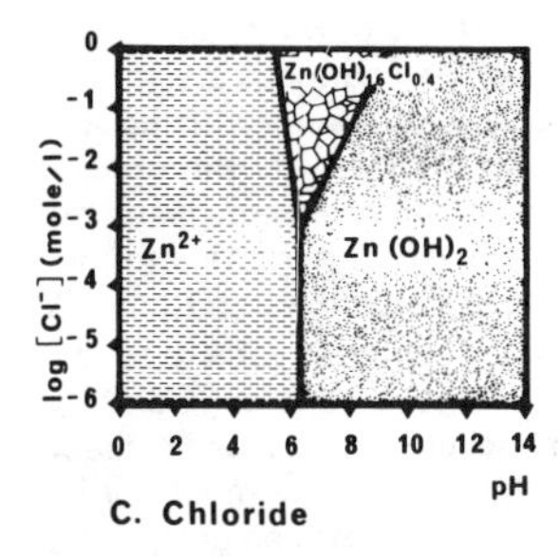

FIGURE 13 — Stability domains of zinc carbonate, basic zinc sulfate, and basic zinc chloride in aerated aqueous solutions with varying anion content and pH value; 25 C, 10^{-1} M Zn.

SO_2 because then the pH value of the water film will probably be too low for such formation. The relatively high corrosion rate on the skyward side found by Legault and Pearson in the Chicago atmosphere (Figure 10) might be due to the surface being exposed to acid rain.[24] Groundward surfaces are not exposed to such rain. In marine atmospheres, rain washing may lower the Zn^{2+} content on the skyward side, which reduces the width of the stability domain for basic chloride. This may explain the higher corrosion rate on the skyward side under marine conditions.

Application Aspects

A few decades back, galvanized steel was considered a corrosion resistant material for structures in outdoor atmospheres. The corrosion rate was only 0.5 to 1.0 μm/y. With increasing pollution by SO_2, the corrosion rate of zinc has increased, in many urban or industrial areas up to 5 μm/y or more, and the life of galvanized steel structures has then sometimes become unacceptably short. Nowadays, it is often necessary to give galvanized steel an additional protection primarily by anticorrosion painting.

As the protection of zinc in outdoor atmospheres largely depends on the formation of a carbonate containing coating, a sufficient supply of CO_2 to the surface is a condition for effective protection. In certain cases, the CO_2 supply is insufficient, *e.g.,* in the crevices between stacked semiproducts, like strips or sections. When moisture collects in the crevices, corrosion takes place primarily under the formation of $Zn(OH)_2$. As no protective carbonate coating can form and the $Zn(OH)_2$ has poor protection properties, the corrosion rate is relatively high and large amounts of loosely adherent "white rust" are formed. If the white rust is exposed to the atmosphere, the zinc hydroxide is converted to carbonate.[28] This type of damage is most frequent during rainy seasons. White rust can also form on moist zinc surfaces covered with mineral wool, which prevents the access of air. This situation may occur in heat insulated wall structures enclosing galvanized steel members.

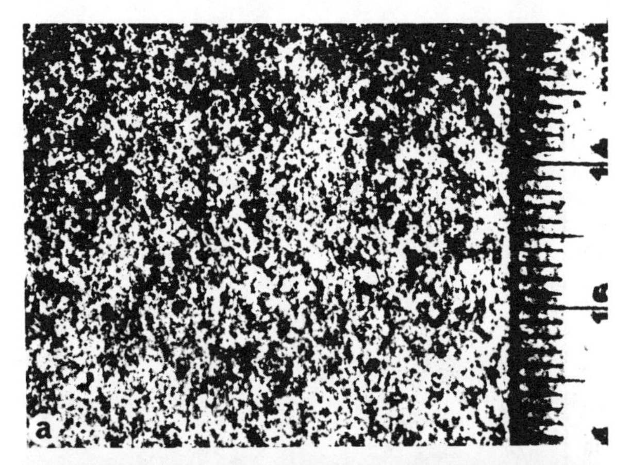

FIGURE 14 — Pitting in aluminum after 10 years exposure in industrial atmosphere:[29] (a) surface appearance; and (b) section.

Aluminum

Corrosion Behavior

In the atmosphere immediately after production, aluminum becomes covered with a thin, dense oxide coating, which is protective. In clean outdoor atmospheres, aluminum will retain its shiny appearance for years, even under tropical conditions. In polluted outdoor atmospheres, small pits develop which are hardly visible to the naked eye (Figure 14). The pits become covered with crusts of aluminum oxide and hydroxide.[29]

The corrosion rate is largely the same for unalloyed aluminum and for most conventional aluminum alloys, except those which are high in Cu. These will generally show higher corrosion rates. The average penetration depth determined from the weight loss is very low and approaches asymptotically a limiting value. As shown in the following table, the mean corrosion rate for AlMn 1.2 during 20 years exposure does not exceed 1 μm/y, not even in polluted atmospheres; in clean atmospheres, it is much lower.[26]

Atmosphere	Mean Corrosion Rate (from weight loss) (μm/y)	Max. Pit Depth After 20 Years Exposure (μm)
Rural	0 to 0.1	10 to 55
Urban	~1	100 to 190
Marine	0.4 to 0.6	85 to 260

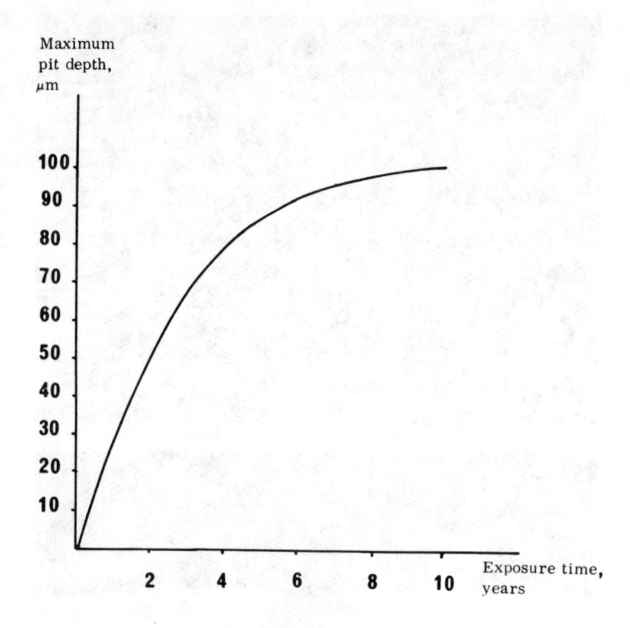

FIGURE 15 — Maximum pit depth vs exposure time for AlMn 1.2 in the urban marine atmosphere of Gothenburgh.[26]

The growth rate of the maximum pit depth is relatively high during the first few years of exposure, but decreases gradually so that the pit depth approaches a nearly constant value (Figure 15).[26] The maximum pit depth rarely exceeds 200 μm after one or two decades of exposure.[30]

Theoretical Background

As shown by the potential-pH-diagram (Figure 16), aluminum is a very base metal, being stable only at low potentials. It can be used in the presence of water only because of its property to develop a protective coating of alumina. Anions like SO_4^{2-} or Cl^- may be incorporated in the lattice forming a variety of basic salts and complexes, which so far are little known. The stability range of the oxide coating extends down to pH 2.5. Thus, the oxide coating is also protective in urban atmospheres with SO_2 pollution, producing a relatively low pH value in the moisture film. In the presence of chloride, however, the oxide coating is more permeable to ions. The chloride ions are believed to migrate into the oxide layer and lower its resistance to outward migration of Al^{3+}.[31]

In the presence of chloride ions, pitting may be initiated. In the propagation stage, aluminum is dissolved anodically to Al^{3+} ions within the pit. The cathodic reaction takes place either outside the pit close to its mouth or inside the pit, and consists in the reduction of oxygen or H^+ ions, respectively. In the former case, it occurs preferentially at surface inclusions with a low oxygen overvoltage, such as segregations of $FeAl_3$ or particles of deposited copper; the passivating oxide layer has a low electronic conductivity. By hydrolysis of the Al^{3+} ions, acid conditions are created within the pit and a cap of $Al(OH)_3$ is formed over the mouth of the pit; the corrosion products finally block the operation of the pit.

TABLE 1 — Basic Copper Salts in Green Patina from Various Atmospheres, Defined by the Anions Ranked with Respect to Content[34]

				Type of Atmosphere			
Reporter	Object	Time of Exposure (years)	Country	Rural	Urban or Industrial	Marine	Mixed Urban-Marine
Vernon & Whitby	Copper roofs; copper conductor in marine atm.	12-300 13	UK	1. SO_4^{2-} 2. CO_3^{2-}	1. SO_4^{2-} 2. CO_3^{2-} 3. Cl^-	1. Cl^- 2. CO_3^{2-} 3. SO_4^{2-}	1. SO_4^{2-} 2. CO_3^{2-}, Cl^-
Vernon	Copper roof on church spire on the isle of Guernsey	33	UK			1. Cl^- 2. CO_3^{2-} 3. SO_4^{2-}	
Freeman Jr	Copper roofs	16-78	USA				1. SO_4^{2-} 2. CO_3^{2-}, Cl^-
Thompson, Tracy, & Freeman Jr	Copper panels from field test	20	USA	1. SO_4^{2-}, Cl^- 2. CO_3^{2-}	1. CO_3^{2-} 2. SO_4^{2-}	1. Cl^- 2. SO_4^{2-} 3. CO_3^{2-}	
Aoyama	Copper conductor for railway		Japan	1. NO_3^- 2. CO_3^{2-}	1. SO_4^{2-} 2. CO_3^{2-}	1. Cl^- 2. CO_3^{2-}	
Mattsson & Holm	Copper-base materials from field test	7	Sweden	1. SO_4^{2-} 2. NO_3^- 3. CO_3^{2-}	1. SO_4^{2-} 2. CO_3^{2-}	1. Cl^- 2. SO_4^{2-}	
Scholes & Jacob	Copper-base materials from field test	16	UK		1. SO_4^{2-} 2. Cl^-, CO_3^{2-}	1. Cl^- 2. SO_4^{2-}	

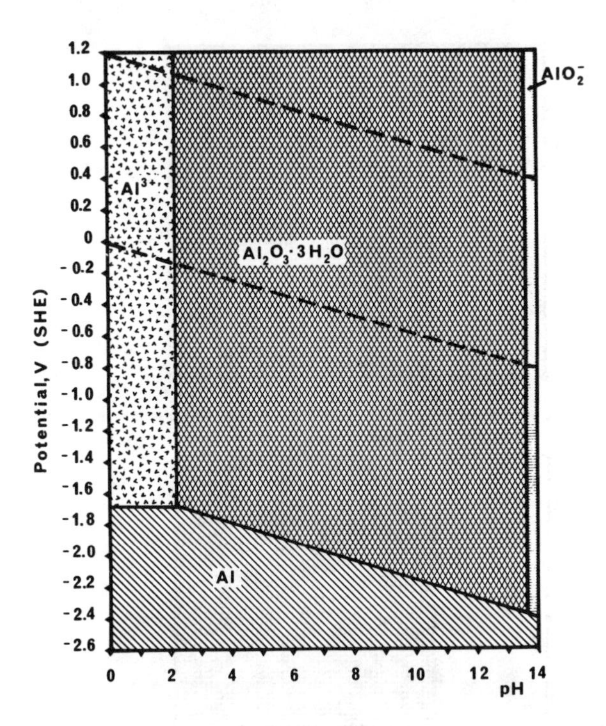

FIGURE 16 — Potential-pH-diagram, Al-H_2O; 25 C, 10^{-1} M Al.[6]

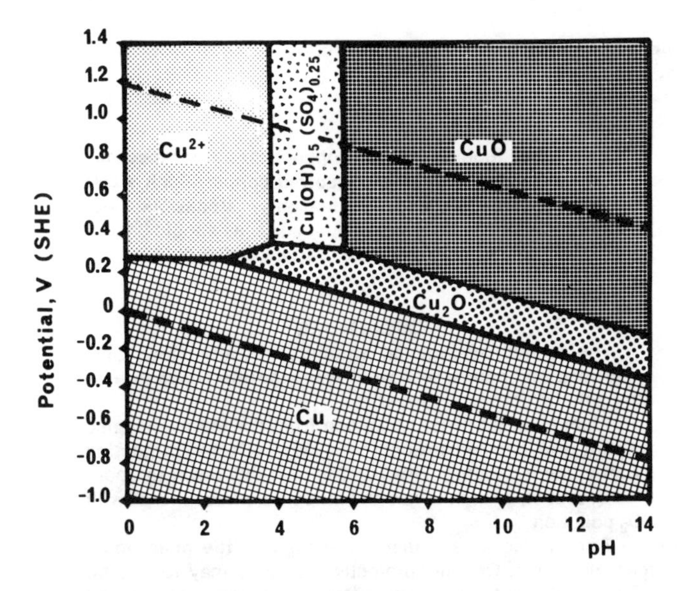

FIGURE 17 — Potential-pH-diagram, Cu-SO_4^{2-} – H_2O; 25 C, 10^{-1} M Cu, 10^{-3} M SO_4^2.

Application Aspects

Due to good passivation properties in outdoor atmospheres, aluminum materials are commonly used for building purposes, *e.g.*, for roofs, facades, window frames, etc. Of particular interest is the relatively good resistance of aluminum in SO_2 polluted atmospheres where most other materials are susceptible to corrosion. Aluminum structures have been reported in excellent condition even after 6 to 7 decades exposure in urban atmospheres.[29,30]

The low penetration rate and the shallow pitting do not generally influence the mechanical properties of aluminum structures, except in excessively polluted atmospheres. The shiny metal appearance will, however, gradually disappear and the surface will roughen under the formation of a grey patina of corrosion products. If the atmosphere contains much soot, this will be adsorbed by the corrosion products and give the patina a dark color. The shiny metal appearance may be retained by anodizing, that is, by anodic oxidation, which strengthens the oxide coating and improves its protective properties. An oxide coating with a thickness of 20 μm will generally preserve the original metal appearance for decades.

In spite of the fact that the oxide coating is more permeable to ions in the presence of chloride, aluminum materials are also useful in marine atmospheres. Under these conditions, however, the aluminum is susceptible to bimetallic corrosion when in contact with a more noble metal like carbon steel or copper.[32] In the presence of chloride, aluminum is also susceptible to crevice corrosion.

Aluminum is used as a protective coating on carbon steel. Due to the high electric resistance of the oxide layer, the aluminum will not give cathodic protection to bare steel surfaces at sheared edges, scratches, etc., so bare steel surfaces may then produce rust staining at least under rural or urban conditions. In marine atmospheres, however, where the oxide coating is more permeable to ions aluminum will offer cathodic protection to carbon steel.[33]

Copper

Corrosion Behavior

If conditions are favorable, copper may develop a characteristic blue-green patina in outdoor atmospheres. The patina has a rather complex composition, varying from place to place (Table 1).[34] The main components are generally copper(I)oxide and one or more basic copper salts; in urban atmospheres, basic sulfate is predominant, and in marine atmospheres basic chloride, while in rural atmospheres, basic sulfate is usually the main component.

The corrosion rate is generally low and decreases somewhat after the first few years of exposure, which indicates the formation of a slightly protective layer of corrosion products.

For copper in various types of atmospheres the following corrosion rates have been determined:[34]

Atmosphere	Corrosion Rate (μm/y)
Rural	~0.5
Urban	1 to 2
Marine	~1

Theoretical Background

The potential-pH-diagram in Figure 17 represents the system Cu-SO_4^{2-} – H_2O at 25 C with a Cu^{2+} content of 10^{-1} mole/L and SO_4^{2-} content of 10^{-3} mole/L. As can be seen, copper metal is stable in a great part of the stability region of water. This is consistent with copper being a noble metal. The diagram also shows a stability domain for basic copper sulfate, $Cu(OH)_{1.5}(SO_4)_{0.25}$. The width of this domain depends on the SO_4^{2-} content; with a decreasing content, the width diminishes (Figure 18 A). The stability domains for $Cu(OH)_{1.5}Cl_{0.5}$, $CuOH(CO_3)_{0.5}$, and $Cu(OH)_{1.5}(NO_3)_{0.5}$ vs corresponding anion contents can also be seen in Figure 18.

From Figure 18 A, it seems reasonable that basic sulfate should be the main component of the green patina on copper in urban atmosphere. It is also reasonable to infer that basic chloride is formed in marine atmospheres (Figure 18 B). As for basic carbonate, Figure 18 C indicates that the conditions in the atmosphere would not favor the formation of this type of patina, for the H_2CO_3 content in the water film would be only

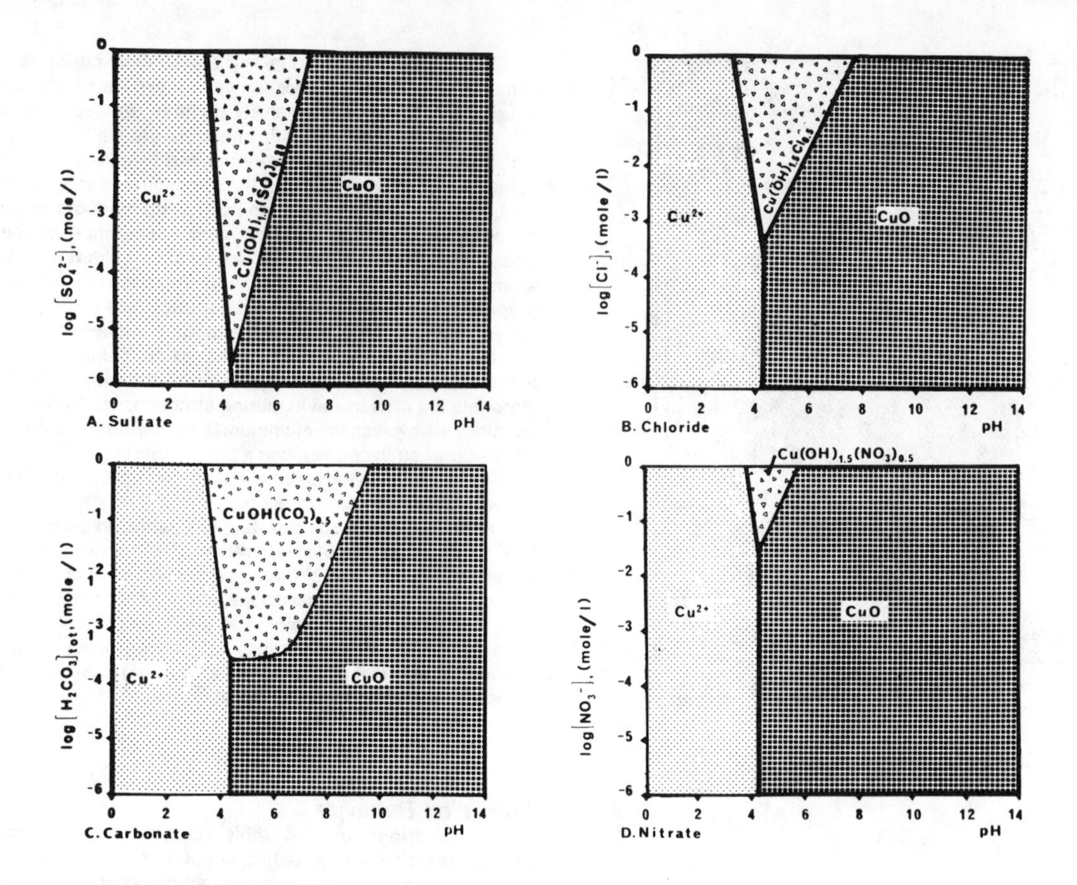

FIGURE 18 — Stability domains of basic copper salts in aerated aqueous solutions with varying anion content and pH value; 25 C, 10^{-1} M Cu.

about 10^{-5} mole/L on equilibrium with the air. In spite of this, basic carbonate is sometimes found in practice, which is remarkable. The presence of basic copper nitrate, also found in some locations, indicates that the water film on the metal surface may contain an appreciable amount of nitrate (Figure 18 D).

Application Aspects

Due to its low corrosion rate and its interesting patina formation, copper has long been used as an architectural material. Many copper roofs on castles and other monumental buildings have lasted for several centuries.

Sometimes it occurs that a user is at first disappointed and complains when his newly laid copper roof shows an ugly, mottled appearance.[35] After 6 to 12 months, however, the surface has usually acquired a uniform dark-brown color. In general, the surface does not develop beyond this stage for a number of years. After five to ten years, a green patina may begin to appear on sloping surfaces under urban or marine conditions. Vertical surfaces generally stay black much longer, as their time of wetness is shorter. In a marine atmosphere, the surfaces facing the sea acquire a green patina sooner than the other ones. This is, of course, due to the greater supply of chlorides by the sea winds. If the atmosphere is low in pollution, the patina may take an extremely long time to form (hundreds of years), due to scarce supply of anions for the formation of basic copper salt. It may even fail to form at all if the temperature is low or the exposure conditions very dry. As shown by the potential-pH-diagram, green patina will also fail to form under acidic conditions, *e.g.*, near chimneys, where acid smoke strikes the roof. Under such conditions, the corrosion products are soluble.

Rain water, running off from copper bearing surfaces, has generally picked up traces of dissolved copper. Such water may cause blue staining on masonry, stonework, etc. Therefore, the rain water from copper surfaces should be properly collected and drawn off with gutters and spouts.

Conclusions

The four most common structural metals—steel, zinc, aluminum, and copper—show interesting differences in their atmospheric corrosion properties.

Aluminum, the least noble one from the thermodynamic point of view, can be used only because of effective protection by an alumina coating. This coating is stable down to about pH 2.5, that is, even in urban and industrial atmospheres with SO_2 pollution.

Zinc is also a base metal, although a little more noble than aluminum. On zinc, protective coatings may form; zinc carbonate (or basic zinc carbonate) in rural atmospheres and basic zinc chloride under marine conditions. When the SO_2 pollution is higher, however, the pH value of the surface moisture will be too low for the stable existence of basic zinc salts. Neither can zinc carbonate form when the supply of CO_2 is insufficient as, for instance, in the crevices between stacked semiproducts of galvanized steel. There, rapid corrosion may take place under the formation of nonprotective white rust.

Iron is still more nobel than zinc. The rust formed on atmospheric corrosion of unalloyed steel, however, is little protective. So, in general, steel structures to be used in outdoor atmospheres have to be given a protective coating of, *e.g.*, antirust paint, zinc, or aluminum. On the so-called weathering steels, however, which have low alloying additions of Cu, P, Cr,

Si, Ni, and/or Mn, a dark-brown rust patina may develop, which is protective.

Copper, the most noble among the metals under consideration, has low corrosion rates in most types of atmosphere. The corrosion products have some protective effect, but are mainly of aesthetic value, especially the pleasing green patina of basic copper salts, which develops after some years in moderately polluted atmospheres.

As shown, potential-pH-diagrams may also form a theoretical basis for discussion of atmospheric corrosion phenomena. Difficulties exist, however, particularly in estimation of the composition of the electrolyte film on the corroding surface. The concentrations will fluctuate extensively due to daily and seasonal climatic variations. Sometimes, the concentrations will reach high values. So, there is need for more knowledge of corrosion in concentrated solutions to gain a deeper understanding of the atmospheric corrosion of metals.

Acknowledgment

The author is grateful to Vladimir Kucera and Jan Gullman for discussions during the preparation of this lecture, to Kaija Eistrat for documentation service, and to Ove Nygren for assistance with the illustrations.

References

1. Barton, K., Bartonova, S., and Beranek, E., Werkstoffe und Korrosion, Vol. 25, p. 659 (1974).
2. Sydberger, T., and Vannerberg, N.-G., Corrosion Science, Vol. 12, p. 775 (1972).
3. Corrosion of Metals, Classification of Corrosion Aggressivity of Atmosphere. Standard of Mutual Economic Aid ST SEV 991-78.
4. Rodhe, L., Stockholm University (private communication).
5. Knotkova-Cermakova, D., and Vlckova, J., Werkstoffe und Korrosion, Vol. 21, p. 16 (1970).
6. Pourbaix, M., Atlas of Electrochemical Equilibria in Aqueous Solutions. NACE, Houston, CEBELCOR, Brussels (1974).
7. Stability Constants of Metal-Ion Complexes. Part A: Inorganic Ligands. Part B: Organic Ligands. IUPAC Chemical Data Series, Pergamon, Oxford (1982, 1979).
8. Ericsson, R., and Sydberger, T., Werkstoffe und Korrosion, Vol. 31, p. 455 (1980).
9. Suzuki, I., Hisamatsu, Y., and Masuko, N., J. Electrochem. Soc., Vol. 127, p. 2210 (1980).
10. Burgmann, G., and Grimme, D. Stahl und Eisen, Vol. 100, p. 641 (1980).
11. Pourbaix, M., Proc. 7th Intern. Congr. Metallic Corrosion, Rio de Janeiro, p. 1181 (1978).
12. Bohnenkamp, K., Burgmann, G., and Schwenk, W., Galvano-Organo, Vol. 43, p. 587 (1974).
13. Krause, J., Korrosion (Dresden), Vol. 6, p. 23 (1975).
14. Barton, K., Kuchynka, D., and Beranek, E., Werkstoffe und Korrosion, Vol. 29, p. 199 (1978).
15. Ericsson, R., Werkstoffe und Korrosion, Vol. 29, p. 400 (1978).
16. Pictorial Surface Preparation Standards for Painting Steel Surfaces. Swedish Standard SIS 05 59 00 (1967).
17. Igetoft, L., Swedish Corrosion Institute (private communication).
18. Gronvall, B., Thureson, L., and Victor, V., Blasted Steel Surfaces—Surface Roughness, Cleanliness and Cold Working (in Swedish). Swedish Corrosion Institute, Bull., No. 67 (1971).
19. Calabrese, C., and Allen, J. R., Corrosion (NACE), Vol. 34, p. 331 (1978).
20. Pourbaix, M., On the Prediction of the Atmospheric Corrosion and Passivation Conditions of Steels. Passivity of Metals. Electrochem. Soc., Princeton, p. 762 (1978).
21. Knotkova-Cermakova, D., and March, V., Die Korrosion niederlegierter Stahle in verunreinigten Atmospharen. Freiberger Forschungshefte, B 189, p. 59 (1976).
22. Schwitter, H., and Bohni, H. J. Electrochem. Soc., Vol. 127, p. 15 (1980).
23. Slunder, C. J., and Boyd, W. K., Zinc: Its Corrosion Resistance. Zinc Institute Inc., New York (1971).
24. Legault, R. A., and Pearson, V. P., ASTM STP 646, p. 83 (1978).
25. Ellis, O. B., Proc. ASTM, Vol. 49, p. 152 (1949).
26. Mattsson, E., Aluminum in Polluted Atmospheres (in Swedish). Teknisk Tidskrift, Vol. 98, p. 767 (1968).
27. Feichtknecht, W., Chimia, Vol. 6, p. 3 (1952).
28. Bird, C. E., and Strauss, F. J., Materials Performance, Vol. 15, p. 27 (1976).
29. Mattsson, E., and Lindgren, S., ASTM STP 435, p. 240 (1968).
30. Godard, H. P., Jepson, W. B., Bothwell, M. R., and Kane, R. L., The Corrosion of Light Metals. John Wiley and Sons, New York, p. 92 (1967).
31. Kaesche, H., Pitting Corrosion of Aluminum, and Intergranular Corrosion of Aluminum Alloys. Localized Corrosion, NACE, Houston, Texas, p. 516 (1974).
32. Kucera, V., and Mattsson, E., Atmospheric Corrosion of Bimetallic Structures. Electrochem. Soc. Monograph on Atmospheric Corrosion (in print).
33. Mattsson, E., and Ericson, W., Aluminum and Zinc Coatings on Steel—A Comparison of Corrosion Resistance in the Atmosphere. AB Svenska Mettallverken, Research Report No. 1274 (1959).
34. Mattsson, E., and Holm, R., Atmospheric Corrosion of Copper and its Alloys. Electrochem. Soc. Monograph on Atmospheric Corrosion (in print).
35. Mattsson, E., and Holm, R., Sheet Metal Ind., Vol. 45, p. 270 (1968).

Source: *Materials Performance*, July 1982, 9-19

Kenneth G. Compton
Consultant
Fort Lauderdale, Florida

SUMMARY

Discusses the many variables that must be accounted for in measuring corrosion. Locations throughout U. S. and Canada were used for the tests described in this article. Discusses what a marine atmosphere is, and why locations with similar atmospheres can have differing corrosion rates.

Factors Influencing the Measurement of Corrosion in Marine Atmosphere

THE MARINE atmosphere extends from areas where sea water directly contaminates a structure to remote areas where the sea salt contamination is carried by the winds. It includes the above-water outdoor atmosphere on ships and offshore structures, as well as piers and bridges connected to the land.

Recently, a team from the American Society for Testing and Materials calibrated a large number of fixed, typical exposure sites that have been classified as "marine." The time of wetness, the sulfur dioxide content of the atmosphere, and the amount of chloride deposition were measured. Quantitative data on the variations in corrosion rate of standard metal specimens could be related in a qualitative manner to the differences in the conditions encountered.

Variables

The amount of sea salt accumulated upon structures depends upon the distance from the ocean and the direction and force and frequency of winds carrying the contamination.

A low, flat stretch of ground with an unimpeded wind flow between the shore line and the specimens or structure will give a maximum contamination. Buildings or bluffs immediately downwind from the site may produce a "dead spot" so that contamination is carried over it.

Another variable is the amount of time that the moisture film from fog, dew, or wet spray remains on the contaminated surface. Low relative humidity or heating by the sun can cause rapid and thorough drying of the surfaces, but on damp, cloudy days, the moisture film can persist for long periods. Frequent heavy rains may wash the contamination off the structure and reduce the rate of corrosion.

Possible contamination from sulfur dioxide and its derivatives is encountered in industrial areas along the ocean and above decks on ships. Sometimes the contamination from sulfur compounds completely obscures the effect of variations in chloride concentration. The rate of corrosion of some material at Sandy Hook, N. J., is possibly influenced in a similar manner.

ASTM Test Results

The ASTM study (Table 1) used corrosion rates of specimens at State College, Pennsylvania as unity and related all other rates to these.

A large variation in corrosion rates occurred between two test sites 720 feet apart at Kure Beach, N. C. The higher rate was encountered at a nominal distance of 80 feet from the waterline and the lower rate at 800 feet. Two variables, the amount of contamination and the duration of wetness from direct ocean spray, were responsible for the differences.

TABLE 1—Relative Corrosivity of Atmosphere at 19 Test Sites Compared to that at State College, Pa.

Location	Desig*	Steel	Zinc
Norman Wells, N. W. T.	R	0.03	0.4
Esquimalt, B. C.	R	0.5	0.4
Saskatoon, SASK.	R	0.6	0.5
Perrine, Fla.	R	0.9	1.0
State College, Pa.	R	1.0	1.0
Ottawa, Ont.	I	1.0	1.2
Middletown, O.	I	1.2	0.9
Trail, B. C.	I	1.4	1.6
Montreal, P. Q.	I	1.5	2.2
Halifax Y. R.*, N. S.	R	1.5	1.6
South Bend, Pa.	R	1.5	1.5
Kure Beach 800 N. C.	M	1.8	1.7
Point Reyes, Cal.	M	1.8	1.8
Sandy Hook, N. J.	M	2.2	1.6
New York (S)*, N. Y.	I	3.1	3.6
Kearny, N. J.	I	3.3	2.6
Halifax F. B.*, N. S.	I	3.8	18
New York (F)*, N. Y.	I	6.0	3.7
Daytona Beach, Fla.	M	7.1	2.6
Kure Beach 80, N. C.	M	13	5.7

* S = Spring, F = Fall, I = Industrial, M = Marine, R = Rural, Y.R. = York Redoubt, F.B. = Federal Building

Another example of the difference in corrosion rates at similar sites was observed on electric power and telephone poles near Astoria, Ore. and in the vicinity of Santa Monica, Cal. Red rust on galvanized hardware at the Oregon site appeared after 35 to 37 years, but at the California site, rusting was heavy within two years. At the first location, heavy rainfall occurred throughout most of the year and washed off the sea salt. At the other, very little rain fell, but heavy dew and fog kept the metal surfaces wet much of the time.

The Sandy Hook site should have shown some contamination from sulfur-bearing compounds because of its proximity to the highly industrial New York City area. However, the prevailing southeast sea breeze and the northeast storms minimized this sulfur contamination.

In another study, a test site at Key West, Fla., was within 100 yards of the Caribbean Sea. The relative rate of zinc corrosion to that at State College was a favorable 0.423, because the prevailing wind was from across the island and the large percentage of time exposed to the heat of the sun.

Larrabee made a study of the boldness and direction of exposure in a mild marine site as they effect corrosion rates.[1] He placed steel specimens at different heights under a roof structure (Figure 1). The specimens faced in four directions and were mounted 30 degrees to the horizontal and in the vertical position.

Two series of test were made because of the unnatural contamination produced by a hurricane on the sheltered specimens during the first test. The test data emphasized the variations that occur from year to year with

★Condensation of a paper titled "Corrosion in the Marine Atmosphere and the Factors Influencing its Severity," presented to the 20th Annual Conference, National Association of Corrosion Engineers, March 9-13, 1964, Chicago, Illinois. Photocopies of the complete, original paper are available from "Photocopies, Materials Protection, 980 M & M Building, Houston, Texas, 77002." Price: $6 per copy.

Figure 1—Specimens exposed in four directions. (Study by C. P. Larrabee, See Reference 1).

varying atmospheric conditions and with angle of exposure.

The behavior of bi-metallic couple and crevice corrosion cannot be discussed in any general way as these situations are highly specific. The only general considerations are the amount of contamination and duration of wetness. In referring to marine corrosion, one must always be very specific as to conditions.

Reference

1. C. P. Larrabee and O. B. Ellis, *Proceedings of the American Society for Testing Materials,* Vol. 59, p. 183, 1959.

Bibliography

H. R. Ambler and A. A. J. Bain, *Journal of Applied Chemistry,* Vol. 5, p. 437, 1955. Also, Tropical Testing Establishment Reports 216 and 247, British Ministry of Supply.

H. P. Godard, *Materials Protection,* 2, No. 6, p. 38, 1963.

KENNETH G. COMPTON is a consultant in Fort Lauderdale, Fla. He is retired from Bell Telephone Laboratories, Inc. He is a past recipient of the NACE Frank Newman Speller Award, the ASTM San Tour Award, the Award of Merit, and the AES Bronze Medal. He has a BS in chemical engineering, an EE, and an MS in physical chemistry. During World War 1, he received two citations in the U. S. Navy. He was a technical consultant.

Source: *Materials Protection,* December 1965, 13-14

some new views on

SOIL CORROSION

Edward Schaschl and Glenn A. Marsh

Research Center
Pure Oil Company
Crystal Lake, Illinois

SUMMARY

Corrosion of steel in soils generally can be explained in terms of long cell action arising from differential aeration. Oxygen plays the dominant role as a cathodic reactant. Cathodic activity is most vigorous when a soil is partially drained, that is, when it contains 50 to 95 percent of the water needed for saturation. Steel in contact with the soil is wetted by a thin film of water and is separated by this film from air in the pores of the soil. When the soil is in this partially drained state, resistivity is generally low enough to permit extensive long cell action. Permeability is important in that it determines how fast the soil will drain, and therefore how long the soil will have from 50 to 95 percent water saturation. So long as a soil is 100 percent water saturated, it provides practically no cathodic activity. But such a soil does provide an anodic region relative to drained soil lying above it. Thus the corrosive nature of a soil is not absolute but depends on its water content and on the state of the surrounding soil.

Introduction

Corrosion of Steel in soil frequently has been blamed on such factors as low pH, electrolysis due to stray currents, bacterial action, or reactive chemicals in the soil. Though these factors may be important in some cases, they should be considered as exceptional rather than common causes of corrosion. A more general cause, differential aeration involving "drained soil," will be the primary topic of this article.

In the differential aeration mechanism, oxygen and water alone are responsible for severe corrosion. Objective of this report is to show the interrelationship of water saturation, resistivity, and draining characteristics and the effects of these variables on the corrosive nature of soil.

To show the behavior of oxygen and water in soil corrosion, some aspects of corrosion in aqueous solutions will be reviewed. The essential point is that, if differences in oxygen availability arise for any reason, long cell action will result.

Soil corrosion will be discussed as a special case of aqueous corrosion. In general, this report deals with well known phenomena, but in several cases the argument will be reinforced by experimental data obtained specifically to illustrate the point in question.

Two laboratory demonstrations are described in the appendix. These can be used to illustrate the pronounced cathodic activity that occurs in drained soil.

Review of Corrosion in Aqueous Solutions

The corrosion rate of steel in neutral aqueous solutions is determined by the availability of dissolved oxygen and by conductivity of the electrolyte.[1] Steel corrodes in an aqueous oxygen-containing electrolyte (natural pH) by local cell action, by long cell action, or by both local and long cell action. In local cell action, available oxygen is reduced at microscopic cathodic sites; iron is oxidied at adjacent microscopic anodic sites. The pH is not a rate determining factor in the range from nine to five.[2]

As conductivity of the electrolyte increases, long cell action becomes more and more possible. In long cell action, the cathodic and anodic sites or zones are separated by a macroscopic distance. The driving force for the long cell action is derived solely from the difference in availability of oxygen between the two zones and is independent of the means by which this difference is achieved. Conductivity of the electrolyte determines how far apart the zones can be for long cell action to occur; in high conductivity situations, the distance can be 100 feet or more.

The dependence of long cell action on differential availability of oxygen can be seen in a variety of corrosion phenomena. An anodic region will be located in a zone where salt content is higher than at a nearby zone; the high salt content is associated with lower solubility of oxygen.[2] An anodic region will be located in a zone where velocity is lower than at an adjacent region even when the bulk oxygen concentrations are the same at both regions. And an anodic region will be located in or under material of low permeability to water such as clay, wood, or rust. The cathodic region will be located nearby where the permeability is higher.

A characteristic of aqueous corrosion involving oxygen (neutral pH) is that insoluble corrosion products are formed. These act as barriers, shielding the anodic areas from oxygen. In this way, the corrosion products themselves stimulate further corrosion by long cell action. Corrosion by long cell action is inherently nonuniform.

Oxygen Mechanism of Soil Corrosion

Soil can be considered as an aqueous electrolyte; therefore corrosion of steel in soil for the most part is a special case of aqueous corrosion. The corrosion rate of a structure buried in water-saturated soil would be expected to be small because of the static nature of the soil. The corrosion rate in dry soil should also be small, in this case due to the high resistivity of the electrolyte. The fact that extremely high corrosion rates are observed in soils indicates

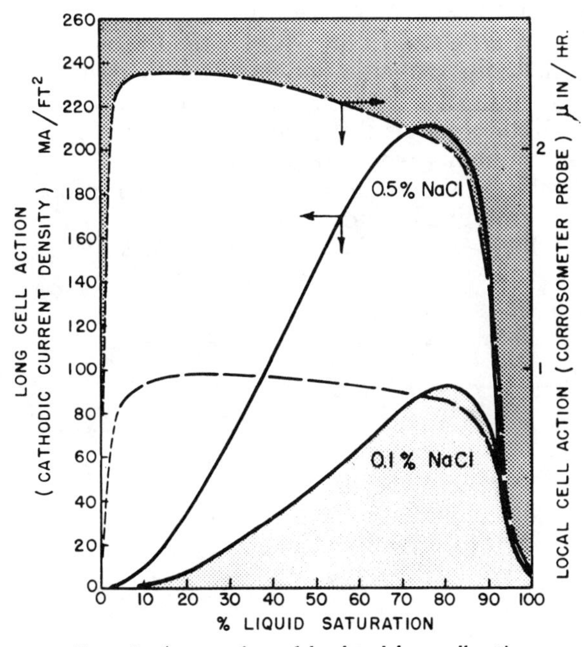

Figure 1—A comparison of local and long cell action as a function of liquid saturation and resistivity.

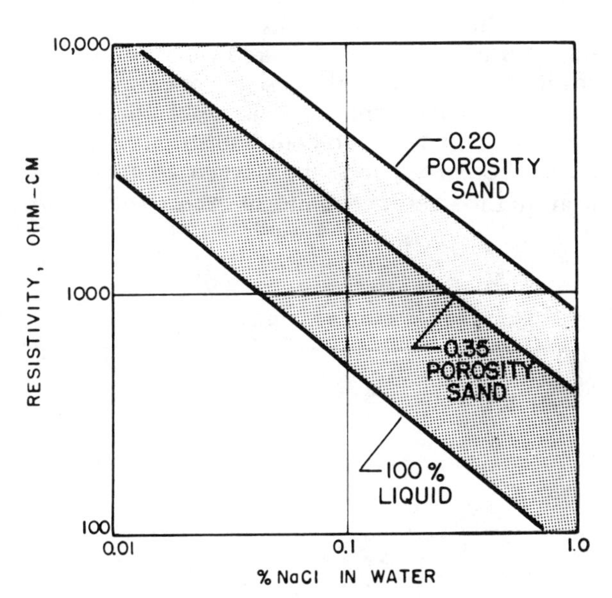

Figure 2—Resistivity of sand having the pore space filled with an electrolyte compared to the resistivity of the electrolyte itself.

that the concept of soil as a homogeneous static electrolyte is not always applicable.

In an effort to explain the high corrosion rates that sometimes occur in what appears to be a noncorrosive environment, exotic mechanisms such as electrolysis and bacterial action have been proposed, perhaps erroneously. Because soil is an aqueous electrolyte, its pH has essentially no effect on corrosion over the range from five to nine. The pH of almost all soils fall into this range. Even where a more acidic soil exists, static nature of the soil makes it difficult for fresh hydronium ions to be brought to the surface of the steel, and so corrosion by the hydrogen mechanism is throttled.

As in the case of aqueous corrosion, both local cell action and long cell action are possible in soil corrosion. Local cell action is not often severe because it results in fairly uniform corrosion. If a steel structure is buried entirely in a soil uniform in water content, corrosion is confined to that arising from local cell action.

In a practical case, there are gross variations in soil composition, not only on a small scale but also over large distances both vertically and horizontally. Even more important, soil can be nonuniform with respect to water saturation and permeability. Thus a steel structure penetrating soil of variable characteristics is exposed to different degrees of availability of oxygen. Long cell action can occur over fractions of an inch or over many feet, depending on the difference in aeration of the zones in contact with the steel and on soil conductivity. Long cell action results in nonuniform corrosion.

An interesting similarity exists between corrosion of steel in soil and corrosion of reinforcing steel in concrete, as discussed by Finley.[3] Often quite permeable to air and water, concrete permits entrance of both these agents, especially in a marine environment. Because of variations in permeability and also nonuniform salt deposition in the pore spaces, the oxygen content adjacent to the steel is nonuniform. As a result, long cell action occurs in a manner analogous to long cell action in soil corrosion.

Pitting in Soil

Pitting, as a result of long cell action, occurs when an impermeable component (e.g. a sand grain) contacts a clean steel surface in an environment which is sufficiently conductive and aerated. Ferrous ions under the sand grain are quickly converted to hydrous ferric oxide. This in turn acts as an adhesive, cementing the sand grain to the surface of the metal and at the same time further restricting the access of oxygen around the sand grain. The anodic area under the sand grain develops more intensely with time as this cementing process continues. In addition, there is a gradual build-up of hydrogen ions in the growing pit which further accelerates the corrosion process.

Long Cell Action Involving Drained Soil

In soils, no agitation is possible to aid the transport of oxygen to the cathodic surface, but the transport can be facilitated by evaporation or draining of the water in the soil.[4] As the water drains, a thin film is left on the surface of the steel exposed to soil. Air in the soil is now separated from the surface of the steel by this thin film of water. In contrast to aqueous solutions, where the diffusion film separates the steel from water that contains a few parts per million of dissolved oxygen, the diffusion film in soils separates the steel from the 20 percent oxygen of the air. Cathodic activity can therefore take place at an extremely high rate.

To explore the drained soil phenomenon experimentally, cathodic activity as a function of water saturation percentage was obtained for several different artificial soils as shown in Figure 1. Experiments were conducted by inserting a steel electrical resistance corrosion meter probe and a zinc anode and a steel cathode into the soil which was maintained at various water saturations. A recording zero-resistance ammeter between the zinc and steel electrodes measured cathodic activity that would be available to drive a long cell. The probe was left isolated, and its corrosion rate provided a measure of local cell action.

In a water-saturated soil (100 percent water saturation), both local cell action and long cell cathodic activity were negligible. As the soil was drained, local cell action and long cell cathodic activity sharply increased with decrease of water saturation. Oxygen of the air was now entering the pores and contacting the steel across the thin water film. By the time the water saturation had decreased to about 70 percent, cathodic activity was at a maximum. Further decreases in water content now introduced additional resistance to long cell action. Local cell action, however, was not affected by further decreases of water saturation until the soil was practically dry, at which point local cell action became negligible.

Implications of these results can be summarized as follows:

1. Long cell action can occur at water saturations between 50 and 95 percent. Local action can occur at water saturations between 5 and 95 percent.

2. Soil resistivity determines the intensity of local cell action at any given water saturation percentage.

3. A soil that does not drain at all but remains 95 to 100 percent water-saturated is noncorrosive regardless of its resistivity.

4. A soil that drains rapidly below the 50 percent water saturation level will sustain some long cell action while draining and also will permit some local cell action. However, such a soil will be relatively noncorrosive regardless of its resistivity.

5. On the other hand, a soil that drains slowly and therefore remains in the region from 50 to 95 percent water saturation for long periods will be corrosive,* especially if its resistivity is low.

6. If steel penetrates soil that is homogeneous except for its water content and if the upper part of the soil has drained to the 50 and 95 percent water saturation level while the lower part is still at 95 to 100 percent, long cell action will occur. Cathodic activity might be so high that no corrosion will occur in the upper, drained area. The observer, noticing intense pitting at the deaerated lower depth, might conclude that the soil is very corrosive in the region, whereas in fact the corrosion depends on the presence of the upper aerated zone.

7. The corrosion rate by long cell action is greatest where least expected on the basis of degree of aeration.

* Such a soil provides a high cathodic activity, and it promotes corrosion by long cell action if a suitable deaerated anodic area is contiguous. The anodic area might be in a nondraining soil as in Item 3.

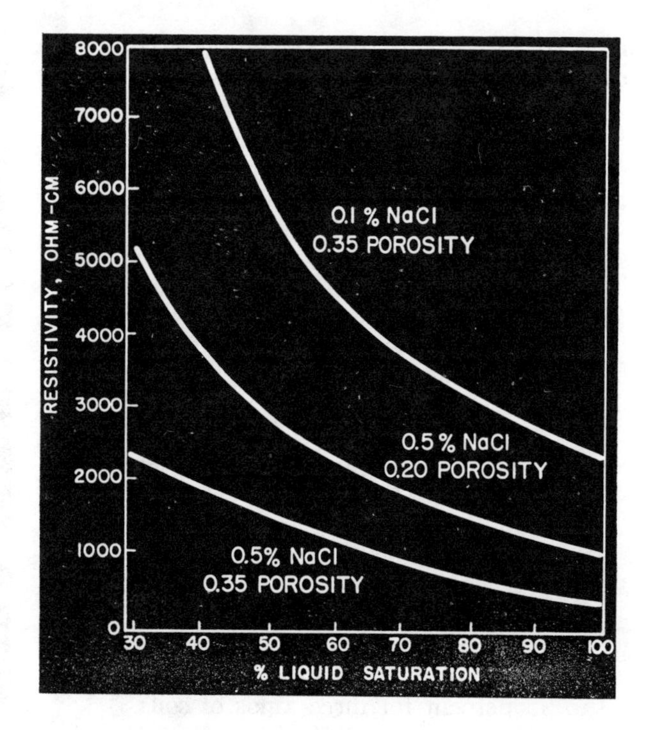

Figure 3—Chart showing soil resistivity as a function of water saturation. Curves indicate the change of resistivity with liquid saturation and porosity.

Resistivity and Draining Charactistics of Soil

From the foregoing sections, the significant factors in soil corrosion are resistivity and draining characteristics. In this section, the interrelationships of these factors will be discussed. For very dilute solutions, resistivity is a linear function of concentration of electrolyte, as shown in Figure 2.[5] In liquid-filled soils, the curves are displaced, the resistivity increasing as the porosity of the soil becomes smaller. Anything that can be done to decrease porosity will therefore increase resistivity.

Soil resistivity is also a function of water saturation, as shown in Figure 3.[6] Different types of soils exhibit different shapes of curves, but in each case the resistivity is higher in dry soil and lower in water-saturated soil. A soil having a high resistivity at intermediate water saturations shows sharp changes in resistivity with changes in water saturation. The distance over which long cell action operates in such a soil will be variable, extending to an appreciable distance at 95 percent water saturation or contracting to local cell action at 50 percent water saturation.

Capillary pressure of soil is just as important as its resistivity. Capillary pressure is expressed as the head of water existing above the free standing water table. Capillary pressure and water saturation are related to each other; with increasing distance upward from the water table (i.e. with increasing head, or capillary pressure) the water saturation decreases to give the curve of Figure 4. Shape of the capillary pressure curve illustrates water distribution above the water table for a particular soil. Coarse sand might yield a curve similar to Curve A while a clay might yield a curve similar to Curve C.[7]

The capillary pressure of a soil also is a measure of the soil's ability to imbibe water, i.e. to soak up by wick action.

Significance of the capillary pressure curve to corrosion in soils can be seen when one considers the critical water saturation range from 50 to 95 percent for three kinds of soil:

1. In a coarse sand with a flat curve as in Figure 4, Curve A, water saturation passes through the 50 to 95 percent zone at almost constant capillary pressure (height above the water table). In theory, steel exposed in such sand would be in contact with the critical water saturation at essentially one line. Cathodic activity would be intense along that line, but the area of metal available as a cathode would be small.

2. In a typical soil with a steep curve as in Curve C (Figure 4) there may be many feet of vertical distance above the water table in which the water saturation is between 50 and 95 percent. Here intense cathodic activity would be expected throughout this vertical distance if steel were exposed in such soil.

3. A practically impermeable clay soil may retain water near 100 percent saturation permanently unless it is allowed to dry out. Such a soil will not drain. Because no air can enter (barring evaporation of water), there is no cathodic activity. Steel exposed entirely to such soil will not corrode, for practical purposes. If the steel contacts a drained soil as well as the impermeable clay, however, the part of the steel in the clay will corrode as the anodic area because of long cell action. An observer finding bacteria in the corrosion products might be tempted to blame them for the nonuniform corrosion observed.

The corrosive nature of a soil is obviously not a fixed absolute value. Before one can talk about a corrosive soil or a noncorrosive soil, one must precisely define (1) the state the soil is in at different locations as points on Figure 1, (2) the orientation or placement of the steel with reference to the zones of various water saturations in the soils, and (3) the nature of the corrosion under consideration, i.e. uniform corrosion arising from local action or nonuniform corrosion arising from long cell action.

Effect of Permeability on Local Cell Action

For a sustained, high rate of corrosion by local cell action, water permeability of the soil must be appreciable.* An illustration of this point was obtained in experiments in which sand or artificial soil of known permeability was used as the corrosive medium for small corrosion probes. Corrosion vessels were constructed of four-inch

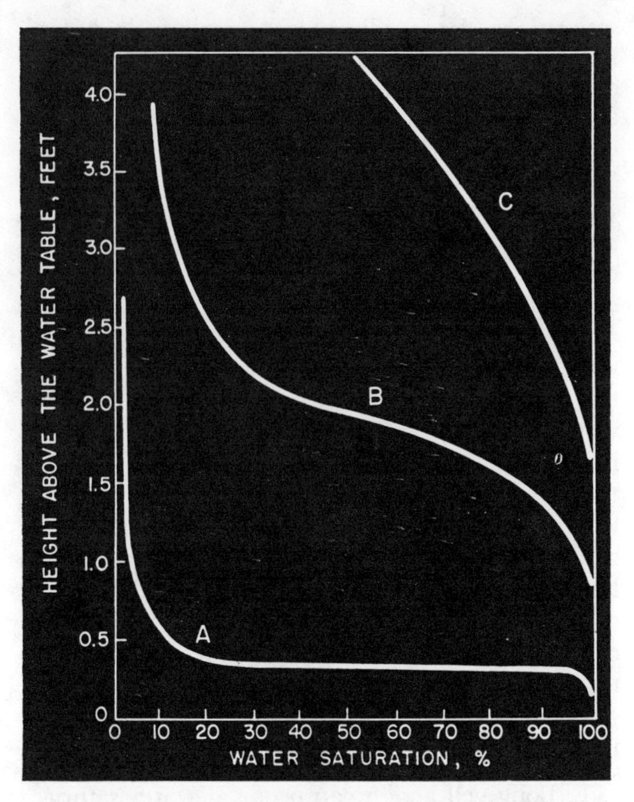

Figure 4—Distribution of water in soil as a function of height above the water table (capillary pressure curve).

lengths of two-inch diameter glass tubing. A corrosion probe was inserted vertically into the bottom stopper which also was fitted with a drain tube. After the probe was inserted, a dry sand mixture was packed into the tube with slight vibration. The vessel was evacuated and then saturated with the aqueous phase.

Permeabilities were determined by timing the flow rates that occurred with application of a known hydrostatic head. The sand mixtures then were drained by applying a vacuum of 20 centimeters of water to the drain tube. Some fine mesh sands and some clay-containing sands did not drain. Under conditions of poor drainage, corrosion rates were low and began to increase only when evaporation permitted air to enter the sand.

Corrosion of steel as a function of permeability in various grades of sand is shown in Figure 5. The sand had been initially saturated with one percent sodium chloride and then drained. To a permeability of about 20 darcys, the corrosion rate was fairly linear, increasing with permeability. Most soil, of course, falls in this permeability range.

The variation of local cell action with permeability (Figure 5) shows that as soils become well packed, less porous, the local cell action decreases. Corrosion due to local cell action should

* Permeability of soil is measured in units of darcys. One darcy is the permeability which will allow the flow of one cubic centimeter per second of fluid of one centipoise viscosity through an area of one square centimeter under a pressure gradient of one atmospheric per centimeter.

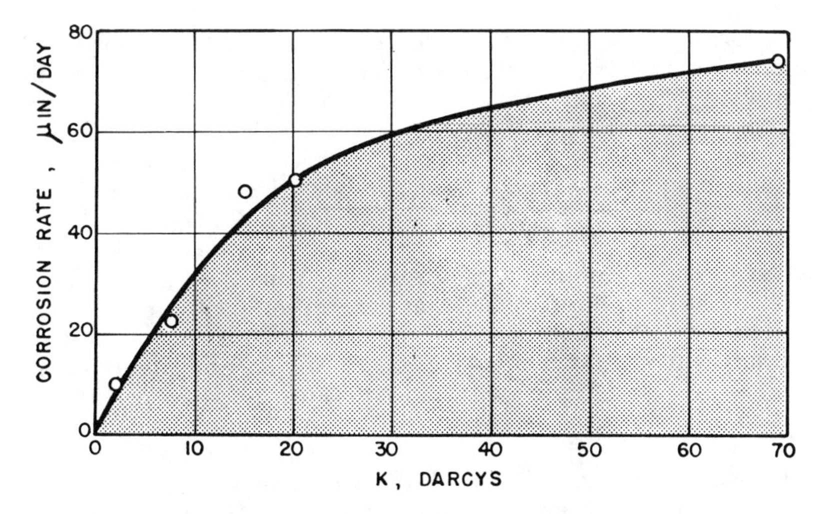

Figure 5—Curve showing corrosion rate of steel as a function of permeability in various grades of sand.

therefore decrease with depth of soil, and this is the case with steel pilings driven into undisturbed soil.[8] But the common experience in disturbed soil is that corrosion increases with depth.[9] Thus, metal loss in deeper soils must be dependent on long cell rather than local cell action. It is usual practice to measure the "corrositivity" of soil by means of soil boxes and small specimens buried in the soil, but such techniques would be expected to measure primarily local cell action.

Effect of Long Cell Action on Local Cell Action

Isolated steel corroding by local cell action in well aerated soil of low resistivity is a potential cathode for long cell action. If such steel is connected to steel in poorly aerated soil, long cell action will occur; the anodic steel will supply electrons to the cathodic steel. The net result is that the steel in the deaerated zone will corrode faster than if it is isolated, and it will cathodically protect the cathodic steel. Thus, if long cell action occurs under this circumstance, the local cell action corrosion rate at the aerated zone is suppressed. This and other effects of long cell action on local cells have been discussed by Pope.[10]

A pebble or a piece of wood in contact with steel in an aerated zone can set up a vigorous long cell in which the adjacent steel is the cathode and the steel area under the foreign object is the anode. With a low anode-cathode area ratio, corrosion rate at the anodic area can be extremely high, particularly if the drained soil mechanism is operating at the cathodic area.

Figure 6 shows the experimentally determined effect of anode-cathode area ratio on the corrosion rate of the anode and cathode of a long cell. Data for this figure were developed for steel in static aqueous solutions, but the principle (if not the numerical result) applies equally well to soils. Probes of various sizes were coupled in a split cell (cathode aerated, anode deaerated) to

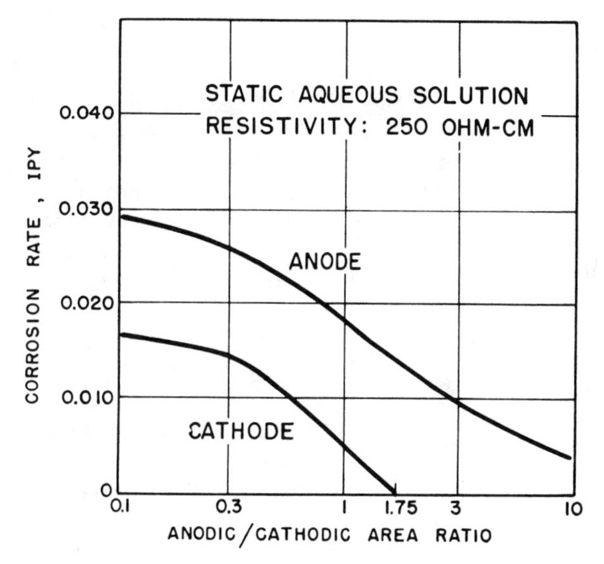

Figure 6—Effect of anode-cathode area ratio on corrosion rate of anode and cathode of a long cell, showing dependence on area ratio.

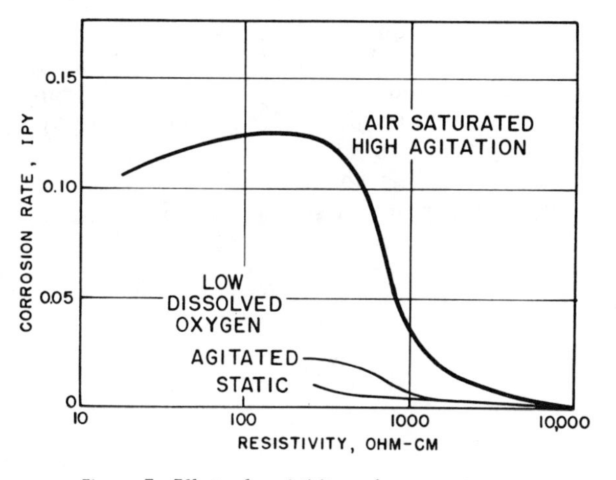

Figure 7—Effect of resistivity and oxygen transport for aerated and agitated solutions on corrosion rate of steel.

Source: *Materials Protection*, November 1963, 8-17

permit measurement of the long cell corrosion and currents. At low anode-cathode area ratios, the small anode corroded rapidly by long cell action, but the cathodic protection effect was not pronounced; therefore the cathode continued to corrode by local cell action. At a one-to-one area ratio, the anode corroded rapidly, but the cathode was small enough that the current density from the anode was almost enough to suppress local action. At an anode-cathode area ratio of 1.75 and greater (for the case studied), the cathode was completely cathodically protected.

A person observing a natural situation where such a ratio exists might conclude that the aerated zone is completely noncorrosive. Strictly speaking, this conclusion is correct, but it ignores the fact that corrosion is occurring nearby as a result of the presence of the noncorrosive zone.

Role of Soil Resistivity

Differences in availability of dissolved oxygen provide the driving force for long cell action in the form of an electromotive force. The maximum open circuit voltage that can be developed by differential aeration of steel is about 0.4 volt. If large areas are involved, even this low voltage can result in the flow of many amperes of current. Each ampere in turn results in a corrosion loss of about 20 pounds of steel per year from the anodic area.

Once the driving force has been estabilished by differential aeration as two separated zones on the surface of a metal, current flow will be determined by resistance in the electrolytic path. Figure 7 shows experimentally obtained local cell action corrosion rates for aerated, agitated solutions as a function of electrolyte resistivity. The observations on the effect of resistivity on local cell action also are applicable to long cell action because local cell action is merely a special case of long cell action. Below resistivity of about 200 ohm-cm in aqueous solutions, further reductions in resistivity have no further accelerating effect on corrosion rate. (The corrosion rate actually decreases slightly with decreasing resistivity because of lower solubility of oxygen in the more saline solutions.) At resistivities above about 400 ohm-cm, the corrosion rate becomes sharply dependent on resistivity.

The dependence of both local and long cell action on resistivity applies to soils as well as aqueous electrolytes. The greater the distance between anode and cathode, the weaker the long cell action. At a constant distance between the anode and cathode, the greater the resistivity, the weaker the long cell action.

So far, no quantitative treatment of long cell action has been made. Tomashov and Mikhailovsky state that penetration depth at the anode is proportional to current density at the cathode divided by the soil resistivity,[11] but the relationship would hold only at a given anode-cathode spacing and area ratio. These same authors believe that the variation of cathodic current density with distance is a measure of probability of long cell action. If this is the case, it should be possible to equate anodic current density (i.e. corrosion rate) with anode-cathode spacing and area ratio, resistivity, and some measure of oxygen availability at each reacting electrode. Such a relationship has not yet been developed.

Variation of Soil Corrosion With Time

Schwerdtfeger[12] has shown that the corrosion rate of a typical buried structure is not uniform with time but varies with soil conditions. In a general case, a long term, moderate average corrosion rate might result from a few weeks of extremely high rate preceded and followed by a long period of almost zero rate. If the conditions are such that the drained soil mechanism can proceed for a long time, the corrosion rate could be sustained at 0.1 ipy or higher, thus leading to rapid failure of the buried structure. Depending on water content, the same soil at different times can be inert or extremely corrosive.

Conclusion

When steel contacts soils of different degrees of aeration, long cell action will result. The greatest corrosion takes place where the soil contains the most water and the least air, contrary to what one might expect.

The most vigorous long cell action occurs when the aerated zone has been water-saturated and then partially drained to 50 to 95 percent water saturation. The less aerated zone can be provided by the same soil at a lower depth where water saturation is from 95 to 100 percent. The smaller the area of steel in the deaerated zone, the more intense the corrosion.

The same soil can be inert or extremely active in promoting corrosion, depending on its water saturation and on orientation or placement of steel relative to the zones of different water saturation.

APPENDIX

Simple Demonstrations of Drained Soil Mechanism

The drained soil phenomenon was demonstrated experimentally in an earlier article.[4] A glass vessel was filled with sand, which in turn was saturated with three percent sodium chloride solution. Corrosion meter probes coupled through a zero reisitance ammeter showed that vigorous long cell action took place when one probe was located in a drained portion of the sand while the other probe was located in a liquid-full, static portion. The probe in the liquid-full zone corroded rapidly.

A demonstration of the drained soil mechanism can be made with the simpler apparatus of Figure 8, in which a vertical series of small steel coupons is placed in a tall jar of 60 mesh sand saturated with a salt solution. The jar is arranged with a siphon tube so that it can be drained. An insulated wire previously soldered to each coupon connects each coupon to a switch box. All the coupons normally are kept short circuited to each other. An ammeter (a zero resistance ammeter gives the best results) is inserted by means of the switch box successively between adjacent coupons. As the sand is drained, the water table is easily detected as the point above which the coupons are cathodic and below which they are anodic. Quantitative distribution of current can be made with this apparatus, as shown in Figure 9. The effects of resistivity, permeability, and time on long cell action are easily measured.

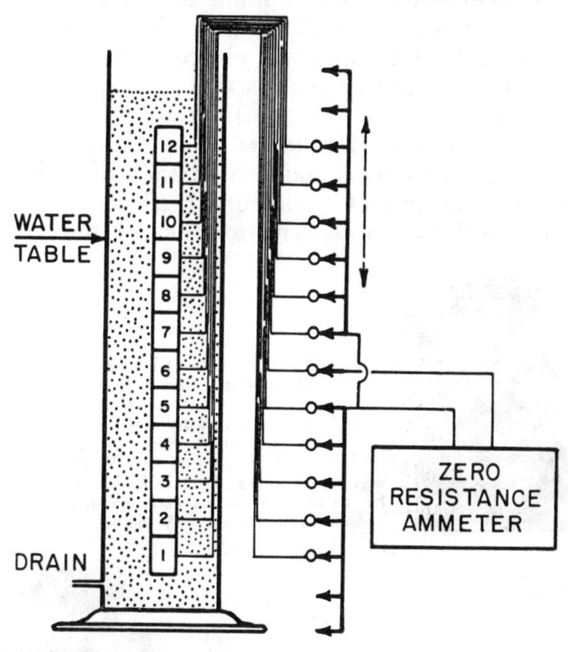

Figure 8—Apparatus for quantitatively measuring long cell currents in drained sand. Vertical series of steel coupons (numbered 1 through 12) are placed in a jar with 60 mesh sand saturated with a solution.

A visual, qualitative version of this demonstration can be set up as in Figure 10, using a two-inch inside diameter glass tube as a vertical container for 60 mesh sand. The tube can be

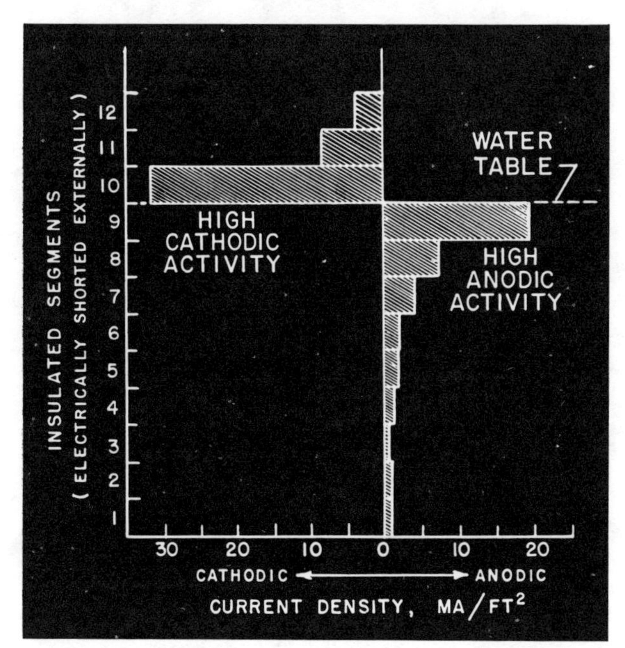

Figure 9—Results of anodic and cathodic current measurements by apparatus shown in Figure 8.

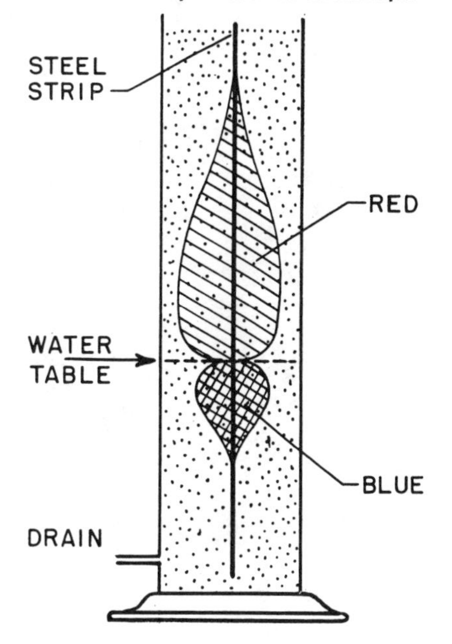

Figure 10—Glass tube demonstration to show distribution of anodic and cathodic areas in a drained sand as determined by a colorimetric technique.

any length over about one foot. A drain tube is placed at the bottom, with a filter paper plug to prevent sand from running out. While the sand is dry, a clean piece of steel shim stock (1 by 12 by 0.01 inch) is placed vertically in the tube so that one edge is visible in the glass tube. Next, a one percent solution of sodium chloride in distilled water is prepared. To 500 cc of this solution is added 3 cc each of a five percent solution of potassium ferricyanide in water and a two percent solution of phenol-phthalein in alcohol. The sodium chloride ferricyanide-phenol-phthalein mixture then is poured into the tube to saturate the sand, then drained so that the liquid level is a few inches from the bottom of the steel strip. As corrosion proceeds, the sand above the water table is stained red by the hydroxyl ions formed in the cathodic reaction. Sand below the water table is stained blue by the presence of ferrous ions which react to form ferrous ferricyanide. In 12 hours, the colors are quite striking. By repeating the experiment with higher and lower concentration of sodium chloride, the effect of resistivity on long cell action can be observed.

G. A. MARSH is a research associate at Pure Oil's Research Center. His work includes corrosion problems pertaining to the petroleum industry and development of corrosion testing methods. He has an MS in chemistry from Northwestern University. An NACE member for 10 years, he is the corrosion division editor for the ECS Journal.

EDWARD SCHASCHL is a technologist at Pure Oil's Research Center, where his work includes corrosion control in all phases of the petroleum industry. He has a BS in chemical engineering from Illinois Institute of Technology and has been an NACE member for 10 years.

References

1. U. R. Evans. Corrosion and Oxidation of Metals. St. Martin's Press, Inc., New York, N. Y., 1960, p. 128.
2. E. Schaschl and G. A. Marsh. Concentration Cells and Aqueous Corrosion. *Corrosion,* **16,** 461t (1960) September.
3. H. F. Finley. Corrosion of Reinforcing Steel in Concrete in Marine Atmospheres. *Corrosion,* **17,** 104t (1961) March.
4. E. Schaschl and G. A. Marsh. Effect of Dissolved Oxygen on Corrosion of Steel and on Current Required for Cathodic Protection. *Corrosion,* **13,** 243t (1957) April.
5. I. C. Calhoun, Jr. Fundamentals of Reservoir Engineering. University of Oklahoma Press, 1953, pp. 135-138.
6. *Ibid.*, pp. 139-140
7. M. C. Leverett. *Trans AIME,* 142, 152 (1941).
8. M. Romanoff. Circular C579, National Bureau of Standards. Underground Corrosion. U. S. Government Printing Office, Washington 25, D.C., p. 70.
9. M. Romanoff. *NBS Journ Research,* 66C, 223 (1962).
10. R. Pope. Effect of Long Cell Action on Local Cells. *Corrosion,* **12,** 169t (1956) April.
11. N. D. Tomashov and Y. N. Mikhailovsky. Corrosivity of Soil. *Corrosion,* **15,** 77t (1959) February.
12. W. J. Schwerdtfeger. *NBS Journ Research,* 65C, 271 (1961).

PART OF RACKS at ASTM's exposure test at Freeport, Texas, on the Gulf Coast. Cast and wrought aluminum and magnesium alloys are being exposed (see References 6 and 7). Two rows of cast tension test bars are shown. These are a portion of some 8000 specimens placed on exposure at this test site alone. Five test series including aluminum and stainless steel alloys, hot-dipped aluminum wire and sheet, and sprayed aluminum panels are included.

Report on

CORROSION and DETERIORATION STUDIES*

conducted by technical committees of the American Society for Testing and Materials

For Over 55 Years, corrosion has been a matter of concern in a larger percentage of American Society for Testing and Materials (ASTM) activities than any other subject within its broad scope of knowledge of materials. Over $2,000,000 have been spent by ASTM member companies to develop, cooperatively, unbiased engineering data on the effects of corrosion, according to a conservative estimate. Presently, there are 34 engineering research projects on exposure at one or more of the 18 ASTM outdoor test sites across the USA.[1] Some of these have been on exposure for periods to 35 years. A few ASTM exposures also are being conducted as private test sites.

Aluminum and Aluminum Coatings

With the new exposures of aluminum coated steel wire and fencing at seven sites sponsored by ASTM Committee A-5 on Corrosion of Iron and Steel, there are now 15 ASTM research programs designed to obtain exposure data of aluminum in one form or another. The oldest exposures of aluminum coatings are those on ferrous hardware of various shapes exposed in 1929 at State College, Pa. No evidence of rust has been reported there.[2] Evidences of rust were reported on similar hardware specimens whose exposure had to be terminated in 1952 at Key West, Fla., and Sandy Hook, N. J.[3] Exposures at Altoona and Brunot Island (Pittsburgh), Pa., had to be discontinued before significant results appeared.[4] A new series of aluminum-coated steel hardware specimens was exposed in 1958.[5] The one-year exposure data have been collected and published.[2] Aluminum coated steel roofing

★Submitted for publication October 30, 1961.

Samuel F. Etris
Staff Secretary
ASTM Advisory Committee on Corrosion
Philadelphia, Pennsylvania

SUMMARY

Survey is given of engineering research projects under way on corrosion and deterioration studies by ASTM. Included are aluminum and aluminum coatings, stainless steel, nonferrous metals and alloys, hot-dip galvanizing and other coatings, electroplated coatings on steel, malleable iron, die-cast metal tests, galvanic corrosion, electrical contacts, atmospheric factors, permanence and corrosivity of plastics, and organic products and coatings. Other projects discussed are water and corrosion products, and exposure and deterioration of wood. Documented references are given to published works.

sheets were first placed on exposure last year.[2]

The most comprehensive engineering research program covering uncoated aluminum alloys was started in 1953 by ASTM Committee B-7 on Light Metals and Alloys, Cast and Wrought. This massive study included 27 aluminum and 7 magnesium alloys, having a total of 10,376 specimens. The 1961 report completes tensile and yield strength and elongation data for five years of exposure.[6,7] These data are of particular interest in the selection of alloys for long-term durability and to establish safe working limits in construction applications.

Samples of aluminum die-casting alloy SC84A (ASTM Specification B 85-60)[8] were prepared with additions of zinc in increments of 0.25, 0.50, 1.0, 1.5 and 2.0 percent. These were cast into bars and exposed at industrial and severe marine locations. Six-year data show that Alloy SC84 undergoes precipitation hardening as evidenced by the increased yield strength in all aged alloys and by the loss of ductility. Alloy SC84 with zinc content from 0.25 to 3.0 percent without any corrosion protective treatment withstands the New York City environment without any appreciable loss of tensile properties. These data also show that the SC84 alloy with zinc content of 0.25 percent affords the best resistance to atmospheric exposure, but that there is no significant difference in resistance to corrosion between alloys in which the zinc content ranges between 0.5 and 3.0 percent.[9]

Two new studies in preparation in this field include specimens for a research project to determine the efficacy of new decorative copper-nickel-chromium electrodeposited coatings on aluminum. The atmospheric corrosion resistance of silver, copper-silver and tin platings on bus bars and accessory electrical hardware is to be investigated in a new project proposed by ASTM Committee B-3 on Corrosion of Nonferrous Metals.

Aluminum also is included in many other programs of broader character that will be mentioned later.

Stainless Steel

The largest non-proprietary exposure study of stainless steels was sponsored by ASTM Committee A-10 on Iron-Chromium, Iron-Chromium-Nickel and Related Alloys in 1958. Over 4000 spot-welded, arc-welded, tension and Erickson cup panels and tension test bars, and 576 wire coils that are static and tension loaded for fatigue testing were exposed at six test sites. A preliminary series of the coils was tested in fatigue after one-year exposure to establish fatigue trends when compared with the three-year specimens that were picked up in 1961.

The seventh inspection of actual installations of stainless steels used in architectural and structural applications in New York, Philadelphia, Chicago, Pittsburgh, Cleveland and Miami appeared this year. This inspection included buildings incorporating AISI Types-202, 302, 316 and 430 stainless steel, the latter near the seacoast, that had not been included before.[10]

Nonferrous Metals and Alloys

In 1931, ASTM Committee B-3 on Corrosion of Nonferrous Metals and Alloys planned an atmospheric exposure program on a vast scale. Over 9000 specimens of 9 by 12 inch panels and tension bars of 24 alloys of aluminum, copper, nickel, lead, tin and zinc were exposed at 12 sites. This program was completed after 20 years' exposure at seven sites.[11]

In the years following the above study it became evident that corrosion data on many new alloys were needed. Therefore, 77 new alloys and high purity metals were chosen and exposed in 1958 at four sites for periods to 20 years.[12] A set of the aluminum alloys in this program was exposed for one year, and a report covering these data has been published.[13]

Hot-Dip and Other Coatings

One of the outstanding reports on corrosion for 1961 was the publication of the "Twenty-Year Atmospheric Corrosion Investigation of Zinc-Coated and Uncoated Wire and Wire Products" by Fred M. Reinhart,[14] National Bureau of Standards, Chairman of ASTM Subcommittee xV on Wire Tests of Committee A-5, and Chairman of the NACE Editorial Review Committee. This report summarizes corrosion data collected over a 20-year exposure including zinc, copper, and lead coated wires at 11 points in the USA.

A sampling of the 21 conclusions shows that the life of zinc coating is proportional to the weight of coating and is linear with time (though at entirely different rates at different locations), and is also independent of diameter of wire. Copper contents of 0.20 to 0.25 percent were no more effective in improving corrosion resistance of the wires than were contents of 0.05 to 0.08 percent. The strengths of the copper and lead coated and stainless steel wires were not significantly affected over this period.

A report on the corrosion of various types of lead, zinc and aluminum coatings on hardware over a 29.6-year period also appeared in 1961. Galvanized roofing sheets at State College, Pa., exposed 32 years ago recorded two new failures (100 percent rust).[15] The last roofing sheet at Altoona failed in 1958.[16]

Electroplated Coatings on Steel

ASTM Committee B-8 on Electrodeposited Metallic Coatings and Related Finishes has intensively studied the many variables that influence the life of decorative coatings on steel. The latest study, which is currently under way, began in 1960. Identified as Program No. 5, the objective of this work is to determine the influence of thickness and types of chromium when plated over duplex type nickel coatings. A report was published in 1961 on plated steel panels exposed over a three-year period to determine the effect of nickel and copper strikes under buffing bright nickel, the effect of type and thickness of chromium, and the effect of bright nickel (Program No. 4).[17] A final report on an investigation of chromate conversion films for electrodeposited zinc coatings on steel also appeared.[18]

Specimens are being prepared for a new exposure program that includes duplex nickel coatings on zinc-based alloy die-castings used in the automotive industry.

A new specification for multi-layer nickel plus chromium on steel (ASTM B 375)[19] has been issued to permit standardization of duplex nickel coatings for outdoor service. Also just published is a series of 207 definitions of terms for electrodeposited metallic coatings and related finishes (ASTM B 374).

ASTM Committee B-8 is developing a standard system of designating plated coatings that will permit labeling of coatings to ensure consumer quality depending on service requirements.

Malleable Iron

The paucity of reliable corrosion data on various production grades of standard (ferritic) malleable and pearlitic malleable irons and their relation to data on the corrosion of steel has long been recognized. Efforts to fill this gap came to fruition in 1958 with the exposure of 1620 specimens of standard malleable, pearlitic, nodular, heat-treated nodular irons, copper-bearing steel, Cor-Ten steel, and AISI 1020 steel at five ASTM test sites for periods to 12 years. One-year exposure weight-loss data published,[20] and the three-year data are being prepared. Some conclusions from the one-year data show that copper additions of about 0.25 percent to steel and 0.5 percent to malleable iron effectively retarded corrosive attack in atmospheres that would ordinarily cause high weight loss in the unalloyed metal. The more highly alloyed steel containing approximately 0.5 percent of both copper and nickel and 1 percent chromium were more resistant to all environments than the mild steel and the copper-bearing steel. Nevertheless, in general, it was not as resistant to environmental corrosion as were the majority of unalloyed cast malleable and nodular irons.

Die-Cast Metal Tests

The need for corrosion data to support the first proposed specifications on light metal die-castings prompted the 1929 study that encompassed 55,000 specimens (20,000 were used for preliminary data and 35,000 were exposed) of 12 aluminum and 9 zinc-base alloys to determine changes in physical properties over exposure periods to 15 years.[21] Data from this study pointed up many changes in then current formulas, and the new alloys were subjected to 20-year exposure at five test sites in the USA and the Canal Zone. In this test, zinc alloys AC41A and AG40A (ASTM Specification B 86), AC43A (obsolete as a result of this study), magnesium alloy AZ91 (ASTM Specification B 94), and three magnesium alloys now obsolete were used. Two reports just issued[22] cover the change in tensile strength, elongation, hardness and Charpy-impact over 20 years' exposure.

Galvanic Corrosion

In 1950, Committee BO3 began a three-part program to develop a standard galvanic couple for atmospheric exposure studies. The first was a disk or 'washer' type,[23] but the results were not satisfactory due to a small exposed area resulting in small losses in weight.

The second study used a threaded bolt wound with a wire of dissimilar metal that gave rapid results.[24] The relative weight loss here was a large percentage, but the specimens did not lend themselves to tension tests.

The third study used a 4 by 6-inch panel with a 4 by ¾-inch panel bolted to it and was designed to have a large crevice loss and its effects measured by tension test specimens cut from the panels. The first year's exposure data from the third study using two magnesium alloys coupled to aluminum alloys, stainless steels, mild steel, monel and brass have been published.[25] The rate of galvanic attack was far greater than expected, and the results indicate the superiority of the panel type specimen. The panels at Kure Beach (80-foot site), N. C., have corroded at such a rate that they will all be withdrawn in less than a three-year period, none lasting the anticipated 10 years. Of particular interest is the publication of a Signal Corps study to develop information to help in the selection of metals, dissimilar metal couples, and protective systems in military communication and associated equipment as a result of a high incidence of corrosion failure of such equipment under environmental conditions of coastal installations and beachheads. The Pitman-Dunn Laboratories designed this study to evaluate galvanic couples of magnesium in electronic assemblies and various protective systems.[26] This study is an excellent supplement to earlier work of ASTM Committee B-3.[23]

Electrical Contacts

ASTM Committee B-4 on Metallic Materials for Thermostats and for Electrical Resistance, Heating, and Contacts has guided the construction of two shelters to contain a series of 720 wires for making 360 static contacts and 100 foils for film growth measurements. The contact metals include copper, beryllium-copper, and silverclad, gold-clad, nickel-clad, and 60-40 tin-lead, clad copper. Foils for film growth measurements include copper, brass, beryllium-copper, tin-lead, silver, nickel, gold and palladium. In addition, each of the two shelters contains certain optional materials that are of interest to specific manufacturers. These include alloys of copper, aluminum, brass and stainless steel plus a variety of single and sandwich plate contact materials. A report on change in resistance of these contacts over a six-month period is in preparation.

The presence of exceedingly small quantities of corrosion products may also significantly degrade the performance of modern electronic devices. Detection of such quantities is being studied intensively in ASTM Committee F-1 on Materials for Electron Tubes and Semiconductor Devices. This committee has sponsored two symposia on the general subject of control of contaminants on surfaces. In the most recent symposium, a paper was given which described a new control technique in which the electrode potential may be used to characterize both the condition of the surface and the bulk purity of the metal.[27] In another paper, studies on electromotive forces are being used to determine the effectiveness of passivating solution used with acid etchants for cleaning iron-nickel alloy electronic parts.[28]

Atmospheric Factors: Apparatus and Exposures

The major factors in inland atmospheric corrosion are moisture, sulfur dioxide and temperature. Marine atmosphere carries chlorides, and the proximity of industry may add other pollutants. Simple effective recording apparatus to measure these factors has been developed by the National Research Council of Canada. ASTM Committee B-3 placed this continuous recording apparatus to measure time-of-wetness,[29,30] sulfur dioxide[31] and temperature at four US sites. The National Council of Canada placed apparatus at three Canadian sites with adjacent specimens of steel and zinc. Placed on exposure in January, a number of specimens was withdrawn each month over a six-month period. Comparison is being made of recorded data and corrosion of the specimens. There appears to be some question about the time-of-wetness measurements at Kure Beach, N. C. High salt concentrations may be affecting the recorded data.

An interesting slant on the role of atmospheric factors at test sites is the determination of relative corrosivity in terms of the corrosion rates of steel and zinc. This approach was instigated by C. P. Larrabee and O. B. Ellis. Under the sponsorship of ASTM Committee B-3, the relative corrosivity of 19 test sites in the US and Canada was established.[32] So much interest in this work was created as a result of this study that a new program was started in 1960 to tie in 39 sites in Canada, England, the Philippines and the US.

Permanence and Corrosivity of Plastics

Determining the corrosivity of molded plastics using a standardized galvanic corrosion cell is being developed through cooperative testing. Results indicate that a recommendation for publication may be made in the near future by ASTM Committee D-20 on Plastics. In this method, a potentiometer registers a galvanic current between a silver plated cathode and a cadmium plated anode which are placed between layers of a filter paper wick covering the plastic and dipping into the distilled water.

A recommended practice for determining light dosage in artificial weathering apparatus has been developed for plastics; final approval by ASTM awaits availability of a supply of standard fading chips. This method has been backed by considerable research effort and has generated considerable interest in this country and abroad.

ASTM Committee D-20 on Plastics has initiated a project to determine if existing pressure tests (ASTM: D 1599) for polyvinyl chloride (PVC) plastic pipe (ASTM Specification D 1785)[19] can be applied also to reinforced plastic pipe used for transporting oil well brines and in chemical processes. A new dimensional schedule and additional performance requirements are being prepared for inclusion in this specification. A draft specification for PVC fittings, both screwed and socket type, is being completed. General agreement has been reached on a specification for polyethylene pipe which is now in process.

There is a great deal of interest in the development of long term strength and working stresses for plastic pipe; one of the problems is the lack of a reliable formula that relates hoop stress to internal pressure. General formulas have been proposed, but none seems entirely satisfactory as yet.

Committee D-20 is considering the biaxial method of evaluating environmental stress cracking. Preliminary results are presented of an experiment involving stress level, environment concentration, temperature, resin density and resin viscosity. Work is also planned on the Lander Stress-Rupture (uniaxial) apparatus.

Organic Products and Coatings

A new method of classifying ferrous surfaces for painting utilizes photographs of steel surfaces cleaned to different degrees of freedom from rust and corrosion products. The results of a questionnaire to interested parties will be used as a basis for the forms in which the standard photographs will appear in the final method.

Another new method in a series of metal preparation methods that include aluminum (ASTM D 1730, D 1733), magnesium (ASTM D 1732) and hot-dip aluminum (ASTM D 1731) surfaces[19] is the preparation of galvanized surfaces for painting. A method for evaluating weathered paint films by infrared analysis technique is being discussed.

A new project covering industrial pitches in ASTM Committee D-8 on Bituminous Materials for Roofing, Waterproofing and Related Building or Industrial Uses has begun cooperative tests to characterize methods of test for softening point, specific gravity and benzene and quinoline insolubles. Plasticity of petroleum-coke-pitch mixtures and methods for ash and coking value are also being developed.

A new Recommended Practice for Atmospheric Exposure of Adhesive-Bonded Joints and Structures (ASTM: D 1828-61T) has been published and explains in detail the procedure for atmospheric testing of this type of material. Specifications have just been published for laminated-wall bituminized fiber drain sewer pipe (ASTM: D 1862) and homogeneous bituminized fiber drain and sewer pipe (ASTM: D 1861) for a variety of uses including industrial waste water.

A new procedure for the detection of the corrosiveness to copper of liquefied petroleum gases (ASTM: D 1838) is available. The test uses the same copper standards as does the 40-year old general petroleum test Method D 130.[19]

Water and Corrosion Products

A new activity of ASTM Committee D-19 on Industrial Water is the development of improved methods for analysis of water formed deposits. Exploratory work is under way to develop a qualitative identification scheme followed by quantitative tests. An X-ray method for deposits is being cooperatively developed.

Tests have been issued for acidity and alkalinity (ASTM: D 1884), chloride ion (ASTM: D 1885), nickel (ASTM: D 1886), suspended and dissolved solids (ASTM: D 1888), turbidity (ASTM: D 1889), hexane extractable matter (ASTM: D 1891), beta particle radioactivity (ASTM: D 1890), and sodium in high-purity industrial water (ASTM: D 1887).[19]

Exposure and Deterioration of Wood

A new specification for pressure treatment of timber products (ASTM: D 1760) has been issued to define treatment of lumber, timbers, bridge ties and mine ties by pressure processes covering 18 species of wood. A method of evaluating wood preservatives by field tests with stakes (ASTM: D 1758) exposed in the ground to the action of wood destroying fungi and insects also has been published. The method describes a preliminary bioassay to prove the presence of decay producing organisms in the soil test plot. The method of testing wood preservatives by laboratory soil-block cultures (ASTM: D 1413)[19] recently was adopted by the National Woodwork Manufacturer's Association in place of their previously used method. An accelerated laboratory method of evaluating natural decay in wood is being considered by ASTM Committee D-7 on Wood.

ASTM Committee D-7 is studying a procedure for evaluating the deterioration of wood piles exposed to sea water by a standard rating system.[33] Plans have been made to develop methods for field evaluation of preservatives in posts and accelerated methods of evaluating wood preservatives against marine borers.

References

1. Report of Advisory Committee on Corrosion. *ASTM Proc*, 58, 229 (1958).
2. Report of Committee A-5 on Corrosion of Iron and Steel. *ASTM Proc*, 61, (1961).
3. *Ibid.*, 52, 118 (1952).
4. *Ibid.*, 44, 92 (1944).
5. *Ibid.*, 58, 116 (1958).
6. L. H. Adam and M. Dougherty. Atmospheric Exposure of Aluminum and Magnesium Sand and Permanent Mold Castings. *ASTM Proc*, 60, 186 (1960).
7. L. H. Adam and M. Dougherty. Atmospheric Exposure of Wrought/Aluminum and Magnesium Alloys. *ASTM Proc*, 61, (1961).
8. 1961 Book of ASTM Standards, Part 2.
9. Exposure Tests of SC84A Aluminum Die-Cast Test Bars With Varying Zinc Contents. *ASTM Proc*, 60, 165 (1960).
10. Report of Task Group on Inspection of Corrosion Resistant Steels in Architectural and Structural Applications. *ASTM Proc*, 61, (1961).
11. Report of Subcommittee VI of Committee B-3 on Atmospheric Corrosion. Symposium on Atmospheric Corrosion of Non-Ferrous Metals. ASTM STP No. 175, p. 3 (1955).
12. Report of Subcommittee VI on Atmospheric Corrosion, 1957 Test Program. *ASTM Proc*, 59, 176 (1959).
13. Report of Subcommittee VI on Atmospheric Corrosion, One-Year Test Data of the 1957 Test Program. *ASTM Proc*, 61 (1961).
14. F. M. Reinhart. Twenty-Year Atmospheric Corrosion Investigation of Zinc Coated and Uncoated Wire and Wire Products. ASTM STP No. 290 (1961).
15. Report of Subcommittee XIV on Field Tests of Atmospheric Corrosion of Metallic Coated Steel Panels, 1960 Program. *ASTM Proc*, 61 (1961).
16. Report of Committee A-5 on Corrosion of Iron and Steel. *ASTM Proc*, 60, 111 (1960).
17. Performance of Decorative Nickel-Chromium Coatings on Steel (Program No. 4). *ASTM Proc*, 61 (1961).
18. Exposure and Accelerated Tests of Supplementary Chromate Treatments for Electroplated Zinc Coatings (Program No. 101). *ASTM Proc*, 61 (1961).
19. 1961 Book of ASTM Standards, Parts 2, 5, 6, 7, 8 and 10.
20. Corrosion Test Results on 15 Ferrous Metals After One-Year Atmospheric Exposure. *ASTM Proc*, 61 (1961).
21. Report of Committee B-6 on Die-Cast Metals and Alloys. *ASTM Proc*, 46, 244 (1946).
22. Atmospheric Exposure Tests Conducted on Die-Cast Test Bars of Four Magnesium Alloys and Atmospheric Exposure Tests Conducted on Zinc Alloy Die Castings AC43A, AC41A and AG40A. *ASTM Proc*, 61 (1961).
23. H. O. Teeple. Atmospheric Galvanic Corrosion of Magnesium Coupled to Other Metals. Symposium on Atmospheric Corrosion of Non-Ferrous Metals. ASTM STP No. 175, p. 89 (1955).
24. Spool and Wire Couple Tests—Part 2 of a Three-Part Program. *ASTM Proc*, 55, 150 (1954).
25. Magnesium Plate-Type Galvanic Couples—Part 3 of a Three Part Program. *ASTM Proc*, 61 (1961).
26. A. Gallaccio and I. Coronet. Report on Marine Atmospheric Exposure of Galvanic Couples Involving Magnesium. ASTM STP No. 255, (1960).
27. D. G. Schimmel. Detection of Inorganic Contamination on Surfaces by an EMF Measurement. Materials and Electron Device Processing. ASTM STP No. 300, p. 46 (1961).
28. D. O. Feder, et. al. Electrode Potential: A Tool for Control of Materials and Processes in Electron Device Fabrication—Part 1: EMF-Time Studies of Clean and Contaminated Electrodes, *Ibid.*, p. 53.
29. P. J. Sereda. Measurement of Surface Moisture. *ASTM Bulletin* No. 228 (TP59) 53 (1958) February.
30. P. J. Sereda. Measurement of Surface Moisture—Second Progress Report. *ASTM Bulletin* No. 238 (TP59) 61 (1958) May.
31. P. J. Sereda. Measurement of Surface Moisture and Sulfur Dioxide Activity at Corrosion Sites. *ASTM Bulletin* No. 246 (TP107) 47 (1960) May.
32. Report of Subcommittee VII on Corrosiveness of Various Atmospheric Test Sites as Measured by Specimens of Steel and Zinc. *ASTM Proc*, 59, 183 (1959).
33. E. V. Dockweiler and H. E. Stover. Evaluation of Wood Piles in Sea Water by Standard Rating System. Symposium on Treated Wood for Marine Use. ASTM STP No. 275, p. 30 (1959).

Causes of Underground Corrosion

By **BERNARD HUSOCK**, PE, Chief Engineer, Harco Corp., Medina, OH

In any discussion of corrosion of underground iron or steel structures, a number of basic truths must be understood and accepted:

- Corrosion of iron and steel is a natural process. Underground corrosion of iron and steel pipes is often viewed as an unusual condition that occurs as a result of unusual circumstances and environments. The question that is often asked is, Will corrosion occur? In reality, corrosion should be expected to occur whenever iron or steel is placed underground. Because iron is not stable in its refined state, it can be expected to become iron oxide (rust) eventually. The energy that was imparted to the metal during the refining process seeks to be released, thus allowing the metal to revert to the ore from which it was derived. Therefore, the question to ask is not, Will corrosion occur? but rather, At what rate will it occur? To put the question in practical terms, How long will it be before the first leak occurs, or before the pipe must be replaced?
- All ferrous metals corrode essentially at the same rate. Tests were performed by the National Bureau of Standards at more than 150 sites nationwide for more than 50 yr. As reported in "Underground Corrosion," Circular C-579, by Melvin Romanoff, these tests show that ferrous metals, including cast iron, carbon steel, wrought iron, and ductile iron, corrode at essentially the same rate underground. The apparent corrosion resistance of cast iron pipe is attributed to the fact that graphitized cast iron can retain its appearance as a pipe even though much of the iron is gone.
- Corrosion is selective and concentrated. The basic corrosion mechanism of iron underground is electrochemical. Furthermore, the corrosion occurs only at anodic areas and is not distributed over the entire metal surface. These anodic areas are relatively small compared to the cathodic or uncorroding areas. Even in severely corroded pipelines with numerous leaks, less than 5 percent of the total surface area is attacked.
- Once leaks start to occur, they continue to occur at a sharply rising rate. When the accumulated number of leaks on an underground pipeline or pipeline network are plotted against time, the resulting curve rises exponentially. If the scale used for the number of leaks is logarithmic, the resultant plot (which can be anticipated on a line without corrosion protection) is a straight line. On a logarithmic scale, the line rises significantly, Fig. 1.

Two basic forms of electrochemical corrosion mechanisms are responsible for underground corrosion: electrolytic and galvanic.

Electrolytic Corrosion—Electrolytic corrosion is a result of direct current from outside sources. These direct currents are introduced into the soil and are picked up by an underground pipe. The locations where the current is picked up are not affected, or at least are provided some degree of corrosion protection.

But there are locations along the pipe where the direct current leaves the pipe to enter the soil. Those locations are driven anodic and corrosion results. An iron or steel pipe under this influence corrodes at the rate of 20 lb per ampere per year. This type of corrosion is often referred to as stray current corrosion. If the outside source of current is a cathodic protection rectifier on a pipeline belonging to others, the corrosion problem is referred to as an interference problem.

The most conspicuous example of electrolytic corrosion is the corrosion of underground pipes that results from stray currents generated by rail transit systems in

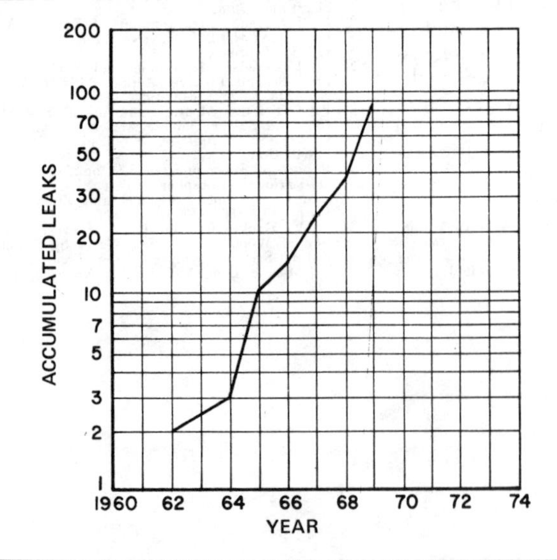

Fig. 1. Leak history of an underground pipeline is illustrated by plotting the accumulated number of leaks against time.

Fig. 2. Galvanic corrosion from dissimilar metals occurrs when bimetal couples are created that corrode one metal with respect to another.

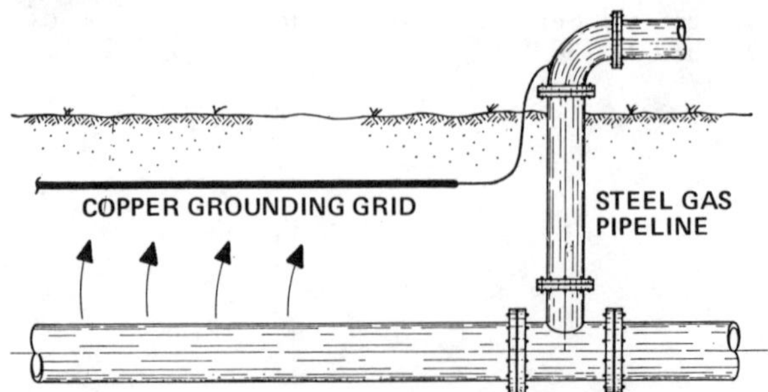

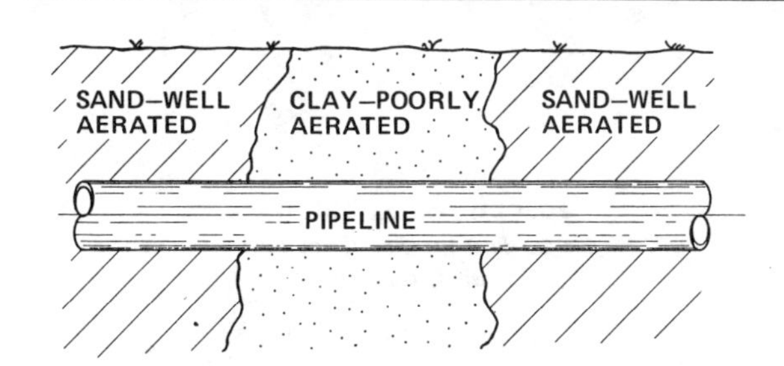

Fig. 3. Galvanic corrosion from differential environments occurs when a single type of metal encounters soils of varying compositions.

Fig. 4. Even if the area around a tank is backfilled with clean, non-corrosive sand, one lump of native clay soil in contact with the tank can result in a corrosion leak because of the differential oxygen concentration.

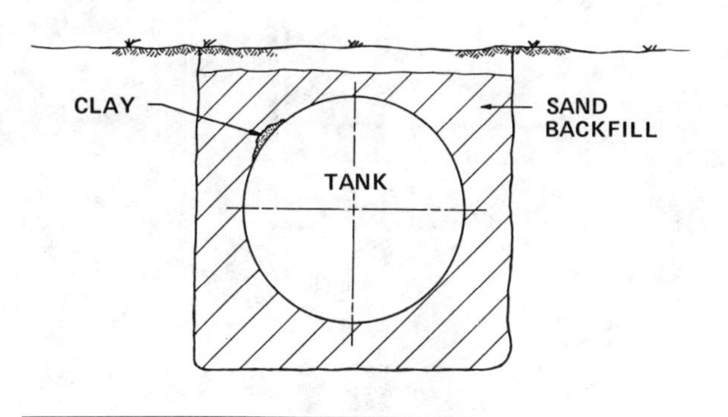

urban areas. Almost all these systems are energized by direct current, and the rails are the negative returns for the electric current. Currents carried back to the substations by the rails can leak into the ground and cause corrosion problems on neighboring underground structures.

Similar problems occur on underground structures in plants as a result of the operation of direct current equipment such as welders, cranes, or plating equipment. Therefore, when sources of direct current are anywhere within the plant, electrolytic corrosion is always a possibility. An investigation of possible causes of corrosion underground must consequently begin with determining whether there are sources of direct current and whether they are adversely affecting underground structures.

Galvanic Corrosion—Although electrolytic corrosion can result in very rapid deterioration and must be dealt with almost immediately, it is encountered far less frequently than galvanic corrosion. The two predominant causes of galvanic corrosion are dissimilar metals and dissimilar environment.

Dissimilar Metals—In most plants, a wide variety of metals are used in underground structures. Water lines are usually cast iron or ductile iron with copper water service lines; gas lines are made of steel; electric conduit is often galvanized pipe; and copper rods are used for electrical grounding. When these metals are connected, bimetal couples are created and one metal corrodes with respect to the other. For example, connecting a steel pipe to a copper pipe or to a copper electrical grounding system results in corrosion of the steel, Fig. 2.

In this type of bimetal couple, a seemingly common sense approach to controlling the corrosion may actually aggravate the problem. Although it may appear sensible to protect the metal with a coating, the result is actually even more concentrated corrosion. Any faults or holidays in the coating make the steel (anode) even more vulnerable to corrosion. It has been shown, in fact, that buried, coated pipe develops leaks in less time than uncoated pipe. For this reason, cathodic protection is necessary whenever coated pipe is used underground.

Galvanic corrosion resulting from dissimilar metals is found at many plants where water lines are connected into the electrical grounding system. At one plant where large-diameter steel pipe was used for process cooling water, leaks occurred within 6 mo even though the pipe was well coated and the trench was backfilled with specially purchased clean sand. However, because of an extensive electrical system, copper ground mats had been installed at approximately 100 ft intervals along the pipelines. They were found to be electrically connected to the water system and the resultant bimetal couple was responsible for rapid and very concentrated corrosion on the cooling water lines. This type of condition is common at many power generating plants.

Dissimilar Environments—A second type of galvanic corrosion occurs when a single type of metal encounters soils of varying composition. For example, if one portion of pipe is in heavy clay soil and another is in well-aerated sandy soil, the portion in the clay will corrode with respect to the portion in the sand, Fig. 3. This corrosion is also galvanic. It occurs because less oxygen is available in the clay than in the sand.

The portion of the pipe that is less accessible to oxygen (the portion in the clay), is the part that corrodes. Because of this characteristic, corrosion leaks are often found on the bottom of a pipe, even when the soils are fairly uniform. Because the bottom of the pipe is less accessible to oxygen than the remainder of the pipe, it is more subject to galvanic corrosion.

The problem of dissimilar environments occurs in many forms. For example, the area around a tank is often backfilled with clean sand, which is usually non-corrosive. Yet, if one lump of native clay is in contact with the tank, Fig. 4, a very concentrated galvanic action can result in a corrosion leak under the clay because of the differential oxygen concentration. Many plants that use buried steel storage tanks experience this problem.

At one plant where volatile and hazardous liquids were stored under these conditions, the problem was aggravated because the tanks were tied into a copper electrical grounding system. The hazard was increased because many of the tanks had been installed directly under a building. Large-diameter tanks are especially vulnerable to corrosion because of the greater differences in environmental conditions between the tops and bottoms of the tanks.

Source: *Plant Engineering*, April 15, 1982, 67-68

Joseph A. Lehmann
Electro Rust-Proofing Corporation
Belleville, New Jersey

Cathodic Protection Fundamentals★

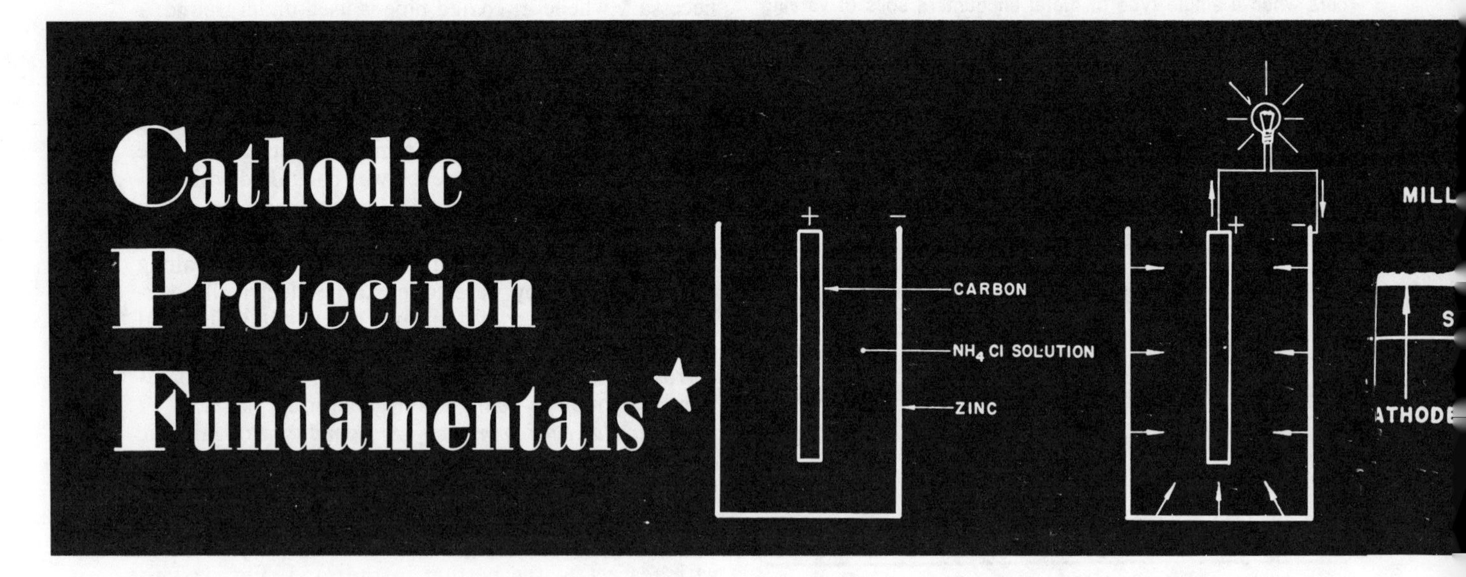

Figure 1—A simple demonstration of the corrosion process is seen in the operation of a dry cell battery. Corrosion occurs when current flows from the zinc casing (anode) through the electrolyte (ammonium chloride solution) to the carbon rod (cathode). Corrosion appears at the zinc casing where current enters the electrolyte. No corrosion occurs at the cathode.

HIGH COSTS of materials and labor demand that engineers use every means at their disposal to prevent material waste. Corrosion of metals is probably the greatest waste incurred by modern civilization. Loss of an automobile muffler, the replacement of a hot water tank, the collapse of a bridge, a gas explosion, or a water main "break" are just some of the corrosion problems man has encountered. No one seems to be immune from its ravages.

Just what is corrosion? Some have called it rust, oxidation, deterioration, chemical attack, and electrolysis. The easiest way to understand the corrosion reaction is to examine a simple application of the corrosion principle—the dry cell battery (galvanic cell), as diagrammed in Figure 1.

Basically, a battery consists of three essential parts: an anode, a cathode, and an electrolyte. The container, usually a zinc casing, is the anode (negative terminal). The cathode (positive terminal) is the center carbon rod. The electrolyte is the compound consisting of ammonium chloride, moisture, and a depolarizing agent such as manganese dioxide which are encased around the carbon. When a light bulb is connected between the positive and negative terminals, an electrical current will flow conventionally from plus to minus because of electrochemical reaction occurring simultaneously at both electrodes. Within the battery (internal circuit) the current will flow from the zinc through the electrolyte to the carbon rod. Corrosion will occur at the anode where the current leaves the metallic surface and enters the electrolyte. No corrosion will occur at the cathode (carbon) where current is collected from the electrolyte.

After the continued use of a dry cell, the container becomes corroded and the electrolyte leaks out. The current flow is caused, essentially, by the difference in electrical potential between the two dissimilar metals. Zinc is more electronegative (less noble) than carbon; therefore, it is anodic to the carbon. Similarly, current can be generated by electrically coupling any two

★Revision of the paper "Fundamentals of Cathodic Protection" presented at North Central Region Conference, National Association of Corrosion Engineers, October 1-3, 1963, Kansas City, Mo.

SUMMARY

Discusses fundamentals of cathodic protection using simple dry cell battery as illustration. Explains application of cathodic protection and suggests methods for maximum efficiency. Describes galvanic and impressed current systems. Emphasis is placed on cathodically protecting pipelines. Discusses current variables for maximum protection.

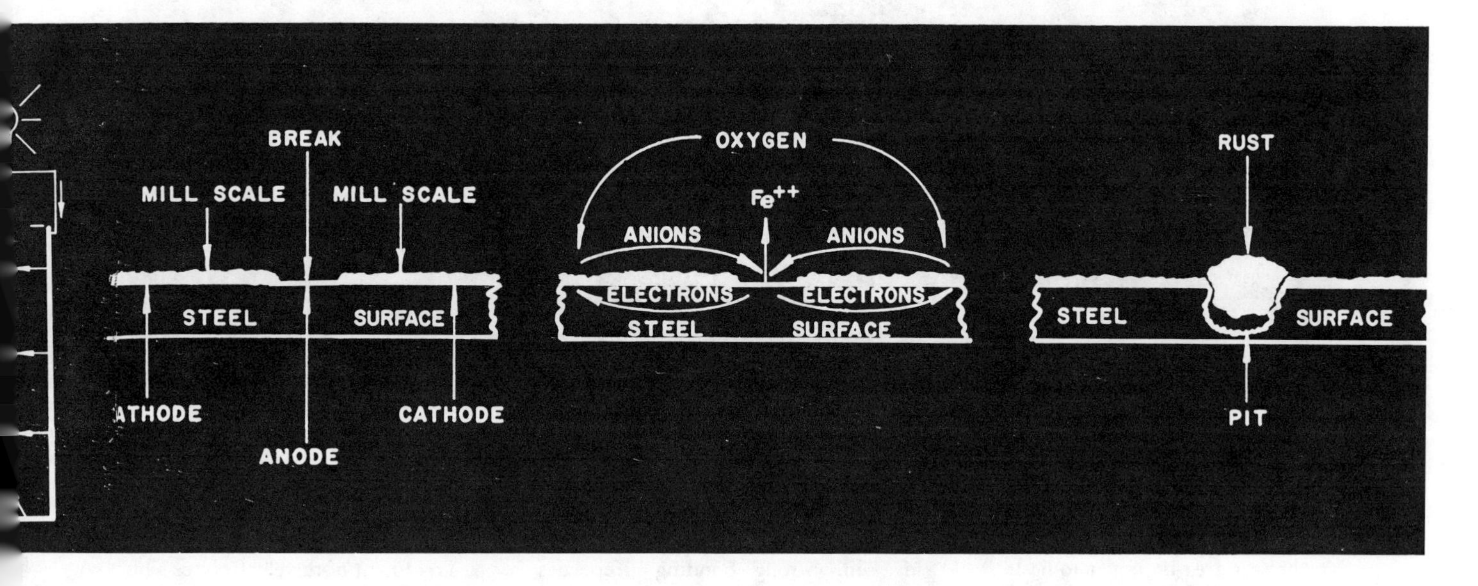

Figure 2—Cause, effect, and result of the corrosion process on a metallic sample exposed to conditions favoring galvanic attack (oxygen, dissimilar metals, and moist atmosphere).

dissimilar metals within an electrolyte. In fact, differences in electrical potential resulting in galvanic (corrosion) cells can occur on metallic surfaces simply because of environmental or physical variations (Figure 2). In addition to corrosion from the inter-connection of dissimilar metals, galvanic cells are created because of variations in ion concentration, oxygen differential, variances in stress, cold worked metal in contact with the same metal annealed, and heterogeneous soil conditions.

Regardless of the conditions causing a corrosion cell, the basic requirements are always the same: an anode coupled to a cathode in an electrolyte. Considering these basic requirements, the following methods of corrosion control are suggested:

1. Electrical insulation of the anode from the cathode.

2. Electrical insulation of both anode and cathode (or either) from the electrolyte.

3. Treatment of the electrolyte so that it will have a high electrical resistance, retarding the flow of corrosion current.

4. Use of nonmetallic materials.

5. Making entire metallic surface (anode-cathode) cathodic to an external anode. This is called "cathodic protection" (Figure 3).

Frequently, corrosion engineers use a combination of two or more of these corrosion control methods when eliminating or retarding corrosion activity. Examples of such applications, as related to pipelines, might be as follows:

1. Electrically insulating anodic area from cathodic area. When new steel pipe is connected to old rusty pipe, generally, the new pipe is anodic to the old pipe. By installing an insulating flange at the junction between the new and old pipe, the long line galvanic cell is "broken up"—the anode (new pipe) is electrically separated from the cathode (old pipe).

Similarly, when a pipeline transverses moist clay areas adjacent to dry loam, steel surfaces

Source: *Materials Protection*, February 1964, 36-41

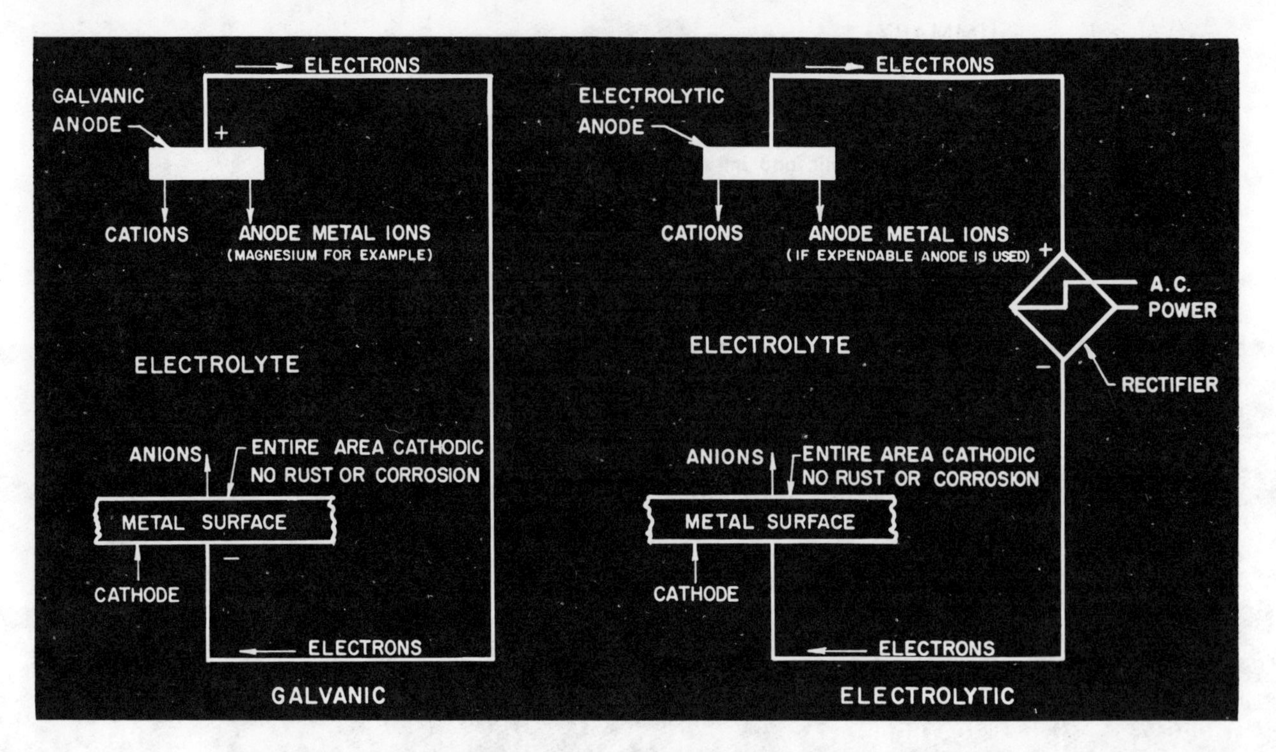

Figure 3—Two types of cathodic protection are the galvanic (sacrificial) system at left and the impressed current system at right. In the galvanic system, a less noble metal (anode) is installed to absorb negatively charged anions. In the impressed current system, an external energy source is used to introduce current into the electrolyte, rendering the pipe cathodic.

in the wet area become anodic to steel surfaces in the dry area. For this reason, corrosion engineers frequently install insulating flanges (or other insulating type fittings) in the pipeline on each side of a swamp to insulate the known anodic area from the cathodic sections.

Insulating fittings also are used to electrically separate dissimilar metals. Insulating couplings are used on gas service lines to insure electrical separation from copper water services (which normally would be connected electrically to the gas line by means of the customer's water heater).

2. Electrically insulating anode or cathode from the electrolyte. This is probably the most common practice used in combatting corrosion. If a structure can be insulated from the electrolyte (soil or water), the metallic surfaces are not affected by the environment; therefore corrosion is eliminated.

In recent years, many types of coatings have been developed for this purpose. These range from heavy asphalt, coal tar mastics, and enamels to thin plastic films. If a coating is to be effective, it should have such properties as high dielectric strength, excellent bonding characteristics, low water absorption properties, and good physical strength. High dielectric strength is required to effectively insulate the pipeline from the surrounding soil. Strong bonding properties are necessary to insure continuous and permanent contact between the steel and the coating itself. If the coating separates from the steel pipe, water can seep in between the coating and the pipe, permitting corrosion beneath the protective coating.

Low water absorption characteristics are necessary to maintain the dielectric strength and physical properties of the coating. This is particularly important when considering the combined use of coatings with cathodic protection. Physical strength is necessary to avoid damage to the coating in shipping, installation, and penetration caused by rock in the pipe trench.

Coatings also must be able to withstand physical damage resulting from soil stress, which can rip the coating from the pipe, especially in certain types of clays that are subject to expansion and contraction because of alternate wetting and drying cycles. Other considerations such as insolubility and inertness are important with regard to pipeline coatings.

Unfortunately it is impossible to achieve perfect coating of a pipeline and, under some circumstances, small pinholes called holidays in the coating intensify the corrosion problem. Penetration occurs more rapidly than it would on a bare line.

3. Treatment of the electrolyte. This method of corrosion control has not been widely used on pipelines. Because of engineering and economic considerations, treatment of the environment has largely been confined to the use of selected sand backfill around a newly installed pipe (usually coated). In some instances, alkalies (lime) have been added to the backfill to create a more favorable environment. This treatment usually is used in conjunction with cathodic protection resulting in protective calcareous deposits at coating failure locations.

4. Use of nonmetallic materials. When corrosion conditions are particularly severe and pressure and soil stress conditions are not too great, the use of plastic pipe (polyethylene or polyvinyl chloride) has been successful. Manufacture of plastic pipe during the past ten years has undergone tremendous technological changes, and there is no doubt that these materials will improve, resulting in even greater use for corrosion control.

5. Cathodic protection. It is important to remember that corrosion always takes place at the anode (where current leaves the metal surface and enters the electrolyte) and that no corrosion or a protective effect is experienced at the cathode (where current is received by the metallic surface from the electrolyte). Therefore, if

an entire metallic surface can be "forced" to become cathodic, it will not corrode—it will be cathodically protected. There are two basic methods of applying cathodic protection to a metallic structure in contact with an electrolyte. These are galvanic cathodic protection and impressed current cathodic protection (Figure 4).

The two types of systems are similar because they both deliver electric (d-c) current to the structure being protected.

Galvanic Cathodic Protection Systems

Galvanic systems produce the required protective current by an electrochemical reaction (the same as a flashlight battery). A metal less noble than the metal to be protected is selected for the sacrificial anode. When protecting steel, cast iron, or copper, the anode material can be magnesium, zinc, or aluminum. The number, size, type, and location of anodes is determined by a detailed survey of the structure to be protected. Such factors as coating resistance, soil resistivity, and interference effects must be considered. In general, galvanic anodes are used when:

1. **Current requirements are relatively low.**
2. **Soil resistivity is low (seldom used in electrolytes having a resistivity of 10,000 to 15,000 ohm-centimeter or above) and voltage requirements not high.**
3. **Electric power is not available.**
4. **Interference problems are prevalent.**
5. **Short life protective systems (low capital investment) are required.**

In most applications the anode is a cylinder or rod connected electrically by a wire to the structure to be protected. Sometimes a calibrated resistor is installed in the anode lead wire to control current output. Under the same conditions, cylinder type anodes have a longer life but lower current output than rod type anodes. Galvanic anode systems are seldom designed for a life of more than 20 years.

Magnesium and zinc anodes in soil usually are surrounded by a special backfill (gypsum-bentonite-sodium sulfate). This backfill is required to insure a homogeneous, moist, low resistivity environment which increases efficiency and extends anode life.

Impressed Current Systems

The use and design of impressed current systems are more flexible than galvanic anode systems. Design requirements are determined by a survey of the structure to be protected.

Basically, the principle is the same for both systems except impressed current systems energize anodes by use of an external energy source. Generally, a rectifier is used to convert available a-c to d-c. The direct current is then introduced into the electrolyte by an anode (or anode bed) especially designed to have a long life under relatively high current, high voltage conditions.

Many different types and sizes of anodes are available for use in impressed current systems. These are graphite, high silicon cast iron, lead-silver alloy, platinum, and even scrap steel rails. Each anode has special design, installation, and operating characteristics.

Impressed current systems are advantageous when:

1. **Current requirements are high.**
2. **Electrolyte resistivity is high, requiring higher voltage than is available from galvanic anodes.**
3. **Long life protective systems are required.**
4. **Fluctuation in current requirements is encountered (systems can be controlled and adjusted automatically).**
5. **Electric power is readily available.**

When impressed currents are used, it is important to consider interference effects on neighboring structures.

Protection of pipe usually involves installation of a transformer rectifier control unit and several anodes (graphite or silicon cast iron rods) wired in parallel. The positive terminal of the rectifier is connected to the anode bed, and the negative terminal is connected to the structure to be protected. In addition, the power unit is usually equipped with indicating ammeter, voltmeter, and overload circuit breaker. Current output of the unit is adjusted by the taps on the secondary winding of the transformer which vary the a-c voltage to the rectifier stacks.

Design and Criteria for Cathodic Protection

In selection of the type cathodic protection system, the most important consideration is that of current requirements. How much d-c is required to protect a given structure? When is protection achieved?

Without going into a great deal of theory regarding electrochemical principles and the solution of metals, it has been determined that corrosion of steel (or iron) is stopped when its potential to a copper-copper sulfate electrode is —0.85 volt or more electronegative with the reference electrode placed close to the electrolyte/structure interface. This potential value is the most significant measurement with respect to corrosion control. It actually is the measurement of voltage drop at the interface of the metallic surface and the electrolyte, the reference cell being one contact terminal and the structure being protected is the other terminal.

In neutral pH range electrolytes, an iron anode will not have a potential more negative than —0.79 volt to a copper-copper sulfate electrode. Corrosion can be controlled if the entire iron surface is raised to —0.85 volt. In other words, the cathode is polarized to the open circuit potential of the anode, and local cathode/anode cells cease to exist. The value of —0.85 volt is generally accepted to compensate for varying conditions, in particular the physical inability to place the reference cell at the surface of the protected structure. Usually, it is necessary to include more voltage drop than should be allowed (in pipeline work, the cell is generally placed on the earth surface, above the line, three or more feet from the metal's surface).

Current required to raise the structure's potential to protected values is a function of the structure's electrical resistance to the electrolyte. On pipelines, this usually is a combination of coating resistance and soil resistivity (or "contact" resistance). For example, a well coated pipeline in high resistance soil may require as little as five microamperes per square foot to achieve cathodic protection; a bare or poorly coated pipe in low resistivity clay may require 1.5 to 2.0 milliamperes per square foot to be polarized to —0.85 volt or more negative. Furthermore, under conditions where depolarizing agents are prevalent, protective current requirements are relatively much greater. For example, in condenser water boxes where water velocity is high (continuously introducing large amounts of oxygen which depolarize the cathode), current requirements to achieve cathodic protection may be 120 milliamperes or more per square foot.

To determine the specific cathodic protection current requirements for a given structure, it is necessary to simulate a protective system. After a series of soil resistivity tests are made, a suitable site (low soil resistance area) is selected for a temporary test ground bed. The test ground bed usually consists of several steel pins driven into the soil on 15 or 20-foot spacing (or 10 feet of aluminum foil submerged in a conveniently located creek.)

The ground bed usually is energized by a portable d-c generator or six-volt "hot-shot" battery. While the applied test current is interrupted, pipe-to-soil potential measurements are made at

Source: *Materials Protection*, February 1964, 36-41

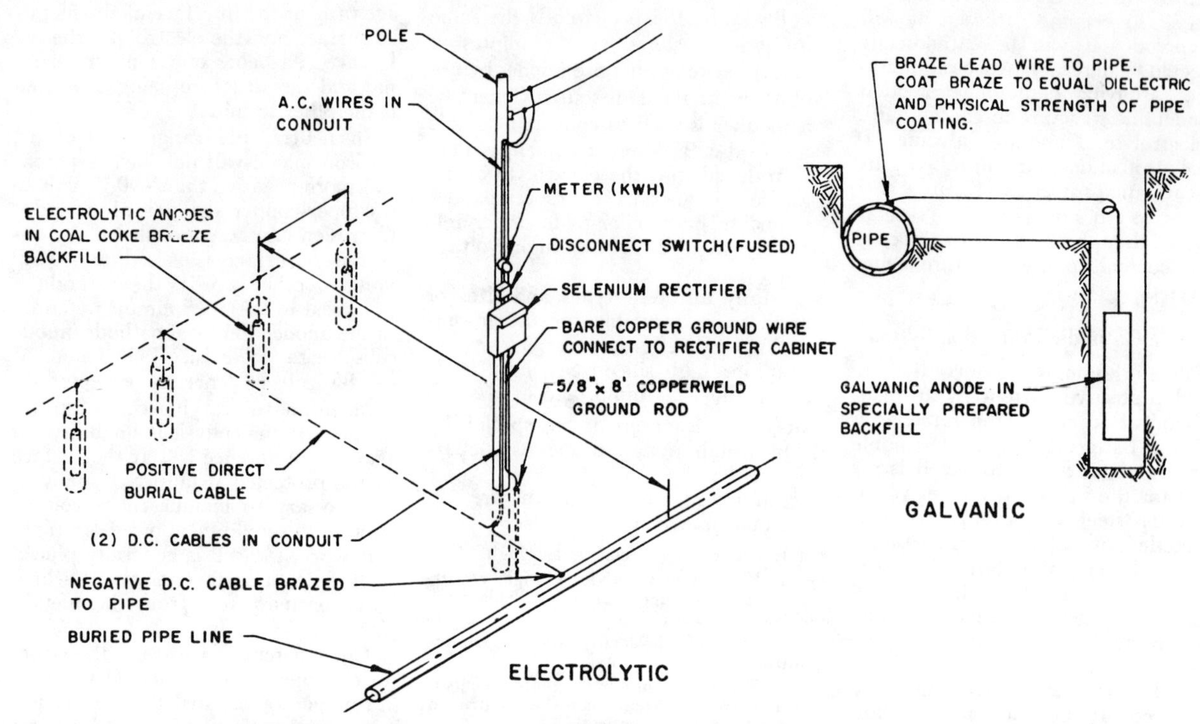

Figure 4—Typical cathodic protection installations. Impressed current (rectifier) system is shown at left and application of a galvanic anode for protection of pipelines at right.

several locations along the line to observe the attenuation effect of the current. Under ideal conditions, the applied test current can be adjusted so that minimum protective potentials (—0.85 volt) are observed at critical (or line end) locations. It is advisable to determine extent of interference effects on neighboring structures while conducting current requirement tests so that allowances can be made for clearing such effects when the permanent cathodic protection system is designed.

Once the current requirement is established, the selection of galvanic or impressed current system can be made and the protective system can be designed.

This brief description may lead one to think that testing and design for cathodic protection is an extremely simple matter. This is not the case. Like other engineering applications, there are many techniques and details which create complex problems. Interference on foreign structures is a major concern. Over protection can literally destroy expensive coatings. Electrically isolating the structure to be protected from other metallic structures and insuring electrical continuity of the protected structure are important considerations. A complete understanding of the associated instrumentation is mandatory, and use of the wrong type meters or misinterpretation of test data can have disastrous results. A complete knowledge of system components (anodes, both galvanic and impressed, rectifiers, backfill materials, and cables) is necessary. Application of cathodic protection is not a simple matter.

During the early history of cathodic protection, systems were crude. Engineers had to use makeshift anodes, rectifier units which were meant to be battery chargers, windmill driven generators, and field test instruments which were designed for laboratory experiments. Today, cathodic protection has come of age, and industry is supplying the corrosion engineer with equipment specifically designed for this work. Recent developments make the use of cathodic protection far more economical than it has ever been—and more effective.

One new "break-through" in this field is the automatic potential control system. This unit makes the use of cathodic protection completely automatic, compensating for any changes which may alter the protective current requirements.

New anode materials now make design more flexible than ever. These include lead-silver alloy anodes, various types of platinum anodes, and new high efficiency aluminum anodes.

It has been said that cathodic protection is the most unique form of corrosion control—"fighting corrosion with corrosion." Indeed, the wide spread use of cathodic protection in almost every industry has been effective in preventing waste due to corrosion on a large scale. Certainly this method of combating corrosion, combined with all the advancements in other corrosion control techniques and materials, can help achieve greater savings by preventing waste.

Other Articles on Cathodic Protection

A. W. Peabody. Pipeline Corrosion Survey Techniques. *Materials Protection.* **1,** 62 (1962) April.

W. K. Abbott and C. M. Schillmoller. Survey of Electrical Grounding Problems. *Materials Protection* **1,** 48 (1962) November.

P. C. Rogers, E. E. Cross, and B. Husock. Cathodic Protection of Underground Heating Lines. *Materials Protection* **1,** 38 (1962) July.

T. J. Lennox, Jr. Characteristics and Applications of Zinc Anodes for Cathodic Protection. *Materials Protection,* **1,** 37 (1962) September.

H. H. Uhlig. The Importance of Corrosion Research. *Corrosion,* **18,** 311t (1962) September.

Shigeo Fukuta, Shin-ichi Kondo, Nubua Usami, and Shigeki Sekimoto. Design for Prevention of Pipeline Corrosion Based on Survey of Ground Potential Distribution. *Corrosion,* **17,** 50t (1961) January.

W. H. Stewart. Aims of T-2 Minimum Requirements. *Corrosion* **16,** 28 (1960) May.

JOSEPH A. LEHMANN, chief engineer for Electro Rust-Proofing Corporation, has been engaged in corrosion engineering since 1949. He attended Central College in Missouri while in the Navy V-5 program and later attended Columbia University and Brooklyn Polytechnic Institute. An active NACE member, he is a past chairman of the Atlanta Section and has presented papers at several region and section meetings. He also is a member of AGA, AWWA, and GES.

Source: *Materials Protection*, February 1964, 36-41

Placement of Reference Electrode and Impressed Current Anode Effect on Cathodic Protection of Steel in a Long Cell

EDWARD SCHASCHL and GLENN A. MARSH, *Union Oil Co. of California, Brea, Calif.*

Experiments showed that cathodic protection of the anodic area of a long cell is most effective when the reference electrode is placed adjacent to the anodic surface and the cathodic protection anode is placed in the vicinity of the anodic area of the cell.

Introduction

THE heterogeneous nature of corroding steel is well known; both local cell action and long cell action often take place simultaneously. In local cell action, the anodic and cathodic sites are microscopic in size, while in long cell action, these sites are larger, sometimes covering many square feet. The anodic and cathodic sites for long cell action are probably regions having different ratios of local cell anodic and cathodic areas. An electrical survey on a buried pipeline, carried out by measuring the flow of current along the pipe when no cathodic protection or stray current is present, will often show appreciable currents which are attributable to long cell action. *Hot spots* where corrosion occurs along a bare pipeline are usually the anodic areas of long cells, the cathodic areas of which may be quite large on either side.[1] Husock has recently shown[2] that hot spots can be located by combining pipe/soil potential surveys with soil resistivity surveys. The accepted treatment is to apply cathodic protection by means of anodes located near the hot spots.

Long cell currents may also occur on a vertical steel structure in sea water when the upper part of the structure is exposed to water more highly aerated than the lower part.[3]

Achieving good current distribution is one of the major problems in cathodic protection. In the case where the structure to be protected is subjected to long cell action, the problem is further complicated, and several questions arise. Since the anodic part of the long cell is supplying electrons to the cathodic part, would it be more economical—in terms of current—to place the cathodic protection anode near the cathodic part of the long cell? One could reason that the cathodic part could then preferentially obtain its electrons and positive ions from the nearby cathodic protection anode rather than from the anodic part of the long cell. Or, on the contrary, is the present practice more economical; here the anode is placed near the *anodic* part of the long cell, and IR drop is thereby minimized through the electrolyte. What effect does placement of the reference electrode have on the current required for cathodic protection?

Laboratory experiments were set up to apply cathodic protection to a long cell under controlled conditions. The objective was to determine quantitatively the effect and significance of anode and reference cell placement. The experiments verified that the best location for placement of the cathodic protection anode is near the anodic part of the long cell. This location requires significantly less current to achieve protection than any other location.

When a potential measurement is made on a cathodically protected structure, ideally the applied current[(1)] should be interrupted just before the potential is measured, in order to eliminate IR drop error. However, if the potential must be measured with the applied current flowing, placement of the reference electrode is critical. Sufficient current must be applied to depress the potential of the *anodic part* of the long cell to the protective level (*i.e.*, -0.85 V vs copper/copper sulfate[(2)]). The reference electrode should be placed adjacent to the anodic part of the long cell for optimum protection. If the potential measurement is made with the reference electrode adjacent to the cathodic part, the anodic part will be over-protected, while if the reference electrode is placed in the electrolyte away from the long cell, the anodic part will be under-protected.

Experimental Setup

A galvanic couple having a well defined separation of anode and cathode was used as the long cell; it consisted of a copper cathode, 50 cm^2 geometric area, and a mild steel anode, 5 cm^2 geometric area. The anode and cathode were placed 30 cm apart and were coupled through a recording zero resistance ammeter. In this cell, there was some cathodic activity on the steel surface, but it was mostly anodic. The copper surface was entirely cathodic, and it was essentially immune to corrosion under the conditions of the experiment. The electrode potential of the copper when uncoupled was governed by the exchange current density for hydrogen on copper and was not affected by any anodic process on the copper.

The electrolyte was a solution of 100 ppm NaCl; the resistivity was about 5000 ohm-cm. The solution was held in a 5 gallon plastic vessel and was aerated but unstirred. The temperature was 25 C (77 F) and the initial pH was 7.2.

Current was applied to the long cell for cathodic protection by means of a platinum anode, a series variable resistor, battery, and microammeter. Electrode potentials were measured with reference to a saturated calomel electrode (SCE). The setup is shown schematically in Figure 1.

The copper/steel couple generated 250 μa of current (Figure 2). Cathodic protection of the steel when uncoupled required 135 μa. Thus, the minimum current that could theoretically achieve protection of the steel when coupled would be 135 + 250 = 385 μa. In the actual experiment, however, far more current was required.

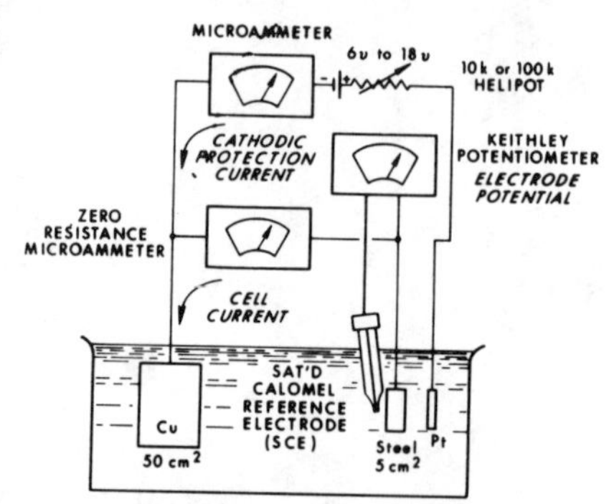

FIGURE 1 — Experimental setup.

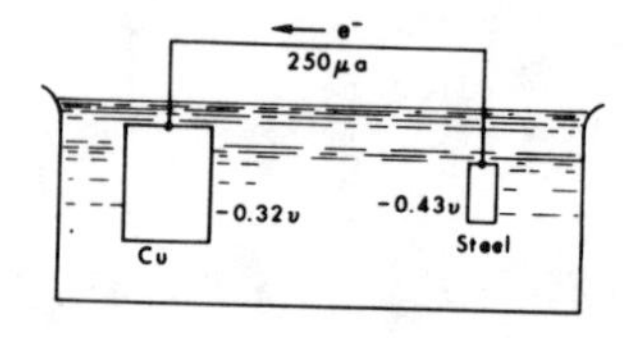

FIGURE 2 — Potentials and current in copper/steel couple.

Effect of Location of Reference Cell on Measured Potential in a Long Cell

In the first experiment, depicted in Figure 3, the steel was held at the protective potential, -0.78 V vs a saturated calomel electrode adjacent to the surface, at which potential the steel was receiving 135 μa of current from a platinum anode equidistant from the steel and copper specimens.

[(1)]Throughout this paper *"applied current"* means current forced onto the structure to which cathodic protection is being applied.

[(2)]Or -0.78 V vs saturated calomel reference electrode which was used in this work.

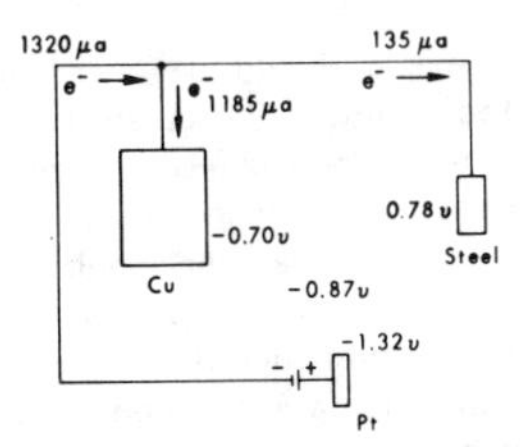

FIGURE 3 — Cathodic protection of couple with impressed current anode equidistant from copper and steel.

With the current and all other conditions held constant, the potential of the couple relative to SCE was measured at various locations as shown in Figure 3. When the potential of the couple was measured near the surface of the copper, a potential of -0.70 V was obtained, indicating a lack of protection of the steel, when in fact the steel was being protected. The reason for the difference in potential when the electrode was placed in different locations was the effect of polarization on the steel and copper surfaces, to be discussed in a later section.

In addition to the polarization effect, IR drop through the solution was also significant, especially when the calomel cell was placed in the vicinity of the platinum anode. Relative to this vicinity, the couple was very negative (-1.32 V). Midway between steel and copper, the IR drop effect was less pronounced but still evident; the potential of the couple was -0.87 V.

If there is a long cell with the centers of the anodic and cathodic area some distance apart, cathodic protection of the corroding (anodic) area can be readily achieved, although a large fraction of the applied current is wasted by going to the cathodic area. The possibility of error in determination of potential is pointed out in the experiment outlined above. If the potential is measured with the reference electrode adjacent to the corroding surface area, the approximately correct current will be applied because IR error will be at a minimum. If the reference electrode is placed adjacent to the cathodic area, there will be some overprotection of the anodic area. But if the reference electrode is placed out in the electrolyte, IR drop will show up as a negative potential and will result in under-protection of the corroding area. The IR drop error can be minimized by placing the reference electrode adjacent to the anodic area of the long cell, and, if possible, using the *current off* criterion for cathodic protection.[3]

Effect of Location of Impressed Current Anode on Cathodic Protection of Steel in a Long Cell

In the experiments described in the previous section, the platinum anode was

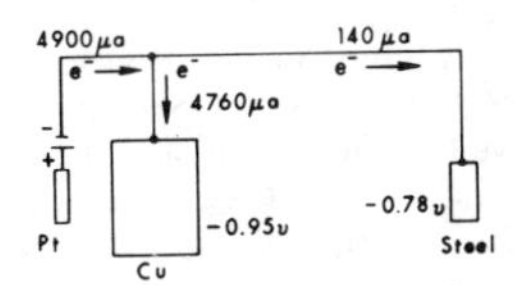

FIGURE 4 — Cathodic protection of couple with impressed current anode near copper.

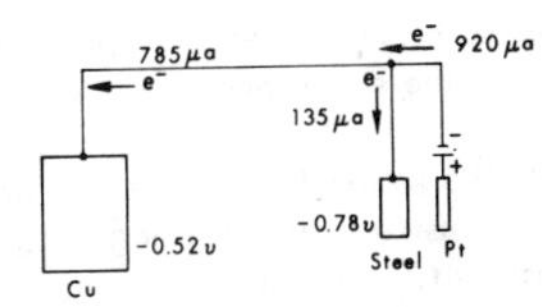

FIGURE 5 — Cathodic protection of couple with impressed current anode near steel.

placed equidistant between copper and steel specimens, as shown in Figure 3. An applied current of 135 μa was required for cathodic protection of the steel, and another 1185 μa flowed to the copper.

Another experiment was then conducted as shown in Figure 4, in which the platinum anode was placed near the copper specimen. A saturated calomel electrode was placed adjacent to the steel and the couple/SCE potential was maintained at the level of protection for steel (-0.78 V).

Placement of the platinum anode near the copper surface resulted in gross wastage of current compared with the current in the experiment of Figure 3, where the anode was placed equidistant from the copper and steel. When the SCE was placed near the copper specimen instead of near the steel specimen, the potential of the couple was -0.95 V. The steel would have been under-protected if the SCE had been placed near the copper, and if the couple/SCE potential had been held at -0.78 V.

In Figure 5, the platinum anode was located near the steel surface, and once again the current flow and potentials were measured.

When the couple was at the protective potential (-0.78 V vs a SCE placed adjacent to the steel specimen), the same current to the steel was required for protection as in the previous cases, namely 135 μa. Now only 785 μa flowed to the copper specimen. While this is still wasted current, it is only about 16% of the current that flowed to copper when the platinum anode was placed near the copper specimen. If the SCE was placed anywhere other than adjacent to the steel surface, however, more current would have been required to maintain the steel/SCE potential at -0.78 V, and the steel would have received more current than needed for protection. The data are tabulated for comparison in Table 1.

Polarization Curves Illustrate Effect of Anode Location

When the currents were measured as described in the previous section, and plotted against electrode potential in the conventional manner, cathodic polarization

TABLE 1 — Cathodic Protection of the Steel Anode of a Steel/Copper Couple (Simulated Long Cell)[1]

From Fig. No.	Location of Cathodic Protection Anode	Current Going to Copper Cathode (μ a)	Location of Ref. Cell	Indicated Potential of Steel Anode (V)	Indicated Degree of Protection of Steel
3	Equidistant between Copper and Steel	1185	At Copper Midpoint At Steel	-0.70 -0.87 -0.78	Under-protected Over-protected Protected
4	At Copper	4760	At Copper At Steel	-0.95 -0.78	Over-protected Protected
5	At Steel	785	At Copper At Steel	-0.52 -0.78	Under-protected Protected

[1] In each experiment, the steel anode was receiving 135 μa of current, sufficient for cathodic protection. However, different required currents were indicated when the reference cell was placed in various positions.

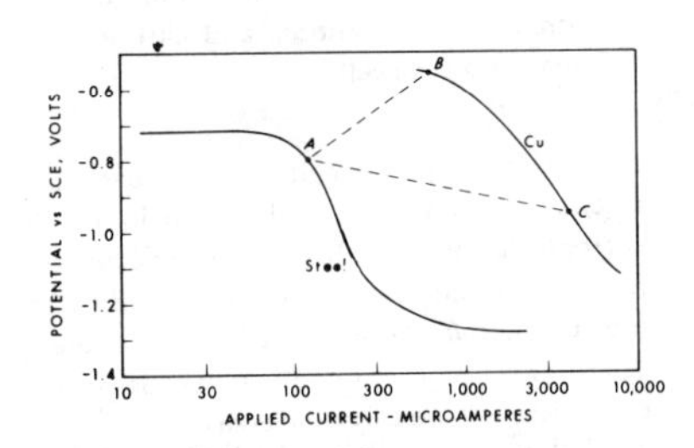

FIGURE 6 — Polarization curves for copper and steel in the couple.

curves (shown in Figure 6) were obtained. The important thing about these polarization curves is that a given potential of steel does not necessarily correspond to the same potential of the copper, even though the steel and copper were electrically connected to each other. For example, when the potential of the steel was at the protective level, -0.78 V (SCE) (point A on Figure 6), the copper potential ranged from -0.52 V (point B) to -0.95 V (point C), depending upon whether the cathodic protection anode was near the steel or the copper. Changes in current density on the copper surface, when the anode was placed in different positions, shifted the potential of the copper along the polarization curve.

It is apparent that the location of the cathodic protection anode relative to the location of anode and cathode in a long cell determines the current density on the anode and the cathode, and the current density in turn determines the potential in accordance with the polarization curve. The shape of the curve is a function of several variables including the availability of dissolved oxygen; so in a practical case, it is not possible to predict quantitatively the effect of anode placement.

Implications for Long Cells Where Both Anodic and Cathodic Areas are Steel

In the general case where a long cell consists of a small mainly anodic area and a large nearby mainly cathodic area, both steel, the cathodic polarization curves would appear as in Figure 7. The distance between the curves (distance FG) would be determined by the area ratio; for a 1:1 area ratio, the curves would be superimposed onto each other. Inasmuch as cathodic and anodic areas merge into each other, the curves as shown would apply to the *centers* of the anodic and cathodic areas. A cathodic

[3] In the *"current off"* criterion, the potential is measured immediately after the cathodic protection current has been turned off.

Source: *Materials Protection*, June 1974, 9-11

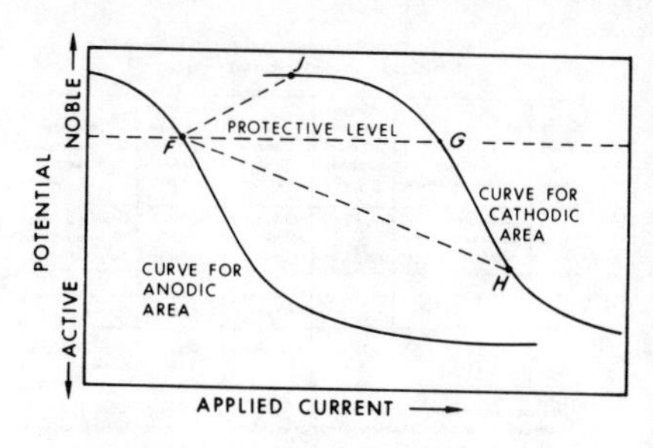

FIGURE 7 — General case, polarization curves for anodic and cathodic areas of a long cell.

protection anode placed near the cathodic area of the long cell would result in a high current density and highly negative potential on the cathodic area corresponding to the protective level on the anodic area. There would be, in other words, over-protection of the cathodic area.

On the other hand, a cathodic protection anode placed near the anodic area of the long cell would protect the anodic area with a minimum of current. The potential of the cathodic area would of course be more positive than required for cathodic protection (point J on Figure 7); so corrosion may still occur at the cathodic area just as it did before cathodic protection was applied to the anodic area.

Considering the case where the cathodic protection anode is placed near the cathodic area, the degree of over-protection will be determined by the amount of anodic activity on the cathodic surface. On the copper cathode used in our experiments, there was no anodic activity. In a long cell where both the anodic and cathodic areas are steel, the degree of over-protection would not be as great as in the case of the copper cathode because there is always some anodic activity on steel. The potential of the steel cathode (with no cathodic protection applied) would be more negative than the potential of the copper cathode.

The resistivity of the electrolyte will also affect the potentials of anodic and cathodic areas. In an electrolyte of low resistivity, the IR drop between the anodic and cathodic areas of a long cell is small; the observed potential difference between these areas will be small for a given degree of long cell action. (The IR drop referred to is the result of the long cell action.) By contrast, in an electrolyte of high resistivity, the IR drop (and hence the observed potential difference) will be appreciable. The effects of placement of cathodic protection anode and reference cell will also be appreciable, as shown in the experimental work described here.

The corrosion behavior is also affected by the resisvitity of the electrolyte. When the resistivity is low (*e.g.*, sea water, $\rho = 22$ ohm-cm), corrosion is controlled primarily by the availability of dissolved oxygen. Long cell action occurs, but it is not pronounced. There can be local variations of several tenths of a volt in the potential of an offshore oil drilling platform, for example.

When the resistivity is high (*e.g.*, soil, $\rho > 10{,}000$ ohm-cm), corrosion is throttled. Long cell action can take place, but it will be serious only if the driving force is very high.

The most common serious case occurs in electrolytes of intermediate resistivity (*e.g.*, soils, ρ = 500-2000 ohm-cm). Local variations in the soil can cause differences in the availability of dissolved oxygen which in turn cause some parts of a structure to become anodic to other nearby parts. Variations in potential as high as 0.5 volt may occur. The local variations in soil may show up also as variations in resistivity from place to place. The more dense, moist areas of soil may have a lower resistivity as well as a lower access to dissolved oxygen than other areas nearby; anodic sites develop on buried pipes passing through such low resistivity areas, and are termed *hot spots.*

Conclusions

Where long cells are involved in corrosion, the reference electrode for cathodic protection testing should be placed as close as possible to the anodic area. This will be the area that is most negative when the cell is corroding freely. *Current off* potential measurements should be made (if practicable) to eliminate IR drop error.

If there is a choice of location for impressed current or galvanic anodes, they should be placed near the anodic area of the long cell. In practice, the anodic area may not be very pronounced, and there may be active corrosion at the cathodic area as well. In such cases, the cathodic protection anodes should be distributed, but a higher current density should be designed for the anodic area.

References

1. Marshall E. Parker. Pipe Line Corrosion and Cathodic Protection (book), 2nd Ed., Gulf Publishing Co., Chapter 8 (1962).
2. B. Husock. A Method for Evaluating Corrosion Activity on Bare Pipelines, paper No. 31 presented at Corrosion/73, March 19-23, Disneyland Hotel, Anaheim, Calif.
3. H. A. Humble. The Cathodic Protection of Steel Piling in Sea Water, Corrosion, Vol. 5, No. 9, p. 292 (1949).

EDWARD SCHASCHL is a senior engineering associate in Union Oil Company of California's research center at Brea, California. He has been employed by the firm since 1948. He received a BS in Chemical Engineering from Illinois Institute of Technology. A member of NACE for over twenty years, Schaschl is an NACE Accredited Corrosion Specialist.

GLENN A. MARSH is a supervisor in Union Oil Company of California's research center at Brea, California. He has been with Union Oil Company of California since 1948. He holds a BS degree in Chemistry from Illinois Institute of Technology and a Masters in Chemistry from Northwestern University. He is a recipient of the NACE Young Authors Award and the Willis Rodney Whitney Award. A member for over twenty years, he is an NACE Accredited Corrosion Specialist.

some successful applications of CATHODIC PROTECTION in BREWERIES★

Edgar W. Dreyman
Petro-Chemical Associates
Leonia, New Jersey

SUMMARY

Because about 50 percent of a brewery's investment in equipment is in tanks and because of the aggressive environments in some of them, this equipment must be protected adequately from corrosion. Although coatings of various kinds are used, cathodic protection systems frequently are required. These may be impressed current or galvanic anode systems, but usually impressed current with inert anodes is necessary because of the need to preserve product purity.

Techniques of applying cathodic protection systems in cold and hot water tanks are described. Several case histories of cathodic protection installations using carbon anodes are given, including special techniques needed because of cleaning procedures and safety precautions provided for energizing currents.

FULLY 50 PERCENT of the capital investment in breweries is in its tanks for holding water, wort and beer. Because beer is a food product destined for human consumption, condition of holding tanks is of utmost concern. Consequently, considerable effort and money are devoted to prolonging the life of this equipment.

Cathodic protection, one of the methods employed to attain these ends, will be discussed here in connection with cold, hot water, fermentation and storage tanks. These may be further defined as plant process water, starting, fermentation, storage, filling and blending tanks.

Cold Water Tanks

Where high water conductivities are encountered as in the middle western states, cold water tanks can be troublesome. Many cement lined tanks fail from gas evolution behind the cement. Blistering of paint and rust tubercle formation are common.

Because these tanks contain aerated water, the problem is agravated. Solutions are good coating systems and/or cathodic protection. Many paint systems have been used successfully for this purpose, with epoxy-phenolic types being most commonly used. The question, "Do we have to sandblast before coating?" is important from the brewer's point of view, because water tanks may be in areas where grain is stored or grinding is taking place.

Although anodes may be made of many materials, in the food field platinum or carbon are used generally. Properly designed, the platinum will last indefinitely and carbon anodes dissolve slowly. In brewery tanks, pure carbon or phenolic-impregnated carbon is used. Linseed oil impregnated carbon should be avoided because oil in brewing water may drastically affect the product.

Long carbon rods hung from insulators attached to the top are used in cold water tanks. Although formulae are employed to determine anode position,[1] anodes must be placed equidistant from each other, the walls of the tank and the floor. Reference cells are used to check current distribution. Anodes are re-arranged for better results if necessary.

Carbon Anodes Used

Carbon anodes of sufficient size to last at least three to five years generally are used. Anodes used in high vertical tanks are assembled like a string of sausages.

This technique was used in a 60-foot high steel malt storage bin which had been converted into

★Revision of a paper titled, "Cathodic Protection in the Brewing Industry," at the 17th Annual Conference, National Association of Corrosion Engineers, Buffalo, N. Y., March 13-17, 1961.

a process water tank. No coating was applied; a current density of 10 ma/sq ft was required. A copper-copper sulfate reference cell reading of about —0.9 volt was obtained in Pittsburgh city water.

Because high current densities can blister paint films, some thought should be given to the paint systems used with cathodic protection. An increase of hydroxyl ions occur at the cathode or tank wall, so most oleoresinous paints will be attacked with saponification of oils. Hydrogen liberated at the cathode surface has a tendancy to blast the coating off.

Considerable work has been done to produce coatings resistant to impressed currents. Certain epoxy-phenolic compositions have been developed which perform well, do not impart odor or taste to the water and are non-toxic.

Cathodic protection in conjunction with these newer coatings is working well in many areas, including hot water tanks in breweries.

Hot Water Tanks

Because large volumes of hot water are consumed in brewing and cleaning operations, the brewery usually has many hot water tanks. The water may be heated directly with steam, a single large heating coil, or with a tube-nest type of heat exchanger.

Hot water tanks have been a major source of trouble, not only to the brewer but also to the coating manufacturer and applicator. Consequently, many breweries have turned to stainless steel or copper tanks which involve high initial cost. Nevertheless, these can be economical in the long run.

The plant with steel tankage must use coatings. Some major coatings applicators recommend cathodic protection in conjunction with linings as a prerequisite to a guarantee for any extended period.

A basic difficulty in hot water tanks is the dissimilar metal contacts which are conducive to galvanic action which is accelerated by the high temperatures. Even in well coated hot water tanks, holes which develop in coatings become small anodic areas coupled with large copper cathodic areas. Depending upon conductivity of the water, the rate of attack will vary, but the end result is the same: deep pitting at breaks in the coating.

The first corrective measure to consider is coating the cathode of the galvanic cell—the heating coil. Because this is not always possible, many plants are using cathodic protection.

By impressing a greater potential on the tank than that produced by the steel and copper couples, the galvanic cell is suppressed. As with cold water tanks, carbon or platinized titanium or tantalum anodes can be used. Platinum coatings on less expensive metals reduce the cost of platinum anodes because as little as a 0.5 mil is required. Their principal drawback is the rather high voltage drop in the anode itself.

Because brewers chemically treat their brewing water in the hot water storage tanks with materials such as lactic and sulfuric acids as well as certain salts to control pH and other factors, these materials may increase the corrosive nature of an otherwise neutral water.

Figure 1—Typical switchboard used for cathodic protection circuits to pasteurizer equipment.

Examples of Cathodic Protection

In a Texas brewery, five water tanks averaging 1000 barrels each were cathodically protected when conventional coatings failed after one year. In this outdoor installation five carbon anodes (each 3 inches by 7 feet) were hung from the top of each tank.

Power was supplied through a switchboard similar to that shown in Figure 1. Five circuits were used, each having its own variable transformer for controlling current output. A float valve switch is provided in each tank to cut off current when the tank is being emptied. In this plant, three years' experience with this equipment has shown it to be most effective in eliminating hot water tank corrosion.

In a Philadelphia brewery, two hot water tanks (500 and 1000 barrel) have been successfully protected with cathodic protection since 1938. During this interval, they have not been coated.

Because a large steam coil is used for heating, disc type anodes 8 inches in diameter by 2½ inches thick are used. They are attached to a porcelain insulator which is attached to the tank wall. This "mushroom" type anode is useful where clearances between the tank wall and coil are limited.

In this plant, anodes last about three years in 5000-ohm water. A light calcareous coating which builds up gradually on tank walls acts as a protective film. In one tank, the only failure observed was limited attack on the internal ladder which was replaced after 15 years.

In another Pennsylvania brewery, a 1000-barrel hot water tank has been under protection for five years, with only the top of the tank coated with a high temperature pitch-type compound. Because cathodic protection is effective only in submerged areas, it is necessary to coat condensate areas, or rusting will take place.

In some hard water areas, such as in Milwaukee, corrosion has occurred in hot water tanks behind loose scale that builds up from precipitation from the water. In one brewery with this problem, cathodic protection was installed in two 1000-barrel tanks. These tanks have propeller type mixers to circulate the water. Carbon rod type anodes, 3 inches by 10 feet were installed with rigid insulators on top and bottom of each anode. Because the tank bottom moves as the tank is filled, the bottom insulator was of a floating design. With the relatively well coated surfaces in these tanks, small current densities were required. This installation has been in operation for about four years with the original anodes still in use.

Many breweries still use square tanks for heating water. While they pose somewhat of a problem in adequately protecting corners, with proper design, this can be attained.

Fermentation Tanks

Coating failures in fermentors are commonplace, so glass lined tanks

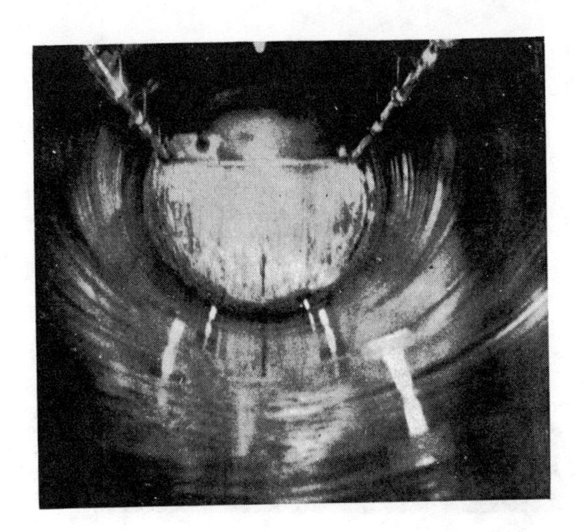

Figure 2—Showing how anodes are mounted in fermentation tank to give preferential protection to bottom.

are preferred for this purpose by most brewers.

Although some organic coatings seem to be performing quite well because breaks in the lining (glass or organic) occur, corrosion can result.

There are many reasons for this. First, beer is an excellent electrolyte, having a resistance of only 80 to 100 ohms and a pH of approximately 4. Due to the high carbon dioxide content, carbonic acid also causes chemical attack.

Because these tanks are cleaned manually after each fermentation cycle, corrosion product which might stifle further attack always is removed and fresh steel is exposed.

Galvanic cells are the result of couples between the steel and copper or brass fittings. Attemperator coils usually are copper. Because of the relatively large cathode areas of copper and brass and small anode areas in breaks in the lining, pits often penetrate half-way through a ⅜-inch steel wall in a year. To reduce this galvanic cell action, coils and fittings should be coated. Attemperator coil design should allow for reduced heat transfer rate due to this coating.

A novel approach used by Rigg and his co-workers at the Olympia Brewery was installation of aluminum fittings which acted as sacrificial galvanic anodes.[2] Other than these experiments, no other practical use of such material is known to the writer.

Where salt brines are used for cooling, attemperators should be insulated from the tank and the plant piping with rubber hose connections. In older plants using direct current, cases have been reported of electrolysis resulting from current carried by the brine.

All practical efforts should be made to protect tanks from mechanical damage caused by improper boots, pails, brushes or staging used during cleaning.

If coatings do not reduce corrosion in fermentors, then cathodic protection should be considered.

Poor Technique Causes Failure

In one large eastern brewery, a cellar of fermentors was pitting badly and coating repair costs were becoming excessive. In this cellar of six tanks, from 1200 to 1800 barrels (horizontal 12 feet by 70 feet) several had been coated with a baked phenolic. Investigation showed that the coating had been sprayed over blasting grit which had not been vacuumed from the tank. Subsequently, as cleaners walked on these sharp particles, the lining was perforated. Rapid pitting resulted. Attempts were made to isolate fittings without significant success.

In 1957 a trial cathodic protection system was installed in one tank. It consisted of six disc-type anodes mounted on insulators about 10 inches from the floor. (See Figure 2.) Several factors favored this design. Because most of the corrosion was on the tank floor, anodes would be close to the bad areas. This type anode would interfere least with tank cleaning operations. Although the pits were numerous, their total area was small. Consequently, polarization was reached with 0.25-ampere at three volts.

Liberation of gases had to be considered, so samples of beer from this tank were analyzed for oxygen and iron. Neither was found at an objectionable level. Before application of cathodic protection, the dissolved iron level had passed the safe limit of 2 to 3 parts per million.

This tank was watched closely for a year. Tests included removing patches of coating which exposed the sandblasted steel surface. Plaster casts of larger pits showed no further metal loss.

Patched Tank is Protected

A second tank was then equipped with anodes. This old glass lined tank had many tin patches, some rather extensive ones on seams. With the additional bare areas, approximately 0.3-ampere at 3.5 volts was required to reach polarization.

Subsequently the remainder of the tanks were cathodically protected. Although current requirements were low and low voltages are in use, several precautionary methods were employed. Each fermentor door is equipped with a waterproof microswitch (See Figure 3) which cuts power to the tank when the door is opened. All lead wires are in insulated conduits and run to a common switchboard (See Figure 4) mounted in a dry area outside the cellar.

Each tank is on a separate circuit with a dummy load on a resistor. When a tank door is opened the dummy load is applied. This maintains a constant load on the rectified power supply.

Manual adjustments are made during the fermentation cycle because the resistance changes as wort becomes beer. Although this could be done automatically, it has not been considered justifiable in view of the additional cost.

Current Density Is Critical

The density of applied current must be kept within a close range. Even a slight increase in potential on the phenolic lined tanks causes blistering and stripping. However, no iron pickup results when additional bare spots are so exposed.

Although glass lined tanks showed little "popping" of the linings, the edges of tin patches were lifted when

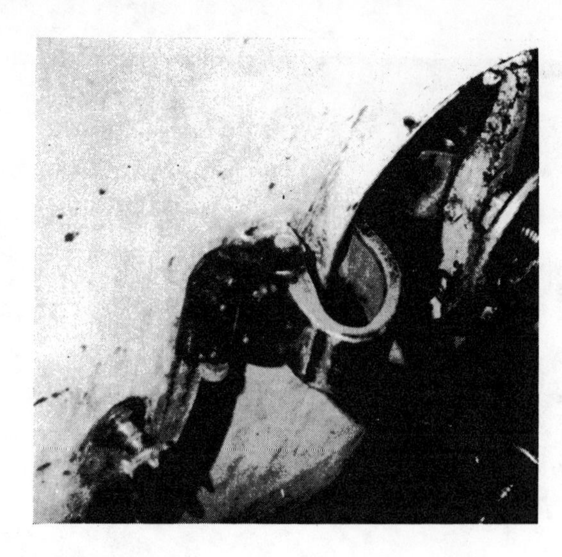

Figure 3—Switch in fermentation tank door which shifts cathodic protection current to dummy load when door is opened.

excessive currents were employed. When glass linings are faulty, cathodic protection is not desirable. However, when there are pores in the glass, a low potential system may be effective. But, if glass is under improper stress, additional hydrogen evolution may unbalance existing stresses with consequent glass chipping.

At a small Pennsylvania brewery, a similar installation was made in phenolic lined tanks of 500-barrel capacity, using carbon rods at about 0.2 amperes per tank. Although some pinholes have developed, this system also has been operating successfully for about four years.

Protection of Pasteurizers

A modern brewery has a large investment in bottle shop equipment. This includes soakers, washers, fillers and pasteurizers. Where high alkalinity cleaning compounds are used, the corrosion problem is negligible, but in pasteurizers the problem is acute.

This bottling equipment consists of a moving belt on which canned or bottled beer moves through sprays of water, which are gradually increased in temperature, until they reach 140 F at which they are held for a specific time. Then they are cooled slowly.

Water circulates from tanks under each different temperature section. These tanks usually have copper or brass heating coils, although some plants use live steam injection. Consequently, an environment results which includes brass, copper and steel in hot water to which generally is added chemicals such as caustics. This, with beer from bottles broken on the chain or walking beams, provides an electrolyte with low resistivity and an aggressive nature.

Because beer is a nutrient, the situation is complicated further by formation of slimes and algaes. These settle to the bottom where concentration cells result. Pits resulting from these cells grow to the size of half dollars and reach a depth of ⅛ inch in a relatively short time.

Because this water does not come into contact with the product, galvanic as well as impressed current anodes can be used. Ambler and Allen have described a type of zinc anode installation which has worked well.[3] More recently, long rod anodes (about 1½ by 1½ by 60 inches) have been used successfully. These are held several inches from the tank floor with cast-in core rods, making it easier to clean the tanks. Several years' operation with this type of anode has given excellent results.

Magnesium anodes have been used to give 12 to 18 months' life. Because replacement costs are high due to the close working areas, longer life zinc generally is preferred. When galvanic anodes having low driving voltages are used, debris should be kept to a minimum because corrosion can take place under it. High purity zinc works best. Maximum iron content in the alky

Figure 4—Switchboard serving tanks in fermentation cellar. It is located outside the cellar for safety reasons.

is maintained here at 0.0006 percent. No evidence of potential reversal at high temperatures has been noted.

Many pasteurizers have been protected adequately with impressed current systems with graphite anodes 3 by 60 inches. Anodes are held on coated saddles with neoprene insulation. Much better current distribution is obtained with them and fewer anodes are required for protection than when galvanic anodes are used. Although inert anodes make for easier cleaning, cost is greater than for a galvanic system of equal performance. However, when waters are very soft, an impressed current system is the only practical solution.

Because several tanks operate at the same water temperatures, a 9-tank pasteurizer may require only five circuits. Variable transformers are used to control input to a rectifier for each circuit.

Reference cells are not mounted in the tanks as a rule. A standard copper-copper sulfate cell is used to check potentials, but where permanent cells are desirable, high purity zinc cells may be used.

Some equipment of this type has been under protection for at least six years. Hybrid systems also are in use where an impressed current system is supplemented by galvanic anodes in areas such as pump boots and circulating channels. For year-to-year comparisons, plaster of paris casts of pits are taken, for checking with a dial gauge.

References

1. H. B. Dwight. Calculation of Resistance to Ground. *Electrical Eng.*, 1319-1328 (1936) Dec.
2. W. D. Rigg. Accelerated Galvanic Corrosion in Beer Tanks in the Presence of Dissimilar Metals. MBAA 47th Annual Convention, New York City, 1954.
3. C. W. Ambler and W. E. Allen. Zinc Galvanic Anodes as a Supplement to Coatings in a Brewery Pasteurizer. Paper presented at Northeast Region Conference, National Association of Corrosion Engineers, October, 1958.

Bibliography

Cathodic Protection Frequently Practical in Corrosion Control of Brewery Vessels. *Corrosion*, 13, 134 (1957) January.

Solving Design Problems for Cathodic Protection of Glass-Lined Domestic Water Heaters. *Corrosion*, 16, 9-17 (1960) September.

EDGAR W. DREYMAN is manager of Petro-Chemical Associates, Inc., Leonia, New Jersey, where his work includes cathodic protection of water immersed equipment and non-destructive testing problems. He has a BS in chemistry from Syracuse University and graduate work in chemical engineering, metallury and corrosion technology at New York University and Stevens Institute of Technology. He has been a member of NACE since 1950.

Source: *Materials Protection*, February 1962, 58-62

CATHODIC PROTECTION OF OFFSHORE STRUCTURES

J. A. Burgbacher, Shell Oil Co., Metairie, La.

SUMMARY

Information in this article is derived from a review of Shell Oil Co.'s cathodic protection of offshore structures in the Gulf of Mexico, offshore Louisiana. Aluminum galvanic anodes and impressed current systems, cathodic protection systems used, are described. Laboratory and field tests to measure output and determine effectiveness of aluminum anodes are presented. Costs of aluminum anodes and impressed current systems are compared for new and existing structures.

ALUMINUM GALVANIC ANODES and impressed current systems are used for cathodic protection of offshore structures by Shell Oil Co. A review of these cathodic protection practices in the Gulf of Mexico, offshore Louisiana, revealed information concerning operation efficiency and costs. This information, presented below, includes: (1) description of aluminum anode systems, (2) results of aluminum anode tests, (3) description of impressed current systems, and (4) discussion of the efficiency and economics of each in new and existing structures.

Galvanic Anode Systems

Aluminum Anodes

Aluminum alloy materials have been available for some time, but it was not until the introduction of the tertiary alloys containing zinc and tin in 1963 that they found widespread usage. Since that time, Shell Oil Co. has installed 2¾ million lbs. of aluminum anodes on new and existing structures offshore Louisiana. Seven hundred lb. anodes are currently specified for new construction; diver-installed anodes on existing structures are limited to 300 to 400 lbs.

Despite the general acceptance of aluminum alloys as galvanic anodes, there was little factual data presented by the anode suppliers to support advertised current outputs. Claims and counterclaims about anode efficiencies were numerous. Limited field testing suggested that at least some of these claims were exaggerated. As a result, many users, including Shell, initiated test programs to evaluate the alloys being offered to the industry.

Laboratory Testing

Shell's test procedure incorporated both the hydrogen evolution and weight loss tests described by Reding.[1] The anodes were driven by an external d-c source at 1000 ma/sq ft or four times the anode density normally used in actual design. The higher density was selected to accelerate the test.

Table 1 summarizes the results. Anode efficiencies at the end of the four-week test varied from 27.3 to 95.8% and current outputs based on weight loss, from 457 to 1309 amp-hr/lb. A second sample of Alloy D was tested when the first sample indicated an output much less than that reported by the supplier on a large number of heats.

Field Testing

Some suppliers of aluminum anodes

TABLE 1—Summary of Laboratory Test Data on Aluminum Alloys

ALLOY	CLOSED CIRCUIT[(1)] POTENTIAL (volts)	HYDROGEN EVOLUTION[(2)] EFFICIENCY OUTPUT (%)	(amp-hr/lb)	WEIGHT LOSS[(3)] OUTPUT (amp-hr/lb)
A	1.05	95.8	1295	1309
B	0.95	54.1	726	988
C	0.95	35.5	454	595
D(1)	0.97	27.8	354	477
D(2)		87.9	1120	1218
E	0.98	27.3	349	457

(1) With reference to a saturated calomel electrode

(2) At the end of four-week test

(3) Four week average

TABLE 2—Comparison of Laboratory and Field Test Data

	CURRENT OUTPUT (amp hr/lb)		
ALLOY	HYDROGEN EVOLUTION	WEIGHT LOSS LABORATORY	FIELD
A	1295	1309	1250
B	726	988	1155
C	454	595	623
D(1)	354	477	495
E	349	457	496

tended to discredit the hydrogen evolution test. Therefore, a 225-day field test was conducted using the remainder of the five anodes from which the laboratory specimens were cut. The anodes were suspended from a large unprotected structure, and current output measurements were made at periodic intervals.

The results are shown in Figure 1. The lower initial output of Alloy B is attributed to its bulkier shape. Note that all of the alloys except Alloy A experienced a reduction in current output of about 50%. Since the structure's potential was not changed and the anode surface areas were not altered significantly by consumption, it was concluded that corrosion products were responsible for this reduction. These anodes were covered with a thick gelatinous layer, probably $Al(OH)_3$, quite similar to that observed in the laboratory.

Evaluation of Results

A comparison of laboratory and field results is shown in Table 2. Agreement between the 30-day laboratory weight loss test and field exposures was good. The data suggest that the hydrogen evolution test gives current outputs somewhat lower than might be expected in the field. Shell is currently purchasing only those aluminum alloys which tested above 1100 amp-hr/lb. Shell design requirements are based on data supplied by the manufacturers on a large number of heats because the Shell program included only a single heat of each alloy.

To assure that only high output anodes are purchased, the supplier must furnish the current output of representative heats as determined from a hydrogen evolution test of at least two weeks duration. A duplicate sample must accompany the order for periodic verification in the Shell laboratory.

Impressed Current Systems

New Structures

The cost of a 20-year aluminum anode system in shallow water represents only a modest initial capital expenditure. As the water depth increases, the capital outlay becomes quite significant and at 340 ft approximates $106,000. In addition to the high first cost, replacement of anodes at these depths is a problem. Currently divers can effectively weld to only 125 ft, and, even with anticipated technological advances, the cost will be prohibitive in the 200- to 300-ft depth range.

It is doubtful whether protection can be maintained on bottom once the original anodes have dissipated and replacement anodes are limited to 125 ft. This leaves the undesirable alternative of suspending the anodes from the surface or changing to an impressed current system. If the impressed current approach is taken, perhaps this method should be considered for the initial installation.

Until recently, most impressed current cathodic protection systems for new structures have been installed in the field. Lead-silver anodes, usually 3-inch x 30-inch, were suspended from a lower deck elevation using either lead wire or plastic rope.[2-5] While many operators find this system acceptable, there are two disadvantages. Storm damage is less severe than with magnesium because of the anodes' heavier density but it still presents an operational problem. Because of the limited current output of this anode size, 25 amps, the system becomes rather cumbersome as the total requirement increases with increasing water depth.

Shell's first permanent impressed current system deviated considerably from the above method in that the anodes were mounted directly to the structure and conductors were brought to the surface in a protective conduit, thus minimizing storm damage. As high output, lead-platinum anodes were used, fewer anodes were needed to satisfy the 600-amp requirement. These were concentrated at one end of the structure and the adjacent steel was coated with dielectric shielding material to a

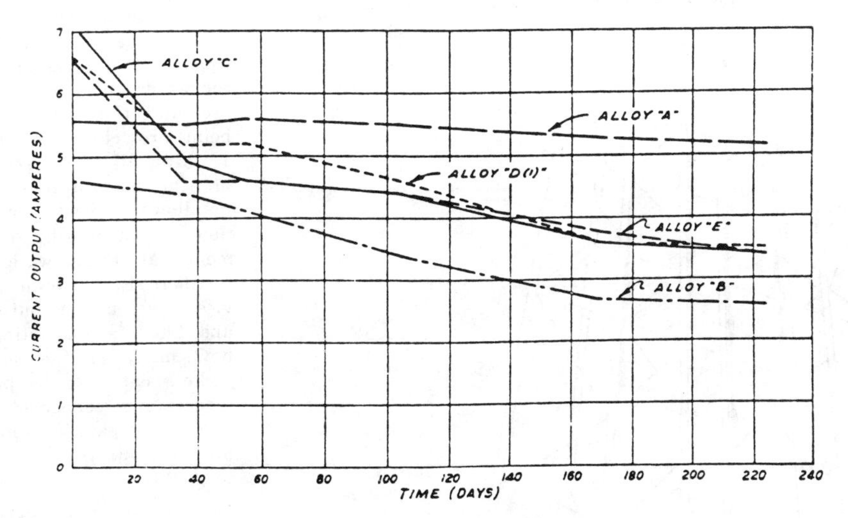

Figure 1—Current outputs field tests.

Source: *Materials Protection*, April 1968, 26-29

distance of ten ft. Benedict had reported that for ships this distance was sufficient to provide a semi-remote anode array.[6] To compensate for a possible poorer current distribution pattern a design criterion of 7 ma/sq ft in the water and 2 ma/sq ft in the mud was used in lieu of the normal 5 and 2 ma/sq ft. A diagram of this installation is shown in Figure 2.

The system was activated on July 19, 1966. All of the structure experienced at least a 200 mv potential shift within the first week and some members had reached a protective potential of –800 mv with respect to an Ag/AgCl reference electrode. After two and one-half months the protective potential had been extended to two-thirds of the structure and a calcareous deposit was observed to a depth of 100 ft on the A-4 leg.

Since this first installation, four additional systems have been designed for platforms varying in water-setting depths from 290 to 340 ft (1000 to 1200 amps). Two of these structures have been launched, but to date only one of the cathodic protection systems has been placed in operation. To improve distribution on these systems, the lead-silver anodes were mounted vertically at two levels on the A-2, A-3, B-2, and B-3 legs as shown in Figure 3. A potential profile on the "A" row taken initially and after 36 hrs and 11 days showed early response on the middle legs to a depth of 150 ft and a lesser potential shift on the outer legs. As the calcareous deposit builds in the immediate vicinity of the anode arrays, the zone of protection should extend to all parts of the structure.

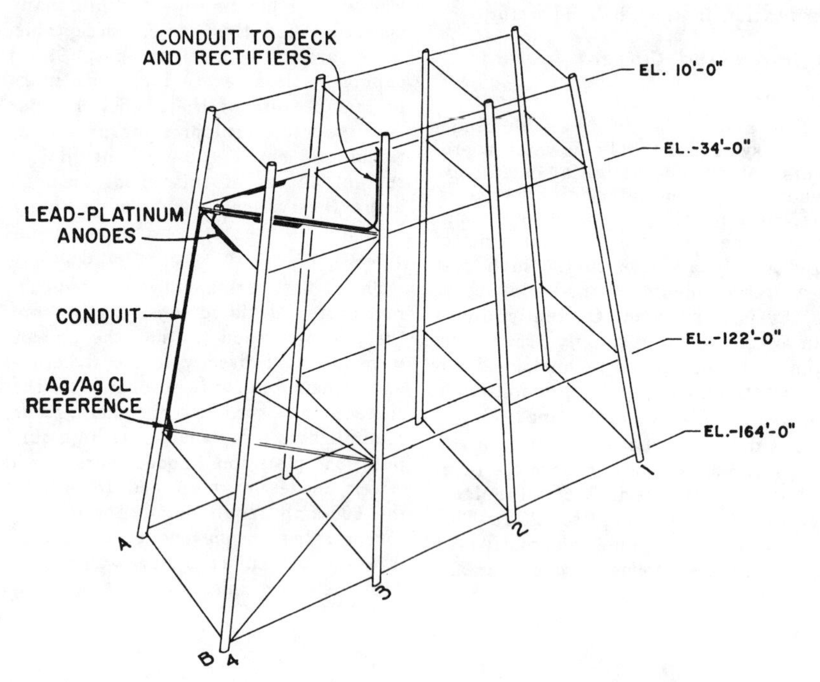

Figure 2—Diagram of impressed current cathodic protection system.

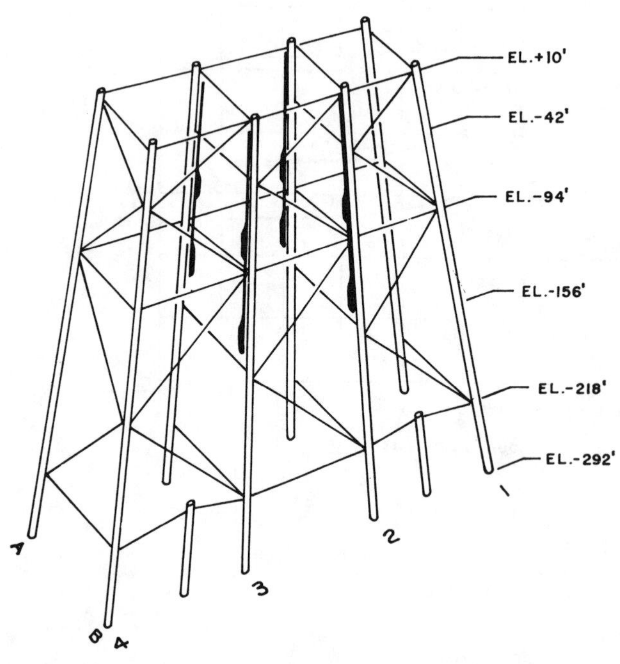

Figure 3—Revised anode location for deep water platforms.

Existing Structures

The concept of fixed permanent anodes has been expanded to include four existing platform complexes of two to four structures each interconnected by walkways. Water depths of these structures vary from 25 to 70 ft. The systems incorporate two 250-amp anode arrays fixed to a 40-ft "dummy" carrier member which in turn is clamped to the structure. The anode location was selected to minimize interference from surrounding members, and the carrier member was coated prior to installation.

In one installation the current output was limited initially to 125 amps from a single anode array because there was considerable difference of opinion as to the current required to protect this complex. This output was equivalent to a design of 5 ma/sq ft in the water and 2 ma/sq ft in the mud, excluding the one structure which was being protected with aluminum. Within 16 days, all of the structures were at or near the protective level. Therefore, no allowance need be made for distribution in calculating current requirements at these water depths. Both rectifiers have been placed in service at a combined output of 500 amps and Shell is attempting to determine how many of the surrounding well jackets can also be protected. These jackets are connected to the central facility by parallel 3-inch, schedule 80 flowlines varying in length from 1000 to 5000 ft.

Economics

New Structures

The capital cost of cathodic protection with aluminum anodes (based on a 20-year life, 1280 amp/lb current output and 64.5¢/lb aluminum, installed) is relatively constant at $87.50/amp required. On the other hand, the initial unit cost of an impressed current system decreases as the requirement increases. In the 1200-amp range, this cost, including a share of the generating equipment, approximates only $27.50/amp.

Figure 4 shows the cost of the two systems at various water depths for 8- and 12-pile structures. The PV cost of power generation and maintenance of the impressed current system is also included. Based on these data, Shell Oil Co. now specifies an impressed current system for any structure set in water depths greater than 150 ft providing that electrical power is installed for other reasons. If the platform is to serve as a terminus for a long pipe line, this type of system will be used regardless of depth.

Existing Structures

The installed cost of aluminum anodes for existing structures varies between $1.00 and $1.10/lb. Thus, a 10-year design installation costs approximately $75/amp. The initial cost of Shell's first four insitu impressed current systems, excluding a share of the generating plant, was in the range of $30 to $35/amp. Based on difficulties experienced during installation, costs of future in-situ systems may be slightly higher.

In addition to the lower first cost, the life of an impressed current system should be extended beyond the time when aluminum anodes are exhausted. If dependable power is available, adequate maintenance can be assured, and the current requirement is sufficiently large to justify the additional engineering, impressed current systems are selected by Shell Oil Co. in preference to aluminum.

Summary

Shell Oil Co. uses both aluminum anodes and impressed current systems to cathodically protect the submerged portion of offshore structures. Aluminum anodes have been found to be more practical for new construction in shallow water. However, in deeper water impressed current systems are more economical than aluminum anodes. On existing structures, impressed current systems are preferred if the current requirement is large enough to justify the additional engineering.

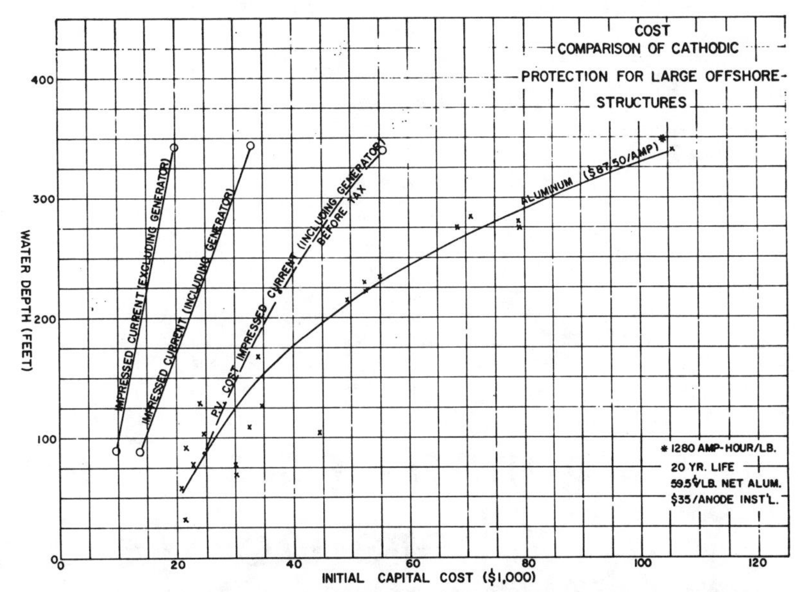

Figure 4—Cost comparison of cathodic protection for large offshore structures.

Acknowledgments

The author wishes to thank W. L. Berry and J. H. Lybarger for their assistance in collecting the data presented. Appreciation is also expressed to Shell Oil Co. for their permission to present this paper.

References

1. J. T. Reding and J. J. Newport. The Influence of Alloying Elements on Aluminum Anodes in Sea Water. *Materials Protection,* 5, No. 12, 15-18 (1966) December.
2. H. R. Hansen. Current Practices of Cathodic Protection on Offshore Structures. NACE South Central Regional Conference, October 17-20, 1966.
3. A. E. Hiller and A. D. Lipps. Cathodic Protection for an Offshore Structure. *Materials Protection,* 4, No 6, 36-39 (1965) June.
4. G. L. Doremus and J. G. Davis. Marine Anodes: The Old and the New. *Materials Protection,* 6, No 1, 30-39 (1967) January.
5. J. L. Andrews. Cathodic Protection Design. *Materials Protection,* 6, No 1, 49-52 (1967) January.
6. R. L. Benedict. Anode Design for Shipboard Cathodic Protection. *Materials Protection,* 4, No 12, 36-38 (1965) December.

Bibliography

J. H. Morgan. Cathodic Protection (1959), 3-4, Leonard Hill Limited, London.

J. D. Baribault. Cathodic Protection for Offshore Structures. *Oil and Gas Journal,* April 8, 1963.

JOHN A. BURGBACHER is staff chemical engineer for Shell Oil Co., Metairie, La., where he is responsible for chemical engineering problems in exploiting and producing crude oil and gas. He received a BS (1948) and an MS (1949) in chemical engineering at Ohio State University. A member of NACE, he is chairman of the New Orleans Section.

Cooling Water Treatment -- Where Do We Stand?

JOHN M. DONOHUE, *Betz Laboratories, Inc., Trevose, Pa.*

More research and engineering studies are being conducted on cooling towers and cooling systems than ever before. Many companies and organizations are involved in the effort to make a science out of cooling water treatment, and the program is yielding good results.

Already developed, tested, and in use are commercially available treatment programs that provide excellent deposit control plus corrosion protection that virtually equals the best chromate programs. Such programs are environmentally acceptable, using no chromates, no zinc where it is undesirable, and some even operate effectively without phosphates.

New biocides continue to be developed that afford the necessary microbiological control within current operating parameters and are amenable to economical removal from effluent.

Any examination of "where we stand" in cooling water treatment must reveal that the changes in recent years have been rapid and are improving the art.

"I KNOW there's more art than science to cooling water treatment. But how do you master the art when the rules keep changing?"

This is the lament of the operator who bears the heavy responsibility for keeping his operating unit on-line. He is expected to keep the cooling system from slowing up the operation, and at the same time prevent costs from rising above their historical pattern.

The rules he refers to are those laid down by Federal legislation, state laws, and even his own company policy as they apply to discharges from his cooling system. Such laws and policies are under alert scrutiny and subject to frequent change because of the swelling public concern for environmental protection. Cooling water systems receive particularly close attention because they handle such huge quantities of water.

Any guidelines for operators must span the major areas of cooling system water treatment. These are identified by the symptoms of the trouble that can occur if proper protective measures are not taken. The three symptoms are corrosion, deposition, and biological fouling.

Corrosion Control

Conventional Treatment

The most widely used corrosion inhibitors are treatments with a chromate base. These highly efficient inorganic inhibitors are difficult to replace economically. Generally, the chromate level (expressed as CrO_4) is maintained at 20 to 30 ppm in cooling tower systems.

Zinc is also used in many treatment programs for its efficient cathodic inhibitory properties. In conjunction with chromate, it gives acceptably complete passivation of metal surfaces. The range of use is 1 to 5 ppm.

Until recent years, these two materials, along with polyphosphate, formed the solid backbone of industrial corrosion inhibitory treatments. But with the increased attention focused on environmental standards, these materials were studied for possible deleterious effects when carried out to streams and lakes as effluent.

A paper by the National Environmental Research Center[1] recently reported on research studies on cooling tower blowdown toxicity. These studies indicated that chromate had no lethal effect on juvenile steelheads in 96 hrs at 31 ppm of CrO_4. But algal productivity (a major source of food to fresh water fish) was decreased at a concentration of 0.14 ppm CrO_4. Zinc was toxic to algae at 0.064 ppm Zn. And the same 96-hr test on juvenile steelheads showed toxic effects at 0.09 ppm. Thus the toxic effects of zinc on both algae and fish occurred in nearly the same concentration.

The report further stated, "these values are for levels that are harmful to the organisms, and stream standards are meant to *prevent* harm, not limit the kill to 50% in 96 hrs." For instance, the recommendation of the US Technical Advisory Committee (1968), for zinc, stipulates that

Reprinted with permission from *Materials Protection*, June 1972, 19-23, © 1972 National Association of Corrosion Engineers

1/100th of the 96-hr Tl_m (LC-50) value is a safe concentration for continuous exposure. Using this factor for zinc and the results of the tests on algae and steelheads, one arrives at "safe" limits of 0.00064 ppm Zn for algae and 0.0009 ppm of Zn for steelheads. For chromium, the Committee recommendation is that "0.02 ppm Cr is safe for salmonid fishes in soft water."

It can be seen, however, from the preceding report that zinc is not acceptable for direct discharge to many streams. Therefore, the newer treatments are based on substituting another cathodic inhibitor or a completely different corrosion inhibiting mechanism.

Both zinc and chromate can be removed or reduced in concentration by waste treatment processes. Many plants continue to use these efficient inhibitors by employing total waste water management. Undesirable constituents of various streams are removed at minimum cost by planned combination of the different streams.

Low Chromate Treatment

In the past few years, techniques have been developed which allow very low chromate levels in the circulating water, 5 to 10 ppm CrO_4, to give good protection. The principal change in technology has been the use of high pH in the range of 7.5 to 9.0 combined with the use of organic phosphorous compounds such as aminomethylene phosphonic acid. The high pH reduces the corrosion load on the system in comparison to the traditional form of treatment. The organic phosphorous compound in many cases eliminates the need of acid feed and results in successful crystalline scale control.

Not all makeup waters can be treated to utilize this system of corrosion inhibition. This form of treatment is limited to mildly alkaline makeup waters with moderate hardness content. Waters that are acidic and low in hardness content require a strong cathodic inhibitor to supplement the low chromate level.

Organic (Nonchromate) Treatments

In the search for acceptable chromate substitutes, numerous organics have been screened, and some have passed laboratory tests successfully. But none have been effective alone. Like chromate, they require supplementary or synergistic agents to stabilize efficiently the high volumes of water used in cooling systems.

The trend to higher pH levels where the corrosion load is lower has continued with the organic approach. Polymers and phosphonates, singly or in combination, have been found to successfully eliminate crystalline scale formation at these high pH levels and supersaturated conditions.

The inhibitors are chosen on the basis of the cooling system's specific makeup water characteristics and the metallurgy of the containing system itself. So in some cases an anodic inhibitor such as polyphosphate, silicate, or lignosulfonate has been found adequate to supplement the cathodic inhibitory effect obtained from the calcium-organic complex of the scale inhibiting mechanism. In other cases, particularly where galvanic couples exist in the cooling circuit, it is necessary to employ a polar organic compound that will be adsorbed on the metal surfaces.

The continuous and intensive research efforts of water treatment companies have yielded several successful alternatives to chromate and zinc treatments. They are generally nontoxic to fish, a criterion that must be satisfied. But beyond this criterion, there is still a great deal of work to be done in the bio-assay realm.

Deposition Control

Environmental regulations have been a major factor in the growing tendency to reduce water losses. Many companies have been tightening up their systems, operating them at higher cycles of concentration, reducing or even eliminating blowdown, and benefiting from the resulting economies.

There has also been a trend to the use of plant effluent as cooling tower makeup. Demonstration studies have been authorized and funded by the Environmental Protection Agency (EPA) so that research data can be available to industry to stimulate water reuse. In all, there are many variations of reuse underway, and many involve the cooling tower system.

Water treatment programs are being adjusted in this new direction as the needs arise, and further research information is thus uncovered. Many plants have set zero effluent as their targets for the near future.

So the need for scientific fouling control continues to grow because of increasing demands on cooling water systems. Such demands include (1) water reuse, (2) the need to use poor quality makeup water, and (3) the thermal pollution control pressures. The nuclear utility industry is one example of the increasing need for control of fouling. This giant in its infancy requires efficient cooling to keep power costs at a reasonable level. But it faces new problems because of thermal pollution regulations which require the use of some method, such as cooling towers, for reducing the high temperatures of their effluent.

The need for fouling control is also evident in view of the greater use of the less expensive metals to decrease capital expenditures.

Fouling was recently termed, "the major unresolved problem in heat transfer" in a paper by members of the Heat Transfer Research Institute.[2] Their conclusion that this situation need not persist is a logical one. The problem is being studied not only from an engineering viewpoint. It is under attack by research in surface chemistry, physical chemistry, water chemistry and other disciplines.

The problem of fouling has recently been greatly reduced by the development of dispersants which prevent particulate and crystalline matter from building up in particle size and settling out in heat transfer equipment. The past several years have seen many research developments dealing mainly in colloidal phenomena. The newest dispersants have even controlled or eliminated the tuberculation of iron pipe, a problem that has plagued many systems for years.

Foulants

Common fouling factors in cooling waters include (1) unclarified surface water–organics, clays, silts, metal oxides, (2) iron fouling from makeup water or corrosion in system, (3) calcium salt deposits–calcium carbonate and calcium sulfate, (4) oil, (5) aluminum, and (6) atmospheric pollutants–dust, pollen, insects and gases.

Dispersants–What They Do

Today's fouling control is described as "dispersing" since the activity is more mechanical than chemical. The basic mechanism includes (1) providing surface protection through affecting particle-surface attractive forces, (2) controlling particle agglomeration and deposit nature through affecting particle-particle attractive forces, and (3) controlling the rate of precipitation and crystal growth rate.

A specialized mechanism identified as "rate control"

Source: *Materials Protection*, June 1972, 19-23

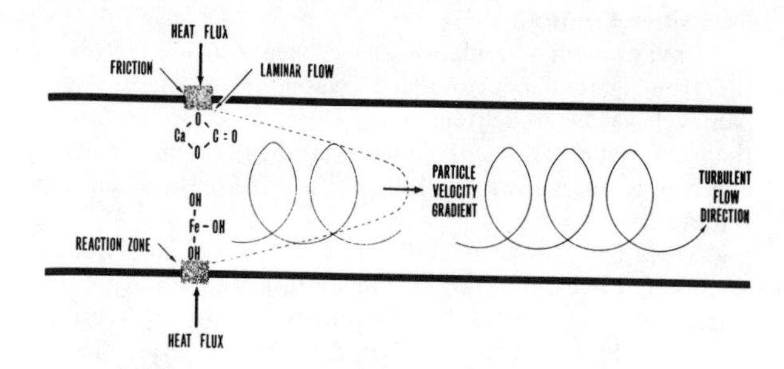

Figure 1 – Growth of crystalline salts such as calcium carbonate increases in higher temperature areas.

involves the formation of a metal-ion complex which controls the growth of a precipitating salt momentarily. This allows time for a polymer to be adsorbed on the scaling nucleus. The electrostatic charge of the polymer then repels particles that could contribute to growth and deposition.

The most sophisticated dispersants have a complex series of activities that do not depend on a single agent, but rather on the successive or "cascading" activities of several agents. They neutralize the greatest area of potential trouble by concentrating their activity at this focal point. This critical area is the layer of water in contact with the heat exchange surface. The reason it is critical is that growth of such crystalline salts as calcium carbonate increases in higher temperature areas (Figure 1). In these interfaces, the concentration of the foulants is greater, the pH is higher, and the activity of most reactions is also faster.

Dispersants–What They Are

Some of the organics used as dispersants include lignosulfonate, polyacrylates, polyacrylamides, phosphonates, and phosphoric acid esters.

A broad-spectrum dispersant might combine two or more primary adsorbents to control the wide range of foulants in a cooling tower. But for a particular foulant, the greatest economy can be obtained by selecting the one which has proved most efficient for the particular fouling problem.

A wetting agent or supplementary dispersant is commonly required to overcome the hydrophobic characteristic imparted to metal and particle surfaces by the presence of oil, grease, and some metal oxides in water. This type of supplementary activity can have a pronounced effect in decreasing fouling and restoring heat transfer efficiency.

The newest developments provide redispersion of soft, amorphous deposits and also prevent further deposition. This activity offers the advantage of employing the dispersant periodically for defouling or using it intermittently at a lower use cost, rather than applying it continuously.

The latest agents have been under study not only for their compatibility with corrosion inhibitors but also to aid directly in the work of corrosion control. Previously, it was sufficient to eliminate deposition, thereby preventing cell-type attack under the deposition. Now the dispersant is expected to film out on the metal surface in some form or another to also minimize cathodic or anodic corrosion.

Biological Fouling

With greater water reuse in cooling towers, the potential for microbiological fouling is more intense than ever. With sewage treatment plant effluents used as makeup to towers and with recycled plant effluents as makeup, the microbiological loading on towers continues to increase. And the trend to high pH cooling waters for corrosion control presents a quite different microbiological complex to the tower operator.

Products are currently available and widely used by industry that can readily control bacteria, fungi, and algae growths in cooling towers, regardless of the makeup water characteristics. The high pH waters and recycled effluents present new conditions where the basic microbiocide of the industry, chlorine, may be less economical and in some cases, even undesirable.

Chlorine combines rapidly with many organics and with ammonia. The reaction products can be less biodegradable and more injurious to the environment than the original organic or ammonia. Even the traditional role of chlorine in disinfecting sewage treatment plant effluents is being questioned because of indications that such treatment is affecting fish life.[3] Chlorine reacts with ammonia, for example, and the chloramine formed by this reaction can be more toxic than free chlorine to warm water fish. As little as 0.4 milligrams per liter can kill adult fish, and 0.05 milligrams per liter is lethal to trout fry.[4]

Chlorination of a phenol-containing tower water can result in the formation of chlorinated phenols which notoriously cause off-taste of municipal water supplies even in trace quantities.

Other agents and combinations of toxicants, however,

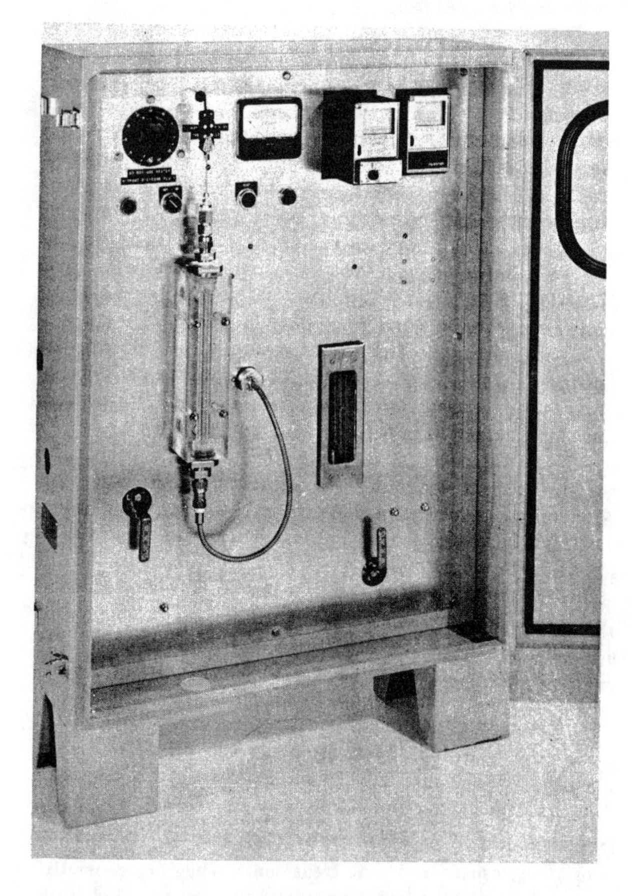

FIGURE 2 – Portable monotube deposition tester (open). Window permits visual check of fouling. Instrumentation records temperature differences to reveal fouling rates.

can take up the slack. Such biocides are often used in an alternating program to insure complete control of slime growths and prevent acclimatization of the organism to a single biocide.

The major projects confronting the water treatment field today are (1) to provide biocides that can be neutralized economically after use, and (2) to determine tolerances for biocides on waste treatment plants. Work in both areas is well underway.

In the first category, such chemicals as acrolein and halogenated organics are available and can be neutralized with sodium sulfite after performing their function.[5] This permits discharge of the treated water without endangering fish life.

In the second category, much data are being gathered from government sponsored projects and from industrial firms. As an example, an EPA report[6] on one of the demonstration programs sponsored by the government states as a conclusion, "Methylene bisthiocyanate is similar to sodium pentachlorophenate in that it passes through the chromate removal system and enters the waste treatment plant. It has no adverse effect on the plant and does not appear in the clarifier effluent."

Another potential problem is created with higher pH circulating waters and the use of efficient dispersants, increasing the possibility of shortening tower wood life through delignification. Interested organizations are now observing and collecting data on this area of operation. The Cooling Tower Institute is just one of such organizations involved and is serving as a reviewing authority.

Figure 3 — This type of respirometer is used for oxygen uptake determinations and toxicity evaluations.

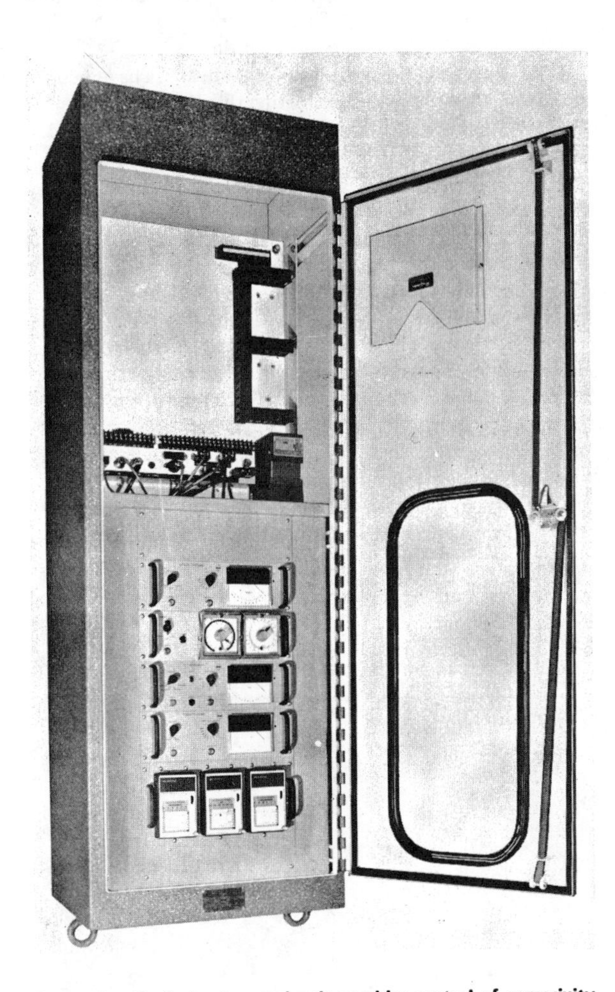

Figure 4 — Typical automated unit provides control of corrosivity, pH, inhibitor, and conductivity.

Monitoring Methods

Corrosion

A standard technique for monitoring corrosion has been the reliable coupon technique. This is still widely practiced. Other tools have been developed by various instrument manufacturers and water treatment companies because of their greater convenience over the coupons and the necessary weight-loss measurements. Instrument methods fall generally into two categories, the electrical resistance technique and the applied potential technique.

With either technique, corrosion measurements are made quickly without removing the sensing device from the installation. Measurements can be continuous, sometimes revealing sudden variations in corrosion rates.

With the electrical resistance technique, the probes can be installed in aqueous and nonaqueous process streams. Generally, the method is more accurate than the weight-loss method, but has disadvantages as well, including complications in reaching the final measurement because of the inherent nonlinearity in the relationship between electrical resistance and cross-sectional area. Any conductive deposits forming on the probe can create misleading results. Also, temperature fluctuations must be compensated for, and pitting characteristics cannot be determined accurately.

The second technique is based on linear polarization at low potential, and the instruments can provide instantaneous corrosion rate data which can be read directly from the instrument face in actual corrosion rate units (mils per year). Instruments are available using the 2-electrode or 3-electrode systsm. Of these, the latter offers the maximum in simplicity and realiability in performance.

Deposition

The monitoring of anti-deposition programs can be accomplished by test heat exchangers, by comparison of operating parameters, and by various differential analyses of water conditions. A specialized type of equipment that gives a relatively quick readout on fouling tendencies is the monotube deposition tester (Figure 2).

Biological

Monitoring methods for cooling tower slime control

Source: *Materials Protection*, June 1972, 19-23

programs have remained virtually unchanged over the past 10 years. The principal method is macroscopic, and is strictly observation of surfaces in the system. In addition, there are microscopic examinations of deposits and of water. The determination of populations of organisms such as algae, sulfate reducing bacteria, and total aerobic organisms is widely practiced both in the field and in the laboratory. One method of monitoring slime control programs is illustrated in Figure 3.

Filter techniques and the use of disposable and injectable tubes have aided in data gathering. Attempts to measure the traces of adenisonetriphosphate (ATP), a substance present in every living cell, have not proven commercially significant in cooling water monitoring.

Instrumentation

More and more, operators of cooling systems are relying on automation. The units insure control of critical variables at the optimum values, thereby safeguarding the system. The close regulation of parameters is also important in minimizing waste disposal considerations.

In the past several years, significant advances have been made in such instrumentation. During this time, equipment has been designed specifically for the unique conditions and requirements of cooling water. Prior to this time, instrumentation was used that had been designed for chemical process operations.

Today, many practical features are built into the instrumentation to achieve maximum dependability. Such features include (1) sensing units located at the cooling tower to minimize response time, (2) preamplification of the signal at the sensing point so it can be transmitted any distance and eliminate interference, (3) instruments located in a control room or other selected position where personnel are available, and (4) dependable solid state electronics and miniaturized recorders. All are characteristics of modern equipment (Figure 4).

Generally, an automation system provides a package to measure and control the inhibitor level (chromate or nonchromate), conductivity for blowdown control, pH, and in many cases, corrosivity. The corrosivity device serves as a backup or as an override on the system[7] to provide an alarm or perform certain functions that are programmed in for rapid restoration of normal conditions.

Chlorination programs can be automated by the use of ORP (oxidation-reduction potential) instruments. Liquid nonoxidizing biocides are frequently injected into cooling systems automatically by the use of timing devices operating proportioning feed pumps.

References

1. R. Garton. Biological Effects of Cooling Tower Blowdown. National Environmental Research Center, Corvallis, Ore.
2. J. Taborek, T. Aoki, R. B. Ritter, and J. W. Palen. Fouling: The Major Unresolved Problem in Heat Transfer. Heat Transfer Research Institute, Inc., Alhambra, Calif. J. G. Knudsen, Oregon State University, Corvallis, Ore.
3. John A. Zillich. Toxicity of Combined Chlorine Residuals to Freshwater Fish. *Journal WPCF,* 44 (1972) February.
4. J. E. McKee and H. W. Wolf. Water Quality Criteria. 2nd Ed., Pub. No. 3-A, Resources Agency of California, State Water Quality Control Board, Sacramento (1963).
5. J. F. Walko, J. M. Donohue, and B. F. Shema. Biological Control in Cooling Systems. Presented at the International Water Conference, Pittsburgh, Pa. (1972) November.
6. Environmental Protection Agency, Water Quality Office, Program No. 12090 EUX. Reuse of Chemical Fiber Plant Wastewater and Cooling Water Blowdown.
7. United States Patents 3,440,525, 3,361,150, and 3,430,129 issued to Universal Interloc, Inc., Santa Barbara, Calif.

Bibliography

G. Hatch and P. Ralston. Aminomethylenephosphonate-Zinc for Control of Oxygen Corrosion. Presented at American Chemical Society meeting (1970) February.

JOHN M. DONOHUE is director of product management at Betz Laboratories, Trevose, Pa. where he coordinates new product development activities in the boiler, cooling, and process fields. Donohue holds a BSc in chemical Engineering from Drexel University. He is a member of the NACE and the API.

Thirty Years' Experience With

Silicate as a Corrosion Inhibitor In Water Systems*

By HENRY L. SHULDENER and SIDNEY SUSSMAN

Abstract

The development of a simple, economical proportional liquid feeding device for distributing sodium silicate in domestic water pipes permitted effective and widespread use of this inhibitor. For the past 32 years it has been used for protection of galvanized iron, galvanized steel, yellow brass, and copper water piping in thousands of buildings in East Coast cities which have corrosive water supplies.

Properly controlled, silicate treatment has eliminated rusty water, maintained satisfactory flow rates, and minimized failures due to pitting and to clogging by corrosion products. Comparative field experiments have demonstrated its effectiveness.

Most cases of poor results when using a silicate inhibitor have resulted from improper feeding by dosing methods, piping defects, poor plumbing design or fabrication, or improper operation of the water system, particularly with respect to lack of hot water temperature control.

Studies of the protective mechanism have shown that it involves formation of a thin film containing both silica gel and an absorption compound of silica and the metal hydroxide. Further basic studies are desirable for understanding the respective roles in protective film formation of alkalinity and silica in natural waters as compared to those in added silicate. 5.8.2

Shuldener Sussman

HENRY L. SHULDENER is President and Technical Director of Water Service Laboratories, New York, N. Y. He holds a 1920 BSChE from New York University's School of Engineering and is a licensed professional engineer in New York. His whole professional life has been devoted to water treatment and corrosion prevention. Mr. Shuldener has invented several chemical feeding devices, published many technical articles and is a member of Association of Consulting Chemists and Chemical Engineers, ACS, NACE, AIChE, AWWA, Electrochemical Society, and other organizations.

SIDNEY SUSSMAN is chief chemist for Water Service Laboratories, New York City. He has a BS from Polytechnic Institute of Brooklyn and a PhD from Massachusetts Institute of Technology in 1937. Before joining his present employer in 1949 he was employed by the DuPont Company, the Permutit Company and Liquid Conditioning Corp. His principal work is in corrosion control in water systems. He has written many technical papers and obtained several chemical patents. Dr. Sussman is a member of NACE, ACS, AWWA and other technical groups and was first president of Metropolitan Water and Wastes Society.

Introduction

SMALL CONCENTRATIONS of sodium silicate were first added to potable waters during the 1920's. Speller[1] suggested this be done to control corrosion of galvanized steel pipe in domestic water systems. Thresh[2] in Great Britain first suggested it as a means for minimizing the health hazard resulting from the solution of lead from lead piping, and later as a means for controlling corrosion of ferrous piping.

Speller's suggestion led National Tube Company to sponsor research by Texter at Mellon Institute[3] and by Russell at M.I.T.[4] which demonstrated the effectiveness of sodium silicate in reducing corrosion rates of steel.

Solid silicate was first used and was applied to the water by means of a drip or by-pass type of feeder.[4,5] With this type of feeder, silicate addition took place whether or not water was being consumed. High concentrations of dissolved silicate accumulated in the vicinity of the feeder at times of low flow and resulted in the development of difficultly removable silicate deposits which caused partial or complete clogging of the piping. These experiences with deposit formation near the point of application caused silicate treatment of water to acquire a poor reputation.

The early workers[3,5] also reported that the silicate treatment was effective for only a limited distance from the feeder. Speller placed this limit at 100 to 200 feet of piping.[6] This conclusion was definitely related to the drip method of feed used initially. The formation of deposits near the point of application depleted silicate available for protection of the piping. Since subsequent developments eliminated this problem, Speller omitted this limitation in later editions of his book.[5] In fact, later studies at Paignton, England, showed that the silicate was protecting a main at a distance 17 miles from the point of application and throughout the town distribution system.[7]

Despite these early handicaps, the method was soon applied in New York, Boston, Pittsburgh, and Great Britain. The primary goal in these installations was the elimination of "red water" (rusty water) and silicates proved most effective for this purpose.

Speller[5] at an early date recognized the limitations of solid silicates for water treatment. These materials were virtually insoluble in cold water, thereby limiting the application of the method to hot water systems. Speller expressed a hope that a convenient means would be found for the controlled feeding of liquid silicates so as to extend the utility of the method.

This was accomplished five years later by the development of the Chemistat,[8] a simple, economical, proportional liquid feeding device which finally made possible the use of liquid silicates for corrosion control in both cold and hot water piping systems in buildings without fear of buildup of silicate deposits. Experience with controlled silicate feeding soon showed that it was possible to avoid formation of such undesirable accumulations while permitting the formation of a thin, self-healing, protective film which did not build up on itself. The application of this feeder in buildings, as well as the use of proportionating pumps in the treatment of large municipal water distribution systems, showed that corrosion could be inhibited at very considerable distances from the point at which the silicate was introduced into the piping.

Although the first use of sodium silicate in water systems was directed at eliminating red water or lead pickup and only indirectly at reducing the corrosion of steel, during the intervening years this additive has also been shown to be an effective corrosion inhibitor for galvanized iron, galvanized steel, yellow brass, copper, and aluminum. It is currently being applied for protection of piping in many thousands of residences, apartment buildings, offices, hospitals, laundries, and hotels. Some municipal water distribution systems in Germany, Great Britain, and the United States are silicate-treated for corrosion control. In this country the method is employed for corrosion control in smaller communities, such as Hollywood, Florida; Lebanon, Pennsylvania[9]; Ossining, New York, and many others. Silicates are not used more often for corrosion control in large water systems because of the economic disadvantage when compared to the use of the less expensive lime.

Industrially, sodium silicate is commonly applied as a corrosion inhibitor to protect piping carrying zeolite-softened water, particularly in commercial laundries, and has been added to oil field brines for protection of piping and to water filled gas holders. There is also some use of silicates for corrosion protection of water tanks on shipboard and of the cooling systems of inboard motors on boats.

Theoretical

Conflicting views have been expressed regarding the mechanism by which

★ Submitted for publication November 30, 1959. A paper presented at a meeting of the North Central Region, National Association of Corrosion Engineers, Cleveland, Ohio, October 20-22, 1959.

sodium silicate minimizes the corrosion in water piping. Early experiences with over-dosing by means of drip feeders led many to believe that the protective layer was a reaction product of the silicate with the hardness of the water and, therefore, that the ultimate buildup of a heavy scale with resultant clogging of pipes must be expected. This view was soon discredited,[6] although not eliminated in the minds of some people. Even to this day despite the many years of contrary experience since proportional feeding of silicate was introduced, this view in some cases still persists.

Some observers have believed that the major effect of the silicate derived from neutralization of dissolved carbon dioxide by its alkali content. In recent years, investigations by Duffek and McKinney,[10] and by Eliassen and co-workers[11] have established that much greater protection is obtained when the pH of a particular environment is adjusted by addition of a silicate, than when it is adjusted to the same final pH by addition of an alkali alone in the absence of any silica.

Fundamental studies by Lehrman and Shuldener[12, 13] have shown that the protective silica film does not begin to form until there are some corrosion products of the metal present, thus confirming a mechanism suspected many years before by Speller.[6] The corrosion products form an adsorption compound with silica, removing the latter from solution. Chemical analyses of the protective layer show that it is high in silica. The silica gel-like structure of the film makes it semitransparent when wet. When dry, it is visible as a thin coating slightly colored, either brown by iron corrosion products or white by zinc corrosion products.

As soon as the thin surface film of adsorption compound forms, further corrosion of the metal surface virtually stops. Because no further corrosion products form, no additional buildup of the film takes place. This mechanism explains why such a film is self-limiting in thickness.

It also explains why the protective film is self-healing when damaged. If the film is removed from a surface by any means, the metal begins to corrode and a fresh surface film forms by reaction with the silica present in the treated water. This, of course, indicates why it is necessary to treat such systems continuously.

Since red water consists of iron oxides in suspension, the fact that the silicate precipitates these in a film on the surface of the piping as fast as formed, explains the great effectiveness of silicate treatment in overcoming red water problems.

Further studies on the mechanism of silicate inhibition are being carried out currently in several laboratories in the United States and Great Britain.

Application

For feeding silicates in large industrial and municipal applications the various types of proportional feed pumps are used.[7, 14]

The widespread successful application of sodium silicate as a corrosion inhibitor in piping systems of apartment buildings, office buildings, hotels and similar structures which are characterized by wide fluctuations in the rate of water use and limited maintenance personnel, would have been impossible without the development of a simple hydraulic-type proportional feeder with no moving parts.[8] This type of feeder, a picture of which is shown in Figure 1, is connected across an orifice inserted in the water line, applying the Venturi principle to feed treatment chemical in proportion to the flow of water.

A cross sectional view of the feeder, shown in Figure 2, indicates how the heavy, viscous sodium silicate solution is present as a separate layer in the lower part of the tank. The water entering the feeder through the inlet impinges upon the water-silicate interface, picking up a small amount of silicate which is carried back into the water line. Tracer studies with a glass hydraulic proportioning feeder and a colored inlet stream show that, following this impingement, the diverted water returns to the top of the tank and flows to the outlet. Thus the silica pickup is independent of the total interface area.

It will be seen from the above description that the amount of silicate solution entering the water line is proportional to the total flow of water so that uniform treatment is maintained under variable flow conditions. When no water is flowing in the line, no treatment chemical enters. The proportioning feeder is always connected below the water line and, therefore, it is impossible for any silicate solution to flow by gravity into the water line.

Because neither inlet nor outlet lines dip into the silicate solution, there can be no pickup of strong chemical from the feeder during any period of low pressure or even negative pressure in the water main. Additional pickup of chemical should there be sudden and heavy reversal of flow in the main is prevented by a baffle inserted just below the outlet line. Independent tests made by the New York City Department of Water Supply, Gas and Electricity demonstrated the correctness of these statements and, as a result, the Department permits installation of this feeder on potable water lines without check valves or vacuum breakers.

In general, it has been found most suitable and economical to utilize the common commercial grade of liquid silicate having a density of 40°—42° Baume and an alkali to silica ratio of 1 Na_2O:3.22 SiO_2. Using this silicate, effective control of red water and of corrosion can be obtained under normal operating conditions by addition of sufficient silicate to increase the silica content of the water by approximately 8 ppm. In some cases it is desirable to increase the initial dosage in order to more rapidly establish a protective film, and in some cases it is possible to reduce the average dosage somewhat and still to maintain the protective film once it has been established.

When applying a chemical to a potable water system, the question of possible overtreatment of the water must be con-

Figure 1—The Chemistat, a hydraulic proportioning feeder, installed across an orifice inserted in a water main for feeding sodium silicate.

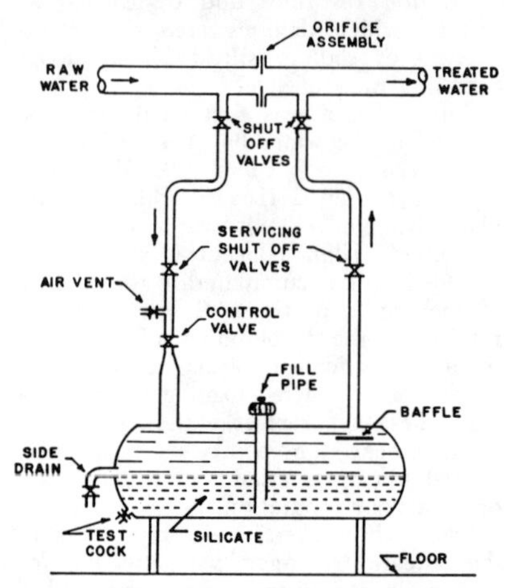

Figure 2—Cross-section of the hydraulic proportioning feeder showing the interface between the treatment chemical and the water layers, and other details of the feeder assembly.

sidered. In New York City,[15] Detroit,[16] and Yonkers, New York, application of chemical treatment to potable waters within buildings is regulated by specific codes, and approval of the chemicals and feeders by the municipal authorities is required.

Field Results

Since 1927 Water Service Laboratories has provided a corrosion control service using sodium silicate as the inhibitor for protection of domestic water piping in large apartment houses, hotels, hospitals, and office buildings in the New York, Philadelphia, and Washington Metropolitan areas. More than 400 large buildings have been treated continuously for more than 10 years and about 50 have been treated continuously for more than 25 years.

Most of the waters being treated are surface supplies, but silicate treatment has also been effective on a number of well supplies which, while much harder

Figure 3—Four inch galvanized iron cold water line from Philadelphia office building. Tuberculated section was in use for two years before silicate treatment of the water was started. The smoother section replaced this when water treatment was started and was in service for an equal length of time.

Figure 4—Orifice disc assembly taken from a 15 year old hydraulic proportioning feeder installation in New York City. Note tuberculation of galvanized iron main on the inlet side where no added silicate was present and absence of significant tuberculation on the outlet side where the added silicate entered the water stream.

and containing more natural silica, are aggressive by virtue of their high carbon dioxide content. Table 1 shows some typical analyses of waters being treated.

Several real estate management companies have kept statistical records for a period of years with regard to the development of leaks and to repair costs on groups of buildings using the same water supply. Some of the buildings were being treated with sodium silicate and others were not. These statistical studies have shown that application of silicate treatment is economically justified on the basis of savings in re-plumbing costs. Numerous other case histories have been cited by Stericker[14] of the effectiveness of silicate treatment in minimizing corrosion.

TABLE 1—Typical Analyses of Waters Silicate-Treated for Corrosion Control*

	LOCATION AND SOURCE				
	Atlantic City N.J.	New York N.Y. (Catskill)	Philadelphia Pa. (Schuylkill)	Jamaica N.Y.	Idlewild N.Y.
PROPERTY	Surface & Well	Surface	Surface	Well	Well
pH	5.5	6.9	6.9	6.8	6.9
Total Hardness (as $CaCO_3$)	13	20	153	308	1650
Calcium (as $CaCO_3$)	6	14	89	124	900
Total Alkalinity (as $CaCO_3$)	6	10	48	218	444
Free Carbon Dioxide (as CO_2)	High**	3	12.5	68	100+
Chloride (as Cl)	10	4	23	20	2150
Sulfate (as SO_4)	11	8	101	95	350
Silica (as SiO_2)	6.5	2.5	8	24.5	30
Total Solids	41	31	257	421	4640

* Analyses before treatment. All values, except pH, are in ppm.
** Includes organic acidic materials.

Housing Development Study

Thirty years ago rusty water and poor flow conditions appeared in the galvanized wrought iron hot water system of an apartment house development consisting of 54 large buildings on Long Island. The insurance company which owned and operated this development carried out a comparative test in three adjacent buildings. After installing long test sections of new galvanized iron piping, the water in the first building was not treated, that in the second building was treated with sodium silicate applied by building personnel by means of a drip feeder, and that in the third building was treated by a regular water treatment service by means of a hydraulic proportioning feeder.

Test pieces of the new pipe were removed at yearly intervals and examined by an independent consulting laboratory. After only one year, test pieces from the first two buildings were coated with a thick, rust-colored coating containing about 14 percent iron oxide and were moderately pitted beneath the coating. The test piece from the third building, at which the silica had been proportionately fed by the hydraulic feeder, was coated with a thin tan coating containing only 4 percent iron oxide and beneath which was found only a slight degree of pitting.

After four years the corrosion rate in the piping of the third building was less than 25 percent of that in the piping of the other two buildings. At this time serious difficulties became manifest in the hot water piping systems of the many untreated buildings and silicate treatment of the hot water was started in all 54 buildings of the development.

Ten years later, serious flow troubles developed in the untreated *cold* water lines and silicate treatment was started after substantial piping replacements were made.

Silicate treatment of the hot water permitted the owners to defer total re-piping until the development was twenty-six years old, although at nine years of age, when silicate treatment was started, the hot water piping was in such condition that normal experience in the area indicated that general re-piping would be required within a few years. Thus, the silicate treatment extended the life of this piping from an estimated additional three or four years without treatment to seventeen years with treatment.

AISI Study

An extensive independent field study of the effectiveness of silicate as a corrosion inhibitor was undertaken by the Committee on Steel Pipe Research of the American Iron and Steel Institute.[17] This was done at another large housing development in Brooklyn, New York, which had been built during World War II with galvanized steel piping. Periodic examinations of test specimens removed from both hot and cold water lines in treated and nearby untreated buildings were made over a period of 17 years. No trouble has been encountered from red water or reduced flow in any of the treated buildings, except for a few short nipples of small diameter supplying kitchen sinks. These difficulties were caused mainly by collected sediment. This was in contrast to normal experience with such pipe carrying New York City water. Rusty water and major pipe replacement in hot water systems are common within five to ten years on this supply.

The tests showed that untreated cold water lines had a greater corrosion rate than treated *hot* water lines. Over the period of the test, pitting was deeper in the untreated cold water lines than in the treated cold water lines.

Qualitative observations made during this test indicated that some tubercles formed fairly rapidly during the early years, but then tended to flatten out. Similar observations have been reported from silicate-treated municipal water systems in Germany.[18] There is, however, no basis to claims sometimes made that silicate treatment of a water will remove existing rust deposits.

Comparative Pipe Specimens

Figure 3 shows sections of cold water line exposed for two years in a Philadelphia office building. The tuberculated section of this 4 inch pipe had been in use for two years with no water treatment. It was replaced by the other section of pipe at the time that silicate

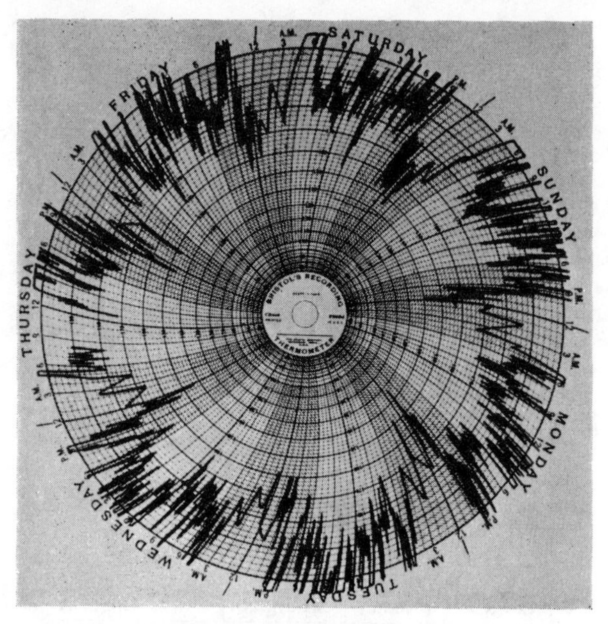

Figure 5—Hot water temperature recorder chart showing poor temperature control at an apartment house. Note the wide and frequent fluctuation, and temperatures as high as 210 F.

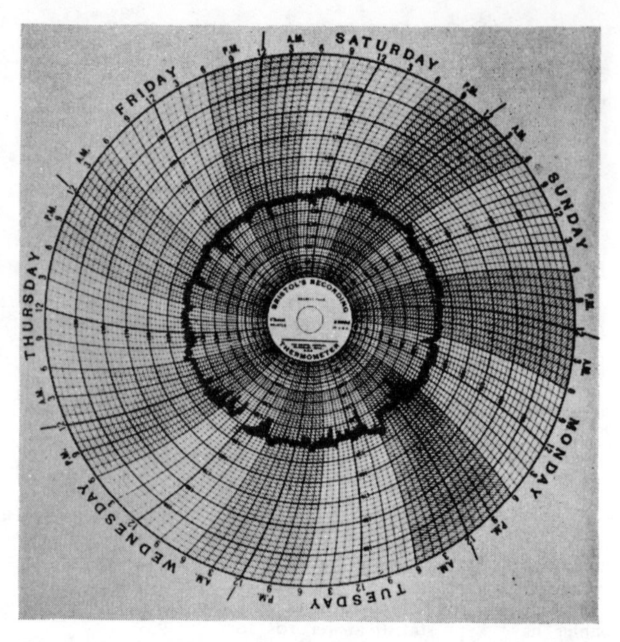

Figure 6—Hot water temperature recorder chart showing very small deviation from 140 F at apartment house with properly adjusted thermostatically controlled mixing valve, and hot water generating and distributing equipment in good working order.

treatment was started and this new section was removed for evaluation after it had been in service for two years. The difference between the tuberculated appearance of the specimen exposed to untreated water and the reasonably smooth coating on the specimen exposed to treated water is made even more apparent by the rust color of the former and the light tan of the latter.

A striking example of the effects of silicate treatment can be seen in Figure 4 which shows an orifice disc which had been in use for 15 years in New York City diverting untreated cold water to the hydraulic proportioning feeder described above. Note the heavy tuberculation of the main at the untreated inlet side and the superficial rust coating at the treated outlet side.

Degree of Protection

Several attempts have been made to evaluate quantitatively the degree of protection afforded to steel pipe by silicate treatment of water. By measurement of the dissolved oxygen content of water entering a test pipe and that of the water leaving the test pipe, Russell[4] estimated that the corrosion rate was reduced 50 percent in cold water lines and 80 percent in hot water lines by silicate treatment.

On the basis of the iron content of water after passing through a test section of steel pipe, Stericker[7] estimated that silicate treatment reduced the corrosion rate of steel by 73 percent.

Hudson and Wormwell[19] reported results of a test carried out in the piping system of an English school. Coils of steel wire were cleaned and weighed before and after insertion in sections of the hot water piping system of one building. In the absence of water treatment, corrosion rates in the range 12 to 30 mils per year were found. When the silica content of the water was increased by 15 ppm, the corrosion rate was reduced to 3.7 mils per year. This corresponds to a reduction of 82 percent in the average corrosion rate of the steel in untreated water, and is a surprisingly good agreement with the above data obtained by other, indirect methods.

Non-Chemical Influences

Poor results obtained with silicate treatment of waters are usually caused by non-chemical factors. These may be related to the original design or construction of the water piping system, to the quality of piping materials, or to operating conditions. This subject has been discussed at some length by Shuldener.[20]

Typical of poor construction was the case of a hot water storage tank which pitted through at the top within six months because a layer of air was trapped in the upper section as a result of the outlet pipe having been screwed down an inch into the tank instead of being flush with the top.

Poor quality of piping materials is illustrated by the common occurrence of skips, flaws, inclusions, and pinholes in galvanized pipe. These help to localize corrosion when water and temperature conditions are such that potentials of the zinc-iron system are reversed.[21, 22]

Probably the most important operational factor tending to counteract the protective effects of silicate addition is excessively high hot water temperature. Not only is 140 F a practical water temperature ceiling from the viewpoint of hazards to people, but higher temperatures accelerate the corrosion process, making it increasingly difficult for any given inhibitor dosage to protect metals as temperature rises.

Temperature fluctuations normally associated with overheating cause expansion and contraction of the piping. This results in leaks at joints due to excessive strains, and flaking off of old rust deposits forming rusty water, developing local stoppages by sedimentation in small diameter branch lines, and accelerating corrosion by exposing fresh metal which behaves anodically. Rusty water thus formed leads users to run large volumes of water to waste in the hope of getting clean water. This can overload the hot water generator and aggravate an already bad operating condition in a continuing vicious cycle. Such temperature fluctuations are shown on a temperature recorder chart from the domestic hot water system of a building in which poor control permitted the hot water to reach 210 F at times (Figure 5). In contrast, Figure 6 shows a temperature recorder chart from a domestic hot water system in which the hot water temperature was thermostatically controlled at about 140 F.

These observations emphasize that silicate treatment of water is not a cure-all and is no substitute for good design, good construction, and good operation of a water piping system.

Summary

Thirty years experience with silicate as a corrosion inhibitor in water systems of large buildings has shown that it provides a simple, safe, and reliable method for eliminating red water and minimizing corrosion in domestic water piping whether it be galvanized, brass, or copper. The experience has also shown that silicate treatment is not a cure-all and cannot be depended upon to overcome the handicaps introduced by poor design, poor construction, or poor operation.

References

1. F. N. Speller. *Journal of the Franklin Institute*, 193, 519-520 (1922).

2. J. C. Thresh. *Analyst*, 47, 459-468, 500-505 (1922).
3. C. R. Texter. *Journal of the American Water Works Association*, 10, 764-72 (1923).
4. R. P. Russell. Publications from M.I.T., Vol. 61, No. 69, Pub. Ser. No. 369, 4pp. (1926) Feb.
5. F. N. Speller. *Corrosion—Causes and Prevention*. 1st Ed., McGraw-Hill Book Co., New York (1926), p. 353.
6. F. N. Speller. *Chemical Age*, 32, 457 (1924).
7. W. Stericker. *Industrial and Engineering Chemistry*, 30, 348-351 (1938).
8. H. L. Shuldener. U. S. Patent 1,796,407 (March 17, 1931).
9. U. S. Public Health Service. Municipal Water Facilities, p. 30 and 118 (December 1956).
10. E. F. Duffek and D. S. McKinney. *Journal of the Electrochemical Society*, 103, 645-648 (1956).
11. R. Eliassen, R. T. Skrinde, and W. B. Davis. Birmingham Technical Meeting, American Iron and Steel Institute, October 15, 1958.
12. L. Lehrman and H. L. Shuldener. *Journal of the American Water Works Association*, 43, 175-188 (1951).
13. L. Lehrman and H. L. Shuldener. *Industrial and Engineering Chemistry*, 44, 1765-1769 (1952).
14. W. Stericker. *Industrial and Engineering Chemistry*, 37, 716-720 (1945).
15. New York City Health Code, Section 141.07 (1959).
16. A. J. Dempster. *Journal of the American Water Works Association*, 45, 81-88 (1953).
17. Committee on Steel Pipe Research, American Iron and Steel Institute, Progress Report PR—104—L.
18. G. Seelmeyer. *Gesundheits Ingenieur*, 75, 69-71, No. 3/4 (1954).
19. J. C. Hudson and F. Wormwell. *Chemistry and Industry*, 1957, 1085.
20. H. L. Shuldener. *Corrosion*, 10, No. 3, 85-90 (1954).
21. R. B. Hoxeng. *Corrosion*, 6, 308-312 (1950).
22. H. L. Shuldener and L. Lehrman. *Journal of the American Water Works Association*, 49, 1432-40 (1957); *Corrosion*, 14, 545t (1958).

Source: *Corrosion*, July 1960, 354t-358t

Chelation compounds as cooling water corrosion inhibitors

A. WEISSTUCH, D. A. CARTER, and C. C. NATHAN, *Betz Laboratories, Inc., Trevose, Pa.*

Investigation of chelates as cooling water corrosion inhibitors has shown that chelating agents can act as accelerators or inhibitors of steel corrosion, depending upon the properties of the chelant. Evidence presented in this paper suggests that improved corrosion inhibition results from the formation of stable, 5 or 6-membered chelate rings between the inhibitor and surface metal atoms.

CHELATING AGENTS such as ethylenediaminetetra-acetic acid (EDTA) and nitrilotriacetic acid (NTA) have become important tools in boiler feedwater treatment and chemical cleaning processes. In these applications, the chelants solubilize metal ions (Mg^{2+}, Ca^{2+}, etc.) which might otherwise form insoluble deposits.[1-6] Much less attention has been directed toward the use of metal chelants as aqueous corrosion inhibitors.[7]

Nature of Chelates[8-9]

By definition, chelates are coordination complexes formed between a metal cation and a molecule (ligand) which contains at least two groups capable of coordinate bonding with the cation. As a result of multiple bonding with the central ion, a ring or claw structure is formed. Chelating agents are classified by the number of co-ordination sites (or dentates) they possess. For example, EDTA is classified as sexadentate, because one EDTA molecule can satisfy metal ion coordinate covalencies as great as six (Figure 1). Several 5-membered rings (or claws) containing the metal ion are formed. Rings of fewer than five members or more than six members are not usually encountered because of their instability. The formation of stable ring structures, the "chelate effect," makes chelate formation preferred over simple complex formation.[10]

Chelating agents generally are organic molecules containing several functional groups. Though amino ($-NH_2$) and carboxyl (-COOH) groups are frequently involved, other functional groups or atoms can supply the pair of electrons necessary to form a coordinate bond with a central atom. Common chelating functionalities are presented in Table 1. The surface activity of the chelant and the aqueous solubility of the resultant chelant determine whether a chelant is used as a descaling agent or as a

TABLE 1 — Common Chelate-Forming Functional Groups

Acid Groups (Replacement of H)		Basic Groups (Electron pair donation)	
-COOH	(Carboxylic)	$\geq$C=O	(Ketone)
$-SO_3H$	(Sulfonic)	$-NH_2$	(1°, 2°, 3° amine)
-OH	(Phenolic)	$\geq$C=N-OH	(Oxime)
-SH	(Mercapto)	-N=	(aromatic hetero)
$-P(O)(OH)_2$	(Phosphonate)	$\geq$C=NH	(imino)
		-O-R	(ether)
		-S-R	(thioether)
		-OH	(alcoholic)

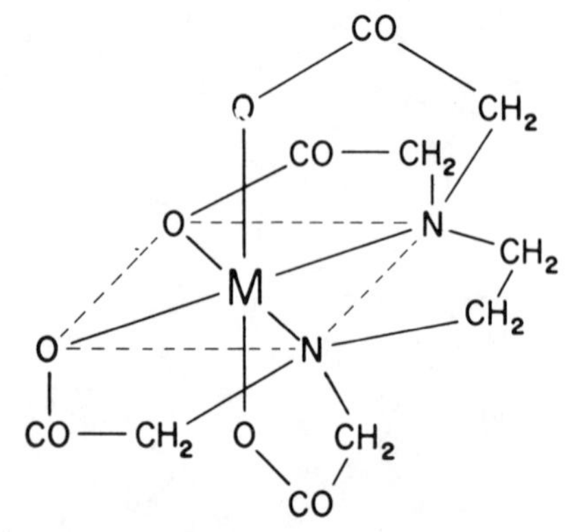

Figure 1 — Sexadentate metal/EDTA complex.

corrosion inhibitor. In general, aqueous solubility[11] is determined by the net charge on the chelate and the hydrophilic/hydrophobic balance of the chelant.[11] This means that neutral chelates and chelates containing large hydrocarbon groups will be less soluble than chelates that are electrically charged or contain polar groups. For example, the charged sulfonic acid group imparts water solubility to 8-hydroxyquinoline/iron chelates (Figure 2).

Chelates and Corrosion Inhibition

Previous papers[7,12-14,20] have suggested that some organic corrosion inhibitors containing more than one functional group may function by chelating the metal ion while it is still associated with its crystal lattice; the metal ion is not solubilized, and the surface chelate acts as a corrosion inhibitor. However, no evidence exists to support this hypothesis.

It is generally accepted that many nonpassivating organic inhibitors function by the following mechanism:[15] Preliminary physical adsorption at the metal surface, characterized by low heats of adsorption, followed by a chemisorption process in which a relatively strong metal-inhibitor chemical bond is formed. The resultant film of chemisorbed inhibitor is then responsible for the observed inhibition either by physically blocking the surface from the corrosive environment or by retarding the electrochemical reactions.

It is in the chemisorption phase that chelation processes can be of importance. Just as the formation of multiple bonds with a metal ion in solution leads to increased stability, multiple attachments to the metal surface by an inhibitor containing several functional groups can lead to stronger chemisorption and better corrosion inhibition. However, chelates used for boiler descaling can cause accelerated rates of metal dissolution.[15] A chelating agent, therefore, can act as an accelerator or an inhibitor of corrosion, depending upon its physical and chemical properties.

Requirements for Corrosion Inhibition by Chelating Agents

Successful chelate corrosion inhibitors require high surface activity and low aqueous solubility, characteristics common to most organic corrosion inhibitors. Surface activity determines to some extent the selectivity of the inhibitor for chelation with surface metal atoms (or ions), relative to solubilized cations. Solubility of the surface chelate is critical in that it is necessary that the chelate remain adherent to the metal surface to form the protective film. Thus, chelants that form soluble chelates (such as EDTA) find application as chemical cleaners and descalers and not as corrosion inhibitors.

These considerations provide the corrosion scientist with several paths toward the development of corrosion inhibitors based on chelation concepts:

1. Existing corrosion inhibitors can be modified so that surface chelation is possible.

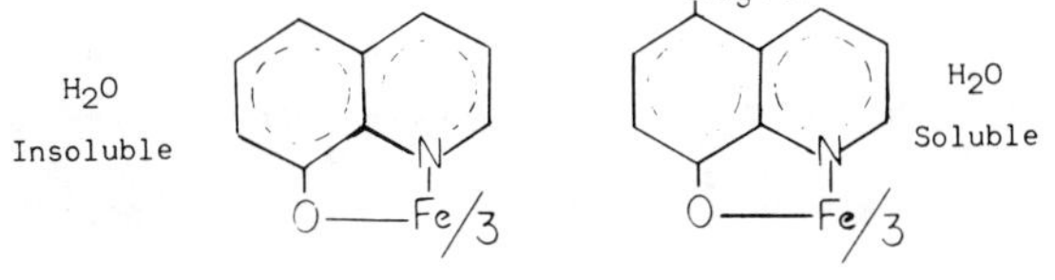

H_2O Insoluble — O—Fe/3

$SO_3^- Na^+$ — O—Fe/3 — H_2O Soluble

Figure 2 — Effect of polar substituents on aqueous solubility of chelates.

2. Entirely new inhibitors containing good chelating groups and high surface activity may be synthesized.
3. Known chelating agents can be structurally modified to increase their surface activity and/or decrease their aqueous solubility.

As an example of the last possibility, the hydrophobicity of EDTA can be increased by alkylation, i.e., addition of one or more long chain alkyl groups.

Experimental

Materials

Most materials were commercially obtainable and were of sufficient purity to be used as received. Sarcosine derivatives were laboratory synthesized and were purified by recrystallization.

Apparatus and Procedure

Three corrosion inhibition test procedures were employed using mild steel specimens: (1) coupon weight loss, (2) extrapolation of the Tafel region of electrochemical polarization curves to zero overpotential, and (3) linear polarization measurements (Stern-Geary method).

All tests were conducted in an aerated simulated cooling water environment at pH 7 and either 30 or 50 C (86 or 122 F). Details are described elsewhere.[16]

Results and Discussion

Survey of Known Chelating Agents

A diversified selection of known chelants was investigated by the electrochemical polarization technique to determine which structural and functional groups provide corrosion inhibition under cooling water conditions. Most of the chelants tested (ketones, diketones, aldehydes, oximes, salicylates, catechols, hydroxyamines, bipyridines, and hydroxyquinolines) do not display high surface activity and low aqueous solubility, and are soluble metal sequesterants rather than metal precipitants. Consequently, very few chelants provided appreciable inhibition; in fact, many of them accelerated corrosion by solubilizing metal surface ions.

Several compounds, however, provided interesting structural information. For example, 2,2 bipyridine, which has free rotation about the C-C bond (Figure 3), gave no inhibition. Whether adsorption occurs with the aromatic rings oriented parallel or perpendicular to the surface, the preferred trans type spatial orientation restricts the probability of surface chelation. On the other hand, 1,10 phenanthroline, which has a rigid structure conducive to surface chelate formation, provided measurable inhibition.

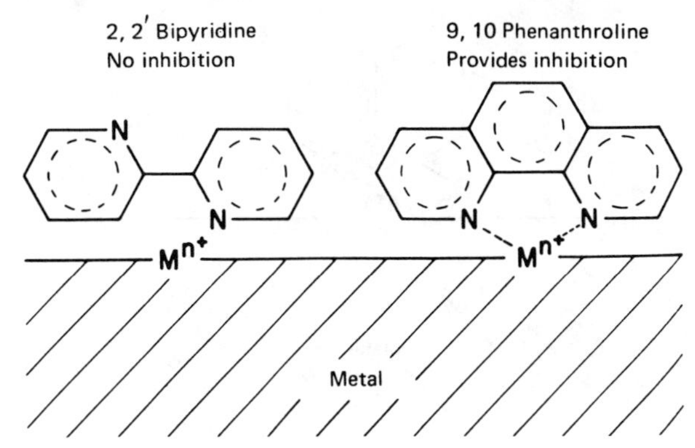

Figure 3 — Steric effects on surface chelation.

This demonstrates the specialized requirements for surface chelation.

8-hydroxyquinoline also provided corrosion inhibition (80% at 100 ppm). As seen in Figure 4, this molecule satisfies the structural requirements for surface chelation, regardless of the orientation of the adsorbed molecule. Although good corrosion inhibition was obtained with this compound, formation of a nonadherent, insoluble chelate in the solution phase was a distinct disadvantage. This demonstrates the need for satisfying the requirement of high surface activity which produces a high surface chelation to bulk chelation ratio.

In another phase of the work, several chelants were alkylated (i.e., a long chain alkyl group was introduced) to decrease their water solubility. This was done with several diketones, phenoxyacetic acids and pyrocatechols. Results of electrochemical polarization tests on pyrocatechol derivatives (Table 2) show that:

1. Decreased solubility as a result of alkylation increases corrosion inhibitor effectiveness, in agreement with the above theory.

2. Straight chain alkyl groups are more effective than branched chain groups, as seen by a comparison of n-butyl and tert-butyl derivatives.

Thus, structural variations can change the operation of a chelant from that of an accelerator to that of an inhibitor of metal corrosion.

Surfactants as Corrosion Inhibitors

The next group of molecules considered as chelating inhibitors consisted of several surface active compounds. Surfactants usually consist of a hydrophobic hydrocarbon group (e.g., a fatty alkyl chain) and a polar, hydrophilic group (e.g., a carboxylic acid group). The polar end of the surfactant is oriented toward the metal surface, with the nonpolar hydrocarbon tail presenting a hydrophobic barrier to the aqueous phase. If the polar group is chemisorbed onto the metal surface, this bonding can be strengthened by multiple attachment to the surface metal atoms.[12] Chelating surfactants should be free of the disadvantages of the nonsurface active, nonspecific chelants mentioned previously. Consequently, enhanced chemisorption through chelate formation should provide a corrosion inhibiting film that is difficult to desorb. Table 3 lists several commercially available surfactants that function as corrosion inhibitors and offer the possibility of surface chelation.

The behavior of the sarcosine derivatives has been studied in some detail. Lauroyl sarcosine, an α-amino acid derivative functions reasonably well as a cooling water corrosion inhibitor (80 to 90% inhibition at 100 ppm). It

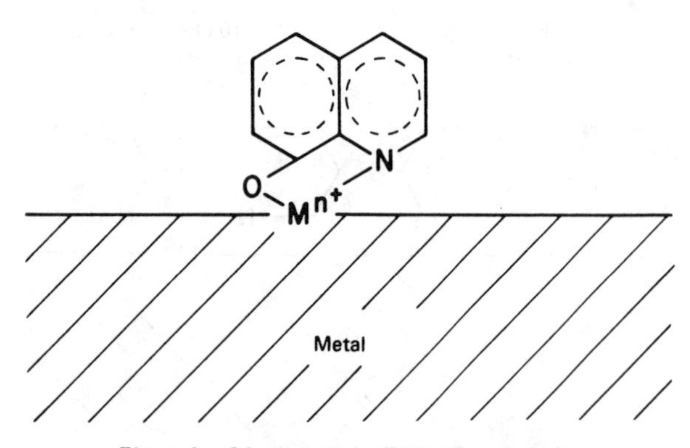

Figure 4 — 8-hydroxyquinoline surface chelation.

TABLE 2 — Cooling Water Corrosion Inhibition by Pyrocatechols

R–(benzene ring)–OH, OH

Compound	% Inhibition	R
Pyrocatechol	-14	-H
4-Methyl Catechol	84	$-CH_3$
4-tert-butyl Catechol	48	$-C(CH_3)_3$
4-n-butyl Catechol	93	$-(CH_2)_3-CH_3$
4-n-hexyl Catechol	96	$-(CH_2)_5-CH_3$

Note: S.C.W.; 30 C; pH 7; linear polarization technique.

can, in principle, form stable, 5-membered chelate rings with surface metal atoms or ions (Figure 5). β-amino acids form less stable 6-membered chelates, and γ, δ, and ε-amino acids do not chelate with metals.[17] In the latter instances, ring sizes of 7, 8, and 9 members, respectively, preclude stable chelate formation. Based on this, a correlation between inhibitor effectiveness and potential chelate ring stability should be possible.

Figure 5 shows that the potential chelate ring size of these molecules can be varied by insertion of additional methylene ($-CH_2-$) groups between the carboxyl and amide moieties. Compounds synthesized included n (= no. of $-CH_2-$ groups) = 1, 2, and 3, corresponding to potential chelate ring sizes of 5, 6, and 7. Chelate stability considerations predict a sharp decrease in inhibition effectiveness when ring size is increased from 6 to 7. Corrosion data obtained with these compounds and similar materials in which the amide methyl ($-CH_3$) group was replaced by hydrogen (-H) are presented in Table 4.

The experimental data confirm the predictions based on chelate stability theory, i.e., there is a slight increase in corrosion rate when n is increased from 1 to 2, and a large increase when n is increased from 2 to 3 (Table 5).

TABLE 3 — Possible Chelating Surfactants

$R = C_nH_{2n+1}$, where $18 \geqslant n \geqslant 11$

Structure	Name
O CH3 ‖ | $R-C-N-CH_2-COOH$	N-Acyl Sarcosine
H | $R-N-CH_2CH_2-COOH$	N-Alkyl - Aminopropionic Acid
O CH3 ‖ | $R-C-N-CH_2CH_2-SO_3H$	N-Methyl - N - Acyl Taurine
O ‖ H $R-S-N-CH_2-COOH$ ‖ O	N-(Alkylsulfonyl) glycine
R–(benzene ring)–$O-CH_2-COOH$	p-Alkylphenoxyacetic acid

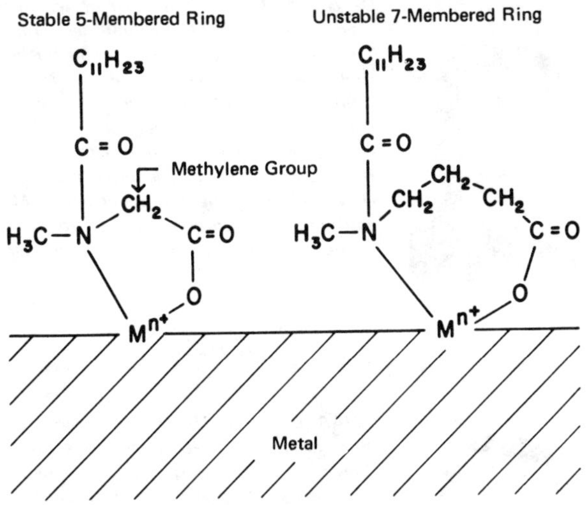

Figure 5 — Sarcosine-type surface chelates.

TABLE 4 — Summary of Corrosion Data Obtained on Sarcosine Type Compounds

$$C_{11}H_{23}\text{-}\overset{O}{\overset{\|}{C}}\text{-}\overset{R}{\overset{|}{N}}\text{-}(CH_2)_n\text{-}COOH$$

n	Chelate Ring Size	Corrosion Rate, mpy R = H	Corrosion Rate, mpy R = CH_3	Remarks
1	5-membered	52	0.5; 2; 5	Can form stable chelate
2	6-membered	12.5; 79; 36	5; 6	Can form moderately stable chelate
3	7-membered	104	42	Chelates are unstable

Note: (100 ppm; pH 7; 120 F; S.C.W.; carbon steel); blank (uninhibited solution) = 132 mpy.

Surface chelation is thus an important factor determining the corrosion inhibition properties of sarcosine-type surfactants. However, all compounds (regardless of ring size) provided some measure of inhibition (Table 4), indicating that surface chelation is not required for corrosion inhibition. Inhibition (at a diminished degree of effectiveness) is possible via simple -COOH chemisorption, as with fatty acid derivatives (e.g., stearic acid). Surface chelation, then, provides an *enhancement* of already existent corrosion inhibition properties, via multiple bonding with the metal surface and formation of stable chelate rings.

Table 4 shows that replacement of the amide $-CH_3$ group by -H causes an increase in corrosion rate. In view of the possibility of surface chelation, this increase may be explained in the following manner: All other factors remaining constant, the ease of chelate formation is determined by the availability of electron pairs of the chelant for formation of coordinate bonds with the metal ion. Because the inductive electron-releasing ability of $-CH_3$ groups is several times greater than that of -H atoms,[18] the differences in corrosion rates can be related to the greater availability of the lone electron pair on the methylated nitrogen atom. As a result, chelation is more likely with the methylated derivatives.

Future Work

The data presented here suggest that surface chelates play an important role in corrosion inhibition. However, *direct* evidence of the actual existence of such chelates is not available at present. Future work should involve the use of multiple specular reflectance infrared spectroscopic techniques[19] to show that both the

$$-COOH \text{ and } -\overset{CH_3}{\overset{|}{N}}-$$

groups are bonded to the metal surface. Another unanswered question: Does the chelation process occur when the surface metal atoms are still associated with the solid lattice or after dissociation and ionization?

Conclusions

The evidence suggests that surface chelation improves the corrosion inhibiting properties of organic molecules due to enhanced chemisorption of the inhibitor. For inhibition to be obtained, the chelant (1) should possess a large hydrophobic group rendering it surface active, and (2) should be of limited solubility.

Much further work is required to substantiate the hypotheses and elucidate the details of surface chelation phenomena.

Acknowledgment

The assistance of Dr. F. Cervi in the synthesis of derivatives used in this study is gratefully acknowledged.

References

1. J. M. Donohue and G. A. Woods. On Stream Desludging of Cooling Systems, *Materials Protection and Performance*, 7, 15 (1968) February.
2. R. B. Coughanour. EDTA Treatment Reduces Boiler Corrosion Problems, *Materials Protection and Performance*, 5, 81 (1966) May.
3. W. E. Bell. Chelation Chemistry–Its Importance to Water Treatment and Chemical Cleaning, *Materials Protection and Performance*, 4, 78 (1965) February.
4. C. N. Loucks. *Ind. Water Eng.*, 4, 11 (1967) February.
5. R. E. Elliot. *Ind. Water Eng.*, 4, 26 (1967) July.
6. D. A. Swanson. *Ind. Water Eng.*, 4, 22 (1967) December.
7. N. Hackerman. *Bull. Indian Sect. Electrochem. Soc.*, 8, 9 (1959).
8. A. E. Martell and M. Calvin. *Chemistry of the Metal Chelate Compounds*, Prentice-Hall, Inc., Englewood Cliffs, N. J. (1952).
9. F. P. Dwyer and D. P. Mellor, editors. *Chelating Agents and Metal Chelates*, Academic Press, New York, N. Y. (1964).

TABLE 5 — Correspondence of Corrosion Rates to Chelate Ring Sizes

No. of methylene groups	1	2	3
Potential chelate ring size	5	6	7
Predicted chelate stability	5	>6	$\gg 7$
Observed corrosion rate	5	<6	$\ll 7$

Source: *Materials Protection*, April 1971, 11-15

10. G. Schwarzenbach. *Helv. Chim. Acta.*, **35**, 2344 (1952).
11. A. E. Martell. *J. Chem. Ed.*, **29**, 270 (1952).
12. R. M. Pines and J. D. Spivack. *Technica I*, Geigy Chem. Co., Ardsley, New York, N. Y., 19-23 (1960) June.
13. F. M. Donahue, A. Akiyama, and K. Nobe. *J. Electrochem. Soc.*, **114**, 1006 (1967).
14. F. M. Donahue and K. Nobe. *J. Electrochem. Soc.*, **114**, 1012 (1967).
15. N. Hackerman. Recent Advances in Understanding of Organic Inhibitors, *Corrosion*, **18**, 332 (1962) September.
16. D. A. Carter and C. C. Nathan. A Polarization Study of Cooling Water Corrosion Inhibitors, *Materials Protection and Performance*, **8**, 61 (1969) September.
17. P. Pfeiffer and E. Lubbe. *J. prakt. Chem.*, **2**, 136, 321 (1933).
18. R. T. Morrison and R. N. Boyd. *Organic Chemistry*, Second Edition, Allyn and Bacon, Boston, Mass. (1966).
19. G. W. Poling. *J. Electrochem. Soc.*, **114**, 1209 (1967).
20. A. Lüttringhaus. *Angew. Chem.*, **64**, 661 (1952).

DAVID A. CARTER is a senior research chemist with Betz Laboratories, Inc., Trevose, Pa., where he is responsible for the cooling water research program. An NACE member since 1965, he is also a member of the Electrochemical Society and an associate of the Royal Institute of Chemistry. He received a BS in chemistry from the University of Sheffield, England, and an MS and PhD in corrosion science from the University of Manchester, Manchester, England.

AARON WEISSTUCH, a research chemist for Betz Laboratories, Inc., Trevose, Pa., is responsible for the electrochemical investigation of corrosion inhibition processes, development of instrumental methods, and techniques for corrosion measurement. He earned a BS from Brooklyn College, an MS from New York University, and a PhD from St. John's University. He is a member of NACE, the Electrochemical Society, and the RESA.

CHARLES C. NATHAN is a research section head with Betz Laboratories, Inc., Trevose, Pa. He manages research activities involving cooling water, boiler treatment, petroleum processing, and additives for prevention of corrosion and fouling. A member of NACE, AIChE, and the New York Academy of Sciences, Nathan is a registered professional engineer in Texas and is recognized by NACE as being qualified in one or more phases of the field of corrosion control. He received a BS in chemical engineering from Rice University, Houston, Texas, and an MS and PhD from the University of Pittsburgh.

B. R. Keeney, R. M. Lasater, J. A. Knox
Halliburton Co., Research Center,
Duncan, Oklahoma

In Oil Field Tubular Steels

New Organic Inhibitor Retards Sulfide Corrosion Cracking

SULFIDE CORROSION CRACKING of oil field tubular steels in sour acid systems can be retarded by use of a new organic corrosion inhibitor. Results of testing indicate this fact. However, exactly how the inhibitor accomplishes the retardation of sulfide corrosion cracking in sour acid systems has not yet been determined.

Investigations of sulfide corrosion cracking of high strength steels in sour brine systems have been conducted by numerous authors.[1-5] Some of these investigations have shown that cracking of high strength, highly stressed steel can occur in uninhibited hydrochloric acid solutions.[1,6] On the other hand, Matsushima and Uhlig state that cracking will often occur in the presence of H_2S or As_2O_3 when highly stressed steel is exposed to acid solutions.[7] However, very little information has been collected concerning sulfide corrosion cracking of high strength steels in hydrochloric acid containing hydrogen sulfide.

Hudson reported that certain acetylenic alcohols decrease the amount of hydrogen absorbed by steel.[8] It was assumed that this decrease of hydrogen in steel would be accompanied by a reduction in cracking tendencies in acid inhibited with acetylenic materials.

Sulfide corrosion cracking of oil field tubular steels during acidizing treatments is of increasing concern to many. Hydrogen sulfide may be present simply because a well is sour; however, another source of hydrogen sulfide is iron sulfide scale that may have formed over a period of time on the tubing. When iron sulfide reacts with hydrochloric acid, sufficient hydrogen sulfide can be liberated to produce cracking.

A series of cracking tests have been conducted to ascertain the severity of sulfide corrosion cracking in relation to varying degrees of steel hardness, exposure temperature, etc. Numerous inhibitors were also investigated to determine their effectiveness for preventing this undesirable and costly phenomenon.

SUMMARY

Hydrogen sulfide cracking of high strength carbon steels in low pH systems can be retarded by use of a new organic corrosion inhibitor. A series of cracking tests were conducted, based on the earlier work of Hudgins, McGlasson, and Rosborough. Tests were conducted on C-ring specimens with hardness from Rockwell "C" 22 to 36 under various conditions. Sour acid systems were used rather than sour brines for these experiments. In every instance, cracking was retarded for the entire test period with a new inhibitor. No other inhibitor evaluated prevented cracking. A detailed discussion of the test methods and procedures is also presented.

Cracking Tests

Procedure

The cracking tests were made with notched C-ring specimens. The C-ring specimens were machined to the dimensions as shown in Figure 1 from three sections of 2-5/16-inch I. D. AISI 4140 low alloy steel tubing.[1] Each section of tubing had been heat treated to different strength levels in accordance with a predetermined standardized procedure.

The three hardness levels selected for evaluation were Rockwell "C" 22, 32, and 36. These hardness levels were obtained by heating the sections of tubing at 1600 F (871 C) for one and one-half hours and then quenching in oil. One section of tubing was then tempered at 1300 F (704 C) for six hours, another at 1150 F (621 C) for one and one-half hours, and the final section at 1000 F (538 C) for one and one-half hours to obtain Rockwell "C" hardnesses of 22, 32, and 36, respectively. Upon completion of the heat treatment, a number of hardness measurements were made to insure uniform treatment.

After machining, the C-ring specimens were cleaned with a mild pumice, dried by dipping in alcohol and acetone and stored in a dessicator until time for use.

Most of the C-rings were prestressed by use of a bolt and suspended in the acid solutions by a copper wire. How-

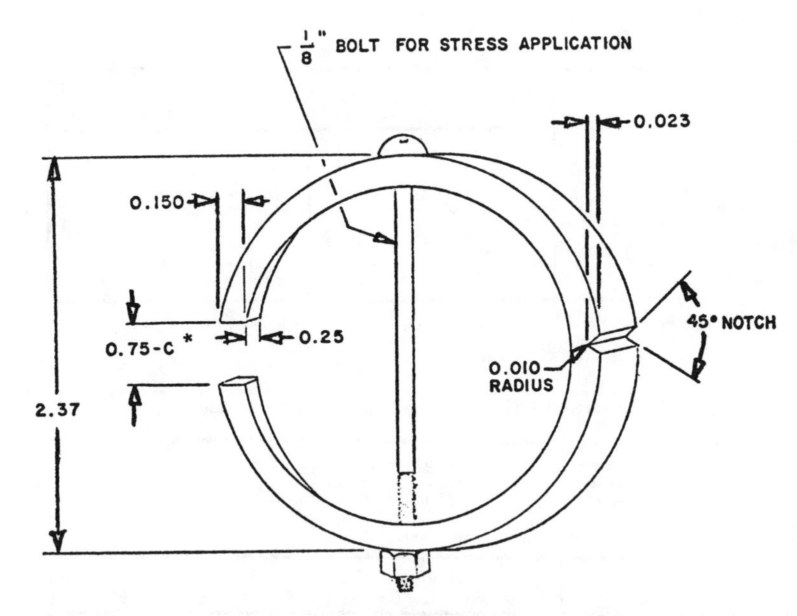

Figure 1—Dimensions of C-ring specimens machined from three sections of 2-5/16-inch I. D. AISI 4140 low alloy steel tubing.

TABLE 1—Hydrogen Sulfide Cracking Tests With No Inhibitor

Test Temperature-----Room Temperature 72 F (22 C)
Corrodent------------15% HCl + 2000 ppm H_2S
% Load---------------Approx. 100% of Yield

Rockwell "C" Hardness	Time to Failure
22	Failed after 2.5 hours.
32	Failed after 1.75 hours.
36	Failed after 1.25 hours.

TABLE 2—Hydrogen Sulfide Cracking Tests With Various Inhibitors

Test Temperature-------Room Temperature 72 F (22 C)
Test Time--------------14 days maximum
Corrodent--------------15% HCl + 2000 ppm H_2S
% Load-----------------Approx. 100% of Yield
Specimen Hardness------Rockwell "C" 36

Inhibitor	Inhibitor Concentration (Vol.%)	Time to Failure
Cracking Inhibitor	0.1%	Intact after 14 days
Cracking Inhibitor	0.2%	Intact after 14 days
Cracking Inhibitor	0.3%	Intact after 14 days
A	0.1%	Failed after 2.5 hours
A	0.2%	Failed after 2.5 hours
A	0.3%	Failed after 2.5 hours
B	0.1%	Failed after 1 hour
B	0.2%	Failed after 2 hours
C	0.1%	Failed after 26 hours
C	0.2%	Failed after 50 hours
D	0.1%	Failed after 1 hour
D	0.2%	Failed after 1 hour
E	0.1%	Failed after 2 hours
E	0.2%	Failed after 2 hours
F	0.1%	Failed between 3-16 hours(1)
F	0.2%	Failed between 3-16 hours(1)
G	0.1%	Failed between 3-16 hours(1)
G	0.2%	Failed between 3-16 hours(1)

(1) The exact failure time of these specimens was not obtained as the C-rings cracked sometime during the night.

TABLE 3—Hydrogen Sulfide Cracking Tests With Various Chemicals

Test Temperature-------Room Temperature 72 F (22 C)
Test Time--------------14 days maximum
Corrodent--------------15% HCl + 2000 ppm H_2S
% Load-----------------Approx. 100% of Yield
Specimen Hardness------Rockwell "C" 36

Inhibitor	Inhibitor Concentration (Vol.%)	Time to Failure
Alkyl Pyridines	0.2%	Failed after 2.5 hours
Ethyl Octynol	0.2%	Failed after 2.5 hours
Ethyl Octynol	1.0%	Failed after 4 hours
Propargyl Alcohol	0.2%	Failed after 2.5 hours
Propargyl Alcohol	1.0%	Failed after 4 hours
Hexynol	0.2%	Failed after 2.5 hours
Rosin Amine ETO(1)	0.2%	Failed after 2 hours
Dibutyl Thiourea	0.2%	Failed after 4 hours

(1) This is an isopropyl alcohol solution (70 percent concentrate) of an adduct of Rosin Amine D and 11 moles of ethylene oxide in admixture with 10 percent free Rosin Amine D.

TABLE 4—Hydrogen Sulfide Cracking Tests At Elevated Temperature

Test Temperature-------150 F (66 C)
Test Time--------------36 hours maximum
Corrodent--------------15% HCl + 2000 ppm H_2S
% Load-----------------Approx. 100% of Yield
Specimen Hardness------Rockwell "C" 32

Inhibitor	Inhibitor Concentration (Vol.%)	Time to Failure
None	None	Failed between 3 and 16 hrs
Cracking Inhibitor	0.2%	Intact after 36 hrs
Inhibitor A	0.2%	Failed between 3 and 16 hrs

ever, in later tests a special holder was designed to allow four specimens to be tested simultaneously. This is shown in Figure 2. This type of holder has an advantage over the bolt method in that the C-ring specimens are insulated from the metal holder by use of glass beads. The C-ring specimens were prestressed to approximately 100% of the yield deformations as determined by load deformation curves.

All the cracking tests were conducted in 15% (wt. %) hydrochloric acid prepared by diluting 20° Bé HCl with tap water. The exact acid concentration was determined by titrating with standard sodium hydroxide.

In a majority of the tests hydrogen sulfide was introduced to the acid solutions by adding 2.5 g ferrous sulfide per 500 cc of acid. This liberated approximately 2000 ppm of hydrogen sulfide. Throughout the entire test each sample was tightly capped with a plastic sheet and examined at regular intervals to determine if failure had occurred. The maximum exposure time for all tests was 14 days. On a few tests H_2S gas was bubbled through the solution for various time intervals to insure saturation throughout the test. The method of introduction of hydrogen sulfide apparently made no difference in the results.

When the acid solutions were inhibited, the inhibitor was added on a volume percentage basis by use of a hypodermic syringe.

Results

The failure times for the uninhibited specimens with Rockwell "C" hardnesses of 22, 32, and 36 are contained in Table 1. The data indicate that the time to failure is directly related to the degree of hardness of the C-ring specimens when other conditions are held as nearly constant as possible. It is quite possible, however, that if a long series of cracking tests were conducted on the Rockwell "C" 22 specimens, many of the C-rings would not experience failure. It is generally agreed that failure will not occur in sour brines below a Rockwell "C" hardness of 22, but Hudgins, et al, suggest that cracking may occur at any hardness level depending on stress, hydrogen sulfide concentration, and degree of cold work.[1] Failure would probably then depend on the uniformity in heat treatment of the test specimens as the degree of hardness may vary over the surface.

Comparison of Inhibitors

With the failure times ascertained for the uninhibited specimens, a series of cracking tests were conducted to compare the effectiveness of various commercially available acid inhibitors to a newly developed inhibitor designed specifically to prevent hydrogen sulfide cracking. These tests were conducted on C-rings with an average Rockwell "C" hardness of 36. The results are listed in Table 2.

The data clearly show the Cracking Inhibitor to be more effective than the other inhibitors evaluated. Of the inhibitors tested, six resulted in coupon failure in less than 16 hours, while with Inhibitor C the specimen failed after 26 and 50 hours at concentrations of 0.1 and 0.2%, respectively. With the Cracking Inhibitor there was no failure after 14 days exposure of the coupon. Since the chemical compositions of the various inhibitors are currently unknown, it is not possible to intelligently discuss the behavior of these materials.

A series of cracking tests were conducted on several well known materials used in acid inhibitor formulations. These data are shown in Table 3. Initial tests were run at a concentration of 0.2% and cracking was observed with all materials tested. Since Edwards, et al,[9] had indicated that acetylenic alcohols can prevent cracking in acid under certain conditions, consideration was given to the possibility that hydrogen sulfide had reacted with the acetylenic alcohols to destroy their effectiveness when low concentrations were used. Therefore, tests were conducted on two of the acetylenic alcohols at a concentration of 1.0% The time to failure was extended slightly at the higher concentration but cracking was observed with these materials within four hours.

Cracking tests were conducted at 150 F (66 C) on specimens with a hardness of Rockwell "C" 32. These data are contained in Table 4. Again, the Cracking Inhibitor prevented failure throughout the test period. Note that failure time for the uninhibited specimen at 150 F (66 C) was longer than the uninhibited specimen at room temperature. This is probably caused by the increased rate of combination of atomic hydrogen to form molecular hydrogen as the temperature is increased.

Inhibitor A was selected for evaluation in this series of tests since it is one of the more widely used commercial inhibitors. Failure was recorded with Inhibitor A at 150 F (66 C) even though the Rockwell "C" 32 hardness C-ring specimen was not severely corroded, indicating that hydrogen sulfide cracking can occur even when the amount of corrosion is very small. Corrosion was extremely low with the Cracking Inhibitor.

Hydrogen sulfide cracking tests were also conducted on lower hardness specimens at 72 F (22 C) using Inhibitor A and the new Cracking Inhibitor. These data are contained in Table 5. Failure was recorded in the uninhibited acid solution; however, both Inhibitor A and the Cracking Inhibitor prevented cracking for the entire test period. As was mentioned earlier, Rockwell "C" 22 is somewhat of a threshold hardness, below which cracking will not normally occur under most load levels. Another possibility is that hydrogen absorption may have been limited sufficiently to prevent cracking in this low hardness steel because the inhibitor is an acetylenic alcohol based material. Acetylenic alcohols will limit the absorption of hydrogen into steel as discussed by Hudson and Riedy.[8]

A photomicrographic study was made of several of the cracked and uncracked specimens. Photomicrograph of one of the cracked specimens (Figure 3) shows that the crack did not start at the notch but rather on the machined surface of the C-ring. No particular area of stress or cold work is apparent where the crack was initiated. The crack branched at several points as it crossed the C-ring. Pitted areas are present in and around the crack.

A photomicrograph also was made at the base of the notch of one of the uncracked C-rings (Figure 4) which had been inhibited with the Cracking Inhibitor. In addition to the typical martensitic structure, an area of cold work can be observed at the notch base. Microhardness tests on this particular specimen showed the body of the C-ring to have a hardness of Rockwell "C" 38 while the cold worked area at the base of the notch was Rockwell "C" 48. Hairline cracks appear to penetrate this cold worked area; however, these were not observed on other uncracked specimens.

A few tests were conducted on Rockwell "C" 58 hardness roller bearings described by Bush and Cowan.[5] In these tests, the time to failure was generally increased; however, no uniform or reproducible data has been obtained. The cracking times for the roller bearings in acid without hydrogen sulfide is less than one minute.

Summary

This work is in its infancy at the present time. Evidence has been submitted which indicates that cracking of high strength steel (Rockwell "C" 36) can be controlled with the Cracking Inhibitor even in hydrochloric acid containing hydrogen sulfide.

Exactly how the Cracking Inhibitor functions is not known; however, research is currently being conducted to determine this mechanism.

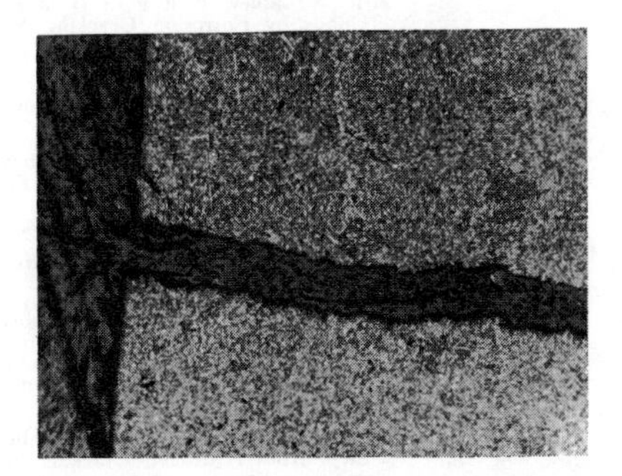

Figure 3—Photomicrograph of one of the cracked specimens.

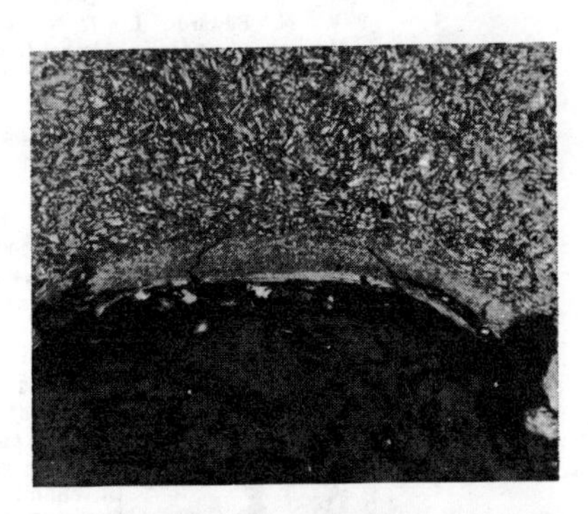

Figure 4—Photomicrograph made at the base of the notch of one of the uncracked C-rings which has been inhibited with the Cracking Inhibitor.

Figure 2—Special holder designed to allow four specimens to be tested simultaneously.

TABLE 5—Hydrogen Sulfide Cracking Tests of Low Hardness Specimens

Test Temperature-------Room Temperature 72 F (22 C)
Test Time--------------14 days maximum
Corrodent--------------15% HCl + 2000 ppm H_2S
% Load-----------------Approx. 100% of Yield
Specimen Hardness------Rockwell "C" 22

Inhibitor	Inhibitor Concentration (Vol.%)	Time to Failure
None	None	Failed after 2.5 hours
Cracking Inhibitor	0.1%	Intact after 14 days
Inhibitor A	0.1%	Intact after 14 days

Source: *Materials Protection*, April 1968, 23-26

References

1. C. M. Hudgins and R. L. McGlasson, et al. Hydrogen Sulfide Cracking of Carbon and Alloy Steels. *Corrosion*, 22, 238 (1966).
2. R. S Ladley. Failure of High Strength Tubular Goods by Corrosion Cracking Mechanisms. Presented at the API Southern District Meeting, Shreveport, La., March 8-10, 1967.
3. Symposium on Sulfide Stress-Corrosion Cracking. *Corrosion*, 8, 325-360 (1952).
4. A. E. Schuetz and W. D. Robertson. Hydrogen Absorption Embrittlement and Fracture of Steel. *Corrosion*, 13, 437 (1957).
5. H. E. Bush and J. C. Cowan. Corrision Inhibitors in Oil Field Fluids. *Materials Protection*, 5, 25 (1966).
6. F. D. Sewell. Hydrogen Embrittlement Challenges Tubular Goods Performance. Presented at the Petroleum Mechanical Engineering Conference of the American Society of Mechanical Engineers, Houston, Texas (1965) September 19-22.
7. I. Matsushima and H. H. Uhlig. Protection of Steel from Hydrogen Cracking by Thin Metallic Coatings. *Journal of Electrochemical Society*, 113, 555 (1966).
8. R. M. Hudson and K. J. Riedy. Limiting Hydrogen Absorption by and Dissolution of Steel During Pickling. *Metal Finishing*, (1964) July.
9. K. N. Edwards, L. J. Nowacki and E. R. Mueller. Acetylenic Alcohol-Inhibited Pickling Bath as a Pretreatment Prior to Lining Steel Pipe. *Corrosion*, 15, 275 (1959).

REGINALD M. LASATER, group leader of the Research Group in the Chemical Services Section, works in the field of chemical stimulation problems. He received a BS in chemistry in 1955 from Central State College in Oklahoma and his MS in chemistry in 1957 from Oklahoma State University. He is a member of ACS, AIME, Clay Minerals Society, and Electrochemical Society.

BILL R. KEENEY is a chemist with Halliburton Co. He conducts research on organic corrosion inhibitors for hydrochloric acid and performs general corrosion studies as related to company problems. He received a BS in chemistry in 1959 from Oklahoma State University.

JOHN A. KNOX is a section supervisor of the Chemical Services Section of Halliburton Co. He has been with Halliburton since 1955. After receiving a BS and MS in chemistry from North Texas State University, he was head of the Science Dept. at East Mississippi State Junior College for two years. He is past Chairman of the Central Oklahoma Section of NACE and is past Director representing corporate members on the NACE Board of Directors. He is also active in AIME, ACS, and ASTM and is a registered Professional Engineer.

Some Vapor Phase Inhibitor Experiments

V. S. Agarwala and K. C. Tripathi, Banaras Hindu University, Varanasi, India

Experiments with vapor phase inhibitors were conducted on four metals: mild steel, copper, brass, and aluminum. Four derivatives of thiourea (allyl thiourea, phenyl thiourea, tetra methyl thiourea, mono and disulphide) were studied for six months in a closed desiccator. Results are shown with microphotographs.

IN RECENT YEARS a new highly effective method of protecting metals from atmospheric corrosion has been developed. Metals are protected in an atmosphere of volatile inhibitors—the so-called Vapor Phase Inhibitors (VPI)—by absorption over the metal surface.[1, 2]

Most known VPI materials are basic nitrogenous compounds of the aliphatic and alicyclic series[3, 4, 5] and have been studied on steel only. To the authors' knowledge no VPI compounds are reported which contain both nitrogen and sulphur. In an earlier article[6] the authors showed the suitability of some compounds containing nitrogen and sulfur as inhibitors of steel corrosion in pickling acids.

Four VPI compounds have been tested for their protective values on ferrous and nonferrous metals; the results are presented here.

Experiments

The tests were conducted with the following metals: (1) mild steel (C 0.20, S 0.06, P 0.08, Mn 0.4%), (2) brass (70/30), (3) aluminum, and (4) copper (99.9% pure). The first three metals were commercially available as sheets of 1.8 mm thickness. Copper sheets were 0.9 mm thick.

Each of the metals were cut to panels of 2 x 3 cms, abraded, and polished with different grades of emery papers down to 4/0. They were degreased with acetone, dried, and finally stored in a desiccator before they were used.

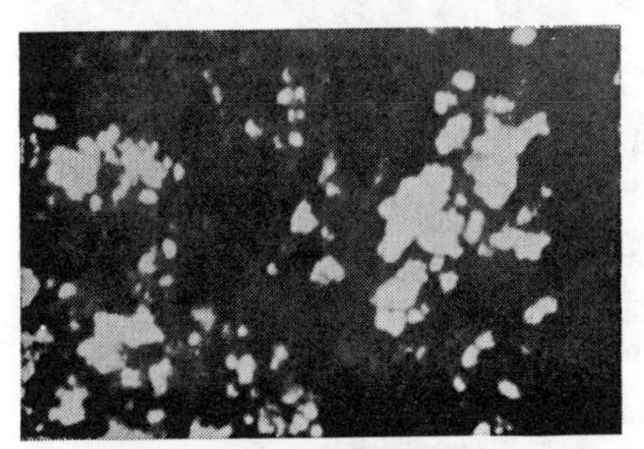

Figure 1—Photograph of mild steel panel after exposure without any VPI action. (Magnification 100x.)

Figure 2—Photograph of mild steel exposed to 4MTU2S (tetramethyl thiouram disulphide) for a period of six months. No protection occurred; corrosion started along grain boundaries and progressed rapidly. Only mild steel was exposed to 4MTU2S because of this severe corrosion.

The following compounds investigated as vapor phase inhibitors were either pure products available commercially or crystallized synthesized products:

(1) Phenyl thiourea (PTU)

(2) Allyl thiourea (ALTU)

(3) Tetramethyl thiouram monosulphide (4MTUS)

(4) Tetramethyl thiouram disulphide (4MTU2S)

Corrosion tests were performed for each inhibitor in desiccators, one-third filled with water. Boats containing the compounds were made to float on the water. The panels were suspended through glass hooks attached to the lid of the desiccator with the help of adhesive tapes.

These units were sealed for six months and the coupons examined periodically. When finally taken out they were photographed through a projection microscope and results compared (see Figure 3). The experiment was so arranged to provide the desired 100% humidity with the amount of oxygen available within the air of the desiccator.

Results and Discussion

The results of the experiment can be seen by the ranking of the VPI in Table 1.

TABLE 1—Ranking of Vapor Phase Inhibitors

	MILD STEEL	COPPER	BRASS	ALUMINUM
PTU	2	1	1	2
ALTU	1	2	2	3
4MTUS	3	3	3	1
4MTU2S	4	-	-	-

Thus, it is clear that no single inhibitor was equally efficient for ferrous or nonferrous metals, although a fair degree of inhibition was obtained by PTU in general. The inhibition by PTU was best for copper, while ALTU was best for mild steel and produced excellent results for them only.

Among these compounds ALTU had the largest vapor pressure while PTU the least; but PTU was a better inhibitor than ALTU in the case of nonferrous metals. This confirmed the view that vapor pressure cannot be the only factor for inhibition. Specific adsorption forces on the metal surface may have been the most dominant point.

Summary

These investigations show that some derivatives of thiourea can be used as vapor phase inhibitors for the four metals. The specimens of mild steel, copper, brass, and aluminum were exposed to a closed 100% humid atmosphere for six months. The following results were obtained:

1. Mild steel was best protected by allyl thiourea, aluminum by tetramethylthiourea monosulphide, copper and brass by phenylthiourea. (The latter compound protected copper better than brass).

2. Phenylthiourea, in general, protected all the investigated metals fairly well up to four months.

References

1. H. L. Bennister. *Research*, 5, 432 (1952).
2. *A New Method of Corrosion Preventive Packaging*. VPI-260, Shell Chemicals Ltd., London, 17 (1952).
3. A. Wachter, T. Skei and N. Stillman. *Corrosion*, 7, 28 (1951).
4. C. A. Rhodes. *Corrosion Prevention and Control*, 4, 4, 37 (1957).
5. I. L. Rosenfeld, V. P. Persiantseva, M. N. Polteva and P. B. Terentiev. *European Symposium on Inhibitors of Corrosion*, Ferrara, Italy, (1960).
6. V. S. Agarwala and K. C. Tripathi. *Trans. Ind. Inst. Metals*, 17, 139 (1964).

VINOD SHANKAR AGARWALA is a graduate student of metallurgy, currently at Massachusetts Institute of Technology, Boston. A research fellow of the department of metallurgy, Banaras Hindu University, Varanasi, India, he conducted the investigation (which this article covers) under the supervision of Dr. K. C. Tripathi during 1962-65. He has a MSc degree from Banaras Hindu Univ., in Chemistry, 1960. He has been a NACE member since 1965.

KRISHNA CHANDRA TRIPATHI is research associate at the Metallurgy Division of Atomic Energy Research Establishment, Harwell, England. At present he is on leave from the department of metallurgy, Banaras Hindu University, Varanasi, India. He received his MSc degree from Lucknow University, India, in physical chemistry, 1952. He received his doctor of science (Dr. Rer. Nat.) from Technical University, Dresden, Germany, 1960. In addition to being a faculty member at Banaras Hindu Univ., he has served as chairman of several technical sessions, including Defense Science Symposium, 1961.

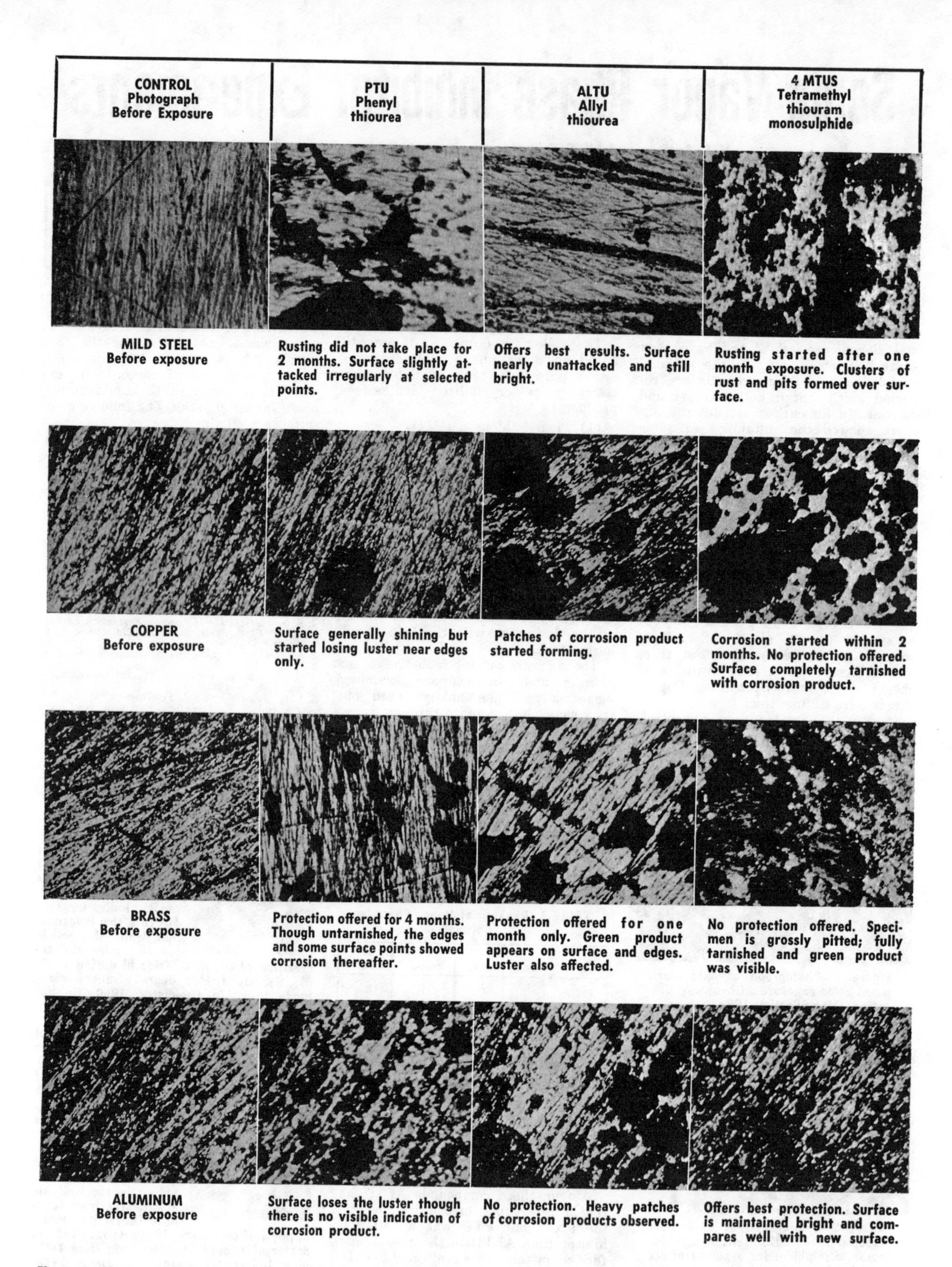

Figure 3—Three vapor phase inhibitors were tested on four metals: mild steel, copper, brass, and aluminum. The results are shown in the microphotographs illustrated here (magnification 100x). Test period was six months, in a closed desiccator containing distilled water. The humidity was 100%, temperature was 32 ± 5 C for all metals (except the mild steel: 25 ± 5 C). Observations are noted under each photograph. Solid dark spots in the microphotographs indicate the pitting.

SUMMARY

A potentiodynamic polarization technique was used to investigate inhibitor effectiveness and mode of action of some inorganic and organic inhibitors on mild steel in a simulated cooling water at neutral pH. The technique gave more reproducible results than did a weight loss technique. Inhibitory performance of inorganic phosphate, chromate, silicate and molybdate and of several pyrophosphate esters of monohydric and polyhydric alcohols was investigated, and it was found possible to relate length of the alkyl chain and the number of hydroxyl groups on the polyol to inhibitor efficiency of the esters.

D. A. Carter and C. C. Nathan
Betz Laboratories, Inc.
Trevose, Pa.

A Polarization Study of Cooling Water Corrosion Inhibitors

CORROSION CONTROL in the corrosive environment of cooling water systems is primarily effected with inorganic inhibitor formulations. The most commonly used materials, phosphate and chromate-based, are known as passivating inhibitors. The mechanism of corrosion inhibition by phosphates,[1-3] chromates,[4-5] and the combination[6-7] has been known for many years.

To evaluate the effectiveness of these inhibitors and others in cooling waters, a test program was initiated. Also investigated during the program were the use and limitations of the potentiodynamic polarization technique.

Testing

Equipment

A potentiodynamic polarization technique was used to evaluate both the effectiveness and the mode of action of inorganic and organic corrosion inhibitors in a cooling water environment. Tests were conducted on a low carbon steel, cylindrical electrode immersed in a simulated cooling water solution with and without inhibitors. A saturated calomel electrode was used as the reference electrode and a platinum screen electrode as the counter electrode.

The experimental instrumentation set-up is shown in Figure 1. A Duffers Model 600 Potentiostat was used as the potential controller. A simply constructed voltage regulator was connected to the potentiostat so that the potential could be changed linearly with time. This enabled the potential of the working electrode to be varied either anodically or cathodically at a rate of 300 mv/hour. The current density between the working and counter electrode was measured on a Beckman 10-inch recorder as a potential drop across a precision 1 ohm resistor. The potential was monitored on a Corning Model 12 pH meter.

Figure 2 shows the experimental cell used for the investigation. The electrode assembly consisted of a mild steel cylindrical electrode surrounded by the platinum-screen counter-electrode. A Luggin capillary placed 3 to 4 mm from the surface of the working electrode was used to measure the potential of the electrode relatively close to the steel surface. A solution bridge from the capillary to another vessel made electrical connection with the reference electrode. The electrode assembly was placed in a beaker containing the solution under study. The solution (pH 6) was constantly aerated and stirred with a magnetic stirrer assembly, and the temperature was controlled at 30 C. The working electrode was constructed from low carbon steel rod cut into 6 inch lengths. The rod was degreased in acetone, pickled in a 33% sulfuric acid solution for 15 seconds, washed with water, cleaned for 2 seconds in a 5 N hydrochloric acid solution, rinsed with water and iso-propanol and dried in a stream of hot air. A 4.7 cm^2 area of the surface was exposed as the working electrode surface. The lower end of the rod was masked off with epoxy resin while the upper end of the exposed surface was masked with epoxy resin set in glass tubing and placed over the rod (Figure 2b).

A simulated cooling water (Table 1) at pH 6 was used for all studies. It represents a typical cooling water used industrially in the Philadelphia area.

Procedure

One anodic and one cathodic polarization experiment were undertaken for each set of conditions. The specimen was immersed in the solution under study for two hours before the commencement of the polarization. The equilibrium corrosion potential attained by the mild steel electrode after this time was termed the initial potential. The initial potential was then changed anodically or cathodically from this potential at a rate of 300 mv/hour for 1½

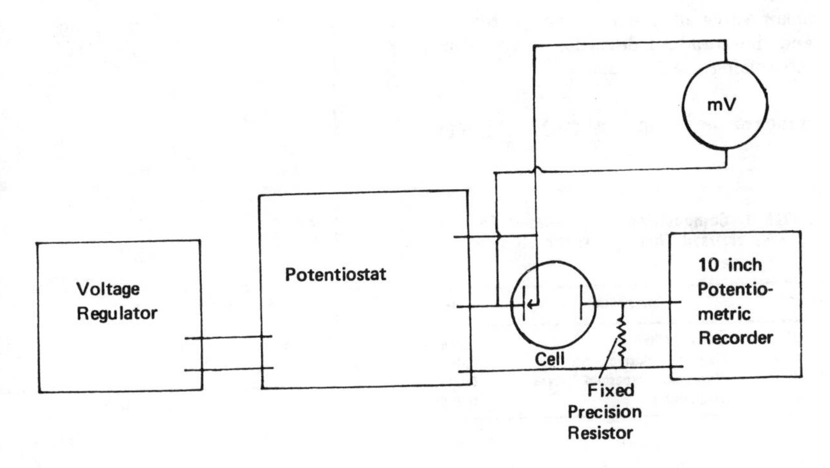

Figure 1—Circuit diagram for the polarization experiments.

hours, the current being recorded continuously during this period.

A control polarization curve in the simulated cooling water alone is plotted in Figure 3. This curve shows the occurrence of the Tafel region between 50 and 150 mv anodic polarization and a limiting oxygen reduction diffusion current upon cathodic polarization. The anodic Tafel slope was extrapolated back to the corrosion potential to give the corrosion current of 2.2×10^{-4} amp/cm², equivalent to 100 mpy. This compared to a corrosion rate of 98 mpy obtained from weight loss data for the same environment. The percent reduction in i_{corr} in the presence of an inhibitor is the percent protection, while the percent reduction in i_{diff} is the cathodic inhibitor efficiency.

Weight loss experiments were conducted on specimens prepared from similar steel rods, using the same surface preparation technique. Rod lengths of 1½ inch were used for the tests. A ⅛ inch diameter hole was drilled through the top of the specimens so that they could be conveniently suspended in a stirred solution maintained at 30 C. The specimens were weighed before immersion and were removed after one and three days immersion. After removal, the specimens were dipped in a 5N HCl solution for two seconds and then rinsed with water. The film was then removed from the metal surface by abrading with a pumice-trisodium ortho-phosphate mixture. The specimens were then washed, dried in hot air, placed in a desiccator, and weighed to constant weight. The difference between the one and three day weight loss was used to calculate the corrosion rate in mpy.

Results

Reproducibility of the Results

The statistical spread of the corrosion rate of mild steel in various cooling water solutions containing phosphate and chromate was determined using the polarization and weight loss techniques. Each set of conditions, representing corrosion rates varying from 0 to 100 mpy, was repeated four times using each technique. The arithmetic mean value of the corrosion rates, $\overline{X}$, and the standard deviation were calculated using the formula:

$$\text{standard deviation} = \sqrt{\frac{\Sigma(X - \overline{X})^2}{N - 1}} \quad (1)$$

TABLE 1—Composition of Simulated Cooling Water Used Throughout the Study[1]

Grams	Compound	Moles
0.829	Sodium Chloride	0.014
0.045	Calcium Chloride	0.004
0.049	Hydrated Magnesium Sulfate	0.0002
0.74	Sodium Sulfate	0.005

[1]These ingredients were made up to 1 liter with distilled water and adjusted to pH 6 and tested at 30 C; ionic strength of the solution = 0.03.

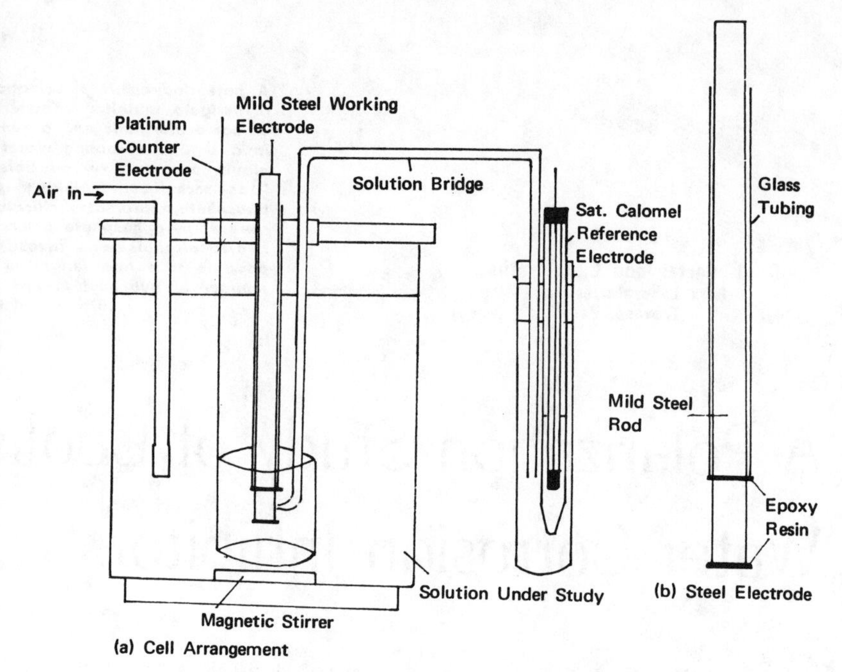

Figure 2—Cell arrangement for polarization studies.

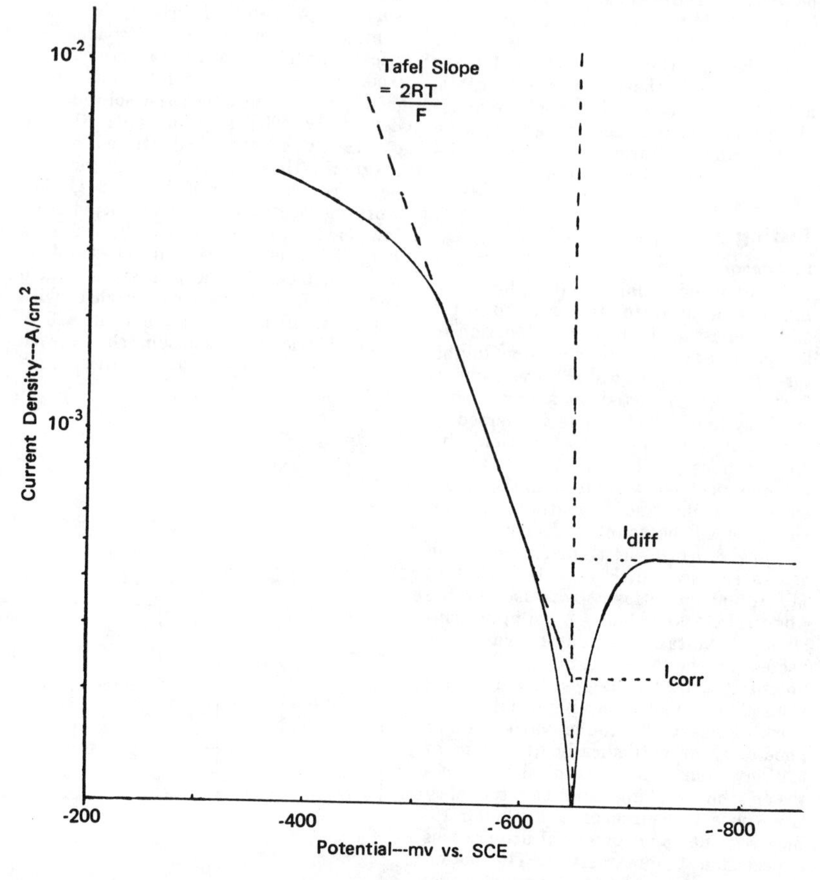

Figure 3—Potentiodynamic polarization of mild steel in a simulated cooling water (pH 6).

where X was the individual corrosion rates measured and N was the number of tests.

The standard deviation was plotted against the corrosion rate using both the weight loss and polarization techniques (Figure 4). The corrosion rate determined from i_{corr} values was converted into mpy so that a direct comparison could be made. It can be seen that the techniques show different behavior. Weight loss data gave maximum scatter at intermediate inhibitory performance while good reproducibility was evident in solutions of very good or very poor inhibition. This trend, also shown in other environments by one of the authors,[8-9] is believed to be due to the nature of the corrosion inhibiting process. Corrosion inhibitors with moderate inhibitory performance may, under the same set of conditions, be more or less effective depending on how the protective film develops. Very good inhibitors always form a protective film on the metal surface, and the scatter is less.

The standard deviation of the corrosion rates using the polarization technique was the same throughout the whole range of corrosion rates. This is surprising, because the behavior exhibited using the weight loss technique should be reproduced. Also the corrosion rate is determined by an intercept on a logarithmic current density axis. If this were the controlling factor, the scatter of the corrosion rate in mpy at high corrosion rates would be expected to be greater than at low corrosion rates.

The difference in behavior between the weight loss and polarization techniques is believed to be a result of the longer equilibration times used with the former technique. The two hours equilibration time with the polarization technique probably did not allow great differences to develop with inhibitors of only moderate performance. Consequently, the scatter of the results was constant throughout the whole range of corrosion rates. The fact that a logarithmic scatter was not evident indicates that the measuring accuracy was not a controlling factor.

The polarization technique, therefore, gives more accuracy in determining inhibitor effectiveness throughout the whole range of inhibitor performance. However, being an accelerated corrosion testing procedure, it gives performance characteristics only over a short period of time and will miss the longer period developments on the metal surface, such as film breakdown and repair, etc.

Inorganic Inhibitors

The polarization technique was used to evaluate the performance of four known inhibitors, molybdate, silicate, polyphosphate and chromate, in the simulated cooling water at pH 6. Results are shown in Figure 5.

Molybdate and silicate were not effective inhibitors at low concentrations. Polyphosphate gave moderate corrosion

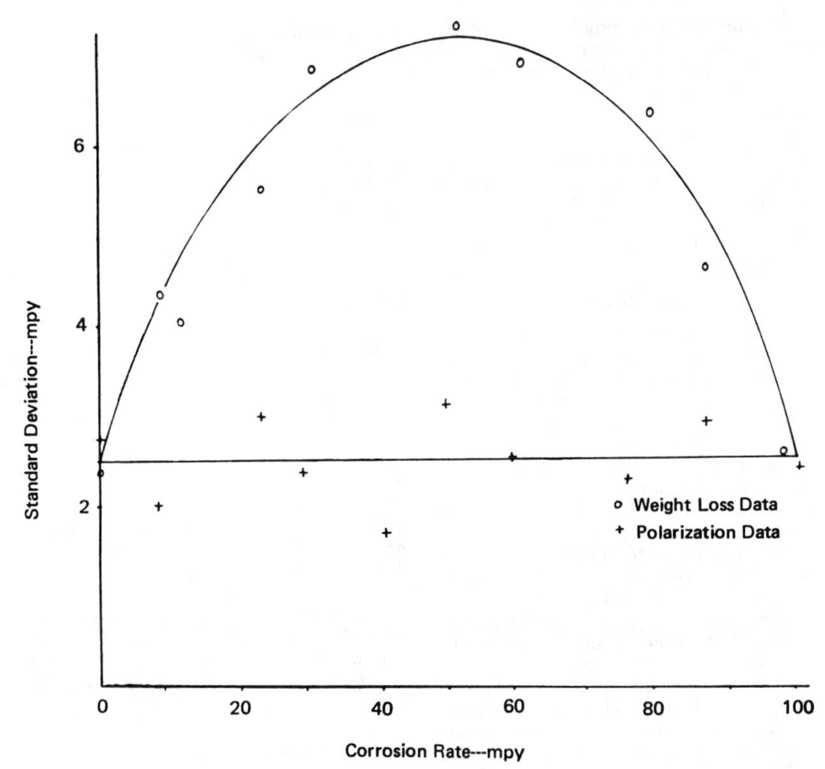

Figure 4—Distributional scatter of corrosion results using weight loss and polarization techniques.

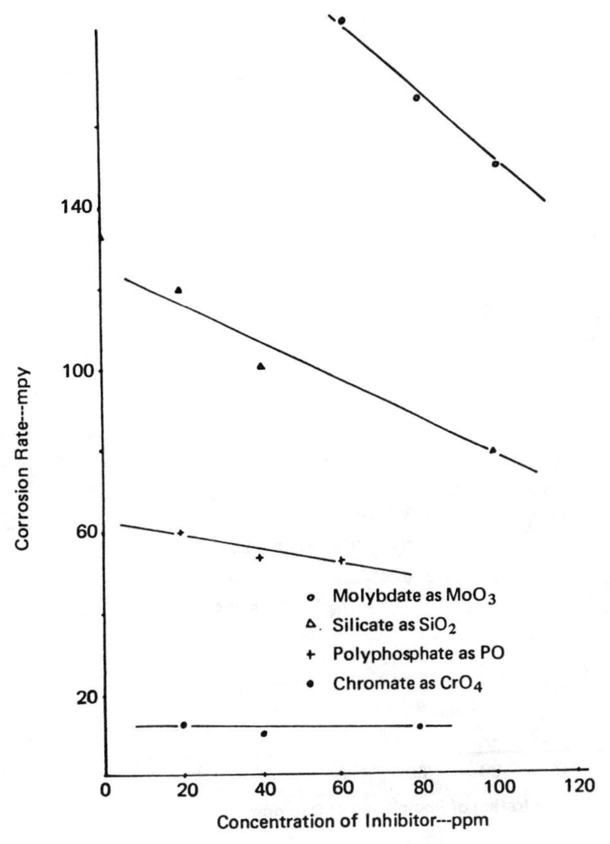

Figure 5—Effect of concentration on the corrosion rate of mild steel in an inhibited cooling water solution (pH 6).

Source: *Materials Protection*, September 1969, 61-66

inhibition, while chromate was very effective. Polyphospate reduced the cathodic diffusion current density from 4.6×10^{-4} amp/cm^2 (cooling water alone) to 2.9×10^{-4} amp/cm^2. Chromate did not noticeably reduce the cathodic diffusion current density, indicating that chromate was an anodic inhibitor.

The cathodic reaction under cooling water conditions is that of oxygen reduction:

$$O_2 + 2H_2O + 4e^- = 4\,OH^- \qquad (2)$$

Under cooling water conditions, the corrosion reaction is under cathodic control; consequently, anodic inhibitors may cause pitting.[4] The above results, therefore, predict a greater pitting tendency with chromate than with polyphosphate. This was evident by microscopic examination of the metal surface after the test and is also evident in the field.

The results also indicate how the polarization technique can be used to predict inhibitor performance as well as the mode of action of inhibitors after only a short time of immersion in the environment under study.

Corrosion inhibitory properties of phosphates alone and in the presence of zinc were studied in more detail. Four phosphates were investigated: sodium dihydrogen orthophosphate (NaH_2PO_4), sodium hexametaphosphate [$(NaPO_3)_n$], tetrapotassium pyrophosphate ($K_4P_2O_7$), and sodium tripolyphosphate ($Na_5P_3O_{10}$).

One of the problems in studying phosphates is the possibility of reversion of one form into another; however, during the short period of investigations, no noticeable reversions were observed. These test results are shown in Figures 6 and 7.

At 40 ppm (as PO_4), the order of effectiveness as corrosion inhibitors of the phosphates alone was hexametaphosphate > tripolyphosphate > orthophosphate > pyrophosphate.

In all cases, inhibitor effectiveness increased with increase in concentration. With the exception of hexametaphosphate, the addition of zinc decreased both the overall corrosion rate and the cathodic diffusion current. Zinc acts as a cathodic inhibitor by reacting with the cathodic reaction product (hydroxyl ions) to form an insoluble precipitate of zinc hydroxide on the cathodic sites. This results in a reduction in both the overall corrosion rate and the tendency for pitting. Hexametaphosphate alone was the most effective cathodic inhibitor, which could account for the lack of noticeable effect with the addition of zinc.

Organic Inhibitors

Corrosion inhibiting properties of several orthophosphate and pyrophosphate esters of organic alcohols were investigated in the simulated cooling water at pH 6. Some of the esters were commercially available and some were synthesized in the Betz laboratories. The structures of the esters are shown below.

$$R(OCH_2CH_2)_xO\overset{OH}{\underset{O}{P}}-O-\overset{OH}{\underset{O}{P}}(CH_2CH_2O)_xR$$

Ethoxylated Pyrophosphates

$$RO\overset{OH}{\underset{O}{P}}-O-\overset{OH}{\underset{O}{P}}OR$$

Pyrophosphates

$$RO{\cdot}PO_3H_2$$

Mono-orthophosphates

$$(RO)_2PO_2H$$

Di-orthophosphates

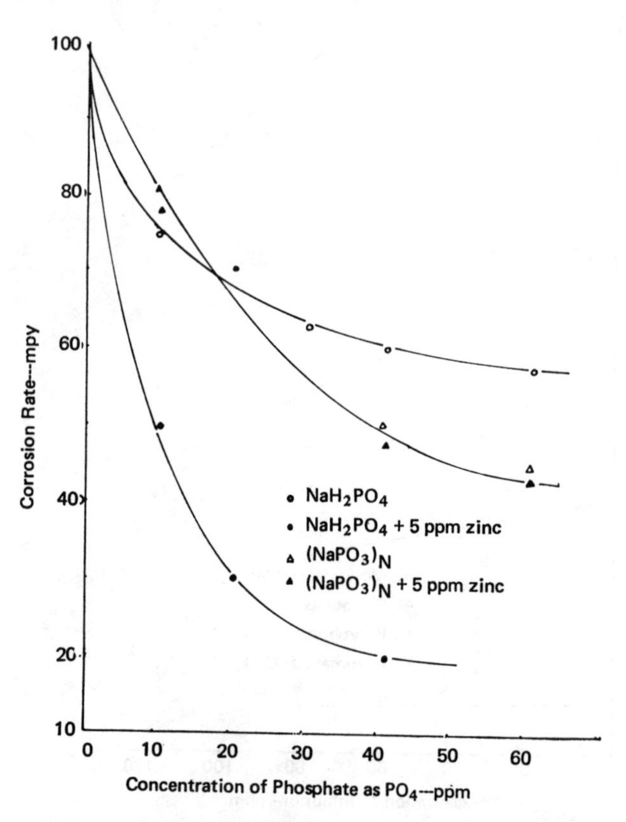

Figure 6—Effect of zinc on the corrosion inhibition properties of ortho- and hexameta-phosphates in a cooling water (pH 6).

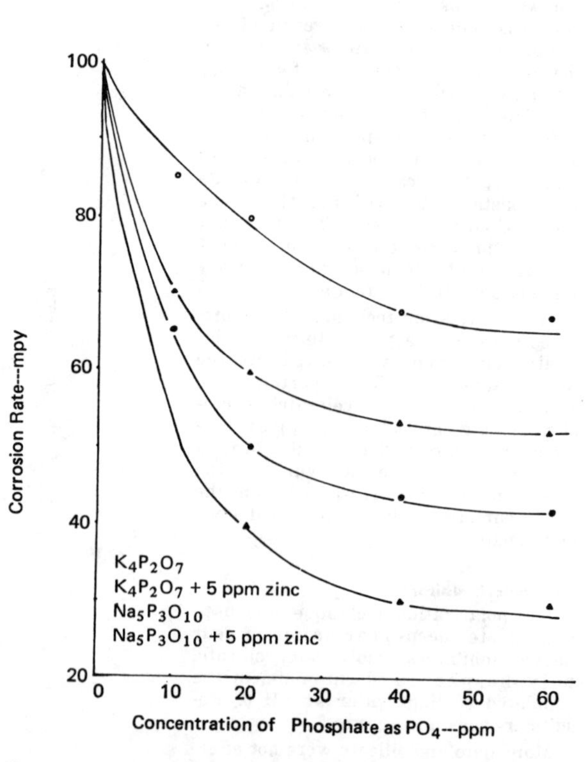

Figure 7—Effect of zinc on the corrosion inhibition properties of pyro- and tripoly-phosphates in a cooling water (pH 6).

All the alkyl groups used in this investigation were straight chained. The ethoxylated pyrophosphates had $_x = 1$. Note that the only ester having the polar phosphate group at the end of the molecule was the mono-orthophosphate. Purity of the synthesized esters was 90% or greater in most cases because extremely pure esters were difficult to prepare due to the complexity of reactions involving phosphates. The esters were evaluated at 30 ppm (based on PO_4 content). Results are shown in Figure 8.

The monophosphate esters gave increasing corrosion inhibition with an increase in the number of carbon atoms in the alkyl group from C_2 to C_5. The diorthophosphate esters showed the same trend but were less effective with the same alkyl substitution (the chain length was the sum of the two alkyl groups). The authors postulate that the phosphate esters function by chemisorption of the phosphate group onto the metal surface. The alkyl group is positioned away from the metal surface, creating a protective organic barrier on the metal surface. As the alkyl group chain length increases, the barrier thickness increases; therefore, the corrosion protection improves. A point is reached where the carbon chain sterically interferes with the adsorption process, but this point is not attained with the orthophosphates. A monosubstituted ester would be expected to be more effective as a corrosion inhibitor than a disubstitute ester, for this would enable closer packing of the phosphates on the metal surface. The intermolecular forces binding the phosphates together would therefore be stronger. The authors' results are consistent with this theory.

The opposite trend was observed with the pyrophosphate esters. Corrosion rates increased from methyl pyrophosphate to hexyl pyrophosphate. In the authors' opinion, the pyrophosphate group is very weakly adsorbed on the metal surface at pH 6, which in part explains the poor performance of inorganic pyrophosphate. Even the presence of short chain alkyl groups on a pyrophosphate should, then, reduce the extent of adsorption of the phosphate group on the metal surface because the alkyl groups will compete with the phosphate for the available metal surface. Therefore, physical adsorption becomes an important factor, and good corrosion protection is not obtained.

The ethoxylated pyrophosphate had one ethylene oxide adduct to each alcohol group. The number of carbon atoms indicated on the abscissa of Figure 8 refers to the number of carbon atoms in the alcohol from which the ethoxylated pyrophosphates would be expected to exhibit greater tendencies for chemisorption than the pyrophosphates would themselves because of the participation of the ether oxygen. The chemisorption resulted in increased inhibition efficiency of the esters from C_3 to C_8. Above C_8 the reverse trend was established because of the steric effect of the alkyl groups.

Several pyrophosphate esters of polyols from ethylene glycol to sorbitol were synthesized and evaluated at 30 ppm PO_4. Results are shown in Figure 9. Inhibitor efficiency of the esters increased from the ester of ethylene glycol to the ester of sorbitol. It is believed that the ester is chemisorbed flat on the metal surface. The larger polyol esters, having more surface coverage per molecule adsorbed, result in a decrease of the corrosion of the steel.

Polarization measurements of the organic phosphate esters showed a maximum of 20% reduction in the cathodic diffusion currents, indicating that the esters were mainly anodic inhibitors and should have a high pitting tendency. Addition of zinc to these esters did increase the cathodic effectiveness as was the case with the inorganic phosphates.

Conclusion

The potentiodynamic polarization evaluation of the performance of corrosion inhibitors in cooling water environments after only two hours equilibrium gave more reproducible results over the whole range of inhibitor performance than did evaluations using a weight loss technique. The technique not only gave values of corrosion rates, but also indicated the mode of action of inhibitors.

Reduction in the cathodic diffusion current when zinc was added to inorganic and organic phosphates indicated that the pitting tendencies were reduced. Several trends were established between inhibitor performance and molecular structure of the organic phosphates. Chemisorption as well as steric factors appear to be important

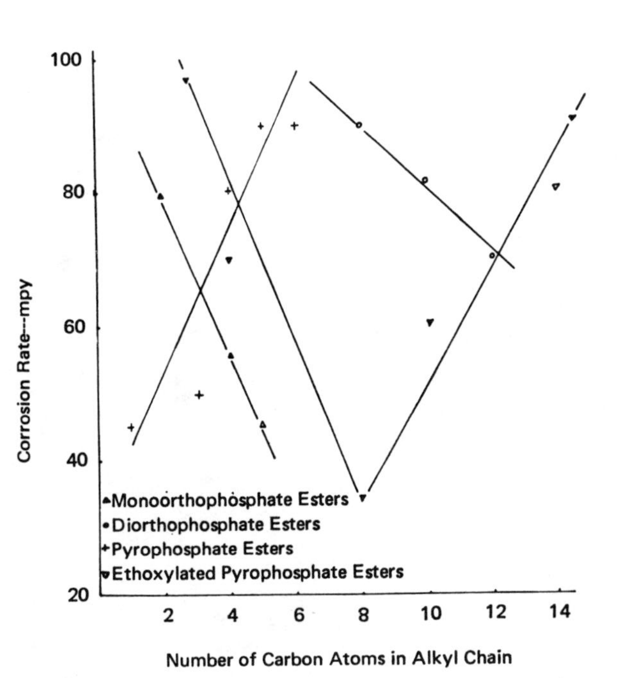

Figure 8—Polarization study of organic phosphate esters in a simulated cooling water (pH 6).

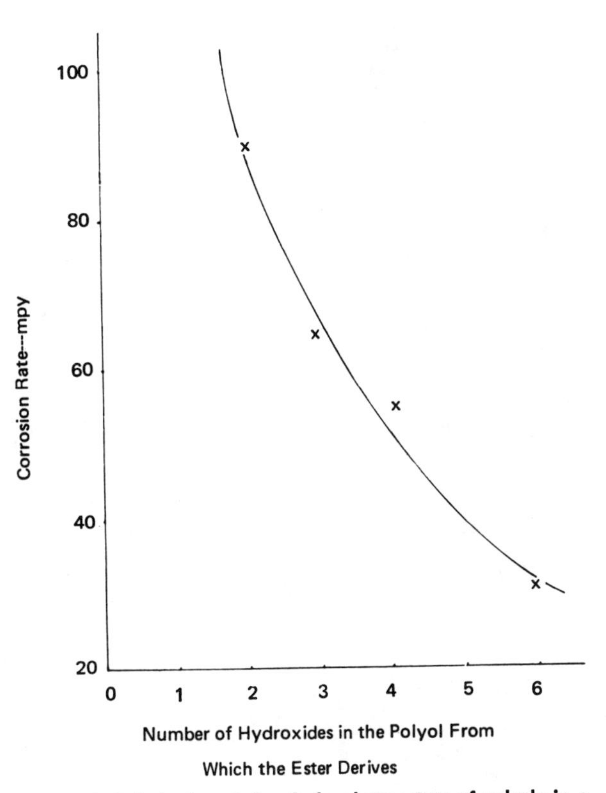

Figure 9—Polarization study of phosphate esters of polyols in a simulated cooling water (pH 6).

Source: *Materials Protection*, September 1969, 61-66

factors governing inhibitor performance.

The result demonstrated the value of polarization technique for corrosion investigations. Being accelerated techniques, they are of limited usefulness for predicting inhibitor behavior over extended periods of time. However, they are valuable tools for the study of corrosion inhibitors.

References

1. G. B. Hatch. Inhibition of Lead Corrosion with Sodium Hexametaphosphate. *J. Amer. Water Works Association,* 33, 1179 (1941); Protective Film Formation with Phosphate Glasses, *Ind. Eng. Chem.* 44, (8), 1775, (1952).
2. G. B. Hatch and O. Rice. Surface-Active Properties of Hexametaphosphate. *Ind. Eng. Chem.* 31, 51 (1939).
3. O. Rice. Brief History of the Use of Glassy Phosphates for the Control of Difficulties Due to Corrosion. *J. Amer. Water Works Association,* 34, 1808, (1942).
4. U. R. Evans, Metallic Corrosion, Passivity and Protection. Chapter X, *Protection by Inhibitive Treatment Water,* 534-596 (1946), Longmans, N.Y.
5. F. N. Speller. Corrosion Causes and Protection. Chapter IX *Prevention of Corrosion Underwater,* 342-383 (1935), McGraw Hill.
6. H. L. Kahler and C. George. A New Method for the Protection of Metals Against Pitting, Tuberculation and General Corrosion. *Corrosion,* 6, 331 (1950).
7. H. L. Kahler and P. J. Gaughan. Protection of Metals Against Pitting Tuberculation and General Corrosion. *Ind. Eng. Chem.* 44, 1770 (1952).
8. C. C. Nathan. Correlations of Oil-Soluble, Water-Dispersible Corrosion Inhibitors in the Oil Field Fluids. *Corrosion,* 18, 282t (1962).
9. C. C. Nathan and E. Eisner. Statistical Concepts in the Testing of Corrosion Inhibitors. *Corrosion, 14,* 4, 193t (1958).

DAVID A. CARTER is a senior research chemist with Betz Laboratories, Inc., Trevose, Pa., where he is responsible for the cooling water research program. An NACE member since 1965, he is also a member of the Electrochemical Society and an associate of the Royal Institute of Chemistry. He received a BS in chemistry from the University of Sheffield, England in 1963 and a MS (1965) and a PhD (1967) in corrosion science from the University of Manchester, England.

CHARLES C. NATHAN is a research section head with Betz Laboratories, Inc., Trevose, Pa. He manages research activities in the fields of cooling waters, boiler treatment, petroleum processing, and additives for prevention of corrosion and fouling. An NACE member for 15 years, he is also a member of AIChE and the New York Academy of Sciences. He is a Registered Professional Engineer in Texas and is recognized by NACE as being qualified in one or more phases of corrosion control. He received a BS in chemical engineering from Rice University, Houston, Texas in 1940 and a MS (1942) and a PhD (1949) from the University of Pittsburgh.

Corrosion Prevention Should Begin on the Drawing Board*

for example: the missile and aircraft industries

Rebecca H. Sparling
General Dynamics/Pomona
Pomona, California

SUMMARY

Ninety percent of the trouble with corrosion in missiles originates in the design. It is not enough for the design engineer to specify anodizing on aluminum parts and plating on steel; he must assure that the assembly will function properly under the worst combination of environmental extremes. To be certain of adequate protection from corrosion, the designer must anticipate conditions during inert life (transportation, storage, rework, and repair) as well as during active operation. Although coatings or maintenance can sometimes counteract mistakes made on the drawing board, maximum emphasis should be placed on preventing corrosion in the initial design.

★Revision of a paper titled "Corrosion Prevention at the Drawing Board" presented at Western Region Conference, National Association of Corrosion Engineers, September 25, 1963, Anaheim, California.

Introduction

DETERIORATION OF MISSILES in the field has led to a great emphasis on corrosion prevention—which should begin at the drawing board. Everyone else—factory worker, lab technician or test engineer—is simply trying to produce what the designer has put on the drawing. If the designer omits protective plating, puts dissimilar metals together, or uses fungus-nutrient materials, it may be impossible later to prevent corrosion. Probably 90 percent of corrosion problems today could be solved in the original design.

The designer's responsibility is (1) to ascertain the environmental extremes likely to be encountered—and (2) to provide corrosion protection against these conditions. This does not imply that the design engineer should be an expert at contract interpretation or a corrosion genius. Highly specialized technology requires full time effort to be proficient in mechanical or electronic design, without additional capabilities. But the designer should ensure the review of his drawings by skilled personnel who can advise just what should be done.

This article considers the environmental requirements and the desired protection. A check list is given of special precautions needed to assure corrosion prevention in missiles and aircraft.

Requirements of Aerospace System

The main requirement of any aerospace system is successful completion of its task, whether it is a weapon, a vehicle, a propellant, or a valve.

Performance must be satisfactory in all kinds of weather and in all locations. The reason for concern over corrosion is that it degrades performance and does so to an unpredictable degree. A loose fastener, an intermittent contact, an electrical short—who can foretell how much malfunction will result?

Exposures to Environments

Natural

Environments anticipated vary with the system which may be missiles launched from aircraft, ground installation, ships or submarines, or may be bombers, fighters, commercial aircraft, or orbiting satellites. All these environments (except satellites) have two things in common: a long period of storage and transportation, and a short time under flight conditions. During these inert and active lives, missiles may be subjected to heat, cold, rain, snow, wind, dust, salt air, dryness, ozone, fungus, thermal shock, mechanical shock and vibration. Satellites have additional requirements of operation under vacuum, solar radiation and meteorite collision.

The major cause of chemical attack is moisture. Without moisture, dissimilar metals do not form a galvanic cell; fungus does not thrive; steel does not rust; other metals seldom corrode. Unfortunately, moisture is always present except in outer space.

Water gets into equipment through openings or by condensation of moisture and collection in sump areas. Two other sources of water intrusion are through hygroscopic materials and capillary action. All this moisture results in probably 85 percent of the corrosion encountered.

Designers must remember that most missiles and many aircraft are planned for use under adverse conditions. Previous wars have been fought in mud, and future wars will probably be the same. Noise, dirt, confusion, darkness, and mud are the natural environments of military action. Air conditioning apparatus is frequently the first casualty in the field. It is inconvenient or impossible to check desiccants there, and the whole atmosphere is different from the controlled laboratory conditions of factory tests.

The designer cannot rely on good intentions. He must provide protection which will assure no degradation under the most adverse conditions.

Induced

In addition to the environments encountered in field service, designers must think of corrosive environments during manufacture and testing. Four areas of particular concern are (1) sequence of manufacturing, (2) residual stress, (3) hydrogen embrittlement, and (4) proof testing.

Test Required

Tests usually required for missiles and aircraft include temperature extremes, temperature cycling or thermal shock, rain, humidity, fungus and salt spray.

Although many of these tests are more severe than the environments, most are accepted as reasonable and realistic. Temperatures of 120 F and higher have been recorded in North America, Australia, India, Pakistan, Iran, Iraq and in many locations in Africa. Similarly, low temperatures of —65 F or less have been measured in Alaska, Canada, both North and South Pole areas, Greenland, Wyoming, Russia and Siberia. The severe thermal shock test is based on a possible check-out of component or sub-system in the tropics followed by high altitude transport at extremely low temperatures.

The salt spray test is almost universally resented. It is "too severe," "not reproducible," "not translatable into service life." Agreed. However, there is a real need for a screening test to weed out inferior finish systems. It is imperative that some kind of accelerated corrosion test be available to give results in 10 days instead of 10 months. Everyone wants a standard test which can be performed by both vendors and purchasers. Even though the salt spray test is rejected by most, it is needed to evaluate corrosion resistance of unknown parts and materials, compared to others which have proved satisfactory. It does not guarantee sufficient life under all corrosive environments and cannot be directly correlated with service.

But the objection is not valid that "this part won't ever be exposed to any salt—it's inside." Condensate anywhere near the ocean contains dissolved salt. All parts not encapsulated or evacuated and hermetically sealed may be exposed to salt attack. Salt probably is the most universal of all corroding agents, which is a good reason for continuing to use the salt spray test.

Specifications for Corrosion Prevention

The design engineer, faced with stringent requirements of world-wide storage and operation, and with tests evaluating the degree of protection, may be discouraged. It is almost impossible to design an efficient light weight part which is completely resistant to all environmental attack. "Light weight" and "RFI (radio frequency interference) shielded" lead the designer one way and optimum corrosion prevention may be 180 degrees in the other direction.

Corrosion protection often cannot be achieved by adding coatings after the design is complete because no space is provided. But a compromise between weight, cost, and protection can be achieved by careful forethought and consideration of corrosion prevention as one of the basic requirements of design.

Guides for the designer are given in government specifications and standards; however, government requirements are necessarily general and can be used only as guides. The slow evolution of government documents and the rapid development of new improved finish systems lead to an appreciable time lag between specified and optimum finishes. For example, epoxy coatings were used successfully on steel, aluminum, and magnesium in 1952, but it took years to include them in specifications.

Additional suggestions can be divided into two main categories: those which protect against nature's corrosion (moisture, fungus, galvanic corrosion), and those necessitated by man-made conditions (manufacturing sequence, stress corrosion, hydrogen embrittlement, corrosive gases, and electrical bonding).

Protection Against Nature's Corrosion

Moisture

The most obvious solution to the mositure problem is to keep equipment dry. Rain can be kept out by gasketed closures. Moisture can be reduced by avoiding hygroscopic materials such as those commonly used for gaskets (felt, leather, cork, asbestos). Cotton wicking in electrical cable also will absorb moisture and should not be used.

Unless sealed, joints can promote water intrusion. Mechanical, crimped, spot or seam welded joints form capillary areas pulling moisture in and should be closed with rubber, epoxy, polyurethane, or another sealant. Oil or grease is not an adequate sealant.

It is important to differentiate between these seals, and the "sealing" of a container or compartment. Except for units which are evacuated, then filled with inert gas and hermetically sealed, sealing of a container is not a good method of keeping moisture out. Mositure will condense inside a sealed container just as it does outside. Also, most seals breathe, permitting moist air to enter the package and resulting in more condensation. Many a sealed electronic package has malfunctioned because of corrosion.

Desiccants should be used in enclosed spaces to maintain humidity levels. Color change will occur if humidity exceeds a specified amount. Desiccants which change color are valuable only when there is periodic inspection and some provision for opening and reconditioning the compartment.

To prevent moisture from reaching specific components, various methods are useful: encapsulation of individual parts, sealants around joints, or protective coatings which form moisture-proof envelopes such as epoxy paints. The designer must select the technique which is best, most practical, and least expensive for his application. Encapsulation is used extensively in electronic design where components are embedded in epoxy, polyurethane, silicone, polyisocyanate, and polysulfide. Encapsulation also protects against mechanical shock and vibration and provides electrical insulation.

Printed circuit assemblies are kept dry by a moisture proof coating, often epoxy or polyurethane. These plastics should be used wherever practical, either by embedding or coating.

Once water gets in, through leakage or condensation, it collects in the low sump areas. These areas are some of the worst contributors to corrosion, and only the design engineer can eliminate them; then he should provide drainage. The designer should not permit any sump areas, unless necessary to design. Also, protective coatings should be applied to any sump areas that cannot be eliminated.

Galvanic Corrosion

Dissimilar metals (military standards define these as more than 0.25 volt potential difference) in an electrolyte will corrode the less noble metal. Three conditions are necessary for galvanic corrosion: (1) dissimilar metals (2) an electron path between the two metals, and (3) an electrolyte (moisture) in contact with both metals.

The design engineer can minimize galvanic attack by any of the following actions:

(1) Reduce voltage by selecting metals which are compatible or by plating with a metal closer than 0.25 volt.

(2) Break the electron path by insulating between the metals, using vinyl tape, an insulating gasket or washer, or organic coatings.

(3) Remove the electrolyte contact by covering the junction with sealant, paint, or plastic.

(4) Reduce the current produced by decreasing the area involved through painting or coating around the joint or by increasing the exposed area of the anodic metal so that corrosion is spread over a larger area and therefore is less severe.

Most problems from galvanic corrosion occur as the result of thoughtlessness or emphasis on other attributes. Electronic designers might consider a junction of gold and aluminum satisfactory because both are electrically conductive and either alone is corrosion resistant. But a gold-aluminum couple can generate 0.9 volt.

Common hardware items can cause galvanic corrosion, too. Dissimilar metal screws, grom-

Source: *Materials Protection*, December 1963, 8-15

mets, and bolts should be installed with wet zinc chromate primer, even when pressfit. Graphite acts as a metal in galvanic corrosion.

If galvanic attack is suspected, a quick test with litmus or polarity paper will establish its presence. Whenever dissimilar metals react in galvanic corrosion, alkali is liberated at the cathode surface. Immerse the dissimilar metal couple in an electrolyte, then place polarity paper on the cathode. Change of color will indicate galvanic action.

Crevice Corrosion

Corrosion of mechanical joints will occur in crevices in stainless steel and nickel which may be either passive (protected by a thin film) or active. The "passive" film does not persist in the presence of electrolyte and the absence of oxygen; therefore, the material in the joint will become active and anodic to the rest of the part. Crevice corrosion occurs even when only one metal is present, for example, a joint of stainless steel and nylon. The designer engineer should take care to use sealants around mechanical joints to keep electrolyte out.

Fungus

Microbiological attack came to widespread notice during World War II when so much damage was done by fungus in the South Pacific. Although they thrive best in warm humid climates, fungi can exist under cold or dry conditions. As more nonmetallic materials are used in electrical and electronic assemblies, increased attention must be devoted to protection from fungal attack. In one large missile system, over 200 items were found to support fungus growth.

Prevention of fungus is a difficult problem but sometimes can be solved by use of materials which will not support fungus growth. However, the designer must also be aware of the fact that fungi can exist on a given material without feeding on it, creating an undesirable film deposit on the material.

Resistance to Man-Made Conditions

In addition to phenomena such as moisture, galvanic action, and fungus, parts frequently have to withstand adverse conditions which are man-made. Improper manufacturing operations, stress corrosion, hydrogen embrittlement, corrosive gases, and electrical bonding techniques all can lead to corrosion failure.

Manufacturing Sequence

When application of corrosion protection before assembly is desirable, the drawings must give adequate notes as to manufacturing procedure. It is not possible to make a blanket rule on the preferred sequence, but in general, the following list is useful:

(1) Protect details by anodizing, plating, or painting before assembly by riveting, bolting, press, or shrink fitting.

(2) Protect the assembly after welding or brazing.

(3) Protect details to some extent before assembly and additionally after assembly when soldering or adhesive bonding.

(4) Protect parts after machining, turning, drilling, forming, grinding, bending, or any other process which may damage the protective finish.

(5) Do not plate or anodize parts which have dissimilar metal bushings in place.

(6) Be careful when match-drilling dissimilar metals on assembly. To avoid galvanic corrosion, note on the assembly drawing that parts should be separated and cleaned after drilling and should be insulated before re-assembly.

(7) Metals can corrode during fabrication; corrosion does not wait until manufacture is complete. Warn the shop by appropriate notes where pitting or corrosion cannot be tolerated (for example, on a rocket motor case with 0.020-inch walls) so that necessary precautions can be taken to prevent corrosion during fabrication.

Stress Corrosion

Material under stress is much more rapidly corroded than material free from stress. The stress may be internally imposed such as that occurring during heat treatment of high strength aluminum alloys. It may be caused by cold working such as deep drawing, or press fitting, and bending on assembly. It also may occur during proof testing. For example, stress corrosion appeared in highly heat treated steel rocket motor cases proof tested with tap water even though the time under stress and corrosive attack was short.

Stress corrosion occurs in almost all metals but is most common in brasses with high zinc content, the heat-treated high strength aluminum alloys of 2000 and 7000 series, wrought magnesium alloys, austenitic stainless steel, and high heat-treat steels. The design engineer cannot always prevent the build up of residual stresses during manufacture by such processes as grinding or forming. But he should be aware of the danger inherent in such processes and should provide either thermal stress relief or extra protective coating. Also, he can specify shot peening which will lower the stress level and improve cracking resistance.

When residual stresses are present, it is important that corrosion protection be more exacting. Anodic coatings, some electroplates (cadmium and nickel), and organic finish systems containing zinc chromate or strontium chromate primer are especially beneficial in retarding stress corrosion.

The designer can minimize such stress corrosion failures through care in selecting metals, in designing so that fabrication will not introduce stresses, and in providing protection against corrosive environments.

Hydrogen Embrittlement

A few years ago, hydrogen embrittlement was considered a problem in electroplated high strength steel. Recent experience has shown that other metals, such as titanium, are also subject to embrittlement and that acid pickling and cathodic cleaning are additional sources of hydrogen embrittlement.

Government specifications require a thermal stress relief of three hours at 375 F after electroplating steel with hardness above 40 Rockwell C (approximately 200,000 pounds per square inch ultimate tensile strength). Present practice is to relieve high heat treat steels with around 240,000 pounds per square inch ultimate strength at 375 F for 24 hours as soon as possible after plating. Shot peening before plating to reduce tensile stresses on the surface is also effective in minimizing hydrogen embrittlement.

There are some plating baths and techniques which produce lower hydrogen than the normal processes, but it is hard for the design engineer to know the methods used. Moreover, there is no nondestructive way to detect hydrogen embrittlement and no sure way to remove hydrogen. Therefore, the design engineer should make every attempt to avoid electroplating high strength steels. Organic coatings should be used wherever possible, and the designer must allow sufficient tolerances to permit the necessary build-up.

The distinction between stress-accelerated corrosion, true stress corrosion, and hydrogen embrittlement is not always clear. The design engineer can let the metallurgists and corrosion experts argue as to which type of failure occurs; his job is to design parts which will not fail. The designer should remember that stress, corrosive environment, time, and temperature are the important factors and that, if the stress is sufficiently high, the part may fracture even though the other factors are low.

Conversely, if the corrosive attack is strong, even low residual stresses can cause failure. The design engineer must avoid or relieve residual stresses, take care not to introduce hydrogen into metal, and protect against corrosive attack by suitable coatings.

Corrosive Gases

Ozone in the atmosphere and phenolic vapor inside equipment are the gases most often responsible for corrosion. Occurring with smog often found in industrial atmospheres, ozone attacks rubber, especially when the rubber is under stress. The designer should use neoprene, which is inherently resistant to ozone. If this is not possible, rubber should be coated with an anti-ozant such as di-octyl-paraphenylene diamine.

Phenolic vapor may be liberated from phenolic insulating varnishes, encapsulating compounds, or uncured phenolic material. Both cadmium and zinc are seriously attacked by phenolic vapors. The design engineer should be careful not to use cadmium or zinc plating in enclosed spaces where phenolic vapors may be liberated.

Some plastic give off chlorine or hydrochloric acid in amounts sufficient to etch aluminum or cause patina on copper when heated to 180 or 200 F. It is important to consider all the materials in the area when selecting protective finishes.

Electrical Bonding

Control of interference, whether electrical voltage or magnetic fields, is one of the big problems in missile system design. Three ways to prevent malfunctions from externally generated interference and to keep the equipment from being a source of interference to other equipment are filtering, shielding, and grounding. Shielding and grounding often involve bonding dissimilar metals. For maximum electromagnetic compatibility (formerly called radio frequency interference), a clean metallic surface bare of any insulating coating is desirable.

From the electrical engineer's point of view, insulating or oxide coatings generally used for protection introduce electrical resistance. This is short-sighted because without adequate corrosion resistance the bare surfaces will oxidize, resistance will increase, and the system performance will suffer.

Such joints as lead-tin solder to gold and copper to aluminum are common where electrical engineers do not consider corrosion problems.

One way to accomplish grounding is by a bus-strap. In cases where this is not practical and where an electrical bond to the structure is required, the treatment before bonding should be as follows:

Aluminum: Mask the bond areas, and anodize remainder, or anodize the structure and spot-face the bond area. The structure also may be chemically film treated, or cadmium or tin plated.

Copper, brass, stainless steel: Bare or plated with tin, cadmium or dipped. Other steels should be cadmium or tin plated or dipped.

When an electrical bond has been established, the joint should receive a complete organic finish or coated with epoxy or other sealant.

Coatings should be investigated which are claimed to be both electrically conductive and corrosion resistant. A simple test in salt spray

Source: *Materials Protection*, December 1963, 8-15

may show less than optimum results.

The designer knows the electrical interference requirements of the equipment, but he must remember that corrosion prevention is a requirement too and that it is even more important for good electrical and electronic performance than for adequate strength. Corrosion may ruin insulation, change electrical properties, cause shorts or loss of contact. Forethought by the design engineer pays off in longer time-to-failure, greater reliability, and better performance.

Check List

To aid the design engineer in preventing corrosion, the following check list highlights a number of possible problem areas.

1. Think of corrosion prevention at beginning of the design.
2. Allow adequate dimensional tolerances for protective coatings.
3. Be sure the environmental requirements are thoroughly understood.
4. Seal joints against moisture intrusion.
5. Use desiccants to maintain desired humidity level in closed compartments—but be sure the desiccant is visible from outside.
6. Do not use hygroscopic or wicking materials.
7. Remember "sealing" a compartment does not always keep moisture out.
8. Use encapsulation to protect electronic components.
9. Avoid dissimilar metal contact by using compatible metals or by insulating between the metals.
10. Remember graphite can be a dissimilar "metal."
11. If dissimilar metal contact is necessary, remove electrolyte by use of desiccants, or coating the contact area.
12. Install dissimilar metal hardware (screws, bolts) with zinc chromate primer.
13. Be alert to the possibility of crevice corrosion in joints.
14. Use fungi-inert materials.
15. Consider manufacturing sequence when calling for protective coatings.
16. Take care to avoid or relieve residual stresses.
17. Use organic coatings instead of electroplate on highly heat-treated steels.
18. Check assemblies to be sure no phenolic or other corrosive gases are given off.
19. Be especially careful about electrical bonding.
20. Always consult a corrosion engineer for help on corrosion problems, from the inception of design.

Conclusion

The designer has prime responsibility for corrosion prevention as well as for the other aspects of mechanical and electrical performance. To meet this responsibility, he does not need to be a chemist or metallurgist or a corrosion specialist; but he does need a realization of the problems concerning corrosion which can arise in any design. He must be willing to get expert help to solve these problems.

Effective design is the result of careful study of requirements by the design engineer with assistance from the corrosion specialist. Such cooperation precludes situations where the close-tolerance joint leaves no room for protective coatings or where the surface to be bonded has been anodized for corrosion prevention. The closer and the earlier that design engineer and corrosion engineer collaborate, the better will be the final design.

Bibliography

Guy V. Bennett. Factors Affecting Resistance of Steel Rocket Motor Cases to Delayed Failures. *Materials Protection,* **2,** 16 (1963) October.

R. McFarland, Jr. Industry, Corrosion, and NACE: A Message to Engineering Management. *Materials Protection,* **2,** 36 (1963) September.

R. J. Koch. Structural Steel Coated Better and Cheaper Before Erection. *Materials Protection,* **2,** 62 (1963) March.

E. H. Phelps and A. W. Loginow. Stress Corrosion of Steels for Aircraft and Missiles. *Corrosion,* **16,** 325t (1960) July.

David Roller and G. H. Rohrback. Development of Thin Metal Film Corrosion Indicators. *Corrosion,* **16,** 399t (1960) August.

Charles W. Holt and R. E. Turley. Cell Apparatus Checks Resistivity of Nitrogen Tetroxide Propellant to Control Corrosion in Missiles. *Corrosion,* **17,** 26 (1961) December.

F. W. Kink and Earl L. White. Corrosion Effects of Liquid Fluorine and Liquid Oxygen on Materials of Construction. *Corrosion,* **17,** 58t (1961) February.

Leo E. Gatzek. Some Corrosion Problems of Missiles in Silo Storage. *Corrosion,* **17,** 28 (1961) February.

C. W. Raleigh and P. F. Derr. Compatibility of Materials With Unsymmetrical Dimethylhydrazine Rocket Fuel. *Corrosion,* **16,** 507t (1960) October.

R. L. Johnson, M. A. Swikert and D. H. Buckley. High Temperature Lubrication in Reactive Atmospheres. *Corrosion,* **16,** 395t (1960) August.

R. M. Krupka and D. E. Taylor. Ablation Behavior of Materials Subjected to Missile Re-Entry Heat Flux Rates. *Corrosion,* **16,** 385t (1960) August.

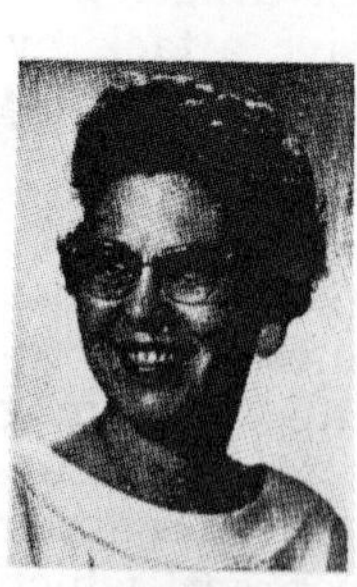

REBECCA H. SPARLING is materials specialist at General Dynamics/Pomona. She has been in the field of materials engineering for over 30 years. She is a graduate of Vanderbilt University with a BS and MS in physical chemistry. In 1957, she received the annual award of the Society of Women Engineers for her contributions to high temperature metallurgy and nondestructive testing. She also was recipient of the Convair/Pomona Management Club Achievement Award. She is listed in "Who's Who of American Women." In private life, she is Mrs. Joseph Sparling, wife of a contract administrator.

DAMAGE to a railroad car which was caused by poor design. Wooden liners inside the car did not reach the ceiling; consequently, "fines" from grain blown into the car settled behind the liner. Within a short period of time, moisture and bacteria caused the above corrosive damage. This problem is now eliminated in railroad cars by extending the liner to the ceiling. In many cases, cars are made of stainless steel.

DESIGNING TO PREVENT CORROSION

PROTECTION OF EQUIPMENT SHOULD BEGIN AT THE DRAWING BOARD

S. K. Coburn, United States Steel Corp., Monroeville, Pa.

The author discusses the importance of initial design to prevent corrosion of equipment in various environments. He briefly introduces some common types of corrosion and explains how they can be minimized by proper design. Among subject areas covered are welding, automobiles, railroad cars, chemical plants, cathodic protection, and protective coatings.

TODAY, in almost every field in which metals are used, the design engineer, the chemical engineer, the mechanical, electrical, missile, and reactor engineers, along with the architect and the do-it-yourself plant engineer must consider the possibility of corrosion. It is not possible to design merely for function, strength, and safety, or even just for durability, trouble-free service, and maintenance for good appearance. Corrosion must also be anticipated; its magnitude must be determined; and its ultimate cost to the organization must be accurately estimated and a minimum cost system designed.

To those with little experience in combating corrosion and to those unfamiliar with its theoretical principles, corrosion can be a mysterious and costly enemy.

This understanding of corrosion phenomena and the means to combat it is the responsibility of the corrosion engineer and/or the materials engineer. Sometimes the same man performs both functions. Whichever prepares materials specifications must include provisions to combat, minimize, and avoid corrosion. (In one large chemical manufacturing organization, the recently formed materials engineering department has been credited with the savings of the better part of $1 million in maintenance costs by close attention to corrosion problems. The department's calculations show that "a dollar saved in replacement materials through the use of good corrosion practices results in a dollar saving in maintenance and labor."[1]

Some engineers claim that it is wise to build a plant from the most expensive materials so as to assure trouble-free operation. Others claim that it is economically practical to operate on the basis of building a low-cost plant, meet troubles as they develop and, at the same time, upgrade the equipment. Consequently, the company can transfer repair and replacement costs to that of operating maintenance and take advantage of another tax category. Still others attempt to live with corrosion by making reasonable modifications in plant and process variables, thus effecting a compromise between high and low-cost operation.

In order to emphasize the importance of design, some theoretical principles of electrochemistry and some of the day-to-day experiences of working engineers will need to be examined.

Corrosion or the deterioration of a metal or of a nonmetallic substance is the result of a physical and/or a chemical reaction between the substance and the environment. The chemical reaction can be direct and rapid, such as that exhibited by sodium in contact with alcohol or water. The reaction also can be electrochemical, and relatively slow, such as the attack of steel by ferric chloride solution, where dissolution of the steel occurs only at discrete points on the surface. Active centers for this type of surface reaction exist at finite distances from one another. Attacked and unattacked areas are identified as anodes and cathodes, respectively. Strong acids and strong alkalis are not required to initiate this reaction. The mere presence of ionic material in a thin aqueous film is sufficient to provide the conductive medium necessary to facilitate the reaction.

Corrosion of some materials occurs in the atmosphere or in the presence of specific gases; it also occurs in liquids and underground. Before corrosion can take place, however, moisture and some ionic material must be present. In many instances oxygen, too, must be available.

Types of Corrosion

Categorically, some writers have identified the main types of corrosion as: general, pitting, corrosion-erosion, dezincification, intergranular, galvanic, stress-corrosion cracking and concentration cell.[2] These represent areas in which distinct theoretical principles are available to explain the mechanism of corrosion and assist in avoiding costly errors in design.

General Corrosion

General corrosion occurs under several circumstances. In can occur in the atmosphere. At the Applied Research Laboratory of the United States Steel Corp., specimens of carbon steel, copper-bearing steel, and high-strength low-alloy steels are exposed on test racks to the industrial atmosphere of the city and to the marine atmosphere of Kure Beach, N. C. After suitable time intervals, the specimens are removed, cleaned and weighed. Examination shows the corrosion on the surface to be fairly uniform. This type of general corrosion occurs on bridges, flag poles, lighting standards, and industrial buildings after paint has failed.

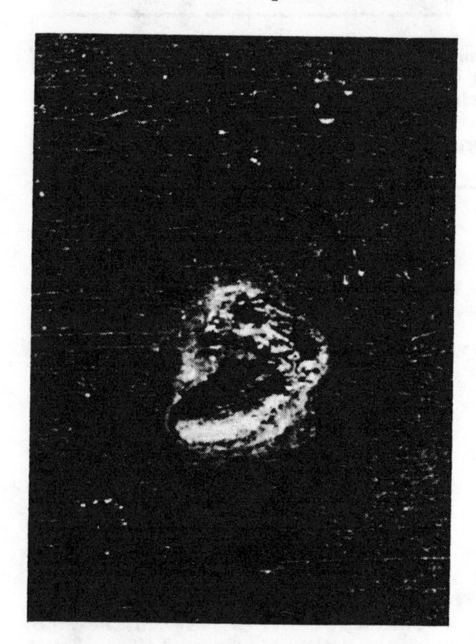

Figure 1—Pitting corrosion. Photograph shows a pit created on a lead plate beneath a glass bead in 1% ammonium chloride solution.

Pitting Corrosion

Pitting is the most insidious form of corrosion because it can render a container, tank, or component piece of equipment unusable through perforation. Pitting varies from that which is crucial to that which can be ignored. It is the characteristic type of corrosion occurring in stainless steels; however, it occurs in all metals. It is always localized, though it may occur at random locations. Ferric chloride, hydrochloric acid, and cupric chloride can initiate pits in stainless steel. Pitting can also occur as a result of cracks in mill scale and beneath certain types of protective coatings. In the cell that forms, pits become anodes and the mill scale acts as the cathode. If the cathode is large and the anode is small, the intensity of pitting at the anode is great.

Pits can form on a piece of lead beneath a glass bead in a 1% solution of ammonium chloride (Figure 1). Pits also can form beneath barnacles on a piece of stainless steel immersed in sea water. Pitting occurs in stagnant areas, such as beneath a wet gasket, in crevices or in areas that are wet and deficient in oxygen. In these cases, crevice corrosion is responsible for the pitting attack. Pitting occurs mainly on metals that depend upon passive films for their protection; however, conditions exist when almost all metals form passive films and thus can pit. Lead, stainless steel, titanium, and aluminum are a few examples.

In addition, pitting can sometimes occur when inhibitors are added to a solution to prevent corrosion and are present in lower-than-optimum concentrations. Pitting occurs when sea water lies stagnant in Monel(1) heat-exchanger tubes; however, no pitting occurs when the water is kept in motion. Pitting that occurs on uncoated underground pipe is related to physical and chemical variations in the composition of the soil. One of the most comprehensive studies in this field has been published by the National Bureau of Standards.[3]

Corrosion-Erosion

The corrosion-erosion attack of metal surfaces takes place when the streamline flow of fluids in a circulating system is interrupted by constrictions or accumulation of debris from suspended solids. Flow rates then are varied, turbulence develops, and voids and vacuum cavities appear. When these cavities collapse, they impose hammer-like blows simultaneously with the initiation of a tearing action that appears to suck away portions of the surface. This phenomenon is called cavitation. Thus, any protective oxide film that exists on the surface of a metal is removed, and active metal is exposed to the corrosive influence of the stream.

Flowing molten paraffin perpetrates this corrosion-erosion type damage. Plastics and glass have been eroded in similar fashion. Pump impellers and piping systems likewise have often been attacked. Also, the cooling systems of some types of diesel engines have been known to require extensive repairs because of corrosion-erosion. Impingement attack might be considered to be a form of corrosion-erosion. Figure 2 shows a copper tube that was attacked by a rapid stream of sea water containing numerous small air bubbles. Protective oxides were removed physically by the abrasive stream, and the surface was depolarized by the oxygen in the water.

Dezincification

Dezincification is a type of selective corrosion that occurs in some copper-zinc alloys. Chloride ions have been associated with this form of deterioration, particularly in marine applications. Dezincification, however, can be avoided for the most part by maintaining the zinc content below 15% or by the inclusion of a small percentage of arsenic or antimony (0.03%).

Intergranular Corrosion

Intergranular corrosion may occur in improperly heat-treated austenitic stainless steels or duraluminum alloys under relatively mild conditions. In both cases, improper heat treatment causes precipitation at grain boundaries. With subsequent exposure to corrosive environments, the depleted solid solution adjacent to the grain boundaries corrodes. Sometimes grains drop out and a lacy network is observed. This problem can be solved

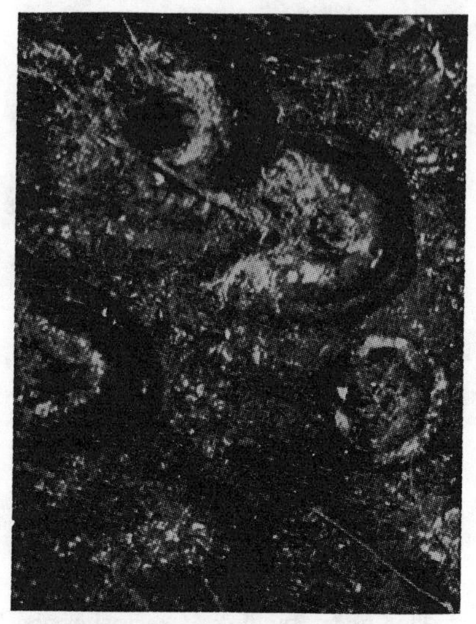

Figure 2—Impingement attack on a copper tubing which was subjected to a rapid stream of sea water containing numerous air bubbles. Protective oxides were removed by the abrasive stream, and the results were islands of unattacked metal in the form of horseshoes. The pits are deeply cut on the downstream side.

(1) Tradename of Huntington Alloy Products Division, Inc., Inco, Huntington, W. Va.

through the use of the stabilized stainless steels (AISI Types 321 or 347) or the low-carbon stainless steels (AISI Types 304L and 316L) or by the use of appropriate heat treatments.

Galvanic Corrosion

Galvanic corrosion, in reality, means the setting up of a galvanic cell. By joining two dissimilar metal strips through a galvanometer and immersing the strips in a conducting medium, current will flow and one metal will corrode at a greater rate than the other. In fact, the second metal may not corrode at all.

Stress Corrosion

A special joint ASTM-NACE Committee, organized in 1956 to study the problem of stress-corrosion cracking, defined stress corrosion cracking as follows: "When some compositions of alloys are exposed in certain environments and are simultaneously under stress remaining after previous cold work or from loads applied in service or some combination of the two, spontaneous cracking may occur after some period of exposure dependent upon the magnitude of the stress and the propensity of the environment to cause cracking."[4]

Stress corrosion cracking often is associated with pitting, galvanic, and intergranular corrosion. The microscopic paths of stress corrosion cracks may be transgranular or intergranular, that is, across the grains or along the grain boundaries; but usually they progress in a direction normal to the applied or residual tensile stress. All common alloys exhibit this characteristic susceptibility in certain environments. The critical environment associated with stress corrosion cracking of austenitic stainless steels is hot, concentrated chloride solutions. Fused caustics also are responsible for such failures. Dissolved oxygen in the presence of chlorides has been known to stimulate this reaction.

Stress corrosion cracking can occur in crevices in heat exchangers and evaporators where localized boiling can lead to a build-up in salt content. Splash zones, localized hot spots, periodic wetting and drying, and tiny leaks in equipment can cause stress corrosion cracking in austenitic stainless steel equipment and should be avoided whenever possible. Stress corrosion cracking also can occur in nonferrous metal systems.

Concentration Cell Corrosion

From an industrial standpoint, concentration-cell corrosion causes considerably more dollar damage than any of the other forms of corrosion. As indicated in the discussion on galvanic corrosion, the joining of two different metals results in a flow of current. The point to be emphasized is that a flow of corrosion current can originate in almost any circuit provided the medium is conductive and a "difference" exists. Some eighteen kinds of environments in which differences can exist that are responsible for generating corrosion currents are listed by Mears and Brown in a paper published in 1941.[5]

Damage sometimes occurs when metals, particularly aluminum and galvanized steel, show humid storage stain in the center of stacked sheets of the metal. At this location, a minimum of air exists and the area becomes anodic, whereas the edges, where the pressure on the sheets is somewhat less and air can circulate, assume the cathodic function. The presence of the stain is an example of concentration cell corrosion.

Chemical-Process Corrosion

In the chemical processing field, several other forms of corrosion should be recognized and understood if equipment is to be designed properly. The following are brief discussions of these types.

Corrosion at Liquid-Vapor Interfaces

Liquid-level corrosion often can occur when liquids are kept at constant levels in tanks or pipes made of materials that are vulnerable to attack at the liquid-vapor interface. For example, a little-used storage tank holding oil at a constant level will corrode at the liquid-vapor interface, or just above it, because the drainage of oil from the wall of the tank leaves the metal exposed to the moist atmosphere in the head space of the tank.

Corrosion Fatigue

Corrosion fatigue may occur when fluctuating loads or stresses are imposed on a system operating in a corrosive environment. Corrosive environments can appreciably lower the ability of a metal to withstand fatigue loads. This type of corrosion is sometimes confused with stress corrosion cracking.

Fretting Corrosion

Fretting corrosion occurs at the interface of tightly fitted, highly loaded metal surfaces. It is characterized by the presence of corrosion products and deep pits in the contacting surfaces where a slight amount of slippage is involved. Highspeed photography can assist in observing fretting corrosion in joint areas.

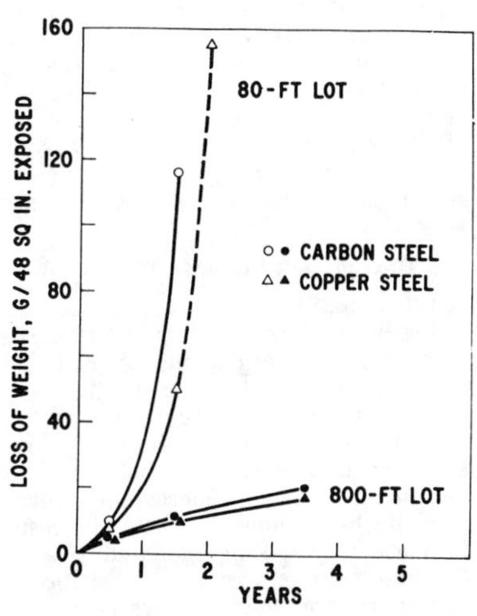

Figure 3—Effect of specimen location on corrosion of carbon steel and copper-bearing steel in marine exposure.

Corrosion in Joints (Gaskets)

Gasket material can be a source of corrosion. For example, gaskets containing graphite or sulfur compounds have caused corrosion in contact areas immersed in or wet by sea water. Gaskets containing ammonium salts or amine-type compounds can initiate stress corrosion cracking of brass flange bolts.

Corrosion Caused by Insulating Materials

Corrosion and cracking of stainless steel piping and tanks can occur as a result of intimate contact with lagging and insulating materials. For example, if lagging becomes soaked with water from a leak, stress corrosion cracking of stainless steel pipe may occur as the result of contact with chloride ion leached from the lagging.

Corrosion in Packages

The Armed Forces have encountered a considerable amount of corrosion of metallic parts stored in wood and plastic containers because of the emission of vapors. For example, certain woods can give off acetic acid vapors. Certain phenolic and urea based plastics can give off formaldehyde vapors, which can form formic acid. Under conditions of high relative humidity, corrosive salts, such as sulfites and chlorides, can be leached from certain wrapping papers.

Many other corrosive situations exist, but these examples will serve to indicate how widespread corrosion is and how conscious of it one must be at all times.

Design Measures to Prevent Corrosion

What then are some of the corrosion preventive measures that can be taken? The corrosion engineer and/or the materials engineer can assist the design engineer by seeking solutions to corrosion problems through one or more of the following approaches:

1. Intelligent selection and use of ferrous and nonferrous alloys and of nonmetallic materials.
2. Use of inhibitors.
3. Control of the environment.
4. Use of cathodic and anodic protection.
5. Flexibility of design in plant and equipment.

This discussion will deal with the last approach; however, the other approaches will be utilized as they relate to the specific problems under consideration.

Atmospheric Corrosion — Plant Location.

Safe, economical plant location is dependent upon the intensity of atmospheric corrosion and the materials selected for exterior construction. A proper appreciation of plant location can go far toward reducing initial capital costs and controlling maintenance and labor costs. The International Nickel Co. operates two atmospheric corrosion test sites at Kure Beach, N.C. One is located 80 ft. from the ocean; the other is 800 ft. from the ocean. Test racks holding specimens face the ocean at the 80-ft. site, but face south at the 800-ft. site. Corrosion rates of the same materials are vastly different at the two sites. Figure 3 gives corrosion rates for structural carbon steel and copper-bearing steel and copper-bearing steel at the two locations. These data indicate that if a decision is made to locate a plant near the ocean, the best location would be at least 800 ft. from the shore line. For a plant or structure located on a peninsula, it would be important to determine the direction of the prevailing winds throughout the year and locate the plant on the far side instead of the windward side of the peninsula (Figure 4.)

Corrosion rates are generally lower in rural areas where the local atmosphere is not seriously contaminated with the exhaust gases from automobiles and industrial smokestacks. Plant location here would be ideal. The location of metal structures in sheltered areas, where relative humidity may be high and the sun and wind do not get a chance to free the metal of condensed moisture, may result in a reduction in service life if the metal is not protected by a well-applied paint or if a poor selection of metal has been made. When several buildings are to be constructed in a complex, it is wise to locate a building that produces corrosive fumes downwind from the other buildings.

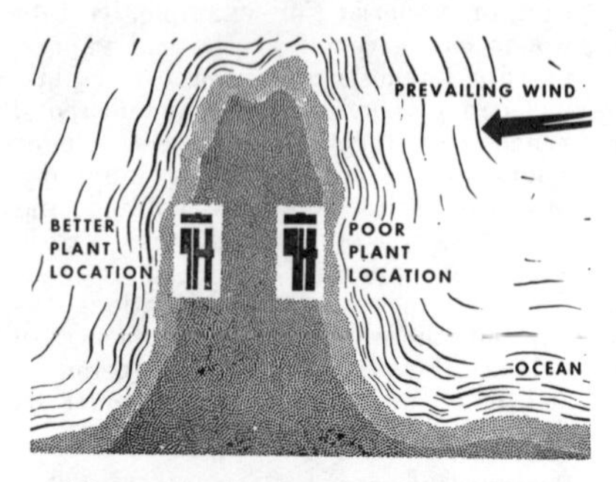

Figure 4—Safe plant location with respect to prevailing winds on a peninsula.

Figure 5—Gate levee after 36 months of service in a river near Pittsburgh, Pa.

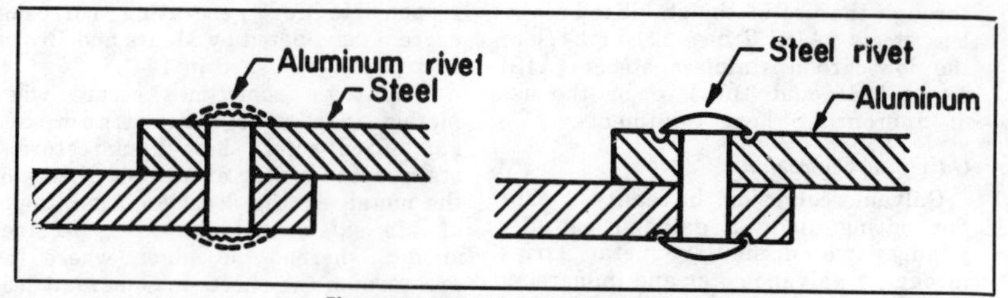

Figure 6—Example of galvanic corrosion.

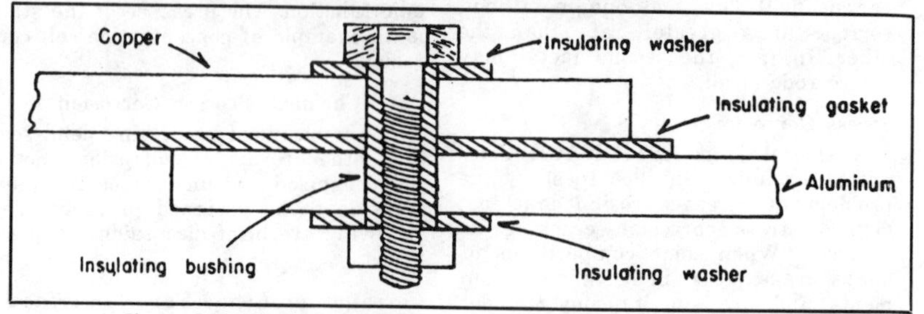

Figure 7—Method of avoiding galvanic corrosion between dissimilar metals.

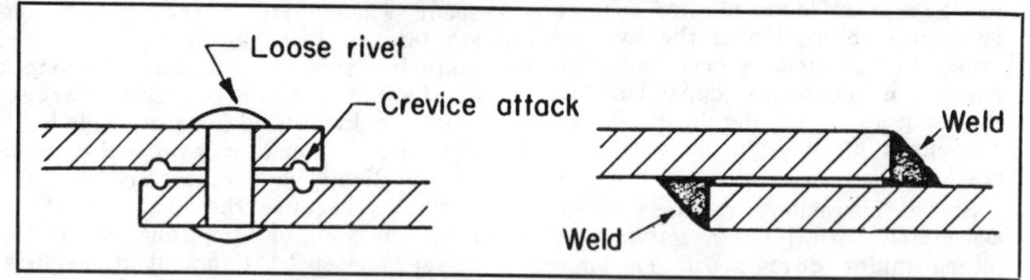

Figure 8—Welding, in many instances, will prevent crevice corrosion.

River Corrosion

Rivers also can be corrosive. Figure 5 shows the reduction in thickness of a 2-inch section of a gate levee constructed of a particular carbon steel composition after 3 years service in one of the major rivers in the Pittsburgh area. The carbon steel gate levee was replaced by cladding AISI Type 304 stainless steel to a structural-carbon-steel backup plate. No evidence of corrosion has been observed after 18 years service.

Design to Eliminate Corrosion

Rivets and Welds

Figure 6 shows how galvanic corrosion occurs in a marine environment. When steel sheets are connected with an aluminum rivet, corrosion of the rivet is likely to occur. Similarly, when steel rivets are used to bolt together aluminum sheets, undercutting corrosion of the aluminum sheet will result in loose rivets, slippage, and possible damage to the structure. The proper procedure for avoiding galvanic corrosion is to apply a nonhardening joint compound in the area where the metal and the rivet or bolt will be in intimate contact. This will serve to insulate the metal.

Another approach is to apply a zinc chromate priming composition to all contacting surfaces, and then coat the primed area with an aluminum paint. Where connections are being made that are not required to resist high stresses, the insertion of plastic bushings, sleeves, shims, or washers, will serve to insulate the points of contact (Figure 7).

Automotive

Some of the types of automotive body corrosion are general, galvanic, crevice and poultice action, concentration cell and oxygen cell, fatigue and stress, and pitting.[6] When a car ages and paint begins to fail, general corrosion occurs. Pitting corrosion may occur on the aluminum and stainless steel trim. Galvanic corrosion sometimes occurs in the cooling system. Fatigue corrosion and stress corrosion occur in some heavy-duty truck applications. However, it has been the crevice, the poultice-action, and the oxygen-concentration-cell types of corrosion that have been most common in accelerating the deterioration of automobile bodies.

When salt is used for deicing purposes, the wheels throw wet, salt-containing dirt into areas capable of trapping the material and thus prolonging the time it remains wet. The area above the headlight is a good example. The result is that corrosion of the steel and perforation can occur in as short a time as two years. Obviously, the solution is to facilitate drainage and ventilation in these areas or seal them off.

The recent adoption of unit-body frame construction may create some problems. The entire body is tied together through a form of box construction. The box sections, which may not be completely closed, can accumulate moisture that ultimately could result in rapid corrosion. Currently, some of these box sections are constructed with steel sheets that are galvanized on one side only. The coated side of the sheet prevents corrosion and perforation from the inside of the box section. It is expected that this product will result in a satisfactory service life. Dip coating of the body frame also has been adopted by some manufacturers to afford internal corrosion protection.

Railroad

An example of corrosion that has cost the railroads considerable sums of money before it was corrected is shown in Figure 9. As grain was blown into the car, some of the "fines" settled behind the wood liner because the liner did not reach the ceiling. In a relatively short time, moisture and bacteria combined to form, by poultice action, oxygen-concentration cells and acid-decomposition products that attacked the lower portion of the side sheets of the car. Extending the liner to the ceiling eliminated this problem. One of the newest approaches to the maintenance problem is the introduction of a stainless steel covered hopper car. Product purity is retained, car maintenance is reduced, and service life is increased.

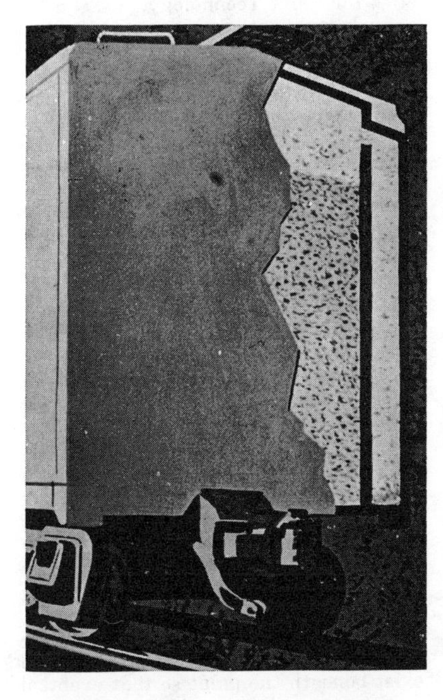

Figure 9—Example of faulty construction of wall liner of a railroad car.

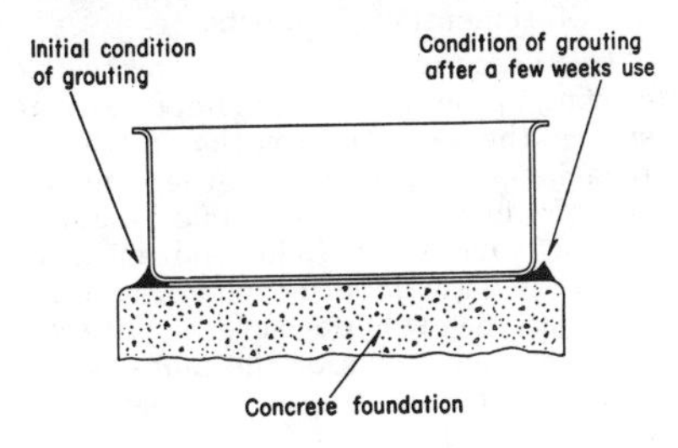

Figure 10—Erosion of grouting on concrete tank base.

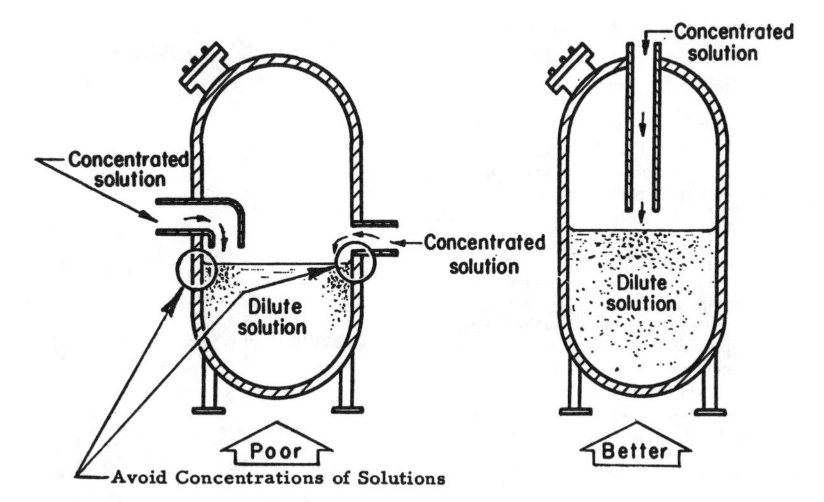

Figure 11—Preferred location of inlet pipe.

Chemical Plants

Where tanks are used for storage, mixing, and chemical reactions, the following illustrations, if adopted in the original design and installation of the equipment, should prove useful in minimizing corrosion.[7] Supporting I beams can be protected by placing the tank slightly beyond the end of the beam to prevent attack of the I beam by drippage. The use of a drip skirt also prevents seepage and attack of the tank base and the beam. In addition, because liquid in manholes tends to become stagnant, the nozzle is vulnerable to corrosion and should be made thick enough to compensate for corrosion during the life of the tank. Tank supports often are a source of trouble, particularly when the tank bottom is flat. Figure 10 shows how grouting on the concrete base is slowly eroded away as a result of drippage. The use of drip skirts and a drain arrangement could eliminate this type of damage. Also, a curved tank bottom facilitates internal drainage, and a smaller curved base prevents drippage from accumulating beneath the tank.

Introduction of a concentrated solution to a tank for dilution purposes can cause generation of corrosion currents along the tank wall because of heat created during the mixing operation, as well as concentration cells. Figure 11 shows a suggested location of the inlet tube. Addition of a cold solution through an inlet tube to a tank containing hot products could lead to severe corrosion of the inlet tube because of condensation on the tube. To avoid this, an inlet tube of greater corrosion resistance should be installed.

Also, location of heaters in tanks can result in localized hot spots that will generate corrosion currents. Figure 12 illustrates proper location of tank heaters. Notice that ledges and crevices which would accumulate solids are largely eliminated. When necessary to heat a tank externally, the heating units should be installed over as much of the exterior surface as possible. Heating over a small area can be responsible for creating conditions that will initiate internal corrosion along tank walls.

Frequently, the sidewalls and roofs of reaction vessels and storage tanks are exposed to the vapors of corrosive solutions and are attacked at excessive rates in comparison with the rate of attack on the immersed portion of the vessel. A parallel situation exists in horizontally placed large diameter pipe in which there is infrequent liquid flow or stagnation. To overcome this type of corrosion, the vessel should be vented to the atmosphere; if this is impossible, provision should be made for vacuum removal of vapor or for a condenser return to the vessel. An alternative approach is to use or apply to the original structure, a more resistant material. This may be a higher alloy material, a nonmetallic substance, a ceramic composition, a clad material, synthetic rubber, or one of the numerous protective coating systems. A thorough evaluation of the material selected should be made before introducing it into the system. Also, turning the pipe 180 degrees will serve to extend the service life of the pipe.

Cathodic protection can be applied to the protection of reaction vessels, provided the contribution of contaminating ions from galvanic anodes is not detrimental to the product. Impressed current systems also can be used to prevent product contamination. In some instances, anodic protection is very effective.

Solving Practical Problems

Joining Old Pipe to New Pipe

Joining of old underground pipe sections to new replacement sections often leads to rapid perforation of the new pipe. The inference is that a reduction in quality is responsible for the short service life. People in the gas transmission field are familiar with the fact that when bare pipe is involved, the rust covered old pipe is cathodic to the new pipe. To overcome this condition, it is necessary to insert insulating flanges at the junction of the new and old pipe. An alternative to this approach is to apply cathodic protection to the new pipe section. Painting the new pipe would reduce the current required for protection.

Shipyard Corrosion

T. H. Rogers, in his work with the Royal Canadian Navy, encountered a rather serious problem in which the urinals of small vessels were made inoperable in a relatively short period.[8] Regardless of the metal used in the 2-inch drain lines, which were flushed with sea water, the formation of a calcareous deposit with resultant severe pitting created an annoying replacement problem. The reason for the difficulty was the chemical complex that formed by the reaction of phosphates in the urine with the calcium and magnesium of the sea water and the metal of the drain. The successful solution of the problem was the replacement of the metal drains with plastic tubing.

Another problem cited by Rogers was the receipt at the shipyards of valves, plugs, and deck fittings that were lubricated with a graphite grease. After a relatively short period in service, many of the fittings involving aluminum-to-aluminum and aluminum-to-stainless steel contact failed because of galling and seizure resulting from the production of voluminous corrosion products of aluminum. A similar type of corrosion occurs on the surface of aluminum sheets that have been marked with a graphite pencil and exposed to the marine atmosphere. Specifications should indicate the need to use an alternative product that does not involve graphite.

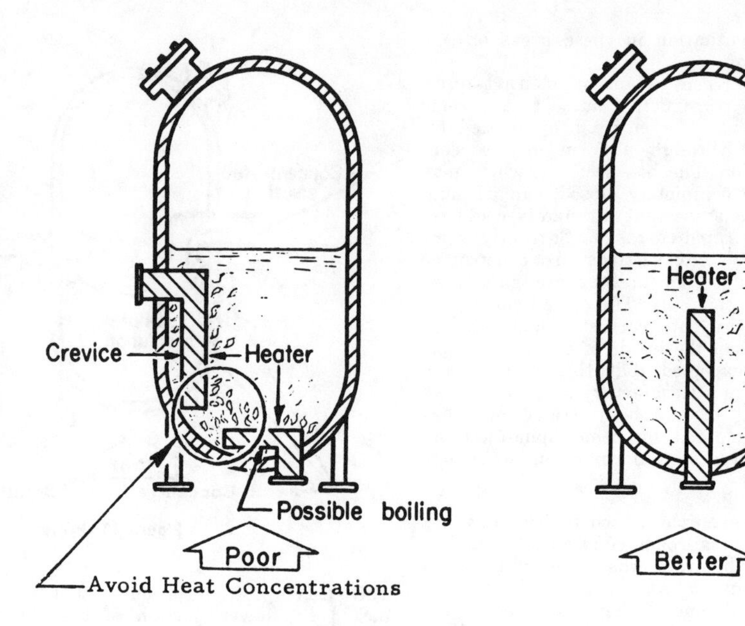

Figure 12—Preferred location of tank heaters.

Protective Coatings

Because it is the surface of metals that is vulnerable to attack, it is appropriate to briefly discuss some of the principles involves in the application of protective coating materials.

Sprayed metal coatings have been found very effective in reducing corrosion. Numerous cases exist of structures in sea water extending from the mud line through the tidal zone into the splash zone. When the corrosion rates are relatively low, it may be more profitable to omit the use of cathodic protection, substitute thicker sections, and coat only the area of the structure exposed in the tidal and splash zones. Sprayed metallic coatings of zinc and aluminum have a performance background that make them worthy of consideration for many applications in which maintenance costs have been excessive. Because these metallic coatings tend to be porous, they should be sprayed with a sealer, such as a clear vinyl coating.

Paint technology has made considerable progress since World War II. One of the most notable achievements of that early period was the development of the wash primer. This material is composed of phosphoric acid, basic zinc tetroxychromate, ethyl alcohol, and vinyl butyral resin. It is applied to clean steel and has a coverage of 300 to 400 sq ft per gallon. It leaves a film of about 0.5 mil and can be applied under conditions of high relative humidity. It has been the major reason for the successful performance of many topcoat formulations.

Also, since the war, several major advances in polymer chemistry were made so that today many new polymer systems are available. They are probably recognized by such names as alkyd resins, acrylic resins, polyesters, polyurethanes, vinyls, and epoxy resins.

Unique protective coating compositions alone are insufficient to assure long service life. In fact, proper surface preparation is critically important to the success of any coating system. Considerable information is available on the proper selection of cleaning procedures and coating materials. Foremost in the field are two volumes entitled *Good Painting Practices* and *Systems and Specifications.*[9] These volumes are available from the Steel Structures Painting Council located at Mellon Institute in Pittsburgh, Pa. In addition, recommended specifications for painting procedures have been prepared by technical committees of the National Association of Corrosion Engineers and by the Federation of Societies for Paint Technology.

Such publications have made clear that the cost of paint is the least expensive factor in the protection of any structure, whether of steel or of nonferrous material. The most costly operation is surface preparation. This problem is one of economics and engineering. Sandblasting, gritblasting, acid cleaning, hand and machine wire brushing, flame cleaning, and natural weathering represents a few of the methods available for preparing the surface.

Once the surface is clean, other factors enter the picture to influence the service life of the protective coating. Welds, crevices, depressions, sharp edges, and ledges can prevent the application of coatings of proper thickness. Sharp points may penetrate the paint film and expose the metal. Welds may contain loosely adhering weld splatter, which will result in lifting of the paint film. Ledges may accumulate water, which will permeate the coating and lead to the formation of blisters. Depressions and ridges may not be completely filled and voids then can exist beneath the paint so that eventual perforation will take place beneath the bridged paint film. The quality of the coating composition will not overcome the aforementioned defects.

In planning an outdoor protective coating program, it is important to secure the best information available. In a large chemical plant where numerous products are being manufactured and a variety of industrial effluents are vented to the atmosphere, it is important to evaluate several different paint systems. One of the more prominent paint consultants has devised a test fixture to assist in the evaluation of paints. The device is a small plate to which is welded a small channel providing a ledge, a weld bead, a broad horizontal and vertical surface, a depression, a scratch, and some sharp edges. Large numbers of these fixtures are bolted to exposure racks in various process areas, and all paints of interest are applied to the fixtures. After a reasonably short exposure period, it is possible to determine which of several paint systems will be most effective in the particular area under consideration.

A word that has come into common usage in recent years is *synergism*. In essence, it means that the union of two substances will result in an effect greater than that achieved by either substance operating independently. The excellent service performance of galvanized and aluminized steel in severe atmospheric exposure is well known. It is well known that vinyl paint systems, too, have an excellent performance record. Neverthless, both systems do have limitations.

The sacrificial metals corrode with time; and painted surfaces require touching up periodically. It appears from current case histories that the combination of a sacrificial metal and a vinyl coating system with an occasional touch-up treatment acts synergistically to give a long and trouble free life.

One example will illustrate the point: A salt-producing plant on the Gulf Coast experienced a service life of 7 to 12 months for uncoated galvanized steel. Vinyl coatings without any touch-up treatments were satisfactory for somewhat less than two years. By combining the two coating systems, there has been no need for touch-up treatment for over three years.

The high strength low-alloy steels represent another case in point. It is well known that such steels exhibit remarkable corrosion resistance in industrial atmospheres.[10] The tight oxide film that forms is hardly more than about one mil thick; and for all practical purposes, corrosion ceases after 18 to 24 months exposure. Also, the application of paint to such steels is far more effective from the standpoint of service life than when the same paint is applied to structural carbon steel.

Summary

The cost of corrosion, in many instances, can be and should be reduced in the design stage. An understanding of the electrochemical theory of corrosion, experience in the recognition of the various types of corrosion, a knowledge of ferrous and nonferrous materials (including plastics, paints, wood, glass, and ceramics), and a familiarity with the technical societies and technical literature dealing with these fields will aid considerably in minimizing corrosion and maximizing the service life of plant and equipment.

NOTE: Figures 10-12 are used with the permission of W. H. Burton. (See Reference No. 7.)

References

1. P. J. Gegner. *Proceedings, Short Course on Process Industry Corrosion*, co-sponsored by Ohio State University and NACE, p. 1 (1960).
2. Mars G. Fontana. *Corrosion: A Compilation*, The Press of Hollenbeck, Columbus, Ohio, p. 29 (1957).
3. M. Romanoff. *Underground Corrosion*, National Bureau of Standards Circular 579, Washington 25, D. C. (1957).
4. F. L. LaQue. ASTM Special Technical Publication No. 264, co-sponsored by ASTM and NACE, p. 1 (1960).
5. R. B. Mears and R. H. Brown. *Industrial and Engineering Chemistry* 33, 1001 (1941).
6. S. K. Coburn, J. W. Stewart and R. B. Mears. *Improved Corrosion Resistance for Automobile Bodies*, paper presented at SAE meeting, (1961) January.
7. W. H. Burton. *Proceedings, Short Course on Process Industry Corrosion*, co-sponsored by Ohio State University and NACE, p. 259 (1960).
8. T. H. Rogers. *1st International Congress on Metallic Corrosion*, Butterworths, London, p. 421 (1961) April 10-15.
9. J. Bigos, Ed. *Steel Structures Painting Manual*, Volumes I and II, Steel Structures Painting Council, Pittsburgh 13, Pennsylvania (1954).
10. C. P. Larrabee. *Corrosion*, 9, 259 (1953).

SEYMOUR K. COBURN is an associate consultant in the Corrosion Technology Division at United States Steel's Applied Research Laboratory, Monroeville, Pa. He has been with the company since September, 1959. He formerly was with the Association of American Railroads in Chicago, Ill. A member of NACE since 1952, Coburn is currently a member of the Publications Committee, the Research Committee, and the Board of Review. In addition, he is chairman of Unit Committee T-3 on General Corrosion. He has a BS from the University of Chicago and an MS from Illinois Institute of Technology.

Classic Blunders in Corrosion Protection*

OLIVER W. SIEBERT
Monsanto Company, St. Louis, Missouri

About 50 examples of blunders of omission, commission, and effects resulting from adventitious influences are related as indicative of the necessity for careful preliminary study of corrosion problems. The examples include references to galvanic, chemical, metallurgical effects, stress corrosion cracking, improper application of cathodic protection, failure of engineers to survey conditions relating to the environment, contents of vessels, details of process contents and operation, stress corrosion cracking, improper selection of materials, erroneous substitutions resulting in damage and numerous other causes. Environments involved are atmospheric, underground, high and low temperature chemicals, fresh and salt water, submerged and vapor space attacks, and others.

THIS PAPER PRESENTS a number of things that have been done in the name of corrosion prevention that should not have been done, and might have been titled: "Big Corrosion Surprises of the Past." One of my compatriots at another chemical company recently told me that, like most practitioners in our business, we have each had our share of goofs. Now, before I give the idea that materials engineering is only the making of mistakes, let us put things in their proper perspective. I prefer to think from a position of relativity, where you're standing when you're looking, It's amazing to me how smart people tend to be, later, when *they* have all the facts. It is also amazing how the facts all seem to come forth *after* the failure, the same facts that were unknown, unavailable, or obviously none of our business, when we were making the original design decision! I grew up in a typical old-world German family—there was a term that used to cover this kind of situation. It was referred to as a "treppenwitz"—loosely translated—"those things/statements which *we* wished *we* had done/said beforehand, especially if something later went wrong."

Before I give you the benefit of the results of a recent survey of corrosion blunders I have conducted within the Chemical Process Industry, I would like to point out that corrosion engineers don't have a corner on blunders. The *American Paint Contractor* magazine surveyed some 100 painting contractors across the United States and asked them to tell about their "biggest professional mistakes."

Question: In your worst dreams have you ever pictured yourself painting the wrong house?

Answer: 34% of the survey group admitted that they had actually painted the wrong house!

For those of you who sometimes feel that you are living under a dark cloud, there may be some consolation in learning that many of your colleagues have already had the same kind of embarrassing experiences you thought were unique with you. I gave correspondents in this survey the opportunity of an identified by-line. Not one asked for credit—they must have felt that there was glory enough for all of us in just knowing that our "successes" will not be forgotten.

People can learn from the experiences of others; it is hoped that you may learn something from the following confessions (of fellow materials engineers).

Case Histories

The first example is truly a classic. After World War II, a book was published reviewing major industrial and military errors of the war effort. Included were the cases of aircraft aluminum propellers, beautifully polished to remove all surface flaws, that failed by fatigue from the stress raisers at the stamping of the manufacturer's name. In this book, it was noted that a mine sweeper was built of stainless steel because, as with standard wooden construction, it too was nonmagnetic—a requirement for operating in mine fields. Sea water is about 3% sodium chloride and a welded ship is highly stressed. Predictably, the ship failed by chloride SCC on its shakedown cruise!

Staying with the fruits of thoughtless expediency practiced during a war, several more ship stories are in order. To get Liberty ships built as quickly as possible, some shipbuilders used steel plate which contained heavy mill scale on the hull bottoms. Some of these ships actually perforated by corrosion during the outfitting period, which was only a matter of a month or so. After a few had been launched this way, the specifications were changed so that the mill scale had to be removed by sandblasting or pickling. The manifestation of a new corrosion phenomenon? No! It had been

*Presented at 8th Annual Corrosion Seminar, NACE and Milwaukee School of Engineering, Milwaukee, Wisconsin, November 22, 1977.

 Reprinted with permission from *Materials Performance*, April 1978, 33-37,

long known that mill scale was strongly cathodic to bare steel, and as such, a very real corrosion hazard. It is worth noting that many storage tanks and much structural steel continues to be erected today without regard to mill scale removal.

Because I hold wood construction in high esteem, it personally hurts to even cite the next bit of history. A wooden salt crystallizer had been in service for over 20 years. Immersion corrosion tests showed that AISI 304 stainless steel was completely resistant, but steel was badly corroded. (I really don't know why anybody would want to replace wood?) A thin AISI 304 stainless steel lining was spot welded into a steel shell and welded up at the seams. Almost immediately, about 1/2 of the spot welds and 1/4 of the weld seams failed by SCC. With brine sandwiched between the dissimilar metals, it is estimated that useful current could have been drawn off the resulting battery! As with the mine sweeper, the replacement vessel was constructed of wood.

Again, to stay with a winner, a large wood blend tank was used successfully for years to blend batches of pigment. The only problem was one of mechanical wear of the bottom and lower sidewall. The process contents were on the acid side with the pH normally below 2. It was suggested that the bottom and lower 1.5 m (5 feet) of the sidewall be lined with acid-proof brick, using a silicate-type mortar, to resist abrasion. Shortly after the brick was installed, and prior to putting the tank back in service, plant production was curtailed due to reduced sales. Because the tank was expected to be out of service for several months, it was decided to fill the tank with water to keep the wood swollen and tight. When the tank was returned to service after approximately 3 months outage, the brick became loose as soon as the agitator was started. Silicate-type mortars have excellent resistance to acid conditions but very poor resistance to water or alkalis!

I would like to give an example of the "emotional" type materials of construction improvement problem. A small wooden 750 gallon tank had been maintenance-free for 18 years. A debottlenecking project was to be easily handled by installing a spare tank (a complete set of wooden stays, hoops, etc.). But the new vessel was made of Hastelloy B because the project engineer was not about to use wood—it wasn't modern enough for him.

Galvanic, bimetallic corrosion has been used to protect steel ship hulls since the days of Sir Humphrey Davy. Cathodic protection is used along with suitable protective coatings. A ship was drydocked for repairs, a coating was applied to the entire hull and a generator placed in the ship. The generator was electrically connected in such a way that the tail shaft and the propeller were not grounded. These items became strongly cathodic with an impressed current going continually up the shaft from the propeller. The ship was only back in service 4 months when perforations of the hull occurred in the bow. Other mechanically damaged areas of protective coating were pitted rapidly, many of these were close to the point of perforation of the hull.

As dramatic as that shipboard impressed current galvanic corrosion example seems, do not overlook the incidental problem resulting from unfavorable area considerations. A steel vessel, with an internal copper heating coil, was only slightly rusted by the moderately corrosive environment. To reduce/control iron pick-up, the tank was lined with a compatible liquid applied coating. In almost no time at all, the tank wall perforated and much of the coating was undercut and lifted off the tank. Since there are *always* some defects/holidays in any thin coating, the small areas of steel exposed at these holidays became highly anodic by action of the large area of the cathodic copper coil.

Somewhat akin to the aircraft propeller manufacturer cited earlier, a 7.5 cm (3 inch) Schedule 80 cast steel fitting was vendor identified by the attachment of a brass medallion. This installation was found on an ammonia barge between the 3 MPa (450 psi) storage tank and the first block valve. The supplier considered the problem of external corrosion and used a long life brass tag. Unfortunately, he did not consider other aspects of corrosion. Over 90% of the fitting wall had corroded away under the brass marker by galvanic and crevice attack.

An underground steel steam condensate return line failed by action of the corrosive condensate. The replacement line was constructed of AISI 304 stainless steel. Leaks occurred in the stainless pipe within 2 years. Failure, by transgranular (chloride) SCC initiated on the exterior (soil side) surface of the pipe, was limited to road crossing locations. Deicing salt in the moist soil supplied a chloride concentration of 0.5%, more than sufficient to cause SCC of the hot stainless steel pipe.

A conventional trailer mounted rubber lined steel "acid buggy" failed by external corrosion from HCl acid piping connection leaks during cleaning of plant equipment. The vessel was replaced with an FRP polyester tank, not subject to the external corrosion problem. Worried about the physical abuse the plastic tank might receive (from personnel used to hauling the steel tank, but not too familiar with limitations of FRP), signs were attached with the warning, CAUTION, GLASS-LINED STEEL. On a rainy weekend, a nitric acid day-tank developed a leak (failures never seem to occur on the day shift, Monday through Friday!). The nitric acid was pumped into the *known* resistant "glass-lined steel" acid buggy. The *nonresistant* FRP tank "burned up" on the spot. The moral of this story is—never try to fool Mother Nature, or your fellow workers.

A large nitric acid liquid/gas scrubber, with a 1 m (42 inch) ϕ stack, was installed to operate in 40% sulfuric acid at 82 C (180 F). The unit was built of asbestos reinforced phenolic. The system's equipment failed after 2 months. A replacement unit of FRP polyester lasted for 6 months. The "ultimate" solution, PVC lined FRP, had a service life of 1 year. A stainless steel stack band had fallen into the scrubber and was observed submerged in the acid solution during an internal inspection of the PVC failure. Instead of it being badly corroded as everyone knows stainless should be in 40% sulfuric acid, it was as bright and shiny as a new dollar. There was sufficient nitric acid present to passivate the stainless against the action of the sulfuric acid. The present AISI 304 stainless steel scrubber has been in maintenance free service for over 12 years. It's not what we don't know that hurts us, it's what we know that isn't so that causes all the trouble!

A cast iron wastewater drain line in a kitchen/service building failed by graphitic corrosion. The metal was replaced with PVC. When the 82 C (180 F) dishwasher rinse water contacted the PVC, it failed; PVC is limited to about 54 to 66 C (130 to 150 F).

Incorrect testing techniques also fit into this story. Chlorides were stripped from a strongly acid mixture with some residual chlorine and a very high salt content. The cold liquor was dropped into a glass-lined steel vessel and heated to about 80 to 90 C to boil off the chlorides. Glass-lined vessels lasted only 1 to 2 years. Laboratory tests of nickel, Inconel, Hastelloys, stainless steels, and titanium found only titanium to be resistant. But in 2 days, a new titanium coil failed [a corrosion rate of 38 mm/y (1500 mpy)!] and the vessel wall lost 0.076 mm (3 mils). In the laboratory test, the iron corrosion products from the alloys tested alongside the titanium were oxidized to the ferric state by the halogenated solution and this provided a most effective inhibitor for the corrosion of titanium. Mixtures of types of metals in laboratory tests should always be avoided. Copper from Alloy 20 could inhibit corrosion of stainless steel, etc.

A flanged and bolted underground stainless pipeline was being replaced with an all welded system because of joint leaks. During the night it rained and submerged the completed portion of the line already in the trench, which had not yet been backfilled (there was the off-shift Murphy at work again!). The next day, stray currents from the welding machine "drilled" thousands of holes through the pipe in the brine environment of wet Gulf coast soil.

Oxygen and carbon dioxide in a steam condensate system was causing high maintenance of carbon steel piping. An FRP epoxy piping system was installed using polyurethane insulation. A steam trap failed and the 182 C (360 F) steam destroyed the pipe and the insulation.

To extract benzene from a product, the wet benzene was distilled and the aqueous phase collected in a separator where the top layer containing the benzene ran back into the still. The water layer was collected in a catchpot. The aqueous layer was tested in the laboratory, and FRP polyester was found to be the recommended replacement for the high maintenance steel tank. Over a longer period than the laboratory environment had "seen", small quantities of benzene collected in the catchpot and cut through the vessel side wall. Know a the conditions of the environment, or as an old army poster warned, "She may look like your sister, but —."

A similar problem arose because it was known that the environment was "only water." The "known" 7% salt solution of the product was to be handled in an FRP polyester column. The specification was modified after the column was delivered because plant people believed there would be excess hypochlorite generated in the system. The unit was lined with PVC and installed. In a few weeks, the PVC was reduced to a pasty mess. Failure was caused by the excess, and unconsidered, morpholine. Stress relieved stainless steel was *believed* to be acceptable, but speed of replacement dictated a "temporary" carbon steel unit. This lasted 2 years and was replaced with a stress relieved stainless column. When replaced 4 years later to install a larger unit, it was found to have suffered SCC. The replacement was not stress relieved and failed in 1 year. Proper *in situ* corrosion tests were conducted, and Incoloy 825 has now been in for 6 years without incident.

External pipe clamps, used to support insulated piping, generally compromise the waterproofing which allows moisture introduction and external pipe corrosion under the wet insulation. A novel approach was to weld a vertical support under the pipe which was then attached to a horizontal shoe for support. This design eliminated the introduction of moisture and stopped the external corrosion. Unfortunately, in an organic acid service, heated to maintain the fluid above the HCl acid condensation dew point, the support acted as a cooling fin. The line corroded from the inside at each support.

Zinc embrittlement has been highlighted widely in recent years. It was decided that galvanized structural steel could be used on a project provided proper safeguards were observed. The contractors agreed. But a subcontractor welded galvanized access ladders directly to a series of 380 ML (100,000 gallon) stainless steel tanks. The hydro test showed nearly every weld to leak. The lesson here is two-fold: not only must the instruction go to all personnel on the job along with adequate field inspection, but don't by-pass the *final* hydro.

As an extension of the earlier example about coating the steel tank containing the copper coil, the oft-told story of beer tanks comes to mind. The original steel tanks, which were coated on the inside with a baked phenolic to reduce contamination, were known to last up to 20 years. Because the coatings in the tank bottoms were subject to mechanical drainage, replacement units were made with stainless steel bottoms and coated steel sidewalls. Within a few months, the tanks started failing by perforation of the steel within a 5 cm (2 inch) band above the stainless steel. As previously noted, all thin, liquid-applied coatings are permeable and have holidays/ defects. This resulted in small steel anodes (exposed through coating defects) coupled to a large stainless cathode. It was estimated that the corrosion rate of steel in that unfavorable area relationship was in the order of 25 mm/y (1000 mpy). The rule is if you must coat a bimetallic couple, be sure to coat the cathode. Just think, all that wasted beer, now that's serious!

Any compendium of "monuments to stupidity" would not be complete without mention of cathodic protection. Because of the possibility of stray current/interference problems, as reviewed before, a steel gas piping system was specified to be protected by a sacrificial magnesium anode system. The customer insisted upon having one of those "good rectifier type systems" (somewhat akin to the "modern" attitude toward wood tanks). Inquiries were made about all neighboring structures. Information was that all water and gas lines (to be protected) were welded steel and there were no underground telephone or electrical cables. Within two months, two underground telephone cables were chewed up. While the telephone lines were on poles, there were three cable "dips" under roads, two right next to the cathodic protection groundbeds. It was also found that there were underground electric cables, the gas lines were not welded but mechanically coupled and the water lines were cast iron bell-and-spigot joints (neither electrically continuous). The original project engineer, the plant representative and the contract cathodic protection engineer had all died. Without information, the system was abandoned. When designing a rectifier system, don't trust information about neighboring structures, no matter how credible the source. Investigate, investigate, and investigate!

Again, an external chloride SCC problem on stainless steel pipe: To keep a replacement, thin-wall stainless pipe from cracking, the pipe was wrapped with aluminum foil, then covered with insulation and weather proofed. All was well until the polymer/tar buildup in the line required removal. Regeneration, the high temperature removal of the tar by burning, melted the aluminum foil. The liquid-metal "corrosion" reaction destroyed the stainless pipe. Serious? Remember the English Flixborough disaster? That explosion was caused by the liquid zinc embrittlement of stainless steel pipe which allowed the release of cyclohexane under pressure into an on-going fire. It was deduced that the stainless pipe was splashed by molten zinc from nearby galvanized steel, heated by a relatively minor fire before the stainless itself became red hot.

We seem to keep coming back to cathodic protection. Today we know that much was misunderstood in times past. Of great consequence was the idea that corrosion of buried pipelines depended *entirely* on resistivity of soil. The rule of thumb was that if the soil resistivity was below 10,000 ohm-cm, the pipes were sure to corrode; 20,000 to 50,000 ohm-cm, maybe; above 100,000 ohm-cm, no worry. Many man-hours were spent taking Sheppard Cane resistivity measurements. In those soils going through 10,000 to 50,000 ohm-cm, the pipes were coated and protected. Above 50,000 ohm-cm, they were left bare and unprotected. On occasion, the skips would be short, but in many cases the coated sections were a small percentage of total. As you know, pipes "protected" as above had a habit of failing. People finally woke up to the fact that abrupt changes in resistivity created corrosion cells, and that the resistivity rule applied only if there was uniformity of resistance. Since there isn't any such animal as uniformity, *ALL* underground metal structures, anywhere in the world, should be properly protected.

An AISI 304 stainless steel Schedule 40 coil was in cold brine service, with occasional hot brine. The installation was duplicated except that this time a thinner AISI 304 stainless steel Schedule 10 coil was installed. The new unit failed from chloride SCC within 6 months; it took 6 years for the original unit to fail. The old "You can always cut a little more" theory doesn't work if the stress level in a SCC system is increased as a result of reduced wall thickness.

Even today the question is still asked by some: "Do we really need a formal Materials Engineering Department?" Some 25 years ago, a process was being developed for a new organic chloride product. The research lab ran some "corrosion" tests by exposing lengths of wire of several candidate metals. It was decided that, based on their untrained visual examination, Nichrome heater wire possessed excellent resistance. At the cost of some several hundred thousand dollars, equipment was fabricated of something that was passed off as Nichrome. All the equipment failed within 2 months.

A series of stainless steel heat exchangers were cooled with recirculating treated water and had been in service over 10 years with no problems. During an emergency, untreated river water was used to cool the units for about 48 hours. Several weeks later, five of the coolers failed by massive chloride SCC. There was a large buildup of dried out mud in the units, which caused reduced heat transfer and higher temperatures, and allowed tube surfaces to become dry. The subject of SCC could support a whole book about failures that were caused by the, 'We can get by with it just one time," and "It's only going to be for a short time," syndrome.

Carbon steel was the accepted material for producing sodium sulfide. Plant exposure tests showed that Inconel was the economically optimum material for construction of a new facility. As predicted, Inconel was an excellent construction material, however, the product was so pure that iron had to be added to it so that customers' processes were not adversely affected. In the past, the customers had adjusted their operations to take care of the iron contamination and, therefore, could not use the new, uncontaminated product.

How many times have each of us been told to stop getting involved in so many of the details, "—just give us a material of construction—." A Hastelloy B pump, good for reducing environments, was specified for 35% sulfuric acid after early failure of Alloy 20. The Hastelloy B unit failed in 2 weeks. It was then discovered that the acid was being pumped through a scrubbing jet which aerated the system. Naturally, oxidation resistant Hastelloy C worked fine.

In another instance, a 380 ML (100,000 gallon) storage tank, fabricated of FRP polyester, collapsed. The correct resin was specified and used, and the tank design and fabrication were not at fault. In fact, the tank was over designed for a specific gravity of 1.2. Unfortunately, the client neglected to tell of a sludge buildup to a depth of 8 to 10 feet which effectively increased the specific gravity to something like 3.3!

Many industrial designs are really the result of "committee action." When one or more design members insist upon independent isolation of their discipline, problems can occur. An over-the-road tank truck was used to haul a very corrosive organic chloride. None of the several metals tried were satisfactory. In cooperation with a glass-lined steel tank fabricator, a trailer was designed with a special suspension/shock absorber system. Other special engineering features were incorporated in the design, and it became a quite costly unit. The vehicle worked beautifully on a straight road, but could not make turns because it had a tendency to tip over.

Remember the coated ship bottom story? An underground bare steel pipeline was corroding and leaks started getting ahead of repairs after about 3 years. The system was replaced with a coated and cathodically protected design. That line has now been in trouble-free service for over 15 years. A companion line was installed 8 years ago and failed in less than 1 year. The dual protection was thought to be excessive, so it was only coated, no cathodic protection. The damaged coating caused the small exposed anodes to corrode more rapidly than the all-bare system. Inadequate protection is worse than no protection.

A large AISI 1020 carbon steel shaft [20 to 25 cm ϕ x 1.5 m (8 to 10 inches ϕ x 5 feet long)] on a rotary kiln failed by fatigue after about 15 years. An AISI 4340 shaft was the replacement. The shaft, heat treated to R_c 32-36, was installed and failed in 3 to 4 months. The present shaft, again of AISI 1020 steel, has been in service for 10 years without difficulty. There were several changes in cross-section which resulted in sharp corners on the shaft. Endurance limits of metals are related to their notch sensitivity. The same notches on a hot-rolled and normalized AISI 1020 carbon steel [UTS 410 MPa (60 ksi)] were not critical, whereas on the AISI 4340 shaft, heat treated to approximately 1.0 GPa (150 ksi) tensile, they were fatal.

Stainless steel, the stuff that's corrosion resistant, was selected for an underground oil line. Approximately a year after installation, the line was being prepared to be put into service. Oil pumped in at one end did not appear at the other end. Investigation revealed numerous holes caused by cathodic protection interference. A major cathodic protection system was in operation to protect carbon steel, the stuff that isn't supposed to be corrosion resistant, but it had not been considered necessary to connect the stainless line into the cathodic protection system.

A 20 year old rubber lined hydrofluoric acid tank, needed replacement. The customer wanted a more modern solution than a like-for-like rubber replacement. He selected a filament wound FRP vessel. The supplier guaranteed the unit for HF service, provided the inside corrosion barrier would be constructed using a synthetic veil and mat reinforcement rather than "C" glass, known to be attacked by HF. The client was told that any defect in the corrosion barrier could be disastrous as would permeation of HF through the internal barrier. Even though the FRP vessel was the one installed, the warnings were sufficiently heeded that he installed the vessel in a diked area, the bottom lined with limestone. Within a year, the vessel ruptured catastrophically at 3 AM (Don't they always?). Luckily, the dike prevented any serious personnel or property damage. The replacement rubber lined steel vessel has now been in service for 10 years.

As is so many times the case, we solve one problem only to generate another. In a pigment-grinding operation, the pigment was mixed with salt crystals and dry ground in a Banbury mixer, the salt being the grinding agent as the mix was churned. The mixer was steel and chilled through a jacket to keep the heat sensitive pigment from heat degradation during grinding. Salt was used because it could be dissolved out of the water insoluble pigment when the operation was finished. Unfortunately, the cold walls of the mixer used to sweat in the humid environment of the plant, whereupon the resulting brine on the walls corroded and roughened the steel. Eventually, the unit was replaced with a stainless steel one which withstood the ravages of the cold brine extremely well. This time, there was no longer any grinding of the pigment—the stainless walls got so smooth and polished that the pigment/salt mix just slid along them without any mixing or grinding. It turned out that about 90% of the grinding was actually being done between the blades and shell and not in the mass. The old corroded steel walls had been serving a function by virtue of being roughened. Finally it was necessary to lay longitudinal weld beads along the inside of the unit to get enough artificial roughening to get the grinding back.

A 61 m (200 foot) long enclosed building housed five externally fired furnaces in which sodium chloride and sulfuric acid were reacted to produce sodium sulfate salt cake, and by-product HCl. There was a considerable amount of overlapping tile roof slabs, no insulation, and much exposed structural steel. HCl fumes collected under the roof, condensed as HCl acid when the roof was cold, and badly corroded the structural steel. Correctly, it was decided to paint the ceiling steel. Incorrectly, it was decided to use only wire brush hand cleaning and one coat of aluminum-pigmented oil paint. Before the far end of the building was reached by the painters, the newly applied coating was hanging down like Spanish moss from the near end.

In spite of the problem, aluminum paint has its place. At another location, a plant perimeter chain link fence was spray painted with aluminum paint. The fence was protected from corrosion for many years—at a $4000 premium. But the wind direction and location of the parking lot resulted in claims for paint damage to employees' cars.

A carbon steel waste gas-induced draft fan showed slight, uniform surface rusting during a pilot plant test. The fan manufacturer "up-graded" the material of construction to one of the controlled rusting low alloy copper containing steels for the full size 1500 kw (2000 hp) operating unit. The waste gas temperature was 82 C (180 F) and contained water droplets. The architectural grade steel corroded rapidly and rust came off in large flakes; the rotor failed mechanically from unbalance problems. The brittle welds of this steel, not designed for machinery construction, added to the unsuitability of this application. The replacement rotor was made of mild steel.

To this collection of problems resulting from corrosion protective actions, must be added those that result for the misguided efforts of kind deeds—as the saying goes, "God save me from my friends." A high temperature gas valve used AISI 310 stainless steel as the structural base with refractory pieces of an inorganic nonmetallic bolted to the shell. The contractor substituted *thick* steel washers for the *thin* AISI 310 stainless steel specified because he didn't think the thin stainless ones were as strong as the thick steel (that he substituted). An expensive failure occurred because the carbon steel washers could not withstand the 810 C (1500 F) temperature.

An AISI 430 stainless steel waste heat boiler averaged about 4 years life with failure due to corrosion from nitric acid condensing on the inside diameter of the tubes. The boiler feed water was demineralized, no chlorides, with very low conductivity (less than mg/L TDS). The unit was replaced with a boiler made of AISI 347 stainless steel because of its known superior resistance to nitric acid. The 6 months "test" installation before failure, by chloride SCC, proved that there is no such thing as "zero" chlorides. The inlet temperature of 300 C (550 F) on the process side, caused local vaporization of water on the outside diameter of the tubes near the inlet tube sheet with extensive SCC on almost all the tubes.

A few years ago, a series of brittle failures occurred in Hastelloy C and C-276 vaporizer shells that operate in high temperature service of about 540 C (1000 F). The units were 1 m ϕ x 2 m long (3 ϕ x 6 feet long) with 6 mm (1/4 inch) thick walls. Embrittlement was caused by μ phase formation, not carbide precipitation. A 1230 C (2250 F) annealing procedure was developed to restore the lost ductility of these units. Periodic cleaning of these vessels involved a burnout cycle at 540 C (1000 F) to remove built up "tar", followed by mechanical scraping to remove residue. Operations decided to skip the "tar" removal and burnout during the annealing. After two units were completely melted down, it was found that the "tar" contained a catalyst (a boron phosphate compound) that causes

Source: *Materials Performance*, April 1978, 33-37

severe fluxing of everything, but platinum, above 1100 C (2000 F). The remaining vessels were "saved" by a lower temperature anneal that transforms the μ phase and causes complete carbide precipitation and spheroidization—a much more stable condition than the normal solution anneal without destruction by fluxing.

A Monel shell-and-tube heat exchanger had been in satisfactory service in an organic acid oil for over 5 years. The product inlet was on the bottom of the removable head section, gasketed to the tube sheet and cover plate. To overcome a fouling problem that was causing concentration cell corrosion on the Monel, the tube velocity was increased by increasing the pump pressure. The head divider was not designed for these conditions; it deflected upward, the gasket blew out, and the product by passed the divider at high velocity. The unit failed by erosion-corrosion.

Several cases of premature equipment failure were found to be from the use of incorrect material. These substitutions were caused by poor materials control in many of the fabricator's shops. The use of a 5 cmϕ (2 inch) piece of Monel in an AISI 304 stainless steel vessel caused a disastrous fire. A new specification was issued that insisted upon paint marking of *all* material. Later, a large vertical, nickel, sodium hydroxide heater failed by massive intergranular corrosion. During heat forming fabrication, nickel and nickel alloys easily alloy at the grain boundaries with low melting materials such as lead and sulfur to form brittle, poor corrosion resistant complex plumbides and sulfides of nickel. The corrosion was at the location of the yellow marking paint. A specification change, made to protect nickel from problems with lead paint, included a prohibition against the use of sulfur bearing lubricants for tube rolling, etc.

Another example of the need for care in testing: Molded reinforced phenolic fume control equipment was failing in 2 to 3 years. Field tests were conducted in this HCl gas scrubber and mist eliminator equipment. Hastelloy C 276 was found to be completely resistant in the scrubber and Alloy 20 in the mist eliminator. These metallic units failed in 2 and 4 weeks, respectively. The corrosion test coupons were exposed to the gas stream, not the scrubbed liquid. The change in acid concentration during scrubbing was not taken into account. The scrubber is now made of FEP-lined FRP, and the mist eliminator elements are now made of FRP.

Internal tank and pipe corrosion is many times corrected by adding external insulation which keeps the inside diameter above the condensation temperature of some corrosive or other. Unfortunately, probably the most frequent, single type of correctional defect found in the CPI is the SCC of stainless steel under wet insulation. Sometimes the insulating material itself contains chlorides, and in others, the chlorides are picked up from outside sources. Improperly weather-proofed or damaged insulation vapor barriers allows water to carry these chlorides to the outside diameter surface.

Prior to 1945, domestic hot water heaters were steel tanks protected with a hot dipped galvanized coating of zinc. Water was generally heated to and stored in these tanks at 50 to 54 C (125 to 130 F). The zinc was anodic to the steel and protected any exposed steel by the process of sacrificial cathodic protection. Historically, such heaters lasted 10 or more years. After the war, automatic washing machines came into general usage. These nonagitated units relied upon the tumbling action of the clothes in 80 to 82 C (175 to 180 F) water to function. The life of zinc coated hot water heaters fell to 6 to 12 months! It seems that above about 57 C (135 F), their rolls reverse and steel becomes anodic to zinc. Combined with the unfavorable area relationship through small defects in the coating, the tanks perforated much faster than would have bare steel tanks. Glass-lined steel and Monel tanks are now the standard for this service.

Sometimes several corrective steps are needed to "improve" the standard of living. The expansion joint portion of a steel shell on a 4 MPa gauge (600 psig) steam heated shell and tube reboiler was always a maintenance problem. A unique torus shaped, double walled, AISI 321 stainless steel joint was specified on a new unit. Intergranular SCC evidenced itself in 6 weeks. The fabricator followed the ASME Code beyond the requirements of stainless steel and subjected the steel shell and stainless steel expansion joint weldment to a 621 to 649 C (1150 to 1200 F) stress relief heat treatment and 10 hour furnace cool. Massive carbide precipitation sensitization of the stainless steel was the failure mechanism. Not to be outdone, the fabricator then used a nonthermal resonant frequency vibration technique to try to stress relieve the shell welds. The failure time of the second unit was almost the same; vibration did not do the job, and the unit failed by chloride SCC. Wouldn't you believe it, the equipment of record is of all-steel construction.

Several all-steel prodction units operating in dry HCl containing organic fluids were monitored with electric resistance Corrosometer probes as an alarm protection against the potential of catastrophic corrosion should the system become contaminated with water. By mistake, water was introduced into one of the feed stocks which caused a massive failure. The Corrosometer system, checked weekly by the instrument technician for calibration, etc., was found to have been shut down and unused by the operators after 3 months. "It was of no value as a control instrument, all the thing ever did was measure zero, day after day, so why keep it running and maintained."

A contract was executed for lining two water tanks. The units were to be blast cleaned, given a wash primer, a vinyl primer, two vinyl intermediate coats, and a vinyl aluminum finish coat. The tank fabricator shop cleaned the plate, applied the wash primer and vinyl primer. The tanks were erected in the field. The coating contractor was obligated to blast the field welds and damaged areas and apply the wash primer and vinyl primer. He knew there were good vinyls on the market that could be applied directly to steel. Even though it violated the contract, to save money he did the first tank without the wash primer using his own purchased material. It went into service. He started on the second tank when an inspector stopped the job. (The question could be asked where was the inspector on the first tank?). He was made to comply with specifications on the second vessel. During removal of the scaffolding, it was found that the coating on the floor would come off in large sheets. Inspection of the incorrectly done first tank found the coating to be very much intact. When the contractor had been forced to redo the job on the second tank, and not to give away the whole fraud, he used the vinyls he had on the job (designed for direct application on steel, but not for topping a wash primer). Had the second job not been inspected, it would have been as good as the first vessel. The tank was redone per specification by another contractor. While the inspector actually caused the failure by trying to do the job correctly, he should have been there for both tanks and should have checked the materials being used on the job.

My last example is not one of corrective action with poor results as has been the theme of the review, but is a followup on the problem of inspection. The roof of a brine storage tank was frequently walked on by operators. Internal corrosion had advanced to the point that complete penetration was visible in several areas. For safety's sake, a request was made to survey the metal thickness of the roof plates with ultrasonics. A corrosion technician (budding corrosion engineer?) was assigned the job. His report indicated essentially original metal thickness everywhere. When questioned how this could be, in light of the obvious holes, his answer was, "Well, many of the readings were unusually low (thin metal), but I just disregarded them as poor readings." After a lecture on proper techniques, another survey was made, which disclosed the actual condition of the roof.

Summary

What have these examples told us, if anything?

The mechanisms of dissimilar metal corrosion, mill scale removal, and the importance of surface preparation, cathodic protection, chloride SCC of stainless steel, corrosion fatigue, erosion-corrosion, etc., are not new. Corrosion engineers already know all of the corrosion concepts noted. But, as the saying goes, "We get so soon old, but so late smart." Too many times we allow the time pressures of the job to control our decisions—which just isn't too smart.

If seeing the wide ranging consequences of the decisions described forces us to think a little longer before we "solve" the next problem, the exercise will have been worthwhile.

SECTION II

Specific Alloys

Effects of atmospheric sulfur dioxide on the corrosion of zinc

F. H. HAYNIE and **J. B. UPHAM,** National Air Pollution Administration,
Raleigh, N. C.

The amount of sulfur dioxide in the air is the major factor in determining the rate of corrosion of zinc. Because little, if any, zinc corrosion would occur in a nonmarine environment if SO_2 were not present, the reduced life of galvanized products can be directly attributed to air pollution.

IT HAS BEEN well established that zinc corrodes more rapidly in highly industrialized areas than in rural areas.[1,2] Sulfur dioxide (SO_2) in industrial atmospheres has long been suspected to be a major contributing factor to this accelerated corrosion because it could react with the protective carbonate film to form a soluble sulfate.

Guttman[3] has shown that there is a direct correlation between the amount of corrosion of zinc panels and the measured atmospheric concentration of sulfur dioxide at a single site. Similar behavior at different sites was confirmed by correlations between zinc weight loss and sulfur dioxide activity as measured by the lead peroxide candle method.[4] This method, however, is not an accurate measure of atmospheric SO_2 concentrations.[3]

In 1961, Continuous Air Monitoring Project (CAMP) stations were established by the U.S. Public Health Service in Philadelphia, Pa., Washington, D. C., Chicago, Ill., Cincinnati, Ohio, New Orleans, La., and San Francisco, Calif. These stations were set up to continuously record atmospheric concentrations of sulfur dioxide, as well as other pollutants. Local agencies operated similar facilities in Los Angeles and Detroit, Mich. The recording of these data offered an excellent opportunity to correlate corrosion rates with atmospheric SO_2 levels. For this reason, a 5-year program to study the atmospheric corrosion of metals was started in 1962 in those eight cities.

Experimental Procedure

A special high-grade commercial zinc (99.9% pure) sheet, 0.06-inch thick, was cut into 4 by 6-inch exposure panels. The panels were scrubbed in a warm detergent solution and degreased in acetone before being weighed on an analytical balance. Teflon coated tongs were used to handle the specimens after they were washed.

The panels were mounted on exposure racks on porcelain insulators facing south and inclined at a 30° angle. A total of 25 panels at each site was initially exposed so that five replicates could be removed after exposure periods of 4, 8, 16, 32, and 64 months. The initial exposure dates and actual exposure times are given in Table 1. After exposure, the panels were returned to the laboratory and the corrosion products were removed by immersing the panels for 5 minutes in a 10% aqueous ammonium chloride solution maintained at 160 to 180 F, followed by scrubbing and rinsing in water and dipping in acetone. The specimens were then reweighed on the same analytical balance.

Exposure sites were selected at or near the eight cities where pollutants were being monitored. With one exception, urban sites were established as close as possible to the CAMP stations. In the San Francisco area, an urban site in Oakland was selected. Rural sites within each area, assumed to have the same general meteorological conditions but lower pollution levels, were selected as a control for the urban sites. Unfortunately, no measurements of pollution levels were made at these rural sites during the time the panels were exposed. The results of the corrosion measurements and subsequent estimates of SO_2 concentrations based on lead candle measurements and diffusion models suggest that the pollution levels at several of these rural sites were much higher than originally anticipated.

Results

Weight-loss data, measured as grams per panel, were converted to corrosion rates (microns per year) by multiplying by a conversion factor of 4.4 and dividing by the length of exposure. Means and standard deviations were calculated for each composite sample of five panels. The results are in Table 2. The corrosion rates were generally

Reprinted with permission from *Materials Protection and Performance*, August 1970, 35-40, © 1969 National Association of Corrosion Engineers

TABLE 1 – Initial Date and Length of Exposure

Site	Initial Exposure Date	No. Months Exposure, Expressed as Years				
		4	8	16	32	64
Chicago						
Urban	9-4-62	0.345	0.674	1.362		
Rural	9-5-62	0.342	0.671	1.362		
Cincinnati						
Urban	7-19-62	0.337	0.674	1.337	2.677	5.340
Rural	7-18-62	0.340	0.677	1.340	2.679	5.340
Detroit						
Urban	9-7-62	0.334	0.666	1.337	2.668	5.326
Rural	9-6-62	0.337	0.668	1.340	2.671	5.332
Los Angeles						
Urban	8-7-62	0.345	0.690	1.345	2.767	5.332
Rural	8-7-62	0.345	0.690	1.345	2.767	5.334
New Orleans						
Urban	7-27-62	0.337	0.666	1.337	2.666	
Rural	7-27-62	0.337	0.666	1.337	2.674	5.340
Philadelphia						
Urban	9-26-62	0.340	0.668	1.337	2.666	5.337
Rural	9-25-62	0.342	0.671	1.340	2.668	5.340
San Francisco						
Urban	1-18-63	0.337	0.666	1.332	2.666	5.323
Rural	1-19-63	0.334	0.679	1.367		
Washington						
Urban	9-20-62	0.337	0.668	1.337	2.701	5.334
Rural	9-19-62	0.340	0.671	1.340	2.704	5.337

TABLE 2 – Average Corrosion Rates of Zinc Panels at Each Site[1]
(μm/yr)

Site	4-Month Exposure		8-Month Exposure		16-Month Exposure		32-Month Exposure		64-Month Exposure	
	$\bar{x}$	±S	$\bar{x}$	±S	$\bar{x}$	±S	$\bar{x}$	±S	$\bar{x}$	±S
Chicago										
Urban	6.554	0.069	9.067	0.086	6.458[2]	0.139	[3]	[3]	[3]	[3]
Rural	1.745	0.063	3.441[1]	0.013	2.432	0.011	[3]	[3]	[3]	[3]
Cincinnati										
Urban	0.987	0.047	1.177	0.052	1.018	0.013	1.376	0.010	1.625	0.022
Rural	0.737	0.355	1.627	0.181	1.308	0.060	1.710	0.145	1.626	0.016
Detroit										
Urban	3.248	0.062	3.939	0.056	3.101	0.062	3.200	0.009	3.460	0.042
Rural	1.363	0.042	1.486	0.082	1.009	0.029	1.241	0.018	1.396	0.023
Los Angeles										
Urban	1.311	0.021	0.938	0.025	0.810	0.012	0.765	0.017	0.874	0.014
Rural	0.619	0.038	0.381	0.015	0.381	0.016	0.393	0.007	0.464	0.029
New Orleans										
Urban	0.835	0.026	0.626	0.019	0.736	0.008	0.816	0.010	[3]	[3]
Rural	2.165	0.159	1.345	0.147	0.916	0.047	0.758	0.049	0.642	0.023
Philadelphia										
Urban	3.767	0.071	3.162	0.070	3.060	0.029	3.367	0.032	3.658[4]	0.019
Rural	3.448	0.293	3.023	0.031	2.986	0.021	3.342	0.015	3.493	0.021
San Francisco										
Urban	1.083	0.051	0.764	0.046	0.771	0.038	0.814	0.018	0.820	0.032
Rural	1.217	0.311	0.753	0.091	0.584	0.104	[3]	[3]	[3]	[3]
Washington										
Urban	2.289	0.035	1.914	0.035	1.782	0.041	2.027	0.041	2.230	0.034
Rural	1.189	0.198	1.070	0.116	0.909	0.055	1.069[2]	0.037	1.175[5]	0.022

[1] Unless otherwise noted, $\bar{x}$ (means) and S (best estimate of standard deviation) were calculated from corrosion rates of five replicate panels.
[2] Four specimens.
[3] Exposure racks destroyed by wind.
[4] Three specimens.
[5] Two specimens.

higher at the urban sites than at the rural sites. At Cincinnati and New Orleans, however, the corrosion rates were initially higher at the rural sites and converged with time. At Philadelphia, the corrosion rates at the rural sites were exceptionally high but not as high as at the urban site.

The length of exposure did not appear to have a significant effect on the corrosion rate. It appeared that a soluble corrosion product was formed. This behavior was somewhat in contrast with the interpretation given by Guttman.[3] He presented weight loss as a function of time of wetness to a power less than 1, indicating a slight buildup of a protective film. A statistical analysis, however, might have shown that the power used (approximately 0.8) was not significantly different from 1.

Guttman[3] showed that the two variables most important in determining the amount of corrosion are time of wetness and atmospheric sulfur dioxide concentration. He also showed that the probability of a specimen being wet was a result of relative humidity. Temperature was not an important variable. This is consistent with a reaction rate controlled by the diffusion of a low concentration species to the reaction site.

If a linear model for corrosion behavior is assumed, the average corrosion rates should be a function of the average relative humidity and the atmospheric sulfur dioxide concentration at each site for each exposure period. These values were calculated from both climatological data obtained at the nearest weather station and CAMP data. Where CAMP data were not complete, trends were used to make estimates. Average sulfur dioxide concentrations reported in parts per million have been converted to micrograms per cubic meter ($\mu g/m^3$) by multiplying by 2620. These results are in Table 3 along with the average corrosion rates at the urban sites.

The following linear model for corrosion behavior is a simplification of two relationships observed by Guttman:

$$Y = K\,(RH - RH_0)\,(SO_2 + B) \qquad (1)$$

where:

Y	= Zinc corrosion rate
RH	= Average relative humidity
SO_2	= Average sulfur dioxide concentration
K, RH_0 B	= Constants

RH_0 has the physical significance of being the relative humidity below which wetting of the zinc was not expected. The constant B allowed for the corrosion of zinc when no SO_2 was present.

TABLE 3 – Corrosion Rates and Climatological Data at Urban Sites

City	Exposure Time (months)	Average Corrosion Rate (μm/yr)	Average Sulfur Dioxide Conc. (μg/m³)	Average Relative Humidity (%)
Chicago	4	6.554	419	66
	8	9.057	479	65
	16	6.448	406	65
Cincinnati	4	0.987	76	69
	8	1.177	97	70
	16	1.018	68	68
	32	1.376	86	69
	64	1.625	79	69
Detroit	4	3.298	139	74
	8	3.939	139	70
	16	3.101	131	67
	32	3.200	121	68
	64	3.460	118	67
Los Angeles	4	1.311	45	79
	8	0.938	45	74
	16	0.810	42	73
	32	0.765	42	70
	64	0.874	39	70
New Orleans	4	0.835	34	76
	8	0.626	29	76
	16	0.736	26	77
	32	0.816	24	77
Philadelphia	4	3.767	191	66
	8	3.162	194	63
	16	3.060	197	66
	32	3.367	194	66
	64	3.658	218	66
San Francisco	4	1.083	10	72
	8	0.764	13	71
	16	0.771	29	74
	32	0.814	34	74
	64	0.820	34	73
Washington	4	2.289	222	64
	8	1.914	207	60
	16	1.782	141	61
	32	2.027	136	62
	64	2.230	126	63

Rearrangement of Equation (1) for linear multiple regression gives

$$Y = b_0 RH \times SO_2 + b_1 RH + b_2 SO_2 + b_3 \quad (2)$$

where:

$b_0 = K$
$b_1 = KB$
$b_2 = KRH_0$
$b_3 = KBRH_0$

The best fit of this equation was obtained using the Abbreviated Doolittle Method.[5] The results, including an analysis of variance, are in Table 4. Only the constants associated with SO_2 and its interaction with relative humidity are statistically significant at the 99% level. Factoring Equation (2), however, into the form

$$Y = 0.00104 (RH-49.4) SO_2 - 0.00664 (RH-76.5) \quad (3)$$

yields an equation with some physical significance. This behavior suggests that, in the absence of SO_2, increasing moisture inhibits corrosion. This is consistent with the formation of a protective carbonate film. It also indicates that zinc is not expected to be wet below a relative humidity of 76.5%. When SO_2 is present, however, the corrosion rate increases with increasing relative humidity, and the surface may be wet down to a relative humidity of 49.4%. This latter behavior is consistent with the formation of a hygroscopic corrosion product.

The results of the analysis of variance suggest that a simpler linear relationship may be just as good. This relationship is:

$$Y = b_0 RH \times SO_2 + b_1 SO_2 \quad (4)$$

where:

$b_0 = K$ and
$b_1 = -KRH_0$.

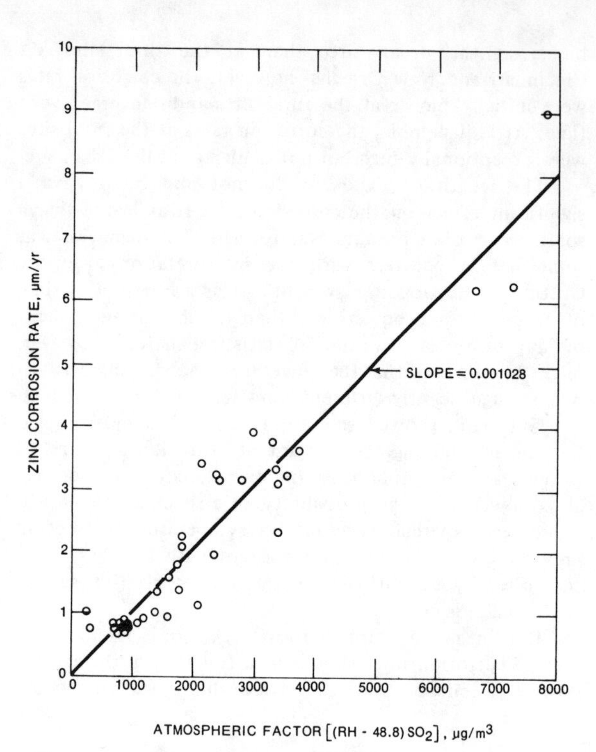

Figure 1 — Effect of SO_2 and relative humidity on the corrosion rate of zinc.

The results of regression and analysis of variance for Equation (4) are in Table 5.

Equation (4) can account for 92% of the variability of average corrosion rates. Equation (3) does not fit the data any better. It is believed that the 8% residual variability does not warrant the extra effort to improve the model by including other variables such as average temperature and average sunshine hours or by using nonlinear relationships.

Equation (4) factors to:

$$Y = 0.001028 (RH - 48.8) SO_2 \quad (5)$$

TABLE 4 — Results of Regression and Analysis of Variance of Four-Constant Equation

Constant	Source of Variation	Degree of Freedom	Sum of Squares	Mean Square	F
$b_0 = 0.001042$	b_0	1	308.780	308.780	989.68(1)
$b_1 = -0.006637$	b_1/b_0	1	1.081	1.081	3.46
$b_2 = -0.05144$	$b_2/b_0,b_1$	1	2.449	2.559	8.20(1)
$b_3 = 0.5078$	$b_3/b_0,b_1,b_2$	1	0.025	0.025	0.08
	Residual	33	10.282	0.312	
	Total	37	322.727		

(1)Significant at 99% level.

TABLE 5 — Results and Analysis of Variance of Two-Constant Equation

Constant	Source of Variation	Degree of Freedom	Sum of Squares	Mean Square	F
$b_0 = 0.001028$	b_0	1	308.78	308.78	1050.27(1)
$b_1 = -0.05019$	b_1/b_0	1	3.66	3.66	12.45(1)
	Residual	35	10.29	0.284	
	Total	37	322.73		

(1)Significant at 99% level.

TABLE 6 — Predicted Useful Life of Galvanized Sheet Steel with a 53 Micrometer Coating at Average Relative Humidity of 65%

SO_2 Conc. ($\mu g/m^3$)	Type Environment	Predicted Best Estimate(1)	Useful Life (yrs) Predicted Range(1)	Observed Range(2)
13	Rural	244	41.0	30 to 35
130	Urban	24	16.0 to 49	
260	Semi-industrial	12	10.0 to 16	15 to 20
520	Industrial	6	5.5 to 7	
1040	Heavy industrial	3	2.9 to 3.3	3 to 5

(1)From Equation (5).
(2)SO_2 concentrations and relative humidity were not observed.

where 48.8 represents the relative humidity below which zinc-wetting is not expected. The results are in Figure 1. The 95% confidence limits on the data are ± 1.082 micrometers per year.

Discussion

The results of this study may be used to predict the useful life of galvanized products. Galvanized sheet steel is normally coated with 53 micrometers of zinc. When this zinc is gone, the steel rusts and, for practical purposes, the useful life of the product ends. Table 6 shows predictions of the useful life of galvanized sheet steel at an average relative humidity of 65% at typical SO_2 pollution levels. These predictions are compared with observed useful life values.[6] These values are consistent with the behavior of 25-micrometer coatings, which have average useful lives of 11 and 4 years in urban and industrial environments, respectively.[7]

Since little, if any, zinc corrosion would occur in nonmarine environments if SO_2 were not present, the reduced life of galvanized products can be directly attributed to air pollution. Thus, if the amount, value, and distribution of these products were known, this fraction of the economic burden caused by air pollution could be calculated using Equation (5), SO_2 concentrations, and relative humidities at sites representative of the product distribution.

Relatively high corrosion rates at rural sites were observed near Chicago, Cincinnati, and Philadelphia; this suggests significant concentrations of sulfur dioxide. Estimates of average SO_2 concentrations at these rural sites, based on a rearrangement of Equation (5), compare favorably with estimates derived from Air Quality Control Region Consultation Reports.[8-10] The results of comparison are in Table 7.

The diffusion model estimates are based on emission sources and are corrected to reflect measured lead candle sulfation levels. Because the lead candle sulfation levels were not measured at these particular sites, it is believed that the SO_2 concentrations estimated by using zinc corrosion rates are more accurate than the values based on the diffusion models.

Conclusions

Atmospheric sulfur dioxide concentration is the major factor in controlling the rate of corrosion of zinc. A simple linear function of average relative humidity and average SO_2 concentration can account for 92% of the variability in average zinc corrosion rates. This function suggests that the zinc surfaces will not be wet, and thus no corrosion will occur, below an average relative humidity of 48.8%. A rearrangement of this function can be used to estimate SO_2 concentrations from average zinc corrosion rates and average relative humidities.

References

1. C. P. Larrabee and O. B. Ellis. Report of Subgroup of Subcommittee VII, on Corrosiveness of Various Atmospheric Test Sites as Measured by Specimens of Steel and Zinc, *Proc. of American Society for Testing and Materials,* **59**, 183 (1959).
2. S. K. Coburn. Chairman of Task Force for report of Committee G-1, Corrosiveness of Various Atmospheric Test Sites as Measured by Specimens of Steel and Zinc. Metal Corrosion in the Atmosphere ASTM STP 435, 360 (1968).
3. H. Guttman. Effects of Atmospheric Factors on the Corrosion of Rolled Zinc. Metal Corrosion in the Atmosphere ASTM STP 435, 223 (1968).
4. H. Guttman and P. J. Sereda. Measurement of Atmospheric Factors Affecting the Corrosion of Metals, Metal Corrosion in the Atmosphere ASTM STP 435, 326 (1968).

TABLE 7 — Comparison of Estimates of SO_2 Concentrations at Three Rural Exposure Sites

Site	Average Relative Humidity (%)	Estimated SO_2 Concentration ($\mu g/m^3$) Equation (5)	Consultation Report Diffusion Models
Chicago	65	145 ± 65	79
Cincinnati	70	76 ± 50	79
Philadelphia	65	155 ± 65	210

5. Bernard Ostle. *Statistics in Research,* The Iowa State University Press, Ames, Iowa, 178 (1963).
6. R. L. Salmon. Systems Analysis of the Effects of Air Pollution on Materials, Final Report Contract CPA-22-69-113 for the National Air Pollution Control Administration by Midwest Research Institute, January 4, 1970, p. 166.
7. H. H. Uhlig. *Corrosion and Corrosion Control,* John Wiley and Sons, Inc., New York, 202 (1963).
8. Report for Consultation on the Metropolitan Chicago Interstate
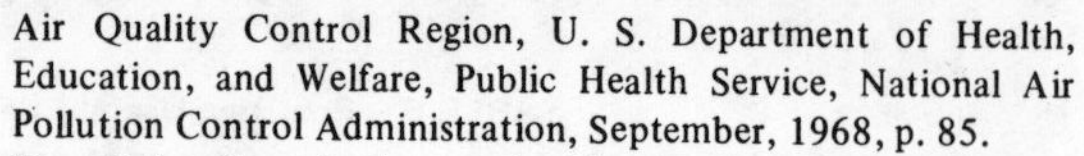
Air Quality Control Region, U. S. Department of Health, Education, and Welfare, Public Health Service, National Air Pollution Control Administration, September, 1968, p. 85.
9. Report for Consultation on the Metropolitan Cincinnati Interstate Air Quality Control Region U.S. H.E.W., PHS, NAPCA, January, 1969, p. 54.
10. Report for Consultation on the Metropolitan Philadelphia Interstate Air Quality Control Region, U.S. H.E.W., PHS, NAPCA, October, 1968, p. 74.

F. H. HAYNIE is Chief, Materials Branch, Division of Economic Effects Research, National Air Pollution Control Administration. He supervises research to determine the effects of air pollution on materials. Haynie holds an MS in chemical engineering from Ohio State University, and is a member of ASTM, AIChE, and NACE.

J. B. UPHAM is a senior chemical engineer in the Materials branch, Division of Economic Effects Research, National Air Pollution Control Administration, Raleigh, N. C. He has studied the effects of air pollution on different materials for the past nine years. He is a graduate of the University of Cincinnati, and he is active in the Air Pollution Control Association.

Corrosion Behavior of Galvanized Sheet in Relation to Variation in Coating Thickness*

C. E. BIRD, *Corrosion Research Division, NCRL, CSIR, Pretoria, South Africa*

Appearance of 5% red rust on commercial galvanized steel at times shorter than the theoretical life of an equivalent thickness of pure zinc is the result of variations in coating thickness. Careful measurements showed that average thickness of 20 g/sq yd (0.7 oz/sq yd or 16.5 g/sq m) zinc was 20.3 microns or 1.87 mils. The thin areas covered somewhat less than half of the total. Tests at 7 sites of widely varying aggressiveness indicated there is no significant difference in the corrosion rate of galvanizing compared to pure zinc. Some evidence indicates that thickness variations are greater as coating thickness increases.

IN A PREVIOUS PUBLICATION,[1] the results of an exposure program designed to establish the corrosion protection of steel by various metallic and plastic coatings at sites in South Africa were reported. During this investigation, it was found that the life expectancy of a pure zinc layer (calculated from the corrosion rate of pure zinc) exceeded the actual life (to 5% red rust) of a galvanized coating of identical thickness up to four times. Various reasons were put forward to account for this apparent discrepancy, among which was the composition of the galvanized coating. To avoid any hasty conclusions regarding inferior corrosion characteristics of a galvanized coating relative to pure zinc, further work was undertaken to clarify the position.

*Voluntary manuscript submitted for publication March, 1976.

Experimental

A galvanized sheet (ex Iscor continuous galvanizing process) of the same size (203 to 127 mm) and grade (commercial), as that used in the exposure program mentioned earlier, was divided into approximately 1000 squares, each 5.0 by 5.0 mm. The specimen was placed on the flat steel table of a Karl Mark precision dial meter, securely clamped into position on the table and the thickness measured in the center of each square to the nearest 1.0 micron.

The zinc coating was then stripped off the measured side of the specimen with inhibited hydrochloric acid, the reverse side being shielded with paraffin wax. The specimen was weighed before and after stripping and removal of the wax. From the weight loss, the

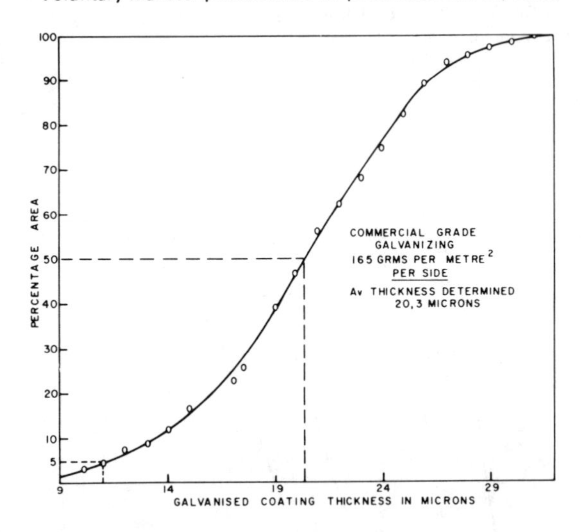

FIGURE 1 — Cumulative distribution curve of thickness vs percentage area.

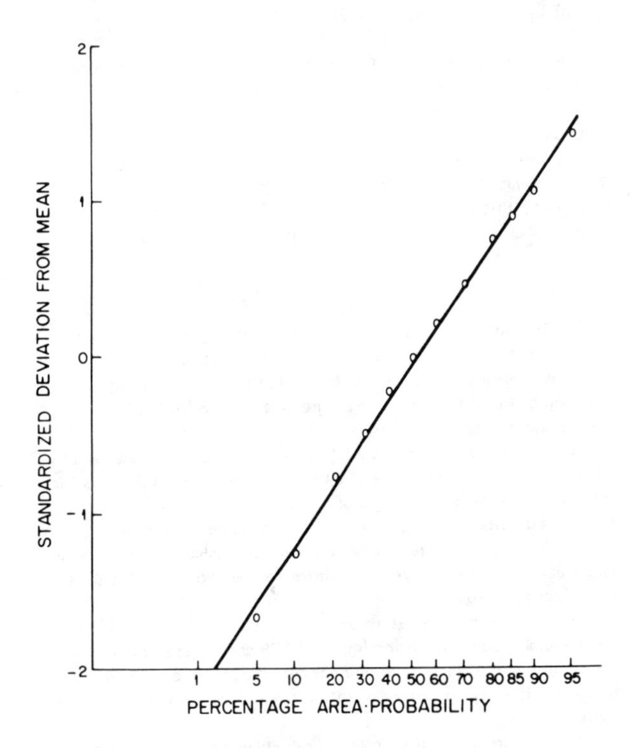

FIGURE 2 — Standard deviation from the mean.

TABLE 1 — Four Year Zinc Atmospheric Corrosion Rates (μm/yr)

Site	Rate
Pretoria CSIR (PTA)	0.51
Pretoria Mamelodi (MAM)	0.76
Cape Town Ysterplaat (CT)	1.78
Durban Sals. Isl (DBN)	3.30
Durban Bluff (Bluff)	11.18
Walvis Bay (WB)	2.34
Rooikop	1.52

average thickness of the coating on the measured side was calculated. The specimen was then replaced on the measuring apparatus as before, with the stripped side facing up, and the thickness redetermined as previously in the center of each square. The difference between the thicknesses determined before and after stripping was then taken to represent the thickness of the coating in that particular square. The average value of the coating thickness determined by weight loss was used to calibrate the instrument and all instrument values were adjusted on this basis.

Results

In Figure 1, a cumulative distribution curve of thickness vs percentage area is shown. This plots near a straight line on normal probability paper. It indicates that on 5% of the specimen surface, the coating thickness was less than 11.0 microns, the average being 20.3 microns.

Figure 2 represents the standard deviation from the mean plotted on normal probability paper. This plot nears a straight line, and with the thickness normally distributed, a calculated mean and standard deviation is a good representation of the thickness. Five percent of the surface will have a coating thickness of less than the mean by 1.62 standard deviation.

Discussion

The average corrosion rates of cold rolled pure zinc at 7 exposure sites during a 4 year exposure program in South Africa are given in Table 1.

If these values are used to calculate the life expectancy of a galvanized steel to 5% red rust (which would represent those areas with a coating thickness of less than 11 μm, Table 2 is obtained. In this table, the actual life of exposed galvanized specimens to 5% area red rust is compared with the theoretical value obtained by calculation from the corrosion rate of cold rolled pure zinc and the cumulative frequency distribution curve.

In the same way, the theoretical life expectancy for progressively higher percentages of red rusted areas may be calculated for the various sites as shown in Figure 3.

In Figure 4, curves obtained by ASTM at Altoona, Pennsylvania[2] may be compared with the shape of those in Figure 3, although the latter curves represent different coating weights instead of different environments. These data plot near straight lines on log probability paper indicating a log normal distribution. Altoona may be compared with Salisbury Island, Durban. From Figure 4, it would appear that the thicker the coating, the greater the variation in coating thickness. Equating life to thickness, the variation is proportional to the mean or the percentage variation is constant.

Conclusions

1. The inherent corrosion resistance of a galvanized coating does not differ to any significant degree from that of pure zinc.
2. Variations in coating thickness due to spangle formation are the main cause of the early and progressive onset of red rust at those areas below average thickness.
3. On the specimen tested (20 μm zinc coating), 5% of the surface had a coating thickness half that of the average thickness for commercial grade galvanizing. For heavier coatings, it seems that the thickness distribution may be even wider, as shown in Figure 4.
4. The normal criterion for the life of a galvanized coating is taken as the number of years exposure to develop a 5% area of the sheet being red rusted (*i.e.*, corrosion of the underlying steel) which is considered the stage at which painting becomes essential. For commercial grade galvanizing (nominal 375 grams per sq meter both sides), this variation in coating thickness reduces the effective life, based on the above criterion, to approximately half its expected value.
5. The formation of spangles on galvanizing and the consequent variation in thickness is a phenomenon common to all galvanized coatings presently produced. Heavier grades of galvanizing (such as 1.50, 2,00, and 2.50 ounces per sq ft) although also being subject to such thickness variations would ensure a proportionally longer life as shown in Figure 4.

TABLE 2 – Life Expectancy in Years to 5% Area Red Rust for 20 μm Galvanized Coating

	Calculated	Actual
CSIR	23	>7
Mamelodi	15	>7
Ysterplaat	7	6
Durban	3.3	2.5
Bluff	1.0	1.0
Walvis Bay	0.5	0.5
Rooikop	8	>7

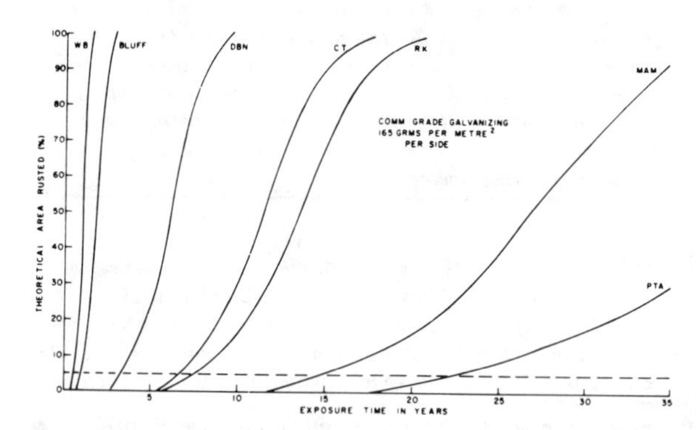

FIGURE 3 – Theoretical life expectancy for progressively higher percentages of red rusted areas.

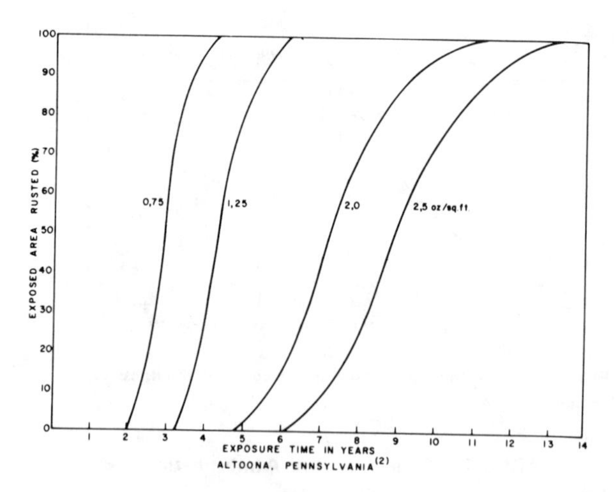

FIGURE 4 – Exposed area rusted vs exposure time in years.

References

1. Joseph, R. M. H., Patrick, G. W. Corrosion Protection by Metals and Plastic Coatings on Mild Steel, Metal Fabrication, Finishing, and Protection. (SA), p. 5-15 (1971) January-February.

2. Am. Soc. Testing Materials, Proc., Vol. 52, p. 113 (1952).

Atmospheric Corrosion Resistance of 55% Al-Zn Coated Sheet Steel: 13 Year Test Results*

H. E. TOWNSEND and J. C. ZOCCOLA
Research Department, Bethlehem Steel Corporation, Bethlehem, Pennsylvania

Results of 13 year atmospheric exposure tests of sheet steel coated with zinc, aluminum, and 55% Al-Zn coatings are reported. Exposures were made in rural, marine littoral, and industrial environments. In all environments, 55% Al-Zn coatings are several times more durable than zinc coatings of equivalent thickness. At long exposure times in industrial and rural environments, corrosion rates of the 55% Al-Zn coating become indistinguishable from those of aluminum coatings. Whereas all these coatings provide sacrificial protection to exposed steel in chloride bearing marine environments, only zinc and 55% Al-Zn coatings provide edge protection in rural and industrial environments. It is believed that the combination of long term durability and cut edge protection afforded by the 55% Al-Zn coating is related to its unique microstructure and the corrosion mechanism. This mechanism is one in which zinc corrosion products from preferential corrosion of the zinc rich portions of cored dendrites are trapped in the interdendritic interstices of an aluminum rich network, thus slowing further attack. Analyses of corrosion products are given.

STEEL SHEET, coated with an alloy comprising 55% Al, 43.4% Zn, and 1.6% Si, is being produced and sold in the United States by Bethlehem Steel Corporation under the name Galvalume, and in Australia by John Lysaght, Ltd. under the name Zincalume. As described in an earlier publication,[1] the 55% Al-Zn composition was selected from a series fo Al-Zn alloys as the optimum coating for the protection of steel sheet. In laboratory and initial outdoor corrosion tests, this coating displayed an ideal combination of the long term durability of aluminum coatings and the resistance to localized corrosion at damaged areas and cut edges of zinc coatings. The coating is applied by a continuous hot dip process similar to that used for Selas type continuous sheet galvanizing.[2] Because the melting point of the 55% Al-Zn coating is low enough to enable hot dip application without causing complete recrystallization of cold worked steel sheet, it is possible to produce high strength (yield strengths in excess of 600 MPa) steel sheet with this coating. A small amount of silicon in the coating is added to the bath for the purpose of controlling the growth of the intermetallic layer that bonds the coating to the substrate.[3]

Nine year results for tests conducted in a variety of environments in the United States have been reported previously.[1] Atmospheric corrosion data have also been reported for tests conducted in other parts of the world, including Australia[4] and Norway.[5] The purpose of this paper is to present quantitative results of the United States corrosion tests after 13 years. These results confirm the long term durability and cut edge protection characteristics previously noted after the shorter test durations.

Materials and Test Procedures

The material tested was 55% Al-Zn coated steel sheet (coating 20 μm thick) that was prepared on a continuous hot dip coating pilot line. Details of the coating process have been described elsewhere.[1,6] For purposes of comparison with other types of coated steels, commercially available G90 galvanized[(1)] and type 2 silicon free, aluminum coated sheet steel were included in the tests. Coating weights, thicknesses, and sheet thicknesses are given in Table 1.

The sheet thickness was not identical for the different types of coated specimens, because the maximum thickness of steel sheet that could be handled on the pilot facility was less than the minimum thickness of commercially available galvanized and aluminum coated sheets. Sheet thickness determines the area of bare steel exposed at cut edges and thus affects the anode to cathode area ratio for coating-substrate galvanic couples. Therefore, comparisons of the degree of edge protection afforded by different coatings are best made with a constant sheet thickness. Accordingly, visual observations are included on coated steels that were later produced commercially on a large scale production facility capable of handling the thicker sheet. (Table 1).

The four atmospheric test sites represent a broad range of environmental conditions, as follows:

1. Severe marine at the INCO test station at Kure Beach, North Carolina, located approximately 25 m from the open Atlantic Ocean.
2. Moderate marine as above, but 250 m from the ocean.
3. Rural at the Pocono Mountain test site located in Saylorsburg, Pennsylvania, about 50 km north of Bethlehem.
4. Industrial at our Bethlehem, Pennsylvania test site, about 3 km from the Bethlehem steelmaking plant.

Test panels were exposed 30 degrees from the horizontal, with skyward surfaces facing South at all locations, except at the severe marine site, where panels faced east toward the ocean.

Details of atmospheric test procedures were given previously[1] but are briefly summarized as follows. Sheet samples 10.2 x 15.2 cm were weighed and, after exposure to the atmosphere for a particular time interval, were cleaned in chromic acid to remove corrosion products and reweighed to determine loss of coating due to corrosion. To facilitate direct comparison of coatings of different densities, average thickness losses were calculated from the weight loss measurements.

Corrosion tests were initiated during the summer of 1964. According to one investigator,[7] corrosion rates of zinc coatings that are first exposed in the summer are lower than those first exposed in the winter. If so, any comparison between the 55% Al-Zn and the galvanized coatings based on the results of our tests is made under conditions that are most favorable to galvanized.

Corrosion products on exposed coatings were analyzed by X-ray diffraction that utilized chromium k_{alpha} radiation.

Results and Discussion

Figure 1 shows the appearance of the coated sheets after a 13 year exposure at the four test locations. The small, localized discolorations at vertical edges of some specimens are points which contacted ceramic insulators in the test racks.

*Voluntary manuscript submitted for publication April, 1979.

(1) ASTM Specification A525, Steel Sheet, Zinc Coated (Galvanized) by the Hot Dip Process, General Reguirements, ASTM Annual Book of Standards, Part 3, ASTM, Philadelphia, Pennsylvania (1973).

TABLE 1 – Description of Coated Steel Sheets Tested

Coating	Nominal Coating Mass (2 sides), gm/sq m^2	Nominal Coating Thickness (1 side), μm	Total Thickness of Coated Steel Sheet (mm)	
			13 Year Tests	6 Year Tests
55% Al-Zn	76	20	0.43	0.48
Galvanized	180	25	0.71	0.46
Aluminum Coated(1)	140	51	0.58	0.64

(1)Nominal coating weights and thickness for aluminum coated sheet are those typical of the Type II aluminum coated product that was available at the start of these tests, whereas presently available Type II coatings are about 20% lighter.

Severe Marine Atmosphere (25 m from Water's Edge)

The galvanized panels, which had started to rust after a 4 year exposure, were now heavily rusted. In constrast, panels with 55% Al-Zn and aluminum coatings were still in good condition, although some corrosion products were starting to creep inward on the faces of panels from cut edges.

Marine Atmosphere (250 m Lot)

All three types of coatings were still in good condition.

Industrial Atmosphere

Most of the galvanized coating was corroded away, and more than 75% of the steel surface was rusted. The Al-Zn and aluminum coated panels exhibited superficial light brown oxide stain due to particulate fallout from nearby steel-making operations but were otherwise in good condition.

Rural Atmosphere

All three materials were in good condition, but there was some rust staining along edges of the aluminum coated panels.

Figures 2 to 5 are corrosion-time curves for the coatings after the 13 year exposure. These results confirm trends found after 9 years of testing.[1] To facilitate comparison of the corrosion resistance of a 55% Al-Zn coating with that of a conventional galvanized coating, the ratio of the 13 year corrosion losses for the two coatings were calculated as shown in Table 2.

Since thicknesses of commercially available coatings were about the same for both the G90 galvanized and the 55% Al-Zn, it can be predicted on the basis of these ratios that the commercially available 55% Al-Zn coating will outlast G90 galvanized by at least two to four times in a wide range of atmospheric environments.

When rusting of galvanized steel occurred in our tests, it generally initiated at cut edges, and then spread across the panel within a few years. After 14 years of testing, initiation of rust at edges of galvanized steel was observed at the remaining two sites. These test results for galvanized coatings are in reasonably good agreement with those of previous work.[8]

Although the zinc coating failed after only 4 years at the severe marine site, samples in test throughout the following years remained intact under an adherent layer of white and red corrosion products. Since bare steel sheet with about the same thickness as the galvanized samples completely corroded within 5 years at this location, the corrosion products from the galvanized coating apparently provide continuing protection to the steel substrate subsequent to what is generally regarded as coating failure. The extra protection afforded by zinc corrosion products is significant for structural integrity, but the surface appearance makes the sheets esthetically unacceptable for many applications.

Corrosion data for a variety of materials, including uncoated steel, nonferrous metals, galvanized, and aluminum coated steels, have previously been shown[9-13] to conform to rate equations of the type

$$C = At^B \tag{1}$$

where C is the corrosion loss, t is time, and A and B are constants.

When written in logarithmic form, Equation (1) becomes:

$$\log C = \log A + B \log t \tag{2}$$

Corrosion data conforming to such a relationship should fall on a straight line when plotted on logarithmic coordinates. When atmospheric data from Figures 2 to 5 are replotted on logarithmic coordinates (Figures 6 to 9), they can be seen to be well represented by a linear relation. (Since the emphasis was on a line characterizing the four longer term results, the data point for the 1 to 2 year data for the 55% Al-Zn coatings at the two marine sites was omitted.) Lines drawn through the data are derived from a least squares fit of the logarithms, and the coefficients are summarized in Table 3. Correlation coefficients for the least squares regression (Table 3) exceed 0.99 in 10 of 12 cases, thus indicating generally excellent fits. The two cases where lower correlation coefficients of 0.973 and 0.854 for aluminum coated steel are observed result from scatter occurring during early exposure when weight losses are low and difficult to measure with precision.

Corrosion rates can be calculated from the constants A and B by taking the time derivative of Equation (1):

$$\frac{dC}{dt} = ABt^{B-1} \tag{3}$$

Rates computed by use of Equation (3) and the constants in Table 2 are plotted in Figures 10 to 13. At the marine sites, corrosion rates for all three types of coatings decreased with time. However, for galvanized coatings initial rates were so high that the coating was consumed before the lower rates observed for Al-Zn and Al coatings could be attained. At the industrial and rural test locations, initially high rates exhibited by galvanized coatings either increased or decreased slightly with time. This finding indicates that the layers of zinc corrosion product formed in these environments are less protective than those formed in marine environments. Indeed, the adherent, thick layers of zinc corrosion products observed in marine environments are not present at the industrial and rural sites.

Corrosion rates for Al-Zn coating tested at the rural and industrial sites decreased with time. As previously described,[1] this can be explained in terms of the unique microstructure and corrosion mechanism exhibited by this coating. The Al-Zn coating has a two phase structure comprising a fine network of aluminum rich cored dendrites and zinc rich interdendritic alloy. The zinc rich interdendritic portion of the coating is preferentially attacked, with the result that the coating behaves much like a zinc coating. That is, the 55% Al-Zn coating: (1) exhibits a corrosion potential similar to that of a zinc coating, (2) provides galvanic protection to steel substrate at exposed cut edges, and (3) has an initially higher rate than an aluminum coating. As corrosion proceeds, protective zinc corrosion products that wash away from ordinary galvanized coatings are mechanically trapped in the interdendritic network, thus slowing further attack. Aluminum coatings without the benefits of protective zinc corrosion products exhibit increasing rates of corrosion with time (Figures 10 to 13). Because of the concomitant decrease in rates of Al-Zn coatings in these environments, rates for the two types of coating converge and are indistinquishable after about a 10 to 15 year exposure.

Compositions of corroded portions of the 55% Al-Zn coatings

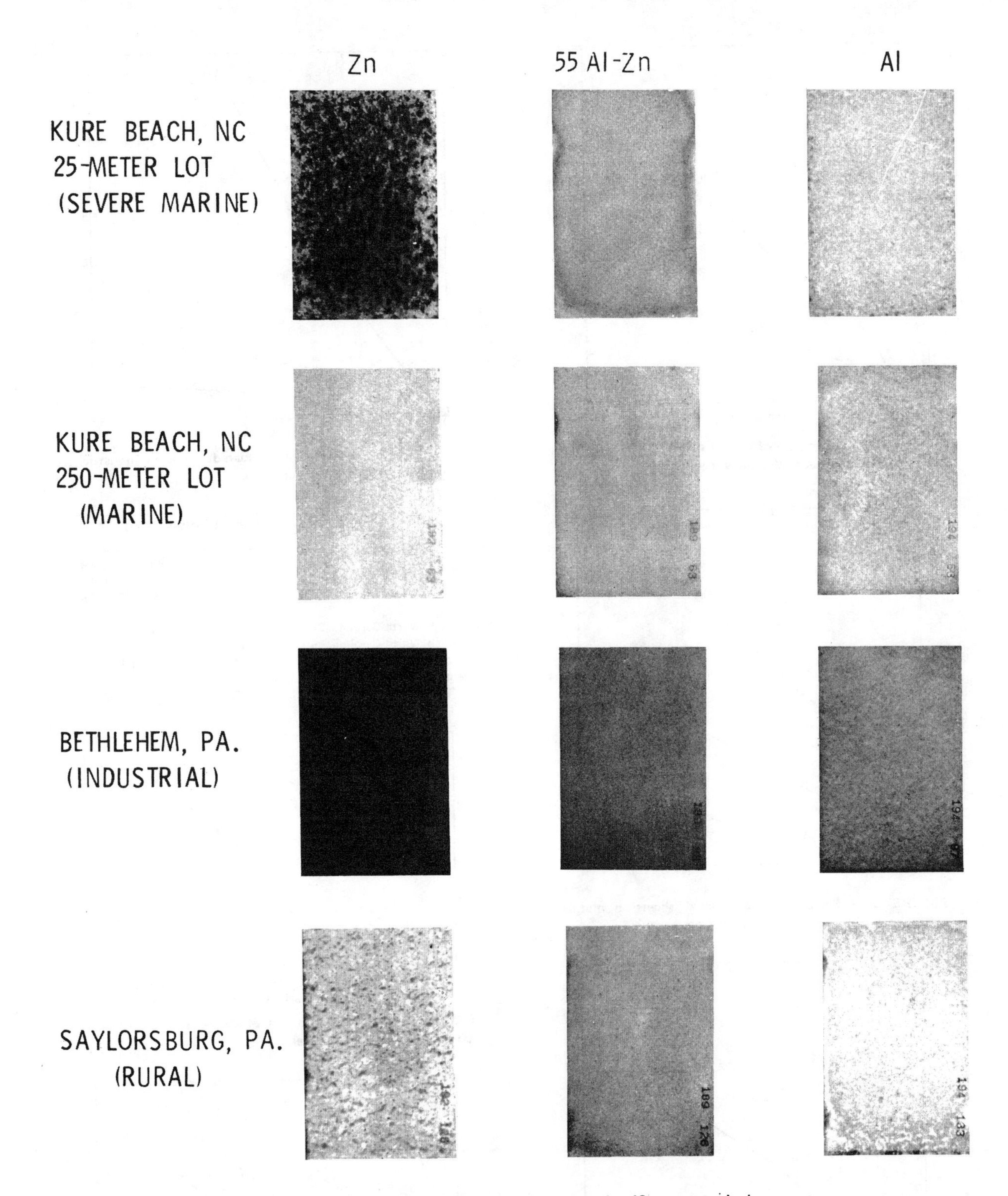

FIGURE 1 – Skyward surface appearance of coated sheets after 13 year atmospheric exposure.

(Figure 14) exhibit a continuation of the previously reported trends.[1] In accordance with the mechanism described above, corrosion of 55% Al-Zn coatings begins with the zinc rich constituents. Passivity of the aluminum rich dendrite matrix in industrial and rural environments is confirmed by the fact that the zinc content of the corroded portion of the coating exceeded 77% even after a 13 year exposure. In the marine environments, where aluminum rich parts of the coating are more active, the composition of the corroded coating was correspondingly higher in aluminum content.

X-ray diffraction analyses indicated that $Zn_5(CO_3)_2(OH)_6$ had formed on galvanized sheet in all environments. In industrial and rural atmospheres, the 55% Al-Zn coating exhibited only this characteristic zinc corrosion product. In marine environments, where the aluminum rich portions of the alloy coating also become active, the $Al(OH)_3$ compound that is characteristic of aluminum

Source: *Materials Protection*, October 1979, 13-20

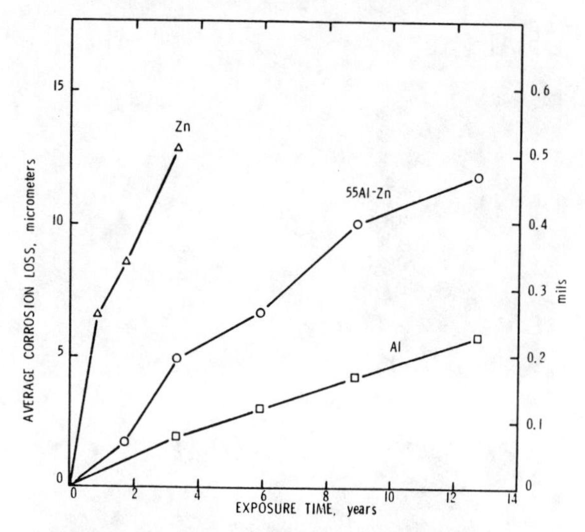

FIGURE 2 – Performance of coated sheets in severe marine atmosphere (Kure beach, 25 meters from ocean).

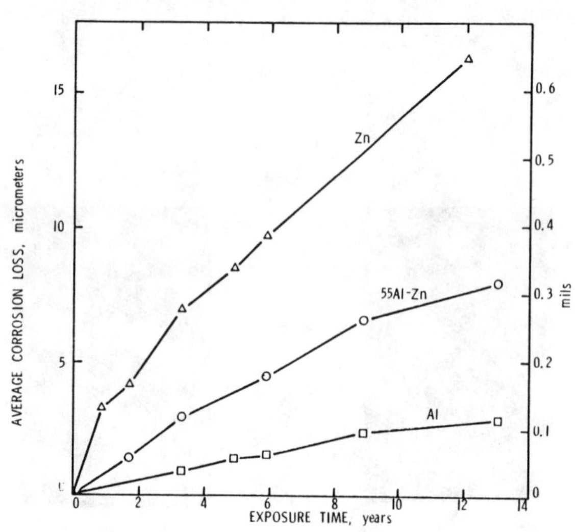

FIGURE 3 – Performance of coated sheets in marine atmosphere (Kure beach, 250 meters from ocean).

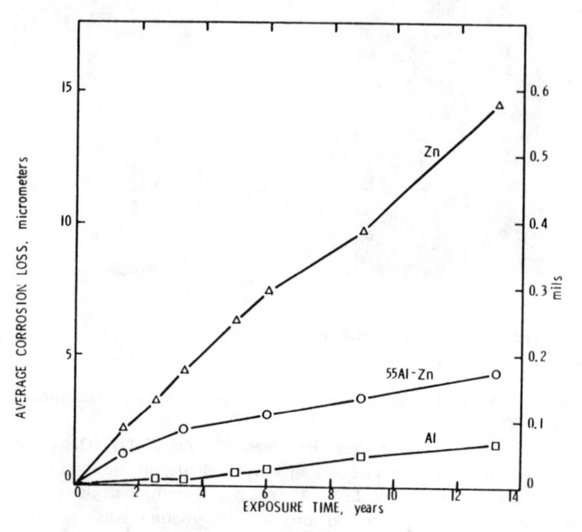

FIGURE 4 – Performance of coated sheets in rural atmosphere (Saylorsburg, Pennsylvania).

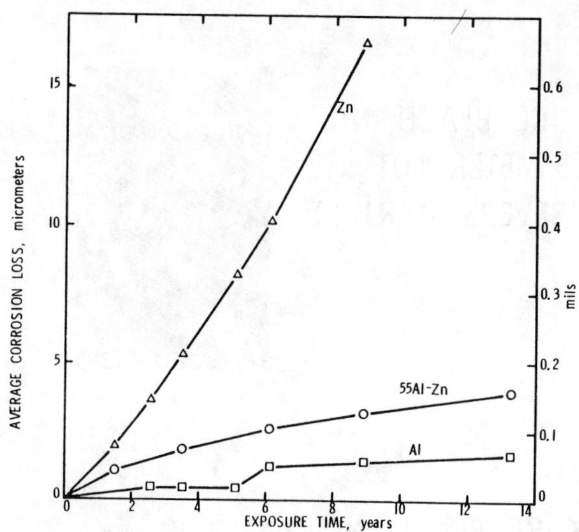

FIGURE 5 – Performance of coated sheets in industrial atmosphere (Bethlehem, Pennsylvania).

TABLE 2 – 13 Year Weight Losses

Site	Ratio of 13 Year Corrosion Losses Galvanized/55% Al-Zn
Kure Beach, North Carolina, 25 m from ocean	4.2
Kure Beach, North Carolina, 250 m from ocean	2.3
Saylorsburg, Pennsylvania	3.4
Bethlehem, Pennsylvania	6.2

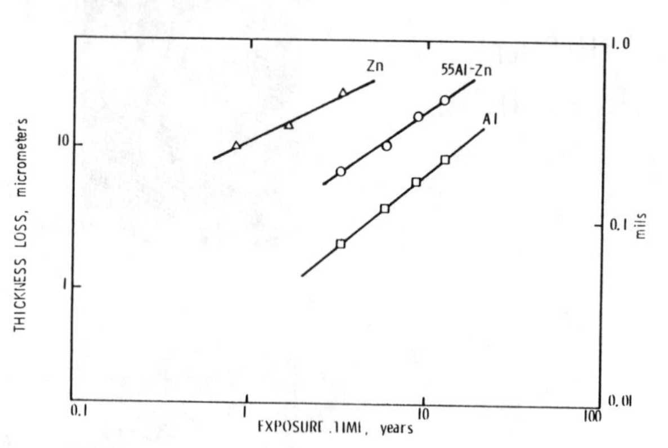

FIGURE 6 – Logarithmic plot of severe marine corrosion data (Kure beach, 25 meters).

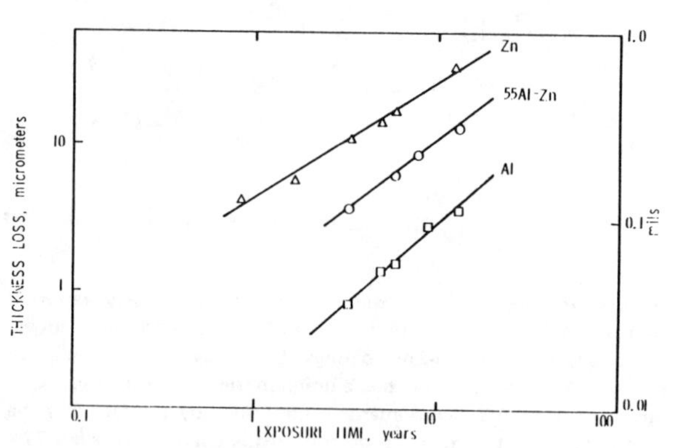

FIGURE 7 – Logarithmic plot of marine corrosion data (Kure beach, 250 meters).

TABLE 3 — Coefficients of Equations for Atmospheric Corrosion Data

	Sheet Coating								
	Zinc			55% Al-Zn			Aluminum		
Test Site	A	B	R	A	B	R	A	B	R
Kure Beach, NC 25 m lot	7.04	0.488	0.992	2.11	0.690	0.990	0.71	0.814	0.999
Kure Beach, NC 250 m lot	3.34	0.613	0.994	1.22	0.749	0.995	0.33	0.856	0.993
Saylorsburg, PA	1.50	0.877	0.999	0.99	0.572	0.997	0.05	1.426	0.973
Bethlehem, PA	1.19	1.198	0.999	0.84	0.611	0.999	0.13	1.020	0.854

Notes: A and B are coefficients of the Equation $C = At^B$, where C is corrosion loss in μm and t is exposure time in years. R is the correlation coefficient of the least-squares regression of the logarithms.

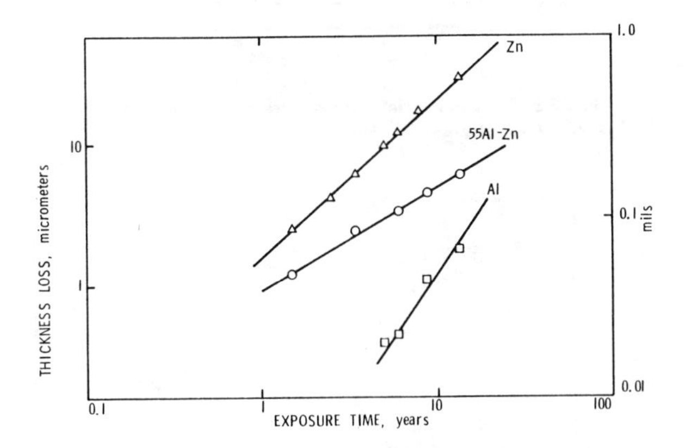

FIGURE 8 — Logarithmic plot of rural corrosion data (Saylorsburg, Pennsylvania).

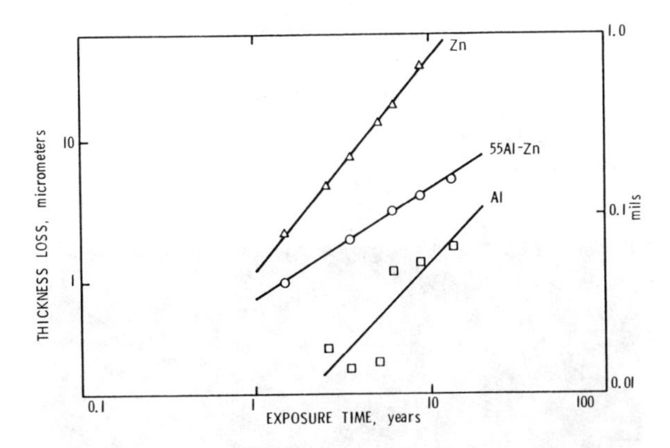

FIGURE 9 — Logarithmic plot of industrial corrosion data (Bethlehem, Pennsylvania).

coating corrosion in marine environments was likewise present on the 55% Al-Zn coating.

The degree of protection afforded by a coating at discontinuities such as pores, scratches, and cut edges is an important characteristic that determines the appearance of weathered coated steel sheet. Cut edge protection depends in part on the method used to cut coated sheet. Shearing of metallic coated sheet tends to wipe some of the coating over freshly exposed cut edges. Although wiping can provide some edge protection, particularly to lighter gage sheet, it is difficult to control and becomes less effective as sheet thickness increases. Nevertheless, sheared edges are always preferable to rough, saw cut edges, which involve exposure of large bare steel areas.

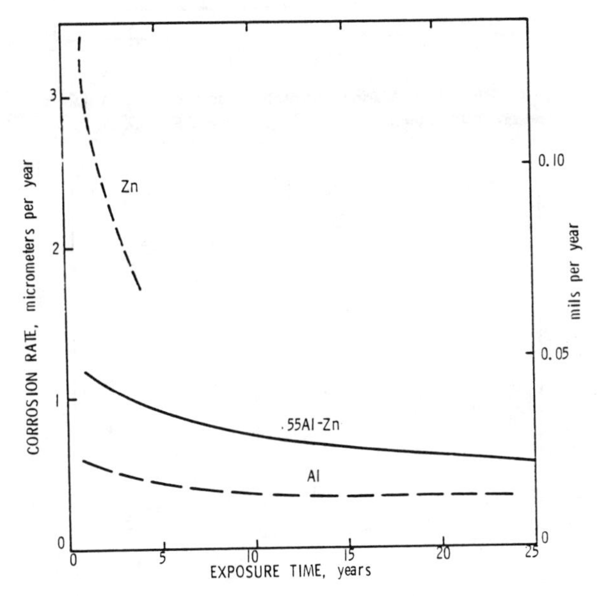

FIGURE 10 — Severe marine atmosphere corrosion rates (Kure beach, 25 meters).

Galvanic characteristics of coating metal relative to the substrate is also important in determining degree of cut edge protection. In most environments, including industrial and marine atmospheres, zinc is anodic to steel and thus provides galvanic protection to steel exposed at cut edges (Figures 15 to 21). The price of sacrificial protection by the zinc coating is accelerated loss of coating and initiation of coating failure at the edges. Aluminum coatings are anodic to steel in marine environments, where chloride ions impair the passivity of aluminum (Figures 15, 16, and 19). In industrial atmospheres, aluminum is more passive and aluminum coatings are cathodic to steel. This property leads to accelerated attack of exposed steel and the growth of rust projections at pores and edges. Rust staining due to bleeding from rust projections is observed at the edges of aluminum coated sheet exposed in industrial and rural environments (Figures 17, 18, 20, and 21).

Given the preferential corrosion of its zinc rich interdentritic constituents, the 55% Al-Zn coating is galvanically similar to a zinc coating and provides protection to steel edges in all environments (Figures 15 to 21). This protection is sufficient to prevent the growth of rust projections and staining observed in the case of aluminum coated sheet. Because the 55% Al-Zn coating is more

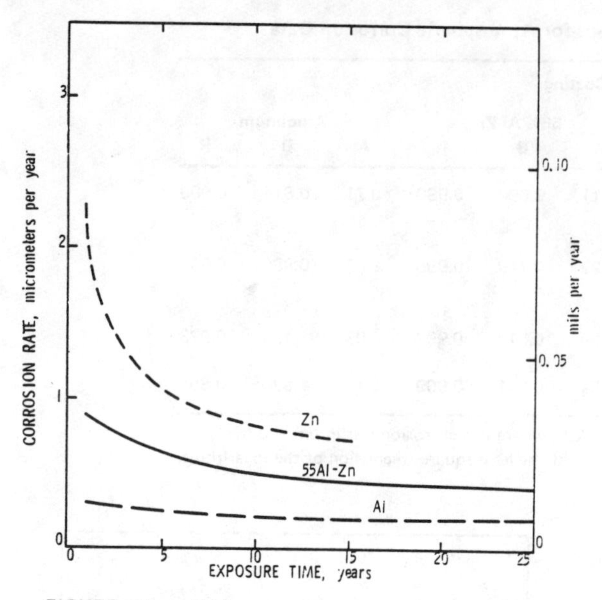

FIGURE 11 – Marine atmosphere corrosion rates (Kure beach, 250 meters).

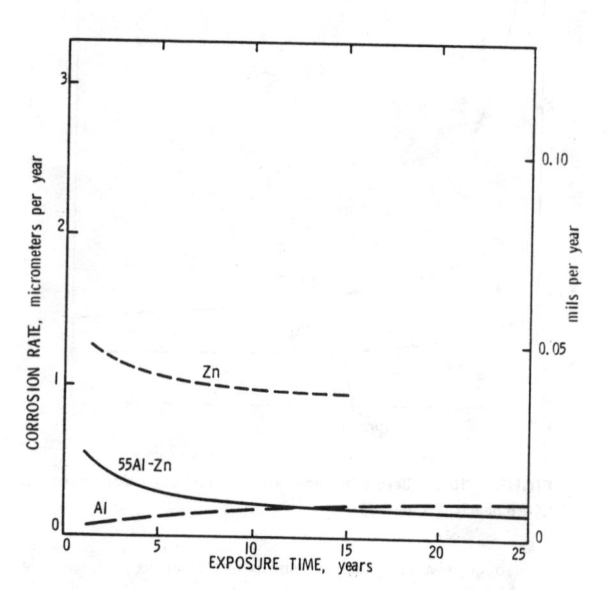

FIGURE 12 – Rural atmosphere corrosion rates (Saylorsburg, Pennsylvania).

durable than galvanized coatings, it provides edge protection for longer times.

Conclusions

The results of the 13 year outdoor corrosion tests of steel sheet protected with 55% Al-Zn alloy, zinc, and aluminum coatings demonstrated that the alloy coating:

1. Provides long term protection to sheet steel in a wide range of environments. Compared to galvanized coatings, 55% Al-Zn of equal thickness is at least two to four times more durable. In the industrial and rural environments, the corrosion rates of 55% Al-Zn coatings become indistinguishable from those of aluminum coatings after about 10 years of testing.

2. Provides cut edge protection to steel sheet. Compared to galvanized coatings, 55% Al-Zn coatings are more durable and thus provide protection for longer times. Compared to aluminum coatings, which provide sacrificial protection only in marine environments, 55% Al-Zn coatings provide galvanic protection at cut

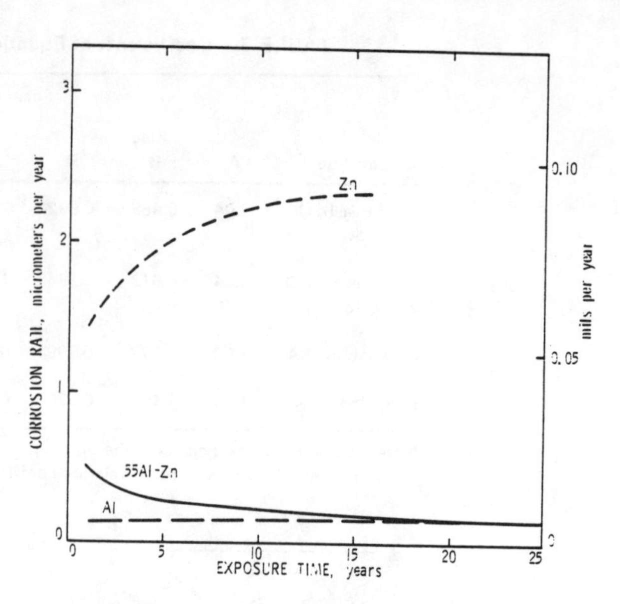

FIGURE 13 – Industrial Atmosphere corrosion rates (Bethlehem, Pennsylvania).

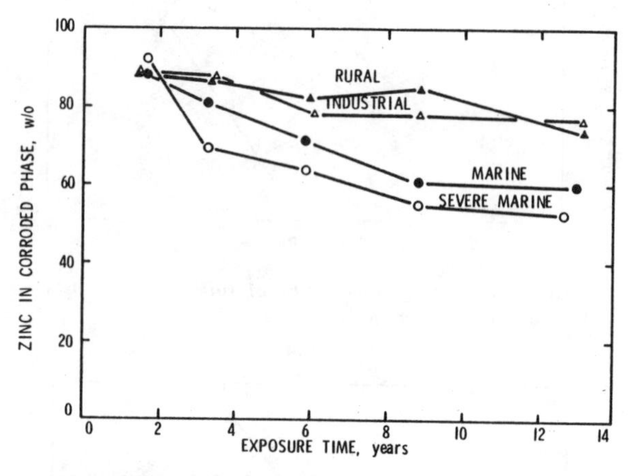

FIGURE 14 – Zinc content of the corroded portion of 55% Al-Zn coatings.

FIGURE 15 – Edge appearance of coated sheets after 13 year severe marine exposure (Kure beach, 25 meters). Top, Al; middle, Zn; and bottom, 55 Al-Zn.

FIGURE 16 — Edge appearance of coated sheets after 13 year marine exposure (Kure beach, 250 meters). Top, Al; middle, Zn; and bottom. 55 Al-Zn.

FIGURE 19 — Edge appearance of coated sheets after 5.5 year severe marine exposure (Kure beach, 25 meters). Top, Al; middle, Zn; and bottom, 55 Al-Zn.

FIGURE 17 — Edge appearance of coated sheets after 13 year rural exposure (Saylorsburg, Pennsylvania). Top, Al; middle, Zn; and bototm, 55 Al-Zn.

FIGURE 20 — Edge appearance of coated sheets after 6 year rural exposure (Saylorsburg, Pennsylvania). Top, Al; middle, Zn; and bottom, 55 Al-Zn.

FIGURE 18 — Edge appearance of coated sheets after 13 year industrial exposure (Bethlehem, Pennsylvania). Top, Al; middle, Zn; and bottom, 55 Al-Zn.

FIGURE 21 — Edge appearance of coated sheets after 6 year industrial exposure (Bethlehem, Pennsylvania). Top, Al; middle, Zn; and bottom, 55 Al-Zn.

Source: *Materials Protection*, October 1979, 13-20

edges in all environments, including industrial and rural.

The characteristics of the 55% Al-Zn coatings represent a favorable combination of the best properties of galvanized and aluminum coatings. This combination is a consequence of the alloy microstructure and corrosion mechanism as described.

Acknowledgment

The authors are grateful to J. B. Horton for interesting discusssions, to E. L. Gehman and D. H. VanBilliard for technical assistance, and to B. S. Mikofsky for editing this manuscript.

References

1. Zoccola, J. C., Townsend, H. E., Borzillo, A. R., Horton, J. B. in Atmospheric Factors Affecting the Corrosion of Engineering Materials, ASTM STP 646, S. K. Coburn, Ed., ASTM, Philadelphia (1978) pp. 165-184.
2. Borzillo, A. R., Caldwell, L. B., Connolly, J. J., Pellatiro, L. P. Proceedings of the Galvanizers Committee, Vol. 65, pp. 41-49, Zinc Institute, New York (1974).
3. Borzillo, A. R., Horton, J. B. U. S. Patent 3, 343, 930, September 26, 1967.
4. Harvey, G. J., Reynolds, J. paper presented at the 6th International Conference on Metallic Corrosion, Sydney, Australia (1975) December.
5. Atteraas, L. Corrosion News, No. 4 (1976) December.
6. Borzillo, A. R., Horton, J. B. U. S. Patent 3,393,089, July 16, 1968.
7. Mansfeld, F. Thirty Months Exposure Study in St. louis, Missouri, I Corrosion Data. NACE National Conference, Houston (1978) Unpublished paper no. 88.
8. Anderson, E. A. Corrosion Resistance of Metals and Alloys, LaQue, F. L., Copson, H. R., Eds., Reinhold Publishing, New York, 2nd Edition (1963) pp. 241-244.
9. Pasano, R. F., The Harmony of Outdoor Weathering Tests, Symposium on the Outdoor Weathering of Metals and Metallic Coatings, Washingtion Regional Meeting of ASTM, March 7, 1934, ASTM, Philadelphia, Pa. (1934).
10. Pilling, N. B., Wesley, W. A. Proc. ASTM, Vol. 40, p. 643 (1940).
11. Horton, J. B. The Composition, Structure, and Growth of Atmospheric Rust on Various Steels, Ph.D. Dissertation, Lehigh University (1964).
12. Legault, R. A., Pearson, V. P. Atmospheric Factors Affecting the Corrosion of Engineering Metals, ASTM STP 646, S. K. Coburn, Ed., ASTM, Philadelphia (1978) p. 83-96.
13. Legault, R. A., Pearson, V. P. Corrosion, Vol. 34, p. 344 (1978).

Jack Blasingame
F. W. Gartner Co.
Houston, Texas

A POPULAR USE of flame-sprayed coatings is in high temperature applications such as on the outside surface of this stack which is in service at a refinery on the Gulf coast. The stack was sprayed with aluminum to a 6-mil thickness. Two sealer coats of silicone were applied to the finished surface. The system will withstand about 1000 F.

Metallizing Takes Back Seat to Flame Spray

THE WORD "METALLIZING" is rapidly disappearing as a term used to describe the repair of worn or corroded metallic surfaces. Today, the industry can hardly limit the process to metal coatings alone because now it is possible to "flame-spray" not only metal but also such materials as ceramics, cermets (mixtures of ceramic and metal powders), plastics, and powders of other materials. Thus, the word "metallizing" is taking a back seat to the more general term "flame-spray."

The process, which involves the spraying of molten material onto a substrate to form a corrosion and wear resistant surface, is generally divided into three groups:

1. **Wire Process.** The wire process incorporates the use of a gun that is capable of spraying metal or ceramic from wire or rod material. The wire is drawn into the gun where it is melted. Compressed air then atomizes and sprays the material onto a previously prepared surface. Normally, acetylene/oxygen mixture is used, although propane and manufactured or natural gases can be used in conjunction with oxygen.

2. **Powder Process.** This process, as the name implies, is used to spray material that is fed into the gun in powder form. These materials include the self-fluxing alloys to produce hard faced coatings, ceramics to provide for thermal insulation, and cermets which are adaptable to a variety of applications. A wider variety of materials can be sprayed with this process than with the wire process.

3. **Plasma Process.** The recently developed plasma flame process is the most sophisticated of the three. An inert gas such as nitrogen, argon, or helium is forced through an electrical arc thereby ionizing and disassociating the molecules of the gas. The energy thus released will produce a gas stream up to 30,000 F; however, the normal operating temperature is around 12,000 to 15,000 F, which is sufficient to melt any known material. The material, in powder form, is introduced into the flame and sprayed with high velocity to produce high quality coatings with no decomposition of the material. While the gas stream is extremely hot, the surface being sprayed does not heat up excessively because of an abrupt drop in flame temperature a few inches from the nozzle of the gun.

Corrosion and Wear Resistant

Flame-sprayed coatings commonly used to provide corrosion resistance for the base metal are zinc and aluminum. These metals are anodic and protect the base by sacrificing themselves. Other flame-sprayed materials are either inert (such as ceramics) to the base or are cathodic and provide protection only when applied in sufficient thickness and properly sealed to form a mechanical barrier.[1]

Ceramics and fusible alloys are often applied in applications which require exceptionally hard surfaces in a corrosive environment. The ceramics are normally sealed with an air-drying phenolic,[1] or air vacuum impregnated with Teflon and other organic coatings.

Although resistance to corrosion is an important factor in the use of flame-sprayed coatings, most flame-sprayed parts are those requiring resistance to wear. Other uses include high temperature protection for rocket engine nozzles, muffler components, melting pots, and wear surfaces of parts in commercial and governmental aircraft.

In addition, electrical products are sometimes flame-sprayed with silver, copper, aluminum, or zinc to provide good conductance of current. Lead can be flame-sprayed to X-ray equipment to serve as a shield. However, probably the most popular use of either metal or ceramic coating is on worn shafts and sleeve wear surfaces under mechanical seals.

The fundamental requirements for adapting a part to the flame-spraying process are accessibility of the part's surface and the ability to prepare that surface for proper bonding of the sprayed material. The term normally used to repair corroded or worn parts with flame-sprayed coatings is "machine element work." However, many companies now have new parts and equipment coated before putting them into service because the sprayed coating will give the part a longer life.

Surface Preparation

The ordinary steps of building up a machine element part are: (1) degreasing, (2) undercutting, (3) masking, (4) blasting with steel grit (aluminum oxide, carborundum), (5) applying a bond coat of molybdenum, an exothermic material, or a nickel-chrome

THE WORKMAN above operates a powder gun used to spray ceramic to a missile component that will be subjected to high temperatures. The component is rotated to insure an even distribution of the ceramic coating.

material, (6) spraying, and (7) final finish—machine or grind.

Cleaning

Vapor degreasing or solvent cleaning is always the first step since worn parts usually contain oil, grease, or other contaminates. These must be removed to obtain an adequate bond. New parts also should be thoroughly cleaned, since they often contain some type of contaminates.

Undercutting

Undercutting is the turning or grinding of a part to provide space for the sprayed material. Thickness of the finish coating is determined by the amount of undercutting.

Masking

Masking is necessary to prevent the sprayed coat from adhering to areas not to be coated. Several good compounds are commercially available for this purpose. Oil should not be used because it may run onto the surface to be sprayed. Masking tape may be used where it is more convenient.

Preheating

Preheating the surface before spraying is desirable because even a small amount of moisture on the base metal will greatly reduce the bond strength and possibly cause shrink cracks. However, it is advisable to use an air cooler at times, especially when spraying small shafts, to keep the part from becoming too hot. Preheating should always be used when a high build-up of a high shrink metal is required.

Preheating alone is seldom used in machine element work because very high temperatures would be required in most cases to secure adequate bond. The use of high temperature heating is generally not advisable on machine element work due to the problems of warpage and oxidation of the surface. A surface temperature of 200 will effectively prevent the formation of condensate.

Bonding Methods

Sprayed material can be bonded to a base metal by roughening the base or by using commercially available bonding materials.

Roughening methods include (1) grooving and roughening with a shaft preparing tool, (2) rough threading, (3) grit blasting, and (4) bonding metal application.

Spray metal has a laminated structure and is made up of many fine, flat particles. The strength of the sprayed material is much greater when parallel to these flat particles and is weaker in a direction perpendicular to the laminations. In other words, it splits parallel to the laminations more easily than it can be broken in another direction. If the laminations of sprayed material are folded up and down over relatively large grooves or bumps, then its strength is improved materially and it cannot split.

Grooves have the effect of restraining any tendency that the material has to shrink and loosen the bond. Thus, if a fairly heavy coating is to be built up on a shaft with a keyway, some form of grooving should be used across the keyway. The keyway requires a strong bond because the ends of the sprayed metal at the keyway will tend to lift as a result of metal shrinkage.

Very heavy coatings, particularly those metals that have high shrink characteristics should also be applied to threaded or grooved surfaces even where there is no keyway. Heavy coatings on flat surfaces require grooving, at least at the edges of the surface, to prevent lifting.

In many instances, considerable time may be saved in both preparing and spraying shafts by deviating from regular methods of preparation involving grooving. The service to which certain types of applications are subjected does not require the high bond strength provided by grooving. Neither does it justify the time and cost entailed in cutting grooves and subsequently filling them in.

This is especially true where comparatively thin coatings are to be applied, such as in restoring press fits, or rebuilding surfaces where maximum allowable wear is only a few thousandths of an inch. Yet, the bond provided by the rotary shaft preparing tool alone may not be sufficient.

Satisfactory and economical compromise between the two methods consists of preparing the surface, after it has been undercut, by first chasing a sharp thread and then applying a bonding material. A regular threading tool can be used and threads cut 40 to 50 to the inch, or 24 to 30 to the inch for a still stronger bond. The nature of the thread and the number of threads per inch can vary considerably without adverse effect on the final job.

Metals used for bonding purposes have almost replaced all other methods, except on parts that require heavy build-up of metal coating. The bonding metals are popular because only a clean surface is necessary for application of the material. They will hold on a highly polished surface. Many job shops, however, still use a blasted surface for application of a bonding metal to insure additional bond strength and also as a method of cleaning the part to be coated. Since bonding metals will stick to a smooth surface, it is necessary to mask areas adjacent to the area to be coated. Two of the most popular bonding metals are molybdenum and Nickel Aluminide.[1]

Molybdenum is an excellent bonding coat for common steels, stainless steels, Monel,[2] nickel, and chrome alloys, cast iron, cast steel, magnesium, and most aluminum alloys. Molybdenum, however, does not provide a bonding base for copper, brass, bronze, or chrome plated or nitrided surfaces.

Molybdenum should not be used in a service of 600 F or higher. Coating thickness is normally 1½ to 2 mils.

Nickel Aluminide is both exothermic and synergistic, and in many applications, it is replacing molybdenum. It is available in powder or wire form. The material can be sprayed from coils easily because it is not stiff, it has a higher bond and temperature resistance than molybdenum, and it can be applied to nitrided steel. Normal thickness is 1½ to 3 mils.

Problems in Flame Spraying

Inside diameters and flat surfaces present a greater problem than out-

[1]Tradename of Metco, Inc., Long Island, N. Y.
[2]Tradename of International Nickel Co., Inc., New York, N. Y.

side diameters, with regard to the shrink of the coating. On outside diameters, the coating shrinks tighter onto the shaft. The only problems with the shaft or similar surfaces are the possibility of cracking and the problem of keyways. On inside diameters and flat surfaces, however, the coatings always tend to shrink in the direction which would loosen them.

For flat surfaces the problem can usually be solved by tying down the edges of the coating so that the coating cannot lift up. This can be done by either notching or grooving the edges, or by spraying over the edge so that the coating actually covers the ends as well as the top of the surface.

Preheating the base surface to 350 F is frequently desirable since preheating will tend to reduce stress in the coating after cooling. Where this is not possible or where very heavy coatings are applied rapidly, cooling with an air blast should also be used during spraying.

Whenever possible, cylinders should be chucked out from the face when spraying to permit through-ventilation. If the base is large and the surface easily accessible, then the regular grooving methods of preparation may be used, if the spraying can be done from both ends.

Machining

After a shaft has been undercut and sprayed, there is usually a ragged edge of sprayed metal over the shoulders at the ends of the undercut section. Where the build-up is heavy, this ragged edge may cause cracking which will penetrate into the main section of the coating. It is good practice to remove this ragged edge immediately after spraying by machining or by grinding before the main finishing is begun.

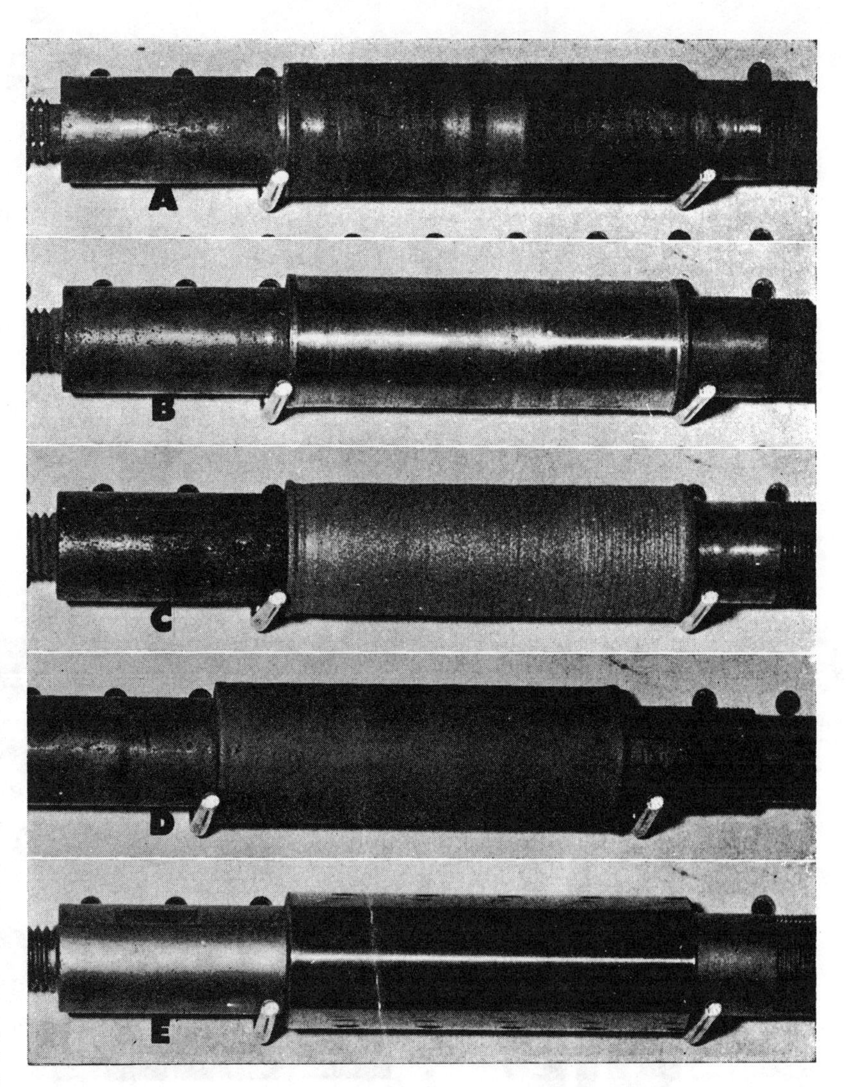

SEQUENCE OF APPLICATION of a flame-sprayed metal coating is shown by the five photographs above. A = worn and corroded shaft; B = machine surface preparation; C = applied bonding coating; D = sprayed metallic coating; and E = finished part ready for service.

Sealing Sprayed Coatings

Although sprayed metal coatings are porous, it is not generally necessary to seal these pores. In service, flame-sprayed coated parts are used where ample lubrication is provided. Thus, the porosity is an advantage because it retains oil; the oil itself tends to seal the pores and prevent entrance of moisture. Stainless or Monel coatings are usually sufficiently thick and relatively impervious to penetration of moisture, even though they are porous.

There are some cases, however, where it is best to seal the pores, such as where thin coatings are used or where the corrosive reagent is quite active. Also, it is sometimes necessary to seal the pores where high pressures will be used, such as in hydraulic press rams.

Conclusion

New developments in alloy and ceramic powder have already begun to extend the use of sprayed coating. These new materials find service in wear, corrosion, and heat resistance fields. There have been major problems to overcome in the development of spraying these materials. Rod or wire forms of alloy or ceramic were not available. A low spray speed, a low deposit efficiency, and greater tendency of the metal sprayed coating to oxidize were factors which kept the powder system running a poor second at the time of development.

The new powder guns, developed for the spraying of alloy and ceramic, have corrected many of the old problems. Deposit efficiency is well above 90% and sufficient heat can be applied for spraying the high melting point materials like alumina and zirconia.[1]

The powder process has two particular advantages. For the production of large numbers of intricately shaped components such as gears, cams, etc., complicated shapes can be produced to close tolerances without requiring subsequent machining. This process also permits the production of uniform dispersions of one metal in another in cases where this cannot be achieved by melting and casting methods due to insolubility effects. Powder metal compacts are a little less dense than wrought materials of the same composition, can retain oil for self-lubrication, and have a high-damping capacity.

References

1. Robert C. Dennison. Flame Spraying, *Machine Design*, (1966) December 8.

Bibliography

Corrosion Tests of Metallized Coated Steel—12 Year Report. American Welding Society, New York, N. Y. (1967).

S. John Oechsle, Jr. Metal Spraying. Paper presented at the NACE Houston Section, November, 1958.

H. S. Ingham. Flame-Sprayed Protective Coating Made From Refractory Materials. *Materials Protection*, 1, 74 (1962) January.

Walter B. Meyer. Flame Sprayed Fusible Alloys. *Materials Protection*. 2, 21 (1963) January.

W. E. Ballard. *Metal Spraying*. Charles Griffin & Company, London, 1963.

F. W. Gartner, Jr. Metallizing Helps Lick Gulf Coast Corrosion. *World Oil*, p. 220, June 1956.

H. S. Hammond. Metallizing for Protection of Bulk Shipments. *Chemical Engineering*, p. 278, June 1956. Reprinted by *Corrosion Forum*.

Source: *Materials Protection*, September 1968, 21-23

Figure 1—Guard rail on this freeway system in downtown Houston, Texas, was galvanized for protection against corrosion. The zinc coating, because of its expected service life of 25 years under atmospheric exposure, permitted the guard rail to be made of thin gauge steel to save tax money on the freeway project. In the background is the 44-story Humble Building, which contains over four million pounds of hot-dip galvanizing.

commercial HOT DIP GALVANIZING of fabricated items*

★Revision of a paper titled "Commercial Hot-Dip Galvanizing of Fabricated Items Used in Construction of Industrial Projects," presented at the 17th Annual Conference, National Association of Corrosion Engineers, March 13-17, 1961, Buffalo, N. Y.

Reprinted with permission from *Materials Protection*, January 1962, 30-39, © 1962 National Association of Corrosion Engineers

Ray Vickers
Smith Industries, Inc.
Houston, Texas

SUMMARY

Describes basic steps in commercial hot-dip galvanizing of large fabricated items. Highway guard rail sections are used as an example. Illustrates the general composition of the alloying layers in galvanizing and explains the different appearances which may be expected. Case histories are given to compare life of galvanized structures. Comparisons also made between cost of galvanizing and other types of protective coatings. Clarifies some of the misunderstandings about galvanizing such as appearance, double-dip and function. Also discusses design factors that should be considered when large structural material is to be galvanized.

COMMERCIAL HOT-DIP GALVANIZING generally is conceded to be an outgrowth of a process patented in England in 1837. Although modernization has been added, the basic process of hot-dip galvanizing today is the same as it was in 1837. This seems strange when one sees the many modern uses of galvanized coatings to protect steel applications such as the highway guard rails in Figure 1.

Galvanizing Process

The zinc hot-dip galvanizing process in theory is relatively simple. The material to be galvanized is thoroughly cleaned, immersed in molten zinc and then withdrawn after a proper length of time. The material so processed has enough molten zinc alloyed as a surface film to give the desired coating for protection from corrosion. The amount of pickling of the material to be galvanized, proper regulation of the zinc kettle temperature, fluxing of the surface of the molten zinc and position of the material as it is lowered into the zinc—all these are important factors. Each of these factors has a definite bearing on the quality of the coating produced.

This galvanizing process is illustrated in Figures 2-7, as sections of the highway guard

Source: *Materials Protection*, January 1962, 30-39

Figure 2—Sections of the highway guard rail are stacked as one load to be immersed in 10 percent alkali at 180 F to remove oil, grease and paint and then pickled in 6 percent sulfuric acid at 180 F to 200 F for removal of mill scale and oxidation. Spacers are placed between each piece of railing so that pickling can be complete on all surfaces.

Figure 3—Load of guard rails are lowered slowly into one of the pickling baths. The sections are light wall, high strength elliptical bridge rails made of domestic steel.

Figure 4—As the load of guard rails are removed from the pickling bath, the operator tilts the load so that the railing can be drained of excess pickling solution. Rail sections have a uniform, blackish appearance, free of scale, oxide, grease and other foreign matter that would interfere with a uniform zinc coating during the galvanizing process.

rail (shown in Figure 1) are galvanized.

Preparation of Base Material

Preparation of the base material to be galvanized involves certain steps depending on the character or condition of the material's surface. In general, there would be three fundamental steps: (1) removal of oil, grease or paint by immersion in 10 percent alkali at 180 F or immersion in an organic solvent, (2) removal of scale or oxide by immersion in 6 percent sulfuric acid at 180 to 200 F, or blasting with sand or grit, and (3) a fluxing step which consists of immersion in a liquid flux or a molten flux.

Primary object of degreasing, pickling and fluxing is to make the steel as chemically clean as possible before it is lowered into the molten zinc. This cleaning and fluxing process is of primary importance to hot-dip galvanizing because the quality and character of a zinc coating is no better than the surface to which it is applied.

Time required to prepare the surface of steel being galvanized varies according to the steel's surface condition. Degreasing of new steel, under normal operating conditions, will take about 30 minutes. Scale and rust removal will take another 30 to 45 minutes. The last step of immersing the steel into dilute hydrochloric acid takes only 3 to 5 minutes. This step is not intended for pickling but is used to make the steel wettable to the molten zinc. Although hydrochloric acid is used in this last step by most commercial galvanizers, some firms substitute a solution of zinc ammonium chloride.

Another important factor in galvanizing is position and angle of the material as it is lowered into the molten zinc. Extreme caution must be taken to eliminate the possibility of air pockets which cause voids in galvanized coatings. Position of the material also must afford proper drainage of excess zinc as the material is withdrawn.

After the steel has been properly cleaned and positioned, it is slowly lowered into the molten zinc through a two-inch blanket of ammonium chloride flux floating on the surface. The steel is left immersed until it has had time to "cook out" and attain the 850 F of the zinc. This usually required seven to nine minutes, depending on the thickness of the steel.

When the material is to be withdrawn, the blanket of flux floating on the surface of molten zinc is skimmed to one end of the kettle so that the steel can be withdrawn through the clean surface of molten zinc.

Composition of Zinc Coatings

Hot-dip galvanizing is not an overlay of coating material like other types of coatings which depend on mechanical bonding. Galvanizing is an alloy with the parent metal. When the molten zinc reacts with the iron, there are several distinct layers of iron-zinc alloy, as shown in Figure 8.

Immediately adjacent to the iron base, a typical galvanized coating will have a thin, adherent gamma alloy layer containing 21 to 28 percent iron. Next to the gamma layer is the delta or palisade layer containing 6 to 11.5 percent iron. Next is the zeta or floating layer with an iron content between 6.0 and 6.2 percent. The outside layer, the eta or pure layer, contains only 0.003 percent iron.[1]

The growth rate of the alloying follows a parabolic curve. The greatest growth develops during the first minute of immersion, with the rate decreasing as the alloys thicken and the diffusion path increases. When a section of structural steel is removed from molten zinc, the alloying process continues as long as the steel retains sufficient heat. This "after" alloying continues at the expense of the outer layer of zinc. If "after" alloying is allowed to progress long enough, the finished coating will appear dull.

Appearance of Zinc Coatings

Varying with different types of steel, the rate of alloying is determined by composition of the base material. Thus, the appearance of galvanizing is affected greatly by the composition of the base steel. Silicon

Figure 5—Three pieces of railing are lowered slowly into the galvanizing kettle which is maintained at 850 F. Items being galvanized are lowered through a liquid flux floating on the surface of the molten zinc. Steam can be seen coming from inside the circular portion of the railing as the molten zinc boils moisture from the railings.

Figure 6—As the guard rails are withdrawn slowly from the kettle, the floating flux (ammonium chloride) has been skimmed to one end of the kettle so that the galvanized item is withdrawn from the molten zinc and not through the flux again. The kettle operator is shown throwing powdered ammonium chloride which makes the molten zinc flow on the steel to give a smooth finish.

Figure 7—After the rail sections have been removed from the molten zinc kettle, each piece is inspected and examined for burrs of zinc which are smoothed off with a file. The discolored portions of the rail were caused by smoke from the kettle as the pieces were withdrawn. This discoloration will be washed off by the first rain after the rails have been mounted on the highway.

and phosphorus—particularly silicon —affect the alloying rate more than most other steel constituents. An increase in silicon content from 0.06 to 0.20 percent will triple the rate of alloying.[2]

Obviously, structural steel having a silicon content of 0.09 to 0.13 percent will tend to have a dull coating. The galvanizing will be even duller when the thickness of the steel is sufficient to retain heat, thus allowing additional alloying.

Unlike production galvanizers who coat only one type item in a continuous line, commercial galvanizers have no time tables regulating the amount of pickling or the time which the material must be immersed in the molten zinc to insure proper alloying. The process of commercial galvanizing is a skill. Few pieces of structural steel being galvanized are the same size, shape or even of the same type material; therefore, quality of the galvanizing depends on the skill of the craftsman performing the process.

ASTM specifications for zinc coating of steel products (A-123-59), under which fabricated steel is classified, states in Section 9a that "The zinc coating shall be adherent, smooth, continuous, and thorough. It shall be free from such imperfections as lumps, blisters, gritty areas, uncoated spots, acid and black spots, dross and flux."

One of the greatest misconceptions of commercial galvanizing is that the coated material must shine and have spangles, such as that on garbage cans or mill sheet galvanizing.

Mill sheet galvanizing is processed in long continuous strips with the appearance being controlled by exit roll gauging, variable air cooling and other mechanical processes. These mechanical controls produce the smooth finish and flowery spangles by which galvanizing is often identified. The commercial galvanizer does not have these controls and must rely on proper drainage to produce a smooth coating. Commercial galvanizing,

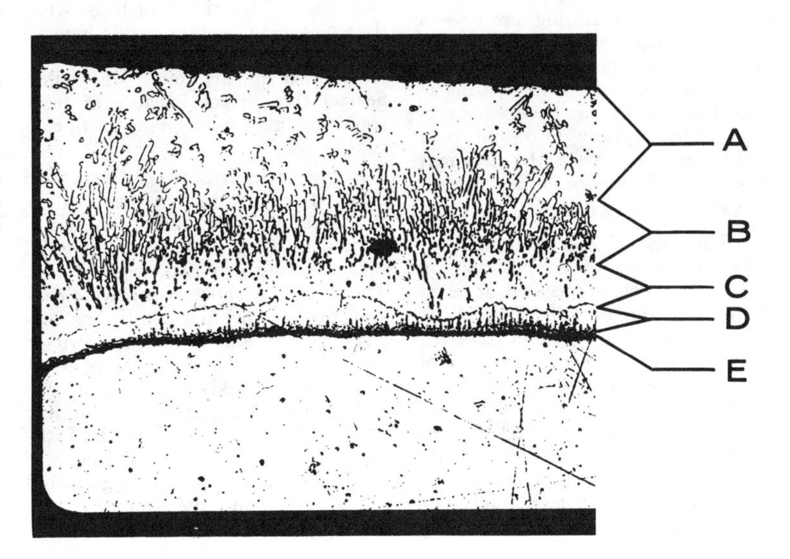

Figure 8—Photomicrograph shows cross section of hot-dip galvanizing and its alloying layers. Labelled sections are as follows: A is the eta or pure zinc layer containing only 0.003 percent iron, B is the zeta diffused layer and C is the zeta compact layer having between 6.0 to 6.2 percent iron, D is the delta or palisade layer containing 6 to 11.5 percent iron, and E is the thin, adherent gamma alloy layer containing 21 to 28 percent iron.

Source: *Materials Protection*, January 1962, 30-39

Figure 9—Rough, dull galvanizing as shown on part of this hand railing does not indicate a poor coating. The type of steel, its surface condition and thickness of various sections all affect the appearance of galvanized zinc coatings. One part of this hand rail was fabricated, probably, of some old steel; the other parts may have been of new steel, thus explaining the different appearance of the zinc coating on some sections.

therefore, is a functional, protective coating—not a decorative one. It has but one purpose: to provide years of low cost corrosion protection.

All too frequently, inspectors have rejected galvanizing because the surface lacked a shiny luster or because the surface was not too smooth. This is the wrong approach in evaluating galvanized material. Inspectors should not overlook the fact that most protective coatings are not necessarily the ones which are most pleasing to the eye.

Several conditions affect the appearance of galvanized structural steel or galvanized pipe. Three of the most important conditions, which could be controlled by persons specifying hot-dip galvanizing, are type of steel, condition of the surface and thickness of various sections.

An example of how galvanizing of a fabricated item may have some portions which appear to be of poor quality because some of the steel may have been thicker or of a different type is shown in Figure 9.

Thickness of Zinc Coatings

The thickness of the coating on galvanized sheets is specified in terms of the total zinc on both sides of the sheet. A two-ounce specification for sheet galvanizing has one ounce of zinc per square foot on each side.

The thickness of commercial hot-dip galvanizing, meeting ASTM specifications, is an average of not less than two ounces per square foot for materials ⅛ and $\frac{3}{16}$-inch thick. For material ¼-inch and heavier, coating weight is an average 2.3 ounces per square foot.

The two-ounce commercially galvanized coating, such as the coating meeting ASTM A-123-59 specifications, will be twice as heavy as two-ounce mill sheet galvanizing processed under ASTM A-93 specifications.

The thickness of the zinc coating is not dependent primarily on immersion time. The major factors in regulating the thickness are heat of the molten zinc and withdrawal speed. Tests of steels of the same composition and thickness have shown that thickness of the outer zinc layer is a function of the withdrawal speed and is almost independent of the immersion time. Material which is withdrawn slowly usually will have a thinner coating. When material is withdrawn quickly, more zinc solidifies on the surface because of more rapid cooling, thus giving a thicker coating.

Actual withdrawal of structural shapes from the molten zinc is done in several steps depending on size of the object being coated. The material is first withdrawn two or three feet. Then, after a short pause which allows excess zinc to run off, the material is withdrawn another two or three feet. If the material is withdrawn too quickly, the excess zinc will solidify and produce a lumpy surface. On the other hand, if the material is held in the molten zinc too long, the coating thickness would be too great and the quality of the coating poor.

Careful consideration of proper drainage for all sections should be taken when designing structural shapes which are to be galvanized. This is especially true with pipe sections. When fabricated pipe is galvanized, molten zinc must enter all sections. Drainage holes must be placed in the pipe so that molten zinc can flow into all areas and on all surfaces. If these holes are not large enough, the withdrawal speed will be greatly reduced, resulting in lowered quality of the galvanizing.

Service Life of Galvanizing

Service life of a galvanized coating is a direct function of its thickness. Sufficient coating thickness must be available for the sacrificial action provided through galvanic protection of the steel.

The corrosion protection offered by galvanizing cannot be expressed in simple terms of "X" number of years. Behavior of the coating depends on a number of conditions and therefore varies with different types of environments.

Being a reactive metal, zinc quickly forms a surface film of corrosion products when exposed to the atmosphere. The resistance of this film and how long this film is allowed to remain on the surface determine the service life of the coating—not the zinc itself.

Numerous tests are being conducted by the American Hot-Dip Galvanizers Association and other groups to determine the life of gal-

Figure 10—Transmission tower in the Gulf Coast area. Hot-dip galvanizing was used to protect these towers from atmospheric corrosion. One utility company has obtained an average 25-year service life from zinc coatings. A study of coating costs showed that the towers could be dismantled, re-galvanized and re-assembled for one-third the cost of painting the towers. Service life on the painting job was estimated at five years although the contractor would guarantee it for only three years.

vanizing. Service life of galvanizing under atmospheric conditions is given in Table 1.

The Houston Lighting and Power Company has transmission line towers throughout the chemical refining area along the Houston ship channel. Most of these galvanized towers have been in service since 1930 and some since 1925. The average service life of galvanized coating on this company's towers is over 25 years. One of these transmission towers is shown in Figure 10.

Tests have shown that galvanizing, like most coatings, does not last as long in sulfur or chloride polluted areas such as chemical refining plants. Atmospheric pollution is the main controlling factor in determining the life of galvanized coatings. Generally, the effects of climate are secondary.

Recently, commercial galvanizers have begun using magnesium in the galvanizing process to extend the life of zinc coatings. Dow Chemical Company for the past six years has conducted extensive tests of galvanized panels containing varying amounts of magnesium in the galvanized coating.

These test results show that an addition of 0.04 to 0.05 percent magnesium will extend galvanized coating life from 20 to 90 percent. This process is offered by several commercial galvanizers at the same price or for a few cents more than the standard type galvanizing.

Galvanizing Costs

The cost of galvanizing has always been expressed in a cwt (hundredweight) price. This means little to most persons who think of coatings in terms of cost per square foot. Numerous published articles compare the cost of galvanizing to various types of paints and coatings. Too often these costs are compared on several specific size beams or columns in convenient lengths. These costs do not include the other items which will be coated on industrial projects such as sag rods, grating, ladders, pipe handrails and other miscellaneous items. Because these items are expensive to coat, they should be considered if a true coating cost is to be estimated.

Initial cost of hot-dip galvanizing is not expensive. On projects containing 15 to 20 tons of structural steel, the cost of galvanizing will be about 20 cents per square foot. Considering the service life of this galvanizing to be 20 years, the cost is one cent per square foot per year.

Three Examples of Galvanizing Costs

Three specific case histories are given to indicate the cost of galvanizing for various types of equipment so that cost comparisons can be made with other types of coatings.

Case Number One

One large chemical company on the Texas Gulf Coast expanded its facilities to include the production of additional chemicals. The price quoted by the fabricator to galvanize about 500 tons of material for this project was only $2.89 per cwt or $57.80 per ton. This price, which covered all types of structural steel, included cost of double dipping plus freight to and from the galvanizing plant. Using the conservative figure of 260 square feet per ton on this particular project, the actual cost of the galvanizing to the customer was 22 cents per square foot.

Case Number Two

A chemical plant on the Louisiana Gulf Coast had 20 tons of structural steel to be coated, including 8 and 10-inch wide flange beams and 4-inch square angles, all to be used for pipe supports. Two types of coatings were considered: hot-dip galvanizing and a coating with the following

TABLE 1—Service Life of Hot-Dip Galvanized Coatings

Weight of Coating*	Years Under Atmospheric Conditions					
	Rural	Tropical Marine	Temperate Marine	Suburban	Urban	Highly Industrial
2.00.......	50	40	35	30	25	15

* Ounce per square foot.

specifications: sandblast, two prime coats, one finish coat of epoxy or vinyl paint. The average cost for galvanizing was 23½ cents per square foot. Average cost of applying paint was 49 cents per square foot. Obviously, the steel was galvanized because of economics.

Case Number Three

In 1956, an electrical utility company in the Houston area conducted an extensive study comparing the cost of painting with the cost of re-galvanizing 250 transmission line towers. This study showed that it is cheaper to set up guyed poles, dismantle the towers, re-galvanize and re-erect the towers than to paint them. The service life of the painting was estimated to be five years although it was guaranteed for only three years.

The cost of re-galvanizing the transmission towers would be only 52.97 percent of the cost of painting the towers if the service life of the galvanizing were estimated at 15 years. With an estimated 20-year life on the galvanizing, the cost of re-galvanizing would be only 39.72 percent of the cost to paint the towers. With a 25-year life, the re-galvanizing cost would be only 31.78 percent of the painting cost for the towers.

Under actual operating conditions, the life of these towers has been over 25 years. Therefore, this study shows that it is three times as expensive to paint the transmission towers than it is to galvanize them.

Designing Equipment to Be Galvanized

When designing equipment or fabricated items that will be galvanized, the engineer should be familiar with the length, width and depth of the galvanizing facilities available so that he will not design items so large that they cannot be immersed in the zinc kettles. Typical size of kettles for commercial galvanizers will be about 30 feet long, 4 feet wide and 3 to 5 feet deep. If the designer is unfamiliar with galvanizing or unfamiliar with the facilities available, a consultation with the commercial galvanizer would be most helpful. Often the galvanizer can make suggestions that will save money on a given project.

The commercial galvanizer is not a metallurgical engineer, however, and can galvanize only whatever material is brought to him. He has no way of knowing the composition of the material unless it is readily recognizable. He should be told when special steels are to be galvanized so that the best quality coating can be obtained.

Galvanizers are asked almost daily to galvanize structural material containing members which have already been galvanized. Many persons believe that their cost should be less on such a job because the material is partly galvanized to begin with. The opposite, however, is true. Such a job is more expensive to galvanize because the zinc on the galvanized areas must be removed. This requires extra pickling. Not only does this add to the cost, but removal of old galvanizing tends to lower the quality of the new galvanizing.

Another factor that should be considered on fabricated items to be galvanized is welds. All weld beads should be sandblasted. Flux used on welding rods is inert to the pickling acids used in the galvanizing process. Hand cleaning methods often will leave a line of residue along the edge of the weld. When hand cleaned material is galvanized, the weld flux is not only unsightly but causes a weak point in the zinc coating.

Galvanizing is a relatively inexpensive coating because large quantities of material can be coated by a standard procedure. Deviation from the galvanizer's standard procedure increases the cost of the coating.

For example, a complete reinforcing bar can be galvanized at less cost than a bar which is to be zinc coated only one foot past the threaded ends of the bar. This type specification requires extra handling and constitutes additional charges.

Double-Dip Process Misunderstood

One great misunderstanding of the galvanizing process is the term "double-dip" used by commercial galvanizers. Material that has been double-dipped does not have a heavier coat-

RAY VICKERS is a special coatings sales engineer for the Smith Industries of Houston, Texas. Since joining the firm in 1956, he has been engaged in the study of hot-dip galvanizing and protective coatings in the sub-tropical environment of the Gulf Coast area. He is on the board of technical advisors for the American Hot-Dip Galvanizers Association and is vice chairman of the NACE Technical Committee T-6B-19 on Hot-Dip Galvanizing as a Protective Coating in Atmospheric Corrosion. He also is a member of other NACE technical committees on protective coatings and is on the board of directors for the Coating Society of the Houston Area.

Figure 11—Stages in Double-Dipping Process

→ A 38-foot truss beam is being partially lowered into the zinc kettle, which is 32 feet long with a five foot depth. Height of the beam is nine feet. → About half the beam is slowly immersed in the molten zinc. → As the first half of the beam is removed from the kettle, the uncoated end is lowered into the molten zinc. The kettle man is shown tossing ammonium chloride onto the galvanized end of the beam. This powdered chemical makes the molten zinc flow, thus insuring a smooth coating. → While the second half of the beam is being galvanized, the kettle man inspects the first half and removes icicles and burrs. → The beam then is removed slowly to complete the galvanized coating. → The galvanized beam is allowed to cool before loading into rail cars. A final inspection is made for burrs which are filed off for a smooth coating job.

ing than a single dipped item. This term is used when the material is longer than the kettle or has a greater vertical dimension than the kettle depth, thus requiring rotation of the item. The stages of double-dipping are shown in Figure 11. These photographs illustrate how a 38-foot truss beam was double-dipped for protection against corrosive environment in an ore storage building.

Conclusions

Hot-dip galvanizing has many advantages over other types of protective coatings. The process itself is not subject to as many human errors as are most of the specialized coatings which are being used today. Material having small segments, angles, thin edges, recesses or inaccessible areas are easily coated by the hot-dip galvanizing process. Many shapes cannot be economically protected by some types of coatings. In the galvanizing process, molten zinc flows freely to cover all surfaces for complete protection with the only requirement that drainage holes be adequate.

The application of hot-dip galvanizing is not regulated by weather or humidity as are most specialized coatings. Also, galvanizing can be applied in less time than other types of coatings.

Galvanizing is not a miracle coating that can accomplish wondrous feats of protection. However, through the past 200 years, it has proved to be a reliable, general purpose protective coating with good resistance to various corrosive environments and to impact and abuse.

References

1. J. J. Sebisty. A Survey of Literature on Hot-Dip Galvanizing 1950-1955. Canada Department of Mines and Technical Survey Mines Branch. Ottawa, Canada.
2. D. W. Fagg. Factors Affecting the Quality of Hot-Dip Galvanized Coatings. American Hot-Dip Galvanizers Symposium, Columbus, Ohio, March 5, 1959.
3. R. M. Burns and W. W. Bradley. Protective Coatings for Metals. Reinhold Publishing Corp., New York, N. Y.

DISCUSSIONS

Question by E. F. Bladholm, Los Angeles, California

Did the speaker assume repainting with a complete system in his comparison with regalvanizing? When we repaint, we use one coat and get 5 to 7 years in a sea coast atmosphere.

Reply by Ray Vickers:

The comparison I gave between painting or regalvanizing was developed by the utility company involved. The only change I made was to use a percentage comparison rather than a dollar comparison. I was asked to do this for obvious reasons.

The utility company received bids from various tower painting contractors on the basis the contractor bid whichever system he considered best. One stipulation in the bidding required the contractors to give the length of time he would guarantee the coating.

It should be taken in consideration that figures or comparison such as these are only accurate for this one bidding at one particular date. Another point is that this study does not plot increased labor and coating cost which will come if our economy follows its present trend. The comparison uses painting cost of 1958 and compares them against galvanizing which will still be good in 1983.

The comparison was made between the regalvanizing and two painting contractors' bids, with each contractor quoting a two-coat system. In both cases, the contractor would guarantee his coating for only three years.

Question by Curtis L. Craig, Omaha, Nebraska

In handling galvanized structural members, a certain amount of nicking or other removal of the galvanizing appears to be inevitable. What methods are available for repairing these spots?

Reply by Ray Vickers:

One of the main features of hot-dip galvanizing is the electrolytic protection which the galvanizing gives to small nicks and scratches. This protection is quite complete and will continue as long as sufficient zinc is available.

If you choose to repair these nicks or void areas which also can be caused by welding after galvanizing, the coating used should be as electrically similar to zinc as possible. First choice for such repairs would be metallizing with zinc. By using Prime Western grade zinc or better, the metallizing and the galvanizing will have the same electro-potential, thereby eliminating the possibility of a galvanic cell.

Second choice would be an inorganic zinc over a sandblasted surface. One manufacturer of inorganic zinc coatings sells galvanizing touch-up kits. That company states that their coatings can be applied over a surface thoroughly cleaned vigorously with a wire brush.

Last choice for repairing galvanized coatings would be a zinc rich organic coating. There are many available which can be dabbed on with little effort.

Comments by John R. Daeson, Park Ridge, Illinois

A discussion of commercial aspects of hot-dip galvanizing should include recognition of the fact that in addition to its well known role of lowest cost protection against atmospheric corrosion, it is becoming increasingly obvious that in exposures where paint systems have to be replaced, hot-dip galvanizing as a priming coat is the ideal solution to the pitting problem which arises from penetration of all paint films by oxygen and moisture.

At costs that compare favorably with that of minimum satisfactory cleaning of the base by sandblasting for painting, the hot-dip galvanized coating remains insurance against pitting and consequent need for expensive sandblasting throughout many applications of renewal paint coats.

Where regular repainting is involved, hot-dip galvanizing, prepared for painting by phosphate treatment, is the controlling element of minimum maintenance cost of painted structures.

Question by Newell W. Tune, Los Angeles Water & Power, North Hollywood, California

What is the deposit or clear coating that occurs on top of hot-dip galvanized sheets that interferes with the adherence of subsequent paint coatings and what is the best way of removing it for painting?

Reply by Ray Vickers:

Mill galvanizers use a number of different coatings over their sheet galvanizing to prevent oxidation or "white rust." These coatings will vary from waxes and oils to vinyls. Dilute phosphoric acid usualy will remove these coatings without too much difficulty.

Question by Forest H. Williams, Pittsburgh, Pennsylvania

In galvanizing structural steel where clip angles are riveted or bolted to the structural member, is it necessary to seal weld the joining surface of the angle to the member? If so, why?

Reply by Ray Vickers:

No, it is not absolutely necessary, but seal welding is preferred to prevent the pickling solutions from entering crevices and in between contacting surfaces. Under normal operations, these surfaces are cleaned by the pickling process, but, because of the heavy viscosity of the molten zinc, the close contacting surfaces remain uncoated. This condition, if not prevented by seal welding, could later become a corrosion problem.

Introduction

CORROSION RESISTANCE of aluminum alloys to marine and urban environments has been analyzed in a long term study begun in 1946, when test panels were exposed at the Naval Air Station, Norfork, Virginia, and at Washington, D.C. Involving exposures of six months, one, two, three and ten years, these studies were designed to show the weathering resistance of typical aluminum alloys.

The corrosion resistance of several aluminum alloys to marine and urban enviroments has been reported previously.[1-3] Long term atmospheric test data at several environments have been published by ASTM.[4-5]

The range of aluminum alloys included in the tests described in this article included the high strength aluminum-copper type, the corrosion resistant and weldable aluminum-magnesium type and a relatively pure aluminum. (See Table 1 for chemical analyses of each alloy.)

Test Procedure

A total of 216 test panels (4 by 14 inches) was prepared for exposure. Panel gauge thickness ranged from 0.019 to 0.051 inches. Each alloy series consisted of ten exposure panels, including an initial control panel and a ten-year control panel.

Exposure racks at Norfork were inclined 45 degrees from vertical and were ten feet above the mean tide level, facing east-southeast. The Washington exposure racks were 45 degrees from the vertical, facing south on the roof of the Northwest Building, National Bureau of Standards. All panels were degreased with acetone before exposure.

After exposure, visual examinations were made of the panels from each exposure period and each condition. Panels were cleaned in chromic-phosphoric acid and nitric acid solutions. Pit depths were measured in mils by a calibrated focus, metallurgical microscope. Averages of the four deepest pits on each side of a panel are shown in Figures 1 and 2. This method of expressing pit depths conforms to the procedure adopted by the ASTM Committee B-3 (ASTM B-3/VI 1957 Test Program).

Effects of long term exposures were evaluated by calculating the changes in tensile strength and elongation over the test intervals for both the aged but unexposed panels (controls) and the exposed specimens. Five specimens were machined from each panel to determine tensile properties of the alloys. These changes in properties were determined by comparing the exposed panel data with the nearest elapsed time control values. All mechanical data were determined by the procedure specified in ASTM E-8 Method for Tension Testing of Metallic Materials.

Electrochemical solution potentials of the test alloys agreed well with the published values for these alloys and shows that these alloys should have the same performance characteristics as similar commercially produced materials. These potentials also indicate the protective nature of the anodic aluminum layers on the duplex alloys.

Aging Effects on Mechanical Properties

Mechanical property changes over the 11½-year exposure were determined by taking the average of five determinations. No significant changes in the tensile or yield strengths of the alloys were indicated. Changes in the elongations of the 1050 and 3003 alloys were apparent, but these were not considered significant because of the magnitude of the values. An increase in the elongation of the full hard 5050 alloy is attributed to the effect of natural aging.

Corrosion Effects on Mechanical Properties

Tensile strength of Alclad 2014 in the annealed (o) condition remained constant over a ten-year exposure period at the marine and urban atmospheric locations. Alclad 2014-T6 and 2014-T3 had constant tensile values over the ten-year period at both exposure sites.

Reprinted with permission from *Materials Protection*, June 1963, 30-36, © 1963 National Association of Corrosion Engineers

William H. Ailor, Jr.
Reynolds Metals Company
Richmond, Virginia

Fred M. Reinhart
Naval Civil Engineering Laboratory
Port Hueneme, California

Ten-Year Weathering Data on Aluminum Alloys★

SUMMARY

Describes procedure used to evaluate corrosion resistance of various aluminum alloys to urban and marine atmospheric exposures. Discusses corrosion effects on mechanical properties of the alloys, pit depths that occurred and concludes that atmospheric corrosion rates on aluminum tends to level off after about two years' exposure and remain relatively constant. Corrosion resistances of various tempers of same alloy showed little variation in these tests.

The aluminum-manganese alloy (strain-hardened to the full hard condition, 3003-H18) had constant tensile properties for the ten-year test at both sites.

Both the annealed and the full hard conditions of the aluminum-magnesium alloy (5050-0 and 5050-H18) showed constant tensile strength values for all exposures to ten years. A slight reduction of elongation was evident during the first year for all exposures of the annealed material. The increase in elongation properties for the full hard alloy could be attributed to the aging of the material because the aged control panels showed a comparable increase in elongation.

No appreciable change in either the tensile or elongation properties for the full hard, 99.5 percent aluminum (1050-H18) or for the full hard 98.7 percent minimum aluminum (8099-H18) was apparent during the test period for both environments.

Corrosion Effects: Pitting

Pit depths were measured with a calibrated focus microscope on the four deepest pits on each side of the test specimen. Some typical results are shown in Figures 1 and 2.

The general ranking of the aluminum alloys with regard to pitting showed Alclad 2014 alloys had the best resistance to both marine and urban exposure environments, attributable to the protective action of the cladding. The higher purity aluminum alloys (1050-H18 and 8099-H18) were intermediate among the alloys tested. Pits were deepest in the aluminum-manganese and aluminum-magnesium alloys.

The Norfolk marine atmosphere was more severe from depth of pitting viewpoint than the Washington urban atmosphere. With few ex-

★ Revision of a paper titled "Ten Years Weathering of Aluminum Alloys" presented at the 18th Annual Conference, National Association of Corrosion Engineers, March 19-23, 1962, Kansas City, Missouri.

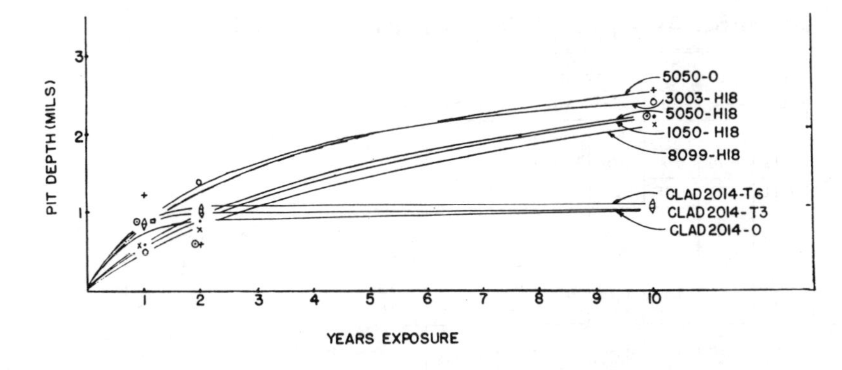

Figure 1—Showing typical aluminum pitting curves derived from atmospheric exposure tests at Washington, D. C. Points represent average of four deepest pits on skyward and groundward sides. (See Table 1 for chemical analyses of various alloys.)

ceptions, pitting depth was greater on the groundward side of the test panels.

Urban Atmosphere

The average depth of the four deepest pits in the urban exposures ranged from about 1.0 mil for the Alclad 2014-0 to a maximum depth of 2.8 mils for the aluminum-manganese and the aluminum-magnesium alloys. The relatively pure alloys (1050 and 8099) had pit depths slightly less than the 5050 alloys.

Maximum pit depth was found on the Alclad 2014 type alloy within a two-year exposure period. All other aluminum alloys showed slightly greater pit depths during the last eight years of exposure. In most cases, these alloys had reached a plateau; little increase in pit depth probably would have been found after a longer period.

The cladding of the Alclad 2014 alloys (five percent of panel thickness on each side) was not penetrated for any exposure period.

Marine Atmosphere

Pit depths of the alloys in the Norfolk marine exposure varied from 1.1 to a maximum of 8.3 mils. With few exceptions, deepest pitting occurred on the groundward side of the panels. For many of the alloys, the rate of pitting decreased rapidly after the first year.

Pit depth in the Alclad 2014 alloy was almost the same after ten years as at the one-year interval. In no instance was the cladding on these alloys penetrated. The aluminum-magnesium type alloys (5050) showed the greatest increase of pit depth with time. The aluminum-manganese alloy (3003) and the higher purity 1050 and 8099 alloys were intermediate between the 5050 alloys and Alclad 2014 alloys in regard to depth of long term pitting.

Conclusions

The good weathering characteristics of aluminum alloys were confirmed by the results of this test program involving exposures to ten years in urban and marine atmospheres. The

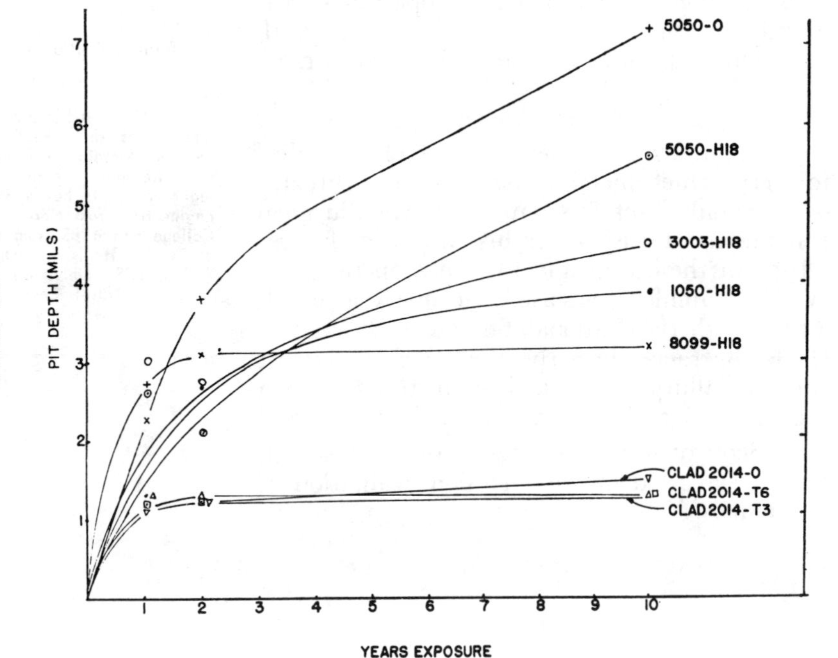

Figure 2—Atmospheric pit depth curves for aluminum alloys exposed for ten years at Naval Air Station, Norfolk, Virginia. Points represent average of four deepest pits on skyward and groundward sides. (See Table 1 for chemical analyses of alloys.)

TABLE 1—Chemical Analysis of Test Alloys

Alloy	Percent							
	Si	Fe	Cu	Mn	Mg	Cr	Ni	Zn
1050-H18	0.14	0.16	0.03	0.01	0.01	....	....	....
8099-H18	0.14	0.70	0.06	0.01	....	....	....	....
3003-H18	0.14	0.29	0.15	1.18	0.00	0.00	....	....
Alclad 2014-T6								
Core	0.98	0.57	4.31	0.86	0.44	0.06	....	....
Cladding	0.72	0.26	0.16	0.43	0.94	0.00	0.00	0.02
Alclad 2014-T3								
Core	0.96	0.53	4.46	0.78	0.40	0.05	....	....
Cladding	0.72	0.18	0.40	0.42	0.95	0.00	0.00	0.02
Alclad 2014-0								
Core	0.91	0.53	4.34	0.84	0.46	0.06	....	....
Cladding	0.71	0.23	0.07	0.44	1.04	0.00	0.00	0.02
5050-0	0.13	0.24	0.02	0.01	1.68	0.03	....	0.05
5050-H18	0.12	0.25	0.02	0.01	1.72	0.07	....	0.05

corrosion in natural environments, measured by pitting depth and changes in mechanical properties, tend to reach a plateau after about two years and are relatively constant thereafter.

The alloy system having magnesium as the major component showed good corrosion resistance in atmospheric exposures.

Bare aluminum-copper alloys (2014 type), one of the higher strength aluminum alloys, generally are not suitable for corrosive environments such as severe marine atmospheres.

Corrosion for most aluminum alloys takes the form of pitting. In this program, a maximum pit depth for all aluminum alloys of about 3 mils was found after ten years in the urban atmosphere and a depth of about 8 mils after ten years in the sea cost environment.

The corrosion resistances of various tempers of the same alloy showed little variation in these tests. Changes in mechanical properties of the aluminum alloys tested over ten years in both environments are not considered significant.

Acknowledgments

The authors are indebted to T. L. Fritzlen, formerly chief metallurgist and now director of Reynolds Metals Company's Metallurgical Engineering Division, for his efforts in the initiation of these tests and for the cooperation of W. H. Mutchler (deceased), formerly a metallurgist with the National Bureau of Standards. H. B. Burrack, formerly Reynolds' assistant chief metallurgist, also assisted in test specimen procurement and preparation. Z. L. Vance and S. B. Scott of Reynolds' Metallurgical Research Division assisted in the corrosion evaluation of test panels.

References

1. R. B. Mears and R. H. Brown. Resistance of Aluminum-Base Alloys to Marine Exposures. *Trans Soc Naval Arch and Marine Eng,* 52, 91-104 (1944).
2. E. H. Dix and R. B. Mears. Resistance of Aluminum-Base Alloys to Atmospheric Exposure. ASTM Symposium on Atmospheric Exposure Tests on Non-Ferrous Metals, February 27, 1946.
3. C. J. Walton, D. O. Sprowls and J. A. Nock, Jr. Resistance of Aluminum Alloys to Weathering. *Corrosion,* 9, 345-358 (1953) October.
4. F. M. Reinhart and G. A. Ellinger. Effect of 20-Year Marine Atmosphere Exposure on Some Aluminum Alloys. ASTM STP No. 175 (1955).
5. C. J. Walton and William King. Resistance of Aluminum-Base Alloys to 20-Year Atmospheric Exposures. ASTM STP No. 175 (1955).

Bibliography

Samuel F. Etris. Report on Corrosion and Deterioration Studies Conducted by ATSM. *Materials Protection,* 1, 10 (1962) April.

E. V. Gibbons. Atmospheric Corrosion Testing of Metals in Canada. *Corrosion,* 17, 318t (1961) June.

Thomas A. Lowe. Aluminum, Aluminized Steel and Galvanized Steel in Severe Coastal Environments. *Corrosion,* 17, 177t (1961) April.

R. E. Brooks. Corrosion Design in the Marine Application of Aluminum Alloy Floodlights. *Corrosion,* 16, 41t (1960), February.

WILLIAM H. AILOR, Jr., has been a member of the Chemical Metallurgy Section, Metallurgical Research Division, Reynolds Metals Company over eight years. He has a BS in chemical engineering from North Carolina State College and a BS from the University of Tampa. He is a member of NACE, ASTM, ECS and American Society of Naval Engineers.

FRED M. REINHART is associated with the Materials Division of the Naval Civil Engineering Laboratory at Port Hueneme, California. Formerly, he had been associated with the National Bureau of Standards, Washington, D. C. He is a member of ASM and ASTM and has been active in NACE, serving as chairman of the NACE Editorial Review Subcommittee for several years.

Corrosion of aluminum alloy balusters in a reinforced concrete bridge

W. J. COPENHAGEN and J. A. COSTELLO, *Council for Scientific and Industrial Research, Cape Town, South Africa*

A bridge built near the sea in the Republic of South Africa was badly damaged by corrosion of the aluminum alloy balusters. Investigation showed the primary cause of corrosion was galvanic coupling of the aluminum balusters and the steel used to reinforce the concrete. Painting the aluminum where it met the concrete reduced the possibility of this type of corrosion.

A REINFORCED concrete highway bridge 180 ft long by 30 ft wide was constructed near the coast in the Republic of South Africa. The bridge was about 100 ft from the sea (Figure 1). The aluminum alloy baluster was a light "H" section.

In 18 months, many of the balusters had corroded at the area of entry into the concrete. A characteristic feature was a collar of foam-like, gelatinous aluminum corrosion products. In many instances, pressure cracks in the concrete were visible (Figure 2). Fragments of concrete from a cracked portion of the bridge were detached, and the corroded aluminum balusters were found to be in contact with the reinforcing steel (Figure 3). The reinforcing steel exposed by removing some of the concrete showed no marked signs of corrosion other than a slight superficial rust film.

Examination of portions of the balusters showed corrosion of the aluminum alloy had been concentrated at the area exposed to the concrete/air interface and took the form of deep pitting to perforation (Figure 4). Analysis of the corrosion product revealed the presence of 2.35% salt as sodium chloride. Analysis of core samples of the concrete taken at random points revealed that sodium chloride was present in concentrations ranging from 0.36 to 1.41% by weight.

Figure 1 — View of bridge over creek 100 ft from the sea.

Figure 2 — Pressure cracking of concrete. In this case, a broken-off section has fallen away.

Reprinted with permission from *Materials Protection and Performance*, September 1970, 31-34,

Figure 3 – Concrete removed from the base of a baluster showing proximity to reinforcing steel.

Figure 4 – Section of corroded baluster showing intensified corrosion at concrete/air interface.

It was necessary to replace all 356 balusters in the bridge, this time ensuring that no contact was made with the reinforcing steel. The embedded portions of the balusters were painted.

Behavior of Aluminum in Concrete

The corrosion behavior of aluminum in contact with building materials such as concrete, mortar, and reinforcing steel has been studied by a number of authors. Wright, Godard, and Jenks[1,2] concluded from a 10-year study that aluminum alloys were not seriously corroded by embedment in concrete, even in the presence of calcium chloride, unless frequent intermittent wetting and drying took place. The alloy involved in the corrosion of the highway bridge was among those tested by these authors. They did state, however, that in conditions of partial exposure (such as the use of aluminum window frames in contact with concrete or mortar), corrosion damage could be greater than in the case of total embedment. Walton, McGeary and Englehart,[3] in studies of the influence of chlorides and galvanic coupling with steel on the corrosion of aluminum in concrete, concluded that corrosion was only superficial when the concrete was cured and exposed under indoor conditions. They indicated that under conditions of continuous or intermittent wetting of the concrete, corrosion appeared to continue, and they predicted that high corrosion rates might occur under such conditions where dissimilar metal contacts and differential aeration conditions in the presence of chlorides were operative.

Wright[4] reported the corrosion of aluminum conduit in concrete in contact with reinforcing steel. The concrete contained calcium chloride additions and the resulting corrosion of the aluminum was severe. Monfore and Ost[5] reported similar failures and high corrosion rates of aluminum conduit, either coupled or uncoupled with steel, in concrete containing calcium chloride additions. Wright and Jenks[6] concluded from laboratory experiments that when aluminum was coupled to steel in concrete containing calcium chloride, the corrosion rate of the aluminum was proportional to the concentration of $CaCl_2$ and to the steel-aluminum area ratio, and inversely proportional to the distance between the coupled dissimilar metals.

Experimental Work

The aim of the investigation was two-fold:

1. An attempt to reproduce, under laboratory conditions, the type of corrosion found on the bridge balustrade.
2. To express the results quantitatively.

Laboratory Reproduction of Bridge Corrosion

A cement mortar mix was used to cast blocks containing metal specimens. A cement/sand ratio of 1:9 by weight was used, with a water/dry mix ratio of 0.16:1, giving a water/cement ratio of 1:6. The aluminum alloy used was of the same composition as those in the bridge balustrade. Four experimental conditions were investigated:

1. Aluminum alloy in mortar made with distilled water.

2. Aluminum alloy coupled with mild steel in mortar made with distilled water.

3. Aluminum alloy in mortar made with sea water.

4. Aluminum alloy coupled with mild steel in mortar made with sea water.

Aluminum strips 3 1/2 inches long, 1/2 inch wide, and 1/8 inch thick were cut from stock of the appropriate alloy. For galvanic coupling with steel, mild steel strips measuring 5 1/2 inches by 3/4 inch by .04 inch were cut from a cold rolled sheet and bolted with steel bolts to the aluminum strips. The coupled and uncoupled strips were embedded in the appropriate mortar mix in 4-inch cube molds from which the mortar blocks were released after two days (Figure 5). The blocks were allowed to stand in the laboratory under ambient conditions for another five days, after which the blocks were broken open and the aluminum samples examined. All four displayed a more or less uniform superficial etching over the entire surface which had been exposed to the mortar. The strip coupled to steel in mortar made with sea water also displayed a zone of intensified pitting attack in a band approximately 1/5 inch wide in the area having been exposed to the mortar/air interface (Figure 6). From these observations, it appeared that aluminum, when coupled to steel in mortar containing chlorides, showed an intensified zone of attack at the mortar/air interface.

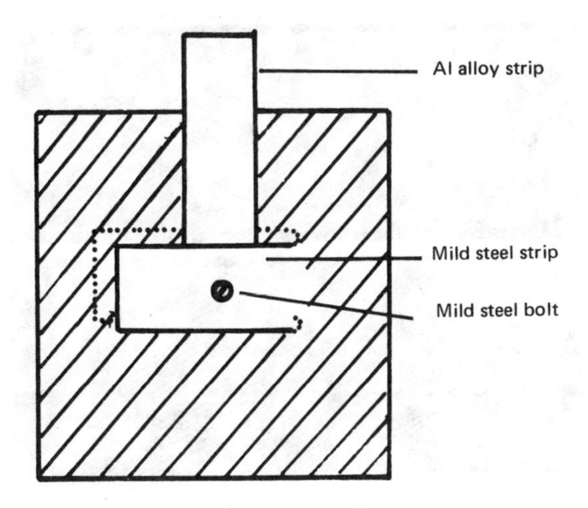

Figure 5 — Embedment of Al alloy strip connected to mild steel strip in mortar block.

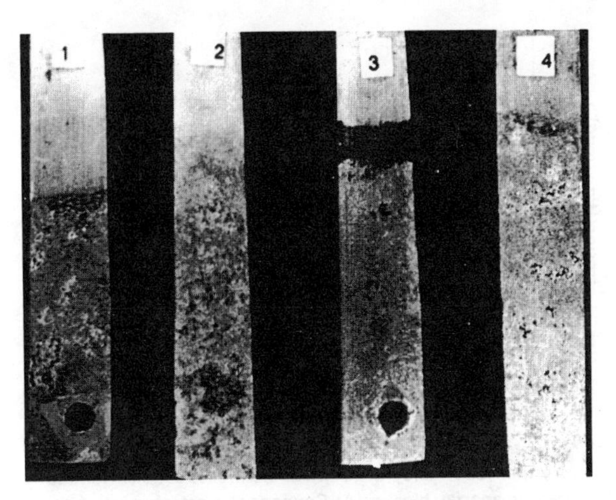

Figure 6 — Specimens used in laboratory reproduction of type of corrosion found on bridge. (1) Al alloy coupled to steel in mortar made with distilled water. (2) Al alloy uncoupled in mortar made with distilled water. (3) Al alloy coupled to steel in mortar made with sea water; shows intensified corrosion at mortar/air interface. (4) Al alloy uncoupled to steel in mortar made with sea water.

Weight Loss Experiments

The object of the weight loss experiments was to express the "mortar line" corrosion demonstrated in earlier experiments quantitatively, and to assess the influence of NaCl concentration on this type of corrosion. In these experiments, an aluminum bi-electrode assembly was used. Both parts of the electrode could be detached and weighed separately, and when assembled, could be short circuited with a common mild steel cathode. Individual electrode components were cut from stock electrode. When all other parts of the electrode were masked with wax, these components had a surface area of 3.5 square inches, making a total exposed area for the bi-electrode assembly of about 7 square inches. Before assembling the electrode and masking with wax, uniform electrode surfaces were prepared by abrasion with successively finer grades of metallurgical grinding papers. The electrodes were degreased with acetone and weighed.

These aluminum bi-electrodes were cast into 4-inch mortar blocks, and were connected externally to mild steel cathodes embedded in the same blocks. The anode-cathode area ratio was 1:2.5. The mix used was a 25% lime in sand (by weight) mix containing the following concentrations of NaCl: 0, 0.1, 1.0, and 3.0% by weight of dry mix. The dry mix-water ratio was 1:0.22 by weight. Triplicate experiments were performed for each NaCl concentration and 12 blocks in all were cast. These blocks were broken open after 13 days in the laboratory and the aluminum alloy electrodes removed. The wax was stripped from the masked portions of the electrodes. The corrosion product was removed from the electrodes by a five-minute immersion in concentrated A.R. nitric acid, followed by drying with acetone and re-weighing. From the weight loss figures, weight losses per unit area and corrosion rates in mdd (mg/sq dec/day) were calculated (Table 1).

The results of Table 1 are plotted graphically in Figure 7. The lines are drawn through the arithmetical mean of the points for each salt concentration. The following may be concluded from these results:

1. NaCl increases corrosion of the aluminum alloy coupled to mild steel. This effect is noticeable both for aluminum at the mortar/air interface and for aluminum immersed in the bulk of the mortar.

2. Of the three NaCl concentrations used, 1% NaCl by weight of dry mix exerts the maximum accelerating effect.

3. Corrosion rates of the aluminum alloy coupled to mild steel in the presence of NaCl are considerably greater at the mortar/air interface than those in the bulk of the mortar. This effect was particularly noticeable at a NaCl concentration of 1%.

During exposure of the aluminum alloy bi-electrodes in the blocks, active corrosion was visible at the mortar/air

TABLE 1 — Corrosion Rates of Components of Aluminum Alloy Bi-Electrode Assemblies in Lime Mortar Containing Different NaCl Concentrations

Percent NaCl in dry mix (by weight)	Sample No.[(1)]	Corrosion Rate mdd
0	A1	10.00
	A2	5.38
	B1	6.92
	B2	4.62
	C1	11.55
	C2	6.92
0.1	D1	136.1
	D2	83.1
	E1	71.6
	E2	17.7
	F1	144.0
	F2	7.68
1.0	G1	212.0
	G2	52.30
	H1	253.9
	H2	71.60
	J1	195.50
	J2	38.50
3.0	K1	146.80
	K2	38.21
	L1	102.50
	L2	50.34
	M1	112.90
	M2	37.60

(1) 1 and 2 refer to components of bi-electrode at mortar/air interface and in bulk, respectively.

Source: *Materials Protection and Performance*, September 1970, 31-34

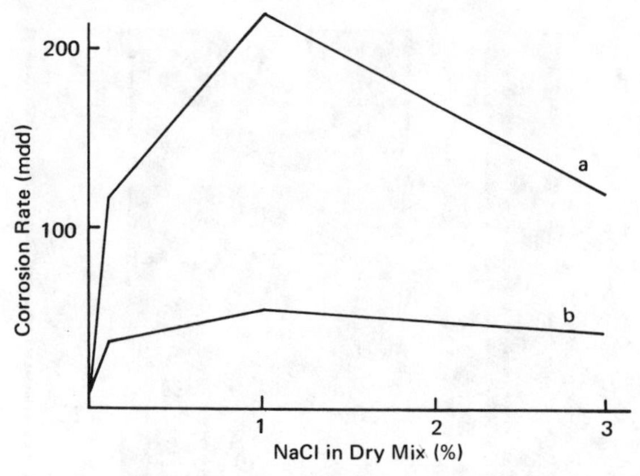

Figure 7 — Plots of corrosion rate vs NaCl concentration in dry mix for Al alloy bi-electrode assemblies coupled to mild steel in mortar blocks. (A) Coupons at mortar/air interface. (B) Coupons in bulk.

interface, and took the form of a bubbling white fluid which became gelatinous on drying. The corrosion pattern at the mortar/air interface was visible after cleaning and took the form of intense pitting corrosion (Figure 8).

Discussion

The phenomenon of intensified pitting at the mortar/air interface on aluminum when coupled to mild steel can be reproduced in the laboratory. The mortar mixes used were either cement or lime mortars. The most aggressive factors in mortar were high pH, porosity, and the chloride ion content. Calcium hydroxide was the main factor which affects pH in mortar, and its solubility was extremely low (less than two grams per liter in water). It was shown that the type of intensified corrosion at the mortar/air interface only takes place in the presence of chloride ion, and when the aluminum alloy was coupled electrically with steel.

The pH situation of the corroding aluminum when coupled to steel in mortar blocks showed an interesting pattern. On breaking open the blocks and treating the mortar adjacent to the corroding aluminum with a pH indicator, it was found that the mortar immediately adjacent to the zone of intensified corrosion at the mortar/air interface had a low pH—probably 4 or lower. This type of pH change in the vicinity of pits on corroding aluminum has been described by Edeleanu and Evans.[8]

Preferential corrosion of a metal at the zone emerging from the waterline medium was a well known phenomenon in aqueous corrosion, and Peers[9] treated this as a type of crevice corrosion which takes place at the meniscus. Peers attributed this to concentration changes at the anode surface accompanying the passage of current, which tended to reduce the anodic overpotential. Hersch[10] has emphasized the importance of areas immediately adjacent to the meniscus (where dissolution is taking place), as catchment areas for oxygen depolarization of local cathodes. Pryor and Keir[7] have shown in studies of the aluminum-steel couple that at high pH values, local cell action on the aluminum makes an appreciable contribution to the total weight loss of aluminum coupled to steel.

Conclusions

An unusual type of necking corrosion of aluminum in concrete takes place at the zone of metal exposed to the concrete/air interface. This type of corrosion appears to take place only with aluminum coupled to the reinforcing steel and in the presence of sea salts. Protective coatings on the aluminum at the concrete/air interface zone would reduce the possibility of this type of corrosion taking place. Galvanic coupling with steel by accident or design should, of course, be avoided at all times.

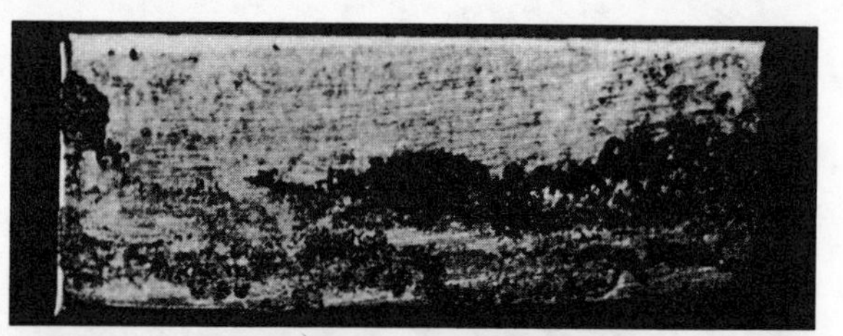

Figure 8 — Al alloy coupons for bi-electrode assembly after coupling with steel in mortar made with 1% NaCl, showing intensified corrosion on upper coupon which was exposed at mortar/air interface.

References

1. T. E. Wright, H. P. Godard, and I. H. Jenks. *Engineering Journal,* 37, 1250 (1954).
2. I. H. Jenks, T. E. Wright, and H. P. Godard. *Engineering Journal,* **45**, 45 (1962).
3. C. J. Walton, F. L. McGeary, and E. T. Englehart. *Corrosion,* **13**, 807 (1957).
4. T. E. Wright. *Engineering Journal,* 38 (1955).
5. G. E. Monfore and B. Ost. *J. Portland Cement Association,* 7, 10 (1965).
6. T. E. Wright and I. H. Jenks. *Journal of the Structural Division, Proc. Amer. Soc. Civ. Engrs.,* 89, ST5, 117 (1963).
7. M. J. Pryor and D. S. Keir. *J. Electrochem. Soc.,* **105**, 629 (1958).
8. C. Edeleanu and U. R. Evans. *Trans. Faraday Soc.,* **47**, No. 346, 1121 (1951).
9. A. M. Peers. *Trans. Faraday Soc.,* **51**, 1748 (1955).
10. P. Hersch. *Nature,* Lond., **160**, 1107 (1957).

W. J. COPENHAGEN is retired. He has an MS from the University of Cape Town and is active in various phases of corrosion research.

J. A. COSTELLO is a research officer, Council for Scientific and Industrial Research, Cape Town, South Africa. He has a BS from the University of Manchester's Institute for Science and Technology.

Corrosion and Heat Resistance of Aluminum-Coated Steel★

R. J. Schmitt
United States Steel Corporation
Monroeville, Pa.
and
Jane H. Rigo
United States Steel Corporation
Cleveland, Ohio

Discusses performance of aluminum and aluminum silicon coated steel in various atmospheres and in water. Tests compare the corrosion resistance of aluminum-coated steel and galvanized steel. Also discusses the hot-dip method for coating steel with aluminum. Gives physical properties and limitations of the coating.

TRENDS toward more durable, lighter, economical, and attractive consumer products have led to the development of many new materials in recent years. In today's competitive markets, products capable of meeting society's stringent demands are being created through the wedding of older, established materials. One example is aluminum-coated steel, which combines the good corrosion-resistance properties of aluminum with the superior mechanical properties of steel. Aluminum-coated steel is being used in items such as home appliances, furnaces, automobile mufflers, fencing, farm equipment, farm buildings, industrial buildings, signs, and insulation jacketing.

Aluminum coatings can be applied to steel by several methods, such as hot dipping, spraying, cementation (calorizing), cladding, electroplating, and vapor phase reaction. This article discusses hot-dipped aluminum-coated steel products and, in particular, continuously hot-dipped aluminum-coated steel sheets and wire.

Aluminum-Coated Steel

Aluminum-coated steel sheets are designated Type 1—Regular or Type 2, according to coating weight. Type 1 sheets are produced with an aluminum-silicon alloy coating; Type 2 sheets are produced with either a commercially pure aluminum coating or an aluminum-silicon alloy coating.*

Type 1—Regular aluminum-coated steel sheet is produced with an aluminum-silicon alloy coating weight of about 0.5 oz/sq ft of sheet. The sheet is coated on both sides, which amounts to about 0.25 oz/sq ft of surface or about a 1-mil-thick coating on each sheet surface. This product is designed for applications requiring heat resistance. Type 2 aluminum-coated steel sheet is produced with an aluminum-silicon alloy coating weight of about 1 oz/sq ft or about a 2-mil thick coating on each sheet surface, and is designed to resist atmospheric corrosion.

Hot-Dipping Process

Aluminum-silicon coatings on steel obtained by hot-dipping are composed essentially of two layers. These are a layer of aluminum-iron-silicon alloy next to the steel base and an outer layer of aluminum-silicon alloy having approximately the same composition as the molten aluminum bath. The amount of aluminum-iron-silicon alloy formed has an important influence on ductility, adhesion, uniformity, smoothness, and appearance. Thickness and hardness of the alloy layer are controlled by adjusting the silicon content of the molten aluminum bath to provide optimum coating adherence and formability.

Aluminum-coated steel wires with minimum coating weights of 0.4 oz/sq ft of surface or less are produced by passage through a molten bath of pure aluminum or silicon-alloyed aluminum. Depending upon the manufacturer, steel wires may or may not be chemically fluxed after cleaning and before dipping in the molten coating bath. At US Steel, the process for aluminum-coating steel wire is a fluxless aluminum-coating method using a vibratory technique on precleaned wire.[1]

Table 1 shows that minimum coating weights will vary from 0.25 to 0.40 oz/sq ft. When coating steel wire, coating weights are increased by increasing the speed at which the wire travels through the molten coating bath. Silicon additions of 7% or less to the bath will reduce the weight only slightly.

In wire processing, operating temperatures of the coating bath range from 1150 to 1300 F (621 to 704 C), as compared with an operating range of 850 to 900 F (454 to 482 C) for a galvanizing bath. The higher the coating temperature, the greater will be the reduction in physical properties. Thus, to produce an aluminum-coated wire with physical properties comparable to a galvanized wire, certain precautions must be taken. These are to coat steels of

TABLE 1—Minimum Coating Weight for Aluminum-Coated Steel Wire.

ITEM	NOMINAL DIAMETER inches	MINIMUM COATING WEIGHT oz/sq ft
Regular Manufacture	0.080 or greater	0.25
	0.076 to 0.0625	0.20
Woven Wire Fence, Barbed Wire	--	0.25
Chain Link Fabric (ASTM A491)	0.192 to 0.1483	Class I 0.32, Class II 0.40
	0.1205	Class I 0.28, Class II 0.35
Steel Wire Strand (ASTM A474)	0.062, 0.065	0.25
	0.080	0.26
	0.093	0.28
	0.104, 0.109	0.30
	0.120	0.32
	0.145, 0.165	0.34
Steel Core Wire (ASTM B341)	0.0500 to 0.0599	0.23
	0.0600 to 0.0749	0.25
	0.0750 to 0.0899	0.26
	0.0900 to 0.1039	0.28
	0.01040 to 0.1199	0.30
	0.1200 to 0.1399	0.32
	0.1400 to 0.1799	0.34
	0.1800 to 0.1900	0.38

★Condensation of the paper, "Aluminum-Coated Steel—A Product with Corrosion Resistance and Heat Resistance," presented at the 21st NACE Annual Conference, March, 1965, in St. Louis. Photocopies of the complete original paper, with 13 figures and 14 tables, are available from: Photocopies, Materials Protection, 980 M & M Building, Houston, Texas, 77002. Price: $9.00 per copy.

* Aluminum-coated steel sheets produced by US Steel Corp. are made with aluminum-silcon alloy coatings. Silicon (5 to 10%) is added to the molten bath to obtain maximum coating adherence.

higher carbon content and/or to use bright process wire of greater total cold reduction than would be used for a galvanizing process. A proper combination of steel chemistry, cold work, bath temperature, and coating speed can yield tensile strengths exceeding 200,000 psi in an aluminum-coated steel wire. The range in physical properties for aluminum-coated steel wire products is illustrated in Tables 2 and 3.

TABLE 2—Tensile Strength Range for Aluminum-Coated Steel Fencing and Barbed Wire.

PRODUCT CLASSIFICATION	NOMINAL DIAMETER, inch	TENSILE STRENGTH, psi
Soft Fence Wire, Barbed Wire	0.177 to 0.0625	60,000 to 75,000
Hard Fence Wire	0.177 to 0.067	75,000 to 100,000
Chain Link Fabric	0.192 to 0.1205	80,000 min
Ranger, Barbed Wire	0.086	90,000 to 115,000

At the present, a variety of aluminum-coated steel wire and wire products are available on the market: barbed wire, chain link fabric, spring coil tension wire, woven farm field fence, core wire, telephone and telegraph wire, and wire strand.

Forming and Joining Aluminum-Coated Steel Products

Aluminum-coated steel sheets will withstand moderate forming and drawing operations. In more severe drawing operations, particularly those involving considerable compression, Type 2 does not perform as well as galvanized steel. Because the coating is twice as thick, the adherence of Type 2 is not as good as that of Type 1 coatings. Type 1 and Type 2 coated steel sheets will withstand a 180-degree cold bend over two sheet thicknesses and corrugating (circular, V-crimp, and rectangular formations) without peeling or flaking the coating. Slight crazing (hairline cracks) may occur in the coating on the Type 2 in V-crimps, and rectangular corrugations may occur if the bends and amount of galling of the coating by forming dies are severe.

Mechanical Joining, Welding

Joining of sheets may be accomplished by metal fasteners, interlocking, and entrapment, or with adhesives. Aluminum-coated steel sheets can be joined by electric-resistance spot or seam welding practices. They can also be metal-arc welded, butt-flash welded, oxyacetylene welded, and brazed. Electric-resistance spot welding will result in some expulsion of the aluminum-silicon coating contacted by the welding electrodes. However, proper spot welding will result in alloying of the coating in the immediate area where welding electrodes contact. Data on corrosion performance of aluminum-iron-silicon and aluminum-silicon alloy coatings are given in Table 4. Although aluminum-iron-silicon alloy is slightly less resistant than the aluminum-silicon alloy, weight losses for both alloys are extremely low; thus, both alloys can be expected to offer excellent corrosion protection to steel.

Corrosion Mechanism

Protection of Steel by Aluminum Coatings

The following is a partial galvanic series of metals in sea water: magnesium, zinc, aluminum, carbon steel, cast iron, copper, nickel, silver, and titanium. Upon joining any two of these materials, the metal nearer the front of this series will be anodic and will experience accelerated corrosion, while the one nearer the rear will be cathodic and will receive some galvanic protection. Therefore, from this series aluminum would be expected to behave similar to zinc and provide electrochemical protection to steel. However, this is usually the exception rather than the rule when an aluminum-iron couple is exposed to most media, particularly atmospheric environments. In the atmosphere, a protective oxide film forms on the surface. This film tends to passivate the aluminum, thus restricting electrochemical protection of base iron at cracks or damaged areas in the coating. If the film is destroyed, the aluminum will provide sacrifical protection to steel, with extent of protection depending on the area of steel exposed and the environment. Because chlorides tend to destroy the film, aluminum becomes effective as a sacrificial coating in providing protection to steel in marine environments.

Aluminum offers little or no electrochemical protection to steel except in chloride media, and its value as a protective coating is its inherent corrosion resistance. In general, the protective value of aluminum coatings on steel is a function of the coating thickness.

A theory introduced by Evans is that pores in aluminum-coated steel become filled with rust shortly after exposure, but in time the oxide film breaks down because of anodic action of the aluminum. This causes a reduction of the rust already formed (possibly to magnetite) and prevents further attack.[2] However, this mechanism has been disputed by Higgins.[3] Whatever the mechanism, all observers agree that corrosion of the steel at discontinuities in aluminum coatings ceases soon after exposure to the atmosphere.

Atmospheric Corrosion

Aluminum-Coated Steel Sheet

Extensive corrosion tests have been conducted to determine the relative performance of aluminum-coated steel, aluminum-silicon coated steel, and galvanized steel sheets. These rack tests, which started as far back as 1933, have included exposure to rural, industrial, and marine atmospheres. More recently, service tests have been conducted on industrial roofs and farm buildings. It should be recognized that sufficient time has not elapsed since introduction of aluminum-coated steel to obtain quantitative data, such as years to first perforation. Therefore, several statements in this article on atmospheric performance of aluminum-coated steel are based on visual observations to date.

TABLE 4—Corrosion Performance of Aluminum-Iron-Silicon Alloy vs Aluminum-Silicon Alloy in Various Atmospheres.

MATERIAL	TIME OF EXPOSURE YEARS	LOSS OF WEIGHT, GRAMS PER YEAR*		
		INDUSTRIAL ATMOS.	SEMI-RURAL ATMOS.	MARINE ATMOS.
Aluminum-Iron-Silicon Alloy	8	0.118	0.079	0.254
Aluminum-Silicon Alloy	2	0.072	0.031	0.016
Carbon Steel	7.5	6.6	4.1	5.6

*Data obtained from 4-inch by 6-inch specimens.

Field Tests

The corrosion resistance of aluminum-coated steel was first observed with samples of 30-gauge steel sheet clad with thin aluminum sheet by roll bonding, which were exposed at South Bend, Pa., and at Kearny, N. J. in 1933. Thickness of the coating was 0.9 mil per surface or approximately 0.4 oz/sq ft of sheet. After three years, these specimens were covered with a dark red stain, obviously iron oxide that leached out from the steel base through cracks in the coating. To date, however, and

TABLE 3—Minimum Mechanical Properties of Several Grades of 7-Strand Aluminum-Coated Steel Wire.

WIRE DIAMETER in.		COMMON		SIEMENS MARTIN		HIGH STRENGTH		UTILITIES		EXTRA HIGH STRENGTH	
STRAND	WIRE	TENSILE STRENGTH ksi	ELONGATION IN 10 in. %	TENSILE STRENGTH ksi	ELONGATION IN 10 in. %	TENSILE STRENGTH ksi	ELONGATION IN 10 in. %	TENSILE STRENGTH ksi	ELONGATION IN 10 in. %	TENSILE STRENGTH ksi	ELONGATION IN 10 in. %
1/2	0.165	52	10	85	8	132	5	176	4	189	4
7/16	0.145	52	10	85	8	132	5	164	4	189	4
3/8	0.120	57	10	92	8	144	5	153	4	204	4
5/16	0.109	--	--	--	-	---	-	97	10	---	-
	0.104	57	10	95	8	144	5	---	-	---	-

at inspections during the subsequent 33 years, no further corrosion of the steel has been noted.

For comparison, a 1-mil-thick galvanized specimen exposed at the same time showed first rust after 3½ years at Kearny and after 7½ years at South Bend. Complete rusting of the skyward side of a galvanized steel sample was observed after six years at Kearny and after 12½ years at South Bend. Figure 1 shows the appearance of the groundward side of galvanized samples (1.25 and 2.5 oz/sq ft) and aluminum-clad steel (1.0 oz/sq ft) after 31 years exposure to the semirural atmosphere of South Bend.

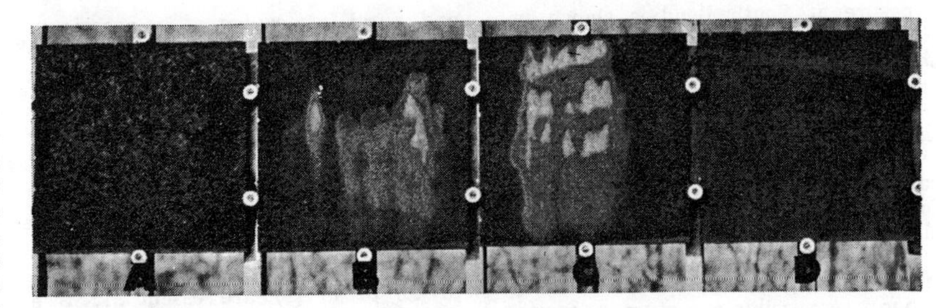

Figure 1—Groundward side of aluminum-clad and galvanized steel samples after 31 years exposure in semirural atmosphere (South Bend, Pa.) Panel A is galvanized steel (1.25 oz/sq ft); B is galvanized steel (1.50 oz/sq ft); C is galvanized steel (2.50 oz/sq ft); and D is aluminum-clad steel (1.0 oz/sq ft).

Comparative Data

Results of 20-year atmospheric tests conducted by the American Society for Testing and Materials (ASTM) shown in Table 5 permit a comparison of the corrosion resistance of aluminum and zinc.[4,5] In rural atmospheres, zinc corroded 7 to 16 times faster than aluminum. In industrial atmospheres, this rate was about six, and in marine atmospheres, it ranged from 2.4 to 6.4. Thus, for equivalent coating thickness, an aluminum coating can be expected to remain intact several times longer than a zinc coating.

Results of corrosion rack tests conducted with 4 by 6-inch specimens of aluminum-silicon alloy coated sheet and galvanized sheet exposed in various atmospheres for periods to four years are shown in Table 6. The data are in agreement with ASTM data discussed earlier—that aluminum-silicon alloy coatings are five to six times more corrosion resistant than galvanized coatings of equivalent thickness when exposed in industrial, semirural, and marine atmospheres.

On the basis of available data, it is conservatively estimated that aluminum coatings have a life expectancy three to five times that of galvanized coatings of equivalent thickness.

Corrugated Aluminum-Coated Steel vs Galvanized Steel

Results obtained with the 4 by 6-inch specimens have been confirmed with 28 by 30-inch corrugated aluminum-coated and galvanized steel sheets exposed for periods up to 16 years in industrial, semirural, and marine atmospheres. For example, Figure 2 shows aluminum-coated steel sheet exposed for five years in the industrial atmosphere at Newark, N. J., to be in excellent condition, whereas the galvanized steel sheet is rusted at the edges and center of the sheet. After 16 years exposure in the semirural atmosphere at South Bend, the aluminum-coated steel sheet, although discolored by dirt, was in excellent condition; however, the galvanized steel sheet was rusted extensively. In another test, both aluminum-coated sheet and galvanized steel sheet exposed for five years to the marine atmosphere at Kure Beach, N. C., were in good condition.

The skyward-exposed surface of a 1.25 oz/sq ft galvanized sheet developed small yellow spots after seven years exposure at Kure Beach and showed appreciable rusting after 15 years. The zinc coating is still furnishing protection to the groundward-exposed surface. Although the skyward-exposed surface of a 0.9 oz/sq ft aluminum-coated steel sheet was unrusted after 15 years, perforation from the groundward side had occurred. Thus, it is apparent that aluminum-coated steel will outperform galvanized steel when rain removes sea water spray at the exposure site. When the surface cannot be washed by rain, galvanized steel probably will be superior in marine atmospheres.

Tests with Cut or Sheared Edges

There is a significant difference in the behavior of cut or sheared edges of galvanized steel and aluminum-coated steel. The edge appearance of a 1.7 oz/sq ft galvanized steel sample and an 0.9 oz/sq ft aluminum-silicon alloy coated steel sample after 17 years

TABLE 5—Summary of ASTM Results on Atmospheric Corrosion of Zinc and Aluminum Alloys.

TEST SITE	TYPE OF ATMOSPHERE	RATE OF PENETRATION,* MILS PER YEAR		RATIO OF RATES, Zn/Al
		ZINC**	ALUMINUM***	
State College, Pa.	Rural	0.041	0.0025	16.4
Phoenix, Ariz.	Rural	0.007	0.001	7.0
Altoona, Pa.	Industrial	0.24	0.040	6.0
New York, N. Y.	Industrial	0.22	0.038	5.8
Sandy Hook, N. J.	Marine	0.064	0.010	6.4
Key West, Fla.	Marine	0.022	0.0035	6.3
LaJolla, Calif.	Marine	0.069	0.029	2.4

* Rate of penetration based on weight losses in 10- and 20- year exposure tests.
** Average results for three different zinc alloys (Prime Western, High Grade, and Special High Grade).
*** Average results for four different aluminum alloys (1100-H14, 3003-H14, 6051-T4, 2017-T3.

TABLE 6—Corrosion Performance of Aluminum-Silicon Alloy Coatings vs Galvanized Coatings in Various Atmospheres.

TYPE OF ATMOSPHERE	TEST SITE	RATE OF PENETRATION,* MILS PER YEAR		
		GALVANIZED COATING	ALUMINUM-SILICON ALLOY COATING	RATIO OF RATES, Zn/Al
Industrial	Kearny, N. J.	0.159	0.027	5.9
Semirural	South Bend, Pa.	0.074	0.016	4.6
Marine	Kure Beach, N. C. (800-ft lot)	0.061	0.010	6.1

* Rate of penetration based on average weight losses of coating metal from 4 by 6-inch specimens exposed for two and four years

Figure 2—Corrugated galvanized and aluminum-coated steel sheets after five years exposure in industrial atmosphere (Newark, N. J.). Panel A is galvanized steel (1.3 oz/sq ft) and Panel B is aluminum-coated steel (Type 2, 1.0 oz/sq ft).

exposure at South Bend is shown in Figure 3. With galvanized coatings, the zinc is anodic to steel and galvanically protects the steel at the cut edge. In time, however, this effect results in the zinc being progressively corroded away from the edge, exposing the steel surface to rusting and ultimate perforation. As discussed previously, the electrochemical behavior of aluminum is such that it does not cathodically protect the cut edges in most atmospheres. A possible exception is marine atmosphere.[9] This lack of protection is an advantage because the coating remains intact at the edges and continues to protect the steel surface. Although some corrosion at the cut edge does occur, it practically stops after a time. Little or no drainage stain is

Figure 3—Appearance of the groundward surface of galvanized and aluminum-coated steel samples after 17 years exposure in a semirural atmosphere (South Bend, Pa.). Panel A is galvanized steel (1.7 oz/sq ft) and Panel B is aluminum-coated steel (Type 2, 0.9 oz/sq ft).

caused by this minor amount of edge corrosion.

Effect of Coating Thickness

Other tests (exposures to 16 years) indicate that corrosion resistance of aluminum-coated steel depends on the thickness of the coating rather than on whether the coating is commercially pure aluminum or aluminum-silicon alloy. Silicon contents from 0.1 to 14% in the aluminum coating had no visual detrimental effect on the atmospheric corrosion resistance. Tests at Kearny and at South Bend show the coatings to be equally corrosion resistant. Aluminum producers have confirmed that aluminum containing silicon has the same good atmospheric-corrosion resistance as commercially pure aluminum.[7] Aluminum-silicon coatings are greyer than pure aluminum coatings as produced and will tend to develop a darker discoloration after prolonged exposure to uncontaminated atmospheres.

Effect of Discontinuities

Atmospheric exposure tests indicate that the development of iron-rust stains on aluminum-coated sheets in industrial and rural atmospheres is associated with discontinuities in the coatings. Coatings thinner than about 0.85 mil (0.4 oz/sq ft) are susceptible to staining after excessive temper rolling or after severe roller leveling, corrugating, or bending. Staining tendency decreases with increased coating thickness, and Type 2 aluminum coatings which are 1.5 to 2.0 mils thick (0.75 to 1.0 oz/sq ft) are essentially free from staining after normal amounts of temper rolling and roller leveling.

In any event, these stains have no effect on serviceability of aluminum-coated sheets in weather-exposure applications. In industrial atmospheres, stains may become obliterated by deposits of air-borne dirt. In rural atmospheres, stains tend to be washed away by rain. In both atmospheres, corrosion product formed on the exposed base metal prevents further corrosion and staining of the coating. In marine atmospheres, iron-rust stains do not occur, although a uniform thin white corrosion product usually develops on the aluminum coating.

Emissivity Valves

Emissivity measurements made on various aluminum-coated steel sheets, galvanized steel sheets, and aluminum before and after one, two, and four years exposure to the semirural atmosphere at South Bend are shown in Table 7. Before exposure, all sheets had about the same emissivity values. After four years, emissivity of galvanized steel increased to 95%, approaching that of a black body which would absorb all thermal radiation. In contrast, the emissivity of aluminum-coated steels and aluminum ranged from about 40 to 50% in the same exposure period. Since these values indicate that 50 to 60% of the thermal radiation would be reflected, the interior of buildings made of these products would be cooler during summer months than that of buildings made of galvanized steel.

The overall good corrosion performance of aluminum-coated steel in the atmosphere has resulted in aluminum-coated steel roofing and siding being used in factories, warehouses, power plants, and agricultural buildings such as barns, garages, machinery sheds, and silo roofs.

Aluminum-Coated Steel Wire

In a current test the relative corrosion resistance of bare, zinc-coated, and aluminum-coated steel wires is being determined visually in terms of percentile loss in breaking strength. Table 8 records the percentile loss in breaking strength of 0.1483-inch diameter wires observed after eight years exposure in industrial, marine, and semirural atmospheres. According to these results, the bare steel wires have lost a significant portion of their original breaking strength. The percentile loss in breaking strength for the metallic-coated wires is negligible. Slight gain in breaking strength is attributed to aging. It should be noted that aluminum coatings in these tests are thin compared with presently established minimums for aluminum coatings for steel wire. Also, the coating thickness for these aluminum coatings is comparable with that of light-weight zinc-coated wire specimens. The metallic-coated wires, particularly those with the light coating weights, have developed some visible signs of corrosion. This is especially true in the case of the 0.46 oz/sq ft zinc-coated wire, which was 100% rusted at the end of three and five years exposure in marine and industrial atmospheres, respectively.

Barbed Wires, Welded Fabric, Woven Fence

Another current test involving barbed wire, welded fabric, and woven field fence demonstrates the superiority of aluminum coatings over conventional

TABLE 7—Emissivity Values for Aluminum-Coated Steel, Galvanized Steel, and Aluminum Specimens Before Exposure and After 1, 2, and 4 Years Exposure in a Semirural Atmosphere (South Bend, Pa.)

MATERIAL	FINISH	EMISSIVITY, %*			
		BEFORE EXPOSURE	AFTER 1 YEAR	AFTER 2 YEARS	AFTER 4 YEARS
Aluminum-Silicon Alloy Coated Steel (0.5 oz/sq ft)	Not temper rolled	9.5	16.1	22.5	45.2
Aluminum-Silicon Alloy Coated Steel (1.0 oz/sq ft)	Not temper rolled	9.9	17.6	22.6	49.2
Aluminum-Coated Steel (0.5 oz/sq ft)	Temper rolled	14.4	11.6	20.1	43.1
Galvanized Steel	Not temper rolled	5.8	37.9	77.9	95.0
Aluminum (3003)	Rolled plain	4.6	15.8	10.7	40.4
Aluminum (3003)	Rolled and embossed	4.8	15.0	10.2	43.8

* The total emissivity is expressed as a percentage of black-body radiation, 100-Reflectivity = Emissivity.

TABLE 8—Loss in Breaking Strength of Aluminum-Coated and Galvanized 0.1483-Inch Diameter Wires After Eight Years Exposure in Various Atmospheres.

	COATING WEIGHT, oz/sq ft	CALCULATED COATING THICKNESS, mils	LOSS IN BREAKING STRENGTH, % ATMOSPHERE MARINE (800-FOOT LOT KURE BEACH, N. C.	INDUSTRIAL (CLEVELAND, OHIO)	SEMIRURAL (SOUTH BEND, PA.)
Bare Steel 1008	--	--	34	21.4	17.1
Bare Steel 1015 Si	--	--	37.6	23.8	25.9
Aluminum Sample A* 1008	0.20	0.88	1.9	+2.3	1.1
Aluminum Sample B* 1008	0.17	0.75	2.95	+1.5	2.0
Galvanized Sample A 1012 Si	0.46	0.78	+1.0	7.25	5.5
Galvanized Sample B	1.15	1.95	+2.7	+5.6	3.0

* Aluminum Sample A - 5.1% Si, 1.8% Fe;
Aluminum Sample B - 5.3% Si, 1.5% Fe.

zinc coatings in industrial atmospheres.[1] After 10 years exposure, aluminum-coated steel samples are intact, showing only heavy soil and rust stain. No attack on the base metal has been observed on chemically cleaned sections.

In contrast, galvanized steel products concurrently exposed with the aluminum-coated steel products are heavily corroded. As shown in Table 9, the time to 100% heavy red rust for the galvanized products ranged from 1.4 to 3.5 years for a coating weight variation of 0.25 oz/sq ft for barbed wire to 0.37 oz/sq ft for field fence.

TABLE 9—Time to 100% Red Rust in Industrial Atmosphere (Cleveland, Ohio).

PRODUCT	GALVANIZED ORIGINAL COATING WEIGHT	YRS TO 100% RED RUST	ALUMINUM-COATED ORIGINAL COATING WEIGHT	YRS TO 100% RED RUST
Barbed Wire	0.25	1.4	0.28	OK at 10 yrs
Field Fence	0.37	3.5	0.22	OK at 10 yrs
Fabric	0.29	2.5	0.24	OK at 10 yrs

Embedment in Cinders, Concrete

In a 3-year embedment test of chain link fabric in cinders and concrete, three types of 9-gauge, 2-inch-mesh chain link were evaluated. These were aluminum-coated, galvanized-after-fabrication, and aluminum alloy 5052. The merit of brush coating chain link fabric with a coal-tar paint prior to contact with concrete also was investigated. Table 10 shows the degree of etching and pitting observed on the base metal after cleaning and, where applicable, after removal of the remaining metallic coating from the base metal.

These data indicate that the aluminum-coated steel was more severely attacked by cinders periodically contaminated with rock salt as a deicing agent than were the galvanized-after or the aluminum fabrics. In concrete, aluminum-coated steel was slightly more corrosion resistant than the other two test fabrics. Coating the fabric with a coal-tar paint prior to contact with concrete was beneficial to all three fabrics, and especially to aluminum alloy fabric.

Fabrics adjacent to cinders and concrete were less severly attacked than the fabrics directly exposed to concrete or to salt-contaminated cinders. Salt-contaminated cinders was the most corrosive environment for aluminum-coated steel; consequently, coating the fabric with a suitable organic finish should be beneficial in such a service.

Chemically Contaminated Atmospheres

The good corrosion resistance of aluminum-coated steel has led to its use in areas where the atmosphere is contaminated by chemical fumes that promote corrosion of other metals. Contaminated atmospheres encountered most often contain hydrogen sulfide, carbon dioxide, ammonia, sulfur dioxide, moisture, and dirt. Tests indicate that aluminum is highly resistant to corrosion in these atmospheres. In high humidity, however, certain chemicals such as calcium chloride, ferric chloride, potassium permanganate, and sodium fluoride can cause corrosion. Tests in a coal-chemical environment involving insulation jacketing made from aluminum-coated steel and aluminum indicate that aluminum-coated steel is as corrosion resistant as aluminum. Both materials performed satisfactorily when exposed to moderately dry contaminated atmospheres. However, these materials were severely corroded by chemical sprays containing chlorides and sulfates.

High-Temperature Oxidation

Aluminum-coated steel sheets resist fire damage in building components better than solid aluminum sheets because aluminum-coated steel can withstand high temperatures which would melt aluminum. Also, aluminum-coated steel maintains a higher degree of mechanical strength than aluminum at elevated temperatures.

Up to about 950 F (510 C), the aluminum coating protects the steel base against oxidation without discoloring. The appearance of Type 1 aluminum-coated steel and carbon-steel specimens exposed to air at 950 F (510 C) for 28 days is shown in Figure 4. The oxide film on carbon steel is loose and has flaked off in large pieces, whereas Type 1 aluminum-coated steel shows no effect of high-temperature oxidation. At temperatures between 950 and 1250 F (510 and 677 C), excellent protection is afforded the steel base by aluminum, although staining may occur because of iron-aluminum alloy formation. In actual use, Type 1 aluminum-coated steel sheets have successfully withstood repeated exposure for short periods to temperatures as high as 1650 F (899 C). Rapid alloying of the aluminum coating and the steel base occurs at high temperatures, but the alloy formed is extremely heat resistant.

Use in Automotive Mufflers

Since 1944, galvanized steel has been the standard construction material for automotive mufflers. Failure of mufflers is the result of corrosive attack by acidic constitutents that condense out of exhaust gases. In recent years, outer shell corrosion of muffers also has become a serious corrosion problem because of the widespread use of road salts for controlling snow and ice on streets and highways throughout much

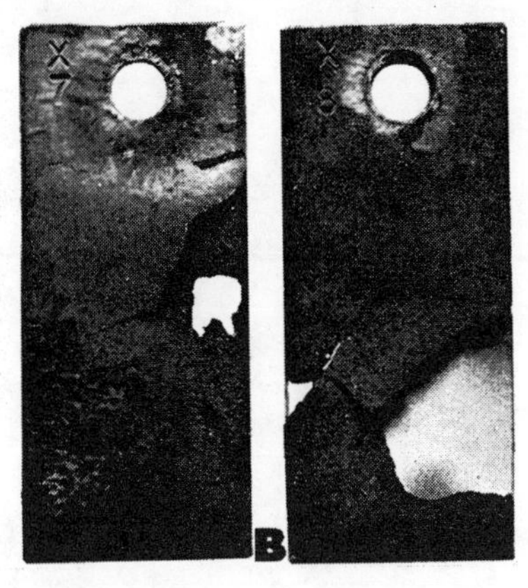

Figure 4—Aluminum-coated and carbon steel samples exposed in air at 950 F (510 C) for 28 days. Panels A are aluminum-coated steel (Type 1, 0.5 oz/sq ft) and Panels B are carbon steel.

TABLE 10—Etching and Pitting of Base Metal of 9-Gauge 2-Inch Mesh Chain Link Fabric Samples After Three Years Embedment in Cinders and Concrete.

	ALUMINUM-COATED (2.6 % Si 0.4 oz/sq ft)		GALVANIZED AFTER FABRICATION (2.4 oz/sq ft)		ALUMINUM ALLOY 5052	
	Etch	Pit	Etch	Pit	Etch	Pit
Cinders - Links	OK	100H	10L	35M	10L	TrH
Cinders - Barbs	OK	50H	10L	10L	10L	15M
Concrete - Links	OK	8M	35L	TrL	50L	50L
Concrete - Barbs	10L	TrL	40L	TrL	50L	50L
Links (Coated with coal-tar paint)	5L	TrL	25L	TrL	5L	5L
Barbs (Coated with coal-tar paint)	5L	TrL	25L	OK	TrL	TrL

* Numbers indicate the percentage of surface affected (Tr indicates trace); H-heavy, M-medium, L-light indicate the extent of etching and pitting.

of the northern United States.

Corrosion tests have shown that aluminum-coated steel is more resistant than galvanized steel to corrosion by acid condensate. Service tests with galvanized and aluminum-coated steel mufflers have demonstrated the superior performance of aluminum-coated steel in this application. Mufflers subjected to extreme heat and continual quenching with corrosive water from the road give longer service life when the outer shell is fabricated from aluminum-coated steel.

Corrosion by Natural Waters

Although aluminum is generally more resistant to natural waters than zinc, the important factor in corrosion of aluminum-coated steel is behavior of the galvanic couple between aluminum and iron. In sea water and in most brackish waters, aluminum has a lower corrosion rate than zinc; thus, aluminum-coated steel should give better service than galvanized steel in these waters. In fresh water, aluminum is not as anodic to steel as zinc; hence, substitution of aluminum-coated steel for galvanized steel must be made with caution in this environment. For example, aluminum-coated steel should not be substituted for galvanized steel in applications such as hot-water tanks because of the possibility of rapid perforation.

Paint Performance

For most applications, aluminum-coated steel sheets will provide satisfactory service without painting. Aluminum-coated steel may be painted with moderate success without special pretreatments; however, weathering of sheets for six months or longer before painting will improve adherence of paint. To obtain paint adherence and service performance usually required for fabricated articles and finished products, the surfaces of aluminum-coated sheets should be treated in the same procedure generally followed in preparing aluminum surfaces for painting. These practices include treatment with wash primers or proprietary phosphates.

Summary

Aluminum-coated steel is a versatile product providing excellent resistance to corrosion and elevated temperatures, good heat reflectivity, and superior mechanical properties. As a result, this product is accepted in a variety of applications, such as farm and industrial buildings, home appliances, mufflers, and enclosure materials such as fencing, barbed wire, and chain link fabric.

References

1. J. H. Rigo. New Developments in Wire Coatings. *Agricultural Engineering* (1964) February.
2. *Chemistry and Industry*, p. 195 (1956) May 24.
3. *Corrosion Prevention and Control*, p. 28 (1956) July.
4. C. J. Walton. Resistance of Aluminum-Base Alloys to 20-Year Atmospheric Exposure. *ASTM Special Technical Publication No. 175* (1956).
5. E. A. Anderson. The Atmospheric Corrosion of Rolled Zinc. *ASTM Special Technical Publication No. 175* (1956).
6. L. J. Gorman. American Society for Testing and Materials, Procedures 39, 247 (1939).
7. J. D. Sprowl. The Production and Uses of Aluminized Steel. *Iron and Steel Engineer* (1961) October.

R. J. SCHMITT is section supervisor of the Applied Research Laboratory at United States Steel Corp. where his work includes studies concerning corrosion of steels exposed to the atmosphere, natural waters, and underground environments. He has a BS in chemical engineering from the University of Pittsburgh and is a member of NACE, AIChE, and AWWA. Presently, he serves as chairman of NACE's Pittsburgh Section.

JANE H. RIGO is a senior research engineer in the Applied Research Laboratory of United States Steel Corp., Cleveland, Ohio. Her responsibilities include supervising projects involving corrosion, organic coatings, and plastics. She is a member of NACE, ASTM, and ACS, and has an MS in organic chemistry from Case Institute of Technology.

SUMMARY

Properties of low alloy steels are discussed and guidelines are presented for their use in buildings and other structures. Factors that affect the atmospheric corrosion of steel and the influence that certain alloying elements have on improving atmospheric corrosion resistance of steel are discussed. Also results of an examination of rust films on exposed carbon steel and corrosion resistant low alloy steel and electrochemical studies on exposed samples of these steels are detailed.

R. J. Schmitt and W. P. Gallagher
Applied Research Laboratory
United States Steel Corp.
Monroeville, Pa.

For Architectural Applications

Unpainted High Strength Low Alloy Steel

ONE OF THE most significant developments in the construction industry in the last decade has been the use of corrosion resistant high strength low alloy steels in an unpainted condition. Certain low alloy steels contain small amounts of those alloying elements that result in atmospheric corrosion resistance superior to carbon steel. When exposed to the atmosphere, these steels develop tightly adherent protective oxide films that substantially seal the surface against further corrosion; as a result, these steels do not require painting.

The excellent atmospheric corrosion resistance exhibited by these steels was recognized in the early 1930's, and they have been widely used since that time in the railroad, shipbuilding, and construction industries. However, in the past, these steels were primarily used in the painted condition. This was particularly true wherever the steel was to be in public view, because the appearance of rust on steel was considered as evidence that the structure was rapidly deteriorating and required immediate painting.

The first major step toward obtaining public acceptance of the appearance of rusted steel was made in 1958 by the late, famous architect Eero Saarinen, who used an unpainted high strength low alloy steel in his unique design for the John Deere and Co.'s administration building in Moline, Ill. Since then, unpainted steel has been used successfully in other prominant architectural accomplishments. Engineers are now following the architect's lead and the applications for corrosion resistant high strength low alloy steels are expanding. They now include buildings, bridges, light standards, guard rails, transmission towers, chemical plant structures, and roofing and siding.

Because of the growing interest in corrosion resistant high strength low alloy steels, there is a need to disseminate information to the architect and engineer concerning the properties of these steels. This paper will discuss the properties of these steels and present guidelines for their use in buildings and related structures. The factors that affect the atmospheric corrosion of steel and the influence that certain alloying elements within the compositional range of the corrosion resistant low alloy steels have on improving the atmospheric corrosion resistance of steel are discussed. Also, results of an examination of rust films on exposed carbon steel and corrosion resistant low alloy steel and electro-chemical studies on exposed samples of these steels are discussed in terms of explaining the differences in the protective qualities of the oxides that form on the steels.

There are several proprietary corrosion resistant high strength low alloy steels being used in buildings and structures at the present time. The steels discussed in this paper are United States Steel's USS COR-TEN high strength low alloy steels (referred to as low Alloy A steel and low Alloy B steel).

Description and Properties

Low Alloy A steel, the earliest of the contemporary high strength low alloy steels, has been furnished for architectural applications. Hot rolled sections of low Alloy A steel maintain a 50,000 psi minimum yield point in thickness up to one-half inch, inclusive. In certain applications, low Alloy B steel is required where a 50,000 psi minimum yield is desired in thickness over one-half inch and when highly re-

TABLE 1—Composition of USS COR-TEN Steels

Type	Weight Percent, ladle								
	C	Mn	P	S	Si	Cu	Cr	Ni	V
USS COR-TEN A Steel	0.12 max	0.20/0.50	0.07/0.15	0.05 max	0.25/0.75	0.25/0.55	0.30/1.25	0.65 max	(1)
USS COR-TEN B Steel(2)	0.10/0.19	0.90/1.25	0.04 max	0.05 max	0.15/0.30	0.25/0.40	0.40/0.65	(1)	0.02/0.10

(1)Not Specified.
(2)U. S. Patent No. 2,845,345.

Reprinted with permission from *Materials Protection*, December 1969, 70-77,

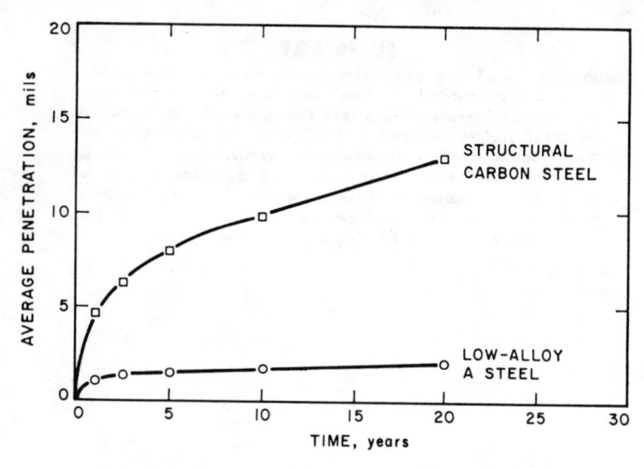

Figure 1—Time corrosion curves of low Alloy A steel and structural carbon steel exposed in an industrial atmosphere.

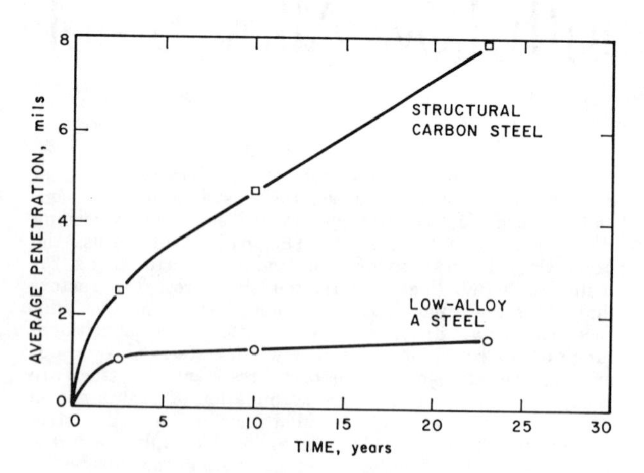

Figure 2—Time corrosion curves of low Alloy A steel and structural carbon steel exposed in a semirural atmosphere.

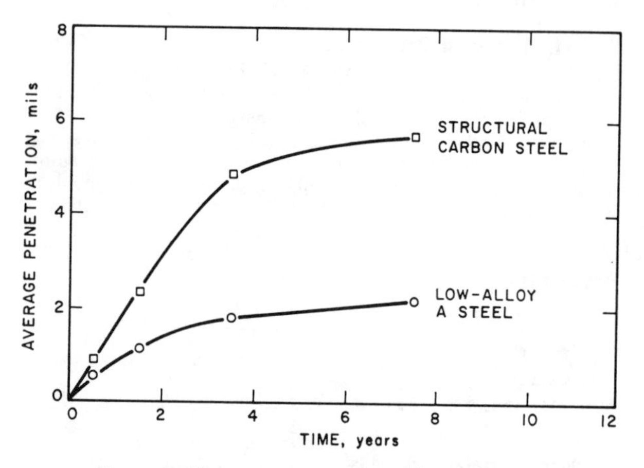

Figure 3—Time-corrosion curves of low Alloy A steel and structural carbon steel exposed to a moderate marine atmosphere 800 feet from the ocean.

strained welds and dynamic loading conditions occur in a structure. The use of both steels in the same structure usually presents no major problems, either in design or aesthetics, because the two steels are closely compatible in performance and appearance. The chemical compositions of low Alloy A and B steels are given in Table 1. Low Alloy A steel meets the requirements of the ASTM A 242 Type 1 and low Alloy B steel meets the requirements of ASTM A 588 Grade A and A 242 Type II. Because most architectural applications for low alloy steel require material that is less than one-half inch in thickness, the discussion hereafter will be concerned primarily with low Alloy A steel.

The atmospheric corrosion performance of low Alloy A steel has been determined over the years in industrial, rural, and marine atmospheres. Atmospheric corrosion tests were conducted by exposing 4 by 6-inch specimens on test racks at an angle of 30° horizontally and facing south (the standard ASTM testing method). Plots of the average penetration in thickness versus time for low Alloy A steel and structural carbon steel exposed in industrial, semirural, and moderate marine atmospheres are shown in Figures 1, 2, and 3. From the slopes of these time-corrosion curves, it can be calculated that low Alloy A steel is five to eight times as corrosion resistant as carbon steel with a low residual copper content. In a moderate marine atmosphere, at a distance of 800 feet from the ocean, the degree of superiority of low Alloy A steel to carbon steel is about the same as it is in industrial atmospheres (Figures 1 and 3). If low Alloy A steel is exposed near the ocean where it can receive recurring wetting by salt spray, its superiority over carbon steel is maintained; however, the corrosion losses for both steels are of several orders of magnitude higher than that obtained in industrial atmospheres. As a result, these steels should not be used in the unpainted condition in severely corrosive marine atmospheres.

Corrosion of the low Alloy A and B steels (and therefore the formation of the oxide film) requires that moisture be present on the steel surface. In addition to rain wetting the surface, moisture may condense on the surface at nights when the temperature of the metal surface drops below the dewpoint of the surrounding air. This moisture will remain on the surface until it is dried by wind or a rise in the ambient temperature. Therefore, it can be expected that the capacity of a steel structure to retain heat may affect the quantity of dew condensed on its surface and thus affect the corrosion rate. Because of this, the amount of corrosion of the low Alloy A and B steels occurring on a large structure, such as a building or bridge, could be less than would occur on small specimens of these steels exposed near the ground on test racks.

The low alloy A and B steels owe their excellent atmospheric corrosion resistance to the formation of tightly adherent protective oxide films. Results of long term corrosion tests have shown that the formation of the protective oxide film on low Alloy A steel results in a loss in thickness of about 2 mils of metal, (Figures 1, 2, and 3). About one half of the initial oxide formed is retained; the balance is lost mainly in the early stages of formation through the eroding action of wind and rain. A portion of the lost metal is in the form of water soluble iron compounds, which can stain certain other building materials.

The texture and color of the oxide film that forms on the low Alloy A and B steels are determined by the alloy content of the steel, the degree of contamination of the atmosphere, and the frequency with which the surface is wet by dew and rainfall and dried by wind and sun. Therefore, slight differences in the appearance of the weathered surface of these steels can be expected in accordance with the type of atmosphere in which they are exposed. The appearance of weathered low Alloy A steel after exposure to an industrial, semirural, and moderate marine atmosphere is shown in Figures 4, 5, and 6, respectively. Low Alloy A steel exhibits a homogeneous dark brown or blue black appearance after two years exposure to industrial atmospheres. In atmospheres with less contamination, the oxides that form on the steel are lighter in color and somewhat rougher in texture. Differences in surface texture are evident when the appearance of low alloy A steel exposed to a semirural atmosphere

(Figure 5) is compared with its appearance in an industrial atmosphere (Figure 4). Low Alloy A steel exposed to a moderate marine atmosphere exhibits a much lighter, brown colored oxide and displays a rougher surface texture than the steels exposed in the other two atmospheres.

The time required for the development of a stable oxide film on the low Alloy A and B steels is also dependent on the atmosphere to which it is exposed. Generally, a stable oxide film will develop more rapidly in industrial atmospheres than in rural atmospheres with little contamination. Views of pickled and blast cleaned low Alloy A steel specimens after atmospheric exposures of up to two years in a semi-industrial atmosphere are shown in Figure 7. After about a year, the steel develops a homogeneous light brown appearance. After about two years, the formation of the protective oxide is nearly complete and the characteristic dark brown color associated with the low Alloy A and B steels has developed. In semi-industrial and rural atmospheres, an additional six months to a year is required for the steels to develop their optimum appearance. Atmospheric exposure tests have also shown that these steels are highly susceptible to rusting on initial exposure, and in fact will develop rust faster than carbon steel on initial exposure. However, after several weeks, the rate of rusting on these steels slows down and becomes less than that for carbon steel.

It is also evident from Figure 7 that the method of removing mill scale (pickling or blast cleaning) has little or no effect on the texture and appearance of the protective oxide film that develops on these steels. However, it is important to recognize that incomplete removal of the mill scale from the steel surface, either as the result of poor pickling or blast cleaning, will cause a non-uniform flaking oxide to develop on the steel surface.

Observations made on both boldly exposed and sheltered low Alloy A steel specimens and test structures indicate that the texture of the oxide film depends on the washing action of rain and the drying action of sunlight. Low Alloy A steel surfaces exposed to the north receive less sunlight than surfaces exposed to the east, south, and west, and therefore are somewhat slower to develop a protective oxide film. Also, surfaces sheltered from the sun and rain tend to develop loose oxide films, whereas surfaces boldly exposed to the sun and rain develop tightly adherent protective oxide films.

White or light yellow streamers or deposits have been observed on low Alloy A and B steels during the early stages of exposure. The light colored deposit is particularly evident on sheltered surfaces. The appearance of low Alloy A steel samples exposed for 18 months in a heavy industrial atmosphere is shown in Figure 8. Note the deposit on the groundward side of the specimens. The light colored deposit that forms on the steel surface has been identified by X-ray diffraction to be ferrous sulfate ($FeSO_4 \cdot 4H_20$). The extent to which the deposit occurs on low Alloy A and B steels varies with the amount of sulfur dioxide (SO_2) in the atmosphere; the higher the amount of SO_2, the greater the amount of ferrous sulfate that forms. Ferrous sulfate is water soluble and as a result is normally washed from boldly exposed surfaces. However, ferrous sulfate will be retained as a hydrated salt when there is insufficient washing action, and this accounts for the light colored deposit that forms on sheltered surfaces of these steels. Streaks of

Figure 4—Appearance of low Alloy A steel after 15 years exposure in an industrial atmosphere.

Figure 5—Appearance of low Alloy A steel after 28 years exposure in a semirural atmosphere.

Source: *Materials Protection*, December 1969, 70-77

Figure 6—Appearance of low Alloy A steel after 15 years exposure in a marine atmosphere.

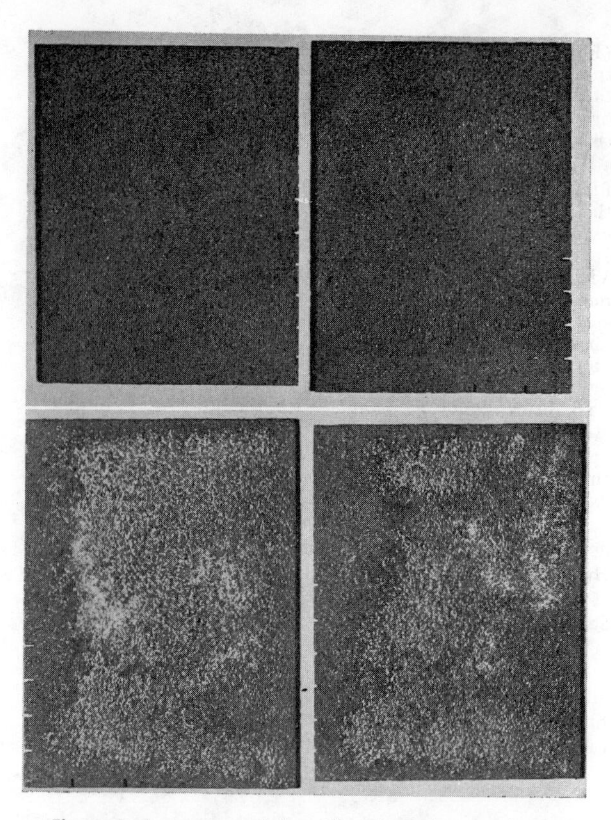

Figure 8—Appearance of low Alloy A steel exposed to a heavy industrial atmosphere for 18 months. Top, skyward surface, uncleaned; bottom, groundward surface, uncleaned.

Figure 7—Appearance of low Alloy A steel after exposures up to 2 years in semi-industrial atmosphere. Top row pickled; bottom blast cleaned.

light colored ferrous sulfate deposits have also been found on boldly exposed steel surfaces of buildings when rainfall has not been sufficient to completely wash the surface. Examination of weathered low Alloy A steel samples has shown that the amount of ferrous sulfate on low Alloy A steel decreases with the time of exposure, and over a period of four to five years, completely disappears. This occurs because the rate of corrosion decreases, and corrosion is the ferrous sulfate generating process.

Guidelines For Effective Use

Results of long term corrosion tests and observations of low Alloy A and B steel structures have provided valuable guidelines for the effective use of these steels. Some of the more important of these are described below.

Cleaning of Exposed Surfaces

To obtain optimum and uniform appearance, regardless of environmental exposure conditions, all low Alloy A and B steel components for exterior architectural applications should be freed of mill scale. Acid pickling and or blast cleaning are satisfactory methods of removing mill scale (Figure 7). Blast cleaning should be done in accordance with provisions of the Steel Structures Painting Council Surface Preparation Specifications, "Near-White Metal Blast Cleaning SSPC-SP 10-638."

Foreign matter such as grease, oil, chalk, crayon, concrete, mortar, and plaster on weathering steel should be removed because its presence will interfere with the normal development of a uniform oxide film. Welding flux, slag, and spatter should also be removed from all exposed surfaces.

Joining

Welding—Low Alloy A and B steels can be welded, with the use of good shop practice, by the shielded metal-arc, submerged-arc, gas metal-arc, and electrical resistance processes. When either of these steels is used for nonexposed or painted structural applications, shielded metal-arc welding may be done with the mild steel electrodes, and gas metal-arc and submerged-arc welding with mild steel wire and wire flux combinations, so that weld metal equivalent to that of the shielded metal-arc electrodes is deposited.

For bare steel applications, when the welded area is required to match or approach the color of low Alloy A and B steels after atmospheric exposure or have atmospheric corrosion resistance similar to that of the base metal, single-pass welds may be made by using the mild steel welding materials, provided that the procedure used insures suitable composition enrichment of the weld metal. In built-up or multiple-pass welds where such a color match is desired, ASTM A 316 E8016-C1 or E8016-C2 electrodes or wires and wire-flux combinations providing 2½ or 3½% nickel weld-metal or a suitable Cr-Si-Cu-Ni composition electrode should be used (Figure 9). The above procedures for multiple-pass welds may also be used for single-pass welds. Further, in multiple-pass butt welds where surface color match is important, consideration should be given to the use of the mild steel welding materials for most of the joint and to the use of the alloy steel welding materials for the completion of the joint. For more detailed welding instructions, the supplier of the steel should be consulted.

Bolting—Low Alloy A steel bolts and nuts are satisfactory for bearing type connections. Friction type connections should be high strength steel bolts that meet ASTM A 325 and should have a chemical composition that will provide adequate atmospheric corrosion resistance.

Staining Of Other Materials

If the iron compounds formed during the process of weathering of low Alloy A and B steels are permitted to evaporate on the surface of a material, staining will result. The degree of staining will depend primarily on the porosity of the material, the amount of iron compounds that can collect on a given surface, and the time of exposure. Some of the more common building materials have been evaluated with regard to staining and to ease of removing stain. A list of materials is in the following table. If there is a question concerning the staining characteristics of a given material, the material should be tested before it is used in a structure.

Materials Suitable

1. Semiglossy and glossy porcelain enamel coatings. (Over 40% - 45° specular gloss.)
2. Washable air-drying and thermosetting organic coatings.
3. Aluminum anodized and unanodized.
4. Stainless steels.
5. Extruded neoprene.
6. Ceramic tile.
7. Common window glass.

Materials to be Avoided Unless Staining is Not Objectionable or Preventive Measures Can be Taken

1. Mortar joints.
2. Cement or concrete.
3. Galvanized steel.
4. Unglazed brick.
5. Matte porcelain enamels (less than 40% - 45° specular gloss.)

Porous and absorbent materials such as concrete will stain readily and are very difficult and sometimes impossible to clean. To follow the progress of staining by the runoff from low Alloy A steel, a concrete base supporting a low Alloy A steel light standard was periodically painted with a proprietary white waterproof paint and the lower part of the standard was painted with a glossy white epoxy paint. After 54 months, corrosion products from the low alloy A steel continue to stain the epoxy paint and the painted concrete base. It can be concluded that staining of other architectural materials by corrosion products of low alloy steels will continue indefinitely, although the amount of stain will decrease appreciably after the oxide film has developed fully. This can be from two to three years, depending on the atmosphere.

It is good practice in designing a building or structure to provide for permanent measures that will drain the runoff from the low Alloy A and B steels away from other materials. If this cannot be done, then consideration should be given to the use of materials that will not develop permanent stains and will be easy to clean or the use of materials that are colored to minimize the difference in appearance caused by staining.

Nonexposed Surfaces

The low Alloy A and B steels, when immersed in water, behave much like carbon steel and require protection by coatings for long term durability. It is not unusual to find conditions in a building that result in the steel becoming wet and remaining wet for long periods of time. For example, a building design may permit water to condense and collect on the back side of the fascia panels. Under these conditions it is necessary to protect these surfaces with appropriate paint systems or galvanized coatings to prevent rapid corrosion of the steel. Moisture can also be trapped in crevices formed when the steel surfaces come in contact with other architectural materials, and therefore the faying surface of the steel should be protected with paint or a sealant. The joint between two faying low Alloy A and B steel surfaces will seal itself if it is tight and fixed. When slight movement or inadequate contact between faying surfaces is foreseen, the surface should be protected. Ledges, crevices, and areas that can hold water should be eliminated or adequate drainage should be provided.

Atmospheric Factors and Steel Composition

The rate of atmospheric corrosion of steel is dependent on (1) the length of time that moisture is in contact with the surface, (2) the extent of pollution of the atmosphere, and (3) the chemical composition of the steel. The extent to which these factors affect the atmospheric corrosion behavior of carbon steel and the low Alloy A and B steels is discussed in this section.

Source: *Materials Protection*, December 1969, 70-77

Results of atmospheric corrosion tests conducted by various ASTM technical committees[1,2] and other societies and associations over the years have shown the importance of the environmental factors. It is recognized from this extensive work that atmospheric sulfur dioxide, atmospheric chlorides, and factors relating to the time of wetness of a corroding surface exert critical effects on the corrosivity of an atmosphere. Copson[3] found from his work that the corrosion rate of steel depends on the quality and quantity of water reaching the steel surface. The most corrosive condition exists when the water contains pollutants, particularly chlorides. Of all the common pollutants, sodium chloride, typified by sea salt, is the greatest destroyer of protective rust films on steel. Sulfur oxides are also corrosive toward steel but not nearly as aggressive as chlorides. When both chlorides and sulfur oxides are almost absent, the atmospheric corrosion of steel is not significant.

The third factor, the chemical composition of the steel, is as important in determining the corrosion rate of steel as are the environmental factors. Since the pioneering work of Buck[4] and the early publications and interpretation of the results of ASTM tests,[5] it has been generally acknowledged that carbon steel containing 0.2% or more copper has twice the atmospheric corrosion resistance of a similar steel containing only 0.01 to 0.02% residual copper. Larrabee and Coburn[6] confirmed these findings and also showed that nickel, chromium, silicon, and phosphorus, singly, are beneficial in improving the corrosion resistance of steel. They further showed that the greatest improvements in corrosion resistance are obtained by the addition of specific combinations of these alloying elements. Such specific combinations can be found in the low Alloy A and B steels.

Although the superior atmosphere corrosion resistance of these steels has been extensively demonstrated over the years, the mechanisms by which the alloying elements improve atmospheric corrosion resistance remain unresolved. However, certain facts have been obtained from the examination of rust films on weathered carbon steel and low Alloy A and B steels and from electrochemical studies on aged samples of these steels, and are worthy of note, since they may aid in future mechanistic investigations.

Investigation Of Rust Films

X-ray diffraction analyses of rust films of various ages on carbon steel and low Alloy A steel have shown that the bulk phases of the films are for the most part the same for the two steels; this is consistent with the findings of Horton.[7] The principal phases present in the rust films are lepidocrocite (γ—FeOOH), goethite (α—FeOOH), and magnetite. The major constituent is generally lepidocrocite but the proportion of goethite increases with the age of the oxide film. Spectrographic analysis of the rust showed that the alloying elements (Cr-Si-Cu-Ni-P, etc) present in low Alloy A steel are subsequently found in the rust film in about the same proportions. However, it was found that the Cr, Mn, and Ni contents increase as the steel surface is approached.

An interesting observation from the X-ray diffraction work is the fact that sulfates are always present in the rust films of carbon steel and low Alloy A steel, and that the percentage of SO_4^{-2} is greater in rust films on low Alloy A steel than in rust films on carbon steel. Since rust on low Alloy A steel contains an appreciably higher percentage of sulfate than carbon steel rust, it is conceivable that the superior performance of low Alloy A steel may be due in part to the manner in which sulfates occur or are incorporated into the rust film.

X-ray diffraction analyses have shown ferrous sulfate tetrahydrate to be present in pockets at the rust/steel interface on low Alloy A steel and carbon steel during atmospheric exposures of six months or less. With longer exposure times, the ferrous sulfate phase and its location on carbon steel remain essentially unchanged, although additional X-ray diffraction lines appear in the patterns after one year. However, on low Alloy A steel after one year exposure, the ferrous sulfate diffraction pattern is no

Figure 9—Appearance of low Alloy A steel (welded with 2½% nickel electrode, right, and Cr-Si-Cu-Ni composition electrode, left,) after 4 years exposure in a semi-industrial atmosphere.

longer present, and a completely new diffraction pattern is obtained. This new phase has a pattern similar but not identical to the basic ferric sulfate, karphosiderite, and gives positive qualitative indentificaton for ferric ions. Although this ferric sulfate phase persists on low Alloy A steel after longer exposure times, the number of pockets decreases until only a few isolated sites are discernible on a 4 by 6 inch panel of low Alloy A steel after four years of atmospheric exposure.

Schikorr[8] has put forth a possible explanation for the role of sulfate in the rusting of steel in terms of the following reaction sequence.

$$Fe + SO_2 + O_2 \rightarrow FeSO_4$$

$$4FeSO_4 + O_2 + 6H_2O \rightarrow 4FeOOH + 4H_2SO_4$$

$$4H_2SO_4 + 4Fe + 2O_2 \rightarrow 4FeSO_4 + 4H_2O$$

In the case of carbon steel, which corrodes freely in air, a self-perpetuating reaction involving water soluble ferrous sulfates possibly develops at the rust/steel interface in the presence of SO_2 in the air. The fact that the low Alloy A steel develops a protective oxide and essentially stops corroding can be understood if a water insoluble sulfate is produced at the rust/steel interface with the help of the alloying elements (Cr-Cu-Ni) in the steel, thereby ending the regeneration sequence proposed by Schikorr. This theory is consistent with the above finding that Cr and Ni concentrations in the rust films are greatest at the rust/steel interface.

Copson[3] has postulated that with low alloy steels containing nickel, copper, or chromium, the sulfates of these elements are formed during the corrosion process. Because the sulfates of these elements are more insoluble than the sulfates of iron, the pores in the rust on low alloy steel become clogged and prevent access of water to the steel itself. Whether the insoluble sulfates of chromium, nickel, and copper merely seem to stop the autocatalytic reaction proposed by Schikorr or actually clog the pores in the rust, as suggested by Copson, it is obvious that either theory is too simple to explain the many effects of various combinations of alloying elements.

Additional evidence of a difference in the two rusts is provided by scanning electron photomicrographs of the surfaces of the rusts on carbon steel and low Alloy A steel that had been exposed to a semirural atmosphere for 18 months. As Figure 10 shows, the carbon steel rust is covered with many small nodules, whereas the low Alloy A steel is relatively smoother with only a few nodules present. These nodules could have been formed by the precipitation of iron ions that have diffused up through the porous carbon steel rust. Horton[7] has pointed out that airborne dust particles are incorporated in rusts, an indication that during the rusting some iron species, probably ferrous ion, diffuses outward through the rust layer to react with water and oxygen, forming rust at the rust/air surface.

The corrosion potentials of rusted carbon steel and low Alloy A steel have been measured by immersing specimens of these steels with their rust films intact in a well stirred electrolyte and recording the potential difference between the steel and a calomel reference electrode. The intent of this experiment was to simulate conditions existing when the steel is wet and corroding. The electrolyte used in these studies was 10^{-4} M H_2SO_4, which has a pH of about 3.8, equal to the lowest measured pH of rainwater collected in a semi-industrial atmosphere. The steel specimens used in this study had weathered for various periods of time in the semi-industrial atmosphere of Monroeville, Pa.

The measured corrosion potentials for low Alloy A steel and carbon steel versus time of atmospheric exposure for the steels are shown in Figure 11. The corrosion potential at any time of exposure is more noble for low Alloy A steel than for carbon steel, and the corrosion potentials of both steels increase with the time of atmospheric exposure of the steels. The rate of increase in corrosion potential decreases with time, and for the steels measured, reaches a limiting value below 200 mv noble to the standard hydrogen electrode. The active corrosion potentials characteristic of the early stages of rusting indicate that the corrosion process is under cathodic control; that is, the anodic process of iron oxidation occurs relatively freely with the rate of corrosion determined by the rate of the cathodic process of

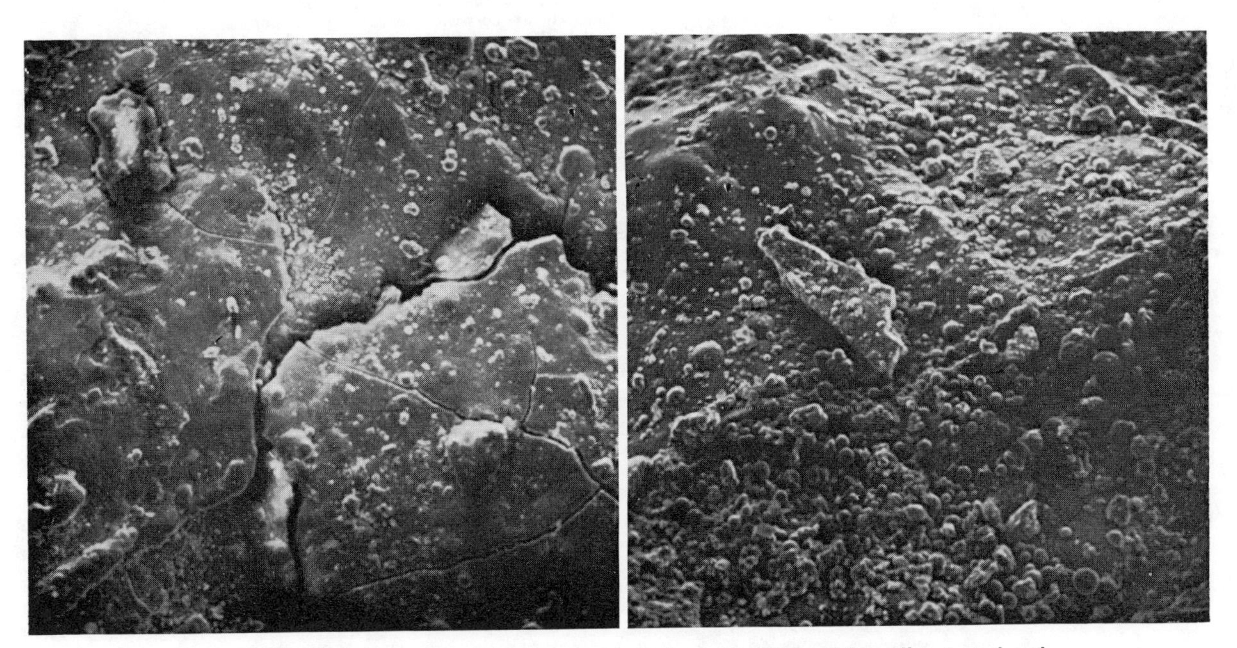

Figure 10—Scanning electron photomicrographs of the rust surface of low Alloy A steel and carbon steel after 18 months' exposure in a semirural atmosphere. Left, low Alloy A steel; right, carbon steel.

Source: *Materials Protection*, December 1969, 70-77

oxygen reduction. With longer exposure times, the decrease in corrosion current, as shown by lower corrosion rates (Figure 1), and the increase in corrosion potential indicate that anodic polarization becomes pronounced with time. The part that the alloying elements in low Alloy A steel plays in making the corrosion potential of this steel more noble than that of carbon steel for identical exposure conditions may be attributed to differences in the physical and electrical properties of the oxide films that form on the steels.

Summary

The successful application of the corrosion resistant high strength low alloy steels in numerous architectural applications has demonstrated that these steels can be used both as an ornamental material and as a structural material. It is hoped that the information presented will promote a better understanding of the use of these steels and the broad range of applications for which they are well-suited.

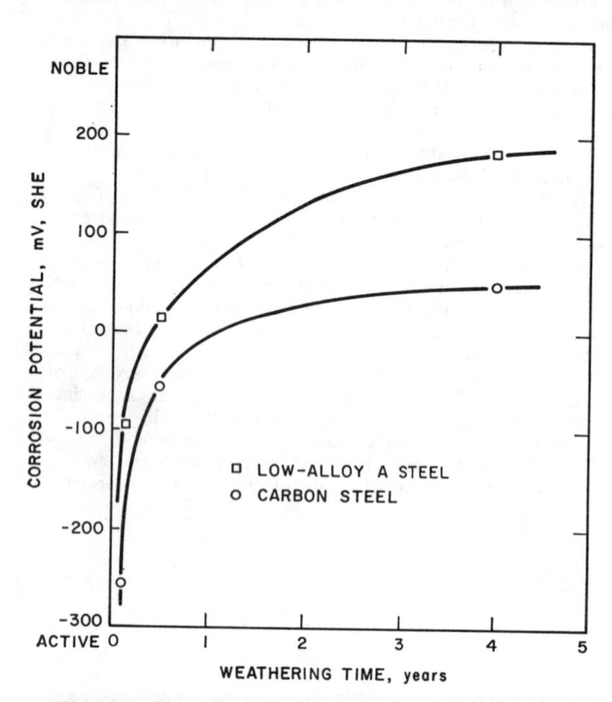

Figure 11—Corrosion potential of weathered low Alloy A steel and carbon steel in dilute H_2SO_4 as a function of time of exposure in a semi-industrial atmosphere.

Acknowledgment

The authors wish to acknowledge the contributions of their co-workers at U. S. Steel's Applied Research Laboratory: C. X. Mullen, who conducted the atmospheric testing program, S. K. Coburn, who provided information on the architectural uses, and M. F. Dean, (formerly of U. S. Steel) who initiated some of the technical investigations.

References

1. *Proceedings of ASTM*, 59, p 183 (1959).
2. H. Guttman and P. J. Sereda. Measurement of Atmospheric Factors, Affecting the Corrosion of Metals. *Metal Corrosion in the Atmosphere*, ASTM STP 435, ASTM, pp 326-359, 1968.
3. H. R. Copson. A Theory of the Mechanism of Rusting of Low Alloy Steel in the Atmosphere. *Proceedings of ASTM*, 45, p 554, 1945.
4. D. C. Buck. *Proceedings of ASTM*, 19, Part 2, 224 (1919).
5. Reports of Committee A-5, *Yearly Proceedings of ASTM*, 1916 to 1951.
6. C. P. Larrabee and S. K. Coburn. *Proceedings of the First International Congress on Metallic Corrosion*, London, 1961 Butterworth, p 283, 1962.
7. J. B. Horton, M.S. Thesis, Lehigh University, 1957.
8. Schikorr. *Werkstoffe u. Korrosion*, 15, 457 (1964).
9. H. R. Copson, op cit p 554.
10. J. B. Horton, op cit.

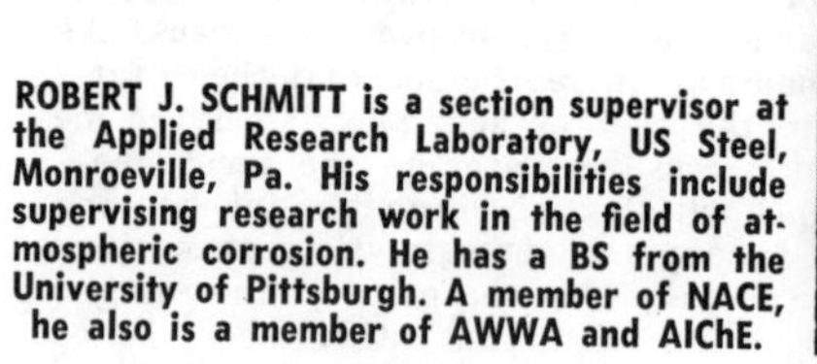
ROBERT J. SCHMITT is a section supervisor at the Applied Research Laboratory, US Steel, Monroeville, Pa. His responsibilities include supervising research work in the field of atmospheric corrosion. He has a BS from the University of Pittsburgh. A member of NACE, he also is a member of AWWA and AIChE.

W. PATRICK GALLAGHER is a senior research engineer at the Applied Research Laboratory, US Steel, Monroeville, Pa. He is involved in research of corrosion occurring in natural media. He has a BS in chemical engineering and an MS and PhD in chemistry. A member of NACE, he also is a member of ACS and AAAS.

The Application of High Strength Low Alloy Steels in the Chemical Industry*

J. B. VRABLE, R. T. JONES, and E. H. PHELPS
Research Laboratory, U. S. Steel Corporation, Monroeville, Pennsylvania.

Resistance to chemical plant atmosphere and other exposures of ASTM A-242 Type 1 (Cor-Ten A) and ASTM A-588 Grade A (Cor-Ten B) high strength, low alloy steels is described. The data are derived from exposures near alumina load, coal, and salt water, waste treatment plants, and limestone stackers. Exposures were made also in exhaust gas condensate, up to 540 C oxidation combustion gases, heat exchangers and condensers for cooling water. The steels performed well unpainted in these environments, acquiring the expected dense, tightly-adhering layer of corrosion products that characterizes steels of the compositions discussed. Alternate wet and dry conditions are necessary for good performance.

HIGH STRENGTH LOW ALLOY (HSLA) STEELS are a group of steels intended for general structural and miscellaneous applications. They have specified minimum yield strengths of 290 MPa (42,000 psi) and above. These steels typically contain small amounts of alloying elements to achieve their strength in the hot rolled or heat treated condition. Some of the HSLA steels have an atmospheric corrosion resistance five to eight times that of carbon steel[1-5] with low residual copper content, along with minimum yield strengths of 345 MPa (50,000 psi). These steels owe their superior atmospheric corrosion resistance to the formation of tightly adherent protective oxide films that reduce the rate of further atmospheric corrosion.[6] However, superior corrosion resistance is seldom if ever observed when the steel is immersed in aqueous and chemical environments. Therefore, in applications where the material is not exposed to the weather and where corrosion is a factor, the low alloy steels should not be assumed to have better corrosion resistance than carbon steel.

Copson and Larrabee[7] reported that field tests and service experience both have shown that paint coatings are more durable on HSLA steels than on carbon steel. The extra durability of the paint is a result of the better atmospheric corrosion behavior of the low alloy steel.

The corrosion properties of these low alloy steels, together with their high strength, have resulted in their wide use to obtain longer service life and increased paint life in many applications. Through the 1960's, only limited use was made of these steels in process plants because of the lack of supporting information on corrosion performance. To develop such information, cooperative corrosion tests were conducted with specimens exposed to different chemical plant atmospheres. The results of the tests showed that USS COR-TEN A[(1)] steel was appreciably more corrosion resistant than carbon steel in the atmospheres tested.[8] Since that time, the HSLA steels have been used for structural applications in a number of chemical plants.

This paper presents the results of monitoring programs to obtain additional quantitative data on the performance of bare low alloy steel structures exposed to the atmosphere in various chemical process plant applications. The steels evaluated in this program were COR-TEN A and COR-TEN B steels, with the chemical compositions shown in Table 1.

COR-TEN A meets the requirements of the ASTM A 242 Type 1 designation in thicknesses up to 12 mm (0.5 inch), inclusive. COR-TEN B meets the requirements of ASTM A 588 Grade A and ASTM A 242 Type 2 in thicknesses to 10 mm (4 inches), inclusive.

Atmospheric Applications

These two HSLA steels have the capability of developing a protective oxide on exposure to the atmosphere and, as a result, have been used extensively in architectural and constructional applications. It is known, however, that this capability to develop a protective oxide is limited to "normal" atmospheres, such as typically found in urban and rural areas. The steels should not be used in the bare condition in atmospheres where high concentrations of strong chemical or industrial fumes are present, unless a thorough evaluation of conditions indicate suitability. The atmospheric environments in and around chemical plants can vary from relatively noncorrosive to quite corrosive depending on the type of plant and operating conditions. In some instances, local small regions within a plant are corrosive, whereas the rest of the plant is not. The satisfactory use of unpainted HSLA steels as structural memebers in chemical plants avoids the initial cost of painting and has the distinct advantage of avoiding maintenance painting, with the attendant high cost of proper surface preparation during the life of the plant.

Case Histories

Alumina Conveyor

In late 1967, a new alumina loadout facility was constructed near Baton Rouge, Louisiana, with COR-TEN A and B steel sturctural members. The steels were erected in the as rolled condition without the removal of mill scale. The structure is about 18 mm (60 feet) high and 520 mm (1700 feet) in length, and runs along the southern extremity of the plant to the Mississippi

*Presented during meeting of ASM, October 24-28, 1977, Chicago, Illinois. It is understood that the material in this paper is intended for general information only and should not be used in relation to any specific application without independent examination and verification of its applicability and suitability by professionally qualified personnel. Those making use thereof or relying thereon assume all risk and liability arising from such use or reliance.

(1)USS and COR-TEN are registered trademarks of United States Steel Corporation.

Reprinted with permission from *Materials Performance*, January 1979, 39-44, © 1979 National Association of Corrosion Engineers

TABLE 1 – Compositions of Carbon Steel and High Strength Low Alloy Steels Investigated

Type of Steel	Composition (Wt%)									
	C	Mn	P	S	Si	Cu	Ni	Cr	Mo	V
Carbon Steel	0.16	0.63	0.012	0.031	0.012	0.01	0.01	0.03	ND	ND
USS COR-TEN A Steel	0.09	0.46	0.075	0.031	0.52	0.31	0.40	0.74	0.025	ND
USS COR-TEN B Steel	0.15	1.3	0.013	0.020	0.26	0.31	ND	0.44	ND	0.05

ND = Not Determined.

riverfront. The basic process at this plant is to convert bauxite ore to alumina (for later reduction to aluminum). The process involves the use of considerable caustic, and a fairly aggressive environment prevails in some portions of the plant. There is also a considerable accumulation of bauxite and alumina dust in many areas. A number of refinery and chemical plants are located across the river and adjacent to the plant. Because this installation represented a new application for the unpainted HSLA steel, a program to monitor corrosion performance was initiated shortly after construction was completed. Since then, the installation has been inspected and thickness measurements have been made periodically (sometimes on a spot check basis).

For the most part, the HSLA steels developed a tight, uniform dark brown oxide. Thickness measurements made after 9 years service indicate, with minor exceptions, no significant change in thickness. No indications of corrosion problems resulting from accumulations of bauxite and alumina dust have been noted. Several areas of flaking oxide were seen during these inspections, the most noticeable being the undersides of a number of horizontal connector plates at the very top level of the conveyor structure. During the two most recent inspections, a thinning of the top edge of several vertical "T" sections directly beneath and sheltered from the washing action of rain by these horizontal connector plates was noted. The corrosion product on these sections was lamellar and typical of that associated with corrosion in chloride containing environments. The affected area showing this type of corrosion was limited to a span of about 60 to 80 mm (200 to 300 feet) near the center protion of the conveyor structure.

A chemical analysis of the lamellar corrosion product showed 0.08 to 0.11% chloride. Based on past experience this level is sufficient to accelerate corrosion. The source of the chlorides is unknown since the plant does not process any chloride containing It is interesting to note that identical "T" sections interspersed among those exhibiting the accelerated attack, but not protected from the washing action of rain by horizontal connector plates above them, did not show appreciable attack.

The inspections also revealed that packout and metal distrotion (as a result of corrosion product formation) had developed at a number of bolted joints. This condition undoubtedly was adversely affected by the high chloride level in the atmosphere. Previous studies have shown that the development of this condition is directly related to wide bolt spacings which do not provide adequate joint stiffness. The previous work has resulted in guidelines for bolt spacings and bolt-to-edge distances to provide adequate joint stiffness to prevent distortion due to packout.[9] Briefly, these guidelines provide that the pitch (spacing on a line of fasteners adjacent to a free edge of plates or shapes in contact with one another) should not exceed 14 times the thickness of the thinnest part, and in any event should not exceed 18 mm (7 inches). The distance from the center of any bolt to the nearest free edge of plates or shapes in contact with one another should not exceed 8 times the thickness of the thinnest part, and in any event should not exceed 13 cm (5 inches). The above requirements are plotted in Figure 1.

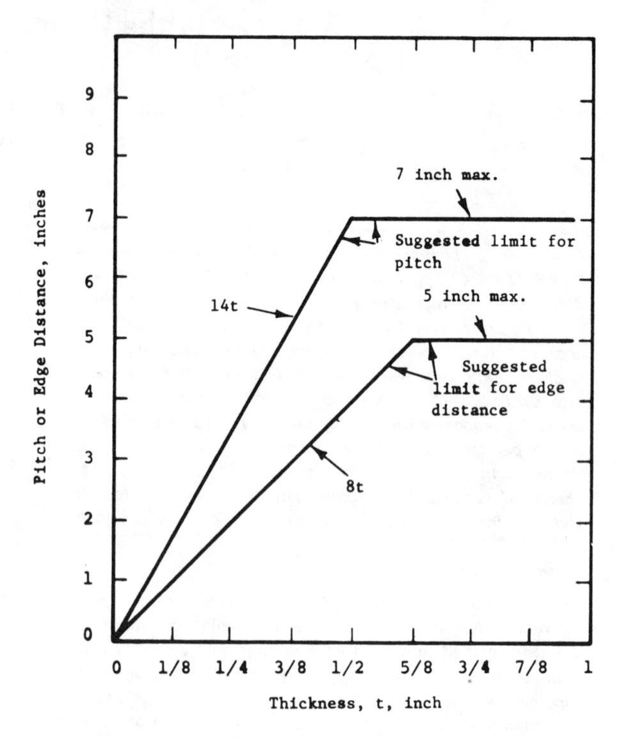

FIGURE 1 – Suggested spacing limits for joints in COR-TEN steel bolted structures.

Paper Mill Conveyor

In 1971, two bare COR-TEN A steel conveyor supports were erected in a paper mill in Georgia. The steel was erected in the as rolled condition without the removal of mill scale. The structures are in the midst of the paper mill and are exposed to the fumes from the processing units and nearby stacks. Shortly after construction was completed, a program was initiated to monitor the performance of the steel in this application.

Thickness measurements after about 4 years service showed that the steel was performing well. Minor pitting was observed under a buildup of wood chips on a few members; however, the pitting has caused no problems and has not affected the structural integrity of the tower. In this type of application, periodic removal of wood chip accumulations by a simple hosing action is advisable.

In most cases, the bolted joints appeared to have been sealed with corrosion products, and only a very slight indication of packout and distortion of the metal was seen. The one exception was a horizontal connection with widely spaced bolts. It is obvious that this bolt spacing did not provide adequate stiffness, and noticeable packout and metal distortion between the bolts occurred as a result. To correct this condition, it was suggested that the existing bolts be removed in a way that would not endanger the integrity of the sturcture, that any existing corrosion products be removed from within the joint, and that additional bolts with spacing to meet the proposed guidelines (as described previously) be installed.

Chemical Plant Structure

In 1971, a large chemical plant complex was constructed near

Baytown, Texas, in which all structural members were COR-TEN A and B steels. Except as required for reasons such as fireproffing, all the steel was erected in the as rolled condition without the removal of mill scale, and was left bare. This complex represents the most extensive use of bare low alloy steel in the chemical industry. Since completion, periodic inspections have been made to determine the performance of the steel throughout the complex.

For the most part, the bare low alloy steel has developed a tight, uniform, dark brown oxide. However, because corrosion was evident in locations near a hydrochloric acid unit, it was necessary to paint the structurals at this location. Evidence of chloride attack on structurals close to the acid unit was indicated by the development of a lamellar type rust.

A highly localized area of scaling and lamellar oxide development similar to that seen in the presence of chlorides was found near a phosgene absorber. The attack was limited to the lower platforms around the absorber column and to a few nearby pipe supports. Periodic inspeciton of these localized areas of attack is continuing to determine whether significant metal loss will occur with additional exposure, and whether it would be advisable to take any corrective action such as cleaning and painting the affected areas.

Waste Treatment Plant Sturcture

In late 1973, a waste treatment facility was erected with bare low alloy steel near New Martinsville, West Virginia. The steel was erected in the as rolled condition without the removal of mill scale. Because this type of facility is a new application for bare low alloy steel, the performance of the structure is being monitored.

The bare low alloy steel has weathered to a dark brown, uniform oxide over most of the structure. Scattered small areas of lighter oxide and some scaling have been observed on a few of the structurals toward the interior of the unit where exposure to the weather is limited. Thickness measurements made on this structure indicat no significant changes in thickness during a 2 year monitoring period.

Coal Conveyor Structure

A coal conveyor of bare low alloy steel was constructed in 1968. This structure consists of two 2.7 mm (9 foot) diameter tubular galleries, each about 55 mm (180 feet) long, housing a conveyor belt which transports coal from the preparation plant to a silo on the Ohio river.

A recent inspection of the conveyor gallery showed that the exterior exhibited a tight, uniform, dark brown oxide typical of that which develops on HSLA steel in most midly corrosive atmospheres. Inspection of the interior showed an accumulation of coal particles and coal dust on the bottom of the gallery throughout its length. There was evidence of moisture in the accumulated coal and dust on the bottom of the tube near the discharge end. Thickness measurements at several points on the tube wall at several locations along the gallery length showed that there has been no loss of metal on the side wall of the tube, but that on the bottom, where moist coal was present, a maximum loss of about 20% of the wall thickness has occurred in the 8 years service. These observations indicate that corrosion is occurring on the portion of the tube where wet coal is in contact with the low alloy steel. To prevent further corrosion of the tube, the bottom area near the discharge end was coated with a 0.30 to 0.38 mm (12 to 15 mil) thick application of a coal tar epoxy paint system.

Coal Preparation Plant

A coal preparation plant was constructed during 1969 to 1971 near Pineville, West Virginia, with all sturctural shapes and plates made of bare COR-TEN steel. The main structures are A-frame and tower support structures for the raw coal, clean coal, and rock conveyor systems. These supports range in height from a meter (several feet) to about 46 m (150 feet). They are constructed of various structural shapes but consist mainly of I-beams (legs) and angles (cross bracing) that are joined primarily by bolting.

Overall, the HSLA steel structural members are in good condition, and there is no evidence of distortion in the bolted joints. Corrosion in hte 0.25 to 0.38 mm (10 to 15 mil) range has occurred at several locations where moist soil, rock, and coal dust were in contact with the steel. These areas were cleaned and painted to minimize further corrosion.

Limestone Stackers

A number of the limestone stackers located near Rogers City, Michigan, are constructed of bare COR-TEN steel. These stackers and associated conveyor system, which were constructed in 1969, are a part of a transfer and storage system for the limestone piled at several locations in the plant.

Overall, the condition of these structures is good after 6 years service. However, 1.0 to 1.3 mm (40 to 50 mil) losses of the bare steel have occurred at locations where calcium chloride is added to the limestone during winter months to prevent the storage piles from freezing. Chemical analysis of a rust sample obtained at the area of salt exposure indicated that the sample contained 1.2% chloride. It was recommended that the areas exposed to the limestone containing salt be painted to minimize further corrosion. It is interesting to note that the inspection showed no significant joint distortion even at locations exposed to the salt. The inspected joints met the bolt guidelines described previously.

Process Side Applications

In addition to the wide application of structurals in chemical plants, COR-TEN steels have been used in a number of specific process side applications involving combustion exhaust gases. As in applications requiring atmospheric corrosion resistance, the degree to which these steels are more corrosion resistant than carbon steel depends on the occurrence of alternate wetting and drying and the percentage of time during which the surfaee is wetted by the corrodent.

Exhaust Gas Condensate Corrosion

For a number of years, COR-TEN A steel has been one of the widely specified construction materials for the sheet gage heat transfer elements of rotary regenerative combustion air heaters. The metallic elements are alternately exposed to hot [260 to 370 C (500 to 700F)] products of combusted fuels where the elements absorb heat from the gases, and then to air at atmospheric temperatures where latent heat from the elements heats the air as it passes by the elements. Sulfur compounds and water vapor from the combustion products condense on the cool elements, forming acid which causes corrosion of the metal surfaces.

A comprehensive test program conducted by the U. S. Bureau of Mines[10] revealed that COR-TEN A steel, although not immune, was substantially more resistant to acid condensate corrosion under the alternate condensing–noncondensing conditions than were carbon steel and several common grades of chromium and nickel-chromium stainless steels. The favorable resistance of HSLA steel to acid condensate corrosion has also been shown by short term tests conducted in the exhaust system of a water cooled furnace fired with high sulfur fuel oil[11] and of a steam generator fired with pulverized coal.[12] Test results are summarized in Figures 2 and 3.

As mentioned above, the degree to which the HSLA steel is more resistant than carbon steel appears to vary considerably with the percentage of time during which the metal is wetted by the condensed acid. Where the metal is continuously wetted, there is little or no difference in resistance between the two materials, as illustrated in Tables 2 and 3.

Both carbon steel and HSLA steel have been used over many years for liners and the wall of exhaust chimneys of boilers fired with coal and fuel oil. Success with the HSLA steel has been somewhat varied. Excellent resistance is obtained when the operating temperature is above the dewpoint of the exhaust gas. Where operating conditions lead to alternate wetting with condensate and drying, HSLA steel can exhibit a substantial advantage over carbon steel. When continuous or nearly continuous wetting with condensate occurs, corrosion of the two steels can be expected to proceed at about the same rate.

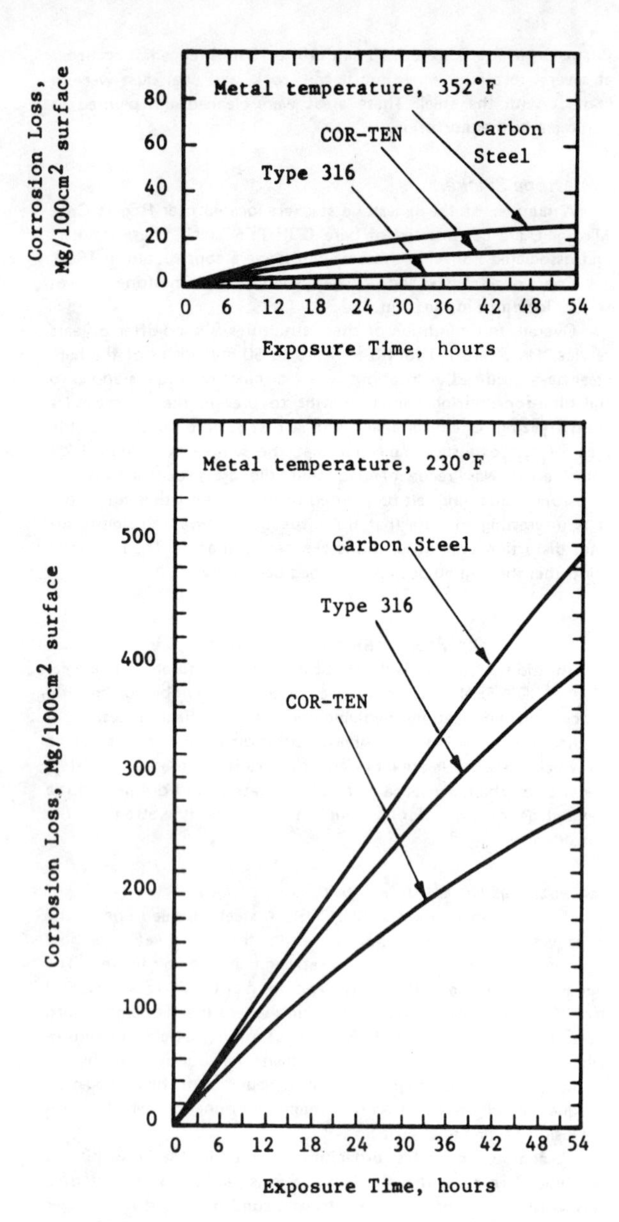

FIGURE 2 – Comparison of corrosion rates under exposure to fuel oil combustion product gas. Dew point of gas is 284 F (from Reference 11).

Elevated Temperature Oxidation

It is generally recognized that carbon steel is suitable for handling air and certain oxidizing combustion product gases at metal temperatures up to about 540 C (1000F). Long term tests in air at temperatures to 650 C (1200 F) have shown that COR-TEN A steel offers an additional degree of oxidation resistance, compared with carbon steel, at and above 540 C (1000 F) Figure 4. Under conditions of severe thermal fluctuations, COR-TEN A steel is much less prone to spalling of the surface oxide which, when intact, controls the rate of continuing oxidation of the underlying steel. This is shown by data given in Table 4. It should be pointed out that in this range of elevated temperatures other design considerations such as strength and ductility may be limiting factors in the use of steel as well as other products, and should be considered along with oxidation resistance in selecting these alloys for elevated temperature service.

Boiler and Cooling Water Service

Carbon steel is used extensively for components of heat

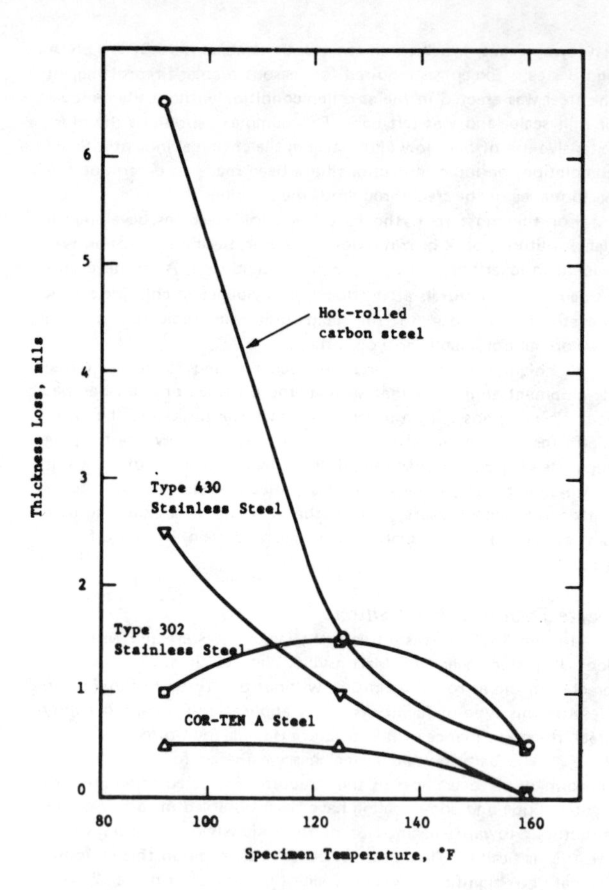

FIGURE 3 – Corrosion resistance of steels exposed to combustion product gases in pulverized coal fired steam generator (from Reference 12).

TABLE 2 – Summary of Data on Corrosion Resistance of SAE 1010 Carbon Steel and COR-TEN A Steel in Natural Gas Combustion Products[(1)]

Exposure Environment[(3)]	Average Corrosion Rate (mpy)[(2)] SAE 1010	USS COR-TEN A
Continuous Condensation at 80 F[(4)]	30	26
Cyclic Between 80 and 400 F[(4)]	38	21
Cyclic Between 80 and 600 F[(4)]	50	20
Cyclic Between 80 and 400 F[(5)]	50	32

[(1)]H. A. Pray, R. S. Peoples, R. S. Dalrymple, and C. T. Sims, Project No. DGR-4-CH, Rept. No. 2 to the AGA, published by the AGA, August, 1949.

[(2)]1 mpy = 0.0254 mm/y.

[(3)]Flue gas composition (Vol% on dry basis): 8.5 CO_2, 5.1 O_2, 86.4 N_2, and 0.0 CO. SO_2 metered in as specified. Total exposure time for all tests was between 500 and 600 hours. Cyclic exposure time was 26 minutes per cycle, which included 7 minutes above 80 F and 19 minutes at 80 F.

[(4)]Sulfur content–25 grains per 100 standard cubic feet of gas.

[(5)]Sulfur content–50 grains per 100 standard cubic feet of gas.

TABLE 3 – Corrosion Rates of Metal Exposed in Sinter Plant Exhaust Gas System

Material	Exposure Period (days)	Corrosion Rate (mpy)(1) Location A	Location B
Carbon Steel	78	127	159
USS COR-TEN A Steel	78	109	159
AISI Type 410 Stainless Steel	78	72	126
AISI Type 430 Stainless Steel	78	83	124
AISI Type 304 Stainless Steel	78	102	102
AISI Type 316 Stainless Steel	78	76	108

(1) 1 mpy = 0.0254 mm/y.

exchangers and condensers where steam and fresh and/or sea water is the heat transfer medium. Particularly in recirculated water systems, some sort of treatment is usually necessary to reduce the corrosivity of the water or a protective coating is used to minimize corrosion. In general, HSLA steel exhibits about the same degree of corrosion resistance as carbon steel in such applications, and the corrosion rate exhibited by both steels is controlled by the same factors, such as water velocity, dissolved oxygen concentration, temperature, and scaling propensity of the water. Table 5 summarized the results of tests in which COR-TEN A steel and carbon steel were exposed in boiler and cooling water systems. The results show that under the conditions noted, carbon steel and the HSLA steel exhibit roughly the same corrosion resistance.

TABLE 4 – Comparison of Oxidation Rates Under Constant Temperature and Cyclic Temperature Conditions (Exposure Period 972 Hours, Except Where Noted)

Material	Average Penetration(1) (mils)(2) Constant Temperature 1000 F	Constant Temperature 1200 F	Cyclic Temperature 1000 F	Cyclic Temperature 1200 F
Carbon Steel	1.2	9.5(3)	1.5	24(4)
USS COR-TEN A Steel	0.33	4.0	0.75	11

(1) Average of duplicate specimens.
(2) 1 mil = 0.0254 mm.
(3) Data for 815 hours exposure.
(4) Removed from test after 815 hours exposure because specimens were perforated.

Summary

Based on the information presented herein, the following conclusions can be drawn concerning the utilization of atmospheric corrosion resistant HSLA steels, in chemical plant applications:

1. Exposure and service tests have shown that these steels can be used in the unpainted condition in many chemical plant sturctural applications. Some areas in chemical plants, however, are sufficiently corrosive that HSLA steels should be used only in the

TABLE 5 – Corrosion Resistance of Carbon Steel and COR-TEN A Steel in Steam and Cooling Water Service

Test Location	Environment	Corrosion Rate (mpy)(1) Carbon Steel	COR-TEN A Steel
Boiler Drum	Steam, 625 F Boiler Water, 625 F $<$5 ppb Dissolved O_2	$<$0.1 $<$0.1	$<$0.1 $<$0.1
Condenser Hotwell–Fossil Fuel Boiler	Boiler Water, 85 F $<$5 ppb Dissolved O_2	4.5(2)	4.3(2)
Deaerating Heater–Fossil Fuel Boiler	Boiler Water, 320 F $>$5 ppb Dissolved O_2	7.7(2)	5.6(2)
Cooling Water System–Blast Furnace	River Water, 65 to 95 F, pH 5.5 to 7.0 Aerated	6.7	6.4
Tar Still Condenser–Tar Plant	River Water, 150 to 180 F, Well Aerated	16	15
Power Plant–Boiler Seals	River Water, 100 F, pH 2.8 to 4.5, Aerated	20	17
BWR-Reactor System(3)	Saturated Water, 550 F 100 ppb Dissolved O_2	0.15	0.15

(1) 1 mpy = 0.0254 mm/y.
(2) Corrosion penetration by pitting.
(3) Vreeland, D. C., Gaul, G. G., Pearl, W. L. Corrosion, Vol. 17, p. 269t (1961) June.

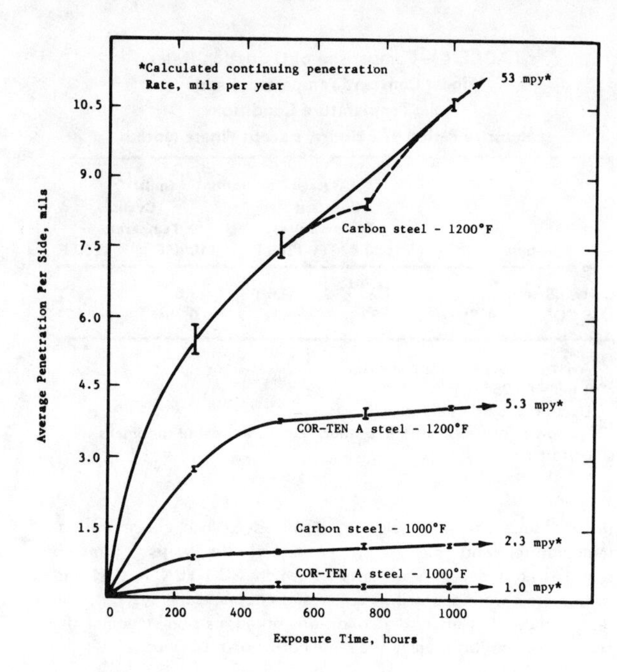

FIGURE 4 — High temperature oxidation tests in air constant temperature conditions.

painted condition, and in this event, an increased paint life can be expected. Atmospheric corrosion tests and monitoring programs on structures should be used to establish suitability. Bolted joints should conform with presented guidelines to provide joint stiffness and minimize packout.

2. In process applications, COR-TEN A steel exhibits substantially better corrosion resistance than carbon steel in applications involving exhaust gas condensates, provided that alternate wetting and drying occurs, and that the percentage of time that the steel is wetted by condensate is not excessive.

3. At high temperatures, COR-TEN A steel exhibits better oxidation resistance than carbon steel.

4. Under immersion or continuously wet conditions, the corrosion resistance of the HSLA steels is about the same as that of carbon steel.

Acknowdedgment

The authors would like to express their appreciation to their coworker, J. D. Swan, who participated in several of the plant inspections described herein.

References

1. American Society for Metals, Corrosion of Metals, p. 30 (1946).
2. Schramm, G. N., Taylerson, E. S., Larrabee, C. P. Railway Age, Vol. 101, p. 780 (1936).
3. Jones, J. A. J. Iron Steel Inst., Vol. 135, No. 1, p. 137 (1937).
4. Pilling, N. B., Wesley, W. A. Proc. ASTM, Vol. 48, p. 610 (1948).
5. Larrabee, C. P. Corrosion, Vol. 9, p. 259 (1953).
6. Copson, H. R. Proc. ASTM, Vol. 45, p. 554 (1945).
7. Copson, H. R., Larrabee, C. P. ASTM Bull. No. 242 (1959) Dec.
8. Schmitt, R. J., Mathay, W. L. Materials Protection, Vol. 6, No. 9, p. 37 (1967).
9. Brockenbrough, R. L., Schmitt, R. J. Paper presented at winter meeting of IEEE, New York, January 26-31, 1975 (Preprint C75 041-9).
10. Barkley, J. F. et. al. Bureau of Mines Report 4996, August, (1953).
11. Haneef, M., J. Inst. of Fuel, Vol. 33, p. 285 (1960) June.
12. Piper, J. D., Van Vliet, H. Trans. ASME, Vol. 80, p. 1251 (1958) August.

TESTS SHOW PERFORMANCE OF LOW ALLOY STEELS IN CHEMICAL PLANT ENVIRONMENTS

R. J. Schmitt
Applied Research Laboratory
United States Steel Corp.
Monroeville, Pa.

W. L. Mathay
Chemical Industry Marketing
United States Steel Corp.
Pittsburgh, Pa.

SUMMARY

The corrosion factors and economic aspects of two high-strength low-alloy steels as materials of construction for chemical plant structures are discussed. The steels are compared in various chemical plant environments with carbon steel, a common material for these structures. The higher cost and better corrosion resistance of the high-strength low-alloy steels are explained in an economic context.

SINCE THEIR development in the early 1930's, high-strength low-alloy steels with a yield point about 1½ times that of structural carbon steel have been produced for specific mechanical properties rather than resistance to chemicals. Some of the low-alloy steels, however, contain small amounts of alloying elements that impart superior atmospheric corrosion resistance.[1-6] These low-alloy steels derive their atmospheric corrosion resistance from tightly adherent oxide films that protect the surfaces from further corrosion.

The superior corrosion resistance often exhibited by low-alloy steels is not usually observed in aqueous and chemical environments. Therefore, under nonatmospheric corrosive conditions the low-alloy steels cannot be assumed to have better corrosion resistance than carbon steel.

Copson and Larrabee[7] report that paint coatings are more durable on low-alloy steels than on carbon steel. Any rust that forms is less voluminous on low-alloy steels, so there is less rupturing of the paint film and less moisture reaches the steel to promote further corrosion.

The corrosion resistance properties of the low-alloy steels, with their high strength, have resulted in their wide use to obtain longer service life and paint life. However, only limited use has yet been made of these steels in chemical plant atmospheres which are generally more aggressive than the environments where low-alloy steels are now being used.

To obtain information on the corrosion properties of low-alloy steels in chemical plant atmospheres, the United States Steel Corp. Applied Research Laboratory initiated tests in one of U. S. Steel's coal-chemical plants. Two years of testing indicated that the corrosion resistance of the low-alloy steel (Cr-Si-Cu-Ni-P composition)[(1)] was at least twice that of carbon steel.

Because of these encouraging findings, a more extensive program was undertaken to obtain complete information on the atmospheric corrosion behavior of low-alloy steels. Various chemical firms cooperated in the program. The low-alloy steels evaluated in this program are of the chemical compositions shown in Table 1 and are referred to as low-alloy A[(2)] and low-alloy B steel. Low-alloy A steel meets the ASTM A 242[(3)] designation of thickness up to ½ inch, inclusive. Low-alloy B steel meets the requirements of ASTM A 242 and ASTM A 441 (modified) in thicknesses to 4 inches, inclusive. ASTM A 242 specifies tensile properties but not the alloying elements required to obtain enhanced atmospheric corrosion resistance. Hence, if a low-alloy steel with atmospheric corrosion resistance equivalent to that of the steels tested is desired, a steel must be used that has a nominal composition similar to those shown in Table 1.

Experiments

Materials

Three corrosion test racks of Type 316 stainless steel were installed at each of the chemical plants to permit removal and evaluation of one complete set of specimens after exposure periods of 6 months, one year, and 2 years. Specimens included unpainted and painted carbon steel, unpainted and painted low-alloy A and B steels and hot-dipped galvanized carbon steel and low-alloy A steel. The chemical compositions of the steels tested are shown in Table 1.

Atmospheric-Corrosion Tests

Prior to exposure the ⅛-inch thick by 4-inch wide by 6-inch long specimens of hot-rolled carbon steel and low-alloy steels were degreased in acetone, pickled in a 20% hydrochloric acid solution then rinsed, dried, and weighed. The galvanized steel speci-

Table 1—Chemical Compositions of Carbon Steel and High-Strength Low-Alloy Steels Used in Chemical Plant Atmospheric Corrosion Studies

	COMPOSITION, PERCENT BY WEIGHT									
TYPE OF STEEL	C	Mn	P	S	Si	Cu	Ni	Cr	Mo	V
Carbon Steel	0.16	0.63	0.012	0.031	0.012	0.01	0.01	0.03	ND	ND
Low-Alloy A Steel(1)	0.09	0.46	0.075	0.031	0.52	0.31	0.40	0.74	0.025	ND
Low-Alloy B Steel(2)	0.15	1.3	0.013	0.020	0.26	0.31	ND	0.44	ND	0.05

ND - Not Determined.

Note: These steel compositions were also used for hot-dipped galvanized specimens and painted specimens.

(1) United States Steel's USS COR-TEN A High-Strength Low-Alloy Steel.

(2) United States Steel's USS COR-TEN B High-Strength Low-Alloy Steel.

(1) United States Steel's USS COR-TEN A High-Strength Low-Alloy Steel.

(2) Low-Alloy A and B steels are United States Steel's USS COR-TEN A and B steels, respectively.

(3) Whereas ASTM A 242 describes the properties for structural shapes, plates, and bars, for welded, riveted, or bolted construction, ASTM A 374 and A 375, respectively, deal with cold-rolled and hot-rolled steel sheets and strips in cut lengths and coils.

mens were hot-dipped to an average coating thickness of about 2 mils. The coated carbon steel and low-alloy steel specimens were sandblasted before the organic coating was applied. The specimens were spray coated with a wash primer and top-coated with 4 to 5 mils (dry film thickness) of a vinyl coating commonly used for chemical-plant structures. A scribe mark (X) was made on the skyward surface of each of the organic-coated specimens to determine whether the resistance of the organic film to undercutting differed with the steel substrate.

The specimens were exposed for periods ranging from 6 months to 2 years. The locations of the test racks and the types of atmospheres encountered are listed in Table 2. At the end of the various periods, the unpainted carbon steel and low-alloy steel specimens were cleaned in a molten bath of sodium hydroxide containing 1 to 2% sodium hydride and were reweighed. The average reduction in thickness of each specimen was calculated on the basis of weight loss. The hot-dipped galvanized and the coated panels were cleaned with mild soap and hot water. The performance of the galvanized coating was evaluated by the degree of roughening of the surface and the amount of red rust. The vinyl coating was judged by the extent of undercutting at the scribe mark.

Table 2—Locations of Corrosion Test Specimens and Types of Atmospheres Encountered

TEST	LOCATION	ATMOSPHERIC CONSTITUENTS
A	Elastomers plant	Chlorine and sulfur compounds
B	Chlor-alkali plant	Moisture, lime, and soda ash
C	Chlor-alkali plant	Moisture, chlorides, and lime
D	Sulfur plant	Chlorides, sulfur, and sulfur compounds
E	Petrochemical plant	Chlorides, hydrogen sulfide, and sulfur dioxide
F	Sulfuric acid plant	Sulfuric acid fumes
G	Chlorinated hydrocarbons plant	Chlorine compounds
H	Petrochemical plant	Ammonia and ammonium acetate fumes
I	Detergent plant	Alkalies and organic compounds
J	Detergent plant	Sulfur compounds

Results and Discussion

Unpainted Steels

The corrosion test specimens after one year exposure in the chlorinated hydrocarbons plant (G) are shown in Figures 1, 2, and 3. The calculated average reductions in thickness for the steels exposed in the 10 chemical atmospheres are shown in Table 3. A plot of the average reduction in thickness against time for the low-alloy steels and carbon steel exposed in the chlorinated hydrocarbons plant is shown in Figure 4.

The data in Table 3 show that the atmosphere in the sulfur plant (D) was substantially more corrosive than the other 9 atmospheres. The atmospheres in the chlor-alkali plant (C) and the chlorinated hydrocarbons plant (G) were the next most corrosive environments, but these atmospheres were appreciably less corrosive than the sulfur plant atmosphere. The least corrosion occurred in the detergent plant (I).

Table 3 shows that the corrosion resistance of the low-alloy steels was superior to that of the carbon steel in all the atmospheres. In fact, the low-alloy A steel was generally at least twice as corrosion resistant as carbon steel. However, the corrosion curve in Figure 4 shows that corrosion rates for the low-alloy steels decreased markedly after 6 months exposure. This indicates that a protective oxide is forming on the steels and that 6-month tests are not sufficient to assess the true merit of these steels. Thus, if the slopes of the curves are considered for all the exposures of long duration, the corrosion resistance of the low-alloy A steel can be calculated as 2 to 13 times that of the carbon steel. Low-alloy B steel was generally superior to carbon steel but was slightly less resistant than low-alloy A steel.

These differences in corrosion performance are due to differences in the oxide films that form on the steels. The oxide films on the low-alloy steels were found to be superior to the film on the carbon steel in all cases. In Figure 2, the oxide film is flaking off the carbon steel specimen, but it is still intact on the low-alloy steel specimens. A more dramatic example of the differences in the protective qualities of the oxides is shown in Figure 5 where specimens of carbon steel and low-alloy A steel were exposed to the atmosphere of a coal-chemical plant for 2 years. A nonprotective lamellar oxide formed on the carbon steel and an adherent protective oxide film formed on the low-alloy A steel.

Galvanized Steel

The corrosion performance of galvanized carbon-steel and galvanized low-alloy A steel specimens exposed to the 10 atmospheres is shown in Table 4. The six-month results indicate that the galvanized coatings on carbon-steel and low-alloy A steel specimens were slightly attacked by seven of the ten

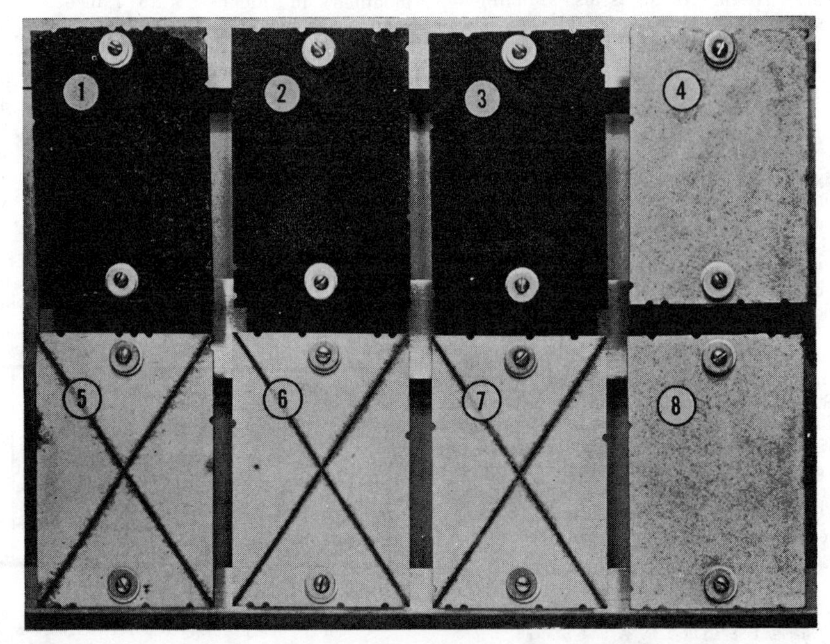

Figure 1—Skyward sides of corrosion test specimens after 12 months exposure in a chlorinated hydrocarbons plant, about 1/5 actual size. (1) structural carbon steel, (2) low-alloy A steel, (3) low-alloy B steel, (4) hot-dipped galvanized carbon steel, (5) painted structural carbon steel, (6) painted low-alloy A steel, (7) painted low-alloy B steel, (8) hot-dipped galvanized low-alloy A steel.

chemical environments (A, B, D, E, F, H, and J). However, the galvanized steels exposed to the atmosphere near the brine tank of the chlor-alkali plant (C), the atmosphere in the chlorinated hydrocarbons plant (G), and the highly alkaline atmosphere in the detergent plant (I) were moderately attacked.

The galvanized coating on the steels continues to provide good protection after 12 months exposure in the chlor-alkali plant (B), sulfur plant (D), petrochemical plant (E), sulfuric acid plant (F), and chlorinated hydrocarbons plant. However, after 2 years exposure in the atmosphere near the brine tank of the chlor-alkali plant, and in the highly corrosive atmosphere of the sulfur plant, the galvanized coating has been severely attacked and does not protect the steel substrate.

Painted Steels

The performance of the painted carbon and low-alloy steels exposed to the 10 atmospheres is shown in Table 5. For 6 months the vinyl coating provided relatively good protection to both carbon steel and the low-alloy steels at 8 of the 10 areas involved. The 8 were the elastomers plant (A), chlor-alkali plant (B), petrochemical plants (E and H), sulfuric acid plant (F), chlorinated hydrocarbons plant (G), and detergent plants (I and J). Further, Table 5 shows that where paint performance differed the performance of the coating was always better on the low-alloy steels than on carbon steel. The locations in which an improvement was noted with the low-alloy steels include the chlor-alkali plant, petrochemical plants, sulfuric acid plant, chlorinated hydrocarbons plant, and the detergent plant. The superior performance of paint on the low-alloy A steel specimens as compared with the carbon steel spcimens after 12 months exposure in the chlorinated hydrocarbons plant is shown in Figure 3.

The longer term exposures suggest that the use of the scribe mark (X) to evaluate the performance of the paint on the different steel substrates may have been too severe a test for the aggressive atmospheres encountered because most of the paint coating deteriorated rapidly after the 6-month evaluation, as shown in Table 5. This was particularly true after 2 years exposure in the highly aggressive atmospheres of the chlor-alkali plant and the sulfur plant where the paint coating was undercut over most of the surfaces of the specimens. In a milder environment that may be more typical of the average chemical plant atmosphere the paint coating continues to provide protection, as indicated by the 2 year specimens from the petrochemical plant (Figure 6). Again, the performance of the coating was better on the low-alloy steels than on carbon steel.

Examination of the painted specimens after 2 years exposure in the chlor-alkali plant and the sulfur plant showed that the paint film was extensively undercut on both the carbon steel and low-alloy steel specimens.

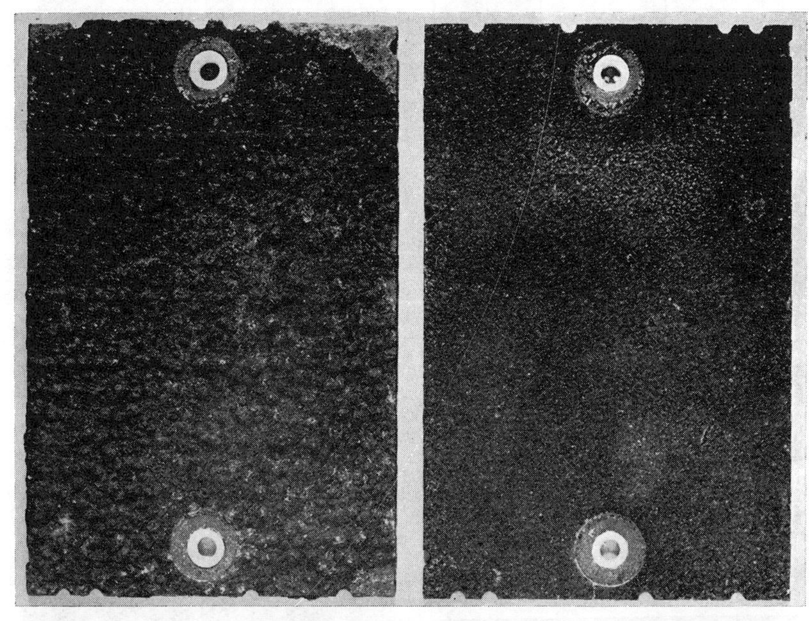

Figure 2—At left is a specimen of structural carbon steel and at right is a specimen of low-alloy A steel. Both have been exposed for 12 months in a chlorinated hydrocarbons plant. About 1/2 actual size.

Figure 3—At left is a specimen of painted structural carbon steel and at right is a specimen of painted low-alloy A steel. Both have been exposed for 12 months in a chlorinated hydrocarbons plant. Note varying degrees of undercutting at scribe marks. About 1/2 actual size.

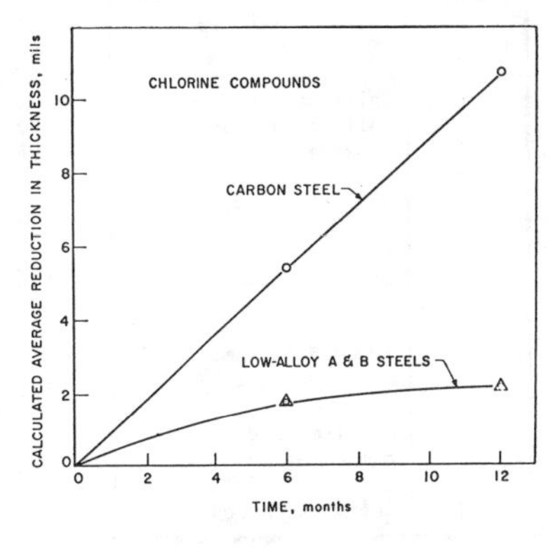

Figure 4—Comparative corrosion rates of low-alloy steels and structural carbon steel exposed in chlorinated hydrocarbons plant.

Source: *Materials Protection*, September 1967, 37-42

Table 3—Corrosion Losses for High-Strength Low-Alloy Steels and Carbon Steel Exposed to Various Atmospheres in Chemical Plants

TEST	ATMOSPHERIC CONSTITUENTS	EXPOSURE PERIOD MONTHS	AVERAGE REDUCTION IN THICKNESS, MILS		
			CARBON STEEL	LOW-ALLOY A STEEL	LOW-ALLOY B STEEL
A	Chlorine and sulfur compounds	6	1.3	0.8	0.9
B	Moisture, lime, and soda ash	6	2.7	1.2	1.3
		12	4.7	1.7	1.8
		24	8.3	2.1	1.9
C	Moisture, chlorides, and lime	6	4.1	2.4	2.7
		12	9.6	3.2	3.9
		24	18.8	5.7	7.4
D	Chlorides, sulfur, and sulfur compounds	6	15.5	7.4	9.4
		12	26.0	10.9	18.5
		24	43.3	20.4	32.4
E	Chlorides, hydrogen sulfide, and sulfur dioxide	6	2.0	0.9	1.2
		12	3.0	1.2	1.6
		24	3.4	1.2	1.9
F	Sulfuric acid fumes	6	3.3	1.8	1.9
		12	4.5	2.1	2.2
G	Chlorine compounds	6	5.4	1.8	1.8
		12	10.7	2.2	2.2
H	Ammonia and ammonium acetate fumes	6	1.5	1.0	1.1
		12	2.3	1.3	1.9
I	Alkalies and organic compounds	6	0.8	0.6	0.6
J	Sulfur compounds	6	1.2	0.6	0.9

Table 4—Corrosion Performance of Galvanized[1] Carbon Steel and High-Strength Low-Alloy Steel Exposed to Various Atmospheres in Chemical Plants

TEST	ATMOSPHERIC CONSTITUENTS	EXPOSURE PERIOD MONTHS	CORROSION PERFORMANCE RATING[2]	
			GALVANIZED CARBON STEEL	GALVANIZED LOW-ALLOY A STEEL
A	Chlorine and sulfur compounds	6	A	A
B	Moisture, lime, and soda ash	6	A	A
		12	A	A
		24	B	B
C	Moisture, chlorides, and lime	6	B	B
		12	B	B
		24	C	C
D	Chlorides, sulfur, and sulfur compounds	6	A	A
		12	B	B
		24	C	C
E	Chlorides, hydrogen sulfide, and sulfur dioxide	6	A	A
		12	A	A
		24	A	A
F	Sulfuric acid fumes	6	A	A
		12	A	A
G	Chlorine compounds	6	B	B
		12	B	B
H	Ammonia and ammonium acetate fumes	6	A	A
		12	A	A
I	Alkalies and organic compounds	6	B	B
J	Sulfur compounds	6	A	A

(1)Hot-Dipped Galvanized--Average Coating Thickness About 2 Mils.

(2)Ratings: A - Slight attack, B - Moderate attack, C - Severe attack.

However, the low-alloy steels showed considerably less corrosive attack under the paint film. Thus, even where paint films on the low-alloy steels are damaged, the superior corrosion resistance of these steels can permit greater freedom in maintenance painting schedules than is practical with carbon steel.

Economic Considerations

The materials engineer is faced with the need to minimize the weight of equipment while obtaining longer service life with lower maintenance cost. For particular applications, the use of reduced weight of the low-alloy steel, combined with its high strength can narrow or eliminate the difference in total cost between these steels and carbon steel. (The mill base price for low-alloy A and B steels is about 2.5 cents per lb more than that for carbon steel.)

Maintaining the same thickness for low-alloy steel as carbon steel may be desirable because the high costs of replacement make service life an important consideration. Although thickness of materials depends on the structural and economic factors involved in any particular application, the combination of superior corrosion resistance and high strength gives low-alloy steels

Figure 5—Above is a specimen of low-alloy A steel and below is a specimen of carbon steel. Both have been exposed for 2 years in a coal-chemical plant.

Table 5—Performance of Painted[1] Carbon Steel and High-Strength Low-Alloy Steel Specimens Exposed to Various Atmospheres in Chemical Plants.

TEST	ATMOSPHERIC CONSTITUENTS	EXPOSURE PERIOD MONTHS	PAINT PERFORMANCE RATING(2)		
			CARBON STEEL	LOW-ALLOY A STEEL	LOW-ALLOY B STEEL
A	Chlorine and sulfur compounds	6	NU	NU	NU
B	Moisture, Lime, and soda ash	6	MU	U	U
		12	SU	MU	MU
		24	CF	CF	CF
C	Moisture, chlorides, and lime	6	MU	MU	MU
		12	SU	SU	SU
		24	CF	CF	CF
D	Chlorides, sulfur, and sulfur compounds	6	MU	MU	MU
		12	SU	SU	SU
		24	CF	CF	CF
E	Chlorides, hydrogen sulfide, and sulfur dioxide	6	MU	NU	NU
		12	MU	U	U
		24	SU	U	U
F	Sulfuric acid fumes	6	U	NU	NU
G	Chlorine compounds	6	MU	U	U
		12	SU	MU	MU
H	Ammonia and ammonium acetate fumes	6	U	NU	NU
		12	U	U	U
I	Alkalies and organic compounds	6	U	NU	NU
J	Sulfur compounds	6	NU	NU	NU

(1)Panels coated with vinyl resin paint system (4 to 5 mils dry thickness) considered suitable for chemical-plant structures.

(2)Ratings: NU - No undercutting at scribe mark.
U - Slight undercutting at scribe mark.
MU - Moderate undercutting at scribe mark.
SU - Severe undercutting at scribe mark.
CF - Complete failure.

Figure 6—Skyward sides of corrosion test specimens after 2 years exposure in a petrochemical plant, about 1/5 actual size. (1) structural carbon steel, (2) low-alloy A steel, (3) low-alloy B steel, (4) hot-dipped galvanized carbon steel, (5) painted structural carbon steel, (6) painted low-alloy A steel, (7) painted low-alloy B steel, (8) hot-dipped galvanized low-alloy A steel.

a definite advantage over carbon steel for chemical plant structures.

Service experience since the introduction of the low-alloy steels has proven their value as structural materials. Steel fabricators should encounter no difficulty in establishing procedures for fabricating low-alloy steels, because these steels, like carbon steel, can be gas cut, sheared, punched, reamed, sawed, milled, drilled, cold-formed, and welded satisfactorily and economically.

A difference between low-alloy and carbon steel is encountered in cold-forming. This difference necessitates the use of more liberal bend radii, slightly increased die clearance, and the provision for greater springback with the low-alloy steels.[8, 9]

The low-alloy steels may be welded by all the usual methods, using good shop or field practice to provide welds with the desired strength and ductility. In single bead welds, because of diffusion of the low-alloy steel base metal with the filler metal, the difference between the corrosion resistance of the weld and that of the base metal is very slight even when a carbon-steel electrode is employed. When built-up welds are made, insufficient diffusion of the base metal into the filler metal could occur and the weld would not be as corrosion resistant as the base metal. Therefore, when corrosion resistance is a critical factor, an electrode similar in composition to the low-alloy steel or an electrode containing 2 to 3% nickel should be used for built-up welds.

Information provided by several fabricators indicates that any extra costs involved in the fabrication of the low-alloy steels will be small (probably less than 10%). Of course, fabrication costs vary with the local labor market, the equipment available, the experience of the fabricator, and the general level of business.

Nevertheless, some of the economic aspects of utilizing low-alloy steels in place of structural grade carbon steel in corrosive chemical-plant atmospheres deserve consideration. It should be noted that in inflationary periods, the trend is to generally favor the use of articles with longer life.

Suppose, for example, that a company plans a new chemical plant requiring 1000 tons of structural steel and considers using unpainted steel structures. The company determines that it can purchase carbon steel structural shapes at a mill base price of $117 per ton and low-alloy A steel structural shapes at a mill base price of $167 per ton. Carbon steel exposed to the aggressive chemical process atmosphere in this particular plant is assumed to maintain its structural integrity for about 4 years, whereas low-alloy A steel is assumed to maintain its structural integrity for about 8 years under the same environmental conditions. This relationship is realistic, because the corrosion studies indicated that the low-alloy A steel was at least twice as corrosion resistant as carbon steel.

The structural carbon steel will cost the company $117,000 initially, and after 4 years a second expenditure of $117,000 for structural materials will be necessary. The low-alloy A steel structurals will cost $167,000 initially, but no further expenditures will be considered necessary for at least 8 years. The company must, therefore, weigh the possibility of investing an additional $50,000 now, as opposed to $117,000 in four years.

Companies, like lending institutions, can evaluate their investment opportunities in terms of an interest rate, or a discounted cash flow rate-of-return on investment. The rate of return on the $50,000 invested now to avoid the future expenditure of $117,000 can be determined by considering the former as a present value (PV) and the latter as a future value (FV). (Future value is defined as the amount to which a sum of money plus its earnings will accumulate by a given date.) Therefore, using compound interest tables and the relationship

$$PV = \frac{FV}{(1 + I)^n}$$

where I is the earnings rate for each period and n is the number of periods or years, the rate of return can be calculated:

$$\$50{,}000 = \frac{\$117{,}000}{(1 + I)^4}$$

$$I = 0.238 \text{ or } 23.8\%$$

Thus, the investment of the additional $50,000 for low-alloy A steel structurals is equivalent to obtaining a rate-of-return of 23.8% before taxes and without considering fabrication costs.

Jelen[10, 11] and Dillon,[12, 13] have pointed out that the factors to be considered in economic evaluations are initial cost, service life, interest rate, tax rate, and depreciation schedule. Hence, a more detailed analysis of the problem has been made with a computer to develop a discounted cash flow rate-of-return recognizing factors such as 1) normal life depreciation schedules, 2) a double-declining/straight line depreciation, and 3) a federal income tax rate of 48%. For purposes of this paper, the fabrication costs for carbon steel have been taken as $240 per ton and those for low-alloy A steel as $264 per ton (10% higher than those for carbon steel). The results of the computer analysis indicate that when the service life of the low-alloy A steel is twice that of structural carbon steel, a rate-of-return of about 45% is obtained.

An analysis of the problem using the discounted cash flow technique as set forth by Dillon[13] has also been made to determine an equivalent uniform annual cost for both structural carbon steel and low-alloy A steel. Equivalent uniform annual cost may be defined as that amount of money which must be invested on a yearly basis for a given period of time to return the same amount of money obtained by investing a lump sum at the beginning of the period. The uniform annual costs based on a service life ratio of low-alloy steel to carbon steel of 2 to 1 are $68,647 for carbon steel and $52,257 for low-alloy A steel. The difference between the two materials in favor of the low-alloy steel is $16,390. Hence, both analyses economically justify the use of the low-alloy steel in place of structural carbon steel.

In this regard, one of the largest chemical plant applications of the low-alloy steels is being made at the new coke-oven gas and anhydrous ammonia facilities of the U. S. Steel Corp. coal-chemical plant mentioned earlier. About 5,000 tons of the low-alloy A and B steels are being utilized in the form of various structurals, as well as in the exposed outer shells for several double-wall low-temperature process units. Most of the facilities will be exposed to the atmosphere and all exposed steel will be painted for longer service life. The decision to use the low-alloy A and B steel structurals in the new plant was based on the results of atmospheric corrosion tests which showed that these steels were considerably more corrosion resistant than carbon steel.

Elsewhere, painted low-alloy A and B steel structurals have been used for the construction of a 4-story open-framework building containing equipment for the production of chlorinated hydrocarbons. This installation was made on the basis of the superior performance exhibited by the low-alloy steels in Test G. A similar structure in the unpainted condition is currently under consideration at another plant. Low-alloy steels are also finding increased use for exterior pipe supports and pipe bridges in both the painted and unpainted conditions and are being evaluated in the unpainted condition for exhaust stacks where the temperature of the stack is above the dew point of the gas.

Summary

The results of these tests show that the low-alloy A and B steels are appreciably more resistant than carbon steel to corrosion by chemical plant atmospheres. Although a protective coating may be required for maximum service life in many of the environments encountered, the use of these low-alloy steels for structurals and plant equipment exposed to the environments appears to have definite advantages both from the corrosion and economic standpoints.

References

1. Corrosion of Metals, Published by ASM, 1946, p. 30.
2. G. N. Schramm, E. S. Taylerson and C. P. Larrabee. Corrosion of Steel Cars by Coal, *Railway Age*, 101, 780 (1936).
3. J. A. Jones, *J. Iron and Steel Institute*, 135, No. 1, 137 (1937).
4. N. B. Pilling and W. A. Wesley, *Proc. ASTM*, 48, 610-617 (1948).
5. C. P. Larrabee. Corrosion Resistance of High-Strength Low-Alloy Steels as Influenced by Composition and Environment, *Corrosion*, 9 (1953).
6. H. R. Copson. A Theory of the Mechanism of Rusting of Low-Alloy Steels in the Atmosphere, *Proc. ASTM*, 45, 554 (1945).
7. H. R. Copson and C. P. Larrabee. Extra Durability of Paint on Low-Alloy Steels, ASTM Bulletin No. 242, December, 1959.
8. C. E. Loos. *Steel Processing*, pp 755-759, December, 1945.
9. H. M. Priest and J. A. Gilligan, Design Manual for High-Strength Steels, 1956.
10. F. C. Jelen. *Chemical Engineering*, Vol. 63, pp 165-169, May, 1956.
11. F. C. Jelen. *Chemical Engineering*, Vol. 64, pp 271-275, September, 1957.
12. C. P. Dillon. *Materials Protection*, 4, 5, 38-45 (1965) May.
13. C. P. Dillon. *Materials Protection*, 5, 6, 47-49 (1966) June.

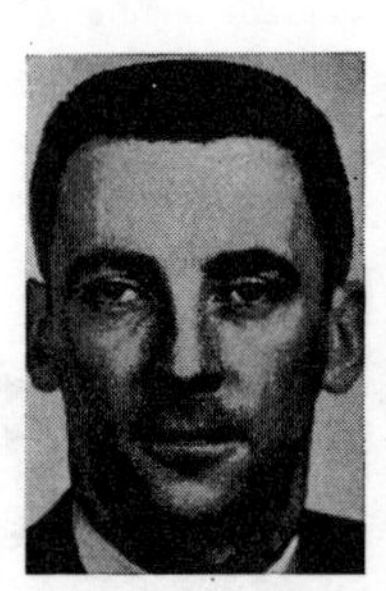

ROBERT J. SCHMITT is section supervisor of the Corrosion Technology Division of the U.S. Steel Corp. Applied Research Laboratory, Monroeville, Pa. He supervises work on steel products for applications in natural environments. He has a BS from the University of Pittsburgh. A member of NACE for 10 years, he is also a member of AIChE and AWWA.

WILLIAM L. MATHAY is manager of the Chemical Industry Marketing Division of the U.S. Steel Corp. Five Gateway Center, Pittsburgh, Pa. He coordinates application development and promotional activities on usage of steel products in chemical process industries. He has a BS in chemistry from Thiel College, has done graduate work at Carnegie Institute of Technology, and has participated in specialized training courses at Ohio State University and Syracuse University. An NACE member for 13 years, he is a member of TAPPI and AIChE and is listed in AMERICAN MEN OF SCIENCE.

Kinetics of the Atmospheric Corrosion of Low-Alloy Steels in an Industrial Environment★

*R. A. LEGAULT and A. G. PREBAN**

Abstract

It has been demonstrated that the natural atmospheric corrosion behavior of low-alloy steels in industrial environments can be accurately described by an equation of the form: $\Delta W = K t^N$. Thus, reliable predictions of long-term behavior become possible, and a vehicle is provided for accurately assessing the effect on atmospheric corrosion behavior of alloying additions, processing variables, and differences in exposure environment.

The reaction of a metal with oxygen is generally conceded to be governed in large part by the manner in which, and the rate at which, the oxide layer grows. It has often been possible to establish precise mathematical relationships between metal oxidation and time. For some of these relationships, it has also been possible to attach a physical meaning to the respective equations and, as a result, make possible an assessment of the relevant mechanisms. An awareness of these would enable the effects of alloying additions and processing variables on oxidation behavior to be predicted.

The atmospheric corrosion behavior of low-alloy steels is also governed primarily by the extent to which the corrosion product oxide, formed during the atmospheric exposure, provides protection to the underlying metal. The quality of this protection can be characterized by the rate of growth of the corrosion-product oxide. Rate equations have rarely, if ever, been determined for the corrosion of metals in natural atmospheric environments. Such determinations require that data be obtained for exposure times of the order of years rather than of minutes. Also, the control of experimental parameters that is enjoyed in the usual laboratory situation is obviously not available, and consequently such a system has commonly, and understandably, been deemed to complex to approach in this fashion.

The purpose of this work was to demonstrate that the natural atmospheric corrosion behavior of low-alloy steels in industrial environments can be accurately described by rate equations of a simple form. As a result, these equations can be used to reliably predict long-term behavior and to assess the influence of alloying additions, processing variables, and variations in the atmospheric environment.

Experimental

Experimental alloys were cast, occasionally from vacuum induction melts, more often from air induction melts, to which the various alloying elements were added at the indicated concentrations. Ingots weighing approximately 50 lbs (23 kg) were cropped, top and bottom, and then furnace soaked for 3 hours at 2250 F (1230 C) in an atmosphere consisting of 300 CFH argon and 10 CFH hydrogen. The material was hot rolled to a thickness of approximately 0.5 inch (1.3 cm), cut into 15-inch (38 cm) lengths, and furnace soaked again for a half-hour under the same conditions as those used in the original soak. Hot rolling was then continued to a final thickness of approximately 0.1 inch (2.5 mm).

The hot-rolled material was then cut into 4 x 6 inch (10 x 15 cm) specimens and pickled in 10 to 20 Vol % HCl at a temperature of 160 F (70 C) for 5 to 10 minutes. Specimens were scrubbed by hand using abrasive pads while rinsing under a stream of cold running water.

The material was then cold-rolled to approximately 0.035 inch (0.9 mm) (20 gauge), degreased in trichloroethylene and dried, prior to annealing for 8 hours at 1250 F (675 C) in a 3% hydrogen-97% argon atmosphere. After a slow furnace cool to room temperature, the experimental steels were recut into 4 x 6 inch test specimens for atmospheric exposure.

Identifying marks were applied to the test specimens, dimensional measurements were made and weighings were obtained. The specimens were then exposed at an angle of 30° from the horizontal with the upper surface facing south. A sufficient number of panels were exposed to provide at least four-fold replication for each exposure-time level. The atmospheric exposure site used was in East Chicago, Indiana, which would be classified as a severe industrial site, comparable in all likelihood to the ASTM site in Newark/Kearny, N. J.

After exposure, test specimens were stripped of corrosion product by immersion into molten caustic[1] which was maintained at a temperature of 700 ± 20 F (370 ± 10 C) and which contained 1.5 to 2.0% sodium hydride. The sodium hydride was introduced into the molten caustic by the addition of 1 ounce chunks of sodium while sparging the molten liquid with gaseous hydrogen. Metal loss during the stripping operation has been demonstrated to be negligible. Specimens were then rinsed in running water and air dried before the final weighing.

★Submitted for publication April, 1974.

*Inland Steel Research Laboratories, 3001 E. Columbus Drive, East Chicago, IN 46312.

TABLE 1 – Full-Factorial Design Used in the Mo-Cu-Ni Series

Alloy	Additive Concentration (%)		
	Mo	Cu	Ni
1FF01	0	0	0
1FF02	1	0	0
1FF03	0	1	0
1FF04	0	0	1
1FF05	1	1	0
1FF06	0	1	1
1FF07	1	0	1
1FF08	1	1	1

TABLE 2 – Cumulative Penetration Depths Averaged Over Specimen Surfaces

Alloy No.	Composition	Average Penetration (mils)			
		3 Months	6 Months	1 Year	2 Years
1FF01	Blank	0.929	1.215	1.656	2.432
1FF02	Mo	0.998	1.324	1.769	2.532
1FF03	Cu	0.851	1.130	1.494	2.096
1FF04	Ni	0.848	1.105	1.471	2.110
1FF05	Mo Cu	0.924	1.189	1.475	2.024
1FF06	Cu Ni	0.823	1.068	1.392	1.908
1FF07	Mo Ni	0.884	1.151	1.489	2.066
1FF08	Mo Cu Ni	0.857	1.083	1.346	1.850

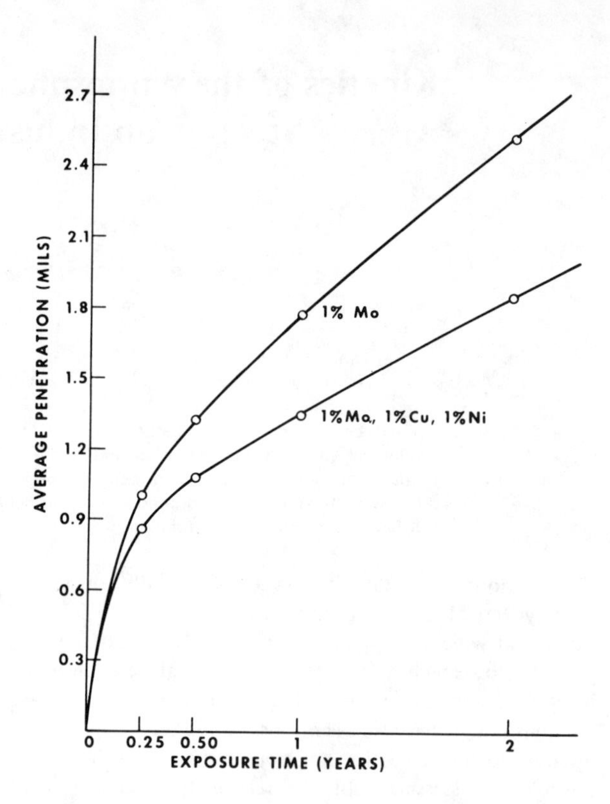

FIGURE 1 – Changes in average cumulative penetration depth with time of exposure.

Results

Atmospheric corrosion data have been obtained on scores of experimental low-alloy steels exposed at an industrial test site for various times ranging from 3 months to several years. A typical example involves corrosion measurements obtained on a full-factorial series of experimental alloys designed to show the effect of 1% additions of molybdenum, copper, and nickel. The alloys in this series were prepared as aluminum-killed, vacuum melts, cast into 50 pound ingots, and all contained approximately 0.04% C, 0.7 to 0.9% Mn, a maximum of 0.015% P and 0.035% S, and a residual to less than 0.035% Al from the killing operation. Table 1 shows the experimental design used to obtain this series of alloys. The atmospheric exposure data obtained on these eight low-alloy steels are shown in Table 2.

All of the experimental steels studied have, without exception, shown exponential weight gain behavior with time. This is illustrated in Figure 1, in which changes in cumulative average penetration are plotted versus exposure time for the same series of experimental alloys. The two curves shown represent the highest corrosion rate (for 1% Mo), and the lowest corrosion rate (for 1% Mo, 1% Cu, 1% Ni). The curves for the other six materials fit within these two extremes.

All of the experimental steels studied have been found to adhere to the following kinetic relationship:

$$\Delta W = K \left(\frac{t_m}{t_1}\right)^N \quad (1)$$

where ΔW = average cumulative penetration depth in mils, t_m = length of exposure, and t_1 = 1 year expressed in the same units as t_m.

When the exposure time is expressed in years, the above relationship reduces to:

$$\Delta W = Kt^N \quad (2)$$

where K and N = constants, and t = exposure time in years.

It is readily apparent that for a one-year exposure:

$$\Delta W = K \quad (3)$$

Equation (2) is readily convertible to a linear expression as follows:

$$\ln \Delta W = \ln K + N \ln t \quad (4)$$

where N is the slope of a straignt line, and ln K is the intercept.

Regression equations of the form $\ln \Delta W = \ln K + N \ln t$ were obtained by computer for each of the eight members of this full-factorial series of experimental alloys. The empirically determined values of ln K and N, thus obtained, are shown in Table 3 together with the corresponding correlation coefficients (R^2) and F-statistics to provide an indication of quality of fit. Additional evidence of the quality of fit is presented in Figure 2, in which calculated values of penetration depth in mils are plotted against the corresponding measured values. It should be noted that although the plotted points for the longest exposure time appear to be further from the ideal 45° line

TABLE 3 – Regression Equation Constants in Kinetic Relationship

Alloy No.	Composition	N	ln K	R^2	F
1FF01	Blank	0.461	0.538	0.993	302
1FF02	Mo	0.445	0.599	0.997	588
1FF03	Cu	0.431	0.425	0.998	935
1FF04	Ni	0.418	0.401	0.998	1101
1FF05	Mo Cu	0.345	0.402	0.999	1831
1FF06	Cu Ni	0.436	0.383	0.988	172
1FF07	Mo Ni	0.357	0.381	0.998	993
1FF08	Mo Cu Ni	0.364	0.336	0.992	258

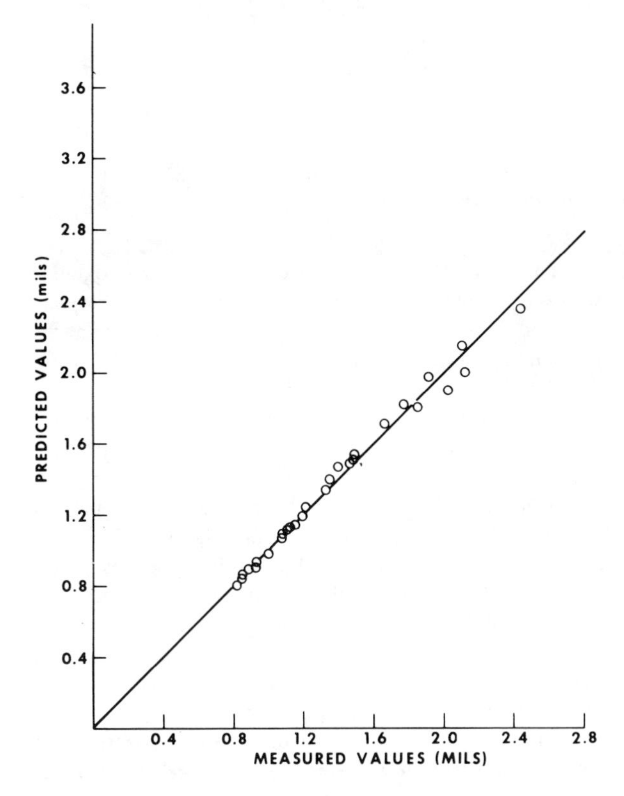

FIGURE 2 – Calculated vs measured penetration values.

than plotted points for the shorter exposure times, the percentage deviations are essentially constant throughout.

Knowledge of ln K and N values for a given experimental alloy allows predictions to be made for behavior at exposure times longer than those for which actual measurements are available. This is illustrated in Figure 3 for the same series of alloys. Again, the alloys not shown here produce curves which fit between the extremes depicted. Extrapolations beyond actual measurements should admittedly be viewed with caution. In this case, it must be emphasized that long-term predictions are assisted by the fact that an approximate steady-state rate of corrosion is often achieved in relatively short exposure times, i.e., about 2 years. This is shown in Figure 4, for the same series of experimental alloys, where the variation in corrosion rate with time of exposure is plotted. Again, only the extremes are shown; the other materials produce curves which fit between the two that are shown.

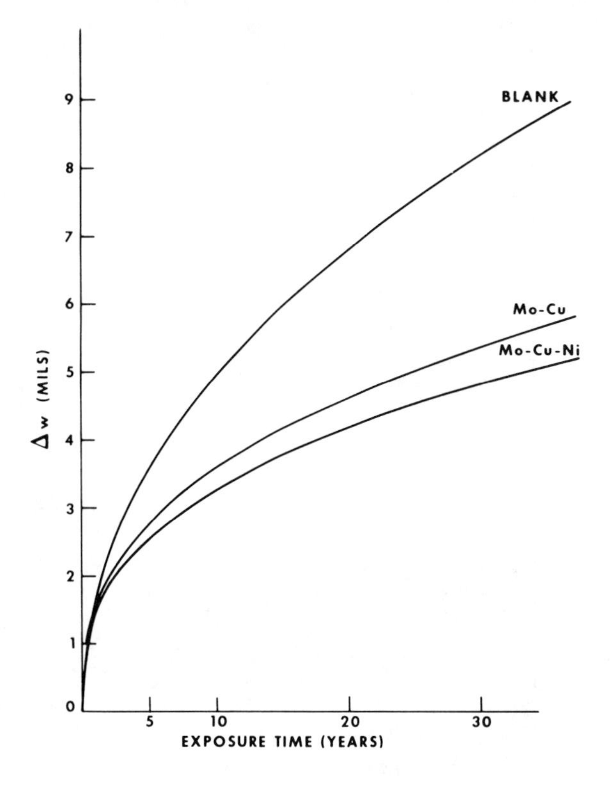

FIGURE 3 – Changes in calculated penetration values with time of exposure.

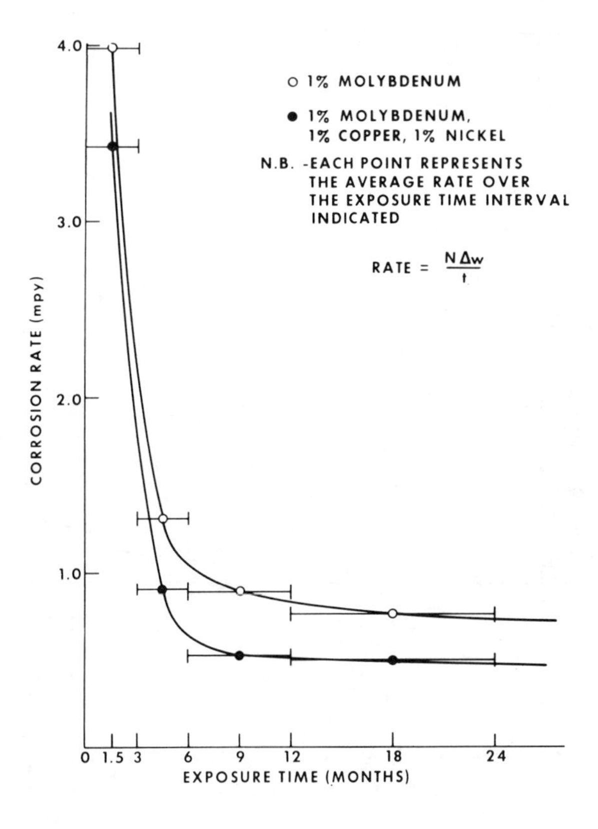

FIGURE 4 – Changes in corrosion rate with time of exposure.

Source: *Corrosion*, April 1975, 117-122

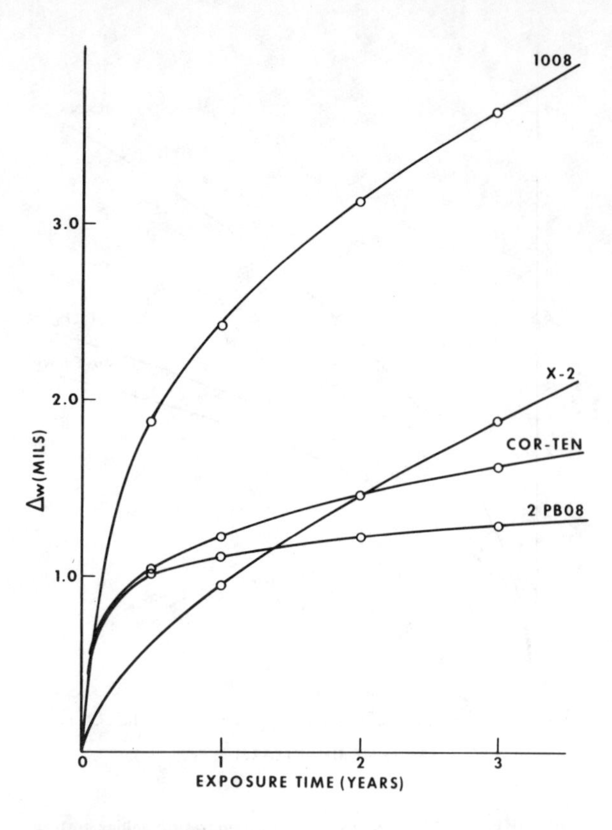

FIGURE 5 – Comparative atmospheric corrosion behavior of several steels.

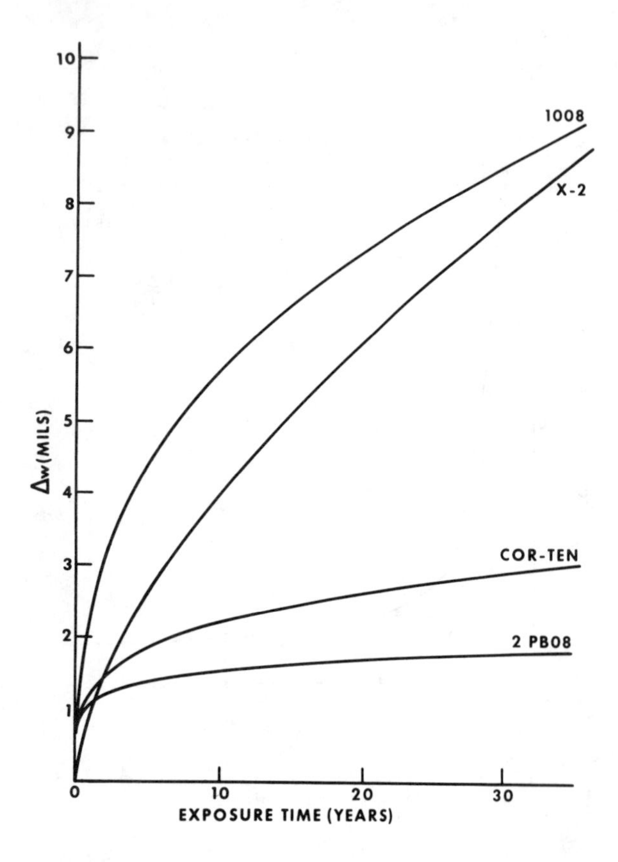

FIGURE 6 – Predicted long-term comparative atmospheric corrosion behavior of several steels.

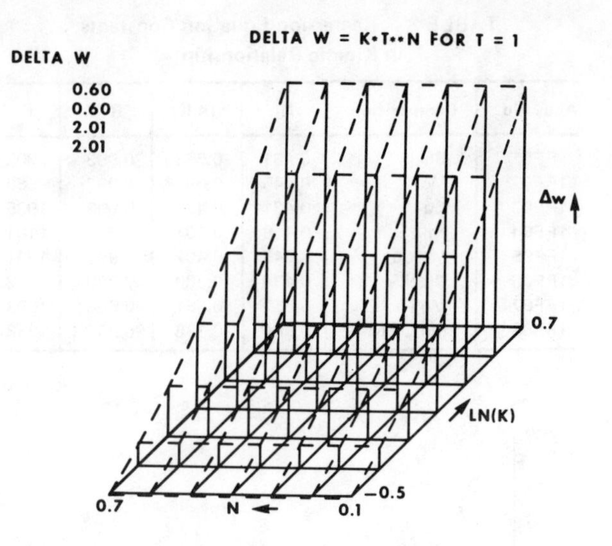

FIGURE 7 – Changes in Δw with changes in both LN K and N for a 1-year exposure.

The relatively rapid achievement of an approximate steady-state rate of corrosion has resulted in a rather widely accepted assumption that short-term exposure data can provide a reasonable indication of long-term behavior. Although it does serve to reinforce the assumption that the kinetic relationship observed during the early exposure will persist, it is dangerous to assume that this implies that the relative corrosion behavior of several steels will also persist. This is illustrated in Figure 5, where 3-year data are plotted for four steels, including a commercial mild-steel and Cor-Ten. The experimental steel, identified as X-2, is clearly superior to the other materials for the very short exposure times, but this advantage is rapidly lost. The extrapolation of these relationships to long-term exposures, as shown in Figure 6, points up how misleading the early advantage noted for a material such as X-2 can be.

The idea of characterizing the atmospheric corrosion behavior of a low-alloy steel in a given natural environment with two parameters is an important contribution to understanding that behavior. The variation in cumulative penetration depth with variation in ln K and N is shown in Figure 7 with a three-dimensional plot for a 1-year exposure. At this time level, the value of N does not contribute to the corrosion behavior. As the time of exposure increases, the value of N becomes more and more important. Figure 8 shows the effect of both ln K and N on the corrosion behavior in a 20-year exposure. This three-dimensional plot shows clearly that the effect of N also increases dramatically with increase in ln K.

Having determined ln K and N values for each member of a full-factorial series makes it possible to write equations for each of these quantities in terms of alloy composition. For the molybdenum-copper-nickel series exposed at the East Chicago test site, for instance, the appropriate equations obtained from regression analyses are as follows:

$$\ln K = 0.54 + 0.06\,(\%Mo) - 0.11\,(\%Cu) - 0.14\,(\%Ni) - 0.08\,(\%Mo\,\%Cu) + 0.10\,(\%Cu\,\%Ni) - 0.08\,(\%Mo\,\%Ni) + 0.06\,(\%Mo\,\%Cu\,\%Ni) \quad (5)$$

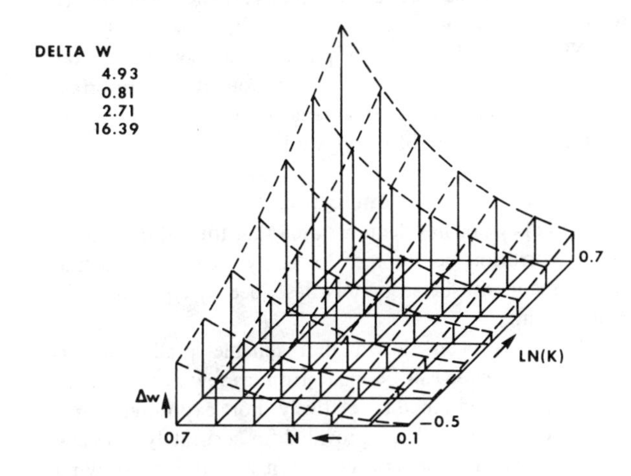

FIGURE 8 – Changes in Δw with changes in both LN K and N for a 20-year exposure.

$$N = 0.46 - 0.02\ (\%Mo) - 0.03\ (\%Cu) - 0.04\ (\%Ni) - 0.07\ (\%Mo\ \%Cu) + 0.05\ (\%Cu\ \%Ni) - 0.05\ (\%Mo\ \%Ni) + 0.06\ (\%Mo\ \%Cu\ \%Ni) \quad (6)$$

An examination of these two equations reveals that whereas the short-term effect of a single addition of molybdenum is to increase the corrosion rate, the long-term effect is to lower it. From the equation for N, it is clear that the largest long-term effect is that of the molybdenum-copper interaction term. Thus, a knowledge of the composition here allows a calculation of the appropriate values for ln K and N, which can then be used in the rate equation to estimate the average penetration depth to be expected at the East Chicago site for a given length of exposure.

All of the scores of steel alloys tested, without exception, fit the rate expression discussed above: $\ln \Delta W = \ln K + N \ln t$. The linearity of the relationship appears to be beyond question. Thus, two points on this line should determine values for ln K and N and accordingly allow long-term predictions of ΔW. From an older exposure site in Porter County, Indiana, we had available replicated data (five specimens for each point) obtained on Cor-Ten in 1-year, 2-year, 4-year, and 7-year exposures. Ln K and N values were determined using only the 1-year and the 2-year data, and predictions for the 4-year and 7-year exposures were obtained using the rate expression given above. Data from the same site obtained on a copper-bearing steel (four specimens for each point) were also available for 1-year, 2-year, 4-year, and 8-year exposures. Again, using only the 1-year and 2-year data, values for ln K and N were determined and were then used to obtain predictions for the 4-year and 8-year exposures. The comparison of predicted versus measured values of ΔW for these two materials are shown in Table 4. Values for the correlation coefficient (R^2) and for the F-statistic are included in this table to indicate the confidence that is warranted for the persistence of linearity out to long exposure times. It should be noted that the reproducibility of the data used to obtain the rate expression for the copper-bearing steel was significantly poorer than normally observed. This apparently accounts for the greater discrepancy between predicted and measured values of ΔW obtained for this material.

TABLE 4 – Comparison of Predicted and Measured Values for ΔW

	Cor-Ten		Cu-Bearing Steel	
	4 Years	7 Years	4 Years	8 Years
Measured ΔW (mils)	1.35	1.56	2.75	3.9
ΔW Predicted from 2 Points	1.34	1.68	2.83	4.2
ΔW Calculated using 4 Points	1.34	1.58	2.7	3.9
R^2 for 4-Point Equation	0.997		0.998	
F for 4-Point Equation	602		983	

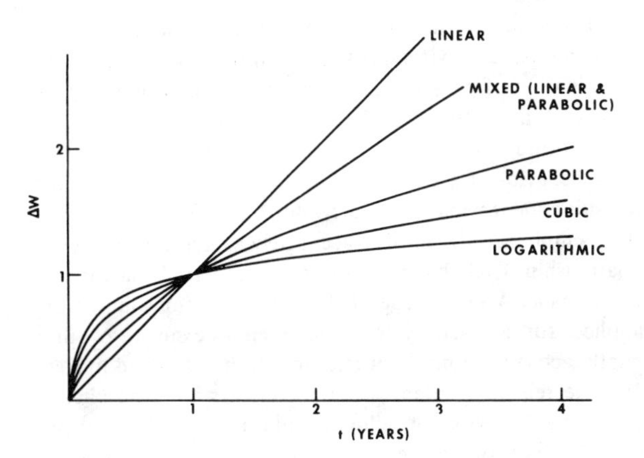

FIGURE 9 – Types of oxidation curves which have been reported.

Discussion

The general rate expression which describes the atmospheric corrosion behavior of alloy steels can be rewritten as follows:

$$\Delta W^M = K_1 t \quad (7)$$

where $M = 1/N$.

Analyzing the expression in this form readily reveals the versatility of this equation. For M = 1, the expression describes a rectilinear relationship; for M = 2, the expression describes a parabolic relationship; for M = 3, the expression describes a cubic relationship, and so on. Thus, the value of N in Equation (2) can serve as a diagnostic tool to indicate the nature of the relationship: N = 1, linear; N = 0.5, parabolic; N = 0.33, cubic, and so on.

These relationships are precisely the relationships which have been reported for the various specific cases of metal oxidation. Figure 9 shows curves representing these various relationships. The curves have been drawn in such a manner as to intersect at one point to facilitate comparison.

The logarithmic relationship has been reported most often for ambient temperature conditions. This is usually explained in terms of deviation from parabolic behavior as a result of changing diffusion conditions as the film grows. Cubic behavior is rare, and rectilinear behavior for steels is usually limited to a very brief period before a significant amount of oxide forms on the surface.

A few alloys have produced N values greater than 0.5.

This behavior can be readily explained as paralinear behavior which is usually attributed to partial control by a boundary reaction. Most of the alloys produce values of N which are smaller than 0.5. The closer the N value comes to 0.5, the more closely the mechanism approaches a uniform diffusion-controlled process. The further the N value is displaced from 0.5 in the direction to lower its value, the larger the contribution from a logarithmic process which implies a more pronounced lowering of the diffusion coefficient as the film grows.

The atmospheric corrosion behavior of low alloy steels in an industrial environment might thus be described as a combination of parabolic and logarithmic behavior with the value of N indicating the relative contribution from each.

Returning to the curves depicted in Figure 9, the very early behavior of a steel is indicated by the value of K, and the subsequent behavior by the value of N. These considerations emphasize the need to accurately establish the kinetic relationship if reliable predictions of long-term behavior are to be made. A knowledge of the kinetic relationship which applied for a given system allows an assessment of the relative contributions from relevant mechanisms and makes possible reliable predictions of the effects on atmospheric corrosion behavior of alloying additions, of processing variables, and of differences in the exposure environment.

The assumptions that have been made in developing this description of atmospheric corrosion behavior include the following:

Environmental conditions remain essentially constant over the exposure period, which is of interest. Whatever progress can be hoped for in the struggle to clean up the environment would tend to lower atmospheric corrosion, and consequently predictions would tend to be conservative.

Seasonal weather variations tend to even out in long exposures. In this regard, it should be emphasized that data obtained in exposures shorter than 1 year should be treated with extreme caution.

No localized corrosion phenomena have been considered in this treatment. These constitute important deviations from the general behavior described here and should be the subject of future study.

Conclusions

1. This work has demonstrated that the natural atmospheric corrosion behavior of low-alloy steels in industrial environments can be accurately described by equations of a simple form.

2. It has been shown that reliable predictions of long-term behavior can be made from as few as two sets of determinations obtained in relatively short exposures.

3. A vehicle has been provided for accurately assessing the effect on atmospheric corrosion behavior of alloying additions, operating parameters, and differences in environment.

4. This work indicates that the atmospheric corrosion behavior of low-alloy steels in an industrial environment might be described as a combination of parabolic and logarithmic behavior, with the value of a constant in the empirically determined kinetic relationship indicating the relative contribution from each. Thus, relevant mechanistic information is made available.

Acknowledgement

The authors gratefully acknowledge the invaluable contribution of Vincent P. Pearson who provided all the atmospheric exposure data used in this study. We also express our appreciation to the Inland Steel Company for making the publication of this work possible.

Reference

1. Annual Book of ASTM Standards, Part 31, G1-72, p. 1082 (1973).

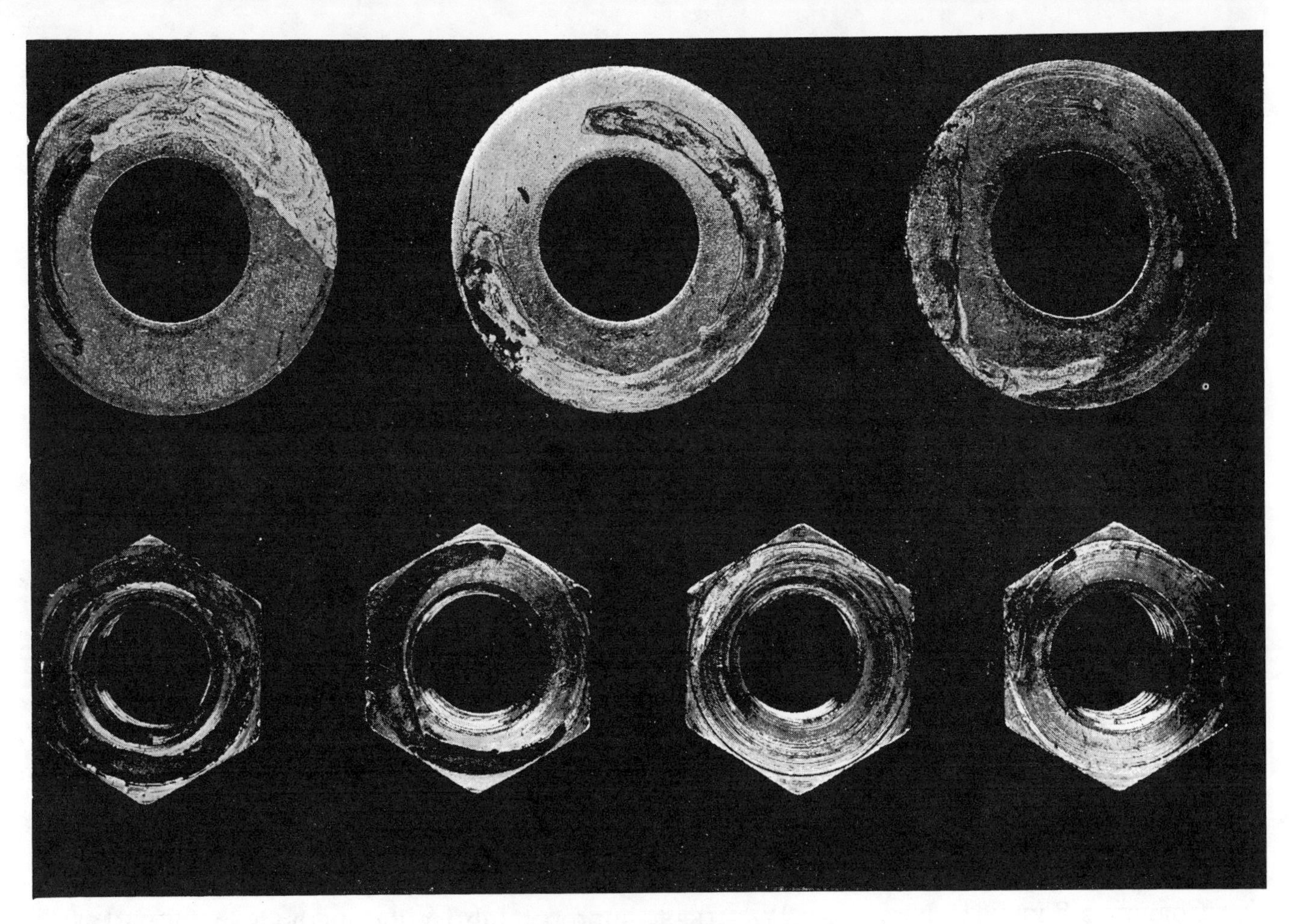

Figure 2—Type 316 stainless nuts and washers after 3 weeks' immersion in quiet sea water. Two right nuts are 0.03 percent carbon; two left nuts 0.06 percent carbon. 1.5X

Improper Fabrication Is Major Cause of Stainless Steel Failures at Two Coastal Steam Stations*

E. F. Bladholm

Southern California Edison Company
Los Angeles, California

★ Revision of a paper titled "Experience with Stainless Steel in Coastal Steam Stations," presented at the 18th Annual Conference of the National Association of Corrosion Engineers, Kansas City, Mo., March 19-23, 1962.

SUMMARY

Experience of Southern California Edison with Types 304 and 316 stainless steels at coastal steam electrical generating stations is described. Case histories are given of failures of wire mesh, nuts, washers and U-bolts mostly because of improper fabrication or heat treatment following fabrication. Apparent need for more stringent specifications is cited in cases of several failures. Experience indicates that extra low carbon grades may be indicated for marine service, especially when equipment is submerged. Beneficial effects of patina from anodic iron "waste plates" on stainless steel strainer mesh is cited. Tests are continuing on components properly annealed before installation to check conclusions. ELC grades also are being exposed under similar conditions.

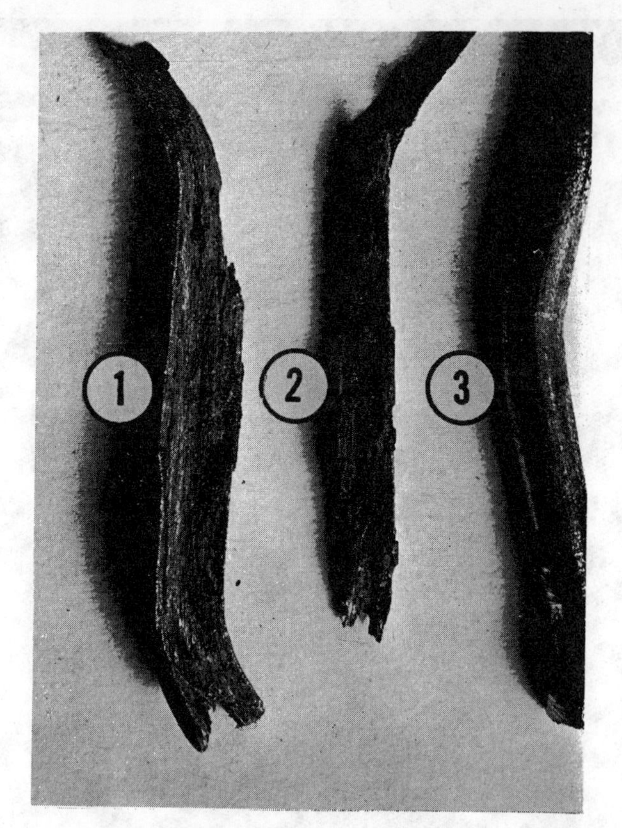

Figure 1—Specimens 1 and 2 are a spring temper hardened Type 316 stainless steel which corroded after two months' service in sea water. Specimen 3 is a medium soft Type 304 stainless after 14 years' service in sea water. 4X

CORRODED TYPE 316 NUTS AND WASHERS, improper fabrication methods in formation of U-bolts, failure of a costly casting from poor heat treatment and lack of knowledge among suppliers of proper techniques for fabricating stainless steel have been experienced at one time or another by Southern California Edison. Experience indicates that extra low carbon grades of stainless may be better in some marine environments, for example.

Especially interesting were examples of baffling failures of stainless steel equipment recently encountered at two of the company's coastal steam stations. Although stainless steels have been used successfully in the sea water circulating systems of Southern California Edison's five coastal steam stations for periods to 13 years, two failures in 1961 prompted an urgent inquiry into the causes. About 18 inches of one strand of wire mesh on the intake vertical slide screens at the Huntington Beach Station virtually corroded away after about two months' service. (Two elements at left in Figure 1). The Type 316 mesh replacement material was perfect elsewhere.

Vertical slide screen hardware corroded after three weeks' immersion at the Alamitos Station. Here new Type 316 wire mesh had been installed late in 1960 as screen replacement for two units. These screens had several rust spots similar to those found also at the Huntington Beach Station. In both cases the wire was from the same supplier.

Eight vertical slide screens for two other units at the Alamitos Station were installed in 1961. Wire mesh and hardware also were Type 316, but wire came from a supplier and mill different from that which supplied the two other units and the Huntington Station equipment. The main frames were of carbon steel plate coated with an epoxy-polyester fiberglass-epoxy coating of about 100 mils total thickness. They were field assembled and lowered into quiet water because the circulating system was not completed.

When these screens were pulled for inspection after three weeks' immersion, it was difficult to believe that stainless steel nuts, studs and washers could corrode so severely in such a short time. (See Figure 3) The wire mesh, however, was perfect.

Reasons for Failure Are Sought

Some reason was sought for the success of some installations and parts of installations and the failure of others. The dangers of crevice corrosion had been considered, but, because of previous successful experience, use of stainless in the applications seemed reasonable.

At the Redondo Beach Station, Type 304 wire mesh and frames and hardware have been corrosion free for 14 years as shown in Figure 3. Four similar screens at El Segundo Station have been corrosion free for seven years. In all, 12 traveling screens, similar to those in Figure 3, with Type 304 mesh and hardware were installed on four separate occasions at two stations, representing 35 years of corrosion free service. Figure 4 shows estimated corrosion rates at the two stations.

Furthermore, as can be seen in Figure 5, Type 316 wire mesh installed at the Mandalay Station in 1960 as replacement material was not damaged after 18 months. Iron waste plates can be seen bolted to the wire mesh when the new material was installed.

Function of Waste Plates

Over the past few years it has been noticed that, where iron oxide stain is found on stainless steel, the film appears to be tough and corrosion resistant; therefore, iron anodes have been attached occasionally to stainless steel. Iron waste plates added to the mesh at Mandalay Steam Station produced a golden brown patina in a matter of weeks.

Figure 3—Traveling screens installed in 1947, Redondo Steam Station. Mesh and frames were fabricated of Type 304 stainless steel.

The initial waste plates were depleted in about six months, partly because of large areas of original bronze parts remaining on the screens. Also, the bronze in contact with the stainless steel wire mesh was bright and was anodic to the passivated stainless steel mesh. Bronze U-bolts were replaced with Type 316 material, and precautions were taken to insulate remaining bronze from the stainless steel.

Small pieces of pipe were used as replacement anodes, and the operation of these screens has been entirely satisfactory. Use of iron waste plates is now advocated for all new stainless steel

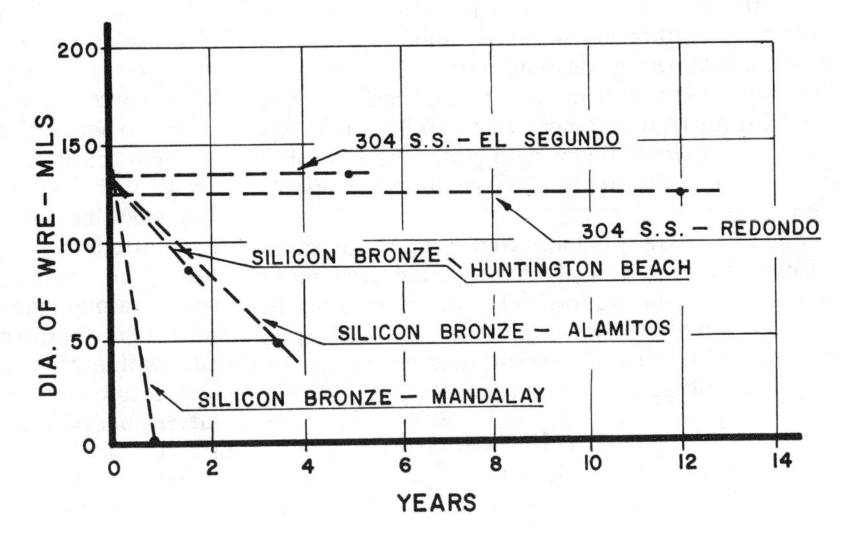

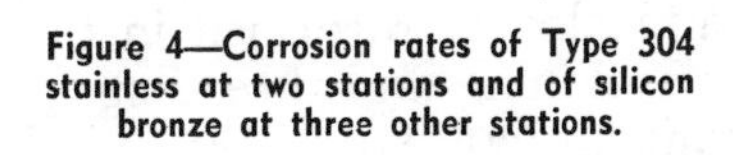

Figure 4—Corrosion rates of Type 304 stainless at two stations and of silicon bronze at three other stations.

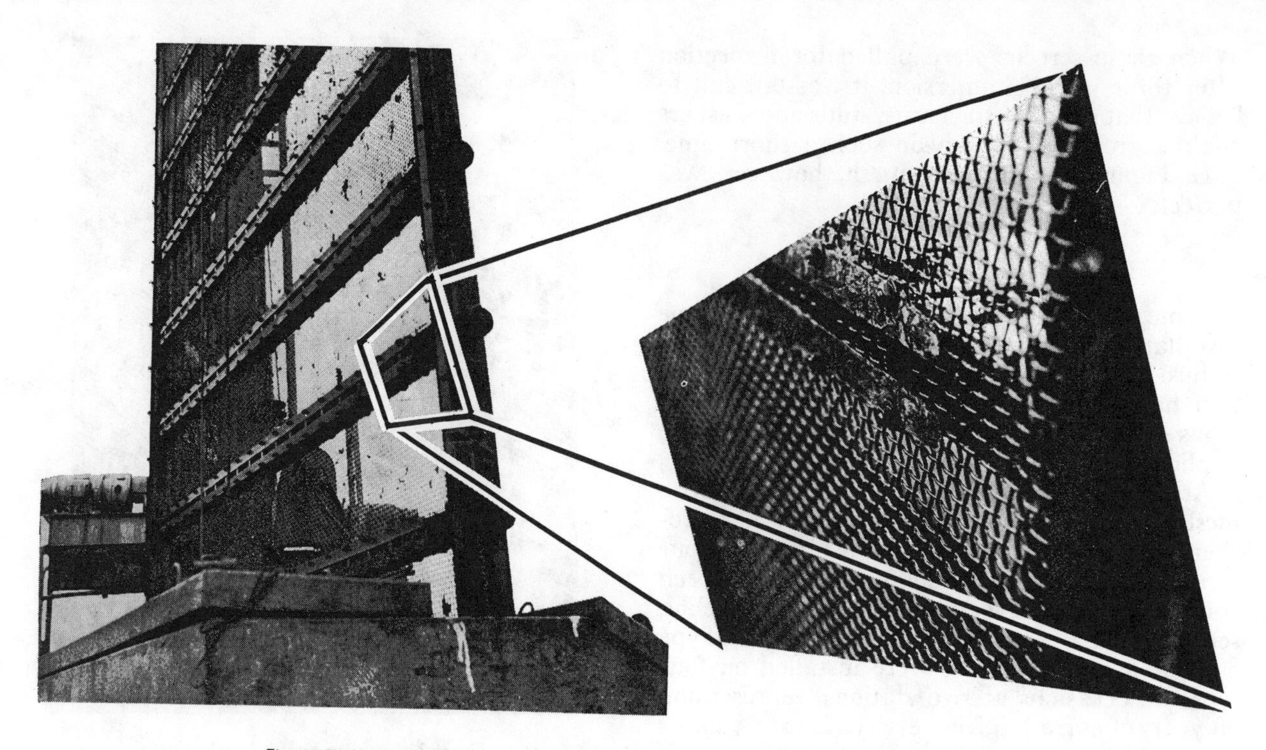

Figure 5—Vertical slide screen at Mandalay Steam Station. After 18 months, corrosion of iron waste plates can be seen. An iron oxide deposit covered the Type 316 mesh.

immersed in sea water, especially where passivation seems desirable. Acid passivation is no longer used.

Results of Investigations

Information from the manufacturer of the corroded wire shown in Figure 1 revealed some interesting facts. Wire supplied to the screen fabricator was of spring temper hardness and was further stressed in the weaving process.

A sample of the corroded wire was analyzed, and photomicrographs were made (Figure 6) to show the grain structure at the bend. Dark area in Figure 6 is center of the wire, and chromium carbide precipitation is clearly shown as is corrosion along the grain boundaries. Transgranular stress corrosion cracking also can be seen progressing from the heavily corroded area. Attention is called to the elongated shape of the grains, probably due to the wire drawing operation.

Figure 7 is a photomicrograph of the same wire as in Figure 6 after annealing treatment consisting of heating to 1950 F and quenching in water. This photomicrograph shows the absence of chromium carbide precipitation noted in the corroded spring temper hardened wire.

Sound Wire is Examined

A piece of Type 304 wire (Specimen 3 in Figure 1) was removed from one of the Redondo screens which had operated successfully for 14 years. A photomicrograph (Figure 8) showing the grain structure was made for comparison with the corroded wire shown in Figures 6 and 7. Grain structure of the 304 wire is comparable in appearance to the 316 wire after annealing. Hardness tests also showed soft condition of the 304 wire (Brinell 68B) as compared to the spring temper hardened condition of the 316 wire.

Reason for Corroded Nuts, Washers

Investigation of the early corrosion of Type 316 nuts, studs and washers on screens at the Alamitos Steam Station revealed some interesting points for consideration. Chemical analysis of a corroded nut showed it had 0.06 percent plus carbon while an uncorroded nut had 0.03 percent carbon. Inspection indicated the material was neither annealed as furnished in the bar stock nor after machining and threading.

Figure 9 shows grain structure of the corroded nut. The corrosion attack apparently is more severe along the concentrated carbon bands where chromium carbide precipitation is evident. An explanation of the stratification or layered appearance of the metal is still needed. This latter condition may have resulted from forming the bar stock or prior rolling at the mill.

Figure 10 shows the same material after annealing by heating to 1875 F and water quench-

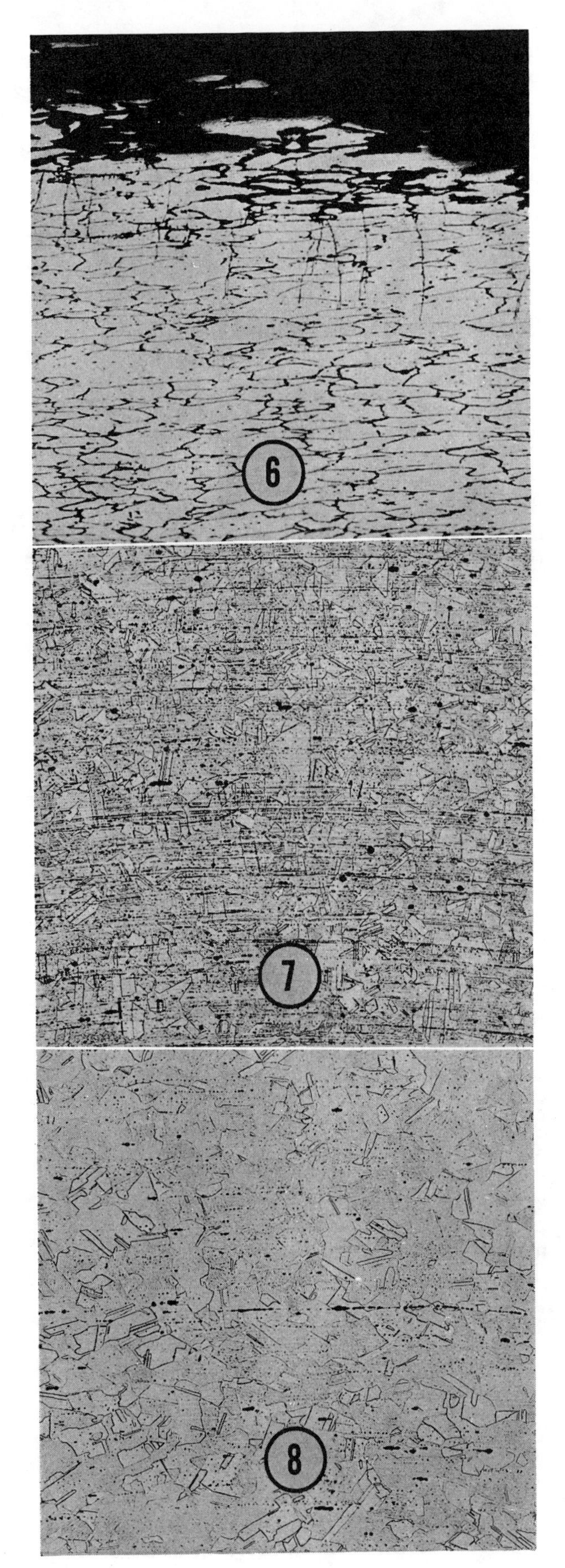

ing. Solution of the chromium carbides appears to have been accomplished although the light and dark bands are still visible. This piece of metal may have come from the top of an ingot where "dirty" material often is found.

Figure 11 shows grain structure of the low carbon nut (top right Figure 2) removed with the corroded nut from the same screen. The light and dark bands are found in this material also. Chromium carbide precipitation is apparent but not to the extent found in the 0.06 percent carbon material. This nut probably did not corrode because the chromium carbide bands may have been pinched off near the surface. Some nonmetallic inclusions appear in both nuts.

U-Bolts Heated With Torch

Figure 12 is typical of some Type 316 U-bolts which corroded near the bends. Although the complete story of fabrication was not obtained, it was concluded that the U-bolts either were cold bent or a torch was played on the local area at time of bending, the latter more probably because of the corrosion pattern. Fabrication was

Figure 6—Type 316 spring temper hardness before weaving. Photomicrograph shows center of wire (dark area) on right side of photograph. Note stress corrosion cracks and intergranular corrosion due to chromium carbide precipitation. 200X 10 percent oxalic etch. Figure 7—Type 316, same as Figure 6, after heat treat at 1950 F and quench. 100X. Figure 8—Type 304 stainless after 14 years' service in sea water. 100X. 10 percent oxalic etch.

not followed by solution annealing. Corrosion stopped when the fabricator removed the corroded U-bolts and solution annealed the lot before re-installing.

In another case where corrosion was experienced on Type 316 U-bolts, the fabricator said, when questioned, that he had heat-treated the bolts after bending, but further inquiry showed that the bolts had been stress relieved at 1150 F. Therefore, they were removed and returned to the fabricator where they were solution annealed at 1950 F and water quenched. There has been no further corrosion of these parts.

Casting Ruined by Stress Relieving

After a large Type 316 stainless casting was factory machined for use on a circulating pump assembly in sea water service, the casting was stress relieved at 1150 to 1200 F by the fabricator. After less than six months' service, it was porous and mealy due to corrosion along the chromium carbide or chromium depleted grain boundaries because stress relieving in the sensitized range had caused chromium carbide precipitation. Correction of this failure was costly to the fabricator.

Low Carbon Content Favored

On the basis of experience and the data presented, low carbon stainless steel grades such as 304L and 316L can be considered for marine use where the material is to be immersed in sea water. In a recent meeting of refinery, aircraft and utility corrosion engineers with factory representatives, advantages were emphasized of the low carbon grades of stainless steel for sea water immersion. Also it was pointed out that low carbon grades (to 0.03 percent carbon) were not as susceptible to weld boundary carbide precipitation.

Type 304 has been found satisfactory by Southern California Edison where used in a coastal atmosphere. Some welded pieces at Redondo Steam Station are not corroded after 12 years' or more exposure to coastal sea air. They were neither annealed after welding nor immersed in sea water.

Are Tighter Specifications Needed?

Apparently strict compliance with specifications is required to avoid the consequences of sloppy performance by some suppliers. For example, a fabricator supplied screens fabricated of double crimp 316 wire. On a similar order with the same specifications, the same fabricator furnished screens with spring temper hardened stainless steel wire further stressed by weaving with an intermediate crimp.

Random corrosion occurred on an occasional stainless steel nut in a marine installation of another local company. An investigation revealed that, though the corroded nuts were Type 304, the uncorroded nuts were Type 316 which had been specified. Inspection of stainless nuts received by Southern California Edison reveals that they appeared to be cold formed or machined and that they were not solution annealed after machining. When questioned, one supplier did not know what solution annealing was. Thus, specifications and inspection apparently should be more rigid. Also, there is a need for further education.

More Testing Underway

To test observations on the use of stainless steel, some of the corroded hardware and screens have been solution heat treated at 1950 F and water quenched. These pieces have been replaced in service in as nearly the same environments as possible. In this manner an attempt will be made to find answers to questions arising from the use of these materials in a marine environment. Added to the testing program are samples of ELC grades of Types 304 and 316.

Acknowledgment

Metal Control Laboratories, Huntington Park, California made analyses and photographs of wire shown in Figures 6-11 inclusive.

S. H. Potter, Bechtel Corporation, for loan of pictures.

DISCUSSION

Comments by C. M. Schillmoller, International Nickel Company, Inc., Los Angeles, Cal.

Eric Bladholm has provided us with a well documented history of nickel stainless steels for sea water intake screens at coastal steam plants. Excellent performance is being obtained in the Redondo Beach, El Segundo and Mandalay Steam Stations while recognition of reasons for failures at Huntington Beach and Alamitos point the way to successful corrosion control.

A word of caution appears in order. The use of Types 304 and 316 stainless steels is highly questionable for sea water intake screens, especially if these alloys should be used throughout so that they would not receive cathodic protection from a substantial area of associated carbon steel with which they would be in electrical contact, or unless they were given deliberate cathodic protection from some installed source of current.

In sea water, one would expect failures of stainless steel to occur, primarily from chloride stress cracking, from intergranular corrosion and from corrosion within crevices formed by overlapping surfaces, and secondarily from pitting under marine organisms, silt and other loosely adherent deposits. Therefore, tighter specifications and procedures as recommended by the author involving stress relief annealing, reduction of hardness of the screen wire and heat treatment of parts to put the chromium carbides back into solution after sensitization has occurred are not the entire answer because they will not eliminate the danger of crevice corrosion. Conversely, if one has to rely on cathodic protection by adjacent steel, either incorporated by equipment design or the deliberate use of iron waste plates, then tighter specifications on type of stainless steel and heat treatment appear no longer economically justifiable.

Several other comments may be of value:

1. We find little difference in corrosion between severely cold worked and fully annealed stainless steel except for an increase in pitting tendency in the hardened material.

2. Weight loss is not an important criteria since corrosion is concentrated in areas of deep

attack. The difference in performance between Types 304 and 316 stainless steel in sea water is related to the probability of pitting and crevice attack rather than the intensity of action. In one test at the Inco Harbor Island Test Station we found, after one year, crevice corrosion on 100 percent of all Type 400 series specimens; 85 percent on all Type 304 specimens and 15 percent on all Type 316 specimens. Molybdenum addition is certainly helpful in a chloride-containing environment.

Figure 9—Photomicrograph of one of corroded nuts (see Figure 2) 100X, Kallings' etch.

3. Type 316 stainless steel impellers for a circulating pump in continuous sea water service are an excellent choice since velocity conditions are high, so that marine organisms and other solids cannot remain attached. Of course, a 2050 F heat treatment followed by quench cooling has to be specified here to prevent intergranular corrosion of the casting.

4. I don't prescribe to the theory that the golden bronze patina caused by iron oxides makes the film on stainless steel tough and increases its protective value. However, it appears a good indicator that cathodic protection is being offered by adjacent steel or iron waste plates so that reliable performance may be expected.

I have two questions:

1. To what do you attribute the excellent performance of stainless steel screens, frames and hardware in the Redondo Beach and El Segundo steam plants? Is there any carbon steel associated with these installations?

2. Your chart on screen performance shows early failures of silicon bronze. Is it correct to assume that their life was related to the frequency of high velocity water jetting during cleaning operations? Also, has replacement of 70-30 Cu-Ni or Monel been considered?

Replies by E. F. Bladholm:

The comparison of the 400, 304 and 316 was very interesting. The following comments and questions may further broaden our base of discussions.

1. Regarding Item 1, we have taken several nuts which were corroded and reinstalled them after annealing. Also, we obtained a new sample of the stressed wire mesh which failed in service. One piece was annealed, and one piece was tested as received. Those pieces were bolted to one of the screens at Alamitos. A duplicate set was exposed at Huntington Beach Steam Station. It will be interesting to see what happens.

4. Your comments under Item 4 are of interest on the value of the iron oxide patina. This idea originated from reading the Corrosion Research Council reports on the tests of metal oxides. Within the last two years they have done some interesting work on the formation and destruction of oxide films on metals.

Figure 10—Photomicrograph of corroded nut in Figure 2 after heating to 1875 F and quench. 100X, Kallings' etch. Figure 11—Photomicrograph of uncorroded nut in Figure 2, Kallings' etch, 100X. Probably explanation of dark bands is that metal came from the top of an ingot which had dirty material in it.

Source: *Materials Protection*, August 1962, 33-39

Question 1: It may be possible that there is some steel present. The lower cast steel sprockets were replaced with 25-20 sprockets after about six months' service. Also it is not known if there are stray currents favoring the screens. It was considered reasonable to assume that the spray wash water would completely aerate the screen mesh and frames.

Question 2: It was considered reasonable to assume that water velocities were too high for the silicon bronze. This material has been entirely successful where used in quiet sea water.

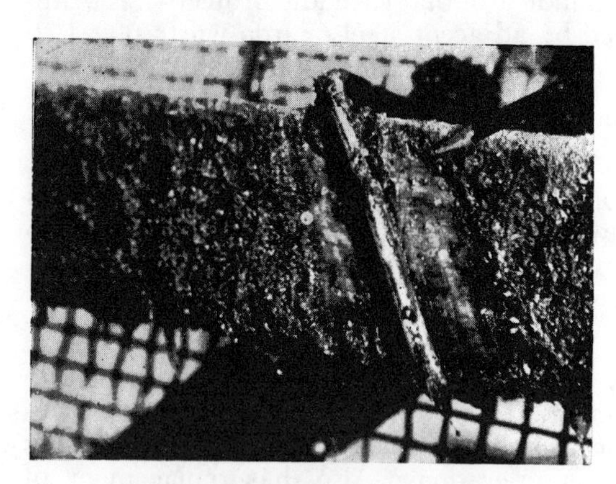

Figure 12—Type 316 stainless U-bolt corroded because it was heated with a torch for bending.

Comments by Frank E. Kulman, Consolidated Edison Company of New York, 4 Irving Place, New York City:

Our experience in the use of stainless steel in sea water confirms the experience of Southern California Edison Company. We intend to continue the use of this material using the same principles outlined by the author.

E. F. BLADHOLM is senior corrosion engineer with Southern California Edison Co., Los Angeles, Cal. He has been involved in corrosion control work with pumps and power plants since 1932. An active member of NACE, he is currently on the NACE Board of Directors as representative for the Western Region. He also has served as chairman of that region.

Field Tests Show Corrosion Resistance of

Stainless Steels in Liquid Fertilizer Service★

SUMMARY

Describes details of a field service test conducted to determine performance and corrosion resistance of several stainless steels for use as liquid fertilizer tanks. Spray-applicator tanks of Types 304, 202, 405 and Tenelon stainless were tested for three years. Tanks of AISI C1010 carbon steel and aluminum alloy 5052 also were included for comparison. Results showed Type 304 stainless will provide satisfactory service with fertilizers and many other farm chemicals. Data include not only corrosion rates but also some test data on stress corrosion cracking of these materials in liquid fertilizer exposures.

T. F. Shaffer, Jr.

Applied Research Laboratory
United States Steel Corporation
Monroeville, Pennsylvania

★Revision of a paper titled "Corrosion Resistance of Stainless Steels in Liquid Fertilizer Service" presented at the 19th Annual Conference, National Association of Corrosion Engineers, March 19-23, 1962, Kansas City, Missouri.

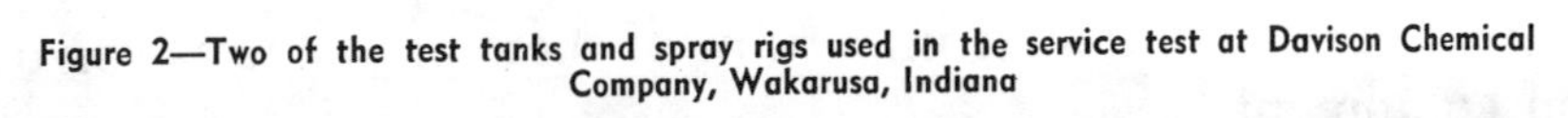

Figure 2—Two of the test tanks and spray rigs used in the service test at Davison Chemical Company, Wakarusa, Indiana

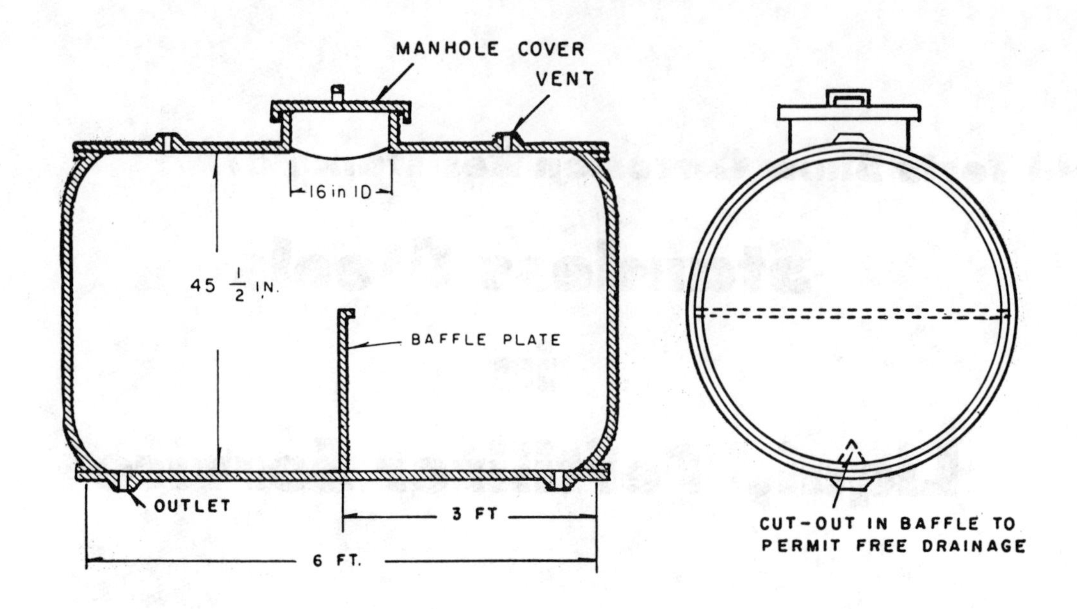

Figure 1—Schematic drawing of a 500-gallon liquid fertilizer tank used in the three-year service test to determine corrosion resistance of various stainless steels.

Introduction

ONE SIGNIFICANT INNOVATION in agriculture is the recent introduction of bulk handling methods. The older methods of handling fertilizers, feed, and grain by sacks were slow, expensive, and laborious. Bulk handling permits one man to do work previously requiring several men and the work itself is considerably easier.

In bulk handling of fertilizers, increased attention is being focused on aqueous solutions of plant nutrients for direct application to the soil. The use of liquid fertilizers has spread rapidly from the west coast, where they were first used, to almost every area in the country. Several problems, however, have hindered rapid acceptance of liquid fertilizers in many areas. One of the most important of these problems is corrosion of storage and application equipment.

Previous Investigations

Previous investigations have been conducted to determine the corrosion resistance of metals in liquid fertilizers. Vreeland and Kalin investigated the corrosion behavior of carbon steel, aluminum, and several stainless steels in liquid fertilizers.[1] In solutions containing only nitrogen as a plant nutrient, several chromium and chromium-nickel stainless steels and aluminum were not attacked, whereas carbon steel was attacked. In mixed fertilizer solutions containing nitrogen and phosphorous or in complete fertilizer solutions containing nitrogen, phosphorus, and potassium, only the chromium-nickel stainless steels were not attacked.

Hatfield and co-workers determined the corrosive effects of mixed and complete fertilizer solutions on carbon steel, AISI Types 316 and 430 stainless steels, and several aluminum alloys.[2] These investigators reported that both carbon steel and the two stainless steels were satisfactory (corrosion rate less than 20 mils per year) with most all combinations of variables tested. Aluminum ranged from unsatisfactory (corrosion rate greater than 60 mils per year) to satisfactory, the rating being dependent on test conditions. However, both these investigations were conducted under laboratory conditions which are not usually representative of service conditions. In addition, these tests did not show what levels in corrosion resistance are necessary to provide satisfactory service performance of materials of construction for equipment used in handling liquid fertilizers.

U. S. Steel's Research

The present investigation by the Applied Research Laboratory of the United States Steel Corporation was designed to determine the service performance of several stainless steels, carbon steel, and aluminum for liquid fertilizer tanks.

Under service conditions, corrosion can affect not only the day-to-day performance of a tank but also the length of service life. Corrosion products, for example, can clog screens and nozzles in the spray system. This clogging requires the immediate attention of the operator or nonuniform fertilizer application will result. Localized corrosion that is not evident in laboratory tests may occur under service conditions.

This investigation, described in this article, considered corrosion rates, types, and areas of corrosion, and service performance of the materials of construction tested. The testing was done in cooperation with W. R. Grace and Company, Davison Chemical Division.

Materials and Experimental Work

Six Tanks Used

Six field spray-applicator tanks were constructed by a commercial tank fabricator. One tank was constructed from each of the following materials: AISI Types 304, 202, and 405 stainless steels, Tenelon stainless, AISI C1010 carbon steel, and aluminum alloy 5052. The sheets of each material were about 0.110-inch thick. Microstructure of each material was normal. Type 308 stainless steel electrodes were used to weld all the stainless tanks. Mild steel electrodes were used for the carbon steel tank. The aluminum tanks was welded by the inert gas, shielded arc welding process, a 5052 filler rod being used where necessary. Tanks constructed of Type 304 stainless and carbon steel had a capacity of 100 gallons each. Tanks constructed of Types 202 and 405 stainless, of Tenelon stainless, and of aluminum had a capacity of 500 gallons each. Figure 1 is a sketch showing the general structure of the 500-gallon tanks; the 1000-gallon tanks were twice as long and had two baffle plates. Tanks were mounted on spray rigs, and the completed tank assemblies (two are shown in Figure 2) were placed in service at the liquid fertilizer plant of the W. R. Grace and Company, Davison Chemical Division, in Wakarusa, Indiana.

Preparation of Test Panels

Weight-loss corrosion test specimens were mounted inside each test tank so that quantitative data could be obtained during the service test. Specimens were made from four-inch square pieces cut from the same sheets as those from which the tanks were fabricated. Each piece was notched for identification, ground on the edges to remove burrs, and cleaned before being weighed to the nearest milligram. Carbon steel panels were vapor-degreased, pickled in inhibited hydrochloric acid, and rinsed in hot water. Aluminum panels were vapor-degreased, dipped in 70 percent nitric acid for ten minutes, and rinsed in hot water. Stainless steel panels were vapor-degreased, dipped in 20 percent nitric acid at 120 F for 20 minutes, and rinsed in hot water.

Specimens were fastened to stainless steel racks by porcelain insulators, and the racks were bolted to the baffle plates inside the tanks. Each rack was insulated from its tank by plastic washers and grommets, to prevent any galvanic action between the tanks, rack, and test specimens. Figure 3 shows a typical test specimen installation.

Two specimen racks were mounted in each tank. One rack contained ten specimens of the same material as the tank in which it was mounted. Two of these specimens were welded and eight were unwelded. The welded austenitic stainless steel specimens also were heated for two hours at 1250 F to ensure a complete intergranular network of chromium carbides which makes the material less corrosion resistant. The other rack contained one unwelded specimen of each material.

The tanks were placed in service during the spring of 1957. Davison kept records of the tank usage and performance during the test. Inspections were made after the spring and fall fertilizing seasons and after winter storage during 1957, 1958, and 1959. At each inspection, one unwelded specimen was removed from the rack in each tank. Observations also were made of the tank's condition.

Specimens removed from the tanks were cleaned and weighed to the nearest milligram. Carbon steel specimens were cleaned in inhibited hydrochloric acid, aluminum specimens in 70 percent nitric acid, and stainless specimens in 20 percent nitric acid.

During the winter of 1959 to 1960, two sections (one by two feet) were cut from each stainless steel tank. One section was taken from the bottom; one section from the side which included the welded shell seam. These sections were examined to determine the type and extent of corrosion. Pit depths were measured by focusing a 250X microscope on the bottom of the pit and on the surface of the metal and by determining the distance between the two points with a micrometer scale on the focusing knob.

Patches of the appropriate type stainless were welded to the tanks, and the tanks were returned to service.

Tank Service and Performance

Service conditions to which the tanks were exposed in this test in northern Indiana are considered to be generally representative, from the standpoint of corrosion, of the conditions that would be encountered in most areas of the United States where liquid fertilizers are used. Most of the fertilizer hauled in the tanks was complete fertilizer, containing the three primary plant nutrients: nitrogen, phosphorus, and potassium.* Davison Chemical Division used ammonia and urea as nitrogen sources, phosphoric acid as a phosphorus source, and potassium chloride as a potassium source for its complete fertilizers. Some nitrogen solutions also were hauled in the test tanks. These were either aqua ammonia or a corrosion inhibited solution of ammonium nitrate and urea in water.

Table 1 summarizes the service of each of the test tanks. A season, as used in Table 1, means either the period from early spring until midsummer or from midsummer until late fall. The tanks were stored during the winter and were not used. The Type 304 stainless tank was used more than any of the other tanks. The aluminum tank was the least used, both in number of seasons used and in times per season.

Some observations made during the test that affected the service performance of the tanks and the test results are as follows:

(1) In fall, 1957, phosphoric acid inadvertently was stored in the Type 304 stainless tank for 12 days. The inside of the tank and the test specimens were corroded by the acid. All unwelded Type 304 test specimens were replaced before the next season.

*Mixed fertilizers are designated by the weight percent of plant nutrients in the fertilizer. A fertilizer designated as 6-18-6 contains six percent available nitrogen (calculated as N), 18 percent available phosphorus (calculated as P_2O_5) and six percent available potassium (calculated as K_2O).

(2) In spring, 1959, the carbon steel tank was damaged in an accident. Test specimens were transferred to a new carbon steel tank.

(3) In spring, 1958, a crack developed near a weld on the Type 202 tank and was repaired by welding. The cause of the crack was not determined.

(4) In spring, 1959, a leak developed in the aluminum tank. Because of past trouble with clogging due to corrosion products in this tank, it was removed from service.

TABLE 1—Summary of Tank Usage

TANK MATERIAL	Number of Seasons Used	Avg Number Times Used per Season	Avg Weight of Fertilizer per 1000-Gal Tank Capacity per Season, tons
Type 304 Stainless	6	33.4	139.7
Type 405 Stainless	5	19.2	67.8
Type 202 Stainless	6	27.0	97.6
USS Tenelon Stainless	6	20.2	82.3
1010 Carbon Steel	6	12.8	41.5
5052 Aluminum	4	10.2	34.9

Weight Loss Determinations

Results of the weight loss determinations made on the specimens removed from the tanks at each inspection are given in Table 2. These results show low corrosion losses during the test for all unwelded stainless steel specimens. Unwelded specimens of carbon steel and aluminum corroded at relatively high rates. Significantly greater corrosion losses were observed on the welded specimens of Type 304, Type 202, and Tenelon stainless than on the unwelded specimens. The high corrosion rates of the welded Type 304 specimens were caused by the exposure to phosphoric acid. The weight loss data on these specimens are not representative of fertilizer service. The corrosion rates of Type 405 stainless, carbon steel, and aluminum were not affected significantly by welding.

The progressive corrosion of carbon steel and aluminum as related to the amount of fertilizer hauled in the tank is shown in Figures 4 and 5. The corrosion rate of carbon steel was not dependent on tank usage. Carbon steel tanks continued to corrode during winter storage. Aluminum specimens corroded appreciably during use periods but did not corrode when the tank was empty.

An analysis of the data of Vreeland and Kalin shows that aluminum's average corrosion rate in several complete fertilizer solutions was about 22 mils per year in continuous immersion laboratory tests.[1] Carbon steel's average corrosion rate in the same test was about four mils per year. The average corrosion rate of carbon steel shown in Figure 4 is about five mils per year, which agrees with results of the earlier laboratory tests. However, the average corrosion of aluminum specimens in Figure 5 during periods when the tank was used is about four mils per year, which does not agree with the results of Vreeland and Kalin.

One difference that might account for this discrepancy in aluminum's corrosion rate in the two investigations is that 3003 aluminum was used in the earlier tests and 5052 aluminum was used in the present service test.

However, other data show that the corrosion resistance of 3003 aluminum and 5052 aluminum in complete liquids is not markedly different.[2] In a 28-day exposure in a solution of 6-18-6 at 72 F, 3003 aluminum had a corrosion rate of 14.6 mils per year and 5052 aluminum 9.1 mils per year. Therefore, the difference in the results of Vreeland and Kalin and the present results is attributed to a difference in test conditions rather than to the difference in alloy composition. In the tests of Vreeland and Kalin, specimens were exposed continuously for 28 days at 135 F. In the present test, corrosion rates are measured over a period of several months during which time the temperature varied considerably and the tanks were empty much of the time even during the seasons when the tanks were being used. Because aluminum corrodes appreciably only when it is in contact with the liquid, the apparent corrosion rate during the entire period between inspections was less than the actual corrosion rate in the fertilizer solution.

Phosphoric Acid Exposure

As mentioned previously, the Type 304 stainless tank was filled with phosphoric acid for 12 days during the test. Because Type 304 specimens were

TABLE 2—Results of Periodic Weight-Loss Determinations

MATERIAL	Penetration, mils							
	3 Months	6 Months	11 Months	16 Months	19 Months	23 Months	27 Months	31 Months
Unwelded Specimens								
Type 304 stainless	<0.01	1.78**	ND*	<0.01	<0.01	<0.01	<0.01	0.02
Type 405 stainless	ND*	<0.01	<0.01	0.02	0.08	0.02	0.02	0.04
Type 202 stainless	<0.01	<0.01	<0.01	<0.01	0.01	ND*	0.01	0.02
USS Tenelon stainless	<0.01	0.01	<0.01	0.02	0.02	ND*	0.04	0.03
Carbon steel	0.57	1.26	2.34	6.9	8.8	10.2	11.1	12.6
5052 aluminum	0.58	1.48	1.70	3.7	3.6	3.8	5.2	6.6
Welded Specimens								
Type 304 stainless	......	......	......	......	......	......	......	1.85** 0.16** 0.06 0.06 0.36 ND* 13.2*** 5.9 5.9
Type 405 stainless	......	......	......	......	......	......	......	
Type 202 stainless	......	......	......	......	......	......	......	
USS Tenelon stainless	......	......	0.19	......	......	......	......	
Carbon steel	......	......	......	......	......	0.29	......	
5052 aluminum	......	......	......	6.9	......	......	......	

* ND = No data.
** Specimens were exposed to 75% H_3PO_4 for 12 days.
*** Specimen was perforated.

corroded severely by this acid, new Type 304 specimens were prepared and placed inside the tank at the 11-month inspection.

Published data indicate that the corrosion of the Type 304 specimens in the phosphoric acid was more severe than would have been expected. Type 304 stainless steel, although not considered suitable for continuous phosphoric acid service, normally is not corroded by phosphoric acid at the rate of 1.8 mils per 12 days (54.7 mils per year). This abnormal behavior prompted the initiation of a laboratory test in which the corrosion rate of annealed Type 304 stainless was determined in reagent-grade, 75 percent phosphoric acid and in reagent-grade, 75 percent phosphoric acid containing five percent by weight potassium chloride.

Tests were conducted for two weeks at room temperature in non-aerated solutions. Rates of 0.01 mils per month occurred in the acid solution, but almost a 2000-fold increase was determined in the corrosion rate of Type 304 in phosphoric acid because of the potassium chloride (1.02 mils per month). This increase indicates that residual potassium chloride from the complete fertilizer solutions could cause higher corrosion rates on a Type 304 stainless tank containing phosphoric acid and probably contributed to the high corrosion rate observed.

In addition to damage to a Type 304 stainless tank because of corrosion in chloride contaminated phosphoric acid, hydrogen would be liberated during the corrosion process. In an unvented system, this hydrogen could create an explosion hazard if a flame source were introduced.[3]

Annealed specimens of AISI Type 202 stainless and Tenelon stainless pitted only at the areas where crevices were present between the specimens and insulators. Welded specimens of Type 202 and Tenelon stainless pitted all over the surface and pitted somewhat more severely near the weld. Type 405 specimens pitted about the same on both annealed and welded specimens. None of the AISI Type 304 specimens had any pits. Corrosion on carbon steel and aluminum specimens began as pitting and developed into uneven general corrosion.

Results of weight loss determinations from the sets of unwelded specimens of the six materials, each exposed in a separate tank, are presented in Table 3. These results include only the sets of specimens exposed in the aluminum tank, which was used slightly, and the Tenelon stainless tank, which received normal use.

The data in Table 3 show that carbon steel corroded almost as much in the little-used aluminum tank as in the normally used stainless tank. This is additional evidence that a carbon steel tank corrodes because of the fertilizer residues and vapors, even when the tank is not in use. In the normally used stainless tank and the little-used aluminum tanks, Types 202 and 304 and Tenelon stainless were equally resistant. Type 405 was the least resistant of the stainless steels.

Inspection of Tanks

Observations of each tank's condition were made at the time of each inspection. General condition of each tank and appearance of the corrosion products and pits were recorded.

Corrosion Patterns

In these inspections, a corrosion pattern for each tank became evident after the first few months' service. In the first inspection, both the carbon steel and aluminum tanks were covered with voluminous corrosion products. These white corrosion products in the aluminum tank caused frequent service failures by plugging screens and nozzles. The carbon steel tank was covered uniformly with corrosion products in the middle and lower portions of the tank and in the head space where the fertilizer solution had not been in contact with the steel. The aluminum tank was pitted in the middle and lower portion but was not corroded in the head space. The Type 202 stainless tank had many small tubercles of corrosion product at the heat-affected zones near welds. The Type 405 tank had a tubercle density of about two tubercles per square foot. Removal of the tubercles revealed pits underneath. The Type 304 tank was not visibly corroded.

Corrosion above and below the liquid line on carbon steel and aluminum tanks was more apparent at the second inspection. In the carbon steel tank, more corrosion product was present above than below the liquid line. In the aluminum tanks, almost no corrosion occurred above the liquid line, but severe attack occurred below it. The only change in tank appearance throughout the remainder of the test was that carbon steel and aluminum tanks collected more corrosion products and that more tubercles formed in Type 405 stainless tank. The effect of the phosphoric acid on the Type 304 stainless tank was apparent after the second inspection.

At the end of the three-year period, sections cut from the four stainless steel tanks were examined at the Applied Research Laboratory.

Figure 6 shows the bottom and side areas from the Type 405 stainless tank. Many tubercles can be seen on each section. More tubercles formed on the Type 405 tank than on any other stainless steel tank. Appearance of the sections shown was the worst of the stainless tank sections.

Pit Depth Measurements

The results of the pit-depth measurements made on these tank sections are shown in Table 4. The five pits in the heat-affected zones of each side section that appeared to be deepest were measured. Average pit depth and maximum pit depth are reported. These results show that the deepest single pit occurred on Type 405 stainless steel. However, the greatest average pit depths occurred on Type 202 stainless. The deepest pit on Tenelon stainless was about equal to the deepest pit on Type 202 stainless, but the

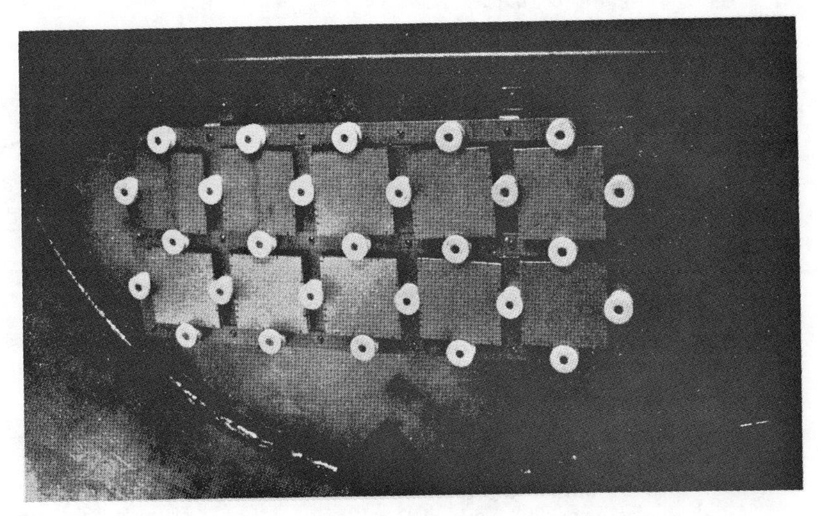

Figure 3—Corrosion test specimens were mounted on the baffle plates inside each test tank. Porcelain insulators and plastic washers and grommets were used to prevent galvanic action between the test specimens and the tank baffle plates.

average pit depth on Tenelon was less than that on Type 202 and about the same as that on Type 405. All the recorded pit depths were between 12.4 and 37.2 mils, which are substantially less than the thickness of the sheets (110 mils).

Metallographic examination of welded sections from each of the stainless tanks showed that areas on each side of the weld on the austenitic stainless steels contained intergranular carbides.

Stress Corrosion Cracking

Stress corrosion cracks were observed in the Type 202 tank section. These cracks ran from the base of the pits in the heat-affected zones near the welds on the inside of the tank. Examination of etched metallographic specimens revealed that the cracks were mostly transgranular, but areas of intergranular cracking also were observed.

Two small cracked areas were observed in the crevice at the weld on the Type 304 stainless steel tank section. Located on the outside sheet of the crevice and also in the weld metal, these cracks had the appearance of stress corrosion cracks. None of the cracks observed penetrated through the sheet.

No cracks were observed in the Type 405 stainless or the Tenelon stainless tanks.

A combination of high residual stresses from welding, a crevice, and carbide precipitation at the grain boundaries in the heat-affected zone near the weld probably provided the conditions under which Type 304 stainless was susceptible to stress corrosion cracking. Top of the baffle plate inside the Type 304 tank was brake-formed at a 90-degree angle to provide rigidity to the baffle. High residual tensile stresses undoubtedly were present at this bend, but no cracks were seen during the inspections. This indicates that annealed Type 304 stainless resists stress corrosion cracking in fertilizer service when there are no crevices in the tank.

This observation also is in agreement with the findings of a previous laboratory stress corrosion cracking test in which annealed and stressed Type 304 stainless steel was resistant to cracking in mixed liquid fertilizers and nitrogen solutions for periods to 122 days. The cracking within the crevice occurred in an area were carbides had precipitated in the grain boundaries. However, cracking did not occur in the areas containing intergranular carbides outside the crevice as it did in the Type 202 tank. All these observations indicate that a crevice and the presence of intergran-

TABLE 3—Corrosion of Annealed, Nonwelded Materials After Three Years

Materials	Average Penetration, mils*	
	Tank With Little Usage	Tank With Normal Usage
Type 304 stainless	0.01	0.03
Type 405 stainless	0.04	1.69
Type 202 stainless	0.01	0.03
USS Tenelon stainless	0.01	0.05
Carbon steel	10.6	13.9
Aluminum	5.52	Specimen lost

* Calculated from weight losses.

Figure 4—Relation between corrosion of carbon steel and tank usage in liquid fertilizer service. Corrosion continued even during storage of the tanks.

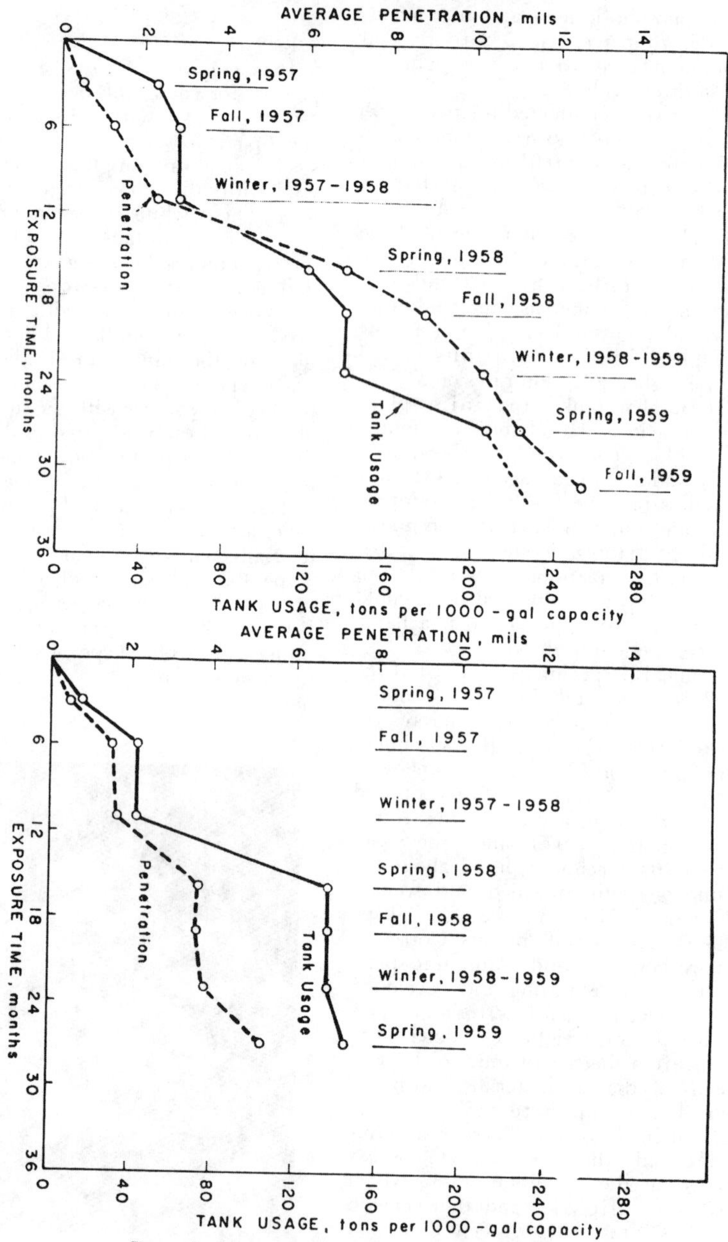

Figure 5—Relationship between corrosion of 5052 aluminum and tank usage in liquid fertilizer service. Aluminum tanks corroded during use periods but not during storage.

TABLE 4—Pit-Depth Measurements on Sections of Stainless-Steel Tanks

Type of Stainless	Avg. Depth of 5 Deepest Pits, mils	Maximum Pit Depth, mils
304	No pits	No pits
405	21.7	37.2
202	25.0	28.0
USS Tenelon	20.2	28.4

ular carbides are necessary prerequisites for stress corrosion cracking of Type 304 stainless steel in fertilizer solutions.

To confirm the theory that both a crevice and intergranular carbides are necessary for stress corrosion cracking of Type 304 stainless in liquid fertilizers, a set of stress corrosion specimens was exposed in the Type 304 stainless tank from April to November, 1961. Specimens were prepared by welding two sheets of Type 304 stainless (6 by 12 by 1/8 inches) together in three ways:

(1) Lapped and welded on one side similar to method used in fabricating test tanks.

(2) Lapped and welded on both sides and on ends of the lapped area to seal the crevice.

(3) Butted and welded on both sides.

One 12-inch square specimen of each of the above three types was stressed by bending the specimen in a jig so that tensile stresses above the steel's yield point and perpendicular to the weld were introduced to the specimen's convex surface. These specimens were mounted as stressed on the tank baffle. Three other specimens, one of each type, also were mounted unstressed.

Only one crack was observed in the six specimens. It was found in the weld metal of the unstressed specimen that had been lapped and welded on one side. The crack began inside the crevice and progressed approximately one-fifth of the way through the weld metal.

This result tends to confirm the theory that a crevice is necessary for stress corrosion cracking of Type 304 stainless in mixed liquid fertilizers and also indicates that crevice-free joints, made by either placing a seal weld over the opening to the crevice or making a butt-welded joint, are not susceptible to cracking. This result also indicates that residual welding stresses can be sufficiently high to cause stress corrosion cracking inside crevices.

The cracking in the Type 304 tank apparently will not cause leaking. After five years' service for the tank, no leaks have been reported in areas similar to the area where the cracks were seen.

Conclusions

Results of the three-year service test showed that carbon steel and aluminum are subject to general corrosion by liquid-mixed fertilizer solutions. Carbon steel corroded at an average rate of about five mils per year and formed voluminous corrosion products. Corrosion in the vapor zone was as severe as corrosion in the liquid zone, or more severe. Carbon steel also corroded when the tank was empty, apparently due to moisture and fertilizer residues trapped inside the layer of corrosion product.

Aluminum showed an average corrosion rate of about four mils per year during test periods when the tank was used. However, there is sufficient evidence to indicate that the actual corrosion rate of aluminum immersed continuously in complete fertilizer solutions is considerably higher. The aluminum tank did not corrode appreciably in the vapor zone, nor did it corrode when empty.

Stainless steel tanks showed no sign of general corrosion, and the weight-loss specimens showed negligible corrosion rates for all the stainless grades tested. Types 202 and 405 stainless tanks and Tenelon stainless tanks were pitted. Type 202 stainless pitted only in the heat-affected zone near the welds and also was subject to stress corrosion cracking. Type 405 and Tenelon stainless pitted at the welds and also at areas not affected by the welds. Type 304 stainless was not susceptible to pitting or to general corrosion by the liquid fertilizers. However, some small stress corrosion cracks, which caused no difficulty in service, were observed in the crevice at the shell-seam weld of the Type 304 tank. Test results with stress corrosion specimens indicate that cracking in Type 304 fertilizer tanks can be prevented by eliminating crevices at welded joints.

Results of the service tests indicate that AISI Type 304 stainless is resistant to corrosion in normal liquid-fertilizer service. This steel has excellent resistance to pitting and general attack in both nitrogen solutions and complete fertilizer solutions and also is resistant to stress corrosion cracking if crevices at welds are avoided. Although resistant to general corrosion, the other stainless steels (Types 202 and 405 and Tenelon stainless) are subject to pitting in complete fertilizer solutions. Carbon steel and aluminum are not resistant to corrosion in service with complete fertilizer solutions. Tanks of carbon steel and aluminum would require considerable maintenance and replacement after a few years' service.

Figure 6—Sections cut from Type 405 stainless steel tank. Top section is from side which included the seam weld. Bottom section is from tank bottom. These were examples of worst appearance of tanks.

References

1. D. C. Vreeland and S. H. Kalin. Corrosion of Metals by Liquid Fertilizer Solution. *Corrosion,* 12, 569t (1956) November.
2. J. D. Hatfield, et al. Corrosion of Metals in Liquid Mixed Fertilizer. *Journ Ag & Food Chem,* 6, No. 7, 524 (1958).
3. J. B. Loew. Corrosion-Generated Hydrogen Explodes in Phosphoric Acid Tank. *Corrosion,* 17, 30 (1961) March.

T. F. SHAFFER, JR., now is senior research assistant at American Steel and Wire Division of U. S. Steel Corp., Cleveland, Ohio. His current work is in general corrosion and stress corrosion of steel wires. Before his April 1 transfer to the wire division, he was a member of U. S. Steel's Applied Research Laboratory, where his work included research on corrosion of carbon, alloy, and stainless steels with special emphasis on pitting and intergranular corrosion of stainless and on applications research directed toward the use of steel products in agricultural equipment and buildings. He is a member of NACE and a graduate of the College of Wooster (Ohio).

Effects of Hydrogen Halides On Anodic Polarization of Stainless Steel★

By OLEN L. RIGGS, JR.*

Abstract

The anodic polarization of stainless steel in 10 moles/liter sulfuric acid with various concentrations of HF, HCl and HBr has been investigated. At sufficiently high concentrations all of these hydrogen halide ions make it more difficult to establish and maintain passivity. They did not change the potential at which passivation was established, but they narrowed the passive potential range by increasing the potential at which breakdown occurred. They all accelerated the decay of passivity when the applied currents were interrupted. Effective protection from corrosion can still be obtained in these systems. In the presence of 1 mole per liter hydrogen fluoride, there was a substantial increase in current to maintain passivity without any corresponding loss of metal.

Introduction

A NORMALLY ACTIVE METAL when placed in a corrosive environment may become chemically inactive. It is then said to be in the passive state. The present article reports on an investigation of the effects of fluoride, chloride and bromide on the passivity of a stainless steel anode in sulfuric acid.

It also presents their effects on the rate of decay of the passive condition. There have been several earlier studies on the effect of halides on the passivity of stainless steels[1, 2] and on the passivity of aluminum.[3, 4, 5, 6] Investigators in Japan[7, 8] have reported that the presence of chloride and bromide makes it more difficult to passivate stainless steel in sulfuric acid. However, the fluoride and iodide did not have this effect. The sulfuric acid concentration employed in the present investigation was too high to permit a comparative study of iodide. Recent anodic polarization measurements show that NaCl additions up to 0.1 N in 10 N H_2SO_4 do not seriously affect the passive behavior of stainless steels.[9] Previous reports[10, 11] showed that low concentrations of chloride did not prevent the passivation of stainless steel in sulfuric acid, but that a higher current density was required to establish passivity.

This study was conducted to determine the limiting and contributing characteristics of hydrogen halides on the anodic passivation of Type 316 stainless steel in 10 moles/liter sulfuric acid.

Experimental Procedure

In all of the experiments, the passivated metal was AISI Type 316 stainless steel (17 Cr, 12 Ni, 3 Mo) immersed in hydrogen halide-treated aqueous 10 moles/liter sulfuric acid. Hydrofluoric, hydrochloric and hydrobromic acids were of reagent grade. No precaution was taken to eliminate air from the experiments.

The stainless steel specimens were scrubbed with a bristle brush, cleaning powder and hot water; following this, they were rinsed in isopropyl alcohol and allowed to dry. They were then cathodically treated in 10 moles/liter sulfuric acid at a current density of 0.25 amp/sq cm for three minutes. The specimens were then ready for the electrochemical cell and subsequent anodic polarization study.

The study was conducted in an electrochemical cell of the design shown in Figure 1. The cell incorporated three electrodes into a circuit, which was completed in a constant potential control (potentiostatic) instrument. The anode was the metal to be studied. It was suspended in the center of the cell from a platinum hook between the cathodes. The cathodes were made from platinum plates, which were 2.4 cm on each side. They were parallel to the anode and 1.25 cm from either side of it. The Type 316 stainless steel anode had an area of 8.0 sq cm. The reference electrode was a saturated calomel half-cell mounted in a vessel remote from the cell. The cell and calomel electrode vessel were joined by a small water-cooled tube. The end of the tube was open into the cell, but contained a porous glass frit closure at the end exposed in the calomel electrode vessel. This permitted the corrosive liquid under investigation to travel the length

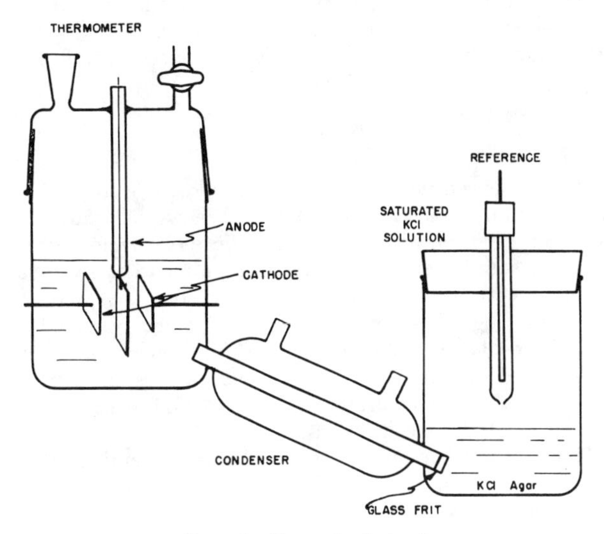

Figure 1—Electrochemical cell.

★ Submitted for publication November 30, 1962.
* Continental Oil Company, P. O. Drawer 1267, Ponca City, Oklahoma.

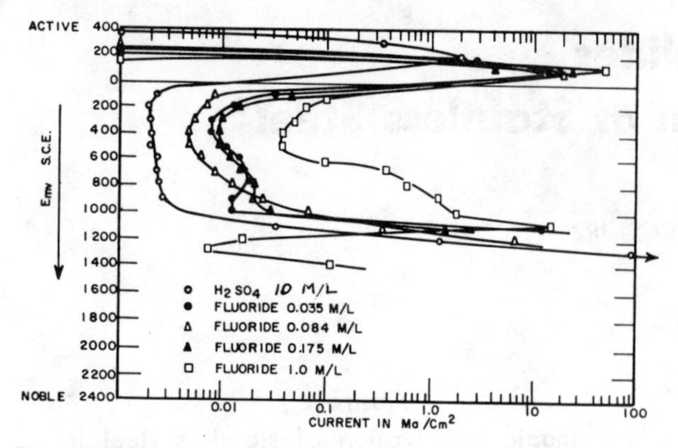

Figure 2—Effect of chloride concentration on anodic polarization of Type 316 stainless steel at 24 C.

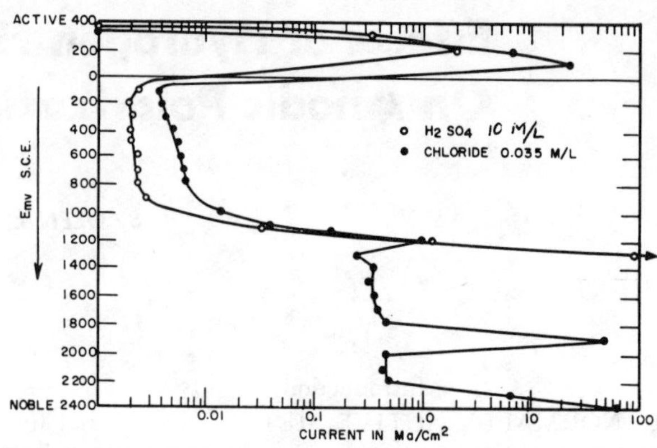

Figure 3—Effect of chloride on extended anodic polarization of Type 316 stainless steel at 24 C.

of the tube and contact the glass frit. In the calomel electrode vessel, the bottom portion was filled with an agar-KCl gel to cover the glass frit. The KCl-saturated water was then poured in over the agar plug.

The polarization measurements were made potentiostatically. The instrument maintained the desired potential value constant within ± 1 millivolt over the range of 0 to ± 2.5 volts. The sensitivity was established as ± 0.5 millivolt. This means the controller went from maximum positive to maximum negative power output with a change of one millivolt in the reference potential. The maximum drift was less than ±1 millivolt over the entire current range. The response time was determined to be less than 100 microseconds. The maximum rated current capacity was 10 amperes anodic or 2.5 amperes cathodic at 10 volts. The instrument had electronic jacks for connecting a 0-100 millivolt recorder to obtain a recording of each full-scale current and potential range. The instrument was capable of manual or automatic operation. This permitted either a time scan integrated polarization curve or one obtained from individually selected potentials and respective time-dependent current values. In the latter instance, the desired potential was set into the instrument. After ten minutes had elapsed, the current was recorded. The corrosion tests at constant potential were determined in much the same manner. The metal specimens were weighed before and after exposure to the system. The constant potential was maintained for the 24-hour exposure interval. The reproducibility was recognizably difficult; however, after an average of five runs, the results were satisfactory.

Results

Effect of Chlorides

Several concentrations of hydrogen chloride were prepared in 10 moles/liter H_2SO_4. Figure 2 shows the relative effect of chloride ion on the anodic polarization of Type 316 stainless steel. The current required to passivate increased with the chloride ion concentration. An apparent narrowing of the passive potential range occurred with successive increases in chloride ion concentration. Pitting was detected in systems containing as little as 0.035 moles/liter of chloride ion. This reaction was intensified at higher concentrations.

The polarization curves indicated that potentials at the breaks do not correspond to the oxidation potential of the chloride ions-to-chlorine reaction (Table 1). Free chlorine was not observed; however, it may have been present. Some purely speculative evidence of this could be the initial change in slope of the passive potential curve at the various chloride ion concentrations. The corrosion reaction could then quite possibly be due to the anodic dissolution process. Hypothetically, the chlorine could combine with iron instantaneously and not be visible as a gas.

TABLE 1—Comparison of Experimental and Calculated Oxidation Potentials of Chlorine and Bromine at 25 C

C_{Cl^-}	E_{Cl_2, Cl^-}	$E_{Calc.}$	$E_{Exp.}$
0.0875	—1.421	1.179	1.400
0.175	—1.403	1.161	1.350
0.350	—1.385	1.143	1.050
0.875	—1.361	1.119	0.450
10.000	—1.299	1.057	0.295
		(0.122)	(1.105)
C_{Br^-}	E_{Br_2, Br^-}	$E_{Calc.}$	$E_{Exp.}$
0.0875	—1.128	0.887	0.495
0.175	—1.111	0.869	0.505
0.350	—1.093	0.851	0.520
0.875	—1.069	0.827	0.545
		(0.060)	(0.050)

NOTE: Figures in parentheses show spread.

An interesting development occurred in the chloride study when the potentials were extended beyond values normally used. In the system of 0.875 moles/liter hydrogen chloride, two additional polarization plateaus developed. The expanded polarization curve is given in Figure 3. Although potentials indicate additional passive regions, corrosion was observed in these regions. This portion of the polarization curve needs further study.

Effect of Bromides

The effects of several concentrations of hydrogen bromide on the anodic polarization study are shown in Figure 4. The hydrogen bromide shortened the passive potential range, which indicated that this range was narrow. However, bromide evolution was observed at a specific potential in each concentration of hydrogen bromide. These observed potentials corresponded to the calculated oxidation potentials of bromine and did not indicate corrosion. The ten-fold increase in the current necessary to maintain passivity was possibly caused by an increase in electrolyte resistance due to the generation of the bromine.

Oxidation Potential Calculations

Although no neat, refined family of polarization curves developed with use of the bromide ion, the curves were reproducible within ± 10 mv. The polarization potential curves suggested that these experimental potentials cor-

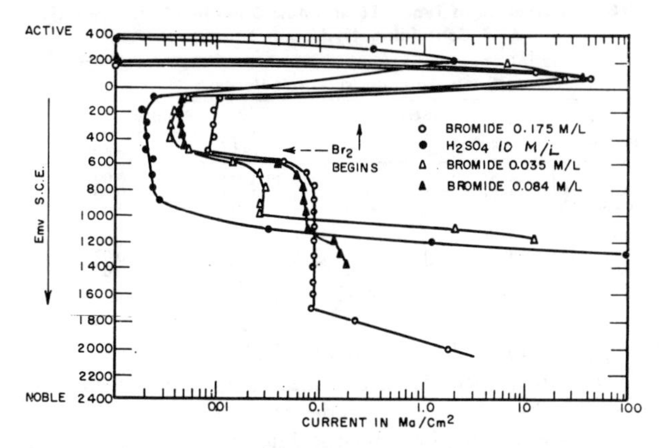

Figure 4—Effect of bromide concentration on anodic polarization.

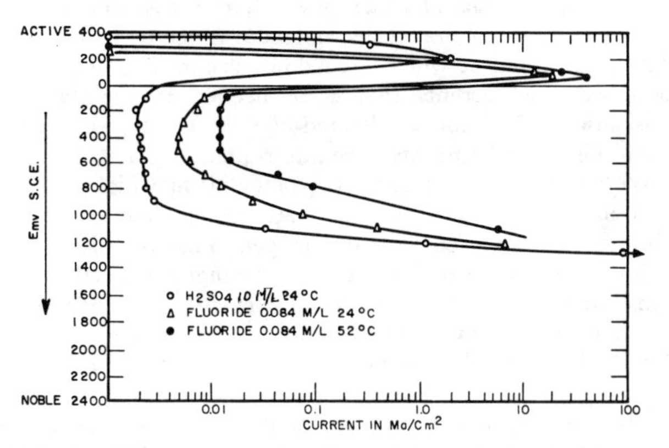

Figure 6—Effect of temperature on anodic polarization of Type 316 stainless steel exposed to 10 M/L H_2SO_4 and 0.084 M/L fluoride.

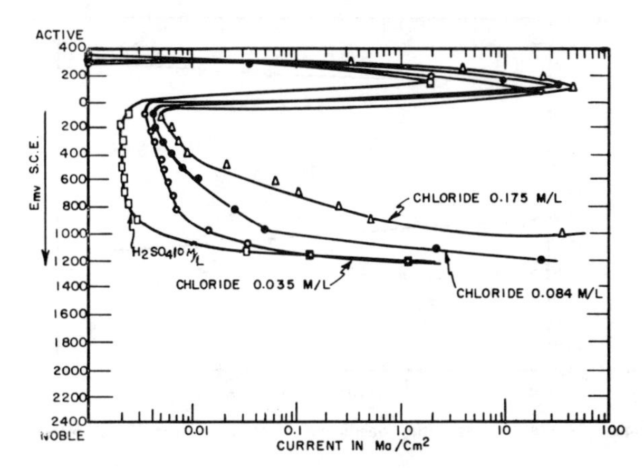

Figure 5—Effect of fluoride concentration on anodic polarization of Type 316 stainless steel at 24 C.

responded to the oxidation potentials of chloride and bromide. Calculations of these potentials were made at each of the concentrations studied using the standard expression for oxidation-reduction electrode potentials:

$$E_{\text{halide}}x = E^\circ_{x_2,x-} - \frac{RT}{2F}\ln a_{x_2} + \frac{RT}{F}\ln a_{x-}$$

It was necessary to know the activity coefficient of chloride in a sulfuric acid system. Such data were not available for these high concentrations of sulfate ions. The calculations assumed an activity coefficient of 1 for all concentrations. At least a constant error existed through all of the calculations. Also, the potentials [$E^\circ_{x_2,x-}$] for chlorine and bromine were calculated on the basis of one atmosphere partial pressure. These potentials, $E^\circ_{x_2,x-}$, were standard hydrogen electrode values. The potentials E_{calc} and E_{exp} were reported as saturated calomel electrode values. These data were summarized in Table 1. The spread in experimental values was in excess of 1 volt. It seemed very unlikely that these experimental potentials corresponded to the oxidation potential of chloride ions. Chlorine was not believed to be a passivating influence in this experiment. If it was formed, it was consumed in the system by complexing either with the iron or the passive film. The net result was that corrosion did occcr. It was possibly due to direct attack by the chloride or due to anodic dissolution.

It seemed probable that this whole effect observed in the hydrogen chloride system could be attributed to the competition between chloride ions and the absorbed passivating substance for the metal surface. This was borne out in part by Cartledge[12] when he used radiotracers to show that halogen ions did concentrate on the iron surface.

The spread of oxidation potentials for the bromide system was only 0.06 volt. The experimental values for bromine had a spread of about 0.05 volt. This, along with the fact that bromine gas was vigorously evolved, indicated that the break in potentials was due to the oxidation of bromide to bromine. It was further observed that no corrosion occurred after this break. The excess current was used in forming bromine gas.

Effect of Fluoride

Fluorine has a very high oxidation potential. The oxidation potential is higher than the potential at which the transpassive region of the polarization occurs. It has been reported[13] that fluoride has a very high capacitance energy of hydration and that it does not behave according to theory at strongly anodic potentials. However, increasing concentrations of fluoride ion cause a respective increase in currents required to passivate. Figure 5 gives the relationship of the anodic polarization curves with the various fluoride ion concentrations. The narrowing of the passive potential range must be attributed to effects other than conversion of fluoride ion to fluorine. Just what contribution the hydrogen fluoride makes is not known. It is not believed to be due to chemisorption,[14] but possibly to some intrabonding ion complex. The possible contribution of silica due to the solution of heat-resistant glass by hydrofluoric acid has been ignored.

At a constant hydrogen fluoride concentration, temperature affects the current requirements in much the same manner as increased concentrations of halide. Figure 6 shows that the current needed to obtain and maintain passivity was increased and the passive range was narrowed.

When the anode potential was plotted against weight loss, as shown in Figure 7, the metal was protected against corrosion over a potential range much broader than was indicated by the current measurements. In the presence of 1.0 mole/liter hydrogen fluoride at anode potentials from 800 to 1200 mv noble to the saturated calomel electrode (SCE), there was a substantial increase in current without any corresponding loss of metal. Table 2 gives the results of weight loss measurements at several fluoride

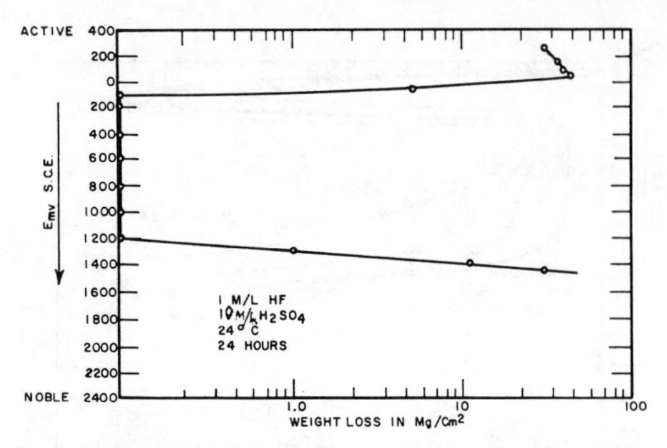

Figure 7—Weight loss during anodic polarization of Type 316 stainless steel.

concentrations and at two different temperatures with and without passivation. At the lower temperature, the lowest concentration of fluoride seemed to protect the metal from corrosion, but at higher concentrations and at all concentrations at the higher temperature, fluoride accelerated attack on the metal. Passivation afforded substantial protection from corrosion under all the conditions investigated.

Effect of Halides on the Breakdown of Passivity

When a stainless steel specimen was maintained in the passive condition at a potential of 500 mv for 50 minutes and the applied current was then disconnected, the potential increased after a few hours to 280 mv and remained constant for 183 days. The surface was purposely made active by contact with a copper-tipped probe. However, when a similarly prepared specimen was transferred to a 10 moles/liter sulfuric acid solution containing appreciable concentrations of fluoride, chloride or bromide, the potential fell very rapidly and in a few minutes it was no longer in the passive range. These observations are shown in Figures 8 and 9. The process was so rapid that the respective halides and two different concentrations of chloride appeared to have approximately identical effects. A trace of chloride (5 ppm) was sufficient to break down the passive condition, although in this case the decay was slower. (Approximately 20 minutes passed before the total decay occurred.)

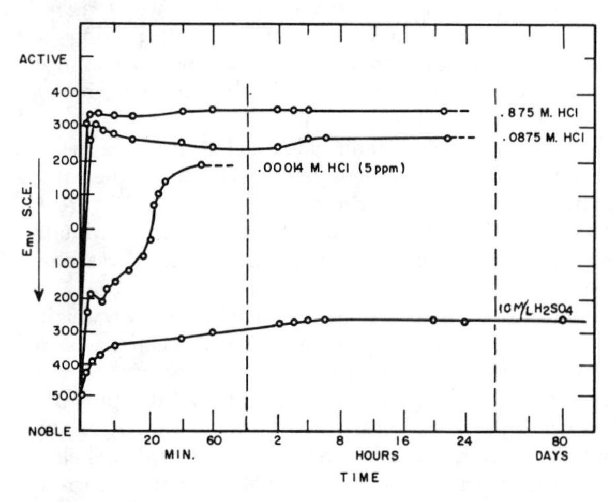

Figure 8—Halide effect on decay time of passive Type 316 stainless steel.

TABLE 2—Corrosion of Type 316 Stainless Steel in 10 M/L Sulfuric Acid Containing Hydrogen Fluoride

	Weight Loss, mg/sq cm/24 Hours			
	Control		Passive	
HF, M/L	24 C	52 C	24 C	52 C
None	0.81	12.50	0.00	0.00
0.035	0.67	9.74	0.00	0.00
0.084	0.12	8.61	0.00	0.00
0.175	0.05	13.96	0.00	0.00
0.875	0.03	15.20	0.00	0.02
1.000	...	....	0.02	0.02

Comparison of Hydrogen Halide Effects

In certain respects, and at sufficiently high concentrations, the three hydrogen halides studied had similar effects on the passivation process. They increased the current necessary to produce passivation. They did not change the potential at which passivation was established, but they narrowed the passive potential range by increasing the potential at which breakdown occurred. They all increased the currents that were needed to maintain passivation. They all accelerated the decay of passivity when the applied currents were interrupted.

With respect to increasing the potential at which passivation breaks down, fluoride had the smallest effect, while bromide and chloride overlapped. That is, chloride had a greater effect than bromide at the higher concentrations but a lesser effect at the lower concentrations. At the lowest concentrations studied, both chloride and fluoride slightly decreased the potential at which passivation broke down.

The increased currents to maintain passivity with fluoride and bromide were not accompanied by loss of metal, but corrosion was observed at passive potentials in the presence of higher concentrations of chloride.

Metallographic Study of Chloride Effects

The effect of chlorides on the passive surface was examined through the use of photomicrographs. Figure 10 shows metallographically the surfaces of anodically passivated Type 316 stainless steel. In the 10 moles/liter H_2SO_4, both 0.03 N and 0.10 N HCl could be tolerated without deleterious effects on the passive state. No pitting was observed throughout the duration of the test. At 0.03 N and 0.10 N HCl, the current density decreased with time.

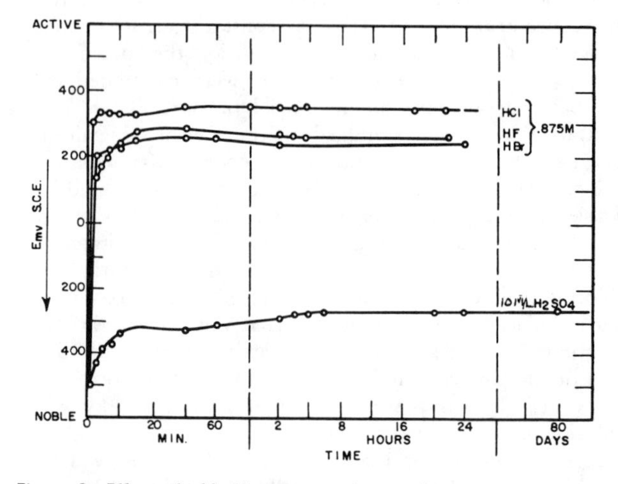

Figure 9—Effect of chloride concentration on decay time of passive Type 316 stainless steel.

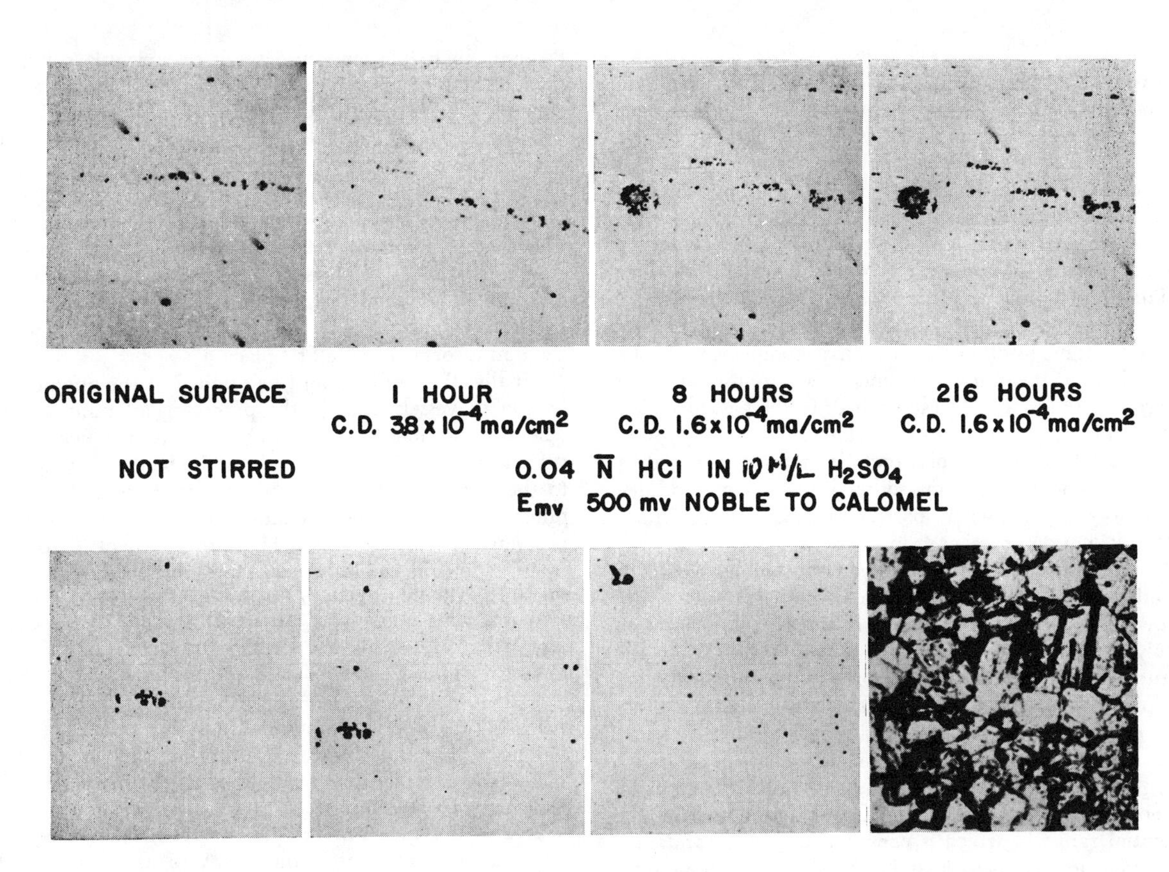

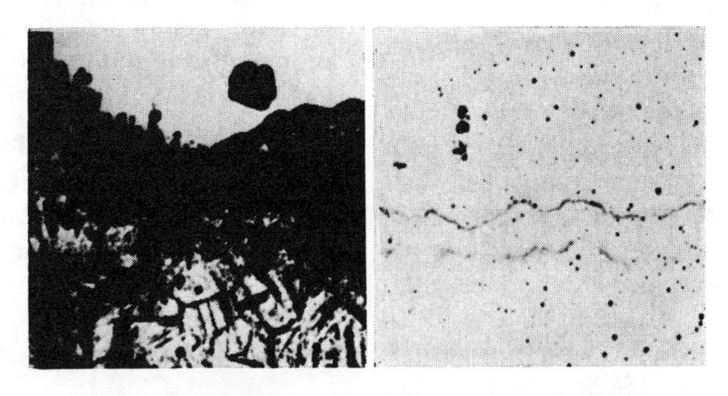

Figure 10—The metallography of Type 316 stainless steel anodically polarized in 10 M/L H_2SO_4 containing hydrochloric acid with temperature at 25 C. (150 X).

In the 0.10 N HCl system, severe etching of the Type 316 stainless steel was observed at the air-liquid interface when no stirring was used. However, the etching was greatly reduced when the system was stirred.

At 0.5 N HCl, passivity was not achieved. After thirty minutes' exposure in this environment, the Type 316 stainless steel surface was severely pitted. Although the limits were not defined carefully, as much as 0.5 N HCl in 10 moles/liter H_2SO_4 could not be tolerated for anodically passivating Type 316 stainless steel. As much as 0.10 N HCl in 10 moles/liter H_2CO_4 offered no problem in anodically protecting this steel.

Discussion

Various theories have been offered regarding the physical or chemical nature of the passive state. It has been ascribed to a layer of oxide, a layer of adsorbed oxygen, the adsorption of some chemical species or reactive sites on the metal surface, or to some modification of the structure of the metal surface itself. The term "film" in what follows refers impartially to the physical product resulting from the change of the active to the passive state.

The film-forming process may occur increasingly rapidly as the current is increased or it may set in abruptly. There may be a film dissolution process that competes with film formation as the current is increased up to the point where the film is actually established. In the passive condition there is presumed to be a relatively inactive dissolution process, the effects of which, if any, are repaired by the current to maintain passivity. When the potential is raised to the point where passivity breaks down, still another dissolution process sets in.

It had been hoped that the hydrogen halides would be found to affect these processes in simple ways that would throw some light on the nature of the chemical reaction involved. The results of adding hydrogen halides to the sulfuric acid, however, have proved to be much too complicated for interpretation as effects on simple processes of film formation and film dissolution. At sufficiently high concentrations, all of the halides appear to be antagonistic to the passive state at all stages. They make it more difficult to establish passivity; they narrow the potential range in which it can be maintained; and they accelerate its decay, at least in the absence of a maintenance current. Looking at the chloride results, only a unified interpretation of this observation is sought, by ascribing to that ion only the ability to accelerate some process of film dissolution. The greater current needed to establish passivity could then be regarded as the result of competition between processes of film formation and film dissolution. The results with fluoride, however, show that this picture is entirely too simple. Since fluoride narrows the passive potential range, increases the applied current, and reduces corrosion effects at low concentration levels, halide ion must affect the processes involved in passivation in more than one way.

The fact that, during the establishment of passivity, fluoride ion on the one hand and chloride and bromide on the other may have different effects, presents a promising lead. It could be related to the conductivity of the solution since, due to the formation of H_2F^+, the introduction of a small amount of hydrogen fluoride may reduce the conductivity, while hydrogen chloride and hydrogen bromide would be expected to increase it. Another possibility is that chloride and bromide produce their effect by complexing ferric ion, since the concentration of the fluoride ion may be extremely small in the solutions to which hydrofluoric acid has been added.

It should be borne in mind that the part played by the halides may not be an active one. It could be that sulfate or bisulfate ion is involved in film formation and any inert action would interfere. The effect that the hydrohalic acids exert on the hydrogen ion activity also cannot be overlooked.

In spite of the several complications, perhaps this conclusion can be made from the experiments. The halides both interfere with the establishment of passivity and accelerate its decay in the absence of a maintenance current. Passivity in halide environments is an unstable condition that eventually disappears. A quite negligible total flow of current, not more than that equivalent to a monolayer of oxygen, is needed to establish the passive potential. When passivity is destroyed on a part of the metal surface, it disappears entirely. Nevertheless, specimens that were maintained in the passive state for 50 minutes retained this condition for an extremely long time. It is possible that some process repairs microscopic chance damage to the film. The halide experiments certainly suggest that this repairing process is similar to the process by which the film is formed in the first place.

Finally, the results might be applied to the anodic protection of vessels used in the processing of mineral ores, where contamination from halides is frequently experienced. Anodic protection could be beneficially applied to the storage vessels for technical grades of sulfuric and phosphoric acids where halides create serious corrosion problems. The vessels used to reprocess spent acids such as sulfuric acid could be anodically protected. Also, it is quite possible that through the use of controlled potentials, bromine might be extracted from sulfuric acid treated brines.

Conclusions

1. The hydrogen halides (HCl, HF and HBr) in 10 moles/liter H_2SO_4 caused an increase in currents needed to establish anodically and maintain the passivity of Type 316 stainless steel.

2. In the absence of the passive potential applied current, halides accelerated the decay of the passive substance. As little as 5 ppm of the chloride ion destroyed passivity. However, in the absence of halide, with no applied current or physical disturbance, a prepassivated specimen of Type 316 stainless steel could remain indefinitely passive to 10 moles/liter H_2SO_4.

3. In the presence of one mole/liter hydrogen fluoride there was a substantial increase in current to maintain passivity without the corresponding loss of metal over the passive 800-1200 potential range.

Acknowledgments

Gratitude is expressed to Dr. Frank H. Dickey for his comprehensive critique of this work, to Dr. William H. Harwood for his guidance and advice, to Mr. Louis H. Wolfe for the photomicrographs, to Mr. Troy Pemberton, and to the management of Continental Oil Company for approval of this study and its publication.

References

1. S. L. Hoyt and M. A. Scheil. *Trans. Am. Soc. Metals*, 27, p 191 (1939).
2. N. A. Nielsen. Symposium on Corrosion Fundamentals. Ed. by Anton de S. Brasunas and E. E. Stansbury, pp 148-69, The University of Tennessee Press (1956).
3. D. J. O'Conner, P. G. Johansen and A. S. Buchanan. *Trans. Faraday Soc.*, 52, 229 (1956).
4. M. J. Pryor and D. S. Keir. *J. Electrochem. Soc.*, 102, 370 (1955).
5. J. P. Clay and A. W. Thomas. *J. Amer. Chem. Soc.*, 60 2384 (1938).
6. H. J. Modi and D. W. Fulistenau. *J. Phys. Chem.*, 16, 640 (1957).
7. N. Ohtani and H. Sugawara. *J. Electrochem. Soc. (Japan)*, 26, No. 4-6, E 65 (1958).
8. S. Morioka and K. Sakiyama. *J. Electrochem. Soc (Japan)*, 25, 191 (1957).
9. S. J. Acello and N. D. Greene. *Corrosion* 18, 8, 286t-290t, (1962) August.
10. J. D. Sudbury, O. L. Riggs Jr. and D. A. Shock. *Corrosion*, 16, 2, (1960) February.
11. D. A. Shock, J. D. Sudbury and O. L. Riggs Jr. *First International Congress on Metallic Corrosion*, April, 1961. Butterworths, Inc., London (1961).
12. K. E. Heusler and G. H. Cartledge. Personal communication (in process of publication).
13. D. C. Grahame *J. Amer. Chem. Soc.*, 76, 4819-23 (1954).
14. D. C. Grahame and B. A. Soderberg *J. Chem. Phys.*, 22, 449 (1954).

SECTION III

Coatings

PROTECTIVE COATINGS

INORGANIC ZINC-RICH COATINGS

How Coatings Protect and Why They Fail*

KENNETH B. TATOR
KTA-Tator Associates, Inc., Coraopolis, Pennsylvania

Mechanisms of coatings failure and contributions of formulation, application and adhesion, and curing are described. Varieties of attacking solutions and particulates are listed and the consequences of inadequate thickness noted.

PAINT COATINGS are used for a variety of reasons. Most homes and office interiors are painted solely for esthetics, and in pastel shades pleasing to the eye and complimentary to the furnishings. The exterior of these same homes and offices are also painted for esthetics—but, in addition, the paint used must also protect the structure from the effects of the weather. Coatings used on automobiles, aircraft, lawn furniture, and many other items, provide a dual function; both appearance and resistance to the atmospheric environment. In these cases for the most part, however, emphasis is on appearance (to sell the customer), with environmental protection serving a secondary role.

For other purposes, however, the roles of the coating are reversed—environmental protection and corrosion resistance are of prime importance, with esthetics assuming a secondary role. Coatings in this category are used predominantly to protect structural steel in more corrosive locations such as chemical plants, tank linings, marine structures such as ships, offshore drilling rigs and coastal bridges; electric transmission towers, above ground pipelines, damgates, and penstocks. The interiors of canneries, meat packing plants, and other food processing areas where cleanliness and hygenics are essential, are painted with smooth tile-like finishes, and nuclear power plants are coated in the primary containment area for both corrosion protection and ease of radioactive decontamination.

It is clear that there are many different environments and services for protective coatings, and accordingly, there are many types of protective coatings. It is certainly safe to say that no one coating will provide the best protection in each of these environments. Accordingly, coatings are formulated to be resistant to specific environments, and may not be the optimum for another set of conditions.

The subject of this paper is to discuss the means by which a protective coating inherently retards or prevents the corrosion process; as well as the mechanism causing it to fail.

How Coatings Protect

Corrosion resistant coatings protect, in almost all cases, by a combination of two mechanisms: (1) by acting as a physical barrier to exclude the environment, and (2) by containing reactive materials (usually pigments) that solubilize or react in a manner that inhibits corrosion, or slows the binder aging characteristics. A classic example of barrier protection would be a clear shellac or varnish coating, and a good example of the reactive type coating would be the sacrificial zinc rich primers, or coatings containing chromate pigments.

*A portion of a paper presented during Corrosion/77, March, 1977, San Francisco, California.

The metamorphosis of a coating from the can to a hardened protective film on a metal surface, proceeds through four steps: (1) packaging in the can, (2) application, (3) fixation, and (4) cure—or final attainment of its protective properties.

Packaging

Obviously, the final properties of any coating film depends upon what is in the can. Coating formulation is a complex technical subject about which much has been written. In most cases, additives to give in-can stability, antigassing, and nonsettling properties, are detrimental to the paint film in its cured state. Thus, coating formulation as it relates to these and other storage problems, while important, is not relevant to the theme of this paper—although one must remember that the formulator must compromise between package stability and application properties vs long term protection after the film has been cured.

Application

Application properties of a coating, like those of storage stability are part of the "formulator's compromise." Organic film forming polymers must be suitably dissolved, emulsified, or made less viscous in order to allow their removal from the can and application by brushing, rollering, or spraying. However, once the coating is applied, the coating material again changes with the evaporation of the now unwanted solvent (water or organic) which, if retained after application, is also detrimental to performance. Thus, while the formulator has to skillfully formulate to allow application of a coating, these considerations are not germane to this paper.

Fixation

As the paint is applied to the surface being coated, it must wet the surface, exhibit initial flow properties in order to form a continuous film—after which it must lose its ability to flow, and be able to effect a transition to the cured, fully hardened, and protective state.

The ability of the protective coating to achieve proper fixation, is extremely important to its ultimate cured properties. This fixation phase is dependent upon a number of factors, including the chemical properties of a coating material itself, application methods and techniques, ambient weather and environmental conditions, surface preparation, etc. Factors other than chemical properties of the coating will be considered elsewhere in this paper. For the moment, let us assume that these other conditions are achieved, and examine only the chemical aspects of the applied coating.

Wetting and adhesion take place during coating fixation, and the presence of polar groups in the coating vehicle markedly affects these properties. The presence of the hydroxyl group, ether oxygen, carbonyl and carboxyl groups, and other polar groups such as the amines, amides, urethanes, nitriles, and ureas, increase adhesion. The adhesive attraction results from a secondary valency bonding involving hydrogen bonds, and van der Waals forces. Since these polar units are determined by the resin—it is clear that wetting, adhesion, and other characteristics of a paint film, directly result from the choice of a resin.

Adhesion at a topcoat-primer interface is much more easily achieved than that between the primer and the substrate. Solvents in the topcoat may be carefully chosen to etch, swell, and even partially solvate the surface layers of the previous coat. This allows the molecules of the topcoat to penetrate and entangle molecules of the previous coat, leading to better adhesion.

Edge coverage of the coating is determined primarily during the fixation process. Surface tension causes the coating to pull thin at edges and other projecting irregularities. This is a particularly acute problem as the paint must flow in order to penetrate, wet, and to provide a smooth, glossy, surface—but should not flow at all if it is to properly cover sharp edges, welds, nuts, and bolts, and other protruding irregularities.

The rate at which the applied paint film "sets up" results from the diffusion and evaporation of volatile solvents and components of a coating, and rheological factors (flocculation and thixotropy).

Solvent release depends upon the evaporation rate (not necessarily the boiling point) of the solvents chosen, and also the diffusion rate by which the solvent can migrate through the coating. The choice of solvents then depends on their solvency for a given resin system, and their rates of diffusion and evaporation from the coating. In general, small molecules will diffuse faster, and polar solvents in a polar resin system will diffuse more slowly and be held longer. This latter aspect is particularly important when formulating vinyls, epoxies, polyesters, and other highly polar coating systems.

Cure

The cure of a coating, in some cases, can be considered an extension of the fixation period. In solvent deposited coatings (such as vinyls, chlorinated rubbers, and all lacquers), there is no change between the fixation and the cure periods—the only difference being the time required for the applied coating to become solvent free.

However, in emulsion coatings and chemically cured crosslinked coatings, the cure period is distinctly separate from the fixation period.

Emulsified particles pack, sinter (at room temperature), coalesce, and even sometimes crosslink as the water evaporates (in water epoxies). The evaporation of solvents from the catalyzed coating allows free radicals and/or polar groups of the co- or homo-polymers to become physically closer, react, and crosslink to form a larger macro-molecule. The type and extent of crosslinking determines both the chemical and mechanical properties of a film. Long chains with high molecular weight, rather loosely crosslinked to other chains, result in greater flexibility and toughness, but usually have lower chemical, solvent, and thermal resistance. An example can be found in the epoxy family of coatings; highly crosslinked amine cured epoxies are the most chemical resistant, but also are the most brittle and inflexible of the epoxies. The polyamide is a much larger molecule, and when crosslinked with the epoxy, spaces the linkages at wider distances resulting in greater flexibility. Epoxy esters result from the esterification of the epoxide and hydroxyl groups with drying oil fatty acids. Epoxy esters dry by solvent evaporation and auto-oxidation in the presence of conventional dryers, and the weaker auto-oxidative crosslinking and the presence of partially unreacted drying oil fatty acids, provide greater flexibility and wetting, but much less chemical and solvent resistance.

How Coatings Fail

Coatings fail for a varied number of reasons. Generally, however, the reasons for failure can be categorized as environmental effects (moisture and oxygen transmission through the film, solvent resistance, and chemical resistance) and application effects (surface preparation and coating thickness).

Moisture and Oxygen Transmission

In order to prevent corrosion of steel, moisture must be prevented from reaching the steel surface. Oxygen is also a corrosion promoter. However, all coatings are permeable to some degree to both oxygen and moisture. As oxygen is always present in atmospheric environments, and the oxygen molecule is relatively small, for all practical purposes, it can be assumed that oxygen will always permeate a paint film. Moisture, however, under good application conditions, is not present in sufficient quantities to adversely affect the paint and subsequently must penetrate to the substrate in order for corrosion to commence. As a result, the penetration of moisture through the coating to the substrate is the controlling factor in the corrosion process. Moisture permeates a coating in a complex and only partially understood manner. Propelling forces are osmotic and electroendosmotic pressures with transport aided by thermally induced molecular movements and vibrations within the polymer.

Hydratable polar groups such as the hydroxyls and other groups that undergo hydrogen bonding may also increase the moisture vapor transmission rate through a paint film. The density of crosslinking within the film decreases the moisture vapor transmission rate.

Besides specific chemical groups and linkages, there are other components of the organic portion of the paint film that are susceptible to chemical solvent or moisture attack. Water emulsion coatings, for example, must be carefully formulated using surfactants, dispersing aids, coalescing aids, and other additives to promote stability in the can, and to provide coalescence and film formation after application. Unfortunately, many of these additives remain in the film after it has cured, and upon prolonged exposure to water or moisture, may swell or degrade, weakening the film properties.

The type of pigment and its packing affect the permeation of moisture. Lamellar pigments (such as "leafing" aluminum flake, mica, and micaeous iron oxide) effectively lengthen the path moisture must travel in order to reach the substrate, while amorphous pigments also provide this function, although to a lesser degree.

Solvent Resistance

Water may attack and dissolve or swell materials with reactive groups having an affinity for water. These include hydroxyl groups, and other groups exhibiting hydrogen bonding. These groups in the paint film will have a stronger affinity for water hydroxyls than for each other, and consequently interchain hydrogen bonds are severed as the polymer hydroxyls form preferential hydrogen bonds with water. This separates the polymer chains and allows penetration of other water molecules between the chains separating them still further and allowing still more water to penetrate. Ultimately the swelling or dissolution of polymer will result. Similar theory accounts for the solubility of polymers in organic solvents. Generally, however, polar resin molecules are not attacked by nonpolar solvents, and vice versa.

Solvent deposited coatings (vinyls and chlorinated rubber) can be readily redissolved in their solvent solution while chemically converted coating systems (epoxies, urethanes, and even alkyds) are less affected because of their crosslinking and the greater number of secondary attractions between adjacent backbone polymer chains.

However, once the coating is removed from the solvent environment, the solvent will rediffuse through the coating to the environment and the coating again will become fully resistant. As a result, solvent attack is undesirable only when it is combined with moisture, chemical attack or some other substance that may more readily attack and penetrate the coating while it is in the weakened solvent softened condition.

Chemical Resistance

Chemical attack, however, is irreversible and results in a chemical breakdown of the primary linkages of the polymer vehicle. Both the organic film forming vehicle and the inorganic pigment are subject to chemical attack.

Pigments are also susceptible to environmental conditions. Reactive metallic pigments shall only be considered here, although inert pigments (titanium dioxide, zinc sulfide, calcium carbonate, and others) provide bulk, reinforcement, and slow the rate of moisture permeation through the film. It is the reactive pigments that are generally used for corrosion resistant coatings. These reactive pigments include red, white and blue lead, lead suboxide, basic lead silico chromate, dibasic lead phosphite, basic zinc chromate, zinc dust, zinc oxide, zinc phosphate, zinc molybdates, strontium chromate, and calcium plumbate. These pigments resist corrosion by a variety of means, singly or in combination, and they:

1. Act as a physical barrier.
2. Are partially soluble in permeating water, and form an alkaline condition to passivate the steel (all above except zinc dust and zinc oxide).
3. React with acids in the decomposing organic film former to form insoluble salts. This reduces moisture permeation and converts the acid to an inert end product (all lead pigment compositions).
4. Absorb ultraviolet light and convert it to less harmful energy forms (zinc dust and zinc oxide).
5. Contribute an oxidizing action which forms a passivating oxidation product in close contact with the base metal (chromate, phosphite, phosphate, molybdate, and red lead).
6. Provide sacrificial galvanic action (zinc dust in high concentrations).

The pigment must supplement the binder's resistance and not react with the attacking chemical reagent. Alkali sensitive pigments such as the lead chromates, zinc yellow, molybdate and chrome orange, chrome green, and pigments containing aluminum should not be used in alkali environments, and acid susceptible pigments (zinc, and some basic pigments that will rapidly react with acids such as calcium carbonate) should not be used in strongly acid environments.

However, premature paint failure usually does not occur as a result of improper selection of coating materials or pigments. For the most part, coating manufacturers test a coating formulation in a laboratory and in the field until they are quite sure that the coating material—if properly applied—will be suitable for an intended service. The key words here are "if properly applied"—and application involving either poor or inadequate surface preparation or deficient coating thickness—is the cause of over 75% of premature coating failures.

Surface Preparation

In outdoor weathering environments, or humid environments, mill scale should be removed prior to coating. In time, moisture will permeate through the coating and through cracks in the mill scale to attack the underlying steel. The mill scale accelerates this attack as it is cathodic to ferrous metal. As corrosion products build up beneath the mill scale, it will spall off removing any coating that has been applied over it and expose more steel to the ravages of environment.

Sometimes it is uneconomical to remove mill scale, or the environment after erection is not severe enough to justify the cost of mill scale removal (such as the interior of a building). Oil-base coatings, pigmented with reactive pigments, will penetrate cracks and discontinuities in the mill scale, sealing them, slowing the corrosion process.

Where rusty steel surfaces must be coated, these same type oleoresinous coatings (oil-base coatings, alkyds, epoxy esters) will thoroughly wet and penetrate any residual rust layer more effectively than most synthetic film formers. Penetration will also be better on porous masonry surfaces (concrete, cement, brick, etc.), but because of hydrolyzable groups in these oil modified coatings, they are susceptible to alkali attack. Unless the masonry is relatively dry, or is aged (usually a year or more) or sealed using an alkali resisting sealer, chemical attack may occur.

Blast cleaning (using either centrifugal wheel or compressed air) is often the best means of surface preparation for the newer more chemically resistant catalyzed coatings as in addition to removing surface contaminants, it roughens the surface. The roughened surface provides a "tooth" or mechanical anchoring for the paint film, but more importantly, it increases the surface area by a factor of from 2 to 10 or more. This greater surface area provides more opportunity for polar attraction to occur between the coating and the clean metal surface. In fact, for adhesion to occur at all, irrespective of its type, the interatomic distances between attracting molecules must be very small [less than 1 nm (10 Å)]. As adhesive forces diminish at the sixth power of this intermolecular distance, it is critically important that the coating achieve an intimate contact with the substrate. To do this, it must thoroughly wet the substrate. Anything that impairs this surface wetting (such as oil, grease, wax deposits) can completely prevent contact between the paint film and the substrate, preventing adhesion. Alternatively, the paint film may adhere to surface contaminants that themselves are not strongly adherent to the base metal. Dirt, chemical deposits, mildew spores and growth, dust from abrasive blasting, and other contaminants will substantially diminish the coating's adhesion, and may accelerate under-film corrosion. Entrapped water soluble contaminants, in immersion, will establish an osmotic cell, leading to blistering failure of the coating.

Large contaminants such as entrapped abrasive or dry spray beneath the primer, or between coats is detrimental since they may be later mechanically dislodged during abrasion or impact, or may not be properly wetted by the coating vehicle, leaving an air bubble or discontinuity within the coating system. Depending upon their makeup, entrapped contaminants may also provide a "wicking" action to accelerate moisture permeation through the coating, and almost always result in poor leveling and a bad appearance.

Coating Thickness

The effect of too thin a coating should be obvious—the coating will fail earlier than that applied at proper thickness. In fact, research work done by the Steel Structures Painting Council[1] concluded that each additional 25 μm (1 mil) of coating above a critical minimum increased protective life 20 months. The study showed there was also a minimum initial paint film thickness above which long term protection was attained. This thickness was not a constant for all types of paint or environment, and was thinner for the more durable or less permeable coatings, and in milder environments. Interestingly, this critical thickness was decreased only slightly by better surface preparation. In moderately severe environments (*e.g.,* Kure Beach, NC 800 foot lot), the minimum initial paint film thickness for oil base paints was 250 μm (9.7 mils), alkyd 130 μm (5.0 mils), phenolic 160 μm (6.4 mils), vinyls 85 μm (3.3 mils), chlorinated rubber 110 μm (4.2 mils), and epoxies 115 μm (4.5 mils). These results were obtained using flat panels cleaned by commercial blast cleaning and coated with a primer and finish coat of each system.

Summary

In summary then, coatings protect by a variety of means, and a skilled coating chemist has a great deal of versatility in devising a protective system. However, the vast majority of premature coating failures are a direct result of conditions not under the control of the coatings chemist—such as surface preparation, mixing and application, and changing or unexpected surface conditions. It is the writer's opinion that the limiting factors in attainment of more durable paint and greater corrosion protection, lie not with a lack of understanding of the corrosion process, coating formulation, coating application techniques, or equipment. Rather, it is human failing attributed to poor coating application inspection, poor workmanship, and inattention to detail (the corrosion process seems to find all imperfections—both major and minor); and, in general, a tendency to blame the other guy (the applicator says it's bad paint, the paint manufacturer says it's poorly applied). Indifference to rectify the problem through training, joint guarantees of responsibility or even better literature disclosing more about the specific properties of the paint or more detailed application literature, has given paint manufacturers and paint contractors a reputation slightly lower than that of a used car salesman, and just barely above politicians. It's time we did something about this—the more obvious problem!

Reference

1. Keane, J. D., Wettach, W., Bosch, W. J. Paint Technology, Vol. 41, No. 533, p. 372 (1969).

Bibliography

Hare, Clive H. System Design in the Modern Coating Industry, Reproduced from Construction Specifier, March, 1973, Courtesy CSI, Washington, DC.

Martens, Charles R. Technology of Paints, Varnishes, and Lacquers,

New York, Chapman-Reinhold, Inc., 1968.
Mattiello, Joseph J., PhD, Protective and Decorative Coatings, Vol. II, London, Chapman & Hall, New York, John Wiley & Sons, 1942.
Myers, Raymond R. and J. S. Long. Treatise on Coatings, Vol. I, Parts I, II, and III, Marcel Dekker, Inc., New York.
Payne, Henry Fleming, Organic Coating Technology, Vol. II, New York, London, John Wiley & Sons, 1961.
Solomon, D. H. The Chemistry of Organic Film Formers, New York, London, Sydney, John Wiley & Sons, 1967.
Turner, G. P. A. Introduction to Paint Chemistry, London, Chapman & Hall, 1967.

Report of Task Group T-6B-8* on Epoxy (Amine Cured). Issued by NACE Technical Unit Committee T-6B** on Coating Materials for Atmospheric Service.

Amine Cured Epoxy Resin Coatings for Resistance to Atmospheric Corrosion

NACE Technical Committee T-6B on Coating Materials for Atmospheric Service was organized for the purpose of assembling and disseminating to the corrosion engineers, factual and quantitative data on the performance and limitations of organic coatings that are successfully used for controlling atmospheric corrosion. Though every attempt is made to be factual in the coverage of application and chemical resistance tables and physical properties, it nevertheless should be recognized that changes in compounding formulations of organic materials for coating enhance or detract from their inherent corrosion resistant characteristics. Under such circumstances, these reports and recommendation lists must be viewed as showing average properties only. The possibility of changes in chemical and physical characteristics as produced by various manufacturers of the materials should be kept in mind.

Definitions

The phrase "epoxy resin based protective coatings" conceivably could cover a broad spectrum of coatings based on this resin. Each type of coating, though being related by the epoxy resin component in the formulation, exhibits properties characteristic of the particular reaction mechanism or composition modification. By definition, the scope of this report will be limited to amine cured bisphenol-A epichlorohydrin resin based coatings. Therefore, the contents of this report should not be construed to apply to polyamide resin cured coatings although many of the characteristics and properties of such coatings are similar or identical to coatings cured with amines or amine adducts.

A second broad category of epoxy resin based coatings used in control of atmospheric corrosion is covered by NACE Task Group T-6B-7 on Epoxy Esters.

Major modifications of amine and polyamide cured epoxy resin coatings will be reported on by additional Task Groups of Unit Committee T-6B.

It is the intention of Technical Unit Committee T-6B to report at this time on only epoxy resin based coatings involving curing mechanisms covered by the Task Groups already established. Interesting new developments in epoxy resin technology such as powdered coatings, foam coatings, emulsion coatings, high bake coatings, and high build coatings applied by other than conventional techniques may be considered by the committee as the use of these specialty types comes into more general use as protection from atmospheric corrosion. Additional Task Groups may be established to report on these types when the consensus indicates that a report is warranted.

Summary of Properties

General

Amine cured epoxy resin coating formulations are unique in providing room temperature cured finishes having many film properties normally associated only with baked coatings. These properties are achieved through reaction of the active hydrogens of aliphatic amines with the epoxy groups of the resin to form complex three dimensional polymers. Several types of aliphatic amines can be used as curing agents for these coatings. The amines can also be modified or reacted by the manufacturer to form adducts which, when used as curing agents, will modify the application and curing characteristics of the coating. The highly stable and chemically resistant ether linkages which predominate are widely spaced, giving unique toughness and flexibility to this thermoset polymer. The polar nature of the molecule contributes to the superior adhesion of these coatings.

Amine cured epoxy resin coatings combine excellent resistance to most corrosive materials with outstanding resistance to mechanical shock and abrasion. A broad range of substrates may be protected by these coatings because of their unusual adhesion. The virtually unsurpassed resistance of these coatings to alkalies has been highly publicized, but it should be remembered that the balance of properties obtained in these films provides excellent resistance to other corrosive media including solvents, dilute mineral acids, and salt water.

Amine cured epoxy resin coatings are not recommended for use in services where continuous exposure to low molecular weight ketones, concentrated organic acids, or oxidizing acids above very dilute concentrations is anticipated. The precautions apply primarily to submergence conditions, and such exposures are not normally encountered in control of atmospheric corrosion.

*H. D. Tarlas, Carboline Co., 328 Hanley Industrial Court, St. Louis, Mo., chairman.
**W. J. Lantz, 63 Springs St., Metuchen, N. J., chairman.

Reprinted with permission from *Materials Protection*, May 1970, 37-42, © 1970 National Association of Corrosion Engineers

Storage Life

Amine cured epoxy resin coatings are supplied as two-package systems. Shelf life of each component normally exceeds 2 years from date of manufacture. Some curing agents become unstable after storage of 2 to 3 years. If excessive settling occurs with some highly filled specialty systems, a three-package system might be preferred. Pot life of the mixed formulation is discussed later.

Precautions

In any discussion of dermatitis or sensitization involving use of amine cured epoxy resin coatings, it should be emphasized that these conditions are not cases of systemic toxicity but merely response to irritation or allergic reaction.

Listed below are the major components of amine cured epoxy resin systems followed by precautionary comments:

1. Epoxy resin. The solid resins are non-irritating and non-sensitizing. Many coatings may contain liquid grade resins and reactive diluents. When in contact with the skin, these latter materials can produce dermatitis in hypersensitive personnel.

2. Solvents. The usual precautions should be observed in respect to fire hazards and toxicity for ketones, alcohols, aliphatic and aromatic solvents. Normal procedures for prevention of solvent burn and for proper ventilation should be observed.

3. Pigments. Pigments normally present no danger. Any exception will be so designated by the coating manufacturer.

4. Curing agents. The sensitizing characteristics of individual curing agents range from non-sensitizing to moderately sensitizing. All amines in the uncombined condition are caustic and will cause dermatitis when in contact with the skin for a sufficient period. Modified amines are materially lower in sensitizing potentialities; amine terminated polyamide resins are essentially non-sensitizing.

Fully cured films of properly formulated epoxy resin based coatings are non-toxic and non-sensitizing.

These comments may seem severe and justification to place limitation on the use of amine cured epoxy resin coatings. Actually these comments are meant only to alert the user to potential industrial hazards. From experience, these coatings can be applied by brush or spray equipment with no more protective measures than is normal in industrial practice. Insistence that good housekeeping standards be maintained during application of any coating pays dividends.

The advice of coating manufacturers is readily available on matters of this nature.

Forms Available

Amine cured epoxy resin based coatings can be classified in two broad categories: solvent containing and solventless. The former is most generally used in atmospheric corrosion control, but the latter is gaining wider acceptance as new curing agents and more versatile equipment are developed which allow practical application techniques.

Amine cured epoxy resin coatings designed for curing at ambient temperature are supplied as two-package systems due to the limited useful working time after the resin component and the curing agent component are mixed.

Effects of Compounding

Several aliphatic polyamines, modified aliphatic polyamines, blocked polyamines and amine-terminated polyamines resins are suitable for reacting with epoxy resins to provide coatings suitable for prevention and control of atmospheric corrosion. A variety of different molecular weight epoxy resins is available by which the properties of the coating can be adjusted to achieve optimum characteristics for specific end uses. The selection of solvents is balanced to achieve the desired application characteristics without detriment to the properties of the cured coating. Other factors influencing the characteristics of these coatings are pigmentation, use of accelerators, other resin modification, and use of thixotroping agents.

Because of the variety and complexity of the epoxy resin formulations commercially available, it is again stressed that this report be used only as a guide, and that specific recommendations and properties be obtained from the coating manufacturers.

Resistances to Fume, Splash, General Atmospheric Conditions

Amine cured epoxy resin coatings exhibit outstanding resistance to almost all corrosive media except strong oxidizing agents as long as the exposure is limited to intermittent splash or fume conditions. The information in this section will be general. More detailed data are given at the end of the report. Because of the unusual balance of properties available in properly formulated amine cured epoxy resin coatings, these systems are some of the most adaptable industrial maintenance finishes commercially available at this time.

Water and Moisture

These epoxy coatings have performed well for several years in ambient temperature distilled and tap water services. Resistance to atmospheric moisture is excellent. Water permeability of clear films of unpigmented amine cured epoxy resin based coatings expressed in milligrams of water per square centimeter of coating per hour is 0.05 to 0.03 for 2- and 5-mil films, respectively.

Inorganic Acids

Amine cured epoxy resin coatings have good resistance to all concentrations of hydrochloric acid, but high concentrations of other inorganic acids will cause film failure. Elevated temperatures accelerate the attack.

Oxidizing Agents

These coatings are not generally recommended for services where continuous exposure to oxidizing agents is anticipated. There are some exceptions to this rule, especially if only very low concentrations of the agent or infrequent exposure is involved and particularly if the primary purpose of the coating is for protection from media other than these agents.

Organic Acids

Some organic acids such as most fatty acids have no effect on these coatings at ambient temperature, but others such as acetic acid will cause film failure. Performance tests should be conducted, or the coating manufacturer's recommendation should be followed when considering protection from these materials.

Alkalies

Amine cured epoxy resin coatings have excellent resistance to most all alkaline materials even at high temperatures.

Salt Solutions

These coatings have excellent resistance to all the salt solutions which have been tested, including sea water. This holds true for all concentrations and temperatures obtainable at atmospheric pressure.

Solvents

Low molecular weight ketone and acetate solvents will soften amine cured epoxy resin coatings, but the films normally will return to their original state and performance upon removal from contact with these solvents. High molecular weight ketone and acetate solvents, alcohols, and aliphatic and aromatic hydrocarbon solvents will not appreciably affect these coatings. Many chlorinated solvents will not lift or soften these films, but when the films are submerged in these solvents at elevated temperatures, evidences of softening become apparent. In fact, several coating strippers contain methylene chloride.

Oils and Fats

None of the fats or oils tested affected amine cured epoxy resin coatings.

Gases

Gases and fumes of materials will have less effect on these coatings than submersion in the various media. Three gases known to attack these coatings are ammonia, chlorine dioxide, and chlorine, particularly in the anhydrous forms.

Properties of Applied Coatings

Temperature Limitations (During and After Application)

At normal room temperature (22 C

or 72 F), amine cured epoxy resin coatings in the prescribed film thicknesses are dry enough to handle in three to six hours depending on the solvent content and composition, and they develop excellent toughness after curing overnight. As temperature is lowered, the curing rate falls rapidly. All the common curing agents are still considered to be applicable at temperatures as low as 13 C (55 F), even though the time to reach a given stage of cure is at least double that required at 24 C (75 F).

Below 13 C (55 F), the curing rate is retarded to such an extent that low molecular weight polyamines are considered unsuitable because of the danger of the amine volatizing from the film before reaction with the epoxy resin occurs, thereby resulting in reduced curing agent concentration and poor ultimate cure.

On the other hand, modified amines and polyamide resins assure good ultimate cure even when such coatings are applied at temperatures as low as 2 to 4 C (35 to 40 F) as long as after a reasonable time the temperature returns to the range where normal curing is accomplished.

Under conditions of high humidity, a curing agent blush may form on the coating's surface when certain amine curing agents are used. This condition can be reduced by the use of modified amine curing agents or by mixing the components about an hour before application. This blush, though somewhat unsightly and thus almost prohibitive for certain decorative applications, does not affect the film's ultimate performance characteristics. If initial appearance is important, the blush is removed easily by washing of the cured coating with water. If additional coats are to be applied, the blush film must be removed by rinsing or water washing to assure good adhesion of the next coat.

The temperature limitations of amine cured epoxy resin coatings in immersion services are normally about 93 C (200 F). In dry exposures, satisfactory experience has been recorded at temperatures as high as 260 C (500 F) for relatively short periods, but a more conservative high temperature service range is 121 to 149 C (250 to 300 F). Above 121 C (250 F), light colored coatings will darken but may remain intact and provide adequate protection.

Abrasion Resistance and Impact

The abrasion resistance of amine cured epoxy resin coatings is excellent. Typical Tabor abrasion resistance of unpigmented films would be in the range of three to four milligrams loss in weight per 100 cycles of a CS-10F wheel supporting a total weight of 1000 grams. Under the same conditions, pigmented films will lose about 10 milligrams per 100 cycles. The reverse impact of unpigmented films normally is greater than 160 inch pounds. Even some pigmented films will pass 160 inch-pounds, but, as pigment concentration increases, reverse impact decreases. High build coatings naturally have lower impact and can be as low as five inch-pounds in highly filled systems. Reverse impact will become lower as the films age but should always remain greater than is sufficient for the intended use.

Hardness

Amine cured epoxy resin coatings have a pencil hardness of 4H to 5H after curing for a week at room temperature. The Sward hardness will range from 35 to 55 for clear and pigmented films. Most formulations will have sufficient flexibility to pass a 180-degree bend on a one-inch conical mandrel. By nature of the resin composition of solventless coatings, they are harder but less flexible than coatings based on higher molecular weight resins. The hardness of both coating types continues to increase on aging, but, if properly formulated, embrittlement does not increase appreciably.

Exposure to high temperatures accelerates the development of hardness and consequent reduction in flexibility. In some cases, where extreme hardness and excellent resistance to solvents have developed in a coating, the coating can present intercoat adhesion problems if topcoated without pretreatment in some way to provide a surface conductive to mechanical adhesion.

Aging and Weathering

Aging and weathering characteristics of these coatings are excellent when correctly applied over properly prepared substrates on exterior exposures. A few months after application, a mild chalk will become apparent. This chalk can be removed and the initial appearance of the film regained by light scrubbing. Although chalking continues for the life of the coating, it is only a surface phenomenon and is not detrimental to the film or of particular hindrance to recoating.

This chalking takes place at a very slow rate of less than one mil per year, and it does not significantly affect the coating's service life.

Most formulations containing pastel or white pigments will yellow slightly on exterior exposure but not to the degree where it is prohibitive for decorative value in most industrial applications. In areas protected from exposure to ultraviolet light, gloss and color retention of all amine cured epoxy resin coatings are excellent.

Toxicity and Odor

Completely cured films of properly formulated coatings of this type are completely non-toxic and odorless.

Weight of Applied Coatings

The weight of formulated coatings of this type will vary from 8.0 to 28.0 pounds per gallon depending on the type and amount of pigmentation. Normally, non-pigmented coatings will be in the 8 to 10 pound per gallon range; pigmented coatings will range between 10 and 13 pounds per gallon. The specific gravity of the applied coating can be calculated from the solids content of the formulations and the film thickness of each.

Electrical Properties

Volume resistivities run under ASTM Method D-257-58 varied from 5.0×10^{12} to 82.3×10^{12} ohms/cm^2/cm. The coatings tested were dried for 24 hours and force dried for six hours at 60 C (150 F). Film thickness ranged from 8 to 10 mils. All the coatings were pigmented. The differences in volume resistivities were attributed to formulation variables.

Adhesion

The excellent adhesion of epoxy resin based coatings can be attributed to the polar nature of the epoxy resin molecule itself. Adherometer adhesion of these films is in the range of 1.0 dyne/mm x 105. Under ASTM tensile adhesion test D-897-49, these coatings had a value of 3000 lb/in^2. It is difficult to evaluate pure adhesion as such, and therefore the practical value of the adherance of a coating to a substrate is of foremost importance. Epoxy resin based coatings repeatedly have demonstrated their superiority over many other coatings.

Gloss Retention

The initial gloss of these coatings is excellent if pigmented in such a manner to optimize this feature. Indoors, the gloss reduction is almost insignificant, but for exterior exposure, the gloss is markedly reduced by the inherent chalking of the films after about three months' exposure in southern latitude and six months in northern climatic exposure. The cleansing action of this chalking phenomenon prevents dirt collection, makes these coatings perform well in this regard, but reduces gloss. Gloss almost comparable to the original can be restored when the chalk is removed by light scrubbing.

Decorative Value

The light color of the resin components of this vehicle permits preparation of almost any color or shade including white and pastels. A coating with almost any desired gloss also can be produced. These coatings have a high indoor decorative value. Due to their chalking and yellowing characteristics on exterior exposure, their use is limited in exterior applications where highest possible aesthetic appearance is mandatory. Seldom would these considerations prohibit the use of these coatings in industrial maintenance.

Application Characteristics

Coverage and Shipping Data

Theoretically, one gallon of any coating will yield a dry film (1-mil dry thickness) coverage of approximately

Source: *Materials Protection*, May 1970, 37-42

1604 square feet, multiplied by the percentage of solids by volume.

Normal solvent-containing amine cured, epoxy resin based coatings range from 35 to 60% solids by volume. The solventless coatings are naturally 100% solids. (It should be noted the definition of solventless epoxy coatings includes both the epoxy and curing agent compounds; however, in some cases it is known that epoxy coatings which are called solventless are of the 100% solids nature in the base component, but the catalyst or curing agent contains a certain amount of solvents.)

Applying the aforementioned formula to solvent containing coatings, the theoretical coverage of one gallon would be 561 to 802 square feet per mil of dry film thickness. In more practical coating terms of two mils dry film thickness per coat, the coverage factor would be 280 to 401, respectively, when based on the above volume solids figures.

The above figures are theoretical values and should be correlated to the percentage waste factor contemplated on specific applications. As might be expected of any surface coating applied under field conditions, waste factors can vary from a low of 5% to as high as 40%

If thinner is added to the coating, the number of gallons before thinning should be used in these calculations because the thinner does not contribute to the dry film thickness. Practical film coverage calculations should always take into consideration the substrate profile. Field film measuring devices measure dried film from the average peak of the profile and, therefore, do not account for the dried film on the surface below the profile pattern.

The solvent containing formulations are shipped to conform with the ICC regulations governing the handling and shipping of flammable materials. No such restrictions apply to the solventless systems.

Flammability and Toxicity

The solvent containing, amine cured epoxy resin coatings normally contain ketones, alcohols, and hydrocarbon solvents. The usual safety measures applicable to other coatings containing these solvents should be observed.

The solventless coatings do not present a fire hazard. Proper protection should be provided to prevent prolonged or excessive skin contact with the coating, and adequate ventilation should be made available to personnel working in closed areas.

The amine cured epoxy resin coatings are no more toxic than any of the commonly used industrial maintenance coatings, but since certain formulations may present more of a hazard in application than others, it is recommended that the precaution on the labels of the containers supplied by the manufacturer be observed and used as applicable. It is recommended that good housekeeping procedures be observed, and the diligent use of soap and water by personnel coming in contact with the uncured coating be established for prevention of dermatitis. It is recommended that solvents not be used to cleanse the skin because the defatting action of these materials could lead to sensitization.

Surface Preparation

Because of their excellent adhesion, amine cured epoxy resin coatings can be applied over a variety of substrates. These substrates must be free of dirt, oil, grease and other contaminates and must be essentially dry. Certain unique formulations containing polyamide curing agents have been developed specifically for application onto clean wet substrates.

In coating over steel, the minimum recommended surface preparation is a commercial sandblast as specified by the National Association of Corrosion Engineers as NACE No. 3* blast cleaning or Steel Structures Painting Council SSPC-SP6-63. A white blast (NACE No. 1*) is preferred, but in certain cases may not be feasible.

If sandblasting is prohibited due to other factors, chemical cleaning and mechanical abrasion are preferred in that order. If an amine cured epoxy resin based primer is selected for use over metallic substrates which previously have been treated with phosphoric acid, precautions must be taken to remove or neutralize all the acid before primer application to avoid any possibility of reaction between the acid and curing agent.

These coatings can be applied over relatively fresh concrete as well as old concrete and other masonry surfaces. Dirty and oily surfaces should be etched with 10% HCl (hydrochloric acid) and then thoroughly flushed with water. The cleaned surface should be allowed to dry. If the surface is greasy, a detergent or strong caustic wash must be used before the acid etch. Because the surface layer of cured concrete is not normally as strong as the body of the concrete, it is good practice to use a light brush sandblast to remove the weaker surface area if other considerations do not prohibit such a procedure.

Plastic surfaces to be coated should be free of any oil, grease or parting agent. If solvents are used, care should be taken that the coating will not attack the plastic material.

Before application over old coatings, the solvent resistance and film strength of the old coating should be considered. If the type is not known, spot testing should be made. The solvent-containing amine cured systems tend to lift prior coatings which have low resistance to active solvents. In such instances, specially formulated coatings may afford an economical approach to this problem and might be used as a barrier coat or a complete topcoat system.

Proper surface preparation is essential to prevent premature failure of any industrial coating. The false economy of inadequate surface preparation invariably will give reduced useful life of the protective coating.

Priming

As with other industrial finishes, amine cured epoxy resin coatings should be applied as a complete system. These systems can range from two to six coats depending on the substrate, service application, and coating formulation. The system may or may not include a primer although primers are generally recommended. Several different solvent resistant primers which are known to adhere well to specific substrates might be used, but for the best over-all performance, an amine cured or a polyamide cured epoxy resin primer is preferred.

Under the special condition of coating zinc or galvanized surfaces, vinyl wash primers and polyamide cured epoxy resin zinc rich primers have shown advantages over other types, but care must be exercised to assure that no residual acid is left on the surface. Special primers may not be required over certain types of inorganic zinc coatings.

Application Methods

Amine cured epoxy resin coatings can be formulated for application by most of the conventional methods. Spraying generally provides the best levelling and is the preferred method over steel and over smooth surfaces, whereas brushing is equally applicable over concrete and other rough surfaces. Conventional spray equipment can be used with most solvent-containing systems.

Airless spray equipment can be used with most systems and under certain conditions will reduce spraying losses. The use of a mastic tip and a pressure pot or airless spray equipment will enhance the appearance and permit greater ease of application of high build and solventless systems.

Some types of solventless coatings require special spray equipment which keeps the base component and the curing agent separated until they reach the nozzle of the gun. At this point the metered components are mixed and ejected toward the substrate. The use of such systems is limited usually to highly specialized applications.

The pot life or usable working life of mixed formulations depends on several factors: type of resin, type of curing agent, solvent system (if any), non-volatile content, quantity of material mixed, and temperature. With the exception of certain specialty solventless systems, most commercial formulations have pot lives ranging from 4 to 36 hours. With the more conventional types of these coatings, no difficulty will be encountered if only as much material is mixed as can be applied during one working day. It is best to follow the coating manufacturer's instructions explicitly on such matters.

---

*NACE Publication 6D165: "Glossary of Terms Used in Maintenance Painting," published in *Materials Protection, 4,* pp 73-80, January, 1965.

Drying Time

At normal room temperature of 22 C (72 F), amine cured epoxy resin coatings are dry to handle in three to six hours and develop excellent hardness in 12 to 16 hours. In coating systems consisting only of amine cured formulations, the topcoat can be applied any time after the previous coat has reached the dry-through stage. This normally is six to eight hours for solvent containing systems and 8 to 12 hours for solventless types. These times are extended as the temperature decreases from 22 C (72 F) or, conversely, times can be reduced as temperatures increase above that point.

Some primers based on newly developed epoxy resins have shorter recoating times.

Some systems have very fast drying properties, and drying time between coats should be limited to a number of hours. Otherwise, sufficient knit may not be obtained, and delamination between coats could occur. Amine cured epoxy resin systems continue to develop hardness, chemical resistance, and solvent resistance for many days after application if normal temperatures are maintained. The type of use intended for these systems is the determining factor as to when they can be placed in that service. For instance, as a coating for dry gas transmission service, the line can be placed in operation as soon as sufficient solvent has been released to assure that the coating will not be moved or disturbed. For the most highly corrosive atmospheres or media, it is best that the coating be allowed to cure for seven days at 22 C (72 F) or higher, or force cured at higher temperatures before being placed in service.

Number of Coats

Multiple coat systems are always recommended. Under no conditions would one coat be expected to perform as well as two or more although satisfactory service has been obtained with one-coat, high build systems in mild atmospheric conditions.

Most conventional systems consist of one prime coat and two finish coats. Another system gaining favor is a primer, a high build coat, and a top seal or decorative coat. Two coat systems of any combination of primer and topcoat will give limited protection and may be satisfactory in mild environments. In applications where regular maintenance programs are observed, a two coat system can be used as a base and successive coats added at one to two-year intervals. Two coats of solventless systems having a total film thickness greater than eight mils have provided adequate protection for most recommended services if application is such that pinholing and other such defects are virtually eliminated.

Film Thickness

Dry film thickness may vary from 1.5 to 30 mils or more per coat, depending on formulation. Seldom is it recommended that total thickness of a multiple coat system be less than five or six mils, but in some unusual applications, the film thickness required might be only one or two mils greater than the blast pattern. For atmospheric protection, the total dry film thickness of an epoxy-amine system is usually in the range of 5 to 15 mils.

Thinner or Solvents

Thinner and clean-up solvents normally consist of a blend of ketones, alcohols and hydrocarbon solvents but may contain other materials known as good solvents for these coatings.

Compatibility

Amine cured epoxy resin coatings can be overcoated with most any type of industrial maintenance coating if applied after the epoxy coating has dried and before it has progressed too far in its cure. The section in this report on drying time can be used as a reference. Overcoating with the same generic type can be done at any time after the first coat has dried, even after a period of years. A solvent type wash of aged amine cured epoxy resin coatings may be recommended before recoating because excessive chalk or surface contamination may prevent good adhesion of the topcoat. To achieve the same result with other types, other special precoating techniques may be required.

Two amine cured epoxy resin finishes may be applied over a variety of primers with the one limitation that the solvent containing types must not be applied over primers which will be lifted by active solvents.

Experience Records

The amine cured epoxy resin coatings are one of the most widely accepted and universally used of any one type of industrial maintenance finish. A vast amount of background and experience data is available from many sources. It is recommended that manufacturers of these coatings be consulted for this type information. Most major raw materials suppliers, coatings manufacturers and corrosion engineers of end-user companies have documented case histories which might also be available upon request.

As a general rule of thumb, the cost of the coatings of this type fall into the range of 0.9 to 2.0¢ per square foot of one mil dry material. This cost is for the coating material only and does not include other expenses such as surface preparation, labor, etc. The material costs are based on the theoretical coverage which can be obtained and do not include any losses which invariably occur due to overspray and spillage. It is recommended that the application of such high performance coatings be closely supervised to assure that efficient techniques are used and that all potential opportunities for loss are minimized.

Chemical Resistance Properties

The recommendation in the table below reflects conditions and temperatures normally encountered in atmospheric corrosion protection services. They include exposures to splash, spillage, and other limited contacts with corrosive liquids as well as fumes and other atmospheric conditions that may prevail in industrial plants and areas. It is not to be construed that these coatings would withstand continuous submergence in the media indicated although in some instances they may be satisfactory. Also, formulation variables cannot be comprehensively covered in such a table. These variables include not only common differences in proprietary coatings, but also the type, i.e. solventless, high build, or solvent containing compositions.

This table is meant only as a guide, and the following coding will be used: R for recommended, LR for limited recommendation, and NR for not recommended. Limited service applications for media rated as NR should be determined by field trials. All concentrations are given by weight. Temperature condition at 22 C (72 F).

Current Membership of NACE Technical Unit Committee T-6B

T. Allan
C. W. Ambler, Jr.
C. R. Anderson
R. H. Bacon
J. L. Barker, Jr.
W. O. Bayer
S. J. Bellassai
M. W. Belue, Jr.
H. S. Bennett
D. M. Berger
A. H. Betley
J. Bigos
C. E. Bixler
J. Blasingame
W. Bosch
J. F. Bosich
W. M. Brackett
F. A. Bristol, Jr.
W. R. Cavanagh
L. L. Christensen
J. H. Cogshall
R. L. Collins
W. F. Connors
J. W. Cushing
J. R. Daesen
J. A. Dittmar
P. R. Dykes
O. H. Fenner
J. R. Fischer
C. E. Fox
K. S. Frasier
N. B. Garlock
F. W. Gartner, Jr.
C. E. Gary
P. J. Gegner
D. H. Gelfer
O. L. Grosz
J. A. Gump
J. P. Halloran
F. P. Helms
W. A. Higgins
P. W. Hill
C. B. Hutchison
R. M. Ives, Jr.
E. D. Jarboe
C. M. Jekot
W. R. Johnston
J. D. Keane
W. E. Kemp
W. J. Lantz
M. R. Leven
H. F. Lewis
P. W. Lewis
A. K. Long
S. L. Lopata
F. W. Luebke
H. McCranie
S. W. McIlrath
R. J. McWaters
V. D. Magat
R. Main
A. J. Marron
J. Meiry
W. B. Meyer
B. Mohr
J. F. Montle
F. C. Morrow, Sr.
C. G. Munger
J. W. Nee
L. J. Nicholas
I. Nicodemus
G. Norman
L. J. Nowacki
S. J. Gechsle, Jr.
P. R. Prokish
J. G. Raudsep
N. F. Reeder
J. E. Rench
G. Repka
F. T. Rice
Z. V. Riders
R. M. Robinson
J. Rodgers
A. H. Roebuck
R. R. Rosenthal, Jr.
S. E. Sankey
E. Saul, Jr.
F. A. Schultz
T. F. Shaffer, Jr.
C. W. Sisler
L. L. Sline
S. Spindel
H. B. Swanson
H. D. Tarlas
R. F. Toma, Jr.
W. R. Tooke, Jr.
N. W. Tune
R. Vickers
G. R. Willemez
V. B. Volkening
D. R. Whiteman
B. C. Wilson
W. A. Wood, Jr.
L. J. Zadra
C. A. Zimmerman

MEDIUM	RECOMMENDATION
Acids	
Acetic acid, 5%	R
Acetic acid, 20%	NR
Acetic acid, 50%	NR
Acetic acid, glacial	NR
Citric acid, 10%	R
Hydrochloric acid, 10%	R
Hydrochloric acid, 20%	R
Hydrochloric acid, 36%	R
Hydrochloric acid, vapor	R
Hydrofluoric, 10%	NR
Lactic acid	R
Linseed acid	R
Nitric acid, 5%	R
Nitric acid, 20%	NR
Nitric acid, 30%	NR
Nitric acid, concentrated	NR
Oleic acid	R
Phosphoric acid, 10%	R
Phosphoric acid, 85%	NR
Sulfuric acid, 10%	R
Sulfuric acid, 25%	R
Sulfuric acid, 50%	R
Sulfuric acid, 70%	NR
Sulfuric acid, 80%	NR
Sulfuric acid, 90%	NR
Sulfuric acid, concentrated	NR
Alkalies	
Ammonium hydroxide, dilute	R
Ammonium hydroxide, concentrated	NR
Calcium hydroxide	R
Sodium hydroxide, dilute	R
Sodium hydroxide, concentrated	R
Fats & Oils	
Animal	R
Mineral	R
Orange peel	R
Vegetable	R
Gases (moist)	
Ammonia	LR
Carbon dioxide	R
Hydrogen sulfide	R
Sulfur dioxide	LR
Halogens (moist)	
Bromine	NR
Chlorine	NR
Iodine	NR
Oxidizing Agents	
Calcium hypochlorite, 5%	R
Chlorine water	R
Chromic acid, 5%	NR
Chromic acid, 40%	NR
Hydrogen peroxide, 30%	NR
Sodium hypochlorite, 5%	R
Sulfur dioxide solutions	R
Salt Solutions	
Alum	R
Ammonium salts	R
Calcium chloride	R
Copper sulfate	R
Ferrous sulfate	R
Sodium acetate	R
Sodium carbonate	R
Sodium chloride	R
Sodium phosphate	R
Sodium sulfate	R
Solvents	
Acetone	LR
Aviation gasoline	R
Butyl alcohol	R
Carbon tetrachloride	R
Esters	R
Ethers	R
Ethyl alcohol	R
Ethyl amyl ketone	R
Jet fuel	R
Methylene chloride	NR
Methyl alcohol	R
Methyl ethyl ketone	R
Methyl isobutyl carbinol	R
Methyl isobutyl ketone	R
Mineral spirits	R
Phenol	NR
Secondary butyl alcohol	R
Toluene	R
Xylene	R
Water	
Distilled	R
Sea	R
Tap	R
Miscellaneous	
Allyl chloride	R
Detergent solution	R
Diethylene triamine	NR
Ethylene dichloride	R
Formaldehyde, 37%	R
Glycerine	R
Sodium chlorate	NR
Sodium chlorite, 25%	R
Sodium methoxide, 40%	R
Sour crude oil	R
Styrene	R

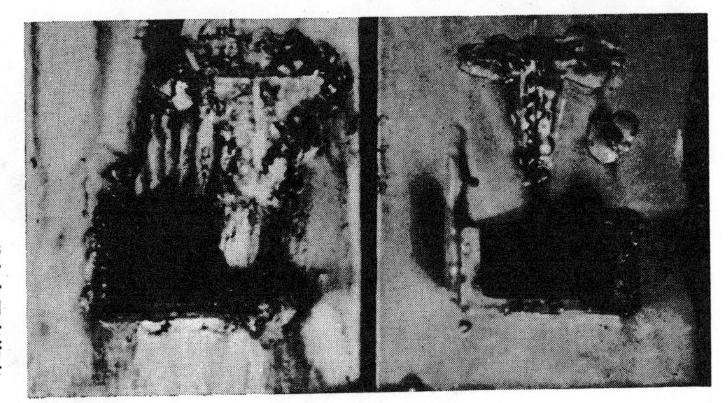

Figure 1—Comparison of test panels with 3-coat epoxy-amine. Panel at left was exposed in a salt fog cabinet for nine months. Panel at right was exposed for 18 months in a severe industrial salt environment. The type and extent of failure on the two panels are almost identical.

Jerry D. Byrd
Carboline Company
St. Louis, Missouri

SUMMARY

Discusses proper test methods for evaluating coatings for particular applications when sufficient case history data are not available. Compares results of laboratory tests to results of field exposures under similar environments. Coatings discussed are epoxies and vinyls in sodium chloride, sulfuric acid, hydrochloric acid, and sodium hydroxide fume environment. Discusses testing coatings for maintenance tanks, floors, and high temperature applications.

how to test coatings for specific applications*

MANY REASONS can be cited for the importance and need of test information showing performance of protective coatings and linings. The need and importance of test data are more easily understood when a coating selection must be made without the time or opportunity to accumulate actual field experience. For example, design of a new plant, process, or establishment of a maintenance program often times must be done without sufficient past records on which to base a decision. In these cases, laboratory testing can aid in prediction of relative performance, give reliable screening of coatings, and indicate which coatings are unsuitable.

Use of Test Panels

As many construction conditions as possible should be incorporated in test panels used for evaluating protective coatings. Some of these conditions would be sharp edges, round edges, welds, crevices, scratches, and impacted areas. A coating which provides excellent protection on plane surfaces but poor edge protection will be demonstrated on these type panels.

Test Panel Size

An argument frequently encountered in testing is the size of the test panel used. Some technicans claim that a test panel must be at least one foot square. This claim is based on the theory that type and extent of failure on a small panel are entirely different from a large panel. However, tests with three different sized panels* painted with three coats of epoxy and exposed for 18 months in an industrial salt environment showed identical failures.

Choice of type and size of a test coupon is best determined by the information which is to be obtained from the test. Frequently, round rods with well rounded ends are used. These rods are dipped in the coating material and then heat cured. Most protective coatings do not give the same type film when dipped as they do when applied by brush or spray. Therefore, if a coatting is to be applied in the field by dipping, lab tests should be conducted with dipped rods so that false information will not be obtained as to chemical and physical resistance.

*KTA panels, supplied by Kenneth Tator Associates, Coraopolis, Pa.

Because most tanks have welds on the interior, tests should include welds. A flat panel of approximately ⅛-inch thickness with rounded edges makes an excellent test piece for simulating weld conditions. This panel should be coated in the same manner as the tank is to be coated.

Accelerated Testing

The purpose of accelerated testing is to duplicate as closely as possible the type of corrosion encountered in industry and to accelerate the time to illustrate failure of the coating.

In all the accelerated testing cabinets, panels were exposed for various times. Results were compared to the same systems exposed in the similar type corrosive environment in the field. The temperature and concentration in the cabinet were altered until the type of failure in the cabinet duplicated that encountered in the field but which occurred in much less time. The acceleration of failure was increased as much as possible without losing the differen-

★Revision of a paper entitled "Accelerated Test Methods for Coating Evaluation" presented at the South Central Region Conference, National Association of Corrosion Engineers, October 15-17, 1963, Oklahoma City, Okla.

Figure 2—Comparison of test panels with 3-coat epoxy system. Panel at left was exposed for nine months in a salt fog cabinet. Panel at right was exposed for four years on the Gulf Coast. Similarity of type and amount of failure are almost identical.

tiating factors between the various coating materials.

The ratio between failure of a coating in the field and failure in the cabinet will vary depending on the severity of the field exposure. This ratio can be as low as 1:1 in severe field exposure or as high as 15:1 in mild exposure. This range, however, is relatively narrow for the average industrial exposures, thus making prediction possible of approximate coating life in an exposure. The ratio for most hydrochloric acid, sulfuric acid, and salt exposures is from 8 to 12:1—that is, one week in the cabinet is equivalent to 8 to 12 weeks in the field.

The design of the accelerated testing cabinets is similar to that of the standard ASTM salt fog cabinet with some modifications.

Environments in accelerated testing cabinets are pure, whereas industrial environments generally are contaminated. The inclusion of small quantities of other corrosives might create a duplication of the plant environment. The type and amount of secondary corrosive will vary with the individual plant. This, however, is difficult to achieve in laboratory testing.

Most accelerated testing cabinets do not allow for weathering. Weathering may or may not be a factor in the actual field exposure; however, where weathering is a factor, results of cabinet tests must be adjusted. An ultraviolet lamp was installed in one cabinet. Preliminary tests with this cabinet indicate, for example, that epoxy systems fail more rapidly than in cabinets without ultra-violet light. Vinyl systems have an increased life in this cabinet. This type testing more closely duplicates failure of the two systems in actual areas where weathering is a factor in corrosive conditions.

The most widely accepted accelerated test cabinet is the ASTM salt fog cabinet. This cabinet uses either a 5% or 20% sodium chloride solution. The 5% salt fog cabinet gives results on maintenance coatings which closely duplicate failure in the field. However, the time required for coating failure is lengthy on all but light duty maintenance systems. The 20% salt fog cabinet gives rapid failure, but the failure is so rapid that some of the differentiating factors are lost.

A combination of the above two cabinets using a synthetic sea salt at 10% concentration provides both sufficiently rapid failure and good field correlation.

Comparison of Cabinet and Field Exposures

Table 1 compares vinyl and epoxy coatings exposed in the 10% salt fog cabinet and field exposures. Failure of vinyl systems is more rapid than epoxy systems. Also, by the time vinyl on the edges of a panel has failed at least 30%, the coating on plane surfaces has begun to fail. This is a condition not normally found in the field. The difference is attributed to aggressiveness of the cabinet environment, which is about four to ten times that of normal field exposure.

Figure 1 shows a comparison of panels coated with epoxy amine exposed for nine months in the salt fog cabinet and for 18 months in a severe industrial salt environment. The type and extent of failure on the two panels are almost identical. Figure 2 shows a comparison of cabinet and Gulf Coast exposures of 3-coat epoxy systems. The similarity of type and amount of failure again is almost identical.

Sulfuric Acid Cabinet

Sulfuric acid corrosion failures are characterized by penetration through porosities or thin spots followed by rapid undercutting (Figure 3). This undercutting is usually accompanied by voluminous deposits of iron sulfate.

Table 2 compares exposure of epoxy and vinyl systems in a sulfuric acid cabinet and the field. Acceleration ratio between cabinet and field exposures is approximately 7:1 for vinyls and 4:1 for epoxies. This acceleration factor will vary depending on severity of the exposure. The type and extent of failure on the two panels were identical.

Hydrochloric Acid Cabinet

Hydrochloric acid is extremely penetrating, with some tendency to undercut a protective coating. However, the extent of undercutting is not as severe as that encountered in sulfuric acid exposure.

The hydrochloric acid cabinet is operated with a 28% solution of hydrochloric acid, which is placed in open pans and allowed to vaporize. Panels are exposed to this vapor for six days. The cabinet is then opened to the atmosphere for one day to replenish oxygen supply. The acid solution is replaced weekly.

Table 3 compares epoxies and vinyls exposed in the hydrochloric acid cabinet and in the field. The acceleration ratio between the cabinet and field in this test was only from 2 to 3:1 be-

Figure 3—Panel exposed in sulfuric acid cabinet. The acid (10% concentration) was atomized with compressed air. Temperature was about 75 F (24 C). Black lines on the panel show the extent of undercutting.

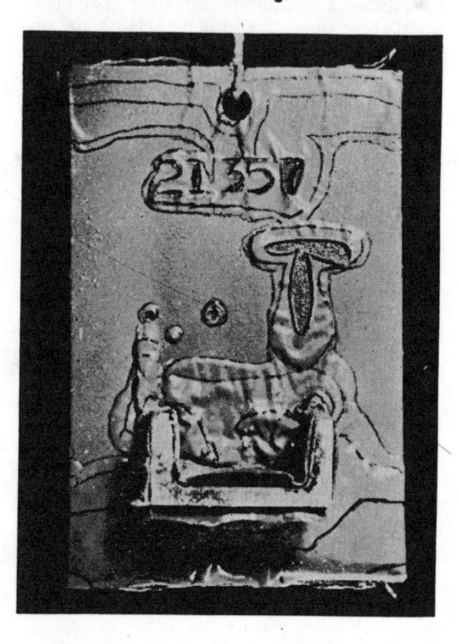

cause this field station was extremely corrosive.

Sodium Hydroxide Cabinet

Corrosion of steel in a sodium hydroxide or caustic environment is not great. Loss of steel due to caustic rusting is negligible because sodium hydroxide has a tendency to passivate the steel. Appearance is about the only reason for coating steel in a caustic environment. The main problems in a caustic environment arise only when coatings are used.

The sodium hydroxide cabinet is operated on a cyclic basis. A 20% solution of sodium hydroxide is atomized with air for three hours. This is followed by three hours in a water fog. Vapor temperature of the cabinet is 95± 5 F (35 C). Cycling is necessary to obtain failure characteristics of coatings obtained in the field.

Failure of a coating in a caustic atmosphere is by loss of bond.

Poor quality epoxy formulations have a tendency to stress crack in a caustic environment. These cracks are tiny at first but grow in width and depth until steel is exposed. Once the steel is exposed, the coating rapidly loses bond.

Coatings containing oil normally does not fail from loss of bond in a caustic environment. The oil in the coating is saponified and washes from the steel. If an oil bearing primer is applied followed by non-oil bearing finish coats, the resultant system will undercut. In essence, the exposed oil bearing primer is saponified and washes from beneath the topcoat, giving the appearance of undercutting or loss of bond.

Immersion Testing

Choice of the panel used for immersion testing is extremely important. If the resistance of the coating alone is desired, then round rods with well rounded bottoms are excellent as are test cubes or free films. However, if the resistance of the coating in conjunction with its performance when applied by methods other than dipping is desired, then flat panels with rounded edges are preferred.

In some instances, the coating is heat cured before testing. Heat curing normally makes a coating much harder and less easily penetrated by corrosives. With thermosetting coatings like epoxies, heat increases the extent and uniformity of cure. With thermoplastic coatings like vinyls, heat insures removal of solvents, thus making them tougher and more resistant.

Acceleration of tests should be by changing temperature, not the concentration of the corrosive. If concentration of the solution is changed, action on the coating may be changed. Caution should be used, however, in increasing the temperature. Temperature increase to accelerate failure should be below the heat distortion point of the coating or poor and misleading results will be obtained. The heat distortion point for thermoplastic materials such as vinyls is low, whereas it is higher for thermosetting materials such as epoxies.

Rating

Rating systems for immersion testing will vary depending on the information desired. The rating system should include:

1. Penetration of the coating as evidenced by blistering.
2. Effect of the corrosive on the coating's surface, evidenced by pitting, surface attack, swelling, and sloughing away of the coating.
3. Discoloration of the coating.
4. Effect of coating on the solution (contamination or color change).

In many instances, discoloration of the solution is critical. To obtain accurate changes in discoloration of the solution, a control sample should be run. Sometimes, the corrosive itself will discolor; unless a control is run, the coating will be blamed.

A large ratio of coated surface to corrosive volume will intensify or magnify any color or viscosity change. This is obtained by using a large panel in a small amount of solution. In normal storage tanks, the surface to volume ratio is approximately one square foot of coated surface per 20 gallons of liquid. If a 1- by 2- by ⅛-inch panel is placed in two ounces of solution, the surface to volume ratio is 2.1 square feet per gallon of corrosive. This is 40 times that encountered in a tank. This is one of the fallacies of accelerated testing. A slight discoloration at this surface to volume ratio may not be detectable when used in a tank. Therefore, interpretation of extent of color change must be considered.

Linings

If a coating is to be considered as a lining material, it should withstand a minimum of 150 days of testing in temperatures and corrosive environments equal to or above those encountered in service. Any penetration, loss of adhesion, swelling, or surface attack in this time period would normally preclude using the coating as a lining material.

Floor Coatings

Floor coatings normally do not receive as severe a chemical exposure as tank linings. Consequently, the testing method for chemical resistance is not as strenuous as that for a lining system. A spot test of a corrosive on floor coatings will normally determine its chemical resistance. This spot test can be conducted by placing a small amount of corrosive on the coating and periodically observing the reaction. The corrosives should be replaced frequently to maintain severity of the exposure.

A watch glass can be placed over volatile chemicals and sealed at edges with wax to prevent evaporation of the chemical.

A rating for the chemical resistance of a floor coating should include (1) softening, (2) surface attack, and (3) discoloration. Film softening can be determined with a knife blade.

Many floor coating failures are caused by loss of bond or cracking after pro-

TABLE 1—Comparison of 10 Percent Salt Fog Cabinet and Field Exposures

Type of Failure	Time to Failure (Weeks)			
	Vinyl		Epoxy	
	Cabinet	Field	Cabinet	Field
10% edge failure	12	72	16	42
30% edge failure	22	93	32	59
10% plane failure	17	109	33	111
30% plane failure	30	114	52	116

TABLE 2—Comparison of Sulfuric Acid Cabinet and Field Exposures

Type of Failure	Time to Failure (Weeks)			
	Vinyl		Epoxy	
	Cabinet	Field	Cabinet	Field
10% edge failure	10	82	14	51
30% edge failure	17	115	24	78
10% plane failure	16	136	21	113
30% plane failure	23	150	36	122

TABLE 3—Comparison of Hydrochloric Acid Cabinet and Field Exposures

Type of Failure	Time to Failure (Weeks)			
	Vinyl		Epoxy	
	Cabinet	Field	Cabinet	Field
10% edge failure	9	30	12	25
30% edge failure	17	44	20	32
10% plane failure	14	57	15	43
30% plane failure	23	73	23	52

longed aging. Aging can be accelerated by heating the coating to 180 F (82.2 C) and cooling it to about 60 F (15.6 C). The coating should be applied to a large concrete block (8- by 16-inch). Small concrete blocks do not have sufficient coating material to cause cracking or loss of bond of the coating. The thermal shock (heating and immediate cooling) will show results on inferior floor coatings in 48 to 72 hours or 36 to 48 thermal shock cycles. Any floor coating which passes 48 thermal shock cycles would be worthy of further testing and consideration for use.

High Temperature Coatings

If the coating is to be applied to a surface which continuously operates at a given temperature, there is no method of accelerating the test.

Failure of high temperature coatings usually occurs by one or a combination of (1) resin degradation at high temperature, (2) loss of adhesion, and (3) penetration and rusting when surface temperature is lowered to the condensation point. During testing, the coating should be thermal shocked from its operating temperature to determine adhesion characteristics. Resistance to penetration and rusting is determined by condensing moisture on the coated surface after exposure at the operating temperature.

Accelerated Weathering

Many different pieces of equipment have been built to give accelerated weathering. One of the best known is the Atlas Weather-O-Meter,* both twin arc and single arc. This equipment uses a carbon arc as its source of light which gives a high concentration of near ultra-violet radiation.

Correlation of actual weathering with these instruments is good in some instances and poor in others. However, these instruments are good for screening materials of similar generic types and for obtaining relative weathering performance.

The Fade-O-Meter* more closely duplicates the long term weathering effect on epoxy, vinyl, and acrylic coatings.

Evaluation of Accelerated Testing

The extent to which a coating is tested will determine its capabilities. However, regardless of the extent of testing, the ultimate worth of a coating can be determined only by its performance in actual service. Testing is simply a method of determining which coating should give good performance and eliminating those doomed to failure. Caution should be exercised to avoid "overtesting" which would not correlate with field conditions thus eliminating from further consideration some promising materials.

Other Articles of Interest

R. Marvin Garrett. How to Choose the Right Protective Coating. *Materials Protection*, 3, 8 (1964) March.

J. J. Madden, Experimenting With Epoxy Resins. *Materials Protection*, 3, 12 (1964) June.

Staff Feature. Laboratory Evaluation of Coatings. *Materials Protection*, 1, 10 (1962) June.

*Products of Atlas Electric Devices, Chicago, Ill.

JERRY D. BYRD is head of the testing department at Carboline Company. He received a BS in chemical engineering from Washington University in 1957 and has been with Carboline since graduation. He is a member of NACE.

Corrosion of Painted Metals—A Review*

HENRY LEIDHEISER, JR.*

Abstract

Seven types of corrosion and precorrosion of painted metals are reviewed: blistering, early rusting, flash rusting, anodic undermining, filiform corrosion, cathodic delamination, and wet adhesion. The importance of the nature of the interfacial region between the coating and the metal is emphasized. Ten unsolved problems are described.

Introduction

Corrosion research has gone through various trends and fashions during the past several decades. The 1950's were marked by emphasis on polarization curves and their applications and new materials of interest to the nuclear energy field, the 1960's by emphasis on stress corrosion cracking and mechanics of metal fracture, and the 1970's by an emphasis on corrosion in unusual environments. The indications are that the major emphasis in corrosion research during the 1980's will be in the area of protective coatings. The severe demands on maintaining integrity of semiconductor devices during service and the fabricating of devices using photosensitive organic coatings are intensifying the interest of semiconductor device manufacturers in protective coatings. Protective coatings include metallic coatings, glass and ceramic coatings, semiconductor coatings, and organic polymeric coatings. It is this latter subject area that this article addresses. The purpose of this presentation is to summarize the present state of knowledge about Corrosion Control by Organic Coatings from the perspective of the author's experience and interests.

The tendency of a coated metal to corrode is a function of three major factors: (1) the nature of the substrate metal, (2) the character of the interfacial region between the coating and the substrate, and (3) the nature of the coating. In many cases, little or no attention is paid to the substrate and the coating is required to compensate for the inadequacies of surface preparation prior to painting. Commercially coated metals, such as the case with automobiles, appliances, and coil coated products, are handled with close attention to every step from the raw material to the finished product. Thus, there are corrosion problems with painted metals that range from improper application procedures to those which are not correctable because of an absence of understanding. It is the intent of this article to focus attention on research aimed at understanding and to stress those aspects where the understanding is inadequate.

An organic coating protects a metal substrate from corroding primarily by two mechanisms: (1) serving as a barrier for the reactants, water, oxygen, and ions, and (2) serving as a reservoir for corrosion inhibitors that assist the surface in resisting attack. The barrier properties of the coating are improved by increased thickness, by the presence of pigments and fillers that increase the diffusion path for water and oxygen, and by the ability to resist degradation. The common degradation mechanisms of the coating include abrasion and impact, cracking or crazing at low or high temperature, bond breakage within the polymer matrix because of hydrolysis reactions, oxidation, or ultraviolet light, and freeze-thaw cycling. The result of such degradation allows access of reactants to the coating/substrate interface without the necessity of diffusion through the polymer matrix. Much work has been published on the degradation of organic coatings and this subject will not be discussed here because it is outside the main thrust of this review.

Sources of Information About the Corrosion of Painted Metals

A novice to the field of coatings science must seek guidance in the library. The important journals in the field that contain articles on corrosion include:

- *Journal of Coatings Technology*
- *Journal of the Oil Colour Chemists Association*
- *Farbe und Lack*
- *Progress in Organic Coatings*

Information on a broad range of coatings types and coatings related issues may be found in a series of pamphlets published by the Federation of Societies for Paint Technology in Philadelphia. These may be purchased collectively in a two-volume set. An excellent summary of paint defects and suggestions for correction may be found in "Hess's Paint Film Defects—Their Causes and Cure."[1] A well organized summary of the properties and resistance to chemical attack of a wide variety of paints may be found in the book, "Design and Corrosion Control."[2] Some outstanding articles on various types of protective coatings will be found in the Encyclopedia of Materials Science and Technology to be published by Pergamon Press in 1983. Two books which include papers given at international symposia are very useful: "Corrosion Control by Coatings"[3] and "Corrosion Control by Organic Coatings" published by NACE.[4]

Types of Corrosion Beneath Organic Coatings on Metals

This review will focus on six specific types of corrosion beneath organic coatings: blistering, early rusting, flash rusting, anodic undermining, filiform corrosion, and cathodic delamination. Each of these will be treated separately. Another type of coating deterioration known as "loss of adhesion when wet," or simply "wet adhesion," which may or may not be related to corrosion, will also be discussed.

---

*This is the first in a series "Current Topics in Corrosion" which are invited reviews by the editor. Submitted for publication March, 1982.

*Center for Surface and Coatings Research, Lehigh University, Bethlehem, Pennsylvania.

Blistering

An up-to-date review of blistering has been given recently by Funke.[5] Blistering is one of the first signs of the breakdown in the protective nature of the coating. The blisters are local regions where the coating has lost adherence from the substrate and where water may accumulate and corrosion may begin. Five mechanisms, operative under different circumstances, are used to explain blister formation which occurs prior to the corrosion process.

Blistering by Volume Expansion Due to Swelling.[6-9] All organic coatings absorb water and those used in corrosion protection are usually in the range of 0.1 to 3% water absorption upon exposure to liquid water or an aqueous electrolyte. Water absorption leads to swelling of the coating and when this occurs locally for any reason, blisters may form and water may collect at the interface.

Blistering Due to Gas Inclusion or Gas Formation.[10] Air bubbles or volatile components of the coating may become incorporated in the film during film formation and leave a void. Such blisters are not necessarily confined to the interface, but when they are, they can serve as a corrosion precursor site.

Electroosmotic Blistering.[11-14] Water may move through a membrane or capillary system under the influence of a potential gradient. Potential gradients, such as may exist with a galvanic couple, have the capability of leading to a blister.

Osmotic Blistering.[15,16] The driving force for osmotic blistering is the presence at the coating/substrate interface of a soluble salt. As water penetrates the coating to the interface, a concentrated solution is developed with sufficient osmotic force to drive water from the coating surface to the interface and a blister is formed. The osmotic mechanism is probably the most common mechanism by which blisters form.

An outstanding example of osmotic blistering was cited by an unidentified discussion participant at the Corrosion 81 meeting in Toronto. A ship was painted in Denmark and made a voyage immediately thereafter across the Atlantic and into the Great Lakes. When it reached port, a blister pattern in the form of a handprint was observed above the water line. Apparently, the paint was applied over a handprint. No blistering occurred during exposure to sea water because of the high salt content of the water, but when the ship was exposed to fresh water, the osmotic forces became significant and the blistering occurred.

Blistering Due to Phase Separation During Film Formation.[17-20] A special type of osmotic blistering can occur when the formulation includes two solvents, the more slowly evaporating one of which is hydrophilic in nature. When the hydrophilic solvent is in low concentration, the phase separation process occurs at a later stage in film formation and may occur at the coating/substrate interface. Water diffuses into the hydrophilic solvent, or into the void left by the hydrophilic solvent, and blisters are initiated. Glycol ethers or esters, which have low volatility, are prone to cause such blister formation.

In all the above cases, the blister provides a locale for collection of water at the coating/substrate interface. Oxygen penetrates through the coating, leaching of ionic materials from the interface or from the coating occurs, and all the constituents are available for electrochemical corrosion. The rate of reaction appears to be controlled by the oxygen permeability of the coating.[19,20] Oxygen is necessary for the cathodic reaction, $H_2O + 1/2O_2 + 2e^- = 2OH^-$, but it is also consumed in the conversion of Fe(II) to Fe(III). The ferric corrosion products tend to concentrate on the inside dome of the blister and at the periphery of the blister where the oxygen concentration is the highest. The cathodic region is at the periphery of the blister and the anodic region is in the center of the blister where the oxygen concentration is the lowest.

Early Rusting[21]

This term is applied to a measles-like rusting that occurs after the coating has dried to the touch. It only occurs after the coated metal is exposed to high moisture conditions. A typical condition under which it is observed has been cited by Grourke.[21] A steel tank was abrasively cleaned and was then painted with an acrylic latex late in the afternoon during the summer. Rust spots were most prominent on the bottom of the horizontally-mounted tank and were observed up to the liquid level within the tank. The top half of the tank exhibited no rust spots.

The three conditions which lead to early rusting are: (1) a thin latex coating (less than 40 μm; (2) a cool substrate temperature; and (3) high moisture conditions. Early rusting can be duplicated in the laboratory under these conditions. The severity of the problem tends to increase as the activity of the steel is increased. For example, early rusting is more severe on panels given a white abrasive blast[22] in comparison with less adequate cleaning using a power tool.

Early rusting occurs with latex coatings because of the mechanism by which they lose water. Film formation occurs through coalescence of the latex particles.[23,24] Particle-particle contact occurs because of water evaporation, particle-particle deformation then occurs as a consequence of surface tension and capillary forces, and finally diffusion of polymer chains occurs among latex particles and the film hardens. Early rusting occurs under those conditions that slow down the rate of drying and allow water soluble iron salts to be leached through the paint film. If moist conditions do not exist during the latex drying process, early rusting does not occur.

In summary, early rusting is a consequence of moist conditions occurring before the latex coating has dried sufficiently. Water ingress and egress occur readily before particle coalescence has been completed and movement of soluble iron salts through the film, followed by water evaporation leads to the rust staining. Success in preventing early rusting has been obtained through the use of soluble inhibitors in the formulation.

Flash Rusting[25]

Brownish rust stains may appear on blast-cleaned steel shortly after priming with a water based primer. This phenomenon is known as "flash rusting." Work done by the Paint Research Association[26] has shown that this defect may be avoided by removing the contaminants remaining after blast cleaning either by careful cleaning or by a chemical treatment before application of the primer.

The steel or ceramic grit on the surface apparently leaves crevices and/or galvanic cells are set up between the steel grit and the steel base sufficient to activate the corrosion process as soon as the surface is wetted by the water based paint. The staining is a result of the soluble corrosion products penetrating the coating and being oxidized to the ferric form within or on the surface of the coating.

The possible adverse effect of steel grit blasting on the performance of paints is a subject worthy of investigation.

Anodic Undermining

Figure 2 shows six planes along which delamination may occur to separate the organic coating from the metal. Anodic undermining represents that class of corrosion reactions underneath an organic coating in which the major separation process is the anodic corrosion reaction under the coating. An outstanding example is the dissolution of the thin tin coating between the organic coating and the steel substrate in a food container. In such circumstances, the cathodic reaction may involve a component in the foodstuff or a defect in the tin coating may expose iron which then serves as the cathode. The tin is selectively dissolved and the coating separates from the metal and loses its protective character.

Aluminum is particularly susceptible to anodic undermining. Koehler[27] cites an example from laboratory studies in which pairs of organic coated aluminum panels were sealed to opposite ends of a cylindrical cell filled with 0.05 citrate solution at pH 3.5 containing 0.5% NaCl. The two panels were connected to a power source and a current of 0.09 μamp/cm^2 was

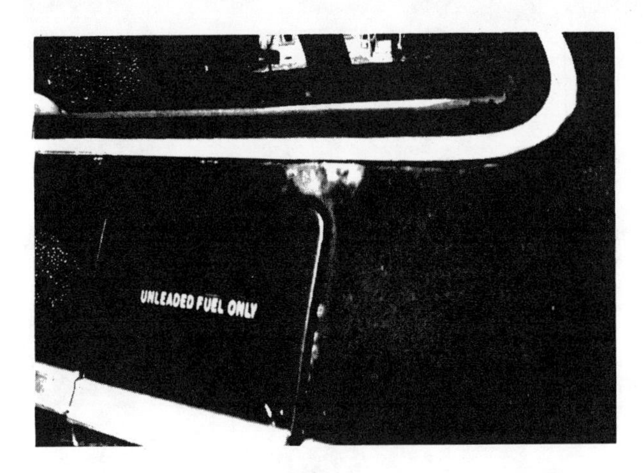

FIGURE 1 — Typical corrosion that was preceded by cathodic delamination. This type of corrosion is common on painted steel which has been damaged by impact or by abrasion. Corrosion is visible as brown blisters along edge to right of gas tank cover and along chrome trim of window at top of picture.

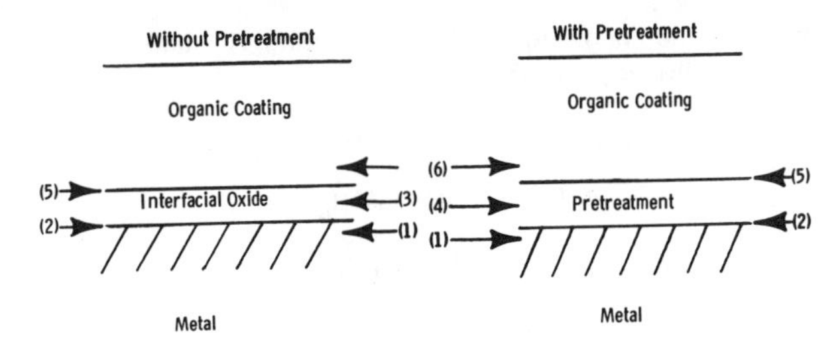

FIGURE 2 — Schematic representation of the six planes along which delamination may occur in an organic coating/metal substrate system. Note that relative thicknesses of oxide and pretreatment layers and the organic coating are not to scale.

passed for one week. Underfilm corrosion was observed on the anode and no underfilm damage occurred on the cathode.

Anodic undermining of organic coatings on steel may occur under circumstances where the steel is made anodic by means of an applied potential. In the absence of an applied potential, coatings on steel fail largely by cathodic delamination.

Anodic undermining has not been studied as extensively as cathodic delamination because there do not appear to be any mysteries. Galvanic effects and principles which apply to crevice corrosion provide a suitable explanation for observed cases of anodic undermining.

Filiform Corrosion

Filiform corrosion is a type of attack in which the corrosion process manifests itself as thread-like filaments. It represents a specialized form of anodic undermining. It generally occurs in humid environments and is most common under organic coatings on steel, aluminum, magnesium, and zinc (galvanized steel). In some cases, filiform corrosion will develop on uncoated steels on which small amounts of contaminating salts have been accidentally deposited. It has also been observed on thin electrodeposits of tin, silver, gold, and under conversion coatings such as phosphate. The most annoying types of filiform corrosion are those which occur under paint films that are designed to retain an aesthetic appearance on metals exposed to the atmosphere and those which occur under the lacquers which protect the interior of food containers.

The threads which form in filiform corrosion exhibit a wide variety of appearances from nodular shapes such as those on aluminum to the very fine, sharply-defined threads observed under clear lacquers on steel. The widths of the filaments are of the order of 0.05 to 0.5 mm and under laboratory conditions grow at the rate of 0.01 to 1 mm/day. The rate of growth of the filaments is approximately constant over long periods of time.[28,29] Filiform corrosion requires a relatively high humidity, generally over 55% at room temperature, but insufficient work has been done to determine an exact lower limit. It can be encouraged to develop on scratching through the coating to the metallic substrate and then maintaining the panel at a relative humidity of 70 to 85%. It develops in the laboratory in some cases simply by putting powders of a salt on the surface of the coating or by putting salt crystals on the metal substrate and applying the coating over the contaminated surface. Preston and Sanyal,[29] for example, obtained filiform corrosion on steels at 99% relative humidity by inoculating the steels with the following powders before applying the coating: sodium chloride, calcium sulfate, ammonium sulfate, sodium nitrate, zinc chromate, flue dust, cinders, iron oxide, and carborundum.

Hoch[30] has made a detailed study of the character of filiform corrosion on steel, magnesium and aluminum substrates. In the case of iron, the very leading edge of the filament had a pH of approximately 1, whereas the liquid immediately behind the leading edge had a pH of 3 to 4, just what would be expected on the basis of hydrolysis of Fe^{++} ions. Magnesium and aluminum also exhibited very low pH's at the leading edge and higher pH values in the liquid adjoining the leading edge. Insoluble corrosion products form in all three cases a short distance back from the leading edge. In the cases of aluminum and magnesium, hydrogen bubbles were observed at the leading edge, indicating that the liquid in the leading edge was low in oxygen and the hydrogen evolution reaction took precedence.

The following mechanism appears to account satisfactorily for the filiform corrosion of aluminum coated with an organic lacquer. First, a highly localized defect forms in the coating. This defect may arise at the edge of a scratch through the coating, at an inclusion in the coating, at a defect in the metal surface, or as a result of the presence of a local high concentration of electrolyte which causes penetration of the coating. In the presence of high relative humidity, water penetrates through the coating. In the presence of electrolyte, which has either penetrated the coating or was inadvertently occluded beneath the coating, a tiny liquid aggregate forms because of the high affinity of ions such as Na^+ and Cl^- for water. Once sufficient molecules are present to have a liquid-like identity, additional water diffusing through the coating is retained because of the low vapor pressure of concentrated electrolyte solutions. Minor corrosion of the substrate occurs yielding additional dissolved ions and promoting further retention of diffusing water species. As the liquid increases in dimension and corrosion occurs, local conditions cause an imbalance in the oxygen supply at some point in the microscopically circular corrosion area. The oxygen-deficient area becomes the anode and the periphery becomes the cathode. The circular droplet then assumes an elliptical shape and the conditions for filamentary growth are present. Once the filament has been nucleated, there is developed an oxygen concentration cell and the propagation of the filament proceeds because of a highly effective anode at the head and a cathodic area present in the areas surrounding the head. Immediately at the interface of the growing head, aluminum is dissolved to yield a highly concentrated Al^{+++} ion solution. Hydrolysis occurs with the following reactions liberating H^+ and thus generating the very low pH:

$$Al(H_2O)_6^{+++} = Al(H_2O)_5OH^{++} + H^+$$

$$Al(H_2O)_5OH^{++} = Al(H_2O)_4(OH)_2^+ + H^+$$

$$Al(H_2O)_4(OH)_2^+ = Al(H_2O)_3(OH)_3 + H^+$$

The latter reaction which occurs some distance from the leading edge results in the precipitation of hydrated aluminum oxide and the pH is reduced to the range of 3 to 4 because of dilution from incoming water.

The oxygen deficiency at the very leading edge along with the low pH also permits the competing cathodic reaction, $2H^+ + 2e^- = H_2$, to occur to a limited extent and a small amount of hydrogen gas is generated.

Koehler[31] has emphasized the importance of the anion in filiform corrosion. He noted that the filaments formed on steel in the presence of sulfate contaminants beneath the coating were less numerous and were much finer than those observed with chloride contaminants. He also noted that the head of the filament contained the anion of the contaminating salt, but not the cation. The anions apparently migrated to provide charge compensation for the ferrous ions formed in the active region at the head of the filament.

The fascinating question about filiform corrosion is why the corrosion occurs in the form of filaments as opposed to circular spots. No complete answer is possible at the present time but it does appear that the limited availability of oxygen, by diffusion through the coating, and the limited availability of water, by diffusion through the coating under relative high humidity conditions, are the determining factors. At very high relative humidities or on exposure to liquid water, filiform corrosion passes over to more general corrosion and the filamentary character is lost.

Much attention has been paid in paint laboratories to reducing filiform corrosion. Phosphate conversion coatings, followed by chromate rinses and distilled water rinses, provide some protection, but do not completely eliminate, the filiform corrosion of iron. The properties of the coating also have an effect on the extent and character of filiform corrosion. Coatings that are highly permeable to water and to oxygen are specially susceptible to filiform attack. Coatings that are very brittle and are ruptured by pressures generated by the corrosion process lose the entrapped moisture and pitting attack often result. In the case of magnesium, filiform corrosion occasionally converts to a virulent pitting attack. More information can be found in reference 17.

Cathodic Delamination

Many coated steel products are subject to scratches or dents with consequent exposure of the steel to the environment. If the coated materials are continuously immersed in an electrolyte, as for example, ships, underground pipelines, and the interior of vessels holding an aqueous solution, it is possible to protect the exposed areas by an applied cathodic potential. One of the undesirable consequences of cathodic protection is that the coating adjoining the defect may separate from the substrate metal. This loss of adhesion is known as "cathodic delamination." This type of delamination may also occur in the absence of an applied potential. The separation of the anodic and cathodic corrosion half reactions under the coating provides regions which are subject to the same driving force as when the cathodic potential is applied externally.

It is generally believed[33,34] that the major driving force for cathodic delamination in corrosion processes in the presence of air is the cathodic reaction, $H_2O + 1/2O_2 + 2e^- = 2OH^-$. When an applied potential is used, the important reaction may be $2H^+ + 2e^- = H_2$, if the driving force is sufficient. Figure 3 shows a typical cathodic polarization curve for steel in 0.5M NaCl saturated with air. The regions of dominance of the two

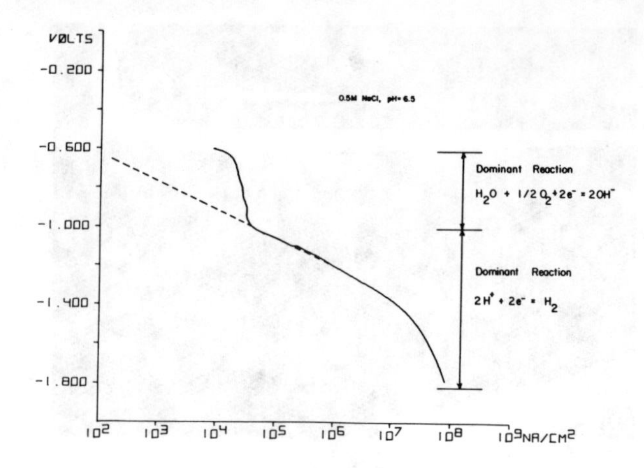

FIGURE 3 — Potentiodynamic polarization curve for steel in aerated 0.5M NaCl. Dotted line represents the extrapolation of the portion of the polarization curve that represents the hydrogen evolution reaction.

cathodic reactions are noted on the figure and the cathodic polarization curve for hydrogen evolution in the absence of oxygen is shown by the dotted line. It is apparent from this figure that at a potential of −0.8 V (vs SCE), the dominant reaction is hydrogen evolution. Polarization at −0.8 V of polymer-coated steel containing a defect in the absence of oxygen leads to no significant delamination from the defect, whereas in the presence of air there is significant delamination.

Studies indicate that the pH beneath the organic coating where the cathodic reaction occurs is highly alkaline, as the cathodic equations indicate. Ritter and Kruger[35] have recently reported that the pH at the delaminating edge is greater than 14 as measured by pH-sensitive electrodes inserted through the metal substrate from the back side. Other studies which integrate the pH over a larger volume of liquid beneath the coating yield pH values of 10 to 12. Cathodic polarization curves on steel in 0.5M NaCl at pH values of 6.5, 10, and 12.5 are approximately the same[36] suggesting that the cathodic behavior beneath the coating may be rationalized in terms of the cathodic polarization curve that is applicable at the exposed defect.

Figure 4 represents the extremes of three types of polarization curves that are observed on metals whose behavior during cathodic treatment has been studied. Point A on each curve represents the potential at the defect and Point B represents the assumed potential at the delaminating front. Curve (1) has the shape of the polarization curve of aluminum in 0.5M NaCl. The surface is not active for the oxygen reduction reaction and the rate of delamination is low as indicated by the location of Point B. Curve (2) is a hypothetical curve somewhat comparable to the polarization curve for tin in 0.5M NaCl. The curve has a steep slope such that the current density falls off greatly with increase in potential. Thus the current density at the delaminating front is low. Curve (3) is typical of iron and copper in 0.5M NaCl. The oxygen reduction reaction is catalyzed over a wide potential range and the current density remains the same over this range. Cathodic delamination occurs at a relatively rapid rate because the current density at Point B is high relative to comparable points on curves (1) and (2).

The cathodic reaction which occurs at the delaminating front generates hydroxyl ions which appear to be the major destructive influence on the organic coating/substrate bond. The value of the pH at the delaminating front is determined by the following factors: the rate at which the reaction occurs; the shape of the delaminating front; the rate of diffusion of hydroxyl ions away from the delaminating front; and buffering

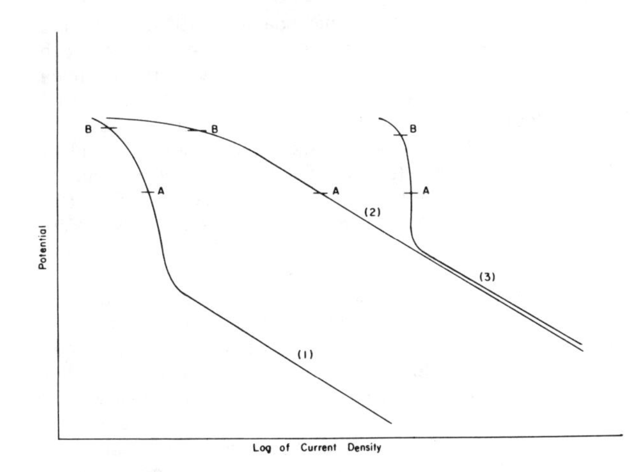

FIGURE 4 — Three types of cathodic polarization curves observed on different metals in aerated 0.5M NaCl. Curve (1) is obtained with aluminum; curve (2) is hypothetical and illustrates the approximate curve obtained with tin; and curve (3) is obtained with steel.

reactions which may involve the interfacial oxide or the polymer.

All the evidence presently available indicates that the cathodic delamination process occurs because of the high pH generated by the cathodic reaction. The real question is what is the consequence of a high pH on the interface. The experimental evidence suggests that the strong alkaline environment may attack the oxide at the interface or may attack the polymer. Attack of the oxide has been seen by Ritter[37] using ellipsometric techniques in the case of polybutadiene coatings on steel and surface analysis techniques in the hands of Dickie and colleagues[38] give clear evidence that carboxylated species are present at the interface as a result of hydroxyl ion attack of the polymer. Dickie[39] has also recently shown that coatings more resistant to alkaline attack exhibited better performance when scribed and submitted to salt spray.

It is proposed that the major mechanism for the delamination process is the solubilization of the thin oxide coating at the interface between the organic coating and the metal. The dissolution of the oxide, as the major delaminating mechanism, has been proposed previously by Gonzalez, Josephic, and Oriani[40] to account for effects observed with food can lacquers on steel when immersed in solutions containing strong complexing agents for Fe^{++} ions. In this process the oxide itself participates in the cathodic reaction by the reaction,

$$\gamma\text{-}Fe_2O_3 + 6H^+ + 2e^- = 2Fe^{++} + H_2O\,,$$

and the complexing agent serves to drive the reaction to the right by complexation of the ferrous ions. In the cathodic delamination process being proposed herein, the dissolution of the oxide breaks the bond between the coating and the substrate metal and the high pH leads to localized attack of the polymer at the interface. The presence of oxidized organic species on the metal surface after delamination has occurred may be the result of *a posteriori* adsorption of oxidized species or may be the result of islands of organic left on the surface. The XPS technique used in reference 38 illuminates a large area and spatial resolution is lacking to determine if the organic material is present over the entire surface or is island-like in nature.

Since the important delaminating process is a consequence of the hydroxyl ions generated by the cathodic reaction, $H_2O + 1/2O_2 + 2e^- = 2OH^-$, the delamination may be prevented by any of the following:

- Preventing reactant water from reaching the reaction site.
- Preventing reactant oxygen from reaching the reaction site.
- Preventing electrons from reaching the reaction site.
- Preventing cation counterions from reaching the reaction site.
- Reducing the catalytic activity of the surface for the cathodic reaction.

Water and oxygen reach the reaction site largely by diffusion through the coating[41] so the rate of reaction may be reduced by reducing the permeability of the coating for these constituents. It appears unlikely on the basis of present knowledge to eliminate completely the diffusion of these constituents through the coating or through defects in the coating that are present when the coating is prepared or occur during normal service. The electrons, however, reach the reaction site through the metal phase and through any interfacial oxide or other film that exists at the metal/coating interface. Any type of metal surface film that is a poor electronic conductor has the possibility of limiting the access of electrons to the reaction site. The low rate of the cathodic delamination from aluminum surfaces, relative to zinc and steel, is probably a consequence of the poor electronic conductivity of the aluminum oxide at the interface.

The cathodic reaction generates anions and there must be available cation counterions to balance the charge locally. Hydrogen ions do not perform this function or else the pH would not rise as dramatically as it does. Available evidence[41] indicates that the major transport medium for the cations is in the liquid layer that forms at the coating/substrate interface, although transport through the coating itself cannot be ruled out.

The cathodic reaction is a catalyzed one and the chemical character of the interfacial region determines whether or not the reaction will occur. It has been shown that cobalt ions have the ability to poison the zinc oxide surface on zinc for the oxygen reduction reaction[42] and it has also been shown that this poisoning leads to a lower rate of cathodic delamination.[43] Some commercial pretreatments possibly include the function of poisoning the surface for the cathodic reaction.

Loss of Adhesion of a Coating When Wet. This phenomenon shows up in a number of ways. The best example is exhibited by the "hot water test" used in characterizing pipeline coatings. A candidate coating on a steel substrate is immersed in water at 80 to 100 C for 10 to 30 days and the adherence of the coating is determined at the end of this time by the use of a strong and sharp knife. A similar test of a duration of approximately 1 hour is used with certain coil coated products. Poor wet adhesion is also exhibited by coatings that dry too rapidly and trap organic solvents at or near the interfacial region. This phenomenon has been discussed by Funke.[44] It also is apparent in the removal of epoxy powder coatings using methyl ethyl ketone. After soaking the coated metal in methyl ethyl ketone for one or more days, the coating may be readily stripped from the substrate if the time after removal from the organic liquid is correctly selected. In too short a time, the coating is gummy and cannot be separated from the substrate. In too long a time, adherence to the substrate is regained. When the correct time is chosen, the coating may be removed in much the same way as a weakly adhering adhesive tape. At the correct time, there apparently is a layer of liquid which is intermediate between the coating and the substrate along with the fact that the coating has a physical state intermediate between a soft clay and a rigid coating.

The major problem in studying wet adhesion is the difficulty in quantitatively defining the degree of adherence. The cross-cut test is used with some degree of success with coatings less than about 100 μm in thickness but is not suitable for very thick coatings. The impact test of Zorll[45] is useful with coatings in the same thickness range as the cross-cut test. Unfortunately, there is no satisfactory test for coatings 100 μm or greater in thickness. A vertical tear off test, commonly used for coating adherence tests, is not applicable to thick, inflexible coatings but it can be used to determine wet adhesion of thin coatings in some cases. Funke[44] has proposed a test which as yet has not received wide acceptance. He suggests that the water absorption of both free films and of the same material on a metal be determined at 90% relative humidity. Loss of adhesion on exposure to high humidity is indicated by a cross-over in the two absorption curves. The cross-over time indicates when the coated metal begins to lose adhesion because of the presence of water at the metal/coating interface. Coatings which exhibit good wet adhesion do not show a cross-over.

The major unsolved problem relating to wet adhesion is the mechansm by which water affects the adhesion. This problem is related to the very basic question as to the factors that control adhesion. The way in which adhesion between two materials is viewed has gone through a number of fashions. Some years ago, adhesion was viewed as an interaction between polar groups; later it became the rule to treat adhesion in terms of dispersion forces. More recently, under the impetus of Fowkes,[46] adhesion of organic coatings is being viewed as an acid/base interaction. This approach has much merit because the acid/base character of a surface can be quantitatively assessed and the adsorption of organic polymers on metals and metal oxides can be studied to determine the appropriateness of this latter viewpoint. The effect of water on the rate and amount of polymer adhesion shows promise of providing the basis for understanding wet adhesion.

Nature of the Interfacial Region

Much research has been carried out with the objective of characterizing the metal surface before the application of the coating. Some recent examples include the work of Schwab and Drisko[47] on the effect of surface profile, the work of Mansfeld, Lumsden, Jeanjaquet, and Tsai on surface chemistry,[48] the work of Zurilla and Hospadaruk[49] on characterizing the oxygen reduction capacity of phosphated surfaces, and the work of Iezzi and Leidheiser[50] on the effect of parameters which control the ability of a steel surface to accept uniform phosphating.

The chemical nature of the intact organic coating/metal substrate interfacial region has been little studied largely because of the difficulty in devising experimental techniques to make such a study. Optical techniques, such as ellipsometry, are useful with thin transparent coatings but they cannot be used with opaque coatings or those which contain pigments or fillers. Ellipsometric techniques can provide information on the thickness of the oxide film at the interface and, with good fortune, the optical parameters may be used to determine the composition of the oxide. In the hands of Ritter and Kruger,[51] it has been found that an oxide exists at the interface and that the optical properties of the interface change when cathodic delamination or corrosion occurs.

Emission Mössbauer spectroscopy provides a technique which may be utilized to identify the chemical nature of the emitting atom. Chemical compounds yield characteristic spectra which can be used as fingerprints to identify the compound. Leidheiser, Kellerman, and Simmons[52] have applied Mössbauer techniques to show chemical changes beneath a coating and, more recently, Leidheiser, Simmons, and Music[53] have been able to show changes in the nature of the oxide film on cobalt when a polybutadiene coating was applied. The simple application of the coating caused a fractional conversion of Co^{+++} to Co^{++} and baking the coating at 200 C resulted in the formation of additional Co^{++}. The emission Mössbauer technique has the severe limitation that it is only applicable with relative ease to cobalt and tin and the complete interpretation of the spectrum is often difficult and/or ambiguous.

Major advances in organic coatings protective against corrosion are dependent on the development of new techniques for studying in a nondestructive way the chemical nature of the interfacial region between the metal and the organic coating. The interfacial region is where the action is, particularly in the case of cathodic delamination. The oxide at the interface is the catalytic surface for the oxygen reduction reaction; it is the medium through which the electrons are supplied for the oxygen reduction reaction; and it provides the bonding which results in the adherence between the coating and the metal.

Commercial systems which provide the maximum resistance against corrosion include an inorganic coating between the organic coating and the metal substrate. This inorganic coating, often called a "pretreatment" or a conversion coating, replaces the normal metal oxide and provides to the organic coating a substrate with different chemical properties—a poorer catalyst for the oxygen reduction reaction, a less conductive interfacial region and, in some cases, a rougher interface that improves organic coating/substrate adherence and resistance to deterioration under service conditions. The more common inorganic coatings include phosphates, chromates and mixed metal oxides. With the exception of the research of Machu,[54] little has been published on the science associated with interfacial inorganic coatings. It is a fertile area for research.

Properties of the Coating

General Comments

Polymers form the matrix of organic coatings. They are the retainers for pigments, fillers, corrosion inhibitors, and other additives present for specific purposes. The polymer selected is based both on end-use requirements and the ability to apply the coating in the desired manner. Emulsion polymers, or latexes, are suspended in an aqueous medium and they form a coating by loss of water by evaporation and coalescence of the individual particles into a continuous film. Some monomers, such as polybutadiene, are applied to the substrate in a solvent and the polymerization process occurs as the solvent is removed. Cross-linking by oxidation occurs at elevated temperature in the case of polybutadiene. Other coatings based on condensation polymers are polymerized *in situ*. A good example is the epoxy-polyamine coating in which the two constituents are mixed just prior to application and the polymerization process occurs over a period of time. Other polymers are dissolved in solvents and the polymer forms the coating as the solvent is evaporated. Production line painting often involves the use of heat or other type of radiation to cause the film-forming process to occur more rapidly.

The properties of polymeric coatings depend not only on the size, shape and chain structure of the individual units, but also on the spatial shape of the polymer molecules. The linking of many carbon atoms and the freedom of rotation about carbon-carbon bonds permits the molecule to assume a variety of spatial shapes such as spirals, coils, and tangles. This wide latitude in shape also leads to a variety of ways in which individual molecules are oriented with respect to their neighbors. Three classes of arrangement are recognized:

1. Segments of the molecule are randomly distributed regardless of whether they belong to the same molecular chain or another chain. Such a material is termed amorphous or glassy and the properties are uniform in all directions.

2. Segments of the molecule possess a degree of lateral order through the folding of individual chains. The volume element over which this occurs may be considered a single crystal. The individual crystals may be randomly oriented or they may be aligned in the same direction. In the latter case, the coating may have physical properties that differ in different directions.

3. Segments of the molecule may show lateral order through the parallel arrangement of extended chains. As in (2),

these parallel arrangements may be unoriented with respect to neighboring volumes or there may be a degree of spatial orientation. Materials of this type are obtained when a polymer melt is solidified while under shear or stress.

An unusual case of corrosion in which the rate of corrosion appears to be related to segmental motion of portions of the polymer chain is shown in some interesting research by Yializis, Cichanowski, and Shaw.[55] These workers observed that the rate of corrosion of aluminum coated polypropylene capacitors in either the dry state or when immersed in a dielectric fluid was a function of the frequency of an applied AC potential. A sharp maximum in corrosion rate occurred at 3.5 $\times$ 10^3 Hz. The following explanation for this phenomenon is offered. On exposure to the atmosphere and during corona discharge prior to metallization, polypropylene dissolves significant quantities of oxygen and water. Appreciable amounts of chloride ion remain in the polypropylene from the manufacturing operation. When the metallized capacitor is exposed to an AC voltage, segmental motions occur in the polypropylene. Over a limited frequency range, the motions of segments of the polymer are such as to allow diffusion of water and oxygen ot occur along special pathways in the polymer. Since the aluminum coating is essentially opaque to the passage of water, chloride ions and oxygen, sufficient reactants accumulate at the aluminum/polymer interface to allow the following reaction to occur:

$$\text{Cathodic: } H_2O + 1/2O_2 + 2e^- = 2OH^-$$

$$\text{Anodic: } Al - 3e^- = Al^{+++}$$

The chloride ion provides sufficient conductivity in the aqueous phase to allow the electrochemical reactions to occur and prevents the formation of a passive film of aluminum oxide at the aluminum/polypropylene interface.

An important characteristic of polymers and organic coatings is known as the glass transition temperature, T_g. It is that temperature at which a discontinuity occurs in a physical property as a function of temperature when the polymer exists in the amorphous condition. Typical physical properties which show discontinuities at T_g include the coefficient of expansion and specific heat. It is interpreted as that temperature above which the polymer has sufficient thermal energy for isomeric rotational motion or for significant torsional oscillation to occur about most of the bonds in the main chain which are capable of such motion. Values of T_g are obtained by many different experimental techniques, the more common of which include dilatometry, dielectric measurements, spectroscopy, calorimetry, and refractive index. Standish and Leidheiser[56] have outlined a simple technique for such determination of T_g of coatings on a metal using dielectric measurements as a function of temperature. The value of T_g is important because physical properties such as water and oxygen permeability and ductility differ above and below T_g.

The mechanical properties of coatings which are not highly loaded with pigments or fillers depend on molecular weight, crystallinity and the three-dimensional arrangement of the branches. An increase in molecular weight makes a polymer harder and stronger. The higher the degree of crystallinity, the stronger the polymer. Chain polymers containing two different groups, R and R', have different mechanical properties dependent upon the arrangement of the branches. Curing or cross-linking causes a polymer to become harder, more brittle, and less soluble.

Polymer coatings are exposed to the environment and thus are subject to degradation by environmental constituents. The main agencies by which degradation occurs are thermal, mechanical, radiant, and chemical. Polymers may also be degraded by living organisms such as mildew. Deterioration takes the form of discoloration, cracking and crazing, loss of adherence to the substrate, or change in a physical property such as resistivity or mechanical strength. The mode of degradation may involve depolymerization, generally caused by heating, splitting out of constituents in the polymer, chain scission, cross linking, oxidation, and hydrolysis. Polymers are subject to cracking on the application of a tensile force, particularly when exposed to certain liquid environments. This phenomenon is known as environmental stress cracking or stress corrosion cracking and there are many analogies to similar phenomena observed with metals.

Zinc-Rich and Zinc-Pigmented Coatings

The term "zinc-rich" applies to coatings which contain up to about 95% metallic zinc in the dry film. They are electrically conductive and protect the metal substrate electrochemically in much the same way as zinc protects steel in galvanized steel. The vehicle is usually a silicate in the case of the so-called "inorganic zincs" and various organic polymers are used to obtain so-called "organic zincs."

Zinc-pigmented paints are usually a mixture of metallic zinc (80%) and zinc oxide (20%), the latter of which is added to reduce the rate of settling of zinc both in the container and after application to a surface. Galvanic protection does not occur in the case of the zinc dust-zinc oxide paints, whose films do not exhibit electrical conductivity.[57]

The galvanic action of the zinc protects the steel at holidays, cut edges, and scribe marks and a similar galvanic action probably applies during the first stages of corrosion beneath the coating. The major action of the zinc, however, appears to be the sealing of the paint film so that it has an improved resistance to penetration by active environmental species. There is evidence also that the zinc pigment and the zinc oxide prevent deterioration of the inorganic and organic binders and assist in maintaining flexibility and desirable mechanical properties of the coating.

Corrosion Inhibitors

It is thought that corrosion inhibitors in an organic coating function in an identical way to those added to a liquid environment. Corrosion inhibitors used in coatings include oxidizing agents such as chromate, inorganic salts that function in the same manner as benzoates, metallic cations of which lead is the most widely used, and organic compounds. All organic coatings are permeable to water and water contents for many coatings at 100% relative humidity are in the range of 0.1 to 3%. Thus, when the coating is wet, some fraction of the inhibitor is solubilized and can be transported to the metal surface. The coating simply serves as a reservoir for the inhibitor.

A recent article has summarized six technical requirements for an ideal corrosion inhibitor to be used in organic coatings.[58] These include the following:

1. The inhibitor must be effective at pH's in the range of 4 to 10 and ideally in the range of 2 to 12.
2. The inhibitor should react with the metal surface such that a product is formed which has a much lower solubility than the unreacted inhibitor.
3. The inhibitor should have a low but sufficient solubility.
4. The inhibitor should form a film at the coating/substrate interface that does not reduce the adhesion of the coating.
5. The inhibitor must be effective both as an anodic and a cathodic inhibitor.
6. The inhibitor should be effective against the two important cathodic reactions, $H_2O + 1/2O_2 + 2e^- = 2OH^-$ and $2H^+ + 2e^- = H_2$.

A critical problem facing the coatings industry at the present time is the need to replace chromates and lead compounds as corrosion inhibitors. Both of these classes of materials have been identified as hazardous and the pressures to remove them from formulations are growing. Adequate proven substitutes have not yet been universally accepted so that there is an increased interest in developing accelerated tests that can be used to select corrosion inhibitors.

Source: *Corrosion*, July 1982, 374-383

Methods for Monitoring the Corrosion of Polymer-Coated Metals

Electrical methods for studying the protective properties of coatings are numerous and many have produced important results. Two reviews of this subject have been published recently[59,60] and selected information will be extracted from these reviews and other published sources. Electrical measurements that provide data which are useful in predicting the lifetime of a coating include DC measurements of coating conductivity;[62,63] of impedance as a function of frequency;[64,65] of equivalent AC resistance at constant frequency;[66] and of the ratio of capacitive to resistive components at constant frequency.[67] The AC properties of a coating have also been used to estimate the amount of water taken up by a coating.[66,68-74] Scanning techniques[75] have proven useful in characterizing the electrical homogeneity of coatings. The rate of diffusion of sodium chloride through coatings has been determined by Kittelberger[76,77] using measured values of the DC resistance and the membrane potential of the film. Radiotracer measurements,[78] however, are much preferred for these types of measurements. Dielectric techniques are also useful in determining the glass transition temperature of coatings, in determining the effects of coating composition and structure, and in the quality control of coating components.[52]

Corrosion potential measurements and their applicability to coated metals have been summarized by Wolstenholme.[79] As a generalization, it can be concluded that movement of the corrosion potential in the noble direction is indicative of an increasing cathodic/anodic surface area ratio and is indicative that oxygen and water are penetrating the coating and arriving at the metal/coating interface. Movement of the corrosion potential in the active direction is indicative that the anodic/cathodic surface area ratio is increasing and that the overall corrosion rate is becoming significant. Increasingly positive potentials with time suggest that alkaline conditions caused by the oxygen reduction reaction are developing locally at the metal/coating interface and that delamination is of concern. Increasingly active potentials indicate rusting (in the case of steel) beneath the coating and represent the signal that the coating lifetime is limited.

Important Unsolved Problems

The unsolved problems related to corrosion and its prevention by organic coatings are legion. Ten of these problems which are considered of special importance are mentioned and briefly discussed. Three of these problems relate to the very practical test development and corrosion monitoring, two relate to a better understanding of practical operations, and five relate to a better understanding of the coating/substrate interface.

1. **The Development of an Accelerated Atmospheric Corrosion Test, the Results of Which Correlate Well with Service Experience.** The most commonly used test for evaluating the corrosion protection by a coating is the salt spray test, used either with an undamaged coating or used with a coating that is scribed in the form of an X. Many workers are dissatisfied with the salt spray test because there is often a lack of correlation with service experience. However, it is a test that is widely used and it is the most likely test to be acceptable both to the supplier of a coating and the potential user of a coating. Other tests involve simulated outdoor exposure with the degradation variables of ultraviolet radiation, temperature, humidity, and cycling taken into account.

It is the author's opinion that the most satisfactory type of accelerated test will be one that provides information about the extent of reaction at the coating/substrate interface on a localized scale and at a time long in advance of visible corrosion on the surface of the coating. Such a test should be capable of being used on test panels of different shapes and upon exposure to different aggressive environments.

2. **The Development of a Satisfactory Test for Screening Inhibitors for Use in Coatings.** Environmental considerations are limiting the use of two satisfactory inhibitors, lead oxides and salts, and chromates. It can be anticipated that other inhibitors will be recognized as harmful and will be unavailable for use by coatings fomulators. No satisfactory test for rapidly screening inhibitors in a coating formulation has yet received wide acceptance. Inhibitors that perform satisfactorily while dissolved in the aggressive medium do not necessarily perform satisfactorily when incorporated in the coating.

3. **The Development of a Corrosion Monitor That Gives Information About the Progress of Corrosion and Allows One to Predict the Lifetime Before Repainting is Necessary.** Many coatings are applied in locations that are not easily accessible. Examples include towers, bridge spans, submerged supports for deep sea platforms, underground pipelines, and hidden parts of metallic structures. It would be very useful to have a cheap device that would periodically sense the coating/substrate system, and transmit the information to a central location so that the system could be followed as a function of time. The most likely type of device would be one that applied an AC potential at a fixed frequency, or perhaps a range of frequencies, and converted this information to capacitive and resistive components. It is difficult to visualize a test that does not depend on an electrical measurement, although long range inspection by fiber optics may be acceptable.

4. **What is the Mechanism of Corrosion or Adhesion Loss of Coated Metals in the Hot Water Test?** Coatings for certain applications are sometimes tested for adhesion by exposing the coating to water at 80 to 100 C for an hour or more and the adhesion is very poor when tested immediately after removal from the hot water, although the coating recovers good adhesion when it is permitted to dry out. What is the mechanism for the loss of adhesion when wet? And why is the adhesion regained when the coating has dried?

Underground pipeline coatings are often subjected to a hot water test for periods of time up to 20 days. Some coatings survive this test and, indeed, show no deterioration of the coating/substrate bond. Other coatings show severe corrosion that appears to correlate with corrosion potential measurements made continuously during the hot water test. These measurements suggest that some coatings are defective at the start of the test. There is need to understand the mechanism of this behavior.

5. **What is the Reason for the Poor Performance of Some Coatings When the Coatings are Applied Over Steel Substrates That are Abrasively Cleaned with Steel Grit?** As stated previously, flash rusting is associated with contaminants remaining on the surface after abrasive cleaning. Work in this laboratory has shown that certain coatings applied over a steel grit blasted surface exhibit poorer performance during exposure to hot water than similar coatings prepared over surfaces blasted with aluminum oxide. Much steel grit remains imbedded in the surface after cleaning as shown by the fact that the steel is released by superficial oxidation at 300 C. Research is needed to determine the essential reasons for the adverse effect of steel grit on corrosion performance.

6. **Development of a Better Understanding Why Cathodically Electrodeposited Organic Coatings Provide Such Good Protection Against Delamination in Chloride-Containing Environments.** Cathodic electrocoats, or E-coats as they are often called, have received wide acceptance in the automobile industry because of the good protection they offer against the delamination process that accompanies corrosion around a break in the coating. There appears to be little understanding of the mechanism by which these coatings operate. One possibility appears to be that the strongly reducing environment at the steel surface coincident with the formation of the coating/steel bond leads to an interface with good integrity. Experiments which seek to characterize the nature of the steel surface during electrocoating, after electrocoating, after baking, and after exposure to an aggressive medium are planned utilizing Mössbauer emission spectroscopy.

7. **Development of a Method for the Detection of Condensed Water at the Coating/Substrate Interface.** Corrosion beneath an organic coating will occur only if there is an aqueous phase to accept the cations formed in the corrosion process and to provide the medium in which the oxygen reduction reaction may occur. Thus, any method for detecting the presence of aqueous-phase water at the interface and the increase in volume of water at the interface as a function of time will have the capability of detecting corrosion at the earliest possible time.

The question is simple but the answer is difficult. Preferably the method should be sensitive to water aggregates of the order of 25 to 100 molecules, but a lower sensitivity would be acceptable. The method should be applicable to opaque coatings and should be able to discriminate between water masses within the coating and condensed water at the interface. Electrical methods or spectroscopic methods to which the coating is not opaque appear to offer the greatest chance of success.

8. **Achieving a Better Understanding of the Potential Distribution Within a Delaminating Region.** It has been shown previously[36,41] that cathodic delamination may be interpreted in terms of polarization curves. It has also been shown in the case of steel that the polarization curve in 0.5M NaCl solution is not strongly a function of pH over the range of 6 to 12.5. The missing information needed for a complete interpretation of the delamination process is the potential distribution radially from a defect when the metal is polarized cathodically. The mathematical treatment is difficult because the following information is not known: the thickness of the aqueous layer at the interface as a function of distance from the defect; the conductivity of the liquid; the amount of charge that passes through the coating radially from the defect; the oxygen concentration gradient across the coating.

9. **Development of Information about Ionic Transport Through the Coating When the Metal Substrate is Polarized Cathodically.** The cathodic reaction that occurs under the coating during corrosion or during cathodic polarization generates OH^- ions. These ions require counterions to maintain charge neutrality. Since the pH rises, the counterions cannot be exclusively H^+ but must be largely alkali metal ions such as Na^+. In one set of experiments carried out in this laboratory, a current of the order of 10^{-11} amp/cm^2 developed across a thin organic coating when the coated metal was coupled to an uncoated metal of equal surface area in 0.5M NaCl. The coated metal was the cathode. It would be desirable to know if the Na^+ ions represent the charge carriers across the coating under the experimental conditions used.

Experiments should be carried out to determine the ability of alkali metal ions to migrate through various coatings under a potential gradient and this ability should be compared with the ability of a coating to resist cathodic delamination. It is essential in experiments of this type to use a coated metal, as opposed to a free film, since the boundary conditions are so different in the two cases.

10. **Development of a Better Understanding of the Chemical Nature of the Bond Between the Organic Coating and the Substrate Metal and How the Bond Changes with Time.** The interfacial region in the absence of a purposely applied chemical conversion coating consists of the metal, a thin oxide coating on the metal, perhaps a water layer, and the organic coating. The bond between the organic coating and the substrate has been variously referred to as a hydrogen bond, a Van der Waals bond, an acid/base interaction, an electrostatic bond, etc. Minimal experimental information is available in specific instances to characterize the bond quantitatively. A subsidiary question is what happens to the nature of the bond when water permeates the coating and becomes available for adsorption or reaction in the interfacial region.

Acknowledgment

This review was written and some of the experimental results reported herein were obtained while the author's research was supported by a major grant from the Office of Naval Research. This support is gratefully acknowledged.

References

1. Hess's Paint Film Defects, Edited and revised by H. R. Hamburg and W. M. Morgans. 3rd Ed., Chapman and Hall, London, 504 pp. (1979).
2. Design and Corrosion Control, V. Roger Pludek, Halsted Press, New York, 383 pp. (1977).
3. Corrosion Control by Organic Coatings, H. Leidheiser, Jr., Editor, Science Press, Princeton, N.J., 500 pp. (1979).
4. Corrosion Control by Organic Coatings, H. Leidheiser, Jr., Editor, Natl. Assoc. Corrosion Engrs., Houston, Texas, 300 pp. (1981).
5. W. Funke. Prog. Organic Coatings, Vol. 9, p. 29 (1981).
6. N. A. Brunt. J. Oil Colour Chem. Assoc., Vol. 47, p. 31 (1964).
7. N. A. Brunt. Verfkroniek, Vol. 33, p. 93 (1960).
8. L. Bierner. Farbe Lack, Vol. 66, p. 686 (1960).
9. D. M. James. J. Oil Colour Chem. Assoc., Vol. 43, p. 391, 658 (1960).
10. J. A. van Laar. Paint Varn. Prod., Vol. 51, No. 8, 31, 88; No. 9, 49; No. 11, 41, 97 (1960).
11. H. Grubitsch and K. Heckel. Farbe Lack, Vol. 66, p. 22 (1960).
12. H. Grubitsch, K. Heckel, and R. Sammer. Farbe Lack, Vol. 69, p. 655 (1963).
13. H. Grubitsch, K. Heckel, and O. Monstad. Farbe Lack, Vol. 70, p. 167 (1964).
14. W. W. Kittelberger and A. C. Elm. Ind. Eng. Chem., Vol. 39, p. 876 (1947).
15. L. A. van der Meer-Lerk and P. M. Heertjes. J. Oil Colour Chem. Assoc., Vol. 58, p. 79 (1975).
16. L. A. van der Meer-Lerk and P. M. Heertjes. J. Oil Colour Chem. Assoc., Vol. 62, p. 256 (1979).
17. W. Funke. J. Oil Colour Chem. Assoc., Vol. 59, p. 398 (1976).
18. C. M. Hansen. Ind. Eng. Chem. Prod. Res. Dev., Vol. 9, p. 282 (1970).
19. W. Funke and H. Haagen. Ind. Eng. Chem. Prod. Res. Dev., Vol. 17, p. 50 (1978).
20. W. Funke, E. Machunsky, and G. Handloser. Farbe Lack, Vol. 84, p. 49 (1978).
21. M. J. Grourke. J. Coatings Technol., Vol. 49, No. 632, p. 69 (1977).
22. According to specifications of Steel Structures Painting Council, Pittsburgh, PA.
23. J. W. Vanderhoff, E. B. Bradford, and W. K. Carrington. J. Polym. Sci., Vol. 41, p. 155 (1973).
24. M. S. El-Aasser and A. A. Robertson. J. Paint Technol., Vol. 47, No. 611, p. 50 (1975).
25. M. J. Grourke and T. H. Haag. Resin Rev., Vol. 24, No. 1 (1974).
26. Paint Research Assoc. Newsletter No. 7 (1978).
27. E. L. Koehler. Localized Corrosion, R. W. Staehle, B. F. Brown, J. Kruger, and A. Agrawal. Editors, Natl. Assoc. Corrosion Engrs., Houston, Texas, p. 117 (1974).
28. H. Kaesche. Werkstoffe Korros., Vol. 10, p. 668 (1959).
29. R. St. J. Preston and B. Sanyal. J. Appl. Chem., Vol. 6, p. 26 (1956).
30. G. M. Hoch. Localized Corrosion, R. W. Staehle, B. F. Brown, J. Kruger, and A. Agrawal, Editors, Natl. Assoc. Corrosion Engrs., Houston, Texas, p. 134 (1974).
31. E. L. Koehler. Corrosion, Vol. 33, p. 209 (1977).
32. ASTM Standard G8-72, Annual Book of ASTM Standards, Vol. 27, p. 869 (1979).
33. H. Leidheiser, Jr. and M. W. Kendig. Corrosion, Vol. 32, p. 69 (1976).
34. R. A. Dickie and A. G. Smith. Chem. Tech. 1980, No. 1, p. 31.
35. J. J. Ritter and J. Kruger. Corrosion Control by Organic Coatings, H. Leidheiser, Jr., Editor, Natl. Assoc. Corrosion Engrs., Houston, Texas, p. 28 (1981).

36. H. Leidheiser, Jr., L. Igetoft, W. Wang, and K. Weber. Proc. 7th Intern. Conf. on Organic Coatings, Athens, Greece (1981); in press.
37. J. J. Ritter. Personal communication, October 1981.
38. J. S. Hammond, J. W. Holubka, and R. A. Dickie. J. Coatings Technol., Vol. 51, No. 655, p. 45 (1979).
39. R. A. Dickie. paper presented at Electrochem. Soc. meeting, Minneapolis, Minn., May 1981.
40. O. D. Gonzalez, P. H. Josephic, and R. A. Oriani. J. Electrochem. Soc., Vol. 121, p. 29 (1974).
41. H. Leidheiser, Jr., W. Wang, and L. Igetoft. Prog. Organic Coatings, accepted for publication.
42. H. Leidheiser, Jr. and I. Suzuki. J. Electrochem. Soc., Vol. 128, p. 242 (1981).
43. H. Leidheiser, Jr. and W. Wang. Corrosion Control by Organic Coatings, H. Leidheiser, Jr., Editor, Natl. Assoc. Corrosion Engrs., p. 70 (1981).
44. W. Funke. Corrosion Control by Coatings, H. Leidheiser, Jr., Editor, Science Press, Princeton, N. J., p. 35 (1979).
45. U. Zorll. Adhasion, Vol. 6, p. 165 (1979).
46. F. M. Fowkes, C-Y. Chen, and S. T. Joslin. Corrosion Control by Organic Coatings, H. Leidheiser, Jr., Editor, Natl. Assoc. Corrosion Engrs., Houston, Texas, p. 1 (1981).
47. L. K. Schwab and R. W. Drisko. Corrosion Control by Organic Coatings, H. Leidheiser, Jr., Editor, Natl. Assoc. Corrosion Engrs., Houston, Texas, p. 222 (1981).
48. F. Mansfeld, J. B. Lumsden, S. L. Jeanjaquet, and S. Tsai. Corrosion Control by Organic Coatings, H. Leidheiser, Jr., Editor, Natl. Assoc. Corrosion Engrs., Houston, Texas, p. 227 (1981).
49. R. W. Zurilla and V. Hospadaruk. Soc. Automotive Engrs. Trans., Vol. 87, p. 762 (1978).
50. R. A. Iezzi and H. Leidheiser, Jr. Corrosion, Vol. 37, p. 28 (1981).
51. J. J. Ritter and J. Kruger. Surface Science, Vol. 96, p. 364 (1980).
52. H. Leidheiser, Jr., G. W. Simmons, E. Kellerman. J. Electrochem. Soc., Vol. 120, p. 1516 (1973).
53. H. Leidheiser, Jr., G. W. Simmons, S. Musić. manuscript in preparation.
54. W. Machu. Werkstoffe Korrosion, Vol. 14, p. 566 (1963).
55. A. Yializis, S. W. Cichanowski, and D. G. Shaw. paper presented at IEEE Meeting in Boston, May 1980.
56. J. V. Standish and H. Leidheiser, Jr. J. Coatings Technol., Vol. 53, No. 678, p. 53 (1981).
57. B. C. Hafford. Zinc Dust Metal Protective Coatings, to appear in Encyclopedia of Materials Science and Engineering, M. Bever, Editor, Pergamon Press. Scheduled for publication in 1983.
58. H. Leidheiser, Jr. J. Coatings Technol., Vol. 53, No. 678, p. 29 (1981).
59. H. Leidheiser, Jr. Prog. Organic Coatings, Vol. 7, p. 79 (1979).
60. Y. Sato. Prog. Organic Coatings, Vol. 9, p. 85 (1981).
61. R. C. Bacon, J. J. Smith, and F. M. Rugg. Ind. Eng. Chem., Vol. 40, p. 161 (1948).
62. E. M. Kinsella and J. E. O. Mayne. Br. Polym. J., Vol. 1, p. 173 (1969).
63. J. E. O. Mayne and D. J. Mills. J. Oil Colour Chem. Assoc., Vol. 58, p. 155 (1975).
64. G. Menges and W. Schneider. Kunststofftechnik, Vol. 12, No. 10, p. 265; No. 11, p. 316; No. 12, p. 343 (1973).
65. H. Leidheiser, Jr. and M. W. Kendig. Corrosion, Vol. 32, p. 69 (1976).
66. R. E. Touhsaent and H. Leidheiser, Jr. Corrosion, Vol. 28, p. 435 (1972).
67. J. M. Parks, M. C. Hughes, and H. Leidheiser, Jr. paper presented at Electrochem. Soc. Meeting, Denver, Colo., October, 15, 1981.
68. D. M. Brasher and A. H. Kingsbury. J. Appl. Chem., Vol. 4, p. 62 (1954).
69. C. P. De and V. M. Kelkar. First Intern. Congr. on Metallic Corrosion, Butterworths, London, England, p. 533 (1962).
70. D. M. Brasher and T. J. Nurse. J. Appl. Chem., Vol. 9, p. 96 (1959).
71. J. K. Gentles. J. Oil Colour Chem. Assocn., Vol. 46, p. 850 (1963).
72. R. N. Miller. Materials Protection, Vol. 7, No. 11, p. 35 (1968).
73. H. C. O'Brien. Ind. Eng. Chem., Vol. 58, No. 6, p. 45 (1966).
74. K. A. Holtzman. J. Paint Technol., Vol. 43, No. 554, p. 47 (1971).
75. J. V. Standish and H. Leidheiser, Jr. Corrosion, Vol. 36, p. 390 (1980).
76. W. W. Kittelberger. J. Phys. Colloid Chem., Vol. 53, p. 392 (1949).
77. W. W. Kittelberger and A. C. Elm. Ind. Eng. Chem., Vol. 44, p. 326 (1952).
78. Y. Sato. Denki Kagaku, Vol. 28, p. 538 (1960).
79. J. Wolstenholme. Corrosion Science, Vol. 13, p. 521 (1973).

Report Prepared by Task Group T-6D-22* on Why Coatings Fail. Issued by Unit Committee T-6D** on Application and Use of Coatings for Atmospheric Service.

Causes and Prevention of Coatings Failures

INDUSTRIAL MAINTENANCE coatings are used mainly to protect metallic structures from corrosion. Deterioration of metals caused by reaction with atmospheric contaminants creates expensive maintenance problems for industry. Coating failures may be associated with the coating itself, application of the coating, or preparation of the surface before coating application.

This report has been prepared to provide information on the various causes of coating failures and the suggested methods of preventing and remedying such failures. Unless otherwise stated, this report will consider the application of protective coatings to steel surfaces. The discussion includes barrier type coatings and does not discuss the sacrificial type zinc-rich coatings.

For a complete list of terms, refer to "Glossary of Terms Used in Maintenance Painting" in the January, 1965 MATERIALS PROTECTION. Reprints available at $2.00 each from NACE Headquarters, as NACE Publication 6D165.

Past experience has indicated that perhaps 70% of all coating failures have resulted from poor or inadequate surface preparation before application of the protective coating. Although this blame is placed on surface preparation, the real cause may be lack of proper specifications between the contractor and the customer or the absence of adequate inspection which is necessary to obtain the results desired in an effective specification. Because so many variables may be involved in a coating failure, this report discusses briefly the types of coating failures and presents some possible methods of prevention or remedy for each type.

* A. K. Long, Celanese Coatings Co., P.O. Box 1863, Louisville, Kentucky, chairman.
** R. R. Rosenthal, Jr., Carboline Co., 328 Hanley Industrial Court, St. Louis, Missouri, chairman.

Types of Coating Adhesion

Adhesion of the coating material to the substrate is one of the many variables that can cause coating failures. Obviously, the coating must adhere to the metallic substrate if the substrate is to be protected from corrosion. Any factor which prevents or reduces coating adhesion will consequently create coating failures.

Three processes can be designated by which a coating can be considered to adhere to a surface: mechanical, polar, or chemical adhesion. The relative importance of each of these three processes has been discussed at great length in the industrial coating field, but most persons experienced in maintenance coating technology agree that mechanical adhesion between the coating and the substrate is the most desirable.

Mechanical Adhesion

Mechanical adhesion is that type of adhesion which depends on surface roughness. The rougher the surface, the better the mechanical adhesion. A surface which has been abrasive blast cleaned or acid pickled under controlled conditions provides optimum surface roughness for mechanical adhesion of coating to substrate. The depth of surface roughness (distance from bottom of depressions to top of peaks) is called "anchor pattern, or anchor profile." Coatings which depend solely on mechanical adhesion require a deeper anchor pattern for proper adherence to the substrate than a coating which has got good physical and chemical adherence. Most coatings in the maintenance field will adhere well over an anchor pattern depth of 1.0 to 2.0 mils, The required coating thickness to provide satisfactory adhesion and to protect the substrate will depend on the nature of the coating and severity of the exposure.

Long-term test data or case history information is required to make this determination.

Polar Adhesion

Polar adhesion or bonding is dependent on the attraction of a resin to a substrate. Each resin acts, in effect, like a weak magnet; thus it is referred to as being the north and south poles of a magnet). The degree of attraction between the resinous binder and elements of the metallic surface determines the amount of polar adhesion. Resins such as the vinyl solution resins used in cocoon or strip coatings exhibit little or no polar bonding and accordingly can be readily stripped in sheets from a substrate. These same materials will have better adhesion to an abrasive blast cleaned or pickled surface (compared to a smooth metallic surface) due to mechanical adhesion only. Resins like the epoxy type exhibit excellent adhesion due to their great polar attraction to a metallic surface. Other resins differ widely in their polar attraction and thus in their ability to stick to a metallic surface.

Though wetting of the substrate has long been known as being essential to

Reprinted with permission from *Materials Protection*, March 1970, 32-36,

adhesion. This is based on the fact that the coating must have intimate molecular contact with the steel to gain ultimate adhesion.

Chemical Adhesion

Chemical adhesion is an actual chemical reaction between the coating and metallic substrate. Such a reaction occurs in the case of vinyl wash primers in which the presence of phosphoric acid initiates a reaction between metal, resin, and inhibitive pigment to produce a tightly adherent, corrosion resistant layer.

In the case of cement-bearing primers for galvanized iron, the presence of cement creates (in the presence of moisture) an alkaline solution which reacts with the galvanized surface to produce a degree of surface etch and consequently good adhesion.

The instances of true chemical adhesion are somewhat limited, but the examples given above indicate at least two types.

Types of Coating Failures

Ten common types of coating failures are discussed, illustrating that some failures are caused by surface preparation, improper application, and improper recommendations or specifications. The prevention of, or remedy for, each type failure is given. There are additional reasons, often obvious, for coating failure which are beyond the scope of this report including mechanical damage, excesively high or low temperatures, erosion, etc.

1. Chalking

Chalking is the term used to describe the formation of a powdery layer on a coating surface exposed to weather. Normally, this powdery layer is white in appearance, hence the designation of chalk. Colored coatings may exhibit this white powdery layer or they may have a colored powdery layer. Chalk is caused by degradation of the coating due to the action of ultraviolet light, moisture, oxygen, and chemicals. Chalk layers will consist of degradation products of the binder, pigment residues, and/or atmospheric deposits of dirt and chemicals.

The chalky layer on a coating is gradually eroded by weather. This factor becomes a definite asset in the production of self-cleaning, white exterior coating. In this case, the rate of chalking is controlled so that erosion will maintain whiteness without too rapid loss of film thickness.

Chalking is a function of the formulation, and most coatings will eventually show this phenomenon to some degree. The rate of chalking is dependent on the type of coating binder, type of pigment, and/or ratio of pigment to binder. Highly pigmented coatings will chalk more readily than underpigmented coatings. An example of the pigment control of chalking may be taken from the best white pigment, titanium dioxide. Titanium dioxide is available in two basic crystalline forms (anatase and rutile). The anatase type causes films to chalk more rapidly; the rutile type tends to resist chalking.

Remedy of this chalking condition lies in the original choice of the coating. Coating should be chosen primarily on the basis of meeting other basic requirements unless resistance to chalking is of primary importance for a given application. An example of this choice is a case where chemical and solvent resistance are the main properties desired.

Because epoxy coatings are designed for chemical and solvent resistance and not necessarily for maximum weathering properties, the user must expect the early chalking and dulling of the sheen of an epoxy base material.

Where chalking of a coating is inevitable, the prime consideration would be the rate of erosion due to chalking. Sufficient film thickness must be applied initially to provide economical service life for the coating.

Heavily chalked surfaces present a problem as far as recoatibility is concerned. This factor will be discussed later.

2. Color Fading or Color Change

Color fading or color change can be caused by reaction of either the binder or the pigment selection. With reference to binder, for example, a white phenolic may yellow whereas an acrylic would stay white longer.

Color fading or color change may be caused by chalk on the surface or by a breakdown of the colored pigment. Pigments can be decomposed or degraded by ultraviolet light (sunlight) or by reaction with chemical environments.

Lead chromate yellow is degraded sufficiently by sunlight to change the color from a bright yellow to a yellow-brown. Iron blue is degraded by its reaction with alkali.

Color fade or color change can be eliminated or diminished by proper formulation of the coating. As far as is consistent with other properties, the choice of binder and pigment can control the color change due to chalking. Fortunately, many pigments are available which are resistant to sunlight (light-fast) or resistant to chemical attack. Again, the original formulation must be balanced to provide the suitable combination of characteristics.

Choice of tinting colors is just as important as the original formulation. A good basic formulation can become a problem in color fading through improper choice of tinting colors.

In chemical environments, protection of the substrate is of greater importance than color fading. This means that color fade in many cases must be tolerated in order to obtain proper protection of a substrate.

3. Cracking, Checking, Alligatoring

Cracking, checking, and alligatoring are coating failures that can be grouped together because they are associated with the aging characteristics of a coating film. Cracking and checking are caused by shrinkage within the film during aging. Eventually the shrinkage strain within the film becomes greater than the cohesion of particles within the film, and film rupture occurs. If this film rupture occurs rather generally over the coating and in a checkerboard pattern, it is called checking. If the failure occurs in long lines, it is referred to as cracking. Either of these conditions is aggravated by excessive film thickness.

Alligatoring is a film rupture which is usually caused by application of a hard, brittle film over a more flexible film. The difference in expansion causes the brittle film to separate in a pattern which resembles the hide of an alligator. Film rupture or alligatoring can sometimes result from inadequate solvent release between coats.

Cracking or checking can be prevented by choosing a coating that is flexible enough to withstand the strain of shrinkage of a given surface in a specific service condition.

Alligatoring can be prevented generally by choosing coating formulations with similar rates of expansion and contraction when two coatings are to be applied as a protective system.

If either of the above failures have occurred without causing any deficiency in adhesion, recoating with more suitable products that will bridge the cracks may be done directly over the old coating. If the failure is associated with poor adhesion, the entire area should be cleaned to bare metal for repainting. Coating adhesion can be determined by scraping with the broad edge of a penknife.

4. Peeling, Flaking, and Delamination

Peeling, flaking, and delamination are coating failures grouped together because they are caused by poor adhesion. Flaking is a failure where the coating comes off in small flakes. Peeling is a failure where the coating comes off in large sections or sheets. When flaking or peeling occurs between coats, this failure is called delamination.

For discussion, these failures will be grouped according to cause: lack of adhesion to original metal substrate or a lack of adhesion to an older coating (delamination).

(A) Peeling and Flaking. Coating failures by peeling and flaking to bare metal are caused by lack of adhesion to the metal. This lack of adhesion may be due to poor surface preparation, poor choice or primer, poor application, too great a thickness of film, or insufficient drying time between coats.

(1) Poor surface preparation: There is no substitute for a clean surface if a good application of a protective coating is to be achieved. Moisture, dirt, oil, grease, mill scale, rust, rust scale, chemicals, and old coatings in various stages of failure can prevent proper adhesion of the prime coat. In order to properly protect a metallic substrate, the primer must be able to reach and adhere to the clean substrate. Loss of adhesion due to inadequate surface preparation may not occur immediately, but loss of adhesion will be related in time to the severity of environment and coating thickness.

One of the more sereve surface contaminants is mill scale, which is a hard, smooth surface layer formed during the hot rolling operation. Mill scale may or may not be tightly adherent and may or may not be subject to surface cracks. If moisture and oxygen penetrate these cracks or get under the mill scale, the resulting corrosion easily forces off the scale and takes the coating with it. Mill scale should always be removed unless exposure is a mild environment with low humidity.

(2) Poor choice of primer: A primer must be chosen to furnish the greatest amount of physical adhesion. Because a primer is the foundation of the coating system, its choice controls the performance of the protective system. Often, surface areas cannot be thoroughly cleaned due to such factors as contamination from nearby equipment. In this case, the chosen primer must be able to penetrate foreign materials remaining on the metallic surface. The primer must have good wettability. Primers are available with good penetration properties, but they are not and must not be considered as a replacement for proper surface cleaning.

In special cases, the surface to be coated may not be dry. In this case the chosen primer must be able to adhere to a moist surface.

Primers must be chosen which will not be softened too greatly by solvents in succeeding coats. Solvents in a topcoat might cause "lifting" of the primer with the consequent loss of adhesion to the metallic substrate.

(3) Poor application: To protect the substrate, a prime coat must be applied in a relatively pinhole-free layer of uniform thickness. Atomizing pressure that is too great at the spray nozzle can cause dry spray (particles which never knit together) to be deposited, thus increasing the porosity of the film. Dry spray can also be produced if the spray gun is held too far from the work or is "fanned" at irregular distances from the substrate being coated. This not only causes excessively dry application but also generally causes considerable overspray and lessened film thickness. Reference to manufacturer's instruction is recommended.

(4) Improper weather conditions: High humidity, low or high temperature, high winds, etc, during application can cause failure either by improper curing conditions or surface contamination.

Use of improper solvent for thinning may cause a partial resin precipitation during the drying period, resulting in poor adhesion.

Application which is not uniform produces too many weak spots where moisture and oxygen can penetrate and undermine the film. In some primers, however, inhibition occurs by reaction of moisture intruding through the topcoat with pigments in the primer.

(5) Film thickness: To yield a proper barrier against corrosion for any given coating system, the specified *film* thickness per coat and for the system should be adhered to. Too little film thickness is a major cause of coating failure.

(6) Too great a thickness of film: As it cures, a coating tends to shrink. This shrinkage builds up stresses within the coating. If these stresses exceed the adhesive power of the film, failure occurs by peeling or flaking. As the film becomes thicker due to successive coats, these internal stresses are cumulative and eventually exceed the force of adhesion.

Usually, excessive coating thickness causes adhesion problems. However, this does not mean that thin coatings are necessarily secure because coating adhesion is a balance between physical adhesion, the opposing shrinkage stresses, and strength of the substrate.

Many companies believe that, if coating failure occurs over more than 25% of the surface area by loss of adhesion, the entire area should be abrasive blast cleaned to bare metal before applying additional coating.

If this coating failure occurs over a minor amount of the surface area, the remaining area should be checked for adhesion by sharp scraping with the broad edge of a penknife. If the coating has a tendency to flake under this test, adhesion is poor; therefore, the entire area should be abrasive blast cleaned to bare metal. If the remaining coating is tightly adherent, the areas of failure may be spot-cleaned and the coating system built up from that point.

(B) Delamination. Loss of adhesion between coats can be caused by incompatibility of the coats, by contamination between coats, by too chalky a surface, by too hard drying of initial coat, as in the case of some converted epoxies, by too great a film thickness, or by dry spray.

(1) Incompatibility between coats: Choice of succeeding coatings which may not properly adhere to each other can cause delamination. This may be caused by lack of polar attraction between the resin in succeeding coats or by the solvent of the top coat not properly "biting" into the previous coat. Many cases of this type of failure result from the indiscriminate use of coatings from two or more suppliers as parts of the coating system. Unless the two coatings are of the same type or unless previous results have indicated compatibility, it is not recommended that mixed systems be used. Such failures can be prevented by following proper instructions from the supplier.

(2) Contamination between coats: Contaminants such as chemicals, dust, and moisture can collect on a coated surface and prevent proper contact between that coating and a succeeding layer. In environments in which chemical fumes and dusts are heavy, the coating system chosen should permit fast recoat, or proper precaution should be taken to prevent or remove the resultant surface contamination. Condensation of moisture on a surface in areas of high humidity is a common occurrence. Either the coating must be compatible with a damp surface or the moisture must be removed. Selection of one or two coat systems, where possible, minimizes this problem.

(3) Chalky substrates: Moderate to heavy layers of chalk on an otherwise sound but aged coating may interfere with or prevent adhesion of any succeeding coating. If the succeeding coating has the power to "wet" (penetrate) the chalky layer, the coating can bind the chalk into a firm substrate. Many coatings cannot perform this binding operation. Emulsion type coatings are not recommended over chalky substrates because they cannot wet the chalky layer; consequently, the emulsion film is anchored only to a weak, powdery outer layer. A special bonding coat should be used, or the chalky material should be removed by scrubbing.

(4) Hard dry of initial coat: For a succeeding coat to adhere to a prior coat, one of two situations must exist: either the solvents must be able to soften the initial coat to some degree or the initial coat must offer sufficient roughness to permit mechanical adhesion. Many coatings dry so slick and hard on aging that they become resistant to ordinary paint solvents. Coatings applied over this type of surface do not attain a proper degree of adhesion. Certain amine-cured epoxy coatings can cure so hard that future coatings do not adhere properly, resulting in delamination. Many structural primers are left uncoated and exposed to weather for periods to one year. These films may become too hard to insure proper adhesion of succeeding coatings. This condition can be prevented by judicious attention to the original formulation or by use of small portions of more active solvents in succeeding coatings.

(5) Too great a thickness of film: Excessive film thickness can cause not only peeling to bare metal but also peeling between coats. The shrinkage stresses can increase until they exceed the force of adhesion between coats. When this happens, the coating peels at its weakest link.

(6) Dry spray: If a succeeding coat has been applied over too much dry spray, that coating becomes a weak layer within the system, and delamination can occur at that point. Sanding to remove dry spray may be necessary.

(7) Chemical attack: If a coating system contains one or more coats which are deteriorated or softened by exposure to environments of chemicals or high humidity, failure by delamination can occur at this point of weakness. This can be prevented by proper choice of coating.

Correction of delamination failures is dependent on many factors. If the delamination occurs only in spots, each area of failure should be scraped down to adherent areas and then spot painted.

Many companies believe that, if delamination covers more than 25% of the surface area or if the knife blade test indicates poor intercoat adhesion, the entire area should be cleaned to sound coating or to bare metal. Depending on the type of delamination and its severity, weak layers often can be removed by a brush-off blast without too much damage to sound, underlying areas. It may sometimes be advisable to waterblast these areas or dry blast with one of the softer abrasive materials such as ground corncobs or walnut shell flour. In some cases it may also be desirable to allow further weathering or exposure for removal of the failed material by natural means.

5. Rusting

Rusting of iron or steel is a common occurrence. Much valuable steel has become useless rust through the action of moisture, oxygen, and chemicals. Corrosion is a broader term which covers the deterioration of all types of surfaces by reaction with their environment.

However, corrosion of metallic surfaces is an electrochemical action similar to that in a storage battery. If a bar of copper and a bar of steel are immersed in a salt solution and then connected externally by a wire, a measurable electrical current will flow just as in a battery. During this action, the iron dissolves first and then reacts with water and oxygen to deposit a film of rust on the copper bar. This electrochemical mechanism will continue until all the iron is dissolved and converted to rust.

When any two different metals are electrically connected, as above, there is a tendency for a minute but measurable current to flow. On the surface of any piece of steel are myraid small areas of varying composition. All that is needed to create a miniature battery is an electrolyte (a solution which will conduct electric current as did the salt solution above). A drop of atmospheric moisture can become this electrolyte to form a microscopic battery. This is repeated many times, and rusting of a surface occurs.

Protective coatings must prevent this corrosion mechanism on metallic surfaces by acting as a barrier to moisture and oxygen, by supplying rust-inhibitive pigments, or by providing a cathodic protection current to the substrate. Even the best coatings have some degree of permeability which permits water vapor penetration.

Failure of a coated surface may appear as (1) spot-rusting at damaged areas, (2) pinhole rusting at minute areas, (3) rust nodules breaking through the coating, and (4) underfilm rusting which eventually causes peeling or flaking of the coating.

(1) Spot-rusting at damaged areas can always occur. Most coatings cannot prevent rust formation at damaged areas. Even to zinc-rich coatings are limited as to the width of the area they can protect from rusting.

(2) Pinhole rusting appears as tiny spots of rust stain either concentrated in an area or scattered. Pinhole rusting can occur when the coating has been contaminated with aggregates such as sand, etc. Pinhole rusting may also be caused by improper application which leaves pinholes or voids in the film, or insufficient film thickness.

(3) Rust nodules breaking through the coating are an advanced stage of pinhole rusting. The expansion of rust products under the film at these small points causes a pinhole breakthrough. This would be a good sign that surface preparation was inadequate or that the primer was not retarding rusting.

(4) Underfilm rusting, flaking, and peeling involve the spreading of rust under the coating, thus creating larger areas of failure. Sometimes this spreading of rust is referred to as "undercutting" of the film. Due to the expansion of rust products, the coating may be forced off in sheets or flakes. When adjoining areas are tested by sharp scraping with a knife blade, rust is often found under the film. This condition can be caused by poor surface preparation, poor primer, low film thickness, or a combintion of these factors.

Where one of these rusting failures covers more than 25% of the surface area, many companies reblast the entire area to a degree consistent with the requirements of the new coating. There is obviously considerable variation in the degree of failure which can be tolerated before initiating repairs, depending on the aggressiveness of the environment.

When the rusting failure covers a small percentage of the surface area, the affected spots can be cleaned and touched up.

6. Blistering

Blistering is a term used to describe a coating failure in which small or large round projections or pimples appear on the coating surface. The following statements cover the various causes of this phenomenon.

(1) Blisters may be caused by exposure of the system to high humidity or water immersion. Moisture vapor passes through the film and condenses at a point of low adhesion. As moisture builds up at this point, the blister enlarges. Its size is dependent on the area of low adhesion and the amount of soluble salts, etc., which may dissolve from the film. Blisters formed in this manner may dissolve some of the rust-inhibitive pigment of the primer and become filled with discolored water. This action may be accompanied by rust if the blister is at the metal surface. If one of the coats in the paint system is sensitive to moisture, it may absorb moisture and swell. Eventually blisters form which may expand to a degree sufficient to cause peeling. Immersion in distilled or deionized water will usually cause more blistering than fresh or salt water.

(2) Blisters may be caused by inadequate release of solvent during the application and drying of the coating system. If the coating is one which forms a surface skin rather rapidly, blisters may be created by entrapped solvents trying to escape from the film. Hot sunlight would possibly exaggerate this condition. Any air, such as an exterior breeze passing over the surface during application and drying also can cause this problem. If the entrapped solvent can absorb any moisture, then high humidity will complicate the blistering.

(3) Blistering can be caused by poor surface preparation which leaves pockets in which moisture can collect and form blisters. This type of blister will eventually break open because of corrosion products formed within the blisters.

(4) Blisters can be caused by poor adhesion of the coating to the substrate or poor intercoat adhesion. In either case, poor adhesion provides potential areas where liquids and gases can collect.

(5) Blisters can be caused by a coat within the paint system which is not resistant to the environment. This coat deteriorates and leaves voids where gases and liquids can collect to form blisters.

(6) Blisters may also form when a relatively fast drying coating is applied over a relatively porous surface. When applied over a porous surface, a coating must displace air from that surface to obtain good adhesion. If the surface is dry or if the tension of the coating surface does not permit the air to pass through and be expelled, blisters can form. This condition is most easily eliminated by using a thinned formulation of the coating for the first coat on porous substrates.

(7) Blistering can be caused by either direct or stray cathodic protection currents. These currents can come from impressed or drainage current systems. In addition, blisters can be caused by the diffusion of hydrogen gas through steel in certain services. This is most often observed when storing hydrocarbon streams which contain hydrogen sulfide.

If blistering becomes serious enough to interfere with the other protective properties of a coating, the entire surface should be abrasive blast cleaned to bare metal and properly recoated. Small areas may be spot cleaned and touched up.

7. Lifting and Wrinkling

Wrinkling is a term used to describe the appearance of a paint surface which has not dried to a smooth layer, but rather into a roughly ridged surface. Normally the phenomenon is caused by rapid surface drying of the coating without a uniform drying throughout the rest of the film. This creates a difference in rates of expansion which causes the wrinkled appearance.

Lifting is the term used to describe a phenomenon which occurs when the solvent of a coat of paint softens the previous coat too readily. This again provides a difference in rates of expansion and may result in a wrinkled surface. If attack on the undercoat is too severe, loss of adhesion to the substrate may result.

In spite of its complacent appearance, a wrinkled surface can still give good protection to the substrate. The wrinkled areas may be sanded off or removed by other mechanical means and repainted.

If lifting has occured, it may be desirable to switch to a "barrier coat" to prevent reoccurrence of that effect. If large areas are affected, the best maintenance policy is to remove the coating and repaint. Abrasive blast cleaning would be effective.

8. Failures Due to Chemical Attack or Solvent Attack

When a coating is not resistant to its chemical or solvent environment, there is apparent disintegration or dissolution of the film. If the primer or intermediate coat is sensitive to the environment, the finish coat will peel off because it is no longer adhering to a strong substrate.

In other cases, failure to protect is the result of chemical attack on the metallic substrate. This is usually caused by pinholes in the film and by low film thickness. This type of failure may take the form of edge failures or spot rusting.

If a coating is being disintegrated or dissolved, the entire area should be abrasive blast cleaned and a new and more resistant system applied.

If the failure is due to pinholes or low film thickness, the procedure depends on the severity of the attack. Many industrial firms believe that, if more than 25% of the surface has been attacked, the entire area should be abrasive blast cleaned to the proper degree and a new coating system of proper thickness applied. If the failure covers a small percentage of the total surface and if the remaining coating is judged to be sound, the affected areas should be spot cleaned and spot painted. The entire area should receive at least one additional coat of material.

9. Failures Around Weld Areas

Weld areas and areas immediately adjacent to welds present considerable coating problems. Weld flux can interfere with coating adhesion and can even accelerate corrosion under the film. Weld spatter presents a relatively large projection and causes possible gaps and cavities which may not be coated adequately with protective materials.

These characteristics of weld areas may cause underfilm rusting, formation of rust nodlues (with consequent breakthrough), and film peeling.

All weld spatter is best removed by scraping or chipping before other cleaning. In severely corrosive environments, the weld area should be ground to a more uniform surface.

Removal of excess flux by blast cleaning is recommended. Abrasive blast cleaning not only removes this flux but also provides "tooth." If blasting cannot be performed in the area, weld areas should be thoroughly scrubbed with water.

10. Edge Failures

Coating failures usually begin at an edge because the coating is normally thinnest at this point. Edge failures usually take the form of rusting through the film and eventual rust creepage under the film.

Tests have shown that the incidence of failure is roughly in the ratio of 100 to 10 to 1 as film thickness over steel is increased from 1 mil to 3 mil to 5 mil.

If a coating lacks the ability to provide maximum build on edges, an extra coat should be applied to all edges.

Existing edge failures can be remedied by spot cleaning and recoating with sufficient coats to obtain the proper thickness.

Current Members of T-6D

S. H. Alexander
R. H. Bacon
J. L. Baker
J. L. Barker, Jr.
R. T. Bell
S. J. Bellassai
M. W. Belue, Jr.
R. L. Benedict
P. J. Bennett
D. M. Berger
Joseph Bigos
C. E. Bixler
D. R. Blue
D. P. Boaz
R. N. Boland
J. F. Bosich
C. C. Boyer
P. C. Briley
J. R. Brock
C. A. Bruno
K. E. Buffington
D. D. Byerley
W. F. Chandler
J. N. Chatfield
C. Christ
D. W. Christofferson
J. C. Coates
W. F. Connors
G. F. Culwell
W. B. Cunningham
J. R. Daesen
J. H. Davis
C. Silvester
E. A. Duligal
P. R. Dykes
H. C. Edwards
H. M. Edwards
O. H. Fenner
F. A. Fonner
J. M. Ford
C. E. Fox
S. C. Frye
R. E. Gackenbach
N. B. Garlock
F. W. Gartner, Jr.
C. E. Gary
P. J. Gegner
D. H. Gelfer
R. D. Goggins
J. A. Gump
R. E. Hall
I. Hamburger
H. R. Hanson
W. B. Harper
L. S. Hartman
F. P. Helms
C. B. Hutchison
E. D. Jarboe
J. R. Jeter
J. D. Keane
K. F. Keefe
T. J. Koenig
J. D. Lady
W. R. Lambert
J. W. Leggett
A. K. Long*
S. L. Lopata*
F. W. Luebke
T. R. McCarroll
H. McCranie
E. McDowell
Q. T. McGlothlin
S. W. McIlrath
G. R. McKinnell
R. W. Maier
C. R. Martinson*
W. B. Meyer
B. T. Mills
J. E. Mitchell
C. G. Munger*
J. K. Nelson
D. A. Niedt*
G. Norman
E. W. Oakes
S. J. Oechsle, Jr.
J. J. Perea
J. A. Peressin
W. S. Quimby
J. W. Redden
N. F. Reeder
D. R. Reedy
J. E. Rench
J. H. Rigdon
J. Rodgers
A. H. Roebuck
R. R. Rosenthal, Jr.
R. Savitt
J. J. Schneider
C. H. Schott
R. L. Seaman
C. M. Simpson, III
W. T. Singleton
C. W. Sisler
L. L. Sline
W. R. Standefer
H. I. Sullivan
H. B. Swanson
H. D Tarlas
A. P. C. Thomson
C. M. Toups
J. W. Tracht
N. W. Tune*
L. S. VanDelinder
R. Vickers
V. B. Volkening*
M. P. Walsh
G. F. Walther
P. E. Weaver
J. Wells
R. L. Whitehouse
D. R. Whiteman
C. L. Wills
J. M. Wise
W. A. Wood, Jr.
J. D. Worrell
J. T. Young

* Members of Task Group T-6D-22 who prepared this report.

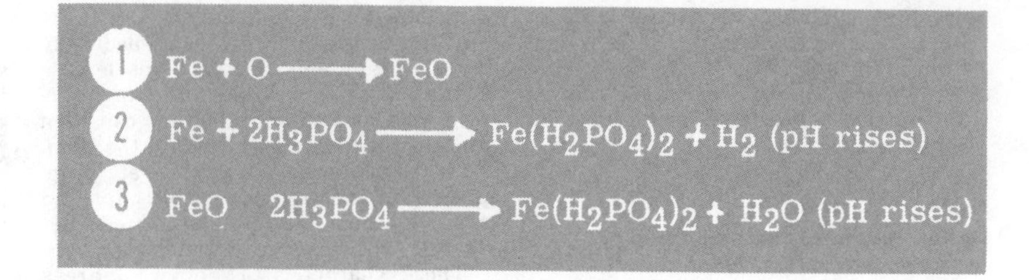

Figure 1—Three preliminary interfacial reactions that occur before precipitation when a phosphate coating is being deposited on metal.

Preparation of Metals by

L. D. Barrett

International Rustproof Company
Cleveland, Ohio

SUMMARY

Describes preparation of metallic surfaces by phosphating. Discusses the importance of cleaning for proper deposition of a quality phosphate coating, and gives four factors to be determined in choosing an effective detergent. An extensive discussion on the four types of phosphate coatings is included. The types, which are iron phosphate, crystalline zinc phosphate, microcrystalline zinc phosphate, and manganese phosphate, are compared as to physical properties, corrosion resistance characteristics, and adhesion. Advantages of using phosphate coatings are discussed.

PHOSPHATE coatings are inorganic crystalline deposits laid down uniformly on properly prepared surfaces by a chemical reaction with the treated base metal. The reaction consists in dissolving some surface metal by acid attack and then causing surface neutralization of the phosphate solution with consequent precipitation of the phosphate coating.

As the phosphate coating precipitates, it coats and blocks the active centers on the steel to present corrosive elements from attacking the metal.

Three preliminary interfacial reactions before precipitation are shown in Figure 1. Two precipitation reactions are given in Figure 2.

Zinc phosphate reacts as shown in the Figures 1 and 2 although other phosphates such as manganese, iron, calcium, or their combinations would react in a similar manner. Thus, a zinc phosphate coating consists of both iron and zinc phosphate in addition to some iron oxide from the first Equation No. 1 (Figure 2).

In an immersion bath, the ferrous iron content in the solution continuously increases in the absence of oxidizing agents, and it is apparent that not all of the iron dissolved in neutralization is formed into the coating.

In spray baths, the iron is continuously oxidized and precipitated as ferric phosphate; in immersion baths, it is allowed to build or is chemically oxidized. This iron phosphate ($FePO_4$) is the primary source of sludge in phosphate baths and cannot be eliminated completely because it is a consequence after reaction.

Metal Cleaners

The phosphate coating depends on the attack of the base metal to form the coating. Consequently, anything which interferes with this attack will influence the coating. This fact explains the great importance of cleaning in the proper deposition of a quality phosphate coating.

In the industrial field today, many complex materials, principally organic, are applied to metal before it reaches the cleaning and phosphating system. Rolling oils, rust proofing oils, paint stripping compounds, heat treating oils, and many others can be applied to metal before phosphate coating.

Most of these materials contain highly polar elements which by nature are strongly adsorbed on metallic surfaces and, as a consequence, are difficult to displace either by cleaners or phosphate solutions.

The variety of materials now being used in these compounds further complicates the problem in that some of them may be incompatible or may react with the metal cleaners used in the coating process.

There are four major factors to be determined in choosing a detergent for a particular soil and, conversely, in selecting the soil (forming compound, etc.). These are: (1) Corrosion effects of the soil on the metal

$$\text{(1)}\quad 3\,Zn(H_2PO_4)_2 \xrightarrow{\text{pH rise}} \underset{\text{coating}}{Zn_3(PO_4)_2} + 4H_3PO_4$$

$$\text{(2)}\quad Fe(H_2PO_4)_2 \xrightarrow{\text{pH rise}} \underset{\text{coating}}{FeHPO_4} + H_3PO_4$$

Figure 2—Two precipitation reactions in the phosphatizing process.

Reprinted with permission from *Materials Protection*, July 1964, 20-23, © 1964 National Association of Corrosion Engineers

PHOSPHATING★

with time, (2) ease of soil removal immediately after application and with time, (3) incompatibility effects of detergent and soil as evidenced by redeposition of soil on the metallic surface after prolonged use of the cleaner, and (4) foaming characteristics.

Some soils, either by redeposition, corrosion, or film formation, can inhibit the phosphate coating or change its character so that it is ineffective. Therefore, the soils present before phosphate coating must be known and must be evaluated carefully before using them in the plant both by themselves and in combination with other soils that might be present. Metalworking compounds should be chosen with care. No changes should be made without careful evaluation of the soils and phosphate cleaners.

Types of Phosphate Coatings

Presently, four types of phosphate coatings are in use. They are iron phosphate, crystalline zinc phosphate, microcrystalline zinc phosphate, and manganese phosphate.

Visually, the coatings are quite different. The iron phosphate is iridescent brown or blue (Figure 3). The crystalline zinc phosphate has a dull finish, usually light gray (Figure 4). Microcrystalline zinc phosphate has a shiny finish somewhat darker than the crystalline zinc coatings (Figure 5). The manganese phosphate tends to be black and coarse unless special techniques are used to control the crystal size (Figure 6).

Iron Phosphate

Iron phosphate is an extremely thin coating of about 50 milligrams per square foot. It is iridescent and is composed principally of iron oxide with small amounts of iron phosphate. Corrosion resistant properties are poor, but flexibility and adhesion are good. The coating is inexpensive and produces lower amounts of precipitate in the solution. The quality is difficult to control by visual examination on the production line. Iron phosphate is used where economy is the prime consideration and corrosion resistance is not too important.

Crystalline Zinc Phosphate

Crystalline zinc phosphate has a coating weight of about 200 milligrams per square foot. Complete coverage of the steel provides good corrosion resistance, but irregular crystal size and structure result in poor adhesion and limited flexibility (Figure 6). Quality control by visual methods on the production line can be attained by careful examination. A further disadvantage of crystalline zinc phosphate is formation of large amounts of precipitate during processing. Crystalline zinc phosphates are used where corrosion resistance rather than adhesion is important and as part of lubrication systems in the forming of metals.

Microcrystalline Zinc Phosphate

Microcrystalline zinc phosphate develops approximately the same coating weight as the crystalline zinc phosphate but is more dense in structure. The actual crystal can be seen only at a high magnification. The uniform, dense structure of microcrystalline zinc phosphate as compared to crystalline zinc accounts for its superior adhesion and corrosion resistance. In addition, the microcrystalline zinc phosphate bath produces much less sludge than that produced by the crystalline zinc phosphate bath and is easier to control visually on the production line. In its adhesion and flexibility properties under paint, it equals or surpasses the iron phosphate.

Manganese Phosphate

Manganese phosphate produces an oil adsorptive coating for mechanically mating surfaces with coating weights to 5000 milligrams per square foot. It is used almost exclusively on wearing and moving ferrous surfaces although it does impart rust proofing characteristics, particularly under oil. The coating has a dark gray to black color. Its functions are to act as a solid lubricant until the two wearing surfaces can form matched mating surfaces, to act as a porous base for liquid lubricants, and to provide corrosion resistance in conjunction with oils.

Chromic Rinses

Chromic rinses as the last stage in the coating process are used to improve the salt spray and detergent resistance of phosphated and painted systems as well as improving the salt spray resistance of unpainted phosphate coatings such as required by government specifications.

In recent years, much research has been spent on chromic rinses, and many improvements have been made in them for specific uses. Although chromic acid is still widely used, the newer modified chrome rinses are being substituted where their special properties are desired. Certain salts and mixtures of salts of chromic acid in proprietary mixtures have shown increased corrosion resistance under paint.

In addition, some of the recent developments in chromic rinses provide a metal treatment which can be followed by a water rinse without loss of the chromic improvement. The painted surface will maintain its improved salt spray resistance and its improved blister resistance to detergents. This effect can be observed with either tap water or deionized water final rinses. Both provide equally satisfactory results. The advantages of such final water rinses are that they eliminate "chrome" problems with paints that are sensitive to chromic acid.

Variables in Corrosion Results

During early laboratory research, reproducibility of corrosion results was difficult to obtain. Researchers working to eliminate variables in the process found that different lots of steel could cause variable corrosion results.

A test was conducted in which samples of eight different lots of steel were phosphated and painted in random at the same time. In testing the panels, the salt spray corrosion varied widely. Creep corrosion differed as much as 3/8 of an inch undercutting from the scribe, such variation due only to the steel lot because this was the only variable.

Comparison of Iron and Microcrystalline Zinc Phosphate

A test was conducted recently to compare iron and microcrystalline zinc phosphate in various field operations on five lots of metals from five steel mills. The results show a variation in salt spray corrosion in both systems. The iron phosphate was much poorer in all cases. Creep corrosion varied from 8.0/32 to 10.5/32 of an inch with iron phosphate but only 0.5/32 to 3.1/32 of an inch with microcrystalline zinc system.

The panels in each case were processed in a single phosphate washer for iron phosphate and a single phosphate washer for microcrystalline zinc.

Following the comparison of the two systems (iron and microcrystalline zinc) on metal from various sup-

★Paper presented at meeting of Cleveland Section, National Association of Corrosion Engineers, Cleveland, Ohio, September 17, 1963.

Source: *Materials Protection*, July 1964, 20-23

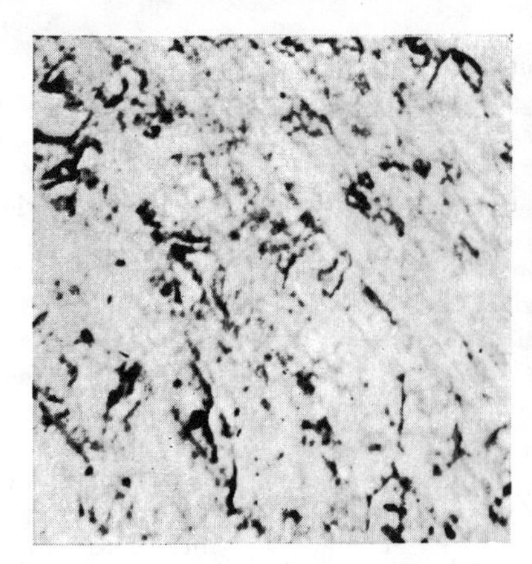

Figure 3—Iron phosphate is an extremely thin coating. It is iridescent, composed principally of iron oxide with small amounts of iron phosphate. Its color is iridescent brown or blue. (250x)

Figure 4—Crystalline zinc prosphate provides good corrosion resistance but has poor adhesion qualities. It is usually gray. (250)

pliers, a test was conducted using a single lot of steel but processing it in a number of different machines using the microcrystalline zinc system and the iron system. The results on the microcrystalline zinc system were uniform regardless of the processing machine; whereas, the iron system varied extremely from line to line. Thus, in processing lines it has been shown that microcrystalline zinc varies much less than iron phosphate irrespective of the source of the metal or the processing equipment.

Why Phosphate?

Why phosphate? The most important answer is to increase the corrosion resistance of the finishing system —that is, its resistance to damage caused by moisture undercutting of the paint film. This corrosion resistance is usually measured in salt spray tests which expose painted specimens to corroding environments. Phosphating can substantially improve the resistance of a finishing system to corrosion.

Secondly, phosphate coatings increase adhesion between the metal and paint as well as supplying a good base for adhesives used to attach rubber, cloth, and plastics to the metal. In this respect, the coatings can be described as universal bonding agents which anchor the paint and prevent blistering, peeling, and flaking. This is accomplished by providing mechanical teeth for the paint as well as a chemical surface of optimum pH for the promotion of adhesion.

Phosphate coatings also provide retained adhesion. A properly phosphated surface will keep the same paint adhesion value after prolonged aging, whereas untreated metal will experience a rapid decrease of adhesion as the painted surface ages. There is a rapid loss of bond between paint and plain steel surfaces. However, paint finishes over phosphate coatings demonstrate strong adhesion over prolonged periods of time.

One of the earliest values of phosphate coating, and still an important one, is the fact that it provides a control which insures clean metal. Microcrystalline zinc phosphate coatings on unclean metal are readily noticed to be imperfect, offering a quick and readily correctable method of detection of unclean surfaces.

In addition to their widespread use as a paint base, phosphate coatings also are used as an aid to the drawing and forming of metals. They provide anti-weld properties to protect against galling and at the same time provide a surface which more easily accepts the forming compound. The forming of innumerable steel products such as auto bumpers, wash tubs, and drawers is improved by the use of the phosphate coatings. In many cases, steels are made formable through phosphating.

Frictional and bearing surfaces such as camshafts, piston rings, valves, and bearings are mated with the help of phosphate coatings to protect against excessive wear and surface seizing during the initial break-in period.

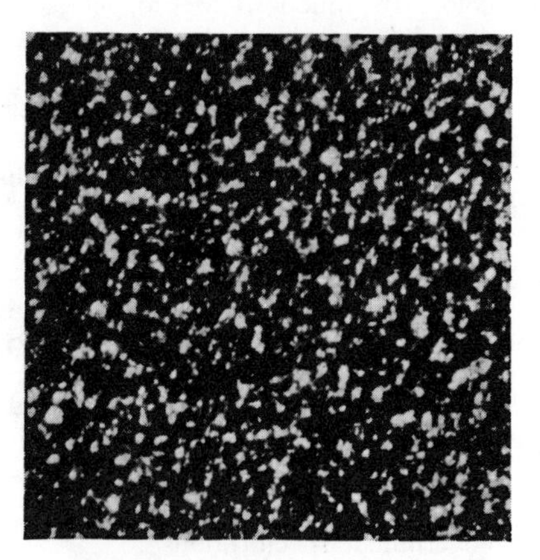

Figure 5—Microcrystalline zinc phosphate is a shiny finish and darker than crystalline zinc coatings. It provides good adhesion to the metallic substrate under paint and in some cases provides better adhesion than the iron prosphate. (250)

Figure 6—Manganese phosphate is used on wearing and moving surfaces, where it functions as a solid lubricant until the wearing surfaces are mated. It provides corrosion resistance in conjunction with oils. It is usually a gray to black color. (250)

Editors Note: Other Articles of Interest

P. D. Neumann and J. C. Griess. Stress Corrosion Cracking of Type 347 Stainless Steel and Other Alloys in High Temperature Water. *Corrosion,* **19,** 345t (1963) October.

Max Kronstein. Methods of Investigating the Characteristics of Reactive Coatings. *Corrosion,* **15,** 508t (1959) October.

Christopher D. Coppard, John H. Lawrence, F. S. Bricknell and Henry W. Adams. Chemical Surface Preparation of Structural after Erection. *Corrosion,* **13,** 141 (1957) November.

H. E. Patterson. Chemical Surface Preparation of Steel Prior to Painting. *Corrosion,* **13,** 61t (1957) January.

L. D. BARRETT is director of field engineering with International Rustproof Company. His corrosion work is primarily in phosphate coatings, electroplating, painting, and rust proof oils. He has a BS and MS from John Carroll University.

Inorganic Zinc Coatings Past, Present, and Future

C. G. MUNGER, *Consultant, Coatings & Corrosion, Fullerton, California*

Reviews early inorganic zinc coatings, their use and reactions. Includes unusual materials and methods of application as well as present day derivations of older products. Some new concepts of the chemistry and reactions of inorganic films with time are discussed including how these may influence future inorganic products.

INORGANIC ZINC coatings have been one of the technological developments of our times which have made a positive impact on society. This impact is not a dramatic one, such as television or space travel, but, nevertheless, a solid impact from the standpoint of preserving scarce materials, eliminating the need for replacement of existing structures, reducing the cost of steel structures, saving manpower, and providing new structures with a substantial increase in life expectancy.

A good example is a refined oil ship. In 1940, the life of such a ship was considered to be 15 or 20 years; the first seven years in refined oil service and the remainder in crude or black oil service. A ship constructed today of the same quality materials but protected inside and out with inorganic zinc is expected to provide 27 to 30 years of service, all transporting refined petroleum.

The Golden Gate Bridge is now coated with inorganic zinc, with a reduction in coating maintenance of several times. Offshore structures originally were reduced to lacework by corrosion in a few years. They now are expected to last the life of the wells without any loss in strength. The corrosion safety factor (increase in steel thickness over what is actually required) has been substantially reduced in the very large super tankers where inorganic zinc is properly used.

Where did all of this start? There are references as early as 1840, Mallet (a French engineer),[1] to the use of zinc for corrosion protection. There are increasing references to zinc in coatings in the early 1900's. However, it wasn't until the 1930's that anyone gave any scientific thought to making a long lasting corrosion-resistant coating from zinc dust. The man was an Australian engineer, Victor Charles Nightingall, who spent the last several years of life in studying ways by which a chemical compound could be made that would duplicate the durability of zinc ores, and, at the same time, provide long-time corrosion protection. It was Nightingall's idea that if he could form a coating which would closely simulate the chemical characteristics of Willemite or Zinc Silicate, he would be able to accomplish his goal. His goal is well stated in one of the opening paragraphs in his U.S. patent on inorganic coatings.[2]

> "Protective coatings for metal surfaces must meet several requirements if they are to be considered effective. Such coatings must be continuous and impervious to corrosive elements, must be sufficiently hard to withstand the normal abrasion and mechanical shock to which metal articles are frequently subjected, and must have a high degree of adherence to the surface of the metal article. If a coating does not adequately meet these requirements, it fails to qualify as a protective coating. This is particularly true with respect to protective coatings for ferrous metal articles where a single flaw in the coating will permit widespread corrosion and even disintegration of the ferrous article beneath the coating."

Nightingall was a man with a purpose and set out to make the zinc silicate approach to the problem work. The following is a quotation from his paper on Dimetallization in 1940.[3]

> "When the right ingredients had been determined by research in one of the zinc-iron silicate ores, it was found that a coating painted or spread on a pickled iron surface would dry in quite a short period of time, usually about half an hour; and that this coating, when dry, would have placed all the chemicals in their right proportions and molecularly closely together, but this was nothing more than a mechanical method of placing the requirements of the chemical reaction in position on the iron surface, and if in this stage, the article was again wetted with water, the coat would dissolve and the zinc washed away. To complete the chemical reaction, it was necessary to place the completed iron goods in an oven and heat to a temperature over and above 180 F (80 C), the reason for this being that to complete the molecular combination of a silica, zinc and iron, it is necessary that this should be done in the presence of a body that will bring about a release of silicic acid when a rapid silication of silica-zinc iron takes place; such a body was found to be carbon dioxide."

The first well-documented field test of this product was initiated in 1942, and was a section of steel pipe in the Woronora pipeline which is part of the water system for the city of Sydney. The line ran above ground, close to the bay and within a few feet of some large oyster beds. The pipe section was inspected in 1950 and found to be in perfect condition. It is still in existence.[4]

The Australian Morgan Whyalla all welded water pipeline, which is the famous 250 mile line in inorganic silicate history was completed in 1944, and carries a 20 year guarantee on the coating. This was done with little more than the Woronora pipe section to go on. Victor Nightingall passed away in 1948, long before the guarantee was up, and before he could see the revolution in coating technology that his ideas had created. His work, and the guarantee was carried on by another engineer with the same drive and single minded approach that he used. This was M. G. McKenzie, whose leadership in the inorganic field helped create the present worldwide use of these coatings. The Woronora pipe and the Morgan Whyalla pipeline were the start of the present era of inorganic zinc coatings.

The Morgan Whyalla waterline ran from a small pumping station at a point called Morgan on the Murray River in South Australia across the Australian bush to the Spencer Gulf and a town named Port Perie. Here, it ran around the Gulf to the site of a new steel mill at Whyalla.

The pipeline was above ground in the Australian bush country and grass lands with its large population of Emu's, kangaroos and sheep. Corrosion here was not severe, however, it was subject to brush fires which are as damaging as any in the world. Both the tall grass and eucalyptus brush are explosively flammable in the dry season. Organic paints do not stand a chance under these conditions. The pipe also runs through a reasonably humid zone along the Spencer Gulf with considerable mileage over salt marshes and salt beds where crystallization of the salt has severely spalled the concrete supports for the pipe. The exterior is subjected to practically all forms of corrosive atmosphere except direct immersion in fresh or salt water.

The coating used was made at the time the pipe was coated. The surface was pickled to remove mill scale, scrubbed with fiber brushes to remove the black pickling deposit, and rinsed with dilute phosphoric acid. As soon as this was complete, the coating

Reprinted with permission from *Materials Performance*, May 1975, 25-29, © 1975 National Association of Corrosion Engineers

was mixed by weighing about 10 pounds (9 kg) of sodium silicate and a small amount of sodium bicarbonate in a bucket. Twenty pounds (18 kg) of zinc dust and about 2 pounds (0.9 kg) of red lead followed, and the whole thing stirred vigorously with a stick. When mixed, it was then applied to the exterior pipe surface with 6 inch (15 cm) brushes. The coating was well worked onto the surface to eliminate holidays. The pipe was then moved in front of a large burner which blew flame, hot air and combustion products into the pipe at one end and out the other. This procedure raised the temperature of the steel pipe to between 300 and 500 F (150 and 260 C) and the time involved was approximately one-half hour.

As the Morgan Whyalla line was to transport water, the interior also required coating. This was done by spinning a concrete lining on the interior of the pipe. The pipe was supported on rubber belts, which merely tended to polish the zinc coating where it spun on the belt. There was apparently no loss of thickness at this point, as the coating, after 30 years exposure, showed no evidence of breakdown in these areas.

The result of the McKenzie conviction, that the zinc coating would be "permanent", was entirely borne out by the guarantee period which was passed in 1965. In 1970, the South Australian government duplicated the original Morgan Whyalla line, using the same exterior coating. A section of the line at Whyalla was inspected firsthand in 1972 at which time the pipe was in perfect condition and showed no evidence of rusting, chalking, or any change due to the long exposure to the atmosphere. This particular area was adjacent to the Spencer Gulf as well as to a steel plant. It therefore was exposed to both a mild marine atmosphere and an industrial atmosphere. Even the field welds, which had been touched up and allowed to air dry, showed no corrosion.

Following inspection of the Morgan Whyalla line in 1949, it was recommended that the Dimetallization process be brought to the United States for use here. It was recognized, that for this material to be entirely effective, the baking step had to be eliminated. The early research then was to find a way in which the coating could be formed without heat, and yet obtain all of the excellent characteristics of the zinc silicate. Many attempts were made to cure the coating by use of various salt solutions. One such attempt was to use the wash primer developed by the U.S. Navy and Union Carbide during the war, both as a cure for the zinc coating and as a primer for organic coatings to follow. Such an attempt was made on a very badly corroded naptha tank at one of the refineries on the east coast of the United States. While not perfect, the coating did prevent corrosion of the tanks for many years in a very corrosive industrial atmosphere. Some pipe was coated here in the United States in a manner similar to that used on the Morgan Whyalla line. This pipe was used as a portable transfer waterline by a construction company and was moved by trucks and bulldozers until finally both the coating and the pipe were worn out by abrasion.

There were some attempts to cure the coating on existing structures by the use of heat. Portable weed burners were used on the coating to bring about a cure. The process was clumsy and dangerous and while the steel temperature never did rise very high undoubtedly water vapor and carbon dioxide in the combustion products did help bring about a cure. This product actually lasted several years before some pinpoint rust started.

Many attempts were made to use various salt solutions. These were primarily fairly concentrated water or alcohol solutions of magnesium chloride, zinc chloride, aluminum chloride, some soluble phosphates, etc. Even washing with sea water was recommended by one manufacturer. These trials had some basis in fact, as this over simplified chemical reaction suggests:[5]

$$2Zn + 2NaCl + 3H_2O \rightarrow ZnOZnCl_2 + 2NaOH + 2H_2$$

The zinc oxy chloride or basic zinc chloride is insoluble. This compound complexed with zinc carbonate, which would surely be part of the reaction product, could provide sufficient insolubility to the silicate matrix to hold it until the zinc silicate reaction could take place. Actually none of these procedures worked satisfactorily except under very limited and controlled conditions.

It was at this point in time that one sales manager, with little imagination or foresight, as it turned out later, issued an order to his sales force to cease selling the inorganic zinc "since it was both practically and theoretically unsound."

Finally, however, it was determined that a solution of dibutylamine phosphate insolubilized the zinc-silicate coating sufficiently that the resulting product had all of the good characteristics of the stoved inorganic zinc originally conceived by Nightingall.

Many other methods have been tried in order to duplicate these original results. These include the use of different sodium silica ratios, potassium silicate, lithium silicate, ammonia silicate, various phosphates, titanates, borates, zinc oxychlorides, magnesium oxychlorides, colloidal silica, silica colloids in solvent, ethylsilicate, cellosolve silicates, and combinations of these above. There was one early coating which used a dry mixture of sodium silicate and zinc powder to which water was added at the time of application.

Much of the research effort since the time of the original coating has been to make a product which will self-cure, and this has been done effectively by many different companies. These materials have been based on potassium silicate, lithium silicate, ethylsilicate, colloidal silica and variations or combinations of these. At the present time, there is one product which is not only self-curing, but is a single package product as well. It is clear that progress has been made over the years, from a product made from individual ingredients just prior to application to the modern ready-to-use product in a can.

The principal raw materials used for the vehicle of inorganic zinc coatings at the present time are sodium silicate, potassium silicate, lithium silicate, colloidal silica solutions and various organic silicates. While this is quite a varied list of products, it is believed that ultimate reaction product as it exists on the steel surface is quite similar for all of these.

How is this possible? Inorganic zinc coatings are composed of powdered metallic zinc in a complex silicate solution. There are many combinations of silicates which are more or less effective depending on the skill in formulation; however, the basic mechanism and formation of the coating are all similar.

The first reaction is the concentration of the silicates by evaporation of most of the solvent. This provides the initial drying and primary deposition of the coating. The second reaction is insolubilization of the silicate matrix by reaction with zinc ions from the surface of the zinc particles and iron ions from the sandblasted steel surface.

Each vehicle (water-soluble or organic silicate) reacts or hydrolyzes to a form of silicic acid with which the zinc reacts to form a silica-oxygen-zinc polymer which is insoluble and forms a strong matrix with zinc powder. At the zinc silicate-steel interface, it is believed that an iron zinc-silicate complex is formed. The reaction creates a complex zinc-silicate cement around the zinc particles and an iron zinc-silicate layer at the interface of the coating and the steel. The cement, or matrix, surrounding all of the zinc globules is very inert to water, sea water, weather conditions, solvents and to many chemicals including mild acids and salts.

A third reaction which takes place in inorganic zinc coatings is one which occurs over a long period of time—in most cases, a number of months or even years. This is the reaction of the zinc and the alkali silicate vehicle with carbon dioxide from the air. Moisture from the air, or actual condensation on the inorganic surface, creates a very mild acid condition which reduces the alkali in the inorganic zinc coating over this long period. This releases additional silicic acid in the silicate polymer and allows a continued reaction with any zinc ions which may be available. This reaction proceeds gradually through the coating to the interface of the steel, increasing the adhesion of the coating to the steel surface and making the coating extremely dense and metal-like. In an ethyl silicate base coating, any alcohol group which remains on the ethyl silicate polymer would also hydrolyze to silicic acid with the zinc insolubilizing the silica polymer in a similar manner to the reactions in an alkali silicate coating. This is a very important reaction, since it increases the effectiveness and durability of the inorganiz coating with age. This is substantially different from the aging characteristics of any organic coating.

Zinc carbonate is also formed on the surface of inorganic zinc coatings and in any porosity which might exist within the film. This

Source: *Materials Performance*, May 1975, 25-29

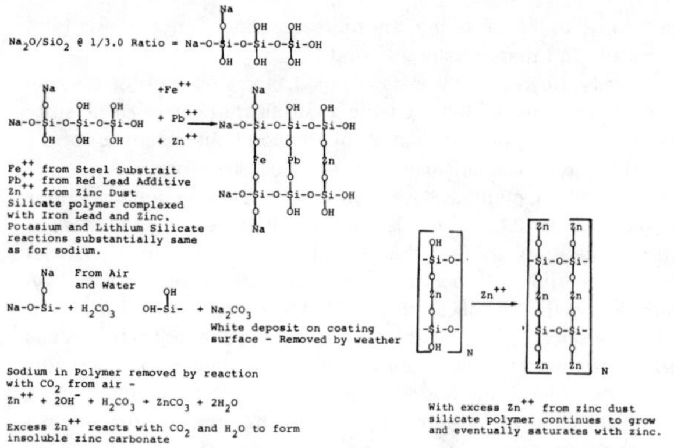

Na_2O/SiO_2 @ 1/3.0 Ratio

Fe^{++} from Steel Substrait
Pb^{++} from Red Lead Additive
Zn^{++} from Zinc Dust
Silicate polymer complexed with Iron Lead and Zinc.
Potasium and Lithium Silicate reactions substantially same as for sodium.

From Air and Water

White deposit on coating surface - Removed by weather

Sodium in Polymer removed by reaction with CO_2 from air -

$Zn^{++} + 2OH^- + H_2CO_3 \rightarrow ZnCO_3 + 2H_2O$

Excess Zn^{++} reacts with CO_2 and H_2O to form insoluble zinc carbonate

With excess Zn^{++} from zinc dust silicate polymer continues to grow and eventually saturates with zinc.

FIGURE 1 — Chemical reactions within a zinc silicate coating.

reaction also aids in increasing the overall chemical resistance of the coating as well as decreasing the possibility of pinpoint rusting.

Figures 1 and 2 indicate the changes that take place within the coating to form an insoluble matrix. No other common metal powders react to form an insoluble silicate polymer in this same way.

One little known chemical reaction which may be a factor in the insolubilizing of the inorganic zinc coatings is one which starts at the formation of zinc dust. Zinc dust is formed by distilling liquid zinc and condensing it in a large chamber. The temperature of the vaporized zinc and the temperature of the condenser controls the size of the zinc particle. The principal basic material used for the manufacture of zinc dust is either zinc dross (waste surface skimmings from the galvanizing process) or scrap die castings. In the galvanizing process, ammonium chloride is used as a fluxing agent and some of this product carried over into the zinc dross. During the distillation process, the zinc and the ammonia gas from the breakdown of the ammonium chloride and possibly nitrogen from the air react to form zinc nitride on the surface of the condensed zinc particles.

This is a little known compound which when subject to moisture can hydrolyze to zinc hydroxide and ammonia. That this actually takes place is indicated by ammonia gas (NH_3) being found in sufficient quantities to bulge tin containers of zinc powder which have been mixed and stored in humid areas at relatively high ambient temperatures [75-95 F (25-35 C)]. The zinc compounds formed are very reactive with the reactivity increased by the complexing tendency of the ammonia and zinc. Much instability of some of the mixed zinc silicate liquid coatings can be attributed to this reaction

$$Zn_2N_3 + 6H_2O \rightarrow 3Zn\ (OH)_2 + 2NH_3$$

The nitride water hydrolysis may also play a role in the original insolubilizing reactions and possibly in the continuing reactions which take place with time. As the zinc nitride hydrolyzes, zinc ions are formed which react with the silicate molecule, polymerizing it and causing it to become insoluble.

As can be seen, the chemistry of these coatings is very complex, and the effectiveness of the end product is due to the skill of the formulator and his addition of minor ingredients that insolubilize the matrix around the zinc particles. It is believed that all of the truly inorganic zinc coatings ultimately have matrices composed of heavy metal silicates. The primary heavy metal being zinc derived from ions dissolved from the zinc particles mixed into the silicate solution, either water or solvent based. Minor quantities of many heavy metals may be reacted into the silicate matrix—lead, magnesium, aluminum, calcium, barium, iron, etc. Immediate insolubility in water is the goal, with the continued long-time reaction of zinc finalizing the insolubility.

In his Dimetallization paper,[3] Nightingall theorized that many of these reactions took place.

"It was at once found on testing this particular mixture that

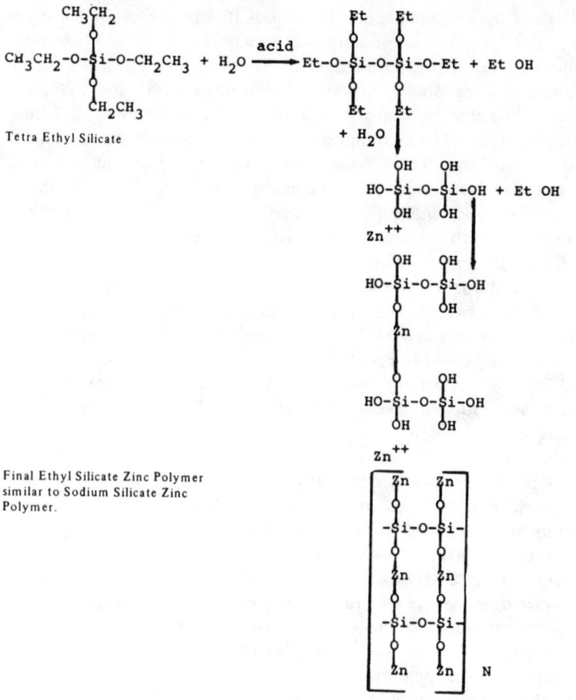

$CH_3CH_2-O-Si(OCH_2CH_3)_2-O-CH_2CH_3 + H_2O \xrightarrow{acid}$ Et-O-Si-O-Si-O-Et + Et OH

Tetra Ethyl Silicate

$+ H_2O$

HO-Si-O-Si-OH + Et OH

Zn^{++}

Final Ethyl Silicate Zinc Polymer similar to Sodium Silicate Zinc Polymer.

FIGURE 2 — Chemical reactions within an ethyl silicate zinc coating.

it laid the foundation for the development and discovery that the artificial iron zinc silica mixture would react on iron surfaces to form a body with similar characteristics to some of the zinc ore bodies found in nature. On testing iron covered with this original mixture, it was found that a time factor starting with four months produced a new phase in which zinc iron silicate and the other chemicals employed, slowly but surely formed a hard chemical combination on the iron surface, which had been covered and that this improved with age."

This reaction appears to be indicated by microscopic examination of the iron zinc silicate interface. The examination shows that the iron surface has very intimate contact with the silicate matrix with no evidence of any cleavage between the two. There also appears to be a reaction layer on the metal surface. There is also practical evidence that such a reaction must take place. Inorganic zinc silicate coatings resist undercutting by corrosion to a very substantial degree. Even where deep anodes are found after long exposure in sea water, there is no evidence of any undercutting of the coating since the zinc silicate surface extends right to the edge of the anode pit. Another indication is where an inorganic zinc coating is blasted from the steel surface. Such a surface, without visible evidence of any coating remaining still will not rust readily, even over a long period of time.

Inorganic zinc coatings have been often compared with galvanizing, however, there are many differences. While both are chemically bonded, the galvanizing by an amalgamation or mutual absorption at the iron zinc interface and the inorganic matrix by a chemical compound of iron and silica. The inorganic zinc coating also has considerably greater general as well as chemical resistance. The reasons are several:

1. The zinc particle is surrounded by and reacted with an inert zinc silicate matrix. This heavy metal silicate is a very inert binder and, except for strong acids or alkalies, is inert to most environmental conditions where coatings would be used.

2. The bond of the inorganic silicate to steel surface is a very tight one, being chemical as well as physical. It prevents the undercutting of the coating by highly corrosive atmospheres, either marine or industrial.

3. The inorganic zinc silicate coating has a controlled conductivity. This has been proven by actual current measurements where zinc was coupled to iron, and where the inorganic zinc silicate coated metal was coupled to iron. The potential of the two coating materials was essentially equivalent, while the current between the zinc and iron was much greater than the current flow between the inorganic zinc silicate coated panel and the steel panel. This indicates, of course, that the metallic zinc would go into solution much faster than the zinc held within the inorganic zinc coating.

4. In many of the inorganic zinc silicate coatings, the film is much harder and much more abrasion resistant than metallic zinc.

These points indicate a longer life span for the inorganic zinc materials as compared to an equivalent thickness of galvanized zinc. This was proven by a practical test where galvanized panels of one, two, and three ounces per square foot (3-9 g/dm^2) were placed in sea water and compared to identical panels coated with an inorganic zinc lead silicate. There were several panels on each rack with one panel remaining above the high tide and one panel fully immersed. Panels between the top and bottom were subject to the rising and falling of the tide. The panels with an inorganic zinc coating showed no evidence of corrosion after 36 months, while each of the three grades of galvanizing showed a rapid and complete breakdown in less than 2 years.

In summary, the advantages of inorganic zinc coatings are as follows:

1. The general and chemical resistance, compared to metallic zinc, is excellent.

2. The coating is unaffected by weathering—sunlight, rain, dew, ultraviolet, wide changes in temperature, bacteria, fungus, etc. Since it does not chalk or dissipate itself due to the above causes, as is usual with an organic coating, the inorganic film remains intact and with essentially the same thickness over many years in the weather.

3. They provide true cathodic protection. The inorganic matrix is conductive and allows the zinc to act in a controlled manner protecting any breaks that may occur in the coating. Being conductive and anodic to steel, the film protects small imperfections in the coating from rust formation. Eventually these minor holidays, pinholes, scratches, and scars heal by the formation of zinc ions and the resultant reaction products of zinc such as zinc carbonate.

4. The inorganic binder reacts with steel chemically in much the same manner as it reacts with zinc powder to form an insoluble zinc silicate. This reaction occurs at the mono-molecular interface of the iron and the silicate, forming a permanent chemical bond between the two.

5. As a result of this bond, the inorganic silicate coatings form a base coating or permanent primer which does not undercut or allow underfilm corrosion. Where topcoats are applied, this property cannot be over-emphasized since most failure of organic films in a corrosive atmosphere is due to undercutting of the coating at breaks or underfilm corrosion through the coating. With the inorganic base, this cannot take place—thus multiplying the effective life of the organic many times.

6. The strong permanent bond to steel surfaces and the rock-like character of the inorganic film form a base which has been proven to have excellent resistance to wear. Therefore, they may be used as a coating for faying surfaces[6] (friction interfaces between structural steel sections) on structural steel buildings, bridges, towers, etc. Faying surfaces have long been a designer's problem since ordinary coatings act as lubricants, and yet without protection, these steel joints can be seriously damaged by corrosion.

7. They are unaffected by gamma ray or neutron bombardment (flux). These coatings have been exposed to radiation up to and beyond 1×10^{10} R without change in properties.[7]

8. The completely cured surface is very hard, metallic, and abrasion resistant.

9. Heat, even at a temperature of 1000 F (540 C), has little effect on inorganic coatings. This is well above the melting point of zinc and yet the coating remains intact and fully protective. The benefits of this characteristic were shown when an oil tank coated inside and out with inorganic zinc caught fire and burned to the point of buckling the plates. The plates were rerolled and rewelded into the tank and still retained their anticorrosive surface.

10. They are unaffected by most organic solvents even including the very strong ones such as ketones, chlorinated hydrocarbons, aromatic hydrocarbons, etc. They are unaffected by petroleum base hydrocarbons such as gasoline, diesel, lube oil, jet fuel or crudes where the neutralization number is below 0.4. This is attested to by the fact that many millions of square feet have been used to line tanks and tankers for petroleum products as well as numerous tanks in the industry which contain the stronger solvents. Some of these have been in service over 15 years.

11. They may be welded without reduction in strength of the joint. This property reduces the cost in steel fabrication and at the same time prevents corrosion from starting at any area other than at the actual weld bead. Battelle Memorial Institute, as well as foreign laboratories, have confirmed this conclusion.

12. They do not shrink on drying or curing as do organic coatings. They wet the steel surface well and because of these two properties, completely follow the configuration of the surface over which they are applied. This is a great advantage in coating rough, pitted, and corroded surfaces.

With such properties, the vision of Victor Nightingall has been confirmed and his confidence justified. All of these, however, did not just happen, as much research and practical experience brought the coatings to this point. The start of the present era of inorganic coating protection actually came about when the process could be applied to existing as well as new structures. The time of in-plant processing and application of the inorganic zinc was passed and the use of these new coatings automatically expanded to massive steel structures, old or new, which required protection from aggressive atmospheres. The applications expanded from individual steel plates, sections of steel pipe, small steel fabrications, sections of steel stacks, to floating roof tanks, offshore drilling structure and production platforms, entire chemical complexes, bridges, ships, and even atomic power containment shells.

Some of the earliest tests on inorganic zinc coatings in the U.S. were located in the 80 foot lot at the International Nickel Company at Kure Beach, North Carolina. Some of the original panels are still there, with the inorganic coating still providing full protection after 23 years. These first test panels were heat cured and were essentially the same composition as the Australian material of the same period. A second test set also exposed 23 years is one of the first trials of a non-baked inorganic coating. Although some rust is now showing on the panel edges, the coating is still providing protection to the steel. The 80 foot lot at Kure Beach is recognized as an aggressive marine test location. Twenty-three years under these conditions illustrates the outstanding resistance of a single coat of inorganic zinc.

As might be expected, some of the early applications were in the marine field and the Gulf Coast with its high temperatures and high humidity provided the need for a new protective coating. Applications were primarily in the oil industry, either just onshore or offshore. Well heads and Christmas trees were some of the first to receive the coating. These pieces of equipment were subject to severe corrosion and yet small enough to make some good tests. Heater-treaters, somewhat larger in size, were next followed by other equipment on the offshore production platforms.

One of the early applications was the first "Mr. Gus". This was a large portable offshore drilling platform and was responsible for many of the offshore wells in the Gulf of Mexico. A later and possibly more spectacular drilling structure was the "Monopod" installed at Cook Inlet, Alaska, where the tides are very high and the ice in the winter continually flows past and against the structure. This platform was coated with inorganic zinc from the mean low tide line up.

The corrosion conditions encountered by offshore petroleum production platforms are the most severe encountered, and many hundreds of drilling and production structures have been coated with inorganic zinc located in the highly humid tropical areas of Indonesia, Singapore, and the Persian Gulf to the United States Gulf Coast and extending into the Arctic areas of Alaska and the North Sea. The inorganic coatings applied alone or overcoated for additional protection and for safety coloration, are providing maximum protection for these essential pieces of equipment.

Bridges, like offshore structures, are extremely vulnerable to corrosion, perhaps more so since many bridge structures are formed from structural steel shapes, with all of the corners, edges, crevices and surface defects inherent in such shapes. One of the very early bridges to be coated is a drawbridge across a tidal river in Florida. This bridge was coated in 1956, with the open grill work being the most difficult part of the structure to fully protect. It is still well protected today by the original single coat of inorganic zinc. Other bridges, such as the Bateman Bridge in Tasmania, which was coated prior to installation; the Golden Gate Bridge; bridges on the original Morgan Whyalla pipeline are all examples of the full protection provided by inorganic zinc over many years of continuous exposure.

There have been many and varied uses of inorganic zinc coatings in the marine field in addition to offshore platforms. One of the major uses has been in the lining of the interior of ship tanker tanks, primarily those transporting refined fuel. One of the oldest documented applications of inorganic zinc coatings is the No. 1 center tank in the *Utah Standard.* This was applied in 1954 to a previously heavily corroded tank surface. This tank was inspected in 1965, after 11 years, and with the exception of holidays or missed areas in the original application, there was no further rust or loss of metal in this tank. It is still in service today, without repair, and is reported to be in very nearly original condition after 20 years of continuous use carrying refined oil products.

The six "Universe" class tankers, 320,000 tons (290,000 T) each, are a good example of ships constructed in Japan protected both on the interior and exterior with inorganic zinc. These ships were first coated with an inorganic preconstruction primer and in the most critical areas with a full coat of inorganic zinc as well. The total footage coated with inorganic zinc in these six vessels was over 18 million square feet (1,700,000 m^2). After five to seven years of service, no corrosion is evident except at severely abraded areas.

As of this time, one of the large shipbuilders in Japan is starting four 470,000 ton (424,000 T) tankers. These are not only very large but because of the cost and environmental problems of sandblasting, they are attempting a new surface preparation procedure. This is followed by manufacturing the steel into large block sections and allowing them to rust. Just prior to coating, the block sections are moved into a large spray building and cleaned with phosphoric acid to remove all of the rust. A very detailed specification covering this cleaning process has been written to specify the quality of surface required. Inorganic zinc coatings are specified in order to provide the necessary adhesion and corrosion resistance. This has been determined by exposure tests of many different products. The fact that the inorganic zinc coatings are chemically reactive to steel has made this different method of surface preparation possible.

Another property of inorganic zinc has been responsible for its use on large surfaces. As previously outlined, it is unaffected by radioactivity or radiation. This being the case, inorganic zinc coatings are used to protect steel in the containment shells at most of the new atomic power plants. These structures are subject to high levels of radiation. From tests and actual exposures, it is expected that the inorganic zinc will protect the containment vessel for its entire design life of 40 years.

Inorganic zinc coatings have come a long way since their conception in the 1930's by Victor Nightingall. Their use is measured in acres rather than square feet and they have proven effective in hundreds of severely corrosive locations. Even with all this extensive use, it is believed that they are only at the starting point. They may indeed provide the means by which most steel surfaces are both protected and decorated in the future. Their qualities and characteristics are proven—it requires only imagination and research to match them with the work to be done.

References

1. Mallet, R., Brit Association Advancement Science, Vol. 10, p. 221 (1840).
2. Nightingall, Victor, U.S. Patent 2,440,969, May, 1948.
3. Nightingall, Victor, Dimetallization for the Prevention of the Corrosion of Iron, Steel and Concrete, Melbourne, Australia, 1940.
4. Munger, C. G., Report of the Inspection of Di-Met Products Used in Australia, November 17, 1949-December 20, 1949.
5. Berger, D. M., Gilbert Associates, Inc., Zinc Rich Coatings Technology, September, 1974.
6. Munse, Walter H., Static and Fatigue Tests of Bolted Connections Coated with Dimetcote, Report, March 10, 1961.
7. Munger, C. G., Coatings for Nuclear Plants, NACE Western Regional Conference, October, 1974.

Bibliography

Munger, C. G., Solvent Service Corrosion in Tanker Ships, Industrial & Engineering Chemistry (1957) July.

Munger, C. G., Interior Tanker Corrosion (1965) November 9.

Munger, C. G., A Revolution in Industrial and Marine Coating, May 22, 1967, Seventh Annual Symposium, Washington Paint Technical Group.

Gelfer, D. H., Comparison of Self-Curing and Post-Cured Inorganic Zinc Coatings as Permanent Primers for Steel. Materials Protection, Vol. 3, No. 3, p. 53 (1964) March.

Cranmer, Walter W., Modern Coatings for Tankership Compartments, 1957 Annual Tanker Conference, American Petroleum Institute.

Munger, C. G., Background Notes on Dimetcote No. 2, October, 1950.

Munger, C. G., Marine Corrosion Prevention with Inorganic Coatings, May, 1972.

New Jersey Zinc Company, Zinc Dust Metal Protective Coatings.

Nightingall, Victor, Aust. Patent 113,946.

Nightingall, Victor, U.S. Patent 2,462,763, February 22, 1949.

Nightingall, Victor, British Patent 505,710, May, 1939.

Effect of Metallic Coatings and Zinc Rich Primers on Performance of Finishing Systems for Steel

M. H. SANDLER, *U. S. Army Aberdeen Research and Development Center, Aberdeen Proving Ground, Md.*

MILITARY VEHICLES are exposed to a wide variety of corrosive climatic environments. Among the more severe exposures are salt atmospheres such as sea coast sites and humid tropical weather conditions. In November, 1965, the Coating and Chemical Laboratory was requested by the U. S. Army Tank-Automotive Command (TACOM) to conduct an exposure program to determine the effect of metallic coated steels and zinc rich primers on the corrosion behavior of finishing systems for automotive steels exposed to severe climatic conditions.

The tropical sites selected were a breakwater marine with very high atmospheric salt content, an open field, and a rain forest located at Fort Sherman, Panama Canal Zone. For temperate zone exposure, the test fence at Aberdeen Proving Ground, Md. was used. Panama is considered representative of most tropical environments, having consistently high but not extreme temperatures, high humidity, and abundant rainfall. The Fort Sherman area averages approximately 130 inches of rainfall a year with monthly means in the rainy season (May to December) from 12 to 22 inches and in the dry season (January to April) from 1.4 to 4 inches. The monthly mean temperatures range from 80 to 82 F with a daily range of 8 to 11 F. The monthly mean relative humidity ranges from 77 to 86%. Although the percentage of cloudiness is high, there are few days without some sunshine. Christobal, just across the bay from Fort Sherman, averages 6.3 hrs/day with monthly totals ranging from about 5 hrs/day in June, July, and November to about 8 hrs/day in March.

The breakwater site is situated at the junction of Limon Bay and the Caribbean Sea. The specimens at this site are exposed to constant spray of salt water with a salt fall for 1 year being calculated as 4514 lb/acre.[1] The open field site is approximately 1/2 mile inland from the breakwater and is subject primarily to rain and sun. The rain forest site is approximately 4 miles inland in the tropical evergreen forest composed basically of 3 tiers of tree growth ranging from 20 to 125 ft in height. The exposure here is primarily humidity and rain.

Note: This article is a revision of a paper of the same title presented at the 1971 NACE Western Region Conference. Copies of the original paper are available for $4.00 from *Materials Protection and Performance,* NACE, 2400 West Loop South, Houston, Texas 77027.

Test Specimens

All test specimens were 4 x 12 inch panels of the following metals:

1. Cold rolled steel, No. 20 gage (Federal Specification QQ-S-698).
2. Hot dip galvanized cold rolled steel, 20 gage, commercial quality, 1.25 oz/sq ft (0.5-1.0 mil zinc/side).
3. Electrolytic zinc coated cold rolled steel with 0.1 mil zinc plate/side. Minimum coating weight 0.10 oz/sq ft.
4. Aluminized steel, Type 1, 20 gage, hot dip coated on both sides with aluminum silicon alloy. Approximate coating weight per side 0.5 oz/sq ft (0.001 inch aluminum/side).

Surface Preparation and Painting

Surface preparation included solvent cleaning, wash primer (MIL-P-15328), and chromate conversion coatings for the plated steels and zinc phosphate (TT-C-490, Type 1), wash primer, and sandblast for the cold rolled steel. Panels with each surface preparation were primed with specification alkyd-phenolic (MIL-P-8585), vinyl (MIL-P-15930), and epoxy (MIL-P-52192, MIL-P-23377) primers and topcoated with specification semi-gloss olive drab alkyd enamel (TT-E-529) and vinyl lacquer (MIL-L-14486), resulting in a total of 8 primer-topcoat systems for each surface preparation. In addition, a group of sandblasted steel panels were primed with a specification organic zinc rich primer (MIL-P-46105) and a proprietary inorganic zinc rich primer for comparison to the plated steels. The zinc primed panels were coated with the two previously mentioned topcoats and vinyl-alkyd enamel (MIL-E-13515).

TABLE 1 – Surface Preparation and Finishes

Surface Preparation	
Solvent clean	1:1 by volume aliphatic naphtha (TT-N-95) – ethylene glycol monoethyl ether (TT-E-781).
Sand blast	SSPC-SP5-52T, blast cleaning to white metal.
MIL-P-15328	Primer (wash) pretreatment (Formula 117 for metals).
MIL-C-5541	Chemical films and chemical film materials for aluminum and aluminum alloys, Type II, Grade C, Class 2.
Chromate conversion	Proprietary for galvanizing.
TT-C-490 (Type I)	Cleaning methods and pretreatment of ferrous surfaces for organic coatings.
Primers	
MIL-P-8585	Primer coating, low moisture sensitivity.
MIL-P-15930	Primer, vinyl zinc chromate type.
MIL-P-52192	Primer coating, epoxy.
MIL-P-23377	Primer coating, epoxy polyamide, chemical and solvent resistant.
MIL-P-46105	Primer coating, weld-through, zinc rich.
Inorganic zinc rich	Proprietary.
Finish Coats	
TT-E-529	Enamel, alkyd, semi-gloss.
MIL-L-14486	Lacquer, vinyl resin, semi-gloss.
MIL-E-13515	Enamel, vinyl alkyd, semi-gloss.

The alkyd and vinyl-alkyd coated panels were prepared with and without a tie coat of wash primer. Surface preparation and finishing systems used are listed in Table 1.

The test panels were given the applicable pretreatments, and the coatings were spray applied using an automatic spray apparatus to assure film uniformity. Wash primer, MIL-C-15328, was applied to a dry film thickness between 0.3 and 0.5 mil; zinc rich primers between 2.0 and 2.5 mils; and all topcoats 0.9 to 1.1 mil except when applied over the proprietary zinc rich primer which required two coats or a thickness of 2.0 mils to obtain a uniform appearance.

Exposure

In July, 1966, the specimens were placed on exterior exposure at the four test sites. The racks at the breakwater face north in the direction of the prevailing trade winds; those in the open field and rain forest face south. All were mounted at an angle of 30 degrees. The racks at Aberdeen Proving Ground face south at an angle of 45 degrees.

Evaluation

At approximately 6 month intervals for up to 40 months and yearly thereafter, the panels were examined for corrosion and/or blistering at the score and for general surface condition and were given a rating from 0 to 5 with 0 being the worst rating. Score ratings were based on none to 1/4 inch corrosion and/or blistering from the score. Surface condition ratings were based on ASTM D 610-43 for corrosion alone and for corrosion accompanied by blistering and on ASTM Method D 714-56 for blistering alone.

In general, ratings of 5 and 4 are considered to provide satisfactory protection. Panel evaluation cannot always be clearly defined by numerical rating, especially when the condition of the specimen falls at the border of two possible ratings; thus, the number assigned is left to the judgement of the evaluator. For this reason, the rating of a specimen was not considered complete until it received the same numerical rating for 2 consecutive rating periods. This is of particular concern in ratings of 4 and 3 since the former is considered satisfactory and the latter unsatisfactory. Therefore, until two consecutive ratings were the same or lower, the specimen was considered to have the higher rating.

Reproducibility among replicate specimens was excellent in most cases. Although there are occasional exceptions to be found in the data, the scope of the program was considered sufficiently broad to show general trends and to provide a meaningful guide for the selection of suitable finishing systems. As expected from previous exposure studies,[2,3] the breakwater is by far the severest site with the major cause of failure being corrosion and/or blistering at the score. This is shown in Table 2 which lists the percent of systems with ratings less than 4 for each of the rating elements.

Test results indicate that hot dip galvanized steel properly finished will offer the most effective corrosion resistant system for severe environments such as salt atmosphere and sea coast site. This is followed in descending order by aluminized steel, zinc rich primer on cold rolled steel, electrolytic zinc, and cold rolled steel. The differences between the metallic coated steels are much less pronounced under less severe exposure with the galvanized steel, aluminized steel, and zinc rich primed cold rolled steel being generally comparable followed by the electrolytic zinc and cold rolled steel.

With regard to metal pretreatment prior to painting, wash primer was more effective with the hot dip galvanized steel than the chromate conversion coating under severe exposure of the sea coast, whereas comparable performance was noted at the other sites. The reverse of this was true for the aluminum coated steel, i.e., MIL-C-5541 chromate film was more effective than wash primer at the sea coast site.

TABLE 2 – Failure Mode After 40 Months' Exposure (Percent of 106 Systems Rated)

	Breakwater	Rain Forest	Open Field	Aberdeen Proving Ground
Score only	58	20	32	36
Score and surface	27	1	2	4
Surface only	8	9	7	4
Total all conditions	93	30	41	44

TABLE 3 – Systems Rated 4 or Better at all Sites, 40 Months' Exposure

Pretreatment	Primer	Topcoat
	Hot Dip Galvanized Steel	
MIL-P-15328	MIL-P-8585	TT-E-529
MIL-P-15328	MIL-P-8585	MIL-L-14486
MIL-P-15328	MIL-P-52192	TT-E-529
MIL-P-15328	MIL-P-52192	MIL-L-14486
Chromate conversion	MIL-P-52192	TT-E-529
	Aluminized Steel	
MIL-C-5541	MIL-P-8585	MIL-L-14486
MIL-P-15328	MIL-P-15930	MIL-L-14486

However, as indicated earlier, the hot dip galvanized substrate provided the most effective performance with 5 systems still rated 4 or better at all sites after 40 months exposure vs 2 systems utilizing aluminized steel (Table 3).

Truck Body Experiment

On the basis of the data obtained after 13 months' exposure, TACOM contracted for the experimental fabrication of three truck bodies (cab and bed) with conventional cold rolled, aluminized, and hot dip galvanized steels. They were finished with a primer topcoat system consisting of MIL-P-8585 zinc chromate primer and TT-E-529 semi-gloss enamel. Metal pretreatment included zinc phosphate TT-C-490 for the cold rolled steel, chemical conversion coating MIL-C-5541 for aluminized steel, and wash primer MIL-P-15328 for the hot dip galvanized.

The vehicles were sent to the Canal Zone and placed in operation in January, 1967. They were used daily in every day transportation of men and materials and were rotated monthly between users. In this manner, the vehicle was used for 1 month on the Atlantic coast of the isthmus where rainfall averages 130 inches/year and the atmosphere is salt laden; for another month on the Pacific side where annual rainfall averages 70 inches; the third month the vehicle was used by a crew of a special project with the vehicle being driven on both sides of the isthmus. The vehicles were used on both improved and unimproved roads and exposed to jungle conditions as well as those of a moderately industrialized, urban area. They were washed weekly with cold water only.

After completion of one year of service, the vehicles were inspected, and the following observations made:

1. The galvanized steel showed some rusting at welded areas where the galvanize had been burned away during welding. The truck bed was in excellent condition with no signs of rust even when the paint film had been damaged.

2. The aluminized steel also showed some rusting at the welded areas. The truck bed was beginning to show some signs of rust where the paint film had been damaged.

3. On the cold rolled steel, the welded areas were rusting the same as the other two vehicles. However, where paint was damaged on the truck bed and other areas, rusting was quite evident. The sill of the driver's door, where the paint had been worn away, also showed noticeable rusting. This was not the case with the other bodies.

The same trends were also noted at the 2-year inspection period.

References

1. W. B. Brierly. Atmospheric Sea-Salt Design Criteria, U. S. Army Natick Laboratories (1965) April.
2. M. H. Sandler. Tropical Exposure of Finishing Systems for Ferrous Metals, CCL Report No. 197 (1966) May.
3. M. H. Sandler and M. Cohen. Tropical Exposure of Finishing Systems for Aluminum and Magnesium, CCL Report No. 233 (1967) September.

MELVIN H. SANDLER is chief of the Paint, Varnish and Lacquer Division, U. S. Army Aberdeen Research and Development Center, Coating and Chemical Laboratory, Md. A graduate of Johns Hopkins University, he has been responsible for the development of many specialized coatings used for the preservation of army material and has served as a consultant to other governmental agencies and industries in this field. He has authored over 50 technical reports and papers and 20 Military Specifications for coatings. He is a member of the NACE, the Federation of Societies for Paint Technology, the Washington Paint Technical Group, and the Surface Preservation Section of the American Ordnance Association.

Source: *Materials Protection and Performance*, August 1972, 26-28

HOW ZINC RICH EPOXY COATING

SUMMARY

How can zinc rich epoxy coatings effectively prevent a structure with only one coat? This question is answered by an explanation of several important factors. The composition, manufacturing process, and dried film construction are discussed. A comparison of organic and inorganic zinc coatings as to performance and protection is presented. Electrical conductivity of the zinc rich coating and crosslinkage are also discussed.

ZINC RICH EPOXY COATINGS have been used as primers on numerous structures for corrosion prevention. A zinc rich primer plus one or more top-coats is required in some situations for adequate protection, but more recently the one-coat system is being used at selected locations.

The single coat zinc rich epoxy material has proven to be effective in corrosive sea water and strong solvents that are part of the environment of offshore structures, sea wall construction, and tank farm lining.

How can a zinc rich epoxy coating protect structures with only one coat? Perhaps an explanation of (1) the composition, (2) the manufacturing process, and (3) dried film characteristics will help answer this question. A comparison of inorganic vs. organic zinc coatings and a discussion of electrical conductivity and cross-linkage will also aid in answering this question.

Composition

The composition of zinc rich epoxy coatings include (in weight percent) zinc, 94%; binder, 5%; and insoluble additives, 1%. The volume figures are 72, 22½ and 5½ respectively.

The zinc is in the form of a powder having a composition of metallic zinc (97% minimum), approximately 2% zinc oxide, and a trace of iron and lead. This material provides the metallic behavior of the coating.

The binder, which consists of cured epoxy resin, holds the zinc and additives tightly in position to form the film and adheres the film to the surface being coated.

The additives are used to control conditions during the various operations of coating manufacture. The additives also provide non-settling, non-gassing and color tint of the packaged product, and aid proper spraying.

Manufacturing Process

During the manufacturing process, the zinc powder and additives are separated into individual particles, and wetted with epoxy resin solution under high speed, high shearing conditions.

The zinc powder, additives, and resin solution are prepared as a slurry in a 200 gallon batch tank. The slurry is turned by an eight inch disk (with a right angle saw tooth outer edge) rotating at 15,000 to 20,000 rpm. A vortex forms around the rotating disk, causing the slurry to assume a cylindrical shape.

As the slurry turns, the top layer folds over the edge of the disk and the resin solution is sheared onto the zinc powder and additive particles. The process continues until the specific batch properties are developed. The zinc rich epoxy is then packaged for storage or delivery.

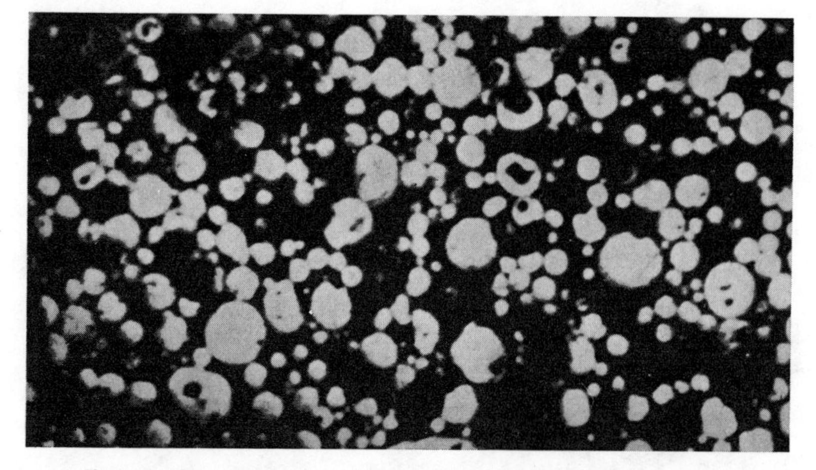

Figure 1—Photomicrograph at 500X showing contact between zinc particles.

The Dried Film

The dried film of the zinc rich epoxy coating is characterized by a network of zinc particles cemented together by a thin veil of epoxy resin that envelopes essentially each particle. The zinc particles are spherically shaped and are packed together in close random arrangement.

There is considerable free space between the closely packed zinc particles. The amount of space has been determined experimentally as being about 39% by volume with one commercial zinc powder product.

In one experiment, zinc dust, having a median particle diameter of eight microns, was stirred into 30 mls of ketone in a test tube. The zinc dust raised the meniscus of the ketone to 50 mls. The test tube was placed in a centrifuge and run until the zinc particles were well settled.

The volume of zinc with ketone filled free space measured 33 mls. The ketone volume above the zinc was 17 mls. Since the initial amount of ketone was 30 mls, 13 mls of ketone filled the free space. The percentage of free space was approximately 39% volume.

The unit volume space occupied in

PROTECT STRUCTURES

E. R. Hinden,
Reliance Universal, Inc., Houston, Texas

the dried film by each phase presented is expanded for review:

PHASES	UNIT VOL.
Insoluble Zinc Spheres......	72
Zinc Free Space............	28
Insoluble Additives	5½
Additives Free Space	Small But Not Measured
Binder	22½

In summation: (1) The binder volume is not great enough to both wet the insoluble particles and fill the free space volume, (2) The insoluble particles touch each other, and (3) The binder functions as an adhesive to wet the insoluble particles, adhere them together, and adhere them to the surface to which they are applied.

As a result the author is of the opinion that: (1) the greater the free space, the greater the coverage rate, and (2) these coatings conduct electricity because the zinc particles are conductive and they touch each other. Figure 1 is a photomicrograph at 500X showing contact between zinc particles.

Inorganic Vs. Organic Zinc Coatings

Protection of Test Panels

Nine coated steel panels, three coated with post-cured inorganic zinc, three coated with self-cured inorganic zinc and three coated with zinc rich epoxy, were used with numerous other panels in a study of the effect of the addition of lead on the rate of zinc sacrifice.

A panel from each of the three types were tested with no additive. Another three were tested with 2% weight additive, and the other three were tested with 5% weight additive. A V-shaped bare window was sandblasted in the center of each panel to evaluate galvanic characteristics and to increase the rate of zinc sacrifice. The panels were immersed for a nine month period in tap water.

The post-cured panel, which was tested with 5% additive, reacted with the environment, developing a cementous surface condition. The reaction was so extensive that galvanic qualities were dissipated and rust developed on the bare window. Also, the passivation qualities of the coating were sacrificed as verified by spot rusting in the general coating area.

The self-cured inorganic zinc coated panels developed an extensive cementous condition. A substantial amount of metallic zinc reacted with the environment to develop the cementous film condition. A small amount of pin-point rusting indicated that some of the passivation qualities of the coating were lost. Pin-point rusting was particularly apparent on the panel with no additive.

The zinc rich epoxy coated panels maintained galvanic protection of the bare window, but slow zinc reactivity was exhibited by the slight white powder, which developed on the surface.

This example demonstrates the progressive loss of galvanic and passivation protection of zinc coatings and the importance of providing a slow controlled rate of zinc reactivity to maintain long-term performance.

Evaluation of Short-Term Performance

The thin film of epoxy resin binder in zinc rich coatings controls both the long and short-term availability of active metallic zinc. Performance under immersion conditions in the Gulf of Mexico has continued for years with galvanic and passivation protection.

An evaluation of the short-term performance of inorganic zinc coatings vs. organic zinc coatings can serve as a basis for understanding the rate of reactivity of zinc epoxy coatings.

Three panels, two coated with inorganic zinc and one coated with organic zinc, were placed in a bottle of tap water for eight days. The panels were prepared with V-shaped bare windows to increase galvanic reaction.

The water was clouded by the reaction products to the extent that it was difficult to recognize the presence of the panels. The panels were removed from the test jar and examined.

Considerable zinc reactivity occurred on the panels coated with inorganic zinc, however, very slight zinc reactivity occurred on the zinc rich epoxy coated panel. All three panels showed galvanic protection of the bare window.

The extensive reactivity of available zinc in the inorganic material is a necessary function of inorganic zinc coatings for proper curing. The zinc rich epoxy is ready for service with overnight curing of the epoxy. No curing or depleting of zinc is required.

Electrical Conductivity

Zinc epoxy coatings conduct electricity. A plot of current conductivity data is shown in Figure 2. To obtain the plotted data, an electrical circuit was constructed using a battery eliminator power source connected to two steel electrodes having a surface area of approximately 16 sq. inches. The electrodes were separated by a distance of approximately 12 inches in an electrolyte consisting of .1 N sodium chloride-tap water solution.

The voltage was held constant at 1.1 v and current measurements were taken using the bare steel electrodes and later, with the cathode coated at several film thicknesses with zinc rich epoxy. Measurements were also plotted when each thickness of zinc epoxy coating

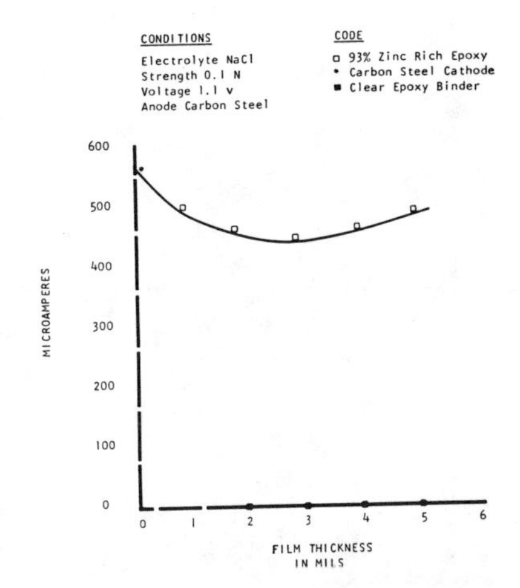

Figure 2—Electrical conductivity data of zinc rich epoxy in sodium chloride solution.

Source: *Materials Protection*, April 1968, 38-40

over two mils was topcoated with an additional one mil of epoxy binder.

The application of up to three mils of zinc epoxy coating on the cathode caused the current conductivity to drop from 580 microamps with no coating to a maximum low of 440 microamps. As more film thickness was applied, readings increased to 500 microamps with 5 mils.

The addition of one mil of cured epoxy binder resin on the surface of the cathode coated with two, three, four and five mils of zinc rich coating resulted in very little current conductivity.

This information indicates that although an insulator binder is used to prepare the coating, electricity can be conducted as a result of the amount, shape and arrangement of the zinc particles.

Current Throwing Capacity

The current throwing capacity of the zinc epoxy coating has been reported of practical value. Steel thermal panes used to heat water to 164 F (74 C) was a corrosion problem at a West Coast refinery. The water contained sulfur and sulfur reducing bacteria. Both are highly corrosive to steel.

After a five week incubation period, corrosion accelerated and within three months, the corrosion rate was 200 to 300 mils per year. Slime and organic scale developed on the panes, thus reducing heat transfer. Rectifier equipment and magnesium anodes only showed a potential drop above –.5 v.

Action was taken to correct this situation, with seven mils of zinc rich epoxy on the panes and with zinc bar anodes connected, the potential was lowered to –.9 v and corrosion was under control. Surface growth was reduced to a light slime after six months of service and the rate of zinc anode consumption was approximately 20 lbs. per 40 sq. ft.

Coating Crosslinkage

Zinc epoxy coatings crosslink so that the coatings are useful for immersion service in strong solvents. For example, a steel probe coated with zinc rich epoxy has been immersed in a test bottle of methanol, half in liquid immersion and half in vapor immersion, since May, 1963. At present, the methanol is clear, the coating is clean, and the steel is still protected.

Other solvents, such as acetone, butyl cellosolve, tap water, 40% formaldehyde, 3% salt water and ethanol (among others), were tested in the same manner and showed no blistering or flaking.

Conclusions

Zinc epoxy coatings protect with only one coat for several reasons. The composition, construction of the dried film, and the manufacturing process contribute to its protective qualities. The zinc rich epoxy coating also conducts electricity well and is being used as a tool in conjunction with cathodic protection to help throw current.

Zinc rich epoxy coating performance and protection surpass that of inorganic zinc coatings. Zinc rich coatings are useful in solvent immersion service because of crosslinkage and zinc content.

E. R. HINDEN is the technical director of the Copon Division of Reliance Universal, Inc., Houston, Texas. He received a BS in Chemistry from Michigan State University in 1955. Hinden was a guest lecturer on epoxy cellular plastics at the University of Michigan Summer Conference. He is an active member of NACE, ACS, and is past-president of the Houston Society for Paint Technology.

Panel Discussion of Inorganic Zinc Primers at Corrosion/75

(The following questions and answers were developed during and after the Toronto symposium on Zinc Coatings that was sponsored by Unit Committee T-6 "Protective Coatings and Linings".)

FOLLOWING A TWO DAY SESSION, at which 16 papers were presented, questions were fielded by the authors and 8 panelists. These and some additional questions received at the symposium are answered below. The moderator was Len Choate of Tennessee Gas Pipeline Company. The affiliations of those answering the questions are shown below:

Berger, D. M., Gilbert Associates, Inc., Reading, PA
Gelfer, D. H., Ameron Corrosion Control Div., Brea, CA
Ginsberg, T., Union Carbide Corp., Bound Brook, NJ
Helms, F. P., Union Carbide Corp., S. Charleston, WV
Hendricks, A. L., Wisconsin Protective Coating Co., Green Bay, WI
Horvick, E. W., Zinc Institute, New York, NY
Keane, J. D., Steel Structures Painting Council, Pittsburgh, PA
Masciale, Mobil Chemical Co., Edison, NJ
Metil, I., Imco Laboratories, Inc., Buffalo, NY
Montle, J. F., Carboline Co., St. Louis, MO
Munger, C. G., Coatings Consultant, Fullerton, CA
Tator, K. B., Kenneth Tator Associates, Coraopolis, PA
Wilhelm, T. P., Glidden-Durkee, Cleveland, OH

Questions and Answers

1. Q. What is the latest on conductive extenders for inorganic zinc?

A. (Metil): Theoretically there are a number of possible conductive extenders, for example: cadmium, graphite, and Ferrophos, but so far, it is the latter that looks promising, as was reported in three papers by Simko, Simpson, and Grove of Hooker Chemical Corporation. I hope to have first hand information on Ferrophos within a few months.

2. Q. In this day and age of material shortages, do you think it would be wise for a buyer to run periodic zinc content checks on large purchases of zinc primers? If so, by what method?

A. (Berger): Definitely yes, both producer and user should use the chemical method in ASTM D521. The Asarco Amalgamation method may also be used by measuring zinc oxide content by atomic absorption spectroscopy or EDTA titration.

3. Q. I've noticed that a porous or pinholed topcoat permits rapid failure of the inorganic zinc primer. Why is this?

A. (Hendricks): In a chemical environment, topcoat porosity permits easy attack of the zinc rich primer. The chemical salts formed are stronger in their reactivity than the adhesive force betwsen the topcoat and the primer. They also frequently form larger particles, *e.g.,* $ZnSO_2 \rightarrow ZnSO_3$. These salts may cause a powdery surface, therefore the topcoat would no longer have a continuous primer film to adhere to. Such failures have been noted in aggressive environments.

4. Q. In evaluating zinc primers, which is considered best; one with a fast sacrificial rate, or one that has a slow sacrificial rate? Why?

A. (Hendricks): In evaluating zinc primers, the sacrificial action is a very important consideration, and yet the test results can be deceiving. One must first consider the testing electrolyte, its conductivity, the temperature, and the frequency of surface wetting, all of which are important factors. The normal test is a comparison test in which a bare spot is left on the steel substrate. An active primer will protect the entire bare area, while a slower reacting primer will protect only the outer edges. As the test continues, a very active zinc primer will completely deplete, resulting in complete failure of the entire system, while the slower reacting primer continues to protect the substrate with the exception of a small area in the center of the bare spot. Consequently, for longer life, a slower reacting primer is preferred. It is possible to control to some degree the sacrificial action of a zinc primer by varying the type of vehicle, and by the use of topcoats. It should be pointed out that to be totally effective, a zinc primer must be sacrificial. A further important consideration in the sacrificial action of zinc primers is in total immersion service. Zinc primers should not be used in an environment where there are products which may be adversely affected by contamination with zinc particles.

5. Q. Do you think that the corrosion potentials of a zinc primer determines its degree of performance? Early test work centered around this property in grading a zinc-rich primer.

A. (Metil): In the past, potential measurements were used for determinations of durability of zinc-rich paints, probably due to work published by Craw in 1957, and Flack in 1963. Unfortunately, there are drawbacks in such methods. In my judgement, the methods are not realistic, and are too complicated. Recently I talked to a European researcher, who four years ago, used the potentiometric method (connected two panels, one protected, one bare steel) and found it impractical. Craw for example, in order to obtain results in a reasonable length of time, accelerated anodic dissolution of the zinc in the coating by impressing a current of 16 mA/ft^2 on the system. The quantity of zinc consumed was then calculated from: (1) The total current passed from the start of the test to the time when the potential of the coated panel had risen to -0.75 volt; (2) the weight of the zinc, in the immersed area of the paint coating; and (3) Faraday's Law. It should be kept in mind that one of the basic requirements for this test is that the paint film be in intimate electrical contact with the substrate. Craw found that the percentage of available zinc was higher when a paint was applied to sandblasted steel than when applied to smooth steel panels. It appears that electrical methods for evaluating zinc-rich coatings in general are not too simple, even when electrical conductivity is used for establishing the amount of zinc needed for galvanic protection. A lot of work has been done by Pass, Elm, and Lantz, who found the minimum resistivity, at different percentages of zinc in the primer: 95, 90, and 85, respectively. Actually, Elm called attention to the fact that there are experimental difficulties associated with performing conductivity tests.

6. Q. **Are there any other coating systems that will work as well as zinc primer systems have in the Gulf of Mexico environment? If so, what are they?**

A. **(Wilhelm):** I have had personal experience with a major oil company operating a refinery in the Beaumont, Texas area, and terminal facilities in Pensacola, Mobil, Tampa, Port Everglades, Savannah, and Jacksonville. This company consistently received 8 to 10 years service with a three coat alkyd system on cone type tanks before touch-up with a refresher coat was required. Surface preparation was commercial or better, the primer was a combination of red lead, zinc chromate, iron oxide in an alkyd vehicle at 1-1/2 mils, and this was followed with two coats of alkyd tank and structural enamel at 2 mils each. Total dry film thickness of the job was 5-1/2 to 6 mils. There were several cases where even longer life (12 years) was received when the same primer was topcoated with a two compartment phenolic aluminum. Where floating roof tanks existed, a three coat vinyl system was utilized. The reason for this kind of performance was no accident. Good inspection was provided by the owner, and only reliable contractors were permitted to bid after a prebidding conference, followed by a conference with the successful bidder before the job started.

7. Q. **What is weldability of steel coated with: (1) 3 mils dry film thickness inorganic zinc? and (2) 3/4 mil dry film thickness inorganic zinc?**

A. **(Gelfer):** (a) Welding through 3 mils dry film thickness inorganic zinc has been satisfactorily done using manual "stick" welding techniques. However, machine welding, particularly submerged arc welding, is not recommended because of the danger of zinc inclusions in the weld, and formation of weld porosities, leading to inadequate weld quality, and (2) welding through 3/4 mil dry film thickness inorganic zinc is accepted good practice using both manual and machine welding procedures, where the zinc coating has been specifically designed to function as a "thin-film" weld-through primer. A number of such products are commercially available; some meet stringent weld quality standards such as the U.S. Navy 0900-000-1000, ASME Welding Qualification, and others.

8. Q. **What is weldability of stainless steel coated with inorganic zinc?**

A. **(Gelfer):** This is not considered proper practice, because of the high probability of weld embrittlement, with resultant reduced corrosion resistance in the weld area.

9. Q. **Which of the following two systems do you think would give the best performance on an offshore platform in the Gulf of Mexico: (1) A 3 mil organic zinc primer (with a 60 to 70% zinc loading) overcoated with 4 to 6 mils of high build epoxy or vinyl, or (2) a 3 mil inorganic zinc primer (with 90% zinc loading) with no topcoat.**

A. **(Gelfer):** If the 3 mil inorganic zinc used without topcoat is a postcured product, then superior performance and longevity can be expected. Case histories demonstrate 15+ years protection from postcured inorganic zinc without topcoat on offshore platform areas above the splash zone. This longevity is obtained without need for maintenance other than to mechanically damaged areas. An organic zinc primer containing 60 to 70% zinc by weight, when topcoated as defined, would be expected to show significant need for maintenance in 3 to 5 years. A longer-life system would consist of a self-curing inorganic zinc primer, at 75 to 80% zinc loading by weight, overcoated with 4 to 6 mils of epoxy or vinyl. Longevity of postcured inorganic zinc is greater than either of these systems, and can be further enhanced by topcoating with high build vinyl or epoxy.

10. Q. **Do inorganic zinc coatings without red lead show pitting in sea water immersion?**

A. **(Gelfer):** Depending on formulation, zincs without red lead can show varying rates of pitting in sea water immersion. Onset of pitting or spot failure is more related to nature of the immersion, *i.e.,* whether it is continuous or intermittent in character. Where red lead-containing inorganics perform well in immersion service, it is considered to be due to the formation of high insoluble lead-zinc silicate complexes that resist attack by the chlorides in sea water.

11. Q. **You stated that epoxies tend to have better adhesion to epoxy zincs than to inorganic zincs. We have found that most epoxies have adhesive strength to inorganic zincs which exceed the cohesive strength of the zinc primer—regardless of overspray, dry spray, etc. Could you be referring to cohesive rather than adhesive strength?**

A. **(Tator):** In my experience, the adhesion of epoxy topcoats to epoxy zinc-rich primers is generally better in all respects than an epoxy topcoat to an inorganic zinc-rich primer. The key words here are "in all respects", as in my opinion, bubbling, blistering, and even to some extent pinholing, are localized ramifications of loss of adhesion—for example, within a blister, the topcoat is not adhering to the underlying surface. Experience also shows that there are fewer problems in the field when applying an epoxy topcoat over an epoxy zinc-rich primer. However, this does not mean that epoxies do not adhere extremely well and tightly to inorganic zinc-rich primers. In fact, as you mentioned, adhesion is excellent, and when breaks or delamination do occasionally occur, it is usually cohesively within the inorganic zinc-rich primer. In the vast majority of cases, adhesion of topcoats to organic or inorganic zinc-rich primers is entirely adequate, and even if pinholes, small blisters, or bubbles should develop, for atmospheric environments, they are generally of no consequence and do not detract from the protective life of the coating system.

12. Q. **Re Union Carbide Program—Are the types of vinyl resins used in coatings "A" and "B" known? Do you feel the problem could just have been due to incorrect resin selection in coating "B"?**

A. **(Tator):** Characterization of the vinyl resin in paint "B" indicated it was probably based on a mixture of VYHH resin and VMCH resins in an 80 to 20 ratio. Paint "A" was probably based on the VYHD type tesin. When the wrong resin combinations are used, the adhesion of a vinyl topcoat applied over a zinc-rich primer is worst shortly after application; it improved somewhat with time. Both paints "A" and "B" showed good initial adhesion, and paint "A" is still adhering. A wash primer applied to the zinc prior to vinyl topcoating would certainly have improved the adhesion of paint "B". However, it is believed the carboxylic modification to the VMCH portion of the resin mixture should be sufficient to provide long-term adhesion to an underlying zinc-rich primer. Furthermore, the mode of failure of paint "B" was not typical of coating incompatibility. Rather, the failure was caused by the formation of crystalline zinc salts (probably zinc sulfide) which was tightly bound to the vinyl paint. It is theorized the formation of these salts at the interface between the vinyl and the zinc-rich primer caused the resulting loss of adhesion—and not incompatibility of the vinyl resins.

13. Q. **Is there any problem of inorganic zinc adhesion over inorganic zinc second coat after 24, 38, 72 etc., hours?**

A. **(Masciale):** If the zinc is fully covered, a "sweep" or "brush" blast is recommended to secure maximum intercoat adhesion. Inorganic zinc does not possess good wettability, and the film must not be under-cured or over-cured if blasting is to be omitted. While there are many instances where good results have been obtained with no special treatment of the first coat surface, there have been many failures. It is always a gamble not to blast between coats.

14. Q. **Do you believe that the zinc dust particles in an inorganic zinc-rich primer are more encapsulated by the binder than in an inorganic system, assuming the zinc dust and its concentration are the same in each?**

A. **(Masciale):** Yes, there is more encapsulation in the case of the inorganic binder. The inorganic vehicle is heavier, and occupies less volume.

15. Q. Are there any other coating systems that will work as well as zinc primer systems in the Gulf of Mexico environment?

A. (Wilhelm): If management is willing to provide sufficient funds for the project, a zinc-rich primer system really can't be beaten for longevity. But, in selecting a coating system, one has to be sensitive to all the concerns of management. The coating system and surface preparation should be matched to the environment within the limitations of the facility and the amount of money management is willing to provide for the project. I have personally seen and have been involved in terminal storage tank painting along the Gulf Coast in Beaumont, Texas, Tampa, Florida, Mobile, Alabama, and Pensacola, Florida, where three coat alkyd systems have provided seven years life before a refresher coat was required. This just didn't happen. Good specifications were written for surface preparation and film thickness, contractors were selected on reputation and inspection was provided to enforce the specifications. The specification included a commercial blast, a multi-pigmented primer at 2 mils, alkyd intermediate at 2 mils, and a self-cleaning white alkyd at 2 mils dry film thickness. We certainly do not wish to eliminate alkyds from the coating engineers stable of protective coatings. They have been the workhorse of the coating industry for years.

16. Q. In combining wash primers with vinyls as topcoats for inorganic zincs, I understand there are possibilities of incompatibility of WP to IZ, and vinyl to WP. How can we assure ourselves of proper materials?

A. (Keane): Empirical experience has shown that compatibility of many topcoats over typical inorganic zinc-rich primers can be aided by the use of primers and other types of tie-coats. Although the WP-1 wash primer is perhaps the most widely used tie-coat, it should not be specified without either the manufacturer's recommendation *or* prior compatibility tests. The best compatibility between wash primer and a vinyl topcoat is obtained when at least half of the vinyl resin is of the hydroxylated type (examples: vinyl resins designated as VAGH, VAGD, etc.). Because of this, some people use vinyl paints such as SSPC Paint 9 which has half of the VAGH and half VMCH primer, permitting it to adhere to the wash primer or, if necessary, to the inorganic zinc or the steel itself. Recommendations should also be followed with regard to weathering and cleaning, if any, between coats.

17. Q. What effect does anchor pattern and surface cleanliness have on the performance of inorganic zinc and how do you measure that profile and cleanliness?

A. (Keane): Surface preparation requirements vary considerably for the various types of zinc-rich paints. The range is discussed in the forthcoming new SSPC specification covering both organic and inorganic coatings. In general, however, the inorganic zincs require a "near white blast" or better (SSPC-SP 10 or NACE 2) although some manufacturers specify "white metal", on the one hand, or a commercial (or even intact millscale!) on the other for their particular products. Among the inorganics, the water thinnable products, strangely enough, seem to be somewhat less tolerant of water on the surface at the time of application, making it less applicable over freshly wet blasted steel. If possible, wet-blasted steel should be dried before priming, and inhibitor treatment is seldom necessary. Many zinc rich plants seem to require both a maximum and a minimum "tooth" in order to develop best adhesion. They tend, however, to be more tolerant of high surface roughness than other paints which do not have the capability of sacrificial protection. It is misleading to specify profile at all unless the method of measuring is specified. The SSPC has developed, and is now drafting, a specification for measuring profile (based on microscopy and alternate methods). In the meantime, however, it is quite satisfactory to use a simple visual profile comparator. (A corrected depth gage or a magnetic gage with a calibration factor may also be used, but most other available methods are subject to serious misuse.) When an approved profile measuring method is specified, a 1 to 3 mil profile height would probably cover the range recommended for most of the zinc-rich products.

18. Q. Compare hot dip galvanizing and inorganic zinc coating as concerns: (1) Longevity in coastal environment, and (2) cost.

A. (Horvick): There are considerable published data on the corrosion behavior of galvanized steel in coastal environments. Unfortunately, zinc-rich coatings do not have comparable test data, but case histories of actual applications show that performance is quite comparable when zinc coatings are of the same order of thickness in the galvanized and in the zinc-rich form. Cost comparison is a bit difficult because galvanizing charges are on a per pound of coated steel basis, whereas zinc-rich coatings are charged on a square foot basis. In general, medium heavy steel structures are more inexpensively coated by hot dip galvanizing, whereas very heavy structures might find zinc-rich coatings less expensive. A number of formulae and charts have been developed to establish relative lines of demarkation showing where the cost advantage lies and can be supplied upon request.

19. Q. What is the influence of zinc particle size on performance?

A. (Horvick): It is the feeling that particle size distribution is of the greatest significance in order to derive best results from zinc-rich coatings. Knowledgeable zinc-rich coatings manufacturers require the particle size distribution to be within a rather narrow range with given curves to define the parameters. The reason for this is that in order to provide maximum particle to particle contact for conductivity, the best geometry is provided by particles of similar dimension. More information can be furnished on this subject if desired, but one suspects that for a Q/A write up, the answer should be brief. There is a difference of opinion as to what the optimum particle of size should be. Some feel that smaller size particles around one micrometer (1 μm), is best, while others feel it should be a large multiple of this. The consensus I have derived is that best results are obtained with a particle size between 5 and 7 μm.

20. Q. What types of inorganic resin are used for nonzinc topcoats with aluminum and stainless steel?

A. (Montle): Our experience has been with the use of a partially hydrolyzed ethyl silicate vehicle; other suppliers may have worked with other vehicles.

21. Q. If you apply inorganic zincs over mill scale, what happens when thermal expansion takes place? You mentioned only that mill scale can be removed by rust, causing it to pop off—there are many more phenomena, vibration, impact, and expansion.

A. (Montle): We are aware of the fact that mill scale can be damaged through vibration, impact abuse, etc., thus causing the mill scale to loosen. Thus, all of the applications we have had to date over mill scale have been in the field and not on shop-coated steel. Thus, the damage from vibration and impact is limited to the normal amount of abuse as what would be encountered in a typical chemical plant. Structures coated would typically be structural steel and tank exteriors. Topcoat use is restricted to thermoplastic coatings or thin epoxies, to avoid excessive stress being placed on the primer from the shrinkage of the topcoat. Under these applications, we have had several years of successful application in typical Gulf Coast and East Coast chemical plants, where the temperature differential from the environment could well vary over 100 F (56 C) from winter to summer. Under this amount of thermal cycling, and typical field abuse, we have observed minimal mill scale loss or corrosion. If the mill scale coated steel were to be subjected to severe thermal shock, it is quite likely some damage would occur. However, under the range of ambient temperatures, since the thermal coefficient of expansion of mill scale is only about 13% more than that of steel (8.5 vs 7-8 x 10^{-6} in/in/F), one would expect little popping of the mill scale under normal atmospheric temperature variation.

22. Q. How many years of successful service do you have over mill scale?

A. (Montle): In actual large scale field applications, our experience is from 8 to 12 years in plants on the East Coast and on the Gulf Coast. All of these are still in excellent condition.

Source: *Materials Performance*, May 1976, 35-38

23. Q. What might be the cause of adhesion loss of a coal tar epoxy applied over inorganic zinc—is this an application problem or possibly due to climatic conditions in northern United States?

A. (Montle): A frequent cause for the loss of adhesion of coal tar epoxy over inorganic zinc is one of excessive thickness of the coal tar epoxy. While there will be considerable differences in the basic adhesion of different coal tar epoxy formulations (with the polyamine cured being far more tolerant and the amine cured less tolerant), high film thickness is the chief cause. If high film thicknesses are coupled with poor air ventilation, as might be encountered in restricted areas, and if further compounded by cool temperatures, difficulty in obtaining adhesion will likely be encountered. Frequently if the coal tar epoxy is applied too thick, under poor air circulation, and low temperatures, adhesion will recover if warm dry air is blown over the surfaces sufficiently long to remove all the residual solvent and complete the cure.

24. Q. On a day-to-day basis, what would be a general recommendation for topcoating inorganic zinc: one high build coating, a mist coat-full coat, or a thinned down first coat?

A. (Montle): My feeling is that the day-to-day basis should be the use of one high build zinc. However, you should be aware of the fact that even with a coating with proven performance, which could well be successful 95% of the time, there will be those surfaces of inorganic zinc which on a particular day, can cause bubbling of topcoat. If bubbling is encountered, the use of a mist coat followed by a full coat will generally eliminate any topcoat bubbling.

25. Q. I have an application of a single coat of inorganic zinc that is now two years old; the environment would be considered moderate (normal-industrial atmosphere weathering, very light concentrations of chemical fumes). Dirt, and I fear, some chemical contamination has collected on the zinc surface and it has taken on a whitish molded appearance. I now wish to topcoat the zinc with a high build vinyl. What surface preparation do you recommend?

A. (Helms): I recommend high pressure water blast—the pressure being determined by the difficulty of removal of the dirt.

26. Q. What are the comparative costs of a zinc paint formulated with all zinc versus one with a ferrophos substitution?

A. (Montle): This, of course, is like answering how long is a piece of rope, as the cost would depend upon the quantity of ferrophos substitution. As a general response, based on a typical amount of substitution, the cost savings would be in the order of magnitude of 10%.

27. Q. Is there documented in service proof that an inorganic zinc-rich coating at 75% zinc in the dried film will perform less satisfactorily than an 85% zinc in the dried film? If so, what is the optimum content of zinc in the dried film?

A. (Montle): I am not really sure what the questioner means by documented proof, so will answer based on our experience. The performance of inorganic zincs at any level is surprisingly quite dependent upon the rest of the formulation. We have found that it is possible to manufacture inorganic zinc coatings with 75% zinc in the dried film that will perform in most applications, equal or superior to those with higher zinc loadings. Because there is less zinc in the dried film, when exposed untopcoated in severe demand areas, there is a performance advantage to the higher zinc loading. However, for exposure in less than the most severe exposures, or when exposed with a protective topcoat, the 75% zinc in the dried film has demonstrated performance equal to or superior to the higher zinc loading. While laboratory data has indicated equivalent or good performance from even lower zinc loadings, we feel that because of the variations in application under field conditions, that the 75% zinc loading is optimum.

28. Q. What pH limitations, topcoated and untopcoated should be recommended with inorganic zincs?

A. (Metil): Since metallic zinc is amphoteric and reacts with both acids and alkalis, zinc rich coatings must be topcoated for environments with a pH below 6 and above 12.5. Obviously, the topcoat does not change the nature of the zinc, so the performance depends on the quality of the topcoat.

29. Q. Has there been any answer to the problem of adhesion loss of some vinyl topcoats over some zinc-rich primers?

A. (Ginsberg): Topcoating zinc-rich with vinyls or *any* other type of generic coating remains one of the thorniest problems in maintenance painting. Not because the variables encountered in the field are so many that the paint technologist cannot cope with all at the same time. A solution to one aspect of the problem may not be compatible with the requirements to another problem. I believe that the question is poorly worded because it implies that it is a universal problem (it is not), and it implies that all problem cases are similar (they are not). The best answer I can give is that there are specific solutions to specific problems, but no one has yet found an infallible panacea.

30. Q. How do we determine the optimum time to topcoat an inorganic zinc-rich coating for atmospheric service? For immersion service?

A. (Berger): Weathering for at least 6 months is recommended, but the surface may have to be cleaned before topcoating. Normally, inorganic zincs should not be topcoated for immersion service.

31. Q. Has any consideration been given to possible performance variations between inorganic zinc on Corten or A.441, and on A.36? Certainly fabricators are finding some difficulty with profile achievement with these corrosion resistant steels, *i.e.,* SP No. 5 or No. 10 are very difficult to achieve at profiles above 3/4 or 1/2 mil?

A. (Munger): Inorganic zinc has been applied to Corten type steels in a number of instances. One that comes to mind is a Corten bridge for unloading coal at a fossil fuel plant in Georgia. The application was satisfactory. Regarding profile—3/4 to 1/2 mil is satisfactory for a number of inorganic zinc coatings. It's my opinion that there are some water base and some solvent base inorganic coatings that can be applied over a surface as smooth as this. In order to make a positive determination as to just which manufacturers would be best, trials should be made.

32. Q. I have heard of a possibility of reversal of zinc-steel polarity under certain conditions of temperature and moisture. What really happens?

A. (Munger): It is pretty common knowledge that this takes place, and it is the reason why magnesium anodes are always used in hot water tanks rather than zinc. The subject is covered in Evans' book, "The Corrosion and Oxidation of Metals" on page 639. He indicated that at high temperatures, the zinc oxide layer on the zinc surface becomes a better conductor and because of this, the potential of the zinc rises above iron at that point. Zinc becomes cathodic and is incapable of affording protection to the iron. The inorganic zinc coatings are not recommended for high temperature water as they would tend to react in much the same way as the metallic zinc.

Zinc-rich Coatings for Marine Application

Richard W. Drisko
Naval Civil Engineering Lab
Port Hueneme, California

SUMMARY

Zinc-rich coatings are generally tough and abrasion resistant. Protection of the substrate from corrosion results as the coating functions as a barrier and as heavy zinc loading imparts cathodic protection. Additional protection may be received by topcoating. Some zinc inorganics harden by post-curing with acid; others self-cure. This paper discusses the characteristics of zinc-rich coatings and methods of application.

THE USE of zinc-rich coatings for protecting steel structures from corrosion has expanded tremendously since these unique coatings were introduced to the United States in the early 1950's. Although they originally received limited application in the marine field, such use is now becoming increasingly important. Because of the many different types of zinc-rich coatings presently available, the consumer is sometimes at a loss to know precisely when use of a zinc-rich system is appropriate and which is the best choice for his particular needs. This paper will discuss the nature of the many types of zinc-rich formulations presently marketed and the mechanism by which they impart protection.

Zinc-rich coatings are generally placed into one of two broad categories, inorganic and organic, depending upon the chemical nature of the binder used to bond the zinc particles to each other and to the steel substrate. Both types require that the coatings be electrically conductive so that cathodic protection can be imparted to the substrate by the reaction $Zn = Zn^{++} + 2e$. The binders of both types do not have good electrical conductivity, and so the coatings must rely upon heavy loading of finely divided zinc particles in contact with each other to provide this continuity. For optimum protection, the highest volume of zinc dust a coating will tolerate and still retain good application properties is desirable. This is usually 85 to 90% for organics and 90 to 95% by weight for dry films of inorganic coatings. Properties of these coatings are more closely related to pigment volume concentrations than to weight loadings.

Conventional protective coatings usually impart much of their protection by acting as barriers preventing the water, oxygen, and electrolyte necessary for corrosion from coming into contact with the underlying metal substrate. Inorganic zinc coatings, however, are quite porous and, at least initially, this porosity may be advantageous in cathodic protection of the substrate. In order to provide satisfactory, long lasting protection in this manner, the coating system must (1) contain sufficient high purity zinc in a condition available to act as a sacrificial anode, (2) have good electrical continuity between individual particles and the substrate, and (3) have sufficient electrolyte present to complete the electrical circuit of the cathodic protection system. Heavy zinc loading fulfills the first two requirements, and the porosity of the coating permits atmospheric (or immersed) water to fulfill the third requirement. As zinc metal is lost in the cathodic protection process, basic zinc compounds resulting from subsequent reactions of the zinc ions form a tough passive film that stifles further loss of zinc and provides a good barrier against corrosion. Should this barrier become damaged by impact or abrasion, the exposed metal will again receive cathodic protection until the barrier film has been healed by the above process.

The first successful zinc inorganic coatings utilized aqueous solutions of alkali (Na, K, Li) silicates. These are sprayed onto a steel substrate and later cured with an acid curing solution. This post-curing results in a hard, abrasion resistant, tightly adhering film. Such curing is rapid and effective under a variety of conditions. Post-cured zinc inorganics have a long history of successful performance.

Later, alkali silicate formulations were devised so that the reaction of zinc, silicate, and water occurred to give a hard, insoluble film after evaporation of the aqueous solvent without the need of an acid curing agent. Curing in this manner does not result in a film identical to that obtained by acid curing. Because of the nature of the reactions involved in curing, temperatures above 45 F (7 C) and relative humidities above 50% are generally necessary for a complete cure.

A later development was a self-curing inorganic zinc coating utilizing the hydrolysis of an alkyl (usually ethyl) silicate as the curing mechanism. Such formulations depend upon evaporation of the organic solvent and further hydrolysis of alkyl groups in the ester to form a completely inorganic binder. Because the rates of reactions involved in curing are dependent upon temperature and humidity, it is sometimes necessary to vary the extent of hydrolysis of the alkyl silicate according to local environmental conditions. Coatings cured in this manner are generally not so hard as alkali silicate coatings.

Self-curing zinc inorganics may be slightly less costly initially than those requiring post-curing because of the extra step necessary for curing, although cost per year of protection may be greater. They also have the advantage of simpler application in the mass production of shop primed structural steel, because an additional curing step is not required. Automatic systems of blast cleaning steel are generally used in conjunction with and prior to application of shop primers. Structural steel for marine use has been shop

"Zinc-rich coatings are generally tough . . ."

primed in this manner and received adequate protection up to five years before being overcoated with a polyamide-epoxy or a high build vinyl system for additional protection of the steel. Such permanently primed steel may also be welded, but the welding must proceed slower than usual and with caution. This permits the zinc metal to be released as a white vapor and results in a smooth weld with little spatter. Since zinc is toxic, good ventilation is required during welding to prevent such effects as zinc shingles to workmen in the area.

Inorganic zinc phosphate coatings are water-based products the curing of which is highly dependent upon temperature and humidity. They are used much less than are alkali and alkyl silicate coatings.

For use in marine environments, it is usually desirable to topcoat inorganic zinc coatings to extend their service lives. Under immersed marine conditions, no more than two years of protection to steel can be expected without topcoating. For post-cured zinc inorganics, it is necessary to remove the curing agent completely before topcoating. Initially, wash primers and tie coats were used extensively as intermediate coats in topcoating systems. They are currently used to a much lesser extent. Because of the textured nature of zinc inorganics, they will accept a variety of topcoats such as epoxies, vinyls, chlorinated rubbers, acrylics, and coal tar epoxies. Topcoated zinc inorganics are extensively used in the atmospheric zone of steel offshore platforms. Navy ships currently utilize two topcoated zinc inorganic systems for topside use. The first consists of a zinc inorganic primer, an epoxy intermediate coat, and an epoxy non-skid coat for decks. The second consists of a zinc inorganic primer, an epoxy intermediate coat, and a decorative topcoat (e.g., silicone alkyd) for vertical surfaces. The Navy does not use zinc inorganics on submerged portions of hull, but Class 3 of MIL-P-23236 covers zinc inorganic coatings for use in steel fuel and salt water ballast tanks. One supplier of a zinc inorganic coating is investigating the possibility of extending its period of protection underwater by use of zinc anodes which are anodic to the coating as well as to the steel. Zinc salts being somewhat toxic to marine fouling organisms, zinc-rich coatings may reduce but not eliminate fouling on marine structures.

Cases of blistering have been reported when self-cured zinc inorganic coatings were topcoated within one to three days of priming. This blistering has been attributed to (1) gas formed in the curing process, trapped by the topcoat and (2) air in the porous inorganic coating displaced by solvent in the topcoat and trapped by the topcoat. Some manufacturers recommend a waiting period of one week at 70 F (21 C) between application of the self-cured inorganic primer and the organic topcoat. This period is not so critical with latex coatings as with those containing organic solvents.

Inorganic zinc coatings are generally packaged in two or three containers with the finely divided zinc in a separate, anhydrous package. The latter minimizes clumping of zinc particles and hydrogen gas formation. A small amount of a water scavenger such as lime is sometimes added for protection from gas formation when the zinc is not packaged separately. Federal Specification TT-P-460 describes two types of zinc dust particles used in zinc-rich paints. The purity of the zinc is at least as important as the particle size. Lead is sometimes incorporated along with zinc in the pigment portion of zinc inorganics, but there is no general agreement among suppliers as to its beneficial effects.

A wide variety of organic zinc-rich coatings is currently marketed. The organic resins used as binders include epoxies (both polyamide and amine cured), epoxy esters, vinyls, chlorinated rubber, polyesters, urethanes, and acrylics. These function not only in providing cathodic protection to steel but also as barrier coatings. Organic zinc-rich coatings are usually tough and have good abrasion resistance but not usually that of the inorganics. Organics are generally easier to topcoat than the inorganics, and topcoats are generally of the same generic type as the binder of the zinc-rich coating. For organic as well as inorganic coatings, it is wise to use a complete coating system recommended by a supplier. Zinc-rich organics may be sold in 1, 2, or 3-package kits.

Zinc-rich coatings require good surface preparation of steel substrate prior to coating application. Surface preparation for most zinc-rich organic coatings is generally less critical than for the inorganics. Many suppliers state that their organic coatings can be applied satisfactorily to new steel with a commercial blast cleaning, while most suppliers of the inorganics state that cleaning to white metal is necessary.

Zinc-rich coatings are usually applied with conventional spray equipment. A pot agitator is necessary to keep the zinc dust suspended during application. It is generally necessary to keep the spray gun and pot at the same level because of the heavy weight imparted by the zinc pigmentation. A brush may frequently be used to touch-up coating work or to reach inaccessible areas.

Bibliography

A. K. Doolittle, "Case History Report on Long Term Coating Performance." *Materials Protection, 2,* 32-39 (1963) January.

C. V. Brouillette. "Field Performance of Zinc-Rich Coatings on Steel Decking." NACE Symposium on Environment Resistant Coatings Research and Technology for Aerospace and Marine Applications, Los Angeles, California. November 3-5, 1969.

R. W. Drisko. "Protection of Mooring Buoys." NACE Western Regional Conference and Symposium, San Diego, California. September 24-26, 1969.

V. L. Flack. "Laboratory Tests of Zinc Filled Coatings." *Materials Protection, 2,* 36-45 (1963) March.

D. H. Gelfer. "Permanent Primers for Steel: Comparison of Self-Curing and Post-Curing Inorganic Zinc Coatings." *Ibid., 3,* 54-61 (1964) March.

E. F. Group, C. L. Uzzell, and C. J. Greene. "Protective Coatings for Offshore Structures." NACE Western Regional Conference and Symposium, San Diego, California. September 24-26, 1969.

J. D. Keane. "Zinc-Rich Coatings: Characteristics, Applications, and Performance." *Materials Protection, 8,* 31-34 (1964) March.

D. G. Mason. "Influence of Zinc Dust Loadings." *Ibid., 6,* 31-34 (1969) March.

Q.T. McGlothline, E. M. Curtis, and J. B. Cox. "Self-Curing Inorganic Zinc Coatings." *Ibid., 5,* 25-28 (1966) February.

J. F. Montle. "Inorganic Zinc Primers: Effect of Porosity on Coverage." *Ibid., 8,* 21-24 (1964) April.

C. G. Munger. "Comparing the Properties of Zinc Dust Coatings in a Variety of Environments." *Ibid., 2,* 3-16 (1963) March.

J. R. Neale, Jr. "Epoxies in the U.S. Navy." NACE Western Regional Conference and Symposium, San Diego, California. September 24-26 (1969).

A. H. Roebuck. "Recent Developments in Marine Coatings." *Ibid.*

J. R. Saroyan. "Zinc-Rich Coatings as Replacements for Galvanizing in Naval Applications." *Materials Protection, 4,* 27-29 (1965) November.

G. R. Sommerville, J. A. Lopez, and J. P. McGuigan. "Accelerated Tests on Zinc Rich Epoxy Coatings for Salt Water Services." *Ibid., 2,* 59-66 (1963) November.

R. Vickers. "Combination Coating System: Hot Dip Galvanizing and Inorganic Zincs Can Give Economical Protection." *Ibid., 3,* 25-31 (1964) August.

H. E. Waldrip. "Testing Zinc-Rich Coatings for Protecting Steel." *Ibid., 5,* 25-26 (1966) May.

RICHARD W. DRISKO is senior project scientist in the materials science division of the civil engineering department of the US Naval Civil Engineering Laboratory, Port Hueneme, Calif. He holds a BS, MS, and PhD from Stanford University. An active member of NACE technical committees, Drisko is also a member of the ACS and ASTM.

Recent Developments in Inorganic Zinc Primers*

JOHN F. MONTLE and MICHAEL D. HASSER
Carboline Company, St. Louis, Missouri

Increased cost of Zn dust for coatings has enhanced interest in applications involving reduced loadings or the use of extenders. Recent tests show Zn at 65 to 85% makes coatings adequate in moderate exposures and elsewhere also when topcoated. Severe exposures or poor surface preparation require heavier loadings. Inorganic Zn primers with less than 65% Zn tend to blister topcoats less in wet exposures but when less than 2 mils is applied, higher loadings are necessary. When less than 85% Zn is used, performance depends on vehicle. Hydrolyzed ethyl silicates tend to tolerate reduced loadings better than others. High loading permits better toleration of application and field variables. Diiron phosphide, an extender for Zn, is electrically conductive. Tests show no difference in surface rust when Zn dust-diiron phosphate ratios were increased from 100:0 to 50:50. Diiron phosphate reduced porosity of welds on weld-thru primer steel. Solvent-based Zn coatings tend to adhere to mill scale and give good service when loading exceeds 86%. Application less than 25 miles from salt water, for immersion or chemical spillage zones is not recommended. Mill scale should be 75% intact. Low stress topcoats are necessary. Colored inorganic zincs do not require topcoats, but color matching depends on Zn dust size. Nonzinc inorganic colored topcoats are successful at 1.5 to 2.5 mils over suitable primers. Single package inorganic zincs are sensitive to topcoat selection. Fast drying inorganic zincs can be topcoated in a few hours. Properties of fast drying polyvinyl butyral coatings are discussed and drying mechanism examined.

THIS PAPER REVIEWS recent trends in the use of inorganic zinc primers. The 6 following subjects are discussed: (1) Reduced zinc level inorganic primers, (2) weld-thru primers, (3) inorganic zinc coatings over mill scale, (4) finish coat trends, (5) single package inorganic primers, and (6) modified, fast dry primers.

Reduced Zinc Level Inorganic Primers

In the last 2 years (1973-74), the zinc dust industry experienced increasing tightness of supplies and a booming demand. The result was rapidly rising prices. In the United States in 1973, the failure to meet this rising demand was aggravated by a previous shutdown of 40% of production capacity (caused in part by environmental demands on producers). Also, price controls on zinc were in effect until December, 1973. These controls resulted in a two tier market, with the United States price held at 21.5 cents/lb for special high grade zinc dust. Of course, this situation was not favorable to increased production in the United States.

Asking prices for foreign and domestic zinc dust suitable for coatings during this period varied from about 20 cents/lb at the beginning of the period, to over 80 cents/lb. Recently, these prices have dropped somewhat. Consequently, manufacturers of inorganic zinc filled coatings examined the effects of a reduction of the zinc content of their coatings for various applications. In this examination, 2 broad areas were covered: (1) The minimum level of zinc in the dry film at which acceptable performance is obtained, and (2) zinc substitutes which enhance certain properties of the coatings.

The problem of zinc content reduces itself to the question: what is the minimum level of zinc required in an inorganic zinc primer to obtain good application properties and excellent performance?

The following questions have a bearing on the answer: (1) What is the service exposure? (2) What range of film thickness will be normally recommended? (3) Will the inorganic zinc primer be topcoated? (4) What is the exact formulation type to be used? and (5) What are the surfaces and how will they be prepared?

Experience to date makes possible the following generalizations:

1. Inorganic coatings with zinc levels of 65 to 85% by weight in the dry film perform equally well in moderate exposures untopcoated, and as maintenance primers (with proper topcoats) for corrosion protection in a wide variety of exposures.
2. In high demand exposures, untopcoated, inorganic zinc primers with higher zinc levels tend to provide longer service life—a trend also noted when coatings are applied over poorly prepared surfaces.
3. Inorganic zinc primers with lower zinc levels (down to 65% by weight in the dry film) tend to have less topcoat blistering in wet exposures, with equal capability of preventing corrosion of the base metal.
4. At low film thickness [less than 50 μm (2 mils)], higher zinc loadings tend to provide longer protection.
5. Performance at moderate zinc levels (less than 85% by weight in the dry film) is very dependent upon the vehicle type. Hydrolyzed ethyl silicates are more tolerant of reduced zinc levels than most silicate types.
6. Higher zinc level coatings have a greater tolerance for application and field variables.

More specifically, extensive laboratory work over the past several years has proven to us that properly formulated inorganic zinc primers with zinc loadings of 60 to 68% by weight in the dry film could perform as well as inorganic zinc primers with zinc loadings of 85 to 87% by weight in the dry film in the following areas of use: (1) Film thicknesses of 64 μm (2.5 mils) minimum, and (2) applied over steel prepared by a minimum surface preparation of a commercial blast, when exposed as follows: (a) as the primer in a maintenance system, or (b) untopcoated exposure in all but the most severe applications.

However, while many field applications suggested that our laboratory conclusions were valid, some problems were encountered. These were primarily due to inadequate preparation of the steel where the coating lacked sufficient tolerance to overcome the deficient surface preparation. Problems were encountered also due to inadequate tolerance of normal field application variables of dry spray, thin areas, and errors due to mixing partial units, etc.

Further work to arrive at a better compromise between the economics of lower zinc levels and the safety factor or "insurance" of higher zinc level has shown us that a zinc content of 75% by weight in the dry film is the optimum level for a reduced zinc content primer.

*Presented during Corrosion/75, April 14-18, 1975, Toronto, Ontario, Canada.

Reprinted with permission from *Materials Performance*, August 1976, 15-18, © 1976 National Association of Corrosion Engineers

TABLE 1 – Characteristics vs % Zinc in the Dry Film

	% Zinc by Weight in Dry Film		
	86%	74%	65%
Performance in various exposures:			
Midwest weathering, untopcoated	10	10	10
Gulf Coast weathering, untopcoated	10	9	7
Tendency to cause topcoat blistering (when exposed to wet environments)	6	9	9
Salt fog, untopcoated	10	9	7
Maintenance primer	10	10	10
Surface preparation tolerance	10	8	3
Field application tolerance	10	9	2
Minimum required film thickness, mils	0.7	1.5	2.5

Note: Ratings are 10 = best, and 0 = poor.

TABLE 2 – Results Obtained in 1146 Hours Salt Fog Exposure

Wt % Fe_2P	Wt % Zinc	Rust in Scribe	Rust on Surface
0	100	10	10
25	75	10	10
50	50	10	10

Note: 10 = No effect.

We must stress that this level is optimum only in a properly formulated hydrolyzed ethyl silicate binder. A summary is shown in Table 1.

Many manufacturers have investigated the use of di-iron phosphide[(1)] pigment-grade as an extender in zinc rich coatings.

The chief advantage of substituting "Ferrophos" for zinc is, of course, cost. The price per pound is approximately 1/3 to 1/4 that of zinc dust.

An important property of Ferrophos is that it is an electrically conductive extender.

It has been found that 25 to 35% by weight of the zinc can be replaced by Ferrophos in marine primers based on silicate binders with a negligible effect on the corrosion resistant properties of the coating after 2500 to 5000 hours salt fog testing.[1]

After 1146 hours in the salt fog (ASTM B117) with lithium polysilicate binder, results presented in Table 2 were observed.[1]

Field testing on the S.S. Baltimore was conducted with 50% of the zinc replaced with Ferrophos in a silicate binder and compared directly to the silicate binder with no Ferrophos substitution for performance. After 2 years exposure, no major differences in corrosion resistance were evident.[1]

Inorganic Zinc Weld-Thru Primers

The use of inorganic zinc primers as weld-thru primers was introduced 10 years ago and is now standard practice.

The advent of automatic centrifugal wheel abrasive cleaning coincided with the introduction of these materials. This combination of automatic blasting and a weld-thru primer has resulted in reduced overall labor costs. Acceptance has been widespread, especially in the marine, nuclear, chemical, and petroleum (pipeline) industry.

Recent developments with these primers have included the substitution of Ferrophos for a certain percentage of the zinc.[2]

TABLE 3 – Weldable Coating Ratings After Salt Fog Exposure

Dry Film Thickness (mils)		Months in Test				
Theoretical By Weight	Measured By Mikrotest	1	2	4	9	15
0.21	1.0	9	6	2	1	1
0.28	1.3	9	7	3	1	1
0.42	1.5	10	10	7	4	4
0.49	1.6	10	10	7	5	3

Coating Ratings After Mild Industrial Exposure in St. Louis

Dry Film Thickness (mils)		Months in Test				
Theoretical By Weight	Measured By Mikrotest	1	2	3	8	15
0.21	1.0	10	10	10	10	10
0.28	1.2	10	10	10	10	10
0.35	1.5	10	10	10	10	10
0.50	1.8	10	10	10	10	10

Notes: Gradings for rust (ASTM D-610) 10 = no effect, and 1 = more than 50% rust. Because of difficulties in measuring accurate thin dry film thicknesses over sandblasted steel, both a weighing technique and a magnetic dry film thickness testor were used.

In continuous arc welding tests performed by Hooker, as the percent Ferrophos increased, porosity of the weld (after breaking) decreased until at 40% Ferrophos, there was no porosity.[2]

Lately, improvements in these primers have made them tougher and faster drying. These materials were always tough and easy to clean. Long term protection is provided at thicknesses less than 25 μm (1 mil) with an ethyl silicate zinc weldable primer at 86% by weight zinc (no Ferrophos) in the dry film (Table 3).

Inorganic Zinc Primers Over Mill Scale

Mill scale itself is an inert protective coating if the scale is continuous. However, breaks in the mill scale (both gross and minute in size) lead to corrosion. The scale is cathodic, and causes corrosion of the underlying steel. As corrosion proceeds, undercutting of the scale occurs and the corrosion products, which occupy a greater volume than the base metal, subject the scale to a "prying" or "lifting" force which causes the mill scale to "pop" off.[(2)] Theoretically, if there were no breaks or porosity in the mill scale, there would be little or no corrosion. Therefore, if this galvanic cell could be insulated or reversed so that corrosion at the break in the mill scale and sub-film corrosion under the mill scale could be prevented, the mill scale would remain intact.

The requirements for a zinc rich coating to achieve this are simple and straightforward: (1) It must be capable of adhering to the smooth mill scale, and (2) it must be capable of providing cathodic protection to the base metal.

If you accept the fact that zinc will cathodically protect bare steel beneath the scale and the fact that the coating will adhere to the smooth scale surface, it follows that an inorganic zinc primer will protect mill scale and prevent "popping".

In general, the coatings industry has always associated inorganic zinc protection with a surface preparation of well-blasted steel. Manufacturers of inorganic zinc primers specify blasting with sand or steel grit and indicate the degree of cleanliness, usually NACE No. 1 or 2. In some cases, steel shot can be used.

There are 3 general types of inorganic zinc coatings: (1) Water-based post-cured, (2) water-based self-cure, and (3) solvent-based self-cure. The nature of the anchor pattern and degree of steel cleanliness recommended will vary with the basic type of inorganic

---

[(1)] Ferrophos is Hooker Chemical Company's trademark for di-iron phosphide.

[(2)] The removal of mill scale by weathering is frequently utilized to facilitate surface preparation.

TABLE 4 – % Zinc in the Dry Film vs Performance Over Mill Scale

Zinc Level (%)	Dry Film Thickness (mils)	Failure[1] After
0	2.5	2 months
20	2.1	2 months
40	2.6	2 months
60	2.3	4 years
70	2.2	5-1/2 years
85	2.5	No failure, less than 1% rust 6-1/2 years
87.5	2.4	No failure, less than 1% rust 6-1/2 years

[1]Greater than 50% rust.

zinc coating used, the service requirements, and the manufacturer's recommendation.

In order to insure proper performance, manufacturers of post-cured water-based materials generally recommend application over a surface cleaned to NACE No. 1, regardless of the service. For this type, correct surface preparation is very critical. Water-based, self-curing inorganic coatings generally require a surface cleaned to NACE No. 1 or 2, having a profile of about 25 to 38 μm (1 to 1-1/2 mils). The solvent-based self-curing inorganic zinc type, in many services, generally has excellent adhesion and performance over a lesser surface preparation such as NACE No. 3 (commercial blasted steel). The solvent-based inorganic zinc will adhere to steel surfaces with little or no profile.

Why does this type coating give useful and economical protection when applied over mill scale-bearing steel, contrary to the long established standards set by the coating industry? Solvent-based, self-curing inorganic zinc coatings have a relatively low surface tension. This property enables the applied material to intimately wet and penetrate clean surfaces and crevices. Consequently, a good bond can be expected, even when used over a smooth surface. On the other hand, most water-based inorganic zinc coatings have a high surface tension with relatively poor wetting properties. This type must depend upon a jagged anchor pattern for mechanical adhesion. For comparison, the surface tensile of water is 72 to 74 dynes/cm^2, while that for typical solvents used in solvent-based inorganic zinc primers is only 22 to 24 dynes/cm^2.

So much for mechanism. Numerous laboratory and field tests have been performed which add credence to the explanation above. A laboratory test to determine the zinc level required for excellent performance over mill scale was started in 1967.

Laboratory data indicate that 85% or greater zinc by weight in the dry film is required to provide excellent performance untopcoated after 6-1/2 years continuous exposure in semi-industrial midwest weathering over mill scale. Increased performance can, of course, be expected if topcoats are used. Results are shown in Table 4. The laboratory test programs were correlated with actual field test exposures over extended periods. A large chemical company field tested this application at 3 of its plants throughout the country. Locations wsre in the Gulf Coast, Mid Atlantic, and New England areas. At each location, test areas of structural steel covered with mill scale were coated with a solvent-based inorganic zinc with 86% zinc by weight in the dry film at a nominal 3 mils dry film thickness. After exposures ranging to 7 years, less than 5% failure had occurred. This failure was concentrated on particular areas of the steel where easy touch-up could be executed and performance was considered excellent. Other paints and coating systems were tested in the same geographic locations over mill scale resulting in mill scale separation with more than 50% failure within 13 months.

Prior to considering an application of inorganic solvent-based zinc coating over power tool-cleaned mill scale steel, the intended service should be carefully analyzed. Then the decision and use must be governed by the following:

1. Service environment is limited to locations more than 40 km (25 miles) away from coastal salt water. Topcoats with a minimum thickness of 100 μm (4 mils) are required in industrial and corrosion environments.
2. Immersion service, such as tank lining, is excluded.
3. Application is not recommended for severe chemical environments where frequent splash or spillage of chemicals occur.
4. There should be at least 75% intact mill scale on the steel to be coated.
5. Heavy duty, high strength topcoats exert severe stress on the substrate. Therefore, low-stress topcoats such as vinyl, acrylic, chlorinated rubber, and nonembrittling epoxy types should be used.
6. Specifications should state that the steel must be solvent wiped or detergent scrubbed to completely remove oil and grease.

Finish Coat Trends

For less severe environments, colored inorganic zinc coatings have been introduced which do not require topcoats.

The coating serves as both a primer and a topcoat. Besides the usual problems associated with color matching and stability, the formulator now must deal with variations in particle size from batch to batch of zinc dust. If this problem is not overcome, using more than one batch of zinc dust will result in a patchwork color effect. Also during application, variation in color may result because of dry spraying. However, because of favorable economics, it is anticipated that obstacles will be surmounted by coating suppliers.

Inorganic colored topcoats (not containing zinc) are another recent development. Single package pigmented inorganic coatings are now available which, when applied at 38 to 64 μm (1-1/2 to 2-1/2 mils) over suitable inorganic primers, provide attractive total inorganic systems which have the usual advantages of any inorganic coating such as solvent resistance, temperature resistance, etc. Two package inorganic coatings containing aluminum or stainless steel are also available. Besides having the advantages listed above, these materials also provide a significant degree of corrosion protection. In addition, in immersion applications, contamination of the liquid medium by zinc is significantly reduced, due to the reduction in surface area, without reducing the basic corrosion protection of the inorganic zinc primer.

Single Package Inorganic Zinc Primers

The main advantage of a single package zinc silicate is, of course, convenience. Instead of blending a zinc filler with a silicate base, one simply remixes the paint as supplied, thins if necessary, and applies the material.

The reaction:

$$Zn + 2H_2O \rightarrow Zn(OH)_2 + H_2$$

precludes the existence of a water-based single package silicate, and is part of the problem when formulating a solvent-based single package.

Three vehicle types are commonly available. The first involves amine-initiated hydrolysis and colloidal suspension of silica in solvent.[3]

The second type consists of partially hydrolyzed ethyl silicate in combination with an alkali metal alkoxide.[4]

The third type vehicle is a polyolsilicate or a polyol hydrocarbon ether silicate, and has patents pending.

One of the main problems that may be encountered with single package zinc silicates is the incompatibility of some topcoats which would normally be compatible with two package solvent-based zincs. This problem exhibits itself through nonadhesion and blistering of the topcoats in certain wet environments, and is due to residual alkaline material present in the dry inorganic zinc film.

Source: *Materials Performance*, August 1976, 15-18

TABLE 5 – Rust Ratings After 1 Year Exposure in Various Environments of One Package vs Two Package Inorganic Zinc Primers

	St. Louis Weathering	Salt Fog	Fresh Water
Two package, 85% zinc in DF	9R	9R	NE
Two package, 75% zinc in DF	NE	NE	NE
One package, 83% zinc in DF	9R	6R	9R

Notes: Rusting = ASTM D-610, and NE = no effect.

We have found performance, untopcoated, over a commercial sandblast, to be similar to what one would expect of a standard two package system at the same zinc loading in the dry film [dry film thickness 51 to 76 μm (2 to 3 mils)] (Table 5).

Stability is, of course, a problem which must be overcome by the formulator. Careful control over manufacturing is required to exclude excessive moisture and prevent gassing upon storage. Careful formulation can provide materials with a minimum of settling and caking of the zinc powder even after long term storage.

Because of the handling of the zinc component during manufacturing, which accounts for most of the weight, the packaged cost is going to be greater for a single package inorganic zinc than a conventional two package at equal zinc content. This disadvantage, plus the disadvantage in topcoatability with most single package zincs under certain conditions, must be weighed versus the convenience of a single package.

Fast Drying Inorganic Zinc Primers

A "fast drying" inorganic zinc can be handled quickly with a minimum of damage and topcoated within a few hours after primer application.

This type of primer has found acceptance in fabricating shops or anywhere a premium is placed on fast handling of the steel after coating. Usually, another requirement is that it be readily adaptable to automatic spraying.

The economic advantage of using a shop-applied approach have already been examined in detail.[5]

Manufacturers have used and investigated various means of obtaining inorganic fast drying primers.

Fast drying inorganic zinc primers can be formulated by carefully balanced solvents, by choosing the basic starting materials carefully, and by rigid control of manufacturing variables such as time and extent of hydrolysis to build suitable large polymers in the vehicle.

Another approach involves the addition of certain organic polymers such as polyvinyl butyral. This approach was demonstrated in United States Patent 3,392,13.

Peace and Mayhan[6] have done work to determine whether a reaction between condensed tetraethyl orthosilicate and polyvinyl butyral does occur, or if the resultant mixture is merely a physical blend. The importance of this study is that if a reaction does take place, then it is likely that the film will retain those properties associated with inorganic primers. If no reaction takes place, and only a physical blend exists, then addition of organic polymers is likely to significantly reduce the properties associated with inorganic primers. Peace and Mayhan concluded:

1. No reaction between tetraethyl orthosilicate and polyvinyl butyral takes place.

2. Increases in viscosity observed during hydrolysis of the tetraethyl orthosilicate in the presence of polyvinyl butyral is explained not by any reaction, but rather appears to be a result of the combined effects of water being formed in the reaction and changes in the silicate structure. In fact, increases in viscosity of a polyvinyl butyral solution in the absence of tetraethyl orthosilicate will be observed if water is added to the solution. Removal of the water with molecular sieves will decrease the viscosity.

There are several general considerations which indicate a low reaction probability between the silicate and polyvinyl butyral. The final reaction mixture contains a very low content of polyvinyl alcohol. Second, the polymer molecules in solution are random coils, not linear chains. Thus, many hydroxyl groups on the polymer are inaccessible for reaction. Third, in order for a reaction to occur between polyvinyl butyral and the silicate, two very large and highly solvated molecules must come together in the proper orientation. This would probably require entry of the silicate into the polyvinyl butyral coils, a very unlikely process. It should be noted that the hydrolyzed silicate is a very large reactant in terms of ordinary chemical reactions.

References

1. Simko and Simpson. Hooker Chemical Corporation, Niagara Falls, New York, A New Class of Conductive Extenders in Zinc Rich Coatings, Presentation—Western Coatings Societies' Symposium and Show, San Francisco, California, March, 1974.
2. Simko and Simpson. Hooker Chemical Corporation, Niagara Falls, New York, The Row of Extenders in Zinc Rich Coatings, Presentation—Interfinish Conference, Basel, Switzerland, 1972.
3. U.S. Pat. 3,615,730 (1972); 3,653,930 (1972); Dutch Pat. 6,600,749.
4. U.S. Pat. 3,660,119 (1972).
5. Schwartz. Carboline Company, St. Louis, Missouri, Shop Versus Field Coating, Presentation—NACE Northeast Regional, Niagara Frontier Section, October 24, 1973.
6. Peace and Mahan. University of Missouri, Rolla, Missouri, unpublished research (1970-72).

Inorganic Zinc Rich Primers – Fact and Fancy*

C. MALCOLM HENDRY
Napko Corporation, Houston, Texas

History of inorganic zinc development is related. Principal properties of inorganic zinc primers, one package advantages and disadvantages, weldability, recommended topcoats, brittleness, shelf life, volume solids calculations, and efficiencies and comparisons with galvanizing are discussed. Test methods to evaluate performance vs % zinc are described. Properties of dry sprayed zinc are considered and consequences of application over mill scale reviewed. Accelerated tests are described.

ZINC RICH PRIMERS have been a great boon to the protection of iron and steel. "Inorganic zincs", as they are called, have been widely acclaimed as the "Cadillacs" of the zinc rich primers. But Cadillacs are expensive, and must be properly used to give satisfactory service. Because of their uniqueness and importance to corrosion resistance, much lore and folklore, fact and fancy, has developed concerning inorganic zincs. This paper will attempt to separate fact from fancy. It is important that the users recognize the differences among types of inorganic zincs, realize their strengths and shortcomings, and know how to use these primers effectively.

History

Beginning with the now legendary Australian pipeline coating of the early 1940's, numerous forms of inorganic zinc rich primers

*Presented during Corrosion/77, March, 1977, San Francisco, California.

have emerged. The Australian (Nightingale) zinc primer contained bicarbonate which was baked to insolubilize the silicate vehicle.

The next important inorganic zinc was the post cured silicate; a curing solution was applied over an aqueous zinc rich silicate. Advantage: film cure was insured. Rains can wash off self curing water reducible primers but can't wash off the post cured variety. Post cured zinc rich silicate primers have performed well for almost two decades of severe service, for example, on offshore drilling rigs. However, application of these post cured primers was problematic. The first coat was water sensitive until the curing solution was applied. The curing solution had to be uniformly applied to insure cure and be washed off before topcoating. After curing, the primer has an unsightly white appearance until topcoated. This white deposit helps fill the pores of the zinc primer, reducing the tendency of topcoats to bubble during application.

A water reducible, self curing inorganic zinc was next developed. Its corrosion resistance approached, but did not reach, that of the post cured zincs. These primers were water sensitive until cured, and their alkaline surfaces were not equally receptive to all types of topcoats.

A major step forward occurred with the advent of self curing, solvent reducible inorganic zincs often classified as alkyl silicates. This type has become the workhorse of the industry. The studies reported in this paper will focus on this type of zinc primer.

A more recent development is "formable zincs" (largely inorganic). The formable zincs are excellent shop primers and, of course, permit forming or bending of the steel *after* application of the primer. Conventional inorganic zincs are friable and do not permit post forming. However, since the formable zincs are not entirely inorganic, their corrosion resistance is slightly inferior to the best self cured or post cured inorganic zincs, and their cost per mil square foot is somewhat higher.

The most recent zincs to be marketed are one package inorganic primers. (The above mentioned inorganic zincs are all at least two package, the zinc being mixed in just prior to application.) The one package inorganics avoid the mixing step, but many are not yet perfected. The one package inorganic zincs are also more expensive than the self curing, two package, zinc rich silicate primers.

Fact and Fancy

Solvent thinned, self curing, zinc rich alkyl silicate primers have become the workhorse coatings for corrosion protection. However, many misconceptions exist about these primers. Some of the "fancies" which have developed about this class of inorganic zincs are the following:

Fancy: "Inorganic zincs are primers".

Fact: Inorganic zincs are more than primers. Untopcoated inorganic zincs offer excellent corrosion resistance from pH 6.0 to 9.5. They do not retain as attractive an appearance as they age as they would if topcoated. The environmental or earth colored zincs maintain their appearance better, and also blend into the landscape. Untopcoated inorganic zincs are often preferred for corrosion resistance.[(1)] They actually become progressively more resistant to the atmosphere as zinc oxide, carbonate, and hydroxides form on the surface. Munger and Berger have cited the performance and economic virtues of untopcoated inorganic zincs.[1-3]

Fancy: "Inorganic zincs are weldable".

Fact: 63 to 75 μm (2.5 to 3.0 mils) films of inorganic zincs on steel cause very slow welding speeds and porous welds. Certain modified inorganic zincs are more weldable; welding speeds and porosity are improved, but are still not comparable to welding unprimed steel, unless accompanied by a significant sacrifice in corrosion resistance. Thin films of primer, *e.g.,* 25 μm (1 mil), interfere very little with welding speed and bead quality.

Fancy: "One package inorganic zincs are the answer."

Fact: One package zincs avoid the mixing time and errors which can occur with two package primers. They do so at some sacrifice in corrosion resistance. They may also be premium priced. Application problems are different from two package zincs. One package versions are often prone to "mud-cracking" or low build, and package instability (gassing).

Fancy: "Inorganic zincs cannot be topcoated until they become hard and resistant to solvent wiping."

Fact: The solvents in these primers may require a week or more to be completely released. They may remain sensitive to solvent wiping until most of the solvent is released. Nevertheless, many silicate primers can be topcoated with a wide range of topcoats within a day after application with no blistering, adhesion loss, or sacrifice in corrosion resistance. Solvent release and "curing" continue after the inorganic zinc primer is topcoated. If the primer is fast drying, it may be prone to dry spray.

Fancy: "Chlorinated rubber topcoats perform well over inorganic zincs."

Fact: This is rarely, if ever, the case. If the chlorinated rubber is applied directly over the zinc primer, blistering almost always occurs in salt fog testing. Chlorinated rubber may perform if an intermediate coat of another generic type insulates the topcoat from the primer, or if another generic vehicle is blended with the chlorinated rubber. Some chlorinated rubbers which are highly diluted with another vehicle are used over zincs. However, these dilutions also compromise many of the chlorinated rubber properties, *e.g.,* chemical resistance.

Fancy: "Inorganic zincs are inherently brittle and friable."

Fact: Normally, yes, but the "formable" (organic modified) zinc rich silicate primers are neither brittle nor friable.

Fancy: "Inorganic zincs have a limited shelf life, posing storage problems."

Fact: Many leading hydrolyzed ethyl silicate primers continue to move toward gelation in the package and may gel in a few months. However, other colloidal silicate zinc primers are quite stable in the package. Certain types have been kept for 7 to 8 years without viscosity buildup and are safe for overseas projects requiring long term storage.

Fancy: "Some zinc rich alkyl silicate primers have 62% volume solids."

Fact: Volume solids is normally a useful parameter to predict coverage and the quantity of paint needed. The measured volume solids by sedimentation,[4] wet film/dry film, and solvent loss (Federal specifications) all give unrealistically high readings, *e.g.,* 53 to 79% volume solids. Calculated volume solids, based on solvent evaporation, for the zinc rich alkyl silicate primers is 41 to 45%. The latter figure is much closer to reality than the claimed 62% volume solids; 45% is a good figure to use when estimating consumption if one also assumes about a 20% loss factor.

Fancy: "Inorganic zinc primers are effective, but not nearly as effective as galvanizing."

Fact: Inorganic zincs are cheaper than galvanizing for almost all large surfaces.[3] In many environments, they will also outlast galvanizing.[2,3] Adhesion of most topcoats to inorganic zincs is excellent. Galvanized steel requires blasting or wash priming before most topcoats can be applied.

Fancy: "Topcoats can be put over inorganic zincs without bubbling."

Fact: All topcoats have a propensity to bubble over all inorganic zincs because they are porous and the air in the pore attempts to escape through the wet topcoat. Bubbling is worse under hot or cool conditions, depending upon the solvents used in the topcoat. Bubbling is less obvious with less glossy topcoats. Mist coating of the topcoat can reduce the bubbling. Some zinc primers are less prone to cause bubbling than others, depending upon the character of the primer surface. Bubbling of topcoats is rarely a serious corrosion risk if the steel surface is properly cleaned and the primer is intact.

Fancy: "High levels of zinc in the dry film (85 to 85%) permit greater application tolerance and insure greater corrosion resistance."

Fact: Montle and Hasser[5,6] have shown that a hydrolyzed ethyl silicate with 86% zinc in the dry film had a far greater tendency toward topcoat blistering and a greater propensity to topcoat bubbling than the same vehicle with 74% zinc in the dry film. Suppliers who previously advocated 86% zinc in the dry film have

---

[(1)]Editor's Note: Most experts recommend topcoating for corrosion resistance.

FIGURE 1 — Cross section of topcoated inorganic zinc over mill scale showing the heaving of the coating due to the corrosion products from the mill scale.

now changed their position. Galvanizing (100% zinc) frequently performance is less than properly formulated zinc primers.[1-3] Further, since the original hydrolyzed ethyl silicates were developed, improvements in silicate chemistry, in the understanding of primer film formation and the mechanism of zinc cathodic protection have occurred. Today's stable inorganic zinc primers perform best with a lower level of zinc than the earlier unstable inorganic zincs. Optimum zinc contents of primers vary depending upon the vehicle type and primer balance. Too high and too low a zinc content is disadvantageous to good film formation and corrosion protection. The following two sections of this paper ("Zinc-in-the-Dry Film" and "Field Application Simulation") present new work on optimum zinc levels in inorganic silicates.

Fancy: "Zinc rich alkyl silicate primers can be applied over tool cleaned mill scale or commercial blast if the zinc content is high enough, *e.g.,* 85%."[5]

Fact: Corrosion cells as seen in Figure 1, will result when inorganic zinc primers are applied over mill scale, with disasterous results, regardless of zinc content of the primer. The authors, Montle and Hasser, who wrote the (Fancy) statement above qualified it with these provisos:

Primers with 85% zinc in the dry film should be used: (1) When the primer is topcoated with a minimum of 100 μm (4 mils) of topcoat; (2) when the location is at least 25 miles from salt water; (3) in the absence of severe chemical environments; (4) excluding immersion service; (5) when no heavy duty, high strength topcoats are used; (6) when at least 75% of mill scale is intact; and (7) if the steel is solvent wiped or detergent scrubbed.

The problem with these provisos is that 75% intact mill scale, solvent wiped, is very unlikely to be achieved. Also, if noncorrosive environments are a requisite, it is hard to imagine why a zinc primer would be chosen at all. Montle and Hasser also are ruling out most conventional topcoats.

This authorization of poorly prepared surfaces is likely to lead the user into misuse of these fine primers. The experimental work presented in the last section of this paper, under "Mill Scale and Rusty Surfaces" shows the hazard of coating over mill scale.

Experimental

Zinc-in-the-Dry Film

Experimental work was undertaken to determine the weight ratio of dry metallic zinc in the dry film required for good corrosion resistance.

Salt fog tests were used as the indicator, because salt fog tests have correlated well with coating performance in aggressive environments. Two different silicate vehicles were chosen, both solvent

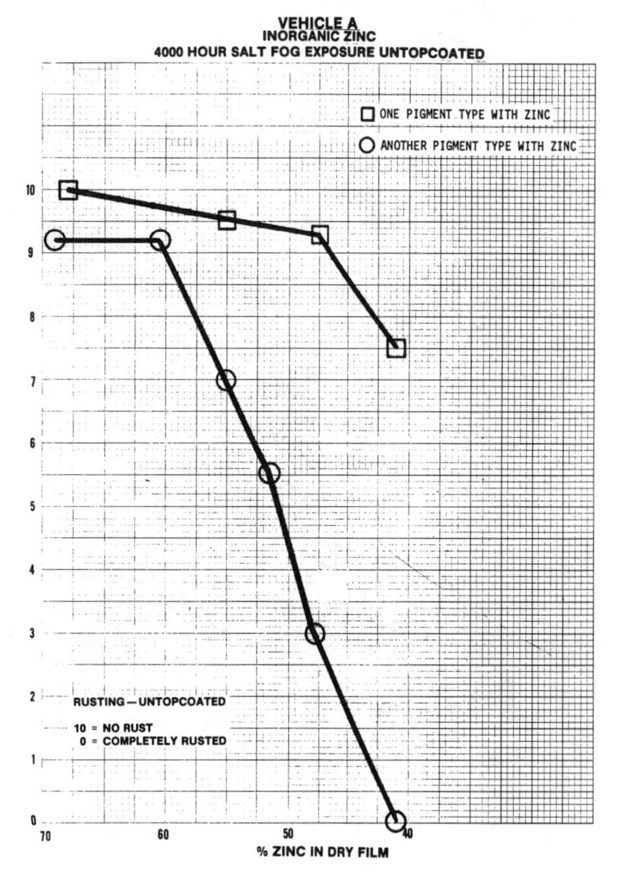

FIGURE 2 — Salt fog tests of primers made from silicate Vehicle A with varying zinc in the dry film.

carried. Primers with various amounts of zinc in the dry film were formulated from these vehicles, the volume solids and pigment volume concentration being held constant by progressively replacing the zinc with other pigments. The zinc content in the dry film of these primers was varied in steps from 71 to 42%. Applied over white blasted steel, the primers, both untopcoated and topcoated with either a high build vinyl or a high build epoxy, were subjected to 4000 hours of salt fog testing. The panels were rated for overall rusting, blistering, and undercutting at the scribe. Figures 2 through 5 show the results of these studies. (The circles and squares represent different pigments to make up the constant PVC as the zinc decreases.)

Figure 2 shows a retention of corrosion resistance, untopcoated, until the zinc content in the dry film of Vehicle A dropped below 60% using the first type of pigment (circles), and no deterioration until the zinc level dropped below 47% using a second type of pigment (squares).

When topcoated with either an epoxy or a vinyl, Vehicle A showed virtually no corrosion deterioration, even with zinc levels as low as 42% (Figure 3). The test panels are pictured in Figures 6 through 9.

Figure 4 shows a deterioration of the corrosion resistance of Vehicle B when zinc content in the dry film dropped below 63% using the first type of pigment (circles) or 57% using the second type of pigment (squares) when untopcoated. Figure 5 shows no loss of corrosion resistance at zinc levels as low as 43% when topcoated with an epoxy or vinyl. The test panels are pictured in Figures 10 through 12.

Neither Vehicle A nor Vehicle B was conventional unstable hydrolyzed ethyl silicate. Vehicles A and B were stable advances over the original silicate primers, and have been in commercial use for many years.

These studies suggest that Vehicle A, properly tormulated,

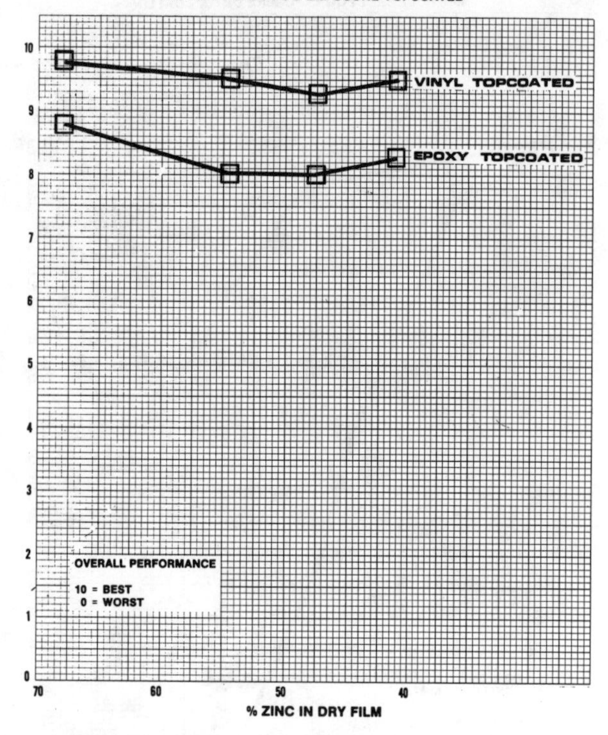

FIGURE 3 – Salt fog tests of topcoated primers made from Vehicle A with varying zinc in the dry film.

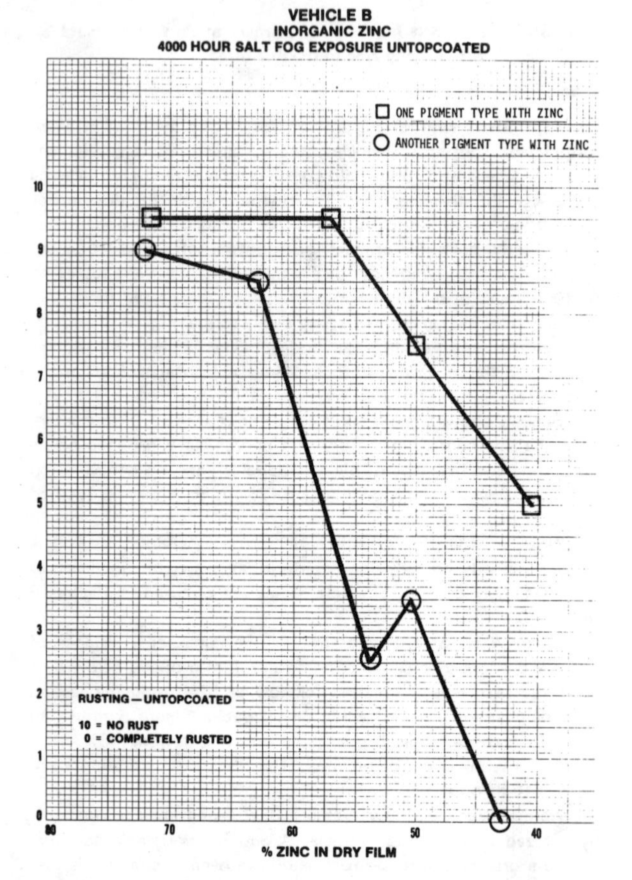

FIGURE 4 – Salt fog tests of primers made from silicate Vehicle B with varying zinc in the dry film.

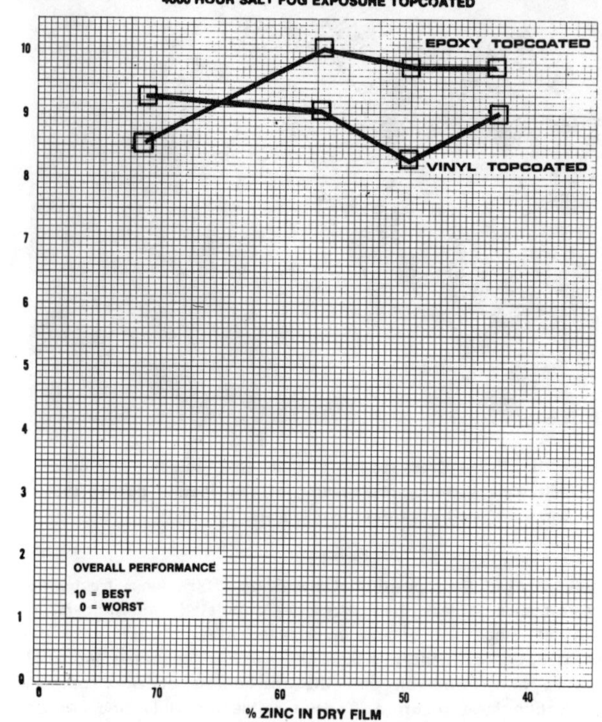

FIGURE 5 – Salt fog tests of topcoated primers made from Vehicle B with varying zinc in the dry film.

FIGURE 6 – Salt fog test panels (4000 hours) of primers made from Vehicle A with zinc in the dry film reduced stepwise from 68% (left) to 41.2% (right). Performance deteriorates below 60.3%.

FIGURE 7 – Salt fog test panels (4000 hours) of primers made from Vehicle A with zinc in the dry film reduced stepwise from 68% (left) to 40.9% (right). Performance deteriorates below 47.7%. The zinc replacement for this series was different from the replacement used in the series shown in Figure 6.

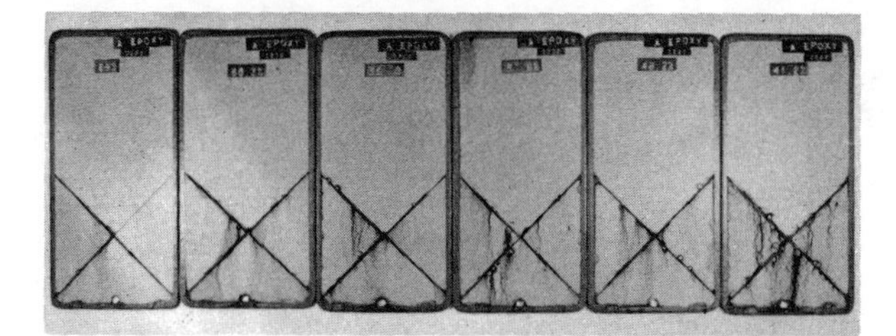

FIGURE 8 — Salt fog test panels (4000 hours) of topcoated primers made from Vehicle A with zinc in the dry film reduced stepwise from 68% (left) to 48.2% (right). The primers were topcoated with 100 μm (4.0 mils) of a vinyl. Performance deteriorates below 51.6% zinc.

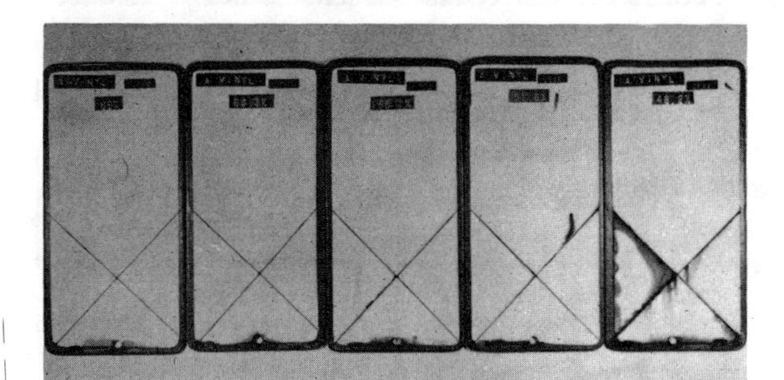

FIGURE 9 — Salt fog test panels (4000 hours) of topcoated primers made from Vehicle A with zinc in the dry film reduced stepwise from 68% (left) to 41.2% (right). The primers were topcoated with 150 μm (6.0 mils) of an epoxy. Performance deteriorates little at 41.2%.

FIGURE 10 — Salt fog panels (4000 hours) of primers made from Vehicle B with zinc in the dry film reduced stepwise from 71.4% (left) to 42.9% (right). Performance deteriorates below 57.1%.

could perform well in salt fog, untopcoated, with zinc levels from 47 to 68%. Topcoated, Vehicle A performed well at zinc levels from 43 to 68%.

Vehicle B was not quite as tolerant. At zinc levels of 57 to 71%, it performed well untopcoated, and performed well topcoated at zinc levels from 43 to 71%.

Field Application Simulation

Further examination of the optimum zinc content in the dry film was undertaken via an accelerated test of zinc primed panels in a sodium chloride/peroxide solution by the procedure described at the end of this paper. This test has consistently identified zinc primers of poor performance expectancy. In this test, rusting is

FIGURE 11 — Salt fog test panels (4000 hours) of topcoated primers made from Vehicle B with zinc in the dry film reduced stepwise from 71.4% (left) to 42.9% (right). No deterioration occurs with reduced zinc content.

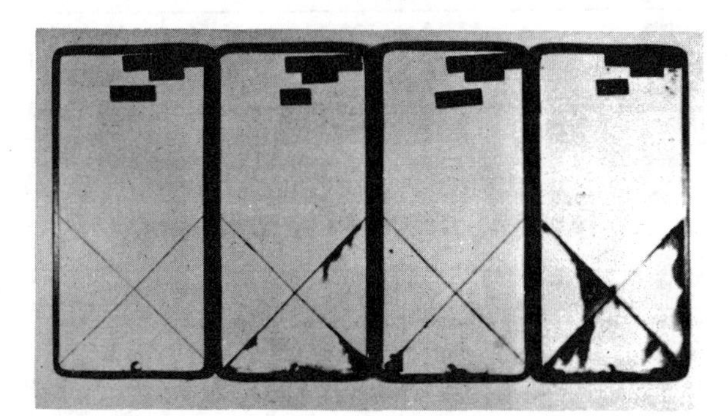

FIGURE 12 — Salt fog test panels (4000 hours) of topcoated primers made from Vehicle B with zinc in the dry film reduced stepwise from 72.1% (left) to 54% (right). Deterioration is modest below 72.1% and levels out.

FIGURE 13 — Steel test panel showing graduated blast pattern from NACE No. 1 (top) to brush blasted mill scale (bottom).

rapidly induced on unprotected or poorly protected surfaces. The degree of rusting was rated from 1 to 20, 1 being best.

Three different vehicles were tested with varying zinc content in the dry film, simulating field conditions. Surface preparation, primer thickness, and dry spraying were varied in the first test. The panel surface preparation was graduated, on the same panel, in four distinct zones as shown in Figure 13. Zone 1 was brush blasted mill scale, zone 2 was commercial (NACE No. 3), zone 3 was near white (NACE No. 2), and zone 4 was white blast (NACE No. 1).

A commercial hydrolyzed fast drying ethyl silicate primer (Vehicle "C") containing 86% zinc in the dry film was tested. It was

TABLE 1 — Film Thickness (mils)

PRIMER AND ZINC IN DRY FILM	DRY SPRAY		WET SPRAY	
PRIMER C-86%	Low Film	Normal Film	Low Film	Normal Film
White	1.3-1.4	3.0-3.1	1.4	3.0-3.1
Near-White	1.3-1.4	2.9-3.1	1.3-1.4	2.9-3.1
Commercial	1.2-1.5	2.9-3.0	1.3-1.5	3.0-3.1
Brush-Blast	1.5	3.0	1.3-1.4	3.2-3.3
PRIMER B-71%				
White	1.5	3.0	1.4-1.6	3.0-3.1
Near-White	1.3-1.5	2.9-3.1	1.6	2.7
Commercial	1.3-1.4	2.9-3.0	1.5	2.7-2.8
Brush-Blast	1.2	3.0	1.4-1.6	2.9-3.1
PRIMER B-86%				
White	1.3-1.4	3.0	1.5	3.0-3.1
Near-White	1.4	2.9-3.0	1.4-1.5	3.0-3.1
Commercial	1.4-1.5	2.9-3.1	1.3	3.0-3.1
Brush-Blast	1.3	2.8-2.9	1.3-1.4	3.1
PRIMER A-68%				
White	1.3-1.5	2.9-3.1	2.8-2.9	1.3-1.4
Near-White	1.3-1.4	2.9-3.1	2.7-2.8	1.3-1.4
Commercial	1.2-1.3	3.0-3.1	3.0-3.1	1.4-1.5
Brush-Blast	1.5	3.0	2.8-2.9	1.4-1.5

SALT/PEROXIDE TEST

TABLE 2 — Rank Order by Performance

PRIMER AND ZINC IN DRY FILM	DRY SPRAY		WET SPRAY	
PRIMER C-86%	Low Film	Normal Film	Low Film	Normal Film
White	20	8	3	1
Near-White	19	7	2	1
Commercial	18	8	3	1
Brush-Blast	17	2	3	1
PRIMER B-71%				
White	5	1	1	1
Near-White	4	1	1	1
Commercial	4	1	1	1
Brush-Blast	1	1	1	1
PRIMER B-86%				
White	4	1	1	1
Near-White	5	1	1	1
Commercial	3	1	1	1
Brush-Blast	2	1	1	1
PRIMER A-68%				
White	14	3	1	1
Near-White	17	3	1	1
Commercial	19	4	1	1
Brush-Blast	14	3	1	1

1 = Best 20 = Worst

SALT/PEROXIDE TEST

applied to the panel, and the cured panel was exposed to the sodium chloride/peroxide test solution. "C" was applied at normal 75 µm (3 mil) film thickness, wet sprayed. It was also applied at 35 µm (1.4 mils), wet sprayed and also dry sprayed at 75 µm and 35 µm (3.0 and 1.4 mils). In the salt/peroxide test, rusting of the 75 µm (3.0 mils) wet sprayed coat did not occur. At 35 µm (1.4 mils), some rusting was evident. At 75 µm (3.0 mils) dry sprayed, more rust was evident. At 35 µm (1.4 mils), dry sprayed, rusting was severe.

Then the same sequence was undertaken with Vehicle A having 68% zinc in the dry film and with Vehicle B having both 71 and 86% zinc in the dry film. Vehicle A and Vehicle B, with both 71 and 86% zinc, were unrusted in the test when wet sprayed at 75 µm (3.0 mils) or 35 µm (1.4 mils). When dry sprayed at 75 µm (3.0 mils), B showed no rust and only a little rust when dry sprayed at 35 µm (1.4 mils). Vehicle A was only slightly rusted after dry spraying at 35 µm (3.0 mils) but showed considerable rust at 35 µm (1.4 mils) dry sprayed.

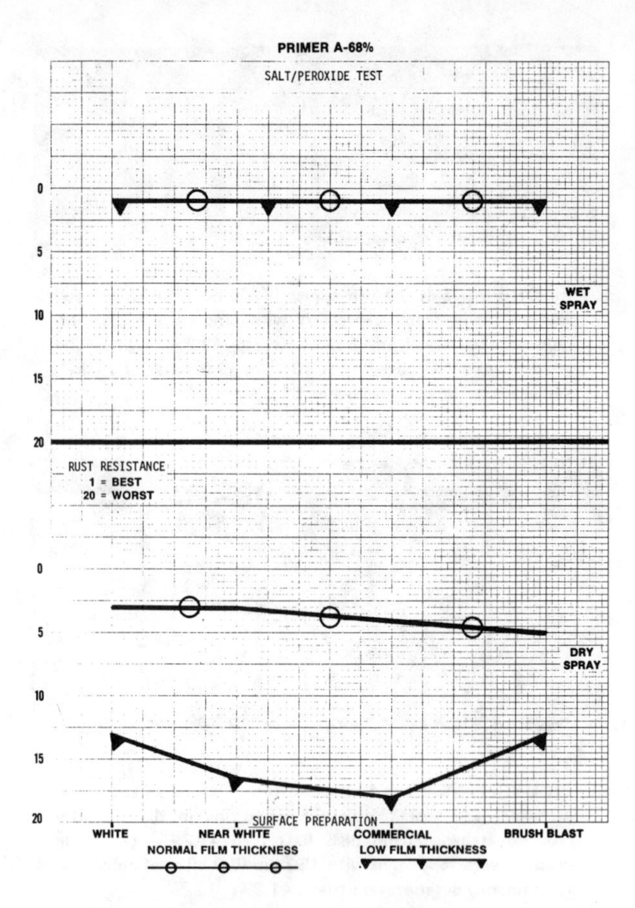

FIGURE 14 — Chart of comparative rusting of primer A with 68% zinc in dry film at normal and low film thickness, wet sprayed (top) and dry sprayed (bottom), after sodium chloride peroxide exposure.

The thickness of the primers on all of the test panels is shown in Table 1 and the corresponding degree of rusting in Table 2. The rust ratings are plotted in Figures 14 through 17, for each of the four primers.

Conclusions: This test series showed no pattern as far as zinc in the dry film is concerned. The commercial hydrolyzed ethyl silicate, Vehicle C with 86% zinc performed well when applied with full wet thickness. When the thickness dropped, performance dropped. The performance of this primer dropped severely when dry sprayed, more than any of the other primers. Vehicle C was very prone to dry spray, perhaps due to very fast drying solvents.

The best overall performers were primers based on Vehicle B, but zinc variations from 71 to 86% made no difference. Primers based on Vehicle A performed between B and C. These data confirm field experience that rust resistance is more vehicle related than zinc content related. The most serious deterioration resulted from dry spraying, especially at low thicknesses.

A second test to simulate possible field occurrences was undertaken to see how serious it would be if part of the zinc package was left out of the inorganic zinc primer. To test this zinc was left out of the primer in 10% increments, using the peroxide/sodium chloride test for rusting. None of the three primers showed any rusting, when applied wet at 75 µm (3.0 mils), until more than 40% of the total zinc had been left out. This was true even of Vehicle C, which had zinc effectively decreased from 86% to 51.6% zinc in the dry film.

These studies suggest that leaving out part of the zinc isn't nearly as critical as the conditions of application, especially dry spraying and low thickness.

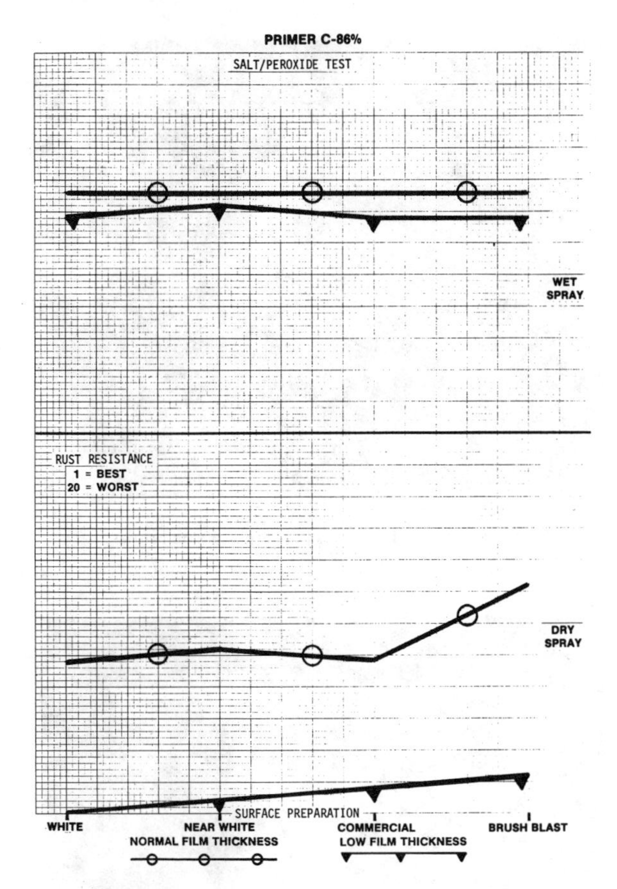

FIGURE 15 — Chart of comparative rusting after sodium chloride/peroxide exposure of primer C with 86% zinc in the dry film at normal and low film thickness, wet sprayed (top) and dry sprayed (bottom).

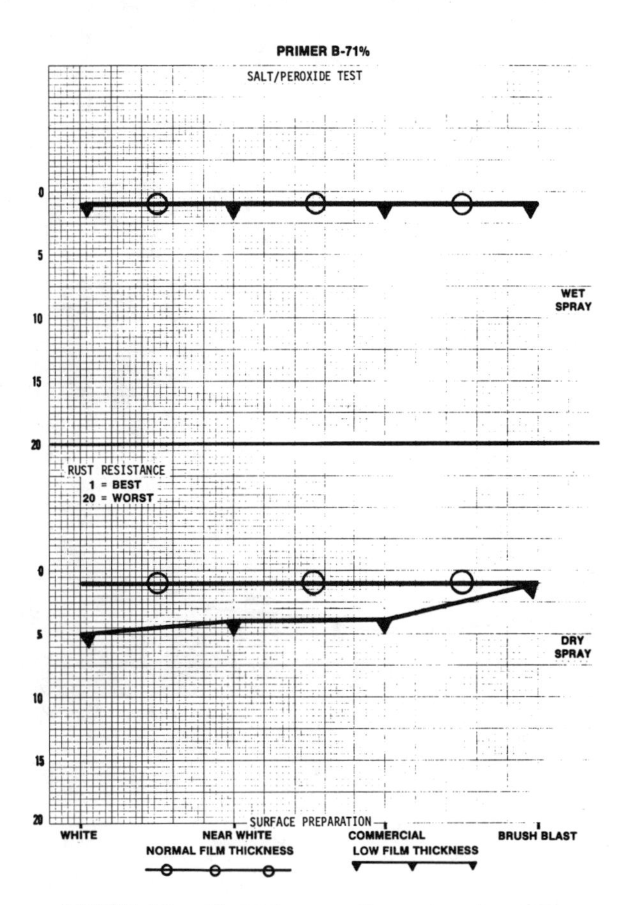

FIGURE 16 — Chart of comparative rusting after sodium chloride/peroxide exposure of primer B with 71% zinc in the dry film at normal and low film thickness, wet sprayed (top) and dry sprayed (bottom).

The Nature of Dry Spraying

The surface of a dry sprayed zinc rich, silicate primer was examined microscopically. Figure 18 shows a normal wet sprayed zinc surface. A dry sprayed primer is shown in Figure 19. Much of the zinc was deposited as huge balls of vehicle plus zinc, attached to the primer surface. Figure 20 is a cross section of a dry sprayed zinc primer when topcoated. Some of the steel surface was unprotected by zinc and much of the zinc was totally encapsulated by the topcoat, eliminating its potential for cathodically protecting the steel surface.

Mill Scale and Rusty Surfaces

When zinc is applied to the surface of steel, it serves as a sacrificial anode which will protect the steel, even in small bare areas. To do so, the zinc must be in contact with the steel and the zinc particles must be in contact with each other. The problem with the application of zinc primers over rust or mill scale is that the zinc clusters are often ineffectual since all the zinc is not in contact with the steel. Additionally, it is a misconception that a paint system consisting of a topcoated inorganic zinc is impervious to oxygen and moisture.

Mill scale is often easily detected under coatings because of the presence of magnetic black magnetite (Fe_3O_4), which forms a continuous hard crystalline film on the substrate iron under hot (mill) conditions:

$$\underset{O_2}{Fe \rightarrow} \underset{\text{non-magnetic}}{FeO} \underset{O_2}{\rightarrow} \underset{\text{magnetite}}{Fe_3O_4} + Fe_2O_3$$ [7]

However, under field exposure conditions, these oxides are further converted to soluble hydrates of iron. As these reactions occur, the firm hard mill scale expands in volume; heaving and splitting of the topcoat may result.

When mill scale is intact, it forms an excellent barrier to the environment, protecting the steel. However, cracks or the partial removal of mill scale lead to the establishment of corrosion cells. The exposed steel becomes anodic to the mill scale and rusting and pitting occur. Unless one could be assured that no loose mill scale is left on the steel and/or that zinc primer could flow into all recesses and cracks, zinc primers are a poor bet for application over mill scale. Figure 21 shows an organic topcoat over inorganic zinc over mill scale. Figure 1 shows a corrosion cell, the expansion of the mill scale and the heaving of the topcoat which will soon erupt, revealing a rusting pit in the steel.

Summary

The following factors are important in the protection of steel with self curing inorganic zinc primers:

1. The zinc content in the dry film has an optimum range which varies with the vehicle and the other components of the primers. Primers have been designed with good corrosion protection with as little as 42% zinc in the dry film. Much of the zinc in conventional primers can be left out or replaced without loss of protection. Some primers with high levels of zinc in the film offer poorer protection than primers with lower (and more optimum) levels of zinc, probably because the excessive zinc interferes with good film formation.
2. The vehicle type is most important.
3. Application of a wet primer coat is a necessity. Dry spraying is disastrous.
4. Film thicknesses near 75 μm (3.0 mils) offer more security

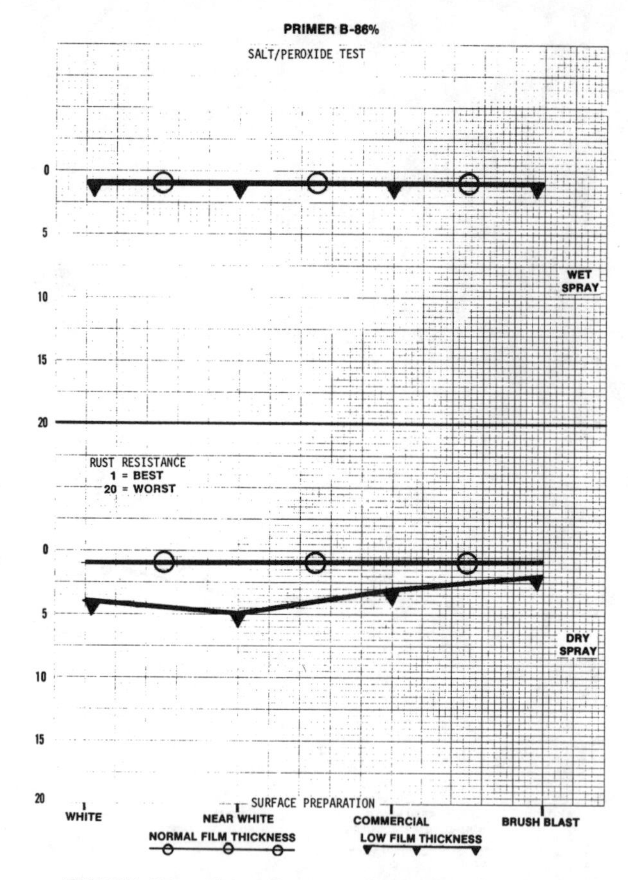

FIGURE 17 – Chart of comparative rusting, after sodium chloride/peroxide exposure of primer B with 86% zinc in the dry film at normal and low film thickness, wet sprayed (top) and dry sprayed (bottom).

FIGURE 18 – Wet sprayed inorganic zinc primer.

than lower thicknesses, but even this amount may not offer protection if the primer is dry sprayed.

5. Application of inorganic zincs over mill scale is not recommended.

Accelerated Test of Primer Protection – Sodium Chloride/Peroxide Test

Scope: This is a useful screening for the detection of film

FIGURE 19 – Dry sprayed inorganic zinc primer.

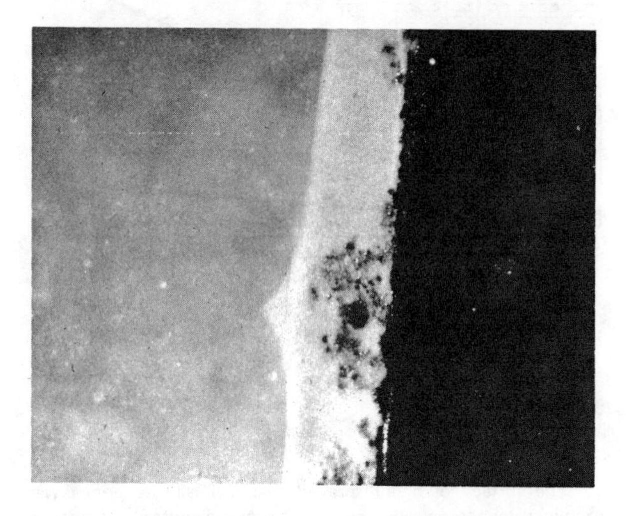

FIGURE 20 – Cross section of dry sprayed inorganic zinc primer showing zinc encapsulated by white topcoat and topcoat in direct contact with metal substrate.

FIGURE 21 – Cross section of paint chip imbeded in resin showing the white topcoat over the inorganic zinc primer over the black mill scale.

defects in organic and inorganic zinc primers which could cause premature failure. It is also useful for establishing the optimum dry film thickness of zinc primers.

Although primers are fully cured before subjecting them to this accelerated corrosion test, no difference in performance has been detected, whether the primer was cured overnight or for several weeks.

Procedure: An aqueous solution of 75 parts water, 15 parts sodium chloride, and 10 parts 3% hydrogen peroxide (by weight) is mixed in a plastic tub. Within 10 minutes after mixing, primed steel panels are placed in the bath at a 45° angle to the horizontal, totally immersed in the solution. Exactly 2 hours later, the panels are removed and washed with tap water. The panels are rated for rust on a scale of 1 to 20, 1 being perfect, (no rust).

Acknowledgment

The work represented in this paper includes contributions from John Eng, Noah Kemp, Howard C. Woodruff, and Joseph E. Rench, all of Napko Corporation.

References

1. Munger, C. G. Inorganic Zinc Coatings, Past, Present, and Future. Materials Performance, Vol. 14, No. 5, p. 25 (1975).
2. Munger, C. G. Environment—Its Influence on Inorganic Zinc Coatings, Materials Performance, Vol. 15, No. 3, (1976).
3. Berger, D. M. Zinc Paint Can Save $500,000 on Your Power Plant, Civil Engineering ASCE, pp. 66-69 (1976) October.
4. Moore, Gilbert. Evaluation of Centrifugal Method for Effective Solids in High Pigmented Inorganic Zinc Coating Systems, Journal of Paint Technology, Vol. 44, No. 565, p. 52 (1972).
5. Montle, J. F., Hasser, M. D. Recent Developments in Inorganic Zinc Primers, Materials Performance, Vol. 15, No. 8, p. 15 (1976).
6. Montle, John. The Minimum Level of Zinc, Carboline Corporation, Newsletter (1975).
7. Suchet, J. P. Crystal Chemistry and Semiconduction, Academic Press, pp. 177-180 (1977).

SECTION IV

Cracking

Designing with Stainless Steels for Service in Stress Corrosion Environments*

GEORGE E. MOLLER
The International Nickel Company, Inc., Torrance, California

Conditions conducive to stress corrosion cracking (SCC) of Al, Mg, Cu, C steel, martensitic and PH hardening stainless, austenitic stainless, Ni-base, and Ti alloys are listed. Compositional and other relevant factors are considered. Case histories are given of failures in AISI 304H tubing in a power station boiler, a test station for 566 C steam, braid covered AISI 321 hoses conducting steam and methane, steam-methane reformers, AISI 321 expansion joints in a salt water cooled power plant condenser, a sour water steam heated reboiler in a petrochemical plant, AISI 432 martensitic bolts in a petroleum refinery vacuum distillation column, AISI 304 heating coil in a paraxylene storage tank, AISI 304 tubes in a coal fired steam evaporator condenser, AISI 304 steam condensate return piping, MEA reboilers, AISI 304 vs polythionic acids in a petroleum refinery, AISI 321 tubes in a diesel hydrofiner, AISI 304 thermocouples in a catalytic reformer, bellows in flue gas expansion joints on oil tankers, AISI 304 in a paper mill pulp washer, and AISI 316L pump in a sea water desalination plant. Preventive measures are summarized.

IT IS WIDELY RECOGNIZED that in spite of their generally good corrosion resistance and heat resistant properties, stainless steels display stress corrosion cracking (SCC) under certain specific mechanical, metallurgical, and corrosive conditions. Table 1 serves to emphasize the fact that SCC is not restricted to stainless steels, but occurs in practically all alloys.[1]

Essentially, SCC is the fracture of a metal or alloy in a corrosive environment, under a tensile stress, either residual or applied. Cracking can be either transgranular, intergranular, or mixed mode, and usually occurs perpendicular to the load. The metallic part generally shows little or no corrosive attack. The risk of SCC occurring can be minimized by appropriate material selection, application of preventive measures, and/or specially designed and executed operating procedures.

This presentation will compare the recognized and well documented underlying factors of SCC with several recent case histories in an effort to assist engineers in a better understanding of the use of stainless steels.

*Presented during Corrosion/76, March, 1976, Houston, Texas.

TABLE 1 — Environment-Alloy Systems Subject to SCC

Alloy	Environment
Aluminum base	Sea water Salt and chemical combinations
Magnesium base	Nitric acid Caustic HF solutions Salts Coastal atmospheres
Copper base	Primarily ammonia and ammonium hydroxide Amines Mercury Sulfur dioxide
Carbon steel	Caustic Anhydrous ammonia Nitrate solutions Carbonate/bicarbonate
Martensitic and precipitation hardening stainless steels	Sea water Chlorides H_2S solutions
Austenitic stainless steels	Chlorides — inorganic and organic Caustic solutions Sulfurous and polythionic acids
Nickel base	Caustic above 315 C (600 F) Fused caustic Hydrofluoric acid
Titanium	Liquid nitrogen tetroxide Sea water Salt atmospheres Fused salt Methanol

Note: Table 1 has been simplified. See Logan, H. L., The Stress Corrosion of Metals, John Wiley & Sons, for a comprehensive list.[1]

Reprinted with permission from *Materials Performance*, May 1977, 32-44, © 1977 National Association of Corrosion Engineers

TABLE 2 – Primary Conditions for SCC

Susceptible Alloy – Metallurgical Condition
Tensile Stress
Damaging Environment
Time

TABLE 3 – Sources of Stress for SCC

Residual
Welding
Shearing, Punching, Cutting
Bending, Crimping, Riveting
Machining
Heat Treating
Service
Quenching
Thermal Cycling
Thermal Expansion
Vibration
Rotation
Bolting
Pressure
Dead Load

Conditions for SCC

Table 2 lists the conditions required for SCC of stainless steels. These are: (1) tensile stress, (2) a specific damaging environment, (3) metallurgical condition, and (4) time. Each of these factors require amplification to differentiate between the various types of stainless steels, to develop important aspects of the physical environments, develop the significant interaction between variables, and point out lower limits of certain variables.

Stress

A major factor in SCC is the presence of a tensile stress. Residual stress from the fabrication of equipment, applied service stress, or a combination of the two, contribute to cracking. Surveys[2,3] indicate that residual stresses are particularly dangerous because they are normally great in magnitude, often approaching the yield point (as is the case in fusion welding) or exceeding the yield point as a result of bending and bolting. Operation stresses, if contributors, can add to the residual stress in equipment. If operational stresses alone cause cracking, they usually have to be extreme in magnitude. Table 3 lists sources of stress in equipment in descending order of importance.

The limiting stress below which a stainless will not crack in a given environment is called the "threshold stress." An example is shown in Figure 1, giving applied stress versus time to fracture for a number of popular AISI 300 series stainless steels tested in boiling 42% magnesium chloride.[4] Relying on a threshold stress can be misleading because its value will vary with alloy, degree of cold work, strain, environment, temperature, and time. Residual stress levels cannot be readily determined. Service stresses can be calculated for ideal circumstances or they can be measured with strain gages, but this is often not practical. Furthermore, these methods do not account for localized submicroscopic stresses. Calculated stresses fail to take into consideration defects which can act as stress raisers.

From a practical standpoint, it can generally be assumed that residual fabricating stresses, especially at welds, are sufficient to cause SCC in as-fabricated susceptible alloys and are almost always present except in the simplest of structures. Residual stresses can be reduced by stress relieving or by the proper recommended heat treatment.

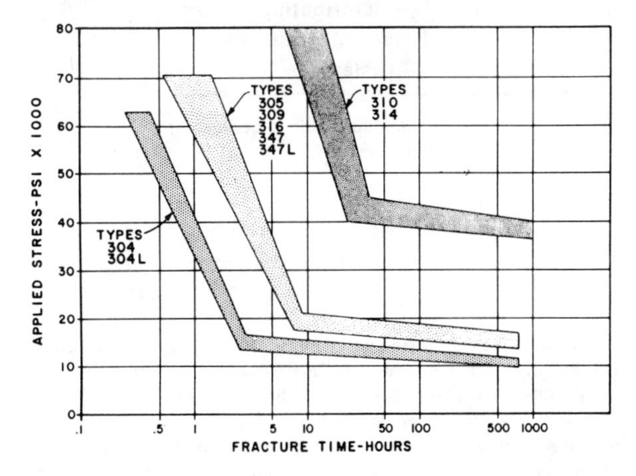

FIGURE 1 – Composite curves illustrating the relative SCC resistance of commercial stainless steels in boiling 42% magnesium and chloride.

TABLE 4 – Common Environmental–Stainless Alloy Systems Susceptible to SCC

Stainless Classification	Metallurgical Condition	Environment	Stress Level	Temperature
Austenitic	Annealed	Aqueous Chlorides	Moderate	Moderate
		Aqueous Hydroxide	Moderate	High
		High Purity Water, Oxygenated	High	Very High
Austenitic	Sensitized(1)	Aqueous Chlorides	Low	Ambient
		Polythionic Acid	Low	Ambient
		Sulfurous Acid	Low	Ambient
		High Purity Water, Oxygenated	High	Very High
		Aqueous Hydroxide	High	High
Ferritic	Annealed	Aqueous Hydroxide	High	High
Ferritic with residual Ni & Cu	Annealed	Aqueous Chloride	High	High
Ferritic	Sensitized(2)	Aqueous Chlorides	Moderate	Moderate
		Aqueous Hydroxide	Moderate	Moderate
Martensitic	>Rc 24	Aqueous Chloride	High	Ambient
		Aqueous Sulfide	High	Ambient
		High Purity Water, Oxygenated	High	High

(1) Chromium carbides precipitated in grain boundaries by holding at 427 to 899 C (800 to 1650 F) for specified time.

(2) Chromium carbides and nitride precipitation in grain boundaries by quenching from above 928 C (1700 F).

A fully austenitic stainless steel as produced in the mill usually has sufficiently low residual stress to be resistant to SCC. Fabricated austenitic stainless steels can be stress relieved by through heating between 843 and 899 C (1550 and 1650 F). This heat treatment, however, can cause chromium carbide precipitation at grain boundaries (sensitization) of regular carbon grades of austenitic stainless steel (AISI 304, 316, 309, and 310) which renders them prone to intergranular corrosion and intergranular cracking at very low values of stress in damaging environments. This rule does not apply to the columbium and titanium stabilized grades or the low carbon grades (AISI 347, 321, 304L, and 316L). These grades can only be sensitized by holding for prolonged periods at temperature below the stress relieving temperature.[5-7]

Therefore, it is advisable to use low carbon or stabilized grades if stress relieving is required, or to employ a full solution anneal at 1065 C (1950 F), or above for stress relief in cases where it is practical.

Environmental Factors

Austenitic Stainless Steels

Table 4 lists conditions which can cause SCC of stainless steels. The metallurgical condition is also shown because it affects susceptibility.

TABLE 5 – Contributing Factors – Chloride SCC Austenitic Stainless Steels

Concentration of chloride (evaporation)
Elevated Temperature
pH > 2
Oxygen
Time

SCC of annealed austenitic stainless steel in aqueous chloride environments is first on the list. This combination represents the most prevalent situation, owing to the broad usage of the AISI 300 series stainless steels and the omni-presence of chlorides. This environment-alloy system is the subject of constant discussion and perhaps the most misunderstood. Although frequent cases arise wherein the AISI 300 type stainless steels suffer from chloride-induced SCC, there are numerous applications where these steels have been used in the presence of chlorides with no cracking.

The contributing factors for chloride-SCC are given in Table 5. The most significant items in Table 5 are concentration of chloride (evaporation) and elevated temperature, which are closely related. From service experiences, some of which will be discussed in the second part of this paper, from constant reference in technical papers,[1,4,8-11] and from the normal way in which chloride-SCC tests are conducted, heat has to be applied to such a degree as to cause evaporation or alternate wetting and drying to concentrate the chloride. Supporting this observation is the fact, also readily apparent in the literature,[1] that austenitic stainlesses do not crack in strong chloride environments at ambient temperature, or even at elevated temperatures under condensing, washing, or diluting circumstances.

For example, Dana and Warren[12] performed laboratory studies to duplicate cracking of hot austenitic stainless steel piping under wet chloride bearing thermal insulation, a condition which exists in some chemical plants. They designed a wick test which results in the concentration of chlorides on the metal surface. The importance of temperature is illustrated in Figure 2 which shows that temperatures of 60 C (140 F) and higher are necessary to achieve cracking of specimens stressed to 90% of their yield strength.

Chloride-SCC is believed to occur at pH levels above 2.0,[13-15] provided the other conditions for cracking exist. At low pH values, the tendency for uniform corrosion or overall general activation of the stainless surface is favored. As the pH value increases towards the alkaline end of the pH scale, the tendency for cracking is reduced, or the time for the onset of cracking is substantially lengthened. However, at relatively high temperatures when free caustic is present and concentrated, AISI 304, 316, and other 300 series stainless steels suffer caustic stress cracking.[1,16-18]

Although the presence of oxygen is not necessary for cracking in boiling magnesium chloride, in NaCl and similar neutral chloride solutions, cracking is observed only if dissolved oxygen is present.[13,19] The low levels or absence of oxygen probably contributes to the excellent service experience of AISI 304 stainless steel tube feedwater heaters so prevalent today in fossil fuel and nuclear power plants. Cracking in concentrated hot alkali solutions can proceed in the absence of oxygen.

Sensitized austenitic chromium-nickel stainless steels, as indicated in Table 4, suffer intergranular cracking at relatively low levels of stress. Sensitization reduces the corrosion resistance of stainless steels to a fairly broad range of corrosive aqueous environments, many of which are readily resisted by the fully solution annealed (nonsensitized) stainless. This topic is covered in detail by the comprehensive publication of the Welding Research Council.[20] Sensitization is particularly important in relation to cracking by polythionic acids at low levels of stress at ambient temperature.[5,6] It is now established that polythionic acids can crack sensitized stainless steel. However, nonsensitized steels are resistant to cracking in polythionic acids.

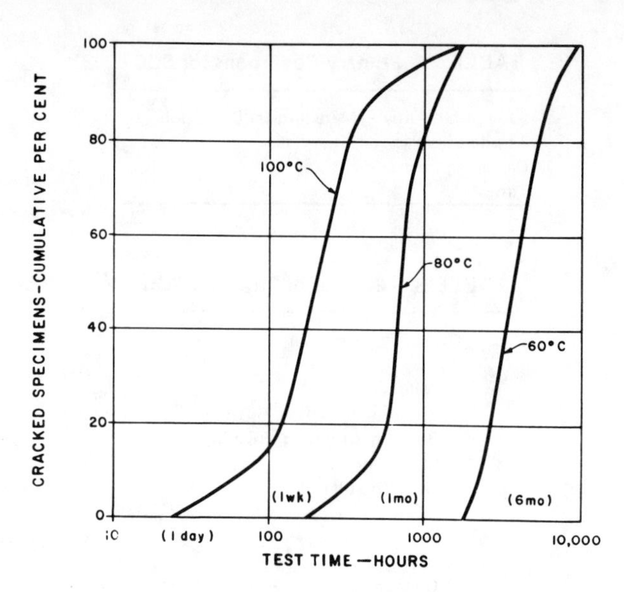

FIGURE 2 – Temperature vs time to SCC for AISI 304 stainless steel in 100 ppm chloride by wick test. U-bend specimens, applied stress 90% of yield strength, tests performed under concentrating conditions.

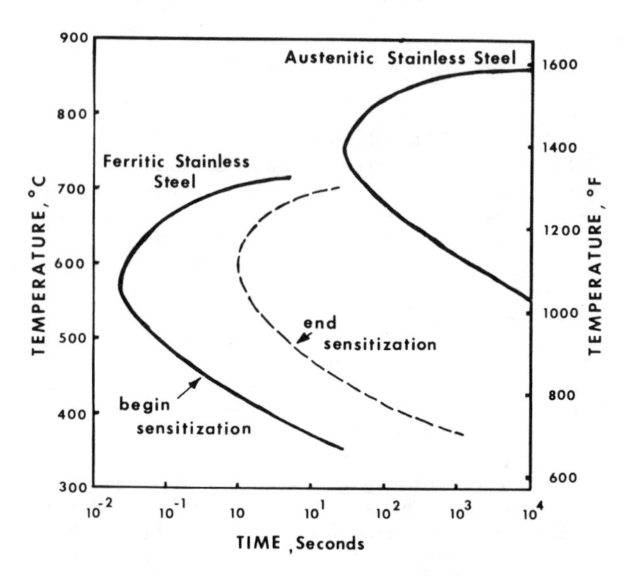

FIGURE 3 – Typical time-temperature-sensitization diagrams for austenitic and ferritic stainless steels.

Ferritic Stainless Steels

High purity ferritic stainless steels have been found subject to SCC in boiling 30% sodium hydroxide in tests exceeding 1000 hours[21] and in boiling 42% magnesium chloride when in a sensitized condition. If ferritic stainless steels have high residual levels of copper and nickel, they become susceptible to cracking in boiling 42% magnesium chloride tests.[22]

When ferritic stainless steels are in a sensitized condition, they become susceptible to intergranular corrosion, and if sufficiently stressed, they become susceptible to intergranular cracking when exposed to hot chloride or other less aggressive environments. The sensitization characteristics of ferritic stainless steels are markedly different than austenitic stainless steels.[23]

Conventional commercial ferritic stainless steels sensitize much more rapidly than austenitic stainless steels of equivalent chromium and carbon contents, as illustrated by the comparative time-temperature-sensitization diagram in Figure 3. The conventional

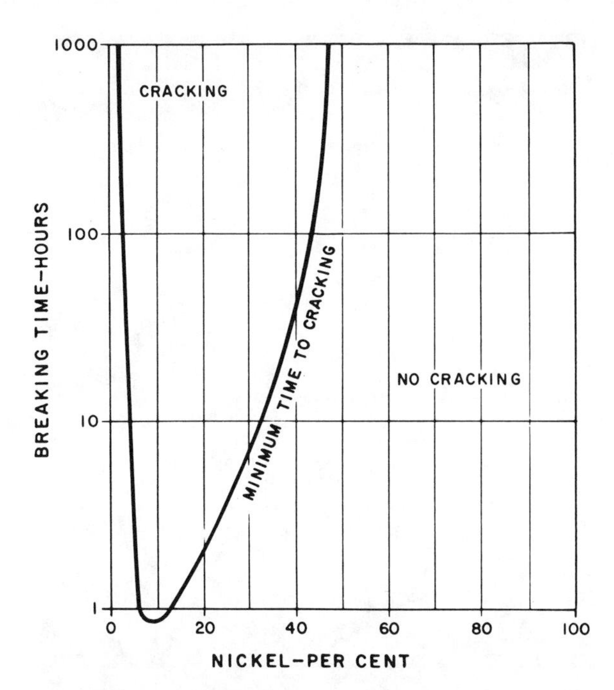

FIGURE 4 — Cracking time for iron-nickel chromium wires in boiling 42% magnesium chloride. Annealed wire loaded to range at 33,000 to 45,000 psi in tension.

TABLE 6 — Results of SCC Tests on Various Stainless Steels and High Alloys Exposed at 100 C (212 F) to 100 mg/l Chloride-Bearing Water[13]

Cracked AISI Type	Crack Resistant
304	AISI 329
304L	AISI 430
316	AISI 446
316L	Fe-Ni-Cr Alloy 20
347	Ni-Cr-Fe Alloy 600
310	Fe-Ni-Cr Alloy 825
202	

TABLE 7 — Summary of Known Compositional Factors to Minimize SCC of Iron-Chromium-Nickel Alloys

1. Increase ferrite content
2. Lower nitrogen content
3. Increase silicon content
4. Increase nickel content

ferritics sensitize even when quenched from 927 to 1149 C (1700 to 2100 F) hot finishing temperatres. Sensitization is removed or reduced through subsequent annealing by holding for sufficient time between 732 and 843 C (1350 and 1500 F). Sensitization can also be removed by holding the material long enough at the sensitizing temperature given in Figure 3 to permit chromium to diffuse back to the grain boundaries.

In view of the possibility of sensitization during mill processing or during the welding of fabricated equipment via carbon and nitrogen pick up, precautions might be deemed advisable. The safeguards are qualification tests according to ASTM A-262 Practice B or D.[24]

Martensitic Stainless Steels

Martensitic stainless steels, specifically the AISI 400 types, strengthened and hardened by quenching from an austenitizing temperature and then tempered to restore some degree of toughness, are susceptible to SCC or hydrogen embrittlement when stressed and exposed to aqueous chlorides or moist sulfide environments.[25,26]

Generally, the martensitic steels are susceptible if they have been tempered at temperatures below 566 C (1050 F) and have hardness values greater than Rockwell C 24. The threshold stress levels appear to be above 276 MPa (40,000 psi). The values given herein for tempering temperature, hardness, and stress vary, depending on the martensitic steel selected.

NACE publication 1F-166 (1973 Revision) entitled "Sulfide Cracking Resistant Metallic Materials for Valves, for Production and Pipeline Service" suggests that AISI 410 stainless steel should be double tempered at 621 C (1150 F) minimum to achieve a hardness level of Rc 22 maximum. Some martensitic stainless, according to this reference can be used at hardnesses above Rc 22 by special permission of the user.

Alloy Composition

A substantial research effort to minimize SCC has been devoted to alloy composition. Copson[27] found, as shown in Figure 4, that the nickel content of stainless steels is very important if a hot aqueous chloride environment is to be resisted. The susceptibility of the chromium-nickel stainless steels is pronounced at nickel contents of 8 to 12%. Decreasing the nickel content, as is done in dual phase alloys such as AISI Types 326 and 329, or increasing the nickel above 32%, increases the resistance of austenitic alloys, making them essentially immune to cracking in most practical situations. For instance, Incoloy Alloy 800,[(1)] Incoloy Alloy 825,[(1)] Carpenter 20-Cb-3,[(2)] Hastelloy G, C, and B alloys,[(3)] and Inconel Alloy 625 are all highly resistant to chloride SCC. Table 6 provides a ranking of several common engineering alloys based on the chloride wick test.[12] The effects of lowering or raising the nickel component of stainless alloys may be noted.

In the stainless steel family, Fontana[28] found by increasing the ferrite contents of cast stainless steels that the threshold stress level for cracking was substantially raised. Alloy Casting Institute chemical resistant grades (the "C" grades) usually contain more ferrite by composition adjustment than the wrought counterparts for ease of foundry practice. It has been observed over the years that fully annealed castings in chloride environments have very seldom displayed cracking. ACI grade CD-4MCu, high in chromium, relatively low in nickel, and containing molybdenum and copper for corrosion resistance, adjusted to be highly ferritic, exhibits a high level of resistance to chlorides as does its wrought equivalent AISI 329.

Uhlig and White[29] have shown that by reducing nitrogen and increasing silicon, wrought chrome-nickel stainless steels become considerably more resistant. This discovery is the basis for proprietary alloy 18-18-2[(4)] which is commercially available. Table 7 summarizes these compositional factors.

Case Histories

With the foregoing brief review of the criteria for SCC of the 3 main families of stainless steels in regular use, a number of recent

---

[(1)] Trademark of Huntington Alloys, Inc., Huntington, West Virginia.
[(2)] Trademark of Carpenter Technology Corporation, Reading, Pennsylvania.
[(3)] Trademark of Stellite Division, Cabot Corporation, Kokomo, Indiana.
[(4)] Trademark of United States Steel Corporation, Pittsburgh, Pennsylvania.

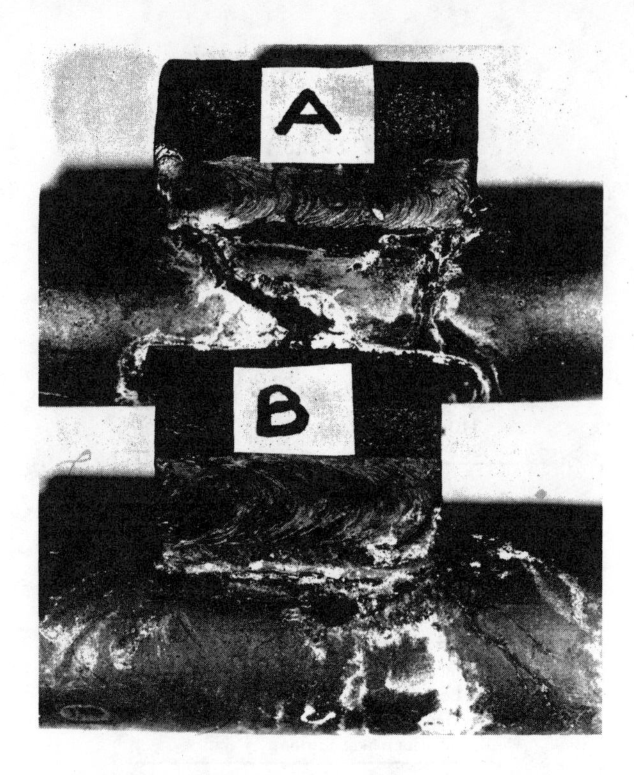

FIGURE 5 – Cracked superheater tubes, AISI 304 H, Case 1.

case histories are presented to illustrate the principles to be considered in alloy selection.

Case 1

A new oil fired subcritical central power generating station unit had certain portions of the superheater constructed of AISI 304H heavy walled tubing. Selection of a large grain austenitic stainless steel is common in superheaters, due to the need of adequate high temperature stress rupture properties.

During a standard alkaline boil out, which is typical for a new steel boiler, some of the NaOH, Na_3PO_4, and Na_2CO_3 aqueous mixture was inadvertently carried into the superheater zone. Tube skin temperatures were recorded at an average of 430 C (806 F) with a maximum of 508 C (948 C). Tubes were found to be cracked from the hot concentrated caustic mixture at points of high stress, which were created by welding spacer lugs on the outside of the tubes with heavy fillet welds. Figure 5 shows 2 tubes with the heavy weldments, cracks originating at the toe of the welds, and white salts on the outside surface. Figure 6 shows typical transgranular cracks of an annealed austenitic stainless steel.

Hot concentrated caustic can cause both intergranular and transgranular cracking of highly stressed austenitic stainless stesls, ferritic steels and even nickel-base alloys at extreme temperatures. Since austenitic stainless steels are necessary for their high temperature strength, the two avenues open to avoid the problem are: (1) avoid caustic carryover, and (2) use bolted spacers rather than welded spacer lugs. External weld attachments to pipe create extremely high localized stresses.

Case 2

A prototype test loop was built for the purpose of testing several designs of steam generators with hot liquid sodium. The system was designed to generate 151 MPa gauge (2200 psig) steam at 566 C (1050 F). The steam generation loop was essentially a once-through system with no steam drum. The high temperature-high pressure portion of the steam system, those portions designed to operate above 4 MPa (600 psi) and 318 C (600 F) were built of AISI 347 austenitic stainless steel, both wrought and cast. The energy in the test facility is dissipated via pressure reduction, desuperheat, extraction, and condensation. The liquid sodium

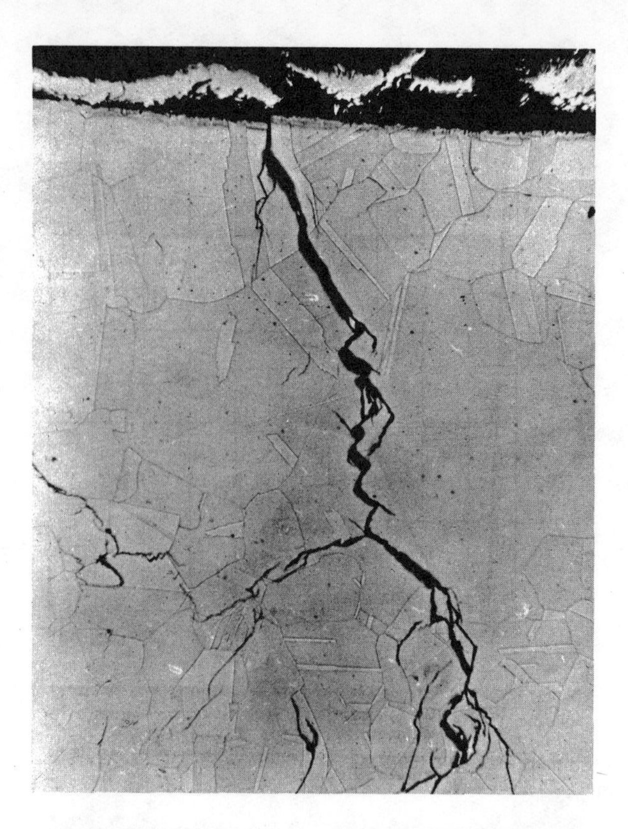

FIGURE 6 – Transgranular cracking of AISI 304 H, oxalic, Case 1. 50X

FIGURE 7 – Dye penetrant crack indications, AISI 347 steam piping, Case 2.

coolant or boiler water is demineralized and deaerated potable water. Chloride residuals vary from 0.5 to 2.0 mg/l.

In the early operation of the facility, highly branching transgranular cracks were found in wrought piping, wrought fittings, and in cast valve components. The facility operated only for two 30 hour periods when it had to be shut down for repair. The facility only achieved the 600 psig pressure and 304 C (580 F).

During start-up, a transient condition occurred during which demineralized water was changed to superheated steam within the piping system. It was found that chlorides, under this transient situation, could concentrate with ease to 486 mg/l and perhaps as high as 5000 mg/l. Cracks occurred only at points of high residual stress as shown in Figure 7. Cracks started on the inside of the pipe,

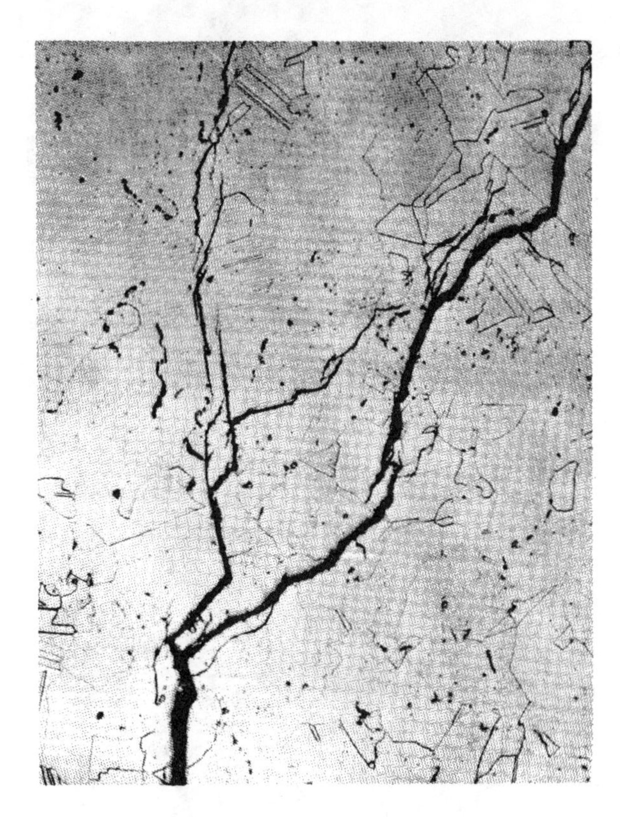

FIGURE 8 — Transgranular cracking in annealed AISI 347 pipe reducer, Case 2. 125X

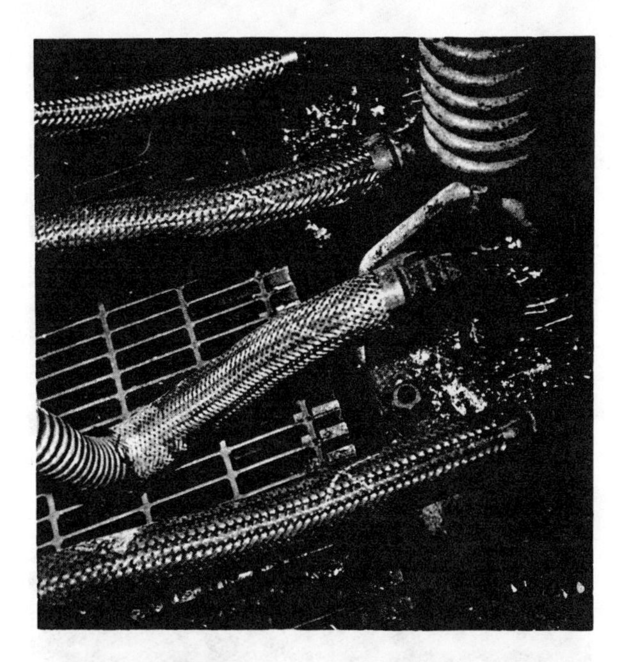

FIGURE 9 — Flexible stainless hoses connecting inlet header to catalyst tubes, Case 3.

opposite the points of stress created by the heavy gusset and reinforcing pad. Figure 8 depicts typical cracking found in piping at points highly stressed from weld attachments. 2-1/4 chromium-1 molybdenum ferritic steel was used for replacement because of its freedom from chloride cracking. Incoloy Alloy 800 and Inconel Alloy 600 were considered as alternatives.

In more conventional steam systems, found in utility fossil fuel and nuclear power plants, AISI 304, 347, and 316 stainless steels are being used successfully for high pressure steam piping. The equipment arrangement in these commercial installations precludes the entry of water or wet steam during operations thereby obviating the salt concentration mechanism required for cracking.

Case 3

Figure 9 shows braid covered AISI 321 stainless flexible hoses for conducting 80% steam and 20% methane to vertical, catalyst filled, centrifugally cast, externally fired, reformer tubes. This step is used in the production of ammonia and hydrogen. Figure 10 shows a hose with a crack in convolution number 5 from the right side of the picture.

Steam at 232 C (450 F) from a boiler treated with the conventional sodium phosphate, caustic for pH control, and sodium sulfite is mixed with methane at 367 C (690 F), then the combined stream is further superheated in a feed effluent exchanger. Total dissolved solids in the steam at 2 mg/l were entrained in steam from the steam drum and collected in the hoses.

Intergranular caustic cracking[19] occurred in a similar manner in 3 separate plants. The hoses, as produced in the factory, have extremely high residual stresses from 25 to 30% cold work, in the order of 690 MPa (100,000 psi). Relative movement of tube versus the inlet header added to the stresses, particularly in plants wherein the hoses were intentionally made short to save money. In Figure 9, the braid has pulled loose in one hose assembly from tremendous tension.

The solution, in new plants and to retrofit older plants, has been to use long flexible arrangements of chrome-moly or carbon

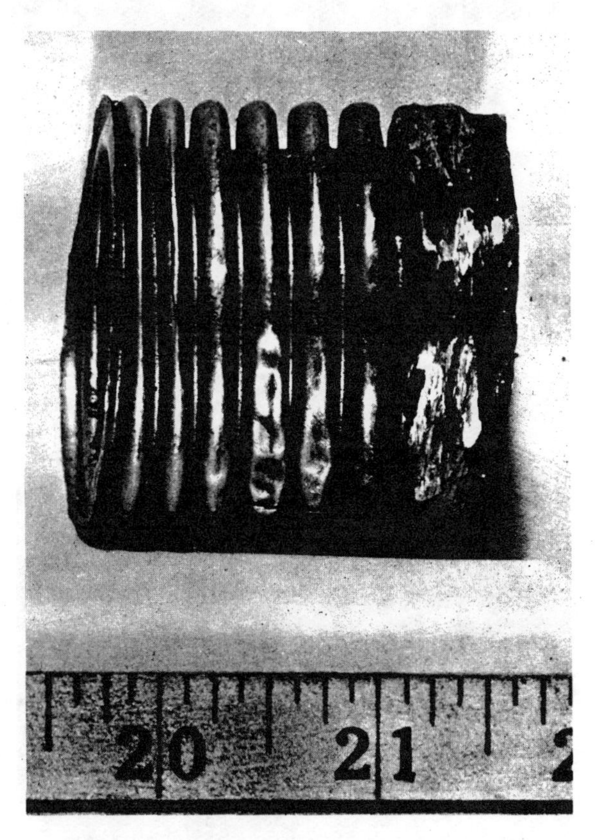

FIGURE 10 — Cracking in fifth convolution from right—intergranular caustic cracking, Case 3.

steel pipe in lieu of flex hoses.

In a number of early steam methane reformers, long flexible hoses were provided at tube inlets. The long hoses provided adequate flexibility to accommodate the relative opposing pipe to tube movements which occur between start up and full operating temperatures. In these cases, in the absence of extreme added

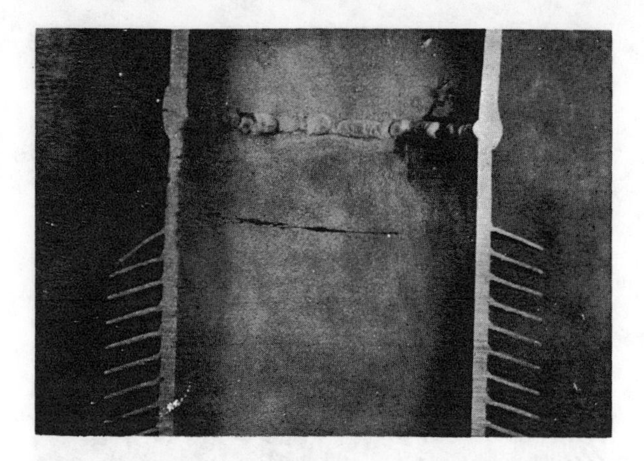

FIGURE 11 – Dye penetrant indication of cracked AISI 304 SMR superheater tube, Case 4. 1X

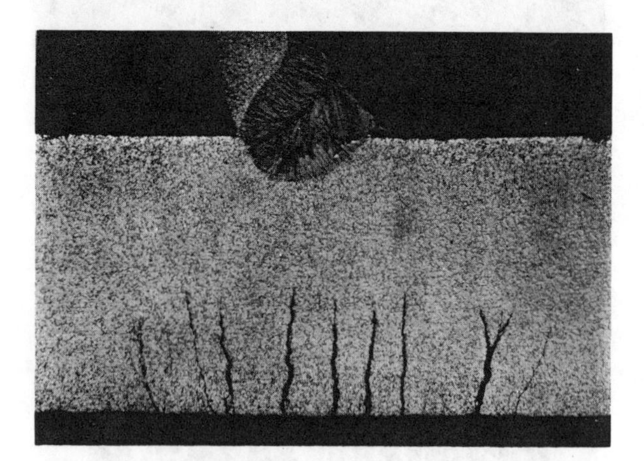

FIGURE 12 – Transgranular cracks starting on inside in high stress area from weldment, Case 4. 10X

FIGURE 13 – Alkaline salt patch on inside of superheater tube, Case 4. 1X

operational loading, austenitic stainless steel flexible hoses have given good service.

Case 4

In another steam methane reformer, steam was preheated prior to mixing with methane in a convection bank superheat coil. Skin temperatures were approximately 483 C (800 F). Figure 11 illustrates a crack on the inside of a finned superheater tube. Figure 12 vividly illustrates that the transgranular cracks are associated with fin attachment welds.

Figure 13 discloses a patch of evaporated alkaline salt material carried into the tubing by entrainment. The deposit, concentrated in the tube, from the more dilute boiler water, contained 5.6% of chloride. The superheater tubing was replaced with Incoloy Alloy 800, an alloy with substantially more resistance to chloride SCC.

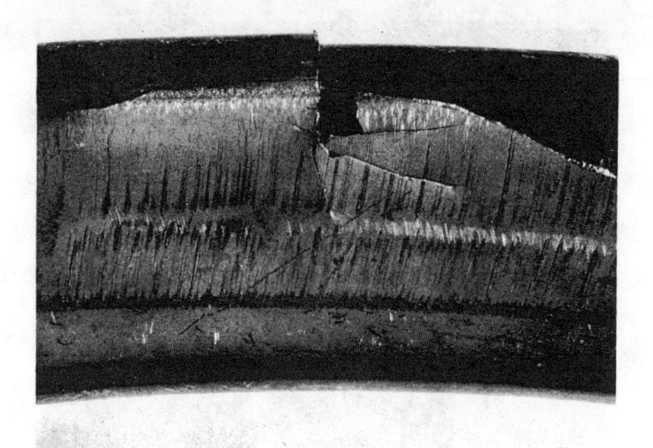

FIGURE 14 – Segment of cracked AISI 321 expansion joint from extraction steam line, Case 5. 1X

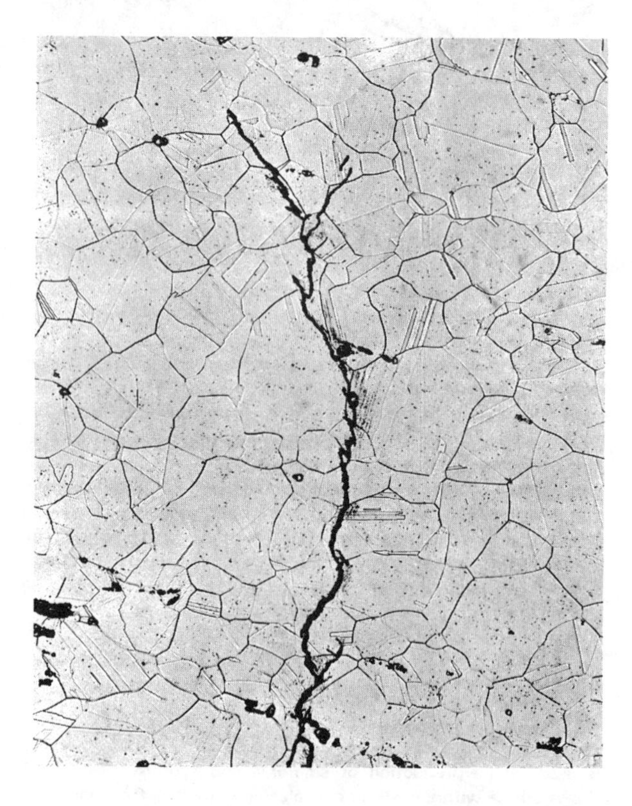

FIGURE 15 – Transgranular cracking of AISI 321, 10% oxalic, Case 5. 250X

Case 5

A segment of a cracked AISI 321 expansion joint is shown in Figure 14. Light microscopy shown in Figure 15 and an SEM

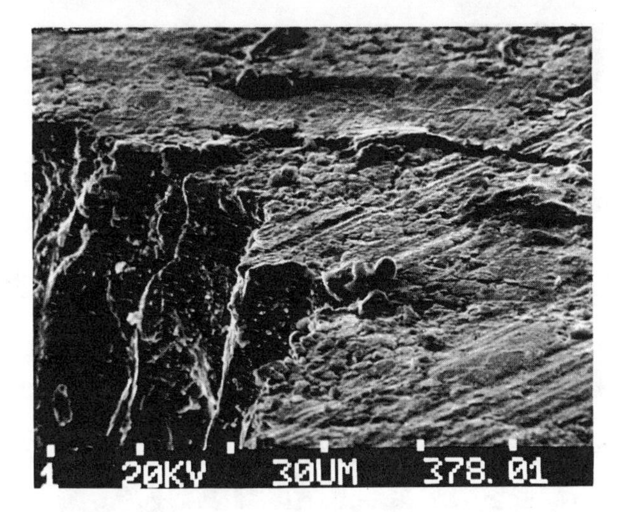

FIGURE 16 — SEM fractograph of surface at vertical transgranular crack with one major and two minor branch cracks. 580X

FIGURE 17 — Intergranular cracking of high purity ferritic oxalic electrolytic, Case 7. 100X

fractograph shown in Figure 16, clearly indicate transgranular cracking.

The expansion joint was taken from an extraction steam line in a supercritical once-through power plant utilizing salt water cooled condensers. An extraction steam line expansion joint can have a temperature differential of 85 C (185 F) across its wall with temperatures up to 160 C (320 F). With a serious condenser leak, chloride salts could permeate the steam system and concentrate on the heavily cold worked and operationally, highly stressed bellows. It has become common to use Inconel Alloy 600 expansion joints in critical locations of power plants to resist chlorides and caustic.

Case 6

A process water shell and tube type fixed tube sheet heater is presently being used in a potassium sulfate plant in the Western United States. Well water, with a pH of 8.2, a calcium hardness of 400 mg/l, and 0.2 Wt% sodium chloride is heated with 827 kPa gauge (120 psig) steam from 10 to 82 C (50 to 180 F).

AISI 304 stainless tubes were installed after carbon steel suffered serious corrosion. The AISI 304 tubes suffered circumferential cracking after 9 months of service and have been replaced with CA-706 (90-10 copper-nickel) which has been in service without failure for 3 years.

Hardness in the water results in scale buildup on the tubes. The tube skins become hotter as scale inhibits heat transfer. The hot tube skins concentrated chlorides on the hot surfaces under the deposits. Stresses were a result of bowing and binding in the baffles of the fixed tube sheet water heater.

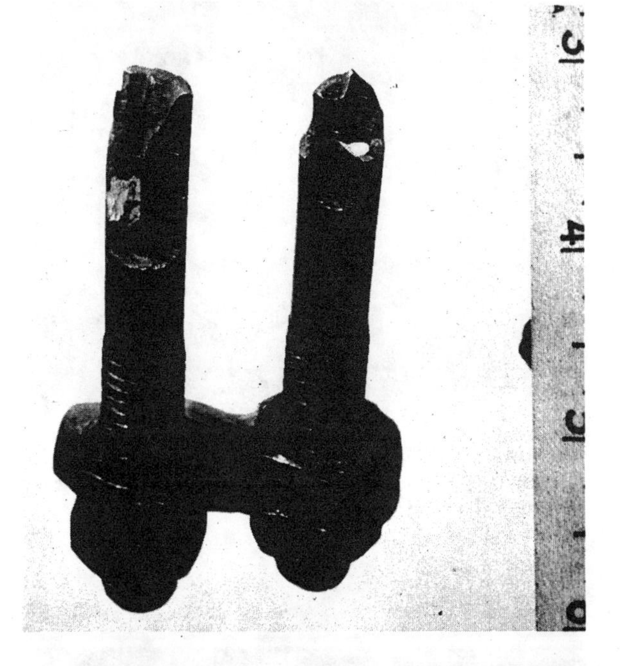

FIGURE 18 — Sulfide cracked AISI 430 martensitic stainless steel bubble cap hold down bolts, Rc 34 and 36, Case 8. 1.5X

Case 7

Figure 17 shows intergranular cracking of a 26Cr-1Mo high purity ferritic stainless steel. The cracking was prevalent in "U" bend areas of 19 mm (3/4 inch), 1.65 mm (0.065 inch) wall (16 gauge) tubing of a sour water steam heated reboiler from a petroleum refinery. Aside from the cracks observed in tube samples, the metal surfaces showed little evidence of surface attack. The "U" bends were not stress relieved or annealed. The 26-1 tubing apparently was sensitized as produced and displayed a "ditched" structure in 15% nitric-methanol etch.

Refinery sour water contains chlorides, ammonia, sulfides, thiocyanates, and cyanides. Sour water reboiler service is unusually arduous because of the boiling action of 345 kPa gauge (50 psig) steam on the inside of the tubes and the presence of contaminated water on the outside of the tubes. The environment is conducive to cracking of susceptible alloys and also pitting under deposits in all but the most corrosion resistant materials such as Inconel Alloy 625, Hastelloy C-276, Hastelloy G, or Alloy 6X.[5]

Case 8

The two cracked AISI 431 martensitic stainless steel bolts in Figure 18 came from an idle petroleum refinery vacuum distillation column. They were exposed to moist sulfides at a pH of 4.5. Hardnesses were Rc 34 and 36. In this vacuum column, 40% of the martensitic stainless steel bubble cap hold down bolts were defective due to hydrogen assisted transgranular cracking in the wrought shank and intergranular cracking at the yoke welds. Replacement bolts were delivered at hardnesses below Rc 25.

Case 9

Figure 19 illustrates cracking which begins in a transgranular mode on the outside of a heavily cold worked and sensitized AISI

[5] Trademark of Allegheny Ludlum Steel Corporation, Pittsburgh, Pennsylvania.

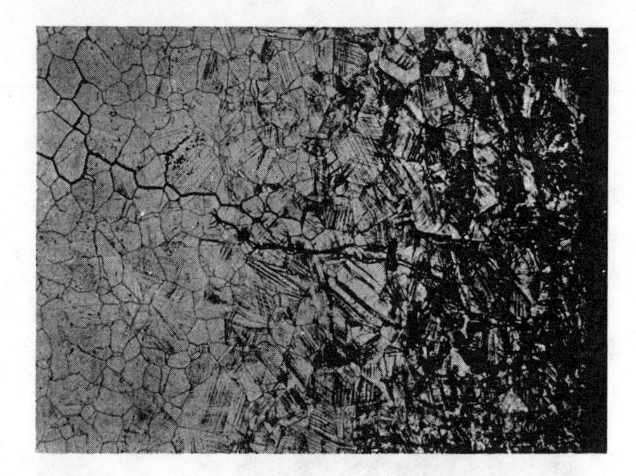

FIGURE 19 – Transgranular and intergranular attack in a single crack in sensitized and cold worked AISI 304, oxalic electrolytic, Case 9. 200X

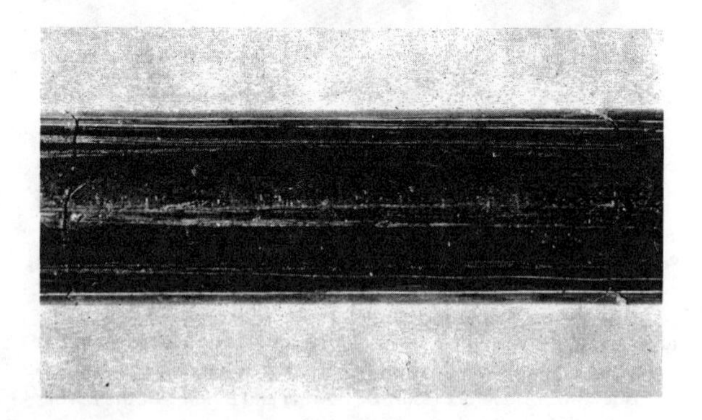

FIGURE 20 – Circumferential chloride cracking in longitudinally stressed AISI 304 evaporator condenser, Case 10. 1.5X

304 pipe and changes to an intergranular mode further in from the cold worked surface.

The equipment in this case was an AISI 304 heating coil in a paraxylene storage tank. Paraxylene is kept at 71 C (160 F) with live steam. The coil was cold bent in fabrication and then given a 843 C (1550 F) stress relief.

Hydrocarbon and petrochemicals contain water and chlorides. Entrained water with the paraxylene stock settles to the bottom of the storage tank and can contact the heating coil if it is not removed frequently from the tank bottom.

Case 10

Figure 20 depicts a circumferentially cracked 19 mm (3/4 inch) diameter, (22 gauge) AISI 304 tube from a coal fired power plant evaporator condenser. On the tube side of this shell and tube low pressure heat exchanger is demineralized boiler feed water at 193 C (380 F), while steam vapors from an evaporator at 205 C (400 F) supply heat by condensing on the shell side. If a vacuum is inadvertently created, salty, concentrated water can be syphoned onto the hot tubes. Evaporators in these units have also been reported to prime or foam, allowing salty water to reach the condenser tubes.

If the demineralizer becomes spent during a massive condenser leak, chlorides can contact the inside of the hot tubes also and concentrate.

Tubes in this unit were found transgranularly cracked from both the inside and the outside. Figure 21 shows a crack from the outside of a tube. Figure 22, an SEM photograph, shows a slight pit at the end of a crack covered with deposits. Chlorides were found in

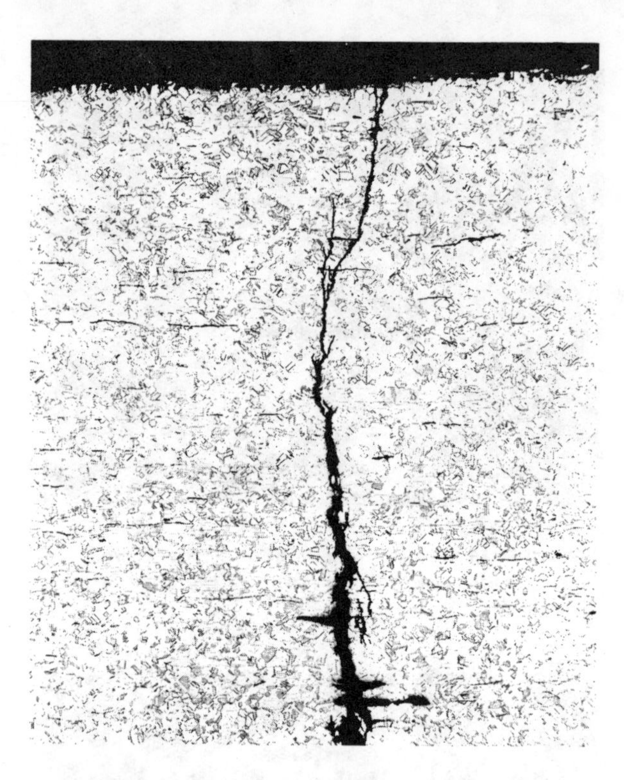

FIGURE 21 – Transgranular crack from outside of AISI 304 tube, evaporator condensate side. 10X

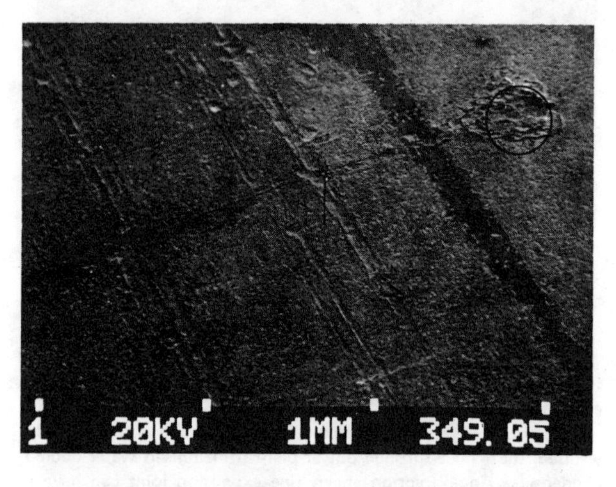

FIGURE 22 – Incipient pit and deposits on outside of AISI Type 304 evaporator condenser tube. Crack starts in pit. Deposits contain chloride. 30X

the deposit by EDX analysis. Residual longitudinal stresses as high as 220 MPa (32,000 psi) were measured in this tubing.

Owing to the hazards of syphon effect of salty water, the choice of a high nickel stainless steel for supercritical units, CA 706 or CA 715 copper-nickel for subcritical units is indicated, since this equipment functions somewhat as a desalination plant. This unit, in this case, was retubed with Incoloy Alloy 800.

Case 11

Figures 23 and 24 show transgranular cracking of slightly sensitized AISI 304 steam condensate return piping which is insulated and buried directly in soil at a University in a farming community. In Figure 23, repair welds can be noticed resulting in additional cracking as shown by heavy dye penetrant indications. It

FIGURE 23 — AISI 304 condensate return piping displaying cracks to dye penetrant test, Case 11.

FIGURE 24 — Transgranular chloride cracks in slightly sensitized AISI 304 pipe, oxalic electrolytic, Case 11. 200X

was felt that chlorides from various sources such as from fertilizer potash, could leach through the insulation and onto the hot stainless steel surfaces. In view of the extent of this piping network, it was suggested that cathodic protection be applied, rather than replacement with a resistant alloy. The chemical industry frequently utilizes silicone base high temperature coatings on austenitic stainless steel piping if there is a danger of salty water contacting hot pipe.

Case 12

In 2 separate plants, MEA reboilers tubed with AISI 304 and 304L failed in periods of approximately 3 months from intergranular SCC. The MEA reboilers were subjected to 843 to 899 C (1550 to 1650 F) stress relief. In the first plant, the exchanger "U"

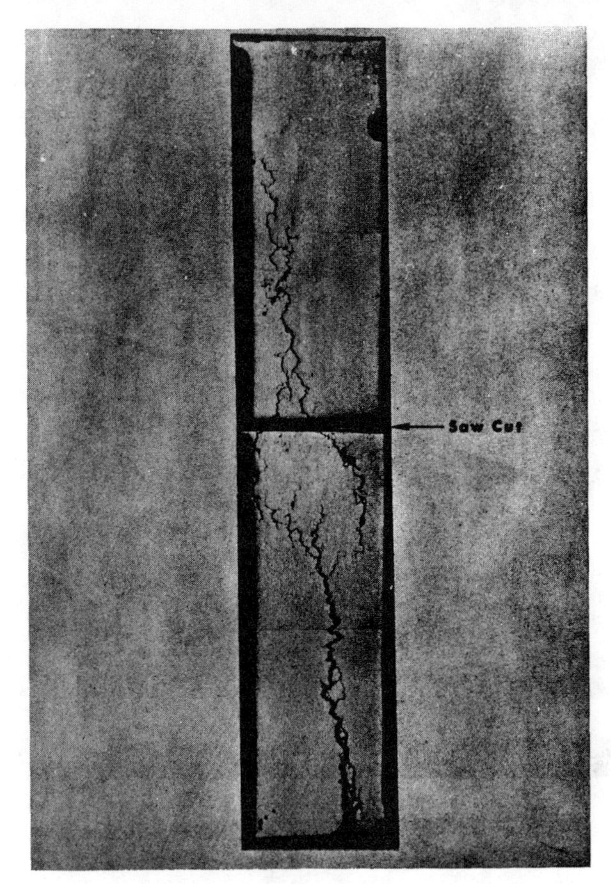

FIGURE 25 — Intergranular cracking in a sensitized forged AISI 304 pipe flange from polythionic acid, no etch.

tube bundle was supposedly held at 899 C (1650 F) for 2 hours. In the second plant, the low carbon AISI 304 dwelled in the sensitizing temperature region for extended periods, 6 hours or longer.

MEA is an alkaline material in a water solution for absorbing H_2S and CO_2 from refinery gases. Chlorides are present in the circulating MEA. Polythionic acid forming compounds are also present in the solution. The MEA reboilers are heated with 1 MPa gauge (150 psig) steam, thereby creating the hot concentrating environment on the shell side. In these instances, it is not certain whether cracking was from chlorides at elevated temperature or polythionic acid during shut down. Both situations could exist.

In Cases 12 and 9, regular grades of austenitic stainless steel became sensitized as a result of a stress relief. Although the opportunity for transgranular cracking was reduced, the austenitic stainless steel became susceptible to intergranular cracking.

The preferred construction would be to employ stress relieved stabilized stainless steel or to solution anneal individual tubes after bending by electrical resistance methods.

Case 13

Figure 25 is an example of polythionic acid intergranular cracking[7] in sensitized AISI 304 in a specimen taken from a high pressure forged flange used in a petroleum refinery hydrodesulfurizer. This plant had operated for 138,000 hours with the flange experiencing temperatures as high as 384 C (720 F). The AISI 304 equipment eventually became sensitized with prolonged exposure at the relatively low temperatures. Eventually, cracking commenced from contact with polythionic acids formed from iron sulfides, moisture, and air during shutdowns.

The AISI 304 equipment has been replaced with AISI 347 stainless steel which does not sensitize at these operating temperatures.

Source: *Materials Performance*, May 1977, 32-44

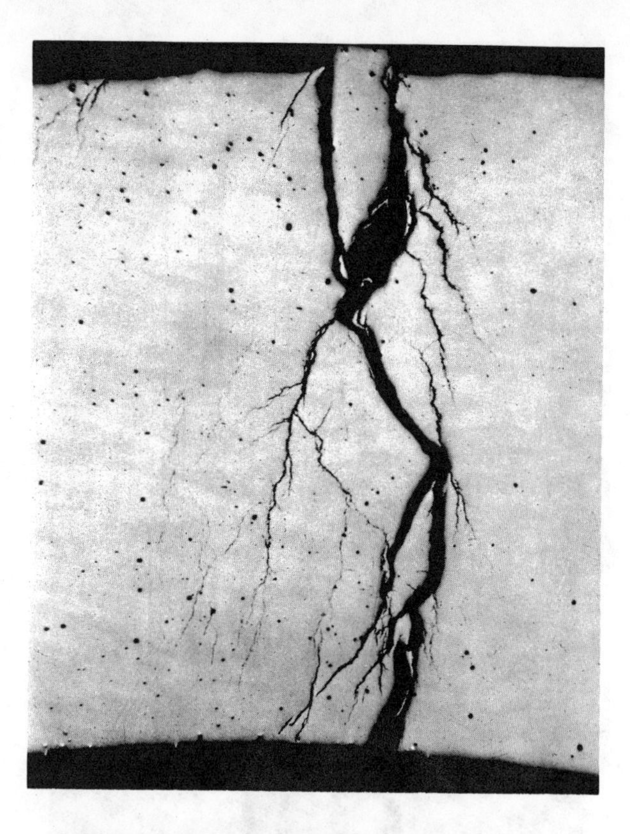

FIGURE 26 – Transgranular SCC in a 3/4 inch diameter AISI 321 feed/effluent exchanger tube, Case 15. 64X

Case 14

In a diesel hydrofiner, 19 mm (3/4 inch) diameter, 1.65 mm (0.065 inch) wall (16 gauge) AISI 321 austenitic stainless steel tubes were found leaking from transgranular SCC after 18 months of service. Figure 26 indicates the extent of cracking in one area in this unetched photomicrograph with cracks starting at the outside surface.

Sulfur containing diesel feed being heated to 288 C (550 F) on the shell side of the "U" tube heat exchanger was receiving heat from reactor effluent on the tube side of the heat exchanger. Reactor effluent enters the tubes at 399 C (750 F) and leaves at 342 C (650 F).

Longitudinal and spiraling cracks were found in straight portions of the tubing. The "U" bend portion of the tubes had been provided with an 843 C (1550 F) stress relief.

Investigation disclosed salt water intrusion into the diesel feedstock by virtue of a salt water condenser leak in the vacuum distillation unit and negligible residence time in the water, gas, hydrocarbon separation vessel, and storage tank between the vacuum distillation section and the hydrodesulfurizer. Present in deposits on the outside of the tubes were 0.3 Wt% of chloride compounds. Feed stock flashes or evaporates on this feed-effluent exchanger, permitting chlorides to concentrate.

Residual hoop stresses from roll straightening of the tubing were considered by the operating company to have been the source of contributory stress. Positive steps have been established to avoid chloride carry through.

Case 15

Sheathed thermocouples are used to register catalyst bed temperatures in catalytic reformer reactors. These thermocouples consist of an AISI 304 6 mm (1/4 inch) outer diameter by 0.71 mm (0.028 inch) wall tube containing the dissimilar thermocouple wires packed in magnesium oxide.

The thermocouple penetrates the 482 C (900 F) catalyst bed inside a thermowell built of 37 mm (1.5 inch) diameter stainless steel pipe. In 3 similar plants in 2 refineries, the AISI 304 sheaths of numerous thermocouples were found to be cracked transgranularly as shown in Figure 27. Moisture had accumulated in the vertically oriented thermowells.

FIGURE 27 – Transgranular SCC in an AISI 304 thermocouple sheath, Case 16. 95X

Chlorides were available since these refineries were located in industrial areas near the seacoast. A hot, alternately wet and dry (condensing-evaporating) condition exists on the surface of the sheaths in the transition zone between 482 C (900 F) interior and outside ambient temperature.

Hoop stresses result from swelling MgO. The succesful remedy has been to coat the susceptible zone of the sheaths with a high temperature silicone coating.

Case 16

The marine division of a petroleum company reported two similar experiences with leaking expansion joints on tankers. The bellows portion of the expansion joints were formed AISI 321 stainless steel sheet. They were located in the flue gas line between the boiler, burning high sulfur fuel oil, and the scrubber, using salt water for the scrubbing media. In both instances, the bellows were not only pitted, but suffering from transgranular SCC.

Stresses are present from cold work and flexing. Chlorides are present in fuel oil, sea spray, and salt water. Heat, necessary to cause concentration in recessed area, is indigenous with flue gas.

For such severe environments, industrial operators and the United States Navy are switching to materials such as Inconel Alloy 625 and Hastelloy C-276 for expansion joints.

Case 17

The use of 0.711 mm (0.028 inch) wall (22 gauge) welded 19, 22, and 25 mm (3/4, 7/8, and 1 inch) AISI 304 stainless steel surface condenser tubing for power plants has become prevalent over the last 14 years. Tubing, in service, primarily in noncoastal areas, served by rivers and lakes, now exceeds 9000 km (300,000,000 lineal feet).

On rare occasions, AISI 316 stainless steel has been employed for condensers using sea water.

No cases of SCC have ever been reported to the author's knowledge. Conditions in surface condensers are not of a severity to cause chloride cracking. Condensing steam at or about 54 C (130 F) is not nearly hot enough to concentrate or boil chloride containing water inside the tubes, even if flow stops or deposits occur.

Case 18

In the kraft process which produces wood pulp for the manufacture of paper products, spent chemicals and organic wood residues, received from final pulp washing at a concentration of 17% solids in water, are concentrated in vertical multiple effect evaporators. The solution is called weak black liquor. The concentrated strong black liquor, at 60% solids by weight, is used as fuel in a steam boiler called a chemical recovery boiler.

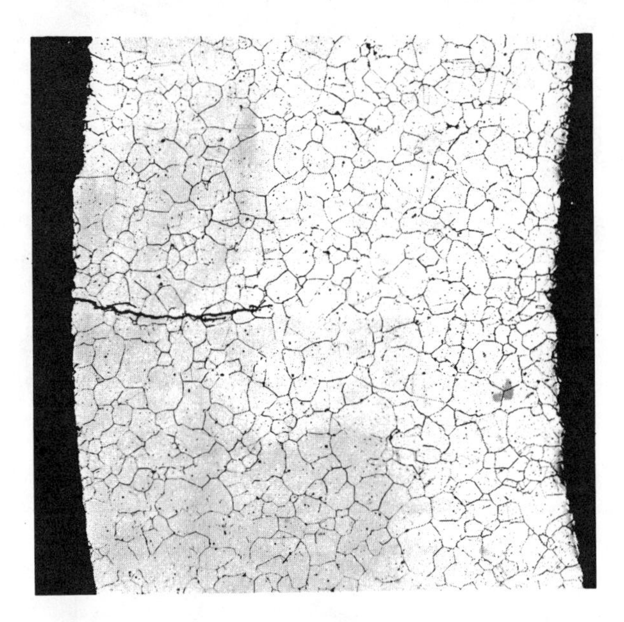

FIGURE 28 — Brine recirculation pump with iron anode attached to manhole cover. Case and impeller are ACI CF-3M (cast 316L), Case 19.

The chemicals present in black liquor are sodium carbonate, sodium hydroxide, sodium sulfide, and organic sodium compounds. Sodium chloride is always present, often at levels exceeding 500 mg/l. The pH level is typically 12.5.

The hottest first effect evaporator is heated with 50 psig steam at a temperature of 147 C (297 F). Strong black liquor leaves the tubeside at the evaporator at 110 C (230 F).

AISI 304 stainless steel has become the standard tube material for the first 2 (hottest) effects to resist hot alkaline corrosion and is often used in all 7 effects. Service experience with AISI 304 has been excellent.

The turbulent condition within the tubing is not conducive to salt deposition and salt concentration. SCC is not normally observed in strongly alkaline solutions, even though chlorides are present, unless the temperature is elevated to the point where caustic can concentrate.

Case 19

A large, single stage scroll type pump which was constructed of ACI CF-3M (cast AISI 316L) stainless steel, except for the bearings, is shown in Figure 28. The pump recirculated brine at temperatures of 38 C (100 F) in a sea water desalination demonstration plant.

The pump, which employed block iron anodes to minimize crevice corrosion and pitting during shutdown periods, was found to be free of SCC or pitting after the 4 year life of the demonstration plant.[30]

In this application, the absence of SCC is attributable to relatively low temperatures, modest service stresses in heavy walled castings, and the presence of ferrite in cast austenitic stainless steel.

Preventive Measures

With scrutiny of each engineering design and process equipment situation, SCC can be prevented or minimized by one or more of the following techniques:[31]

1. Avoid concentration of chloride or hydroxide by equipment design or arrangement.
2. Reduce stresses by stress relief of the proper alloy. Remember that stress relief of nonstabilized or low carbon materials may result in sensitization which renders austenitic stainless steels more susceptible to chemical corrosive attack and intergranular corrosion cracking than if not heat treated at all. Highly stressed areas can be minimized or avoided by good design practice. For example, forged elbows in pipe systems are preferred over mitered elbows. Forged tees are preferable to direct sub-in joints. The use of heavy weld attachments, such as support lugs on piping and other equipment is undesirable. In metal fabrication, avoid poor welding workmanship, such as lack of penetration which will result in stress raisers.
3. Eliminate chloride salts.
4. Remove oxygen or use oxygen scavengers in the case of chloride containing environments.
5. Use corrosion inhibitors.[31]
6. Apply cathodic protection.
7. Apply chemical resistant coatings in appropriate situations to provide a barrier to the corrosive species.
8. Convert tensile stresses to compressive stresses by shot peening.
9. Select a cracking resistant alloy with corrosion resistant characteristics adequate for the environment when none of the foregoing preventive measures will remedy a given situation.

Summary

SCC usually occurs to susceptible alloys only under extreme circumstances wherein 2 or 3 interrelated physical phenomenon exist concurrently. In the case with chloride SCC of austenitic stainless steels, for example, an elevated temperature is necessary to cause the concentration of the chlorides in the presence of a highly stressed component.

A review of the predominant damaging environments and variables for stainless steels has been presented with substantiation from technical literature. The principles involved in engineering decisions to understand and avoid SCC have been illustrated with recent case histories.

References

1. Logan, H. L. The Stress Corrosion of Metals, John Wiley & Sons, New York (1966).
2. Fontana, M. G., Greene, N. D. Corrosion Engineering, McGraw-Hill (1967).
3. ASTM Special Publication No. 264, Report on Stress Corrosion Cracking of Austenitic Stainless Steels.
4. Denhard, E. E., Jr. Corrosion, Vol. 16, p. 131 (1960).
5. Samans, C. H. Corrosion, Vol. 20, p. 256t (1964).
6. Piehl, R. L. Proc. API, Vol. 44, (III), p. 189 (1964).
7. Stickler, R., Vinckier, A. Trans. of ASM, Vol. 54, p. 363 (1961).
8. Logan, H. L., McBee, M. J. Materials Research and Standards, Vol. 7, No. 4, p. 137 (1967).
9. Copson, H. R., Cheng, C. F. Corrosion, Vol. 13, p. 357t (1957).
10. Couper, A. S. Materials Protection, Vol. 8, No. 10, p. 17 (1969).
11. ASTM Standard G36-73, Standard Recommended Practice for Performing Stress Corrosion Cracking Tests in a Boiling Magnesium Chloride Solution.
12. Dana, A. W., Warren, Donald. DuPont Engineering Dept. Bulletin. Dana, A. W. ASTM Bulletin No. 225, October, 1957.
13. Staehle, R. W., Beck, F. H., Fontana, M. G. Corrosion, Vol. 15, No. 7, p. 51 (1959).
14. Rideout, P. S. NACE Conference, Chicago, 1964, unpublished paper No. 67.
15. Scharfstein, L. R., Brindley, W. F. Corrosion, Vol. 14, No. 10, p. 60 (1958).
16. Snowden, R. P. J. Iron Steel Institute, Vol. 194, p. 180 (1960).
17. Swanby, R. K. Corrosion Charts: Guides to Materials Selection, Chemical Engineering, Vol. 69, p. 186 (1962) November.
18. Wanklyn, J. N., Jones, P. Chem. & Ind., p. 888 (1958).
19. Williams, W. Lee. Corrosion, Vol. 13, No. 9, p. 539t (1957).
20. Intergranular Corrosion of Chromium-Nickel Stainless Steels—Final Report, Welding Research Council Bulletin No. 138, February, 1969.
21. Huntington Alloys, Inc., private communication (1976).
22. Bond, A. P., Dundas, H. J. Corrosion, Vol. 24, No. 10, p. 344 (1968).
23. Cowan, R. L., II, Tedmon, C. S., Jr. Intergranular Corrosion of

Iron-Nickel-Chromium Alloys, Advances in Corrosion Science and Technology, Vol. 3, p. 332-337, Plenum Press (1973).
24. Dillon, C. P. Materials Performance, Vol. 14, No. 8, p. 36 (1975).
25. Lillys, P., Nehrenberg, A. E. Trans. of ASM, Vol. 48, p. 327 (1956).
26. McGuire, M. G., Trioiano, A. R., Hehemann, R. F. Corrosion, Vol. 29, No. 7, p. 268 (1973).
27. Copson, H. R. Effect of Composition of Stress Corrosion Cracking of Some Alloys Containing Nickel, Physical Metallurgy of Stress Corrosion Fracture, Interscience Publishers, New York, 1959.
28. Flowers, J. W., Beck, F. H., Fontana, M. G. Corrosion, Vol. 19, No. 5, p. 186t (1963).
29. Uhlig, H. H., White, R. A. Trans. ASM, Vol. 52, p. 830 (1960).
30. Moller, G. E. The Successful Application of Austenitic Stainless Steels in Sea Water, National Association of Corrosion Engineers, Toronto, Canada, 1975, unpublished paper No. 120.
31. Couper, A. S. Materials Protection, Vol. 8, No. 10, p. 17 (1969).

Susceptibility of Aluminum Alloys to Stress Corrosion*

D. O. Sprowls and H. C. Rutemiller
Alcoa Research Laboratories
New Kensington, Pennsylvania

STRESS CORROSION is a complex interaction of corrosive attack and sustained tension stress at metal surfaces resulting in cracking. Stress corrosion cracking involves greater deterioration in mechanical properties through the simultaneous action of static tension stress and corrosion than would occur as a result of separate actions of the two processes. Cracking can result in the brittle failure of an otherwise ductile material. For this to occur there must exist (1) susceptibility of the material to selective corrosion along more or less continuous paths, and (2) a condition of sustained stress in tension on the surface acting in the direction tending to pull the metal apart along these paths.[1] These paths are susceptible to such attack because they are relatively anodic to the rest of the metal. Stress corrosion cracking characteristically occurs at the grain boundaries in aluminum alloys.

All intergranular attack is not stress corrosion. Some heat treated alloys, such as 6061-T6, 2024-T6 and 2024-T8 can corrode intergranularly and still not be susceptible to stress corrosion cracking.

Testing for Stress Corrosion

Though it is relatively easy to produce failure of susceptible metals in an accelerated stress corrosion test, much testing is required to determine whether the material possesses a degree of susceptibility that will hamper its useful-

D. O. SPROWLS has been associated with the Chemical Metallurgy Division of Alcoa Research Laboratories since 1936. He has a BS in chemical engineering from Drexel Institute of Technology. As a research engineer, he is responsible for coordinating the evaluation of wrought aluminum alloys' resistance to corrosion and stress corrosion cracking. Active in NACE, he is chairman of Technical Committee T-5E on Stress Corrosion Cracking of Metallic Materials.

HERBERT C. RUTEMILLER is an instructor in the Department of Statistics at Western Reserve University, Cleveland, Ohio. Formerly with Alcoa Research, he was working in aluminum castings and forgings. He has a BS in physics from Case Institute of Technology and an MS in statistics from Western Reserve University.

*Extracted from a paper titled "Stress Corrosion of Aluminum—Where To Look For It, How to Prevent It" presented at the 18th Annual Conference, National Association of Corrosion Engineers, Kansas City, Missouri, March 19-23, 1962.

Figure 1—Tension bar 0.125 inch in diameter is stressed in aluminum alloy fixture coated with clear strippable material to prevent galvanic corrosion. Micarta bushings are used for the same reason on the C-ring at right. Constant deflection has been used because this load arrangement has been associated with most service failures due to stress corrosion cracking. When small specimens are used, the sampling procedure and loading method must be selected carefully to avoid failures from other causes.

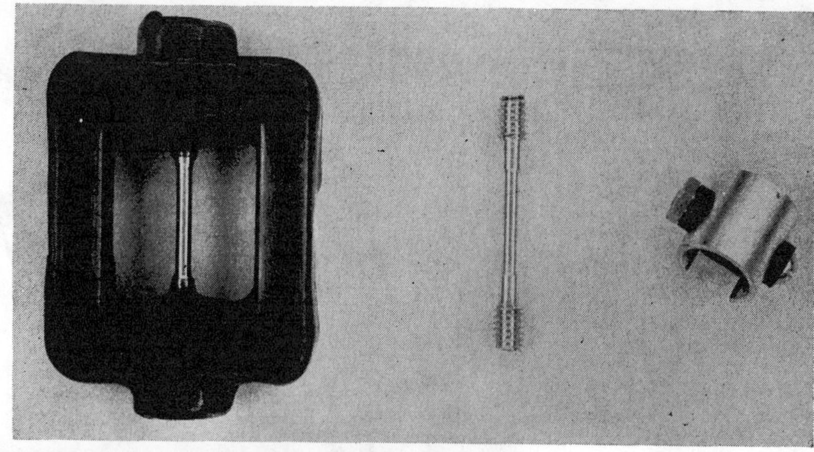

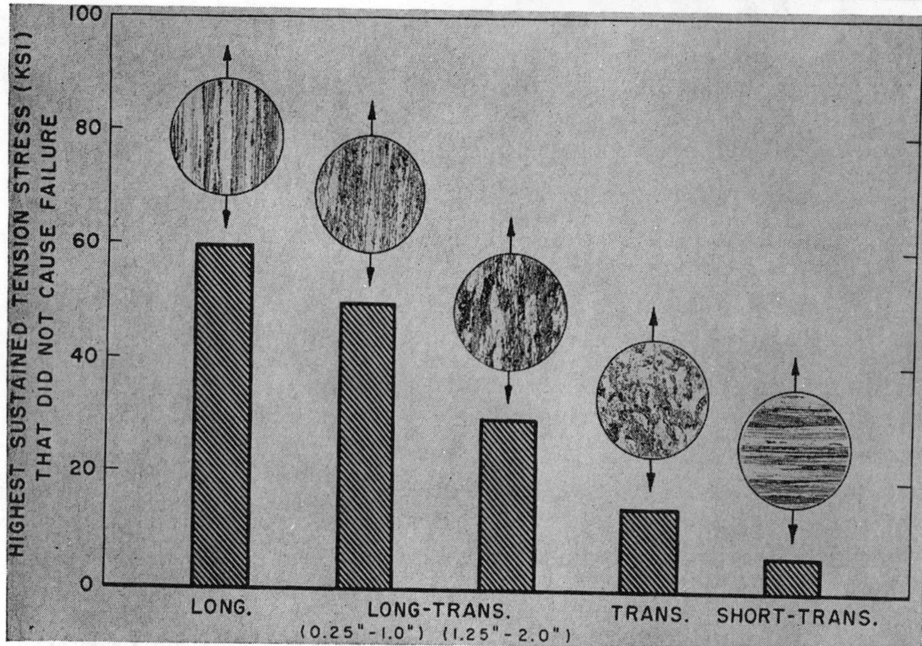

Figure 2—Effect of stress direction on resistance to stress corrosion cracking of 7075-T6 alloy extrusions. Bar graphs represent a summary of stress corrosion cracking tests of specimens exposed to 3½ percent sodium chloride solution under various stresses by alternate immersion. Typical microstructures for the materials are shown by arrows indicating direction of stress relative to grain structure.

Figure 3—Typical example of stress corrosion failure caused by re-adjustment of stress patterns resulting from machining external surfaces of an aluminum alloy bar. Residual stresses caused by quenching are usually compressive at the surface, but are countered by interior tensile stresses. In this case a mechanically stress relieved stock would have prevented this. The spool is machined from 2024-T4 alloy. The cross section through one of the cracks shows its intergranular nature.

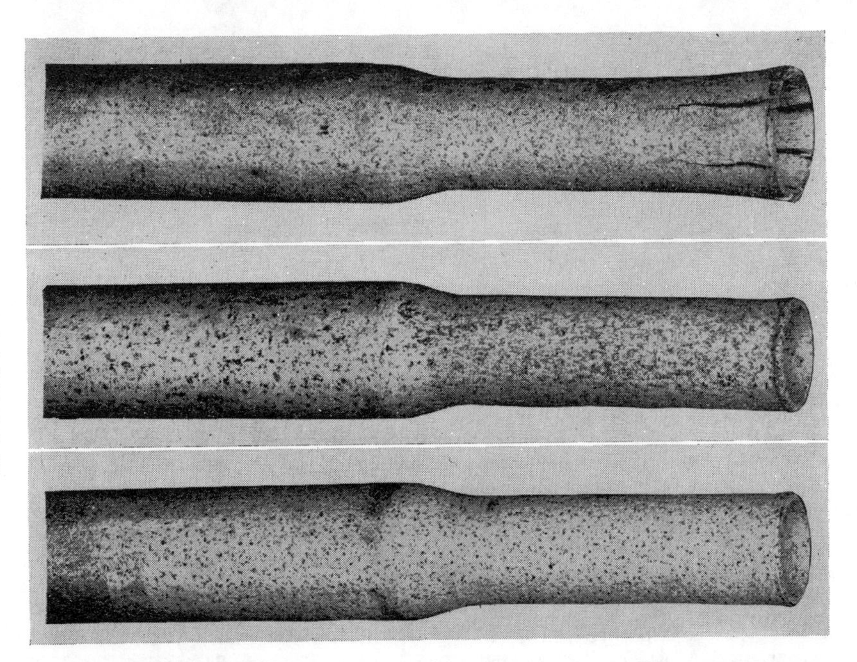

Figure 4—Tube ends of various tempers of 2024 alloy exposed to 3½ percent sodium chloride for 84 days by alternate immersion with corrosion product chemically removed after exposure. Numerous instances of stress corrosion cracking of tubular fittings have shown the inadequate resistance of 2024-T4 or 2024-T351 bar stock for such uses. Superior performance of artificially aged T6 or T851 tempers have been shown repeatedly in laboratory tests and in field applications. Top: swaged 2024-T3. Middle: swaged 2024-T3, aged to —T81 temper (12 hours at 375 F). Bottom: swaged 2024-0, heat treated to—T42 temper. Corrosion products chemically removed after exposure.

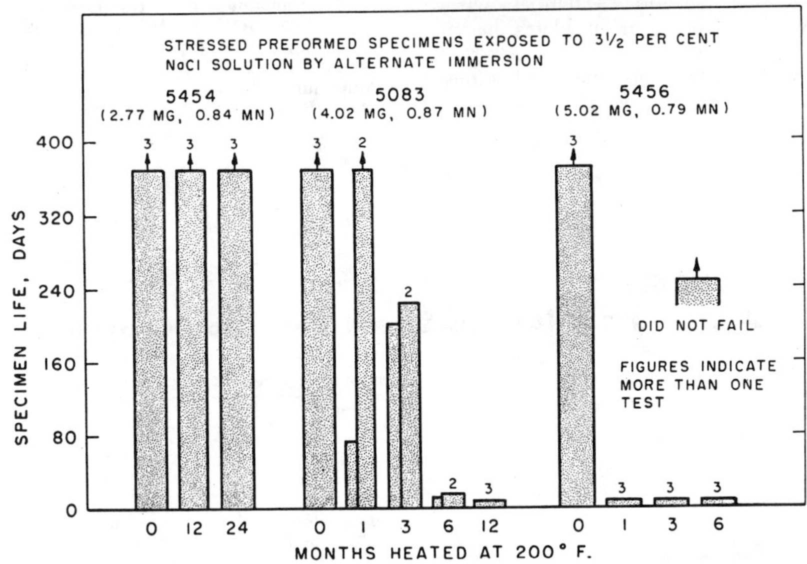

Figure 5—Comparative resistance to stress corrosion cracking of 7075-T6 and 7075-T73 alloys. Bar graphs represent average stress corrosion cracking tests of many specimens exposed under various tension stresses. Arrows indicate that no stress corrosion failures occurred at the highest stress used (75 percent of yield strength).

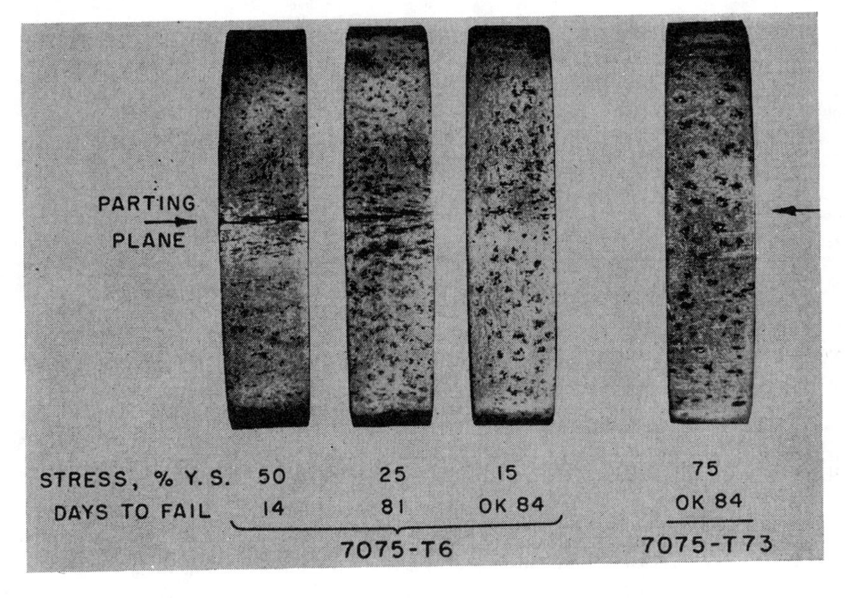

Figure 6—Slightly enlarged photograph of ring stress corrosion specimens exposed to 3½ percent sodium chloride by alternate immersion. Corrosion product chemically removed after exposure. Stress corrosion cracking and directional attack in the parting plane area of 7075-T6 alloy were absent in the 7075-T73 specimen.

Source: *Materials Protection*, June 1963, 63-65

ness. Determining the performance that can be expected from any alloy and temper necessitates using tests that will compare the alloy with other materials for which a service record has been established. Stress corrosion testing must include evaluation of material performance on a comparative basis and under controlled conditions.

Size and shape of material to be tested for stress corrosion will determine which of several types of specimen are most practical. High tension stress applied by dead weight loading can cause stress ruptures and failures that are not related to stress corrosion cracking. Microscopic examination of fractured specimens may be required to identify the cause of failure in borderline cases.

When accelerated tests are used, they must be correlated with longer term atmospheric exposures and service experience. Most results reported were obtained using alternate immersion exposure to 3½ percent sodium chloride solution. This test consists of a one-hour cycle with a ten-minute immersion followed by 50 minutes of drying.[2] Excellent correlation has been established between this alternate immersion test and natural atmospheric exposures.

The relative resistance to stress corrosion cracking of a wide variety of aluminum alloys and tempers is indicated in Table 1.

Prevention of Stress Corrosion Cracking

Stress corrosion problems can be avoided in many cases by careful selection of alloys and tempers and especially by adherence to design and assembly practices based on experience and understanding of stress corrosion. For example, various products of one of the most widely used high strength aluminum alloys (2024) have been in use for many years in the susceptible T3 and T4 tempers with few stress corrosion problems being noted.

Heat treated and quenched alloys in thick sections, especially those with irregular shape, can contain residual quenching stresses. Fabrication methods such as tube sinking and power bending, performed after heat treatment, can introduce high residual stresses. Mechanical stress-relief or controlled quenching in hot or boiling water can be used under some circumstances to reduce stresses to a much lower level. Relieval of surface tension stresses by peening or rolling also may be effective. Thermal stress-relief treatments generally are not used for heat treated alloys because they lower the tensile properties.

Several illustrations of the causes and methods of preventing stress corrosion cracking are shown in the accompanying photographs.

References

1. R. B. Mears, R. H. Brown and E. H. Dix, Jr. A Generalized Theory of the Stress Corrosion of Alloys. ASTM-AIME Symposium on Stress Corrosion Cracking of Metals (1944) pp. 323-339.
2. D. O. Sprowls and R. H. Brown. Resistance of Wrought High Strength Aluminum Alloys to Stress Corrosion. *Metal Progress,* April, May, 1962.

TABLE 1
Relative Resistance of Various Aluminum Alloys to Stress Corrosion Cracking

Type of Alloy	Commercial Examples	Resistance to Stress—Corrosion Cracking: Temper	Rating*
	Wrought, Strain-Hardened		
Unalloyed Al	1100	All	Excellent
Al-Mn	3003	All	Excellent
Al-Mg	5005, 5050, 5154	All	Excellent
Al-Mg	5356	Controlled	Very high
Al-Mg-Mn	3004, 3005, 5454	All	Excellent
Al-Mg-Mn	5086	All	Very high
Al-Mg-Mn	5083, 5456	Controlled	Very high
Alclad	3003, 3004	All	Excellent
	Wrought, Heat-Treated		
Al-Mg-Si	6061, 6262	All	Very high
Al-Mg-Si	6063	All	Excellent
Al-Si-Mg	6151,6351	All	Very high
Al-Cu	2024, 2219	—T4	Moderate
Al-Cu	2024, 2219	—T6, —T8	Very high
Al-Cu	2014	—T4, —T6	Moderate
Al-Zn-Mg-Cu	7075, 7079, 7178	—T6	Moderate
Al-Zn-Mg-Cu	7075	—T73	Very high
Alclad	2014, 2219, 6061, 7075	All	Excellent
	Cast		
Al-Mg	214, 218, Almag 35	As cast	Excellent
Al-Mg	220	—T4	Moderate
Al-Si-Mg	356, A356, 357, 358, 359, Tens 50	All	Excellent
Al-Si-Cu	319, 333, 380	As cast	Very high
Al-Si-Cu	355, C355, X354	—T6, —T61	Very high
Al-Cu-Si	195, B195	—T6	Moderate
Al-Zn-Mg	Ternalloy 7	—T6	Moderate
Al-Zn-Mg	40E	As cast	Moderate
Al-Zn-Mg-Cu	A612, C612	As cast	Excellent

* Ratings are based on exposure conditions at normal outdoor or room temperatures.
Elevated temperature exposure would lower the ratings in some cases.
Excellent = no known instances of failure in service or in laboratory tests.
Very High = no know service failures, laboratory failures only under extreme conditions.

Stress Corrosion Cracking of Brass in an Air Conditioning System

Figure 1—Stressed yellow brass plate exposed in humidifying chamber of an air conditioning unit for 2½ to 3 years. Intergranular crack is shown. 100X.

Herbert H. Uhlig and Joseph Sansone
Corrosion Laboratory
Massachusetts Institute of Technology
Cambridge, Massachusetts

SUMMARY

Describes failures of brass brackets and hooks for tubing support in humidifying chamber of an air conditioning system. Failure was stress corrosion cracking caused by exposure to ammonium salt. Gives laboratory tests to determine cause of failure and explains source of salts.

AN EPIDEMIC of failures in a large office building in Boston, Mass., occurred some years ago in the humidifying chamber of the air conditioning system. Brass brackets or plates (2 by 1½ by 0.08 inch) assembled by brass bolts and nuts, plus a number of brass hooks used to support copper tubing, had completely cracked after 2½ to 3 years' service. The broken pieces could be shovelled up from the floor of the humidifying chamber.

Analysis of some of the material which failed is given in Table 1. By and large, the material corresponded to 60-40 brass or 70-30 yellow brass. Sufficient green corrosion products appeared on the surface to be scraped off for analysis. The mixture of copper-zinc salts included a large percentage of ammonium salts. Photomicrographs of the brass plates showed that the cracks were intergranular (Figure 1). Cracks through the heavily cold worked hooks were both transgranular and intergranular. The failure was diagnosed as stress cor-

TABLE 1—Analysis of Brass Components (Percent)

	Hooks	Bolts	Plates
Copper........	59.90	65.15	67.22
Zinc...........	39.28	34.71	32.67
Tin............	0.71	....	....

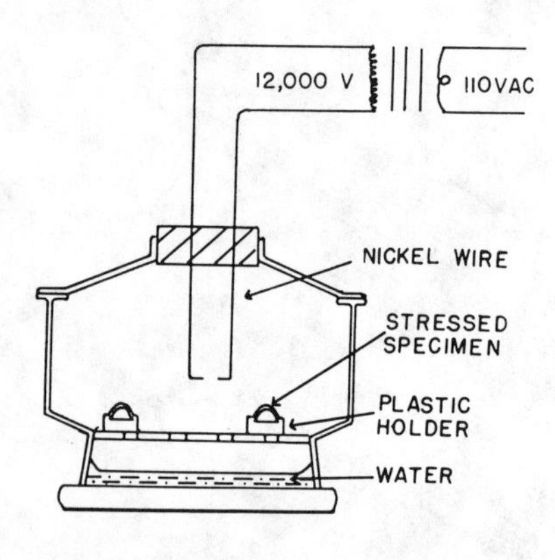

Figure 2—Laboratory arrangement for accelerated test of stress corrosion cracking in moist air to simulate humidifying chamber exposure in an air conditioning unit.

$$3/2\ O_2 + N_2 \xrightarrow[\text{discharge}]{\text{spark}} NO + NO_2 \xrightarrow[+\ O_2]{H_2O} 2\ HNO_3 \qquad (1)$$

$$4\ Zn + 10\ HNO_3 \longrightarrow 4\ Zn(NO_3)_2 + 3\ H_2O + NH_4NO_3 \qquad (2)$$

$$Cu + 1/2\ O_2 + 2\ HNO_3 \longrightarrow Cu(NO_3)_2 + H_2O \qquad (3)$$

$$Cu^{++} + 4\ NH_4^{+} \longrightarrow Cu(NH_3)_4^{++} + 4H^{+} \qquad (4)$$

Figure 3—Typical chemical reactions which in sequence produce cupric ammonium salts. These salts caused the failures of brass brackets and tubing support hooks in an air conditioning unit.

rosion cracking caused by exposure to ammonium salts.

But where did the ammonia come from which caused the failures? Air intake was far above the street level at two opposite sides of the building, and there was no chemical manufacturing in the immediate area. A thorough check disclosed no obvious source either of ammonia or of amines.

It was finally concluded that the indirect source was an electrostatic dust eliminator located ahead of the humidifying chamber, through which all incoming air passed. The precipitator operating at high voltages produced a corona discharge which in turn induced reaction of nitrogen and oxygen of the air to form nitrogen oxides. These oxides combined with water to form acids, including nitric acid, which in turn reacted with the brass to form cupric ammonium salts. Cupric ammonium salts are well known accelerating agents for stress corrosion cracking of brass.

This diagnosis was proved correct by exposing cold rolled brass in the stressed condition to air in which a spark discharge took place. Specimens of 65 percent copper, 35 percent zinc strip 0.02 inch thick were cut into 2.4 by 0.2 inch specimens. They were cleaned by abrading, followed by washing in hot benzene, then bent into the form of a "C" with a span of 1.4 inch and held in this position by notched plastic holders. Several stressed specimens were placed in a 6-liter glass desiccator equipped with nickel wire electrodes (Figure 2). A water layer on the bottom of the desiccator saturated the air within, and a 12,000-volt transformer supplied continuous spark across the electrodes. All specimens cracked in an average of 25 hours' exposure with maximum time to cracking of 30 hours.

On the other hand, annealed strips unstressed showed some slight surface attack but were not cracked. Condensate on the cracked specimens was analyzed using ferrous sulfate and concentrated sulfuric acid for detection of nitrate, and applying Nessler's reagent for the ammonium ion. Both species were present in unmistakably large concentrations.

The water layer at the bottom of the desiccator was found by titration to be equivalent to 0.03 normal acid, indicating that a considerable amount of nitrogen oxides had formed during the tests, reacting with water to form acid.

There was still the question whether cracking could possibly be caused in part by ozone, hydrogen peroxide, or some similar reaction product within the desiccator. This possibility was eliminated by running another series of tests after filling the desiccator with pure oxygen instead of air.

After 3½ days of spark discharge, no cracking had occurred with any of three stressed brass specimens that were exposed. A thin layer of black copper oxide coated each of the specimens, presumably caused by reaction with ozone. After 3½ days, air was admitted to the desiccator and the sparking was continued without disturbing the specimens. After 19 additional hours, all three brass specimens had cracked.

Several additional experiments were run both with air and oxygen to check the above results. In each case, the crack pattern repeated itself as described. Photomicrographs of the failed specimens showed a mixture of transgranular and intergranular cracks through the severely cold-worked strip.

Typical chemical reactions which in sequence produce cupric ammonium salts directly responsible for the failures are shown in Figure 3. Reactions identical or similar to (2), (3) and (4) in Figure 3 probably also account for stress corrosion cracking of a stressed brass chain exposed to a dilute nitric acid laboratory atmosphere, as described by J. P. Fraser.[1]

Conclusions

This experience points to the necessity of avoiding stressed copper base alloys in air conditioning systems using electrostatic dust eliminators. It also points to the surprisingly high concentrations of nitrogen oxides picked up by air passing through electrostatic dust eliminators, sufficient to cause corrosion damage.

Reference

1. J. P. Fraser. Stress Corrosion Cracking of Brass in Nitric Acid Vapor Causes Collapse of Lab Lighting Fixture. *Materials Protection,* **2,** 97 (1963) January.

Other Articles of Interest

N. A. Sinclair and H. J. Albert. Stress Corrosion Cracking of Naval Brass. *Materials Protection,* **1,** 35 (1962) March.

Ken Nobe and S. Tan. Electrical Potential Responses of Silver, Steel and Brass Stressed in Tension in Sodium Chloride Solutions. *Corrosion,* **18,** 391t (1962) November.

T. P. Hoar, Stress Corrosion Cracking. *Corrosion,* **19,** 331t (1963) October.

W. Lee Williams. Stress Corrosion Cracking: A Review of Current Status. *Corrosion,* **17,** 340t (1961) June.

H. H. UHLIG is Professor of Metallurgy at Massachusetts Institute of Technology in charge of the Corrosion Laboratory. He and his students have contributed many papers to the corrosion literature over the past 25 years. He edited the "Corrosion Handbook" first published in 1948. His latest book, "Corrosion and Corrosion Control, an Introduction to Corrosion Science and Engineering," was published in 1963. He is a member of NACE and received the association's Whitney Award in 1951 for his public contribution to the science of corrosion science.

JOSEPH A. SANSONE was a student at MIT (1962-63) during which time he conducted a research project in the Corrosion Laboratory. Results of the project are in part reported in this article.

Source: *Materials Protection*, February 1964, 21-23

A Summary of Stress Corrosion Cracking Tests Applicable to Stainless Alloy Weldments*

R. H. ESPY
Armco, Inc., Middletown, Ohio

Stress corrosion cracking is a mechanism of failure in alloys requiring tensile stress in a corrosive medium. In weldments, the tensile stress occurs by shrinkage of the weld during cooling. The magnitude of stress is a function of the weld deposit size and the natural restraint which any weldment offers to shrinkage. The corrosive media required to produce failure are known to vary with materials. As-welded martensitic stainless steels fail when exposed to aqueous corrosive media which produce hydrogen at the metal surface. As-welded austenitic stainless alloys are susceptible to failure in aqueous media containing chlorides when the temperature of the media is about 160 F (71.1 C) or higher. Polythionic acid at room temperature and high purity oxygenated H_2O at temperatures of 550 F (288 C) also caused failures in austenitic stainless steels when sensitized by welding. Although numerous stress corrosion cracking tests have been run on unwelded wrought alloys, the evaluation of weldments for stress corrosion cracking characteristics is best done by using actual weldments. The type of test selected like circular bead, constant strain rate, or wick test depends on the application involved. The test media may vary from actual service exposure to highly accelerated media. The correlation of laboratory data to service life in an actual application for some austenitic stainless steel alloys has been good but more information is needed.

Introduction

STRESS CORROSION CRACKING is a mechanism of failure in alloys requiring tensile stress in a corrosive medium. In weldments, a residual tensile stress develops on shrinkage of the weld during cooling from the deposition temperature. These tensile stresses are not only present in the weld deposit, but may extend for some distance into the base material. The magnitude of stress is a function of the weld deposit size and the natural restraint which any weldment offers to shrinkage.

The corrosive media required to produce SCC vary with materials. For example, the martensitic stainless steels will fail through the hardened, as-welded heat affected zone (HAZ) of the base metal when exposed to aqueous corrosive media which produce hydrogen at the metal surface. The austenitic stainless alloys are generally considered susceptible to failure in aqueous media containing chlorides when the temperature of the media is about 71 C (160 F) or higher as reported by Scharfstein and Brindley.[1] Polythionic acid at room temperature reported by Samans,[2] and high purity H_2O containing O_2 at levels greater than 0.2 mg/L at temperatures of 288 C (550 F) as reported by Berry, White, and Boyd[3] also caused failures in austenitic stainless steels when sensitized by the heat of welding.

Although numerous SCC tests have been run on unwelded wrought alloys, the suitability of such data for evaluation of weldments is questionable. This is because weld filler metal may differ significantly from the base metal in resistance to SCC. Also, the heat of welding may cause metallurgical changes in the base metal. In addition, stresses are more complex in weldments, and may be the result of both residual stresses from welding and applied service stresses. This report summarizes the state of the art in SCC testing of weldments and suggests areas for improvement.

Weldment Test Specimens

Several types of specimens have been employed to develop actual weldment data for applications where SCC is a potential hazard. They can be divided into three general classes: (1) Specimens that are tested in the as-welded condition; (2) specimens that are welded, and artificially stressed subsequent to welding; and (3) specimens that are welded, formed, and then further stressed. Individual specimens representative of the three classes are shown in the following tabulation:

Class I—(As-Welded)
- Circular Bead on Plate
- Circular Patch on Plate
- Bead on Bar

Class II—(Welded and Artificially Stressed)
- Bent Beam
- C-Ring
- Split Ring
- 4 Notch Tensile
- Constant Strain Rate

Class III—(Welded, Formed, and Artificially Stressed)
- U-Bend

Diagrams of the several test configurations are shown in Figures 1 to 11. A discussion on each specimen, possible modifications, and value of data obtained follows.

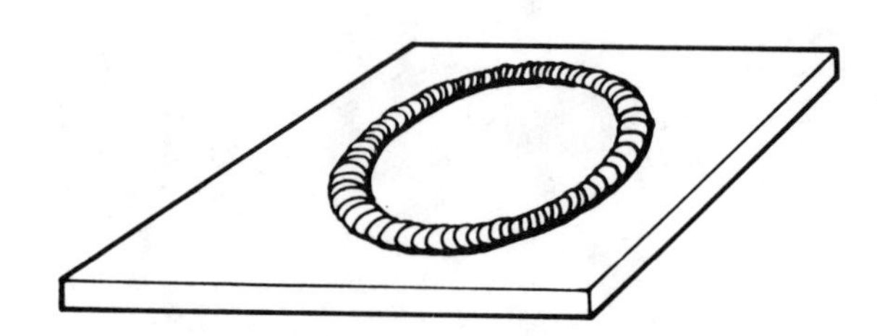

FIGURE 1 — Circular weld bead plate test plate 4 inch square x 3/16 inch thick weld circle 2 inch diameter.

Class I—Circular Bead on Plate. This type of specimen offers a wide variety of test parameters: plate thickness; plate configuration—round or square and dimension—diameter or side length; weld bead diameter; filler added or fusion only; weld travel speed—manual or mechanized and fixturing for added restraint and/or heat removal.

A standard specimen used by the author[4,5] ASTM G58-78 for austenitic stainless steels is 100 mm (4 inch) square x 4.7 mm (3/16 inch) thick on which a 50 mm (2 inch) diameter stringer bead is deposited manually by the SMA (shielded metal arc, commonly

*Voluntary manuscript submitted for publication October, 1978.

Reprinted with permission from *Materials Performance*, November 1980, 20-24,

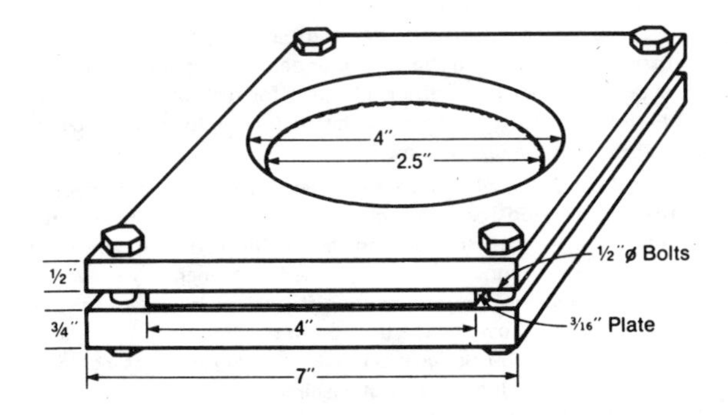

FIGURE 2 — Restraining fixture for circular weld bead plate test material mild steel.

called stick electrode) process using a 4 mm (5/32 inch) diameter electrode. The specimen is placed in a restraining fixture as shown in Figure 2 and is retained in the fixture after welding until the sample and fixture have cooled to room temperature.

The sample dimensions and weld parameters were selected arbitrarily and the presence of sufficient stress to cause cracking was checked by exposure to a boiling 45% $MgCl_2$ media described by Streicher and Sweet[6] (ASTM G-36-73). Crack initiation in AISI 304 stainless steel was detected within one, four-hour exposure and had progressed extensively after a 24 hour exposure (Figure 3).

Denhard[7] showed that stress relief heat treatments at 650 C (1200 F) (38% relief) resulted in a threshold stress of 83 MPa (12,000 psi). It was determined that stress in the as-welded bead on the plate was approximately 172 MPa (25,000 psi).

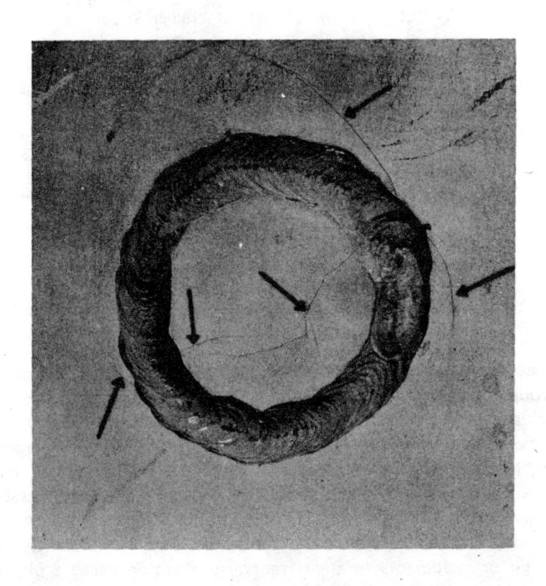

FIGURE 3 — Circular weld bead plate test with AISI 304 stainless steel after 24 hours exposure to boiling $MgCl_2$ solution—arrows show cracks.

In a report published in 1971, Steklov[8] determined stresses in similar type specimens through the use of strain gages. The specimen suggested in this work differed in that full penetration welds were used in a **circular patch configuration**. A disc, rather than a square, was used for the base plate (Figure 4).

With the patch specimen, added flexibility is obtained in that different thicknesses of material may be used for the patch and base plate. The optimum specimen geometry suggested is base plate diameter 130 to 150 mm (5 to 6 inches), weld diameter 20 to 40 mm (3/4 to 1-1/2 inches), and plate thickness 3 to 6 mm (1/8 to 1/4

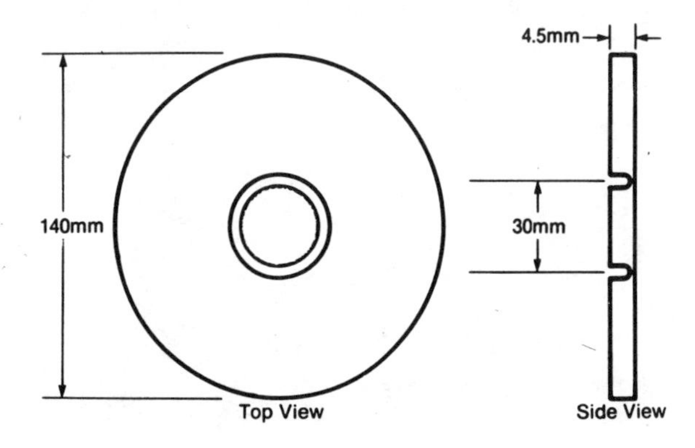

FIGURE 4 — Circular weld patch plate test,[2] plate ϕ—140 mm, weld ϕ—30 mm, plate thickness—4.5 mm.

inch). Using this specimen with KH18N10T alloy—AISI 321 stainless steel—a maximum stress of 470 MPa (68,000 psi) as-welded is reported.

A **bead-on-bar** type of specimen developed by the author[4,5] (ASTM 58-78) has been useful for determining relative resistance for materials to SCC in boiling 45% $MgCl_2$ media. Using a well characterized material such as AISI 304 austenitic stainless steel, the time to crack initiation in a 240 hour exposure becomes a guide in determining weldment performance in hot chloride-containing media.

The sample is prepared by fusing three autogenous or fillerless beads on a 150 mm (6 inch) length of 25 mm (1 inch) diameter bar as shown in Figure 5. A 19 mm (3/4 inch) length is then cut from

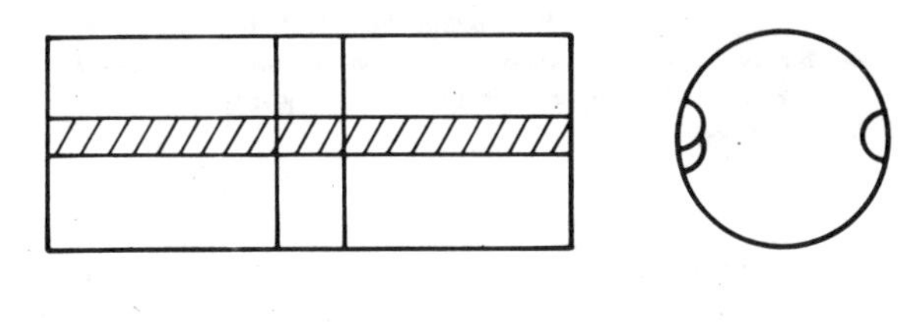

FIGURE 5 — Weld bead on round bar test specimen. Bar size 1 inch diameter x 6 inches long with 3/4 inch long section removed for exposure to corrosion media.

the welded bar, surface ground on the ends, and exposed to the test media. Examination for crack initiation is made after the first four hours and then every 24 hours until full time of exposure has been achieved. If cracking occurs before the 240 hour exposure limit, testing is discontinued.

The effect of increasing Ni content in austenitic stainless steels above the 8% level of AISI 304 for increased resistance to failure in hot aqueous chloride media was reported by Copson[9] and Denhard[7] from their work with stressed base metal samples. The use of weld bead-on-bar specimens confirmed the effect of increased resistance to failure with increased Ni levels as reported by the author.[4,5] In addition, this work shows that austenitic stainless steels having nickel contents less than 8% also exhibit increased resistance to failure. Figure 6 illustrates results obtained by the author[10,11] with a Mn-N modified 3% Ni austenitic stainless steel (Nitronic 33 stainless steel) compared to an AISI 304 stainless steel. The weld bead-on-bar type of specimen has to date not been used in test media other than the boiling 45% $MgCl_2$.

Class II—Bent Beam. A typical bent beam is shown in Figure 7. The idea of applying stress to a weldment for testing is attractive from the standpoint of evaluating a weldment placed under load in service. However, the difficulty of determining the stress level in a complex system where both applied and residual stresses exist

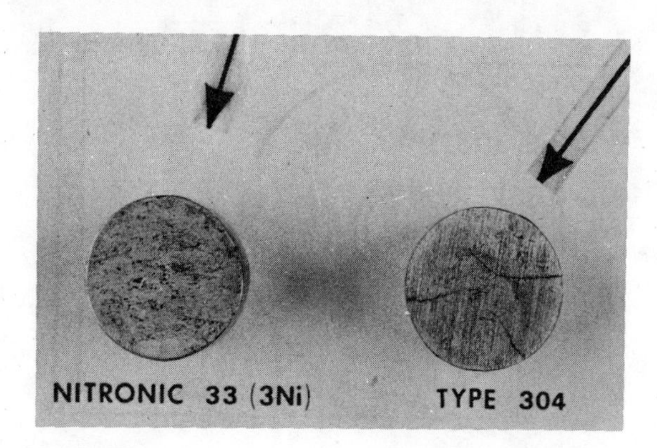

FIGURE 6 — Low nickel stainless steel (Nitronic 33) compared to AISI 304 stainless steel using weld bead on bar specimens. Arrows indicate location of welds. Nitronic 33 showed no cracking after 240 hours exposure in boiling $MgCl_2$—AISI 304 cracked in 4 hours.

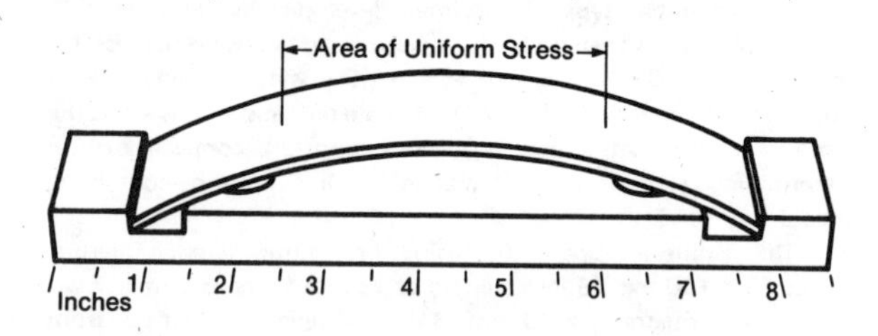

FIGURE 7 — Typical bent beam weld test specimen. The weld may be either transverse or longitudinal to the bend direction. Load clamp should be electrically insulated from test specimen.

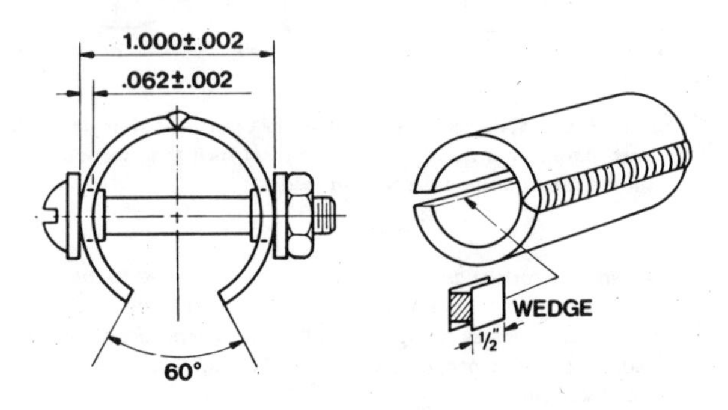

FIGURE 8 — Typical C-ring and split ring weld test specimens. The C-ring (1 inch long) is a method of applying tension stress to the external surface. The split ring represents internal loading of pipe and tubing. Load clamp should be electrically insulated from test specimen.

creates confusion. Therefore, quantitative interpretation of the test results is impossible.

For the mechanics in preparation and use of bent beam stress corrosion specimens, reference is made to the ASTM designation G39-79. The **C-ring** (outer fibers in tension) and **split ring** (inner fibers in tension) tests applied by the Oak Ridge National Laboratory (ORNL) to welded tubing as shown in Figure 8 are similar to bent beams except that they are applicable to tubular or pipe sections. The split ring idea as used by ORNL differs from the ASTM designation G38-73 "Making and Using C-Ring Stress Corrosion Specimens," in that a greater percent of the welded tube's diameter is placed in condition of stress for test. Special attention should be given to selection of materials for the loading device and insulation from the test specimen.

The C-ring and split ring type of test gives useful data because of reasonably good load control and a good simulation of actual conditions in service. Results reported by ORNL to date are based on welded plus annealed plus stressed specimens. The use of as-welded specimens presents similar difficulties with regard to stress level as indicated above for the bent beam specimen.

Use of the ASTM designation G38-72 is suggested as a guide in the preparation and loading of test specimens.

The **4 notch tensile** specimen used by Denhard[7] as shown in Figure 9 is based on a constant load principle with a stress

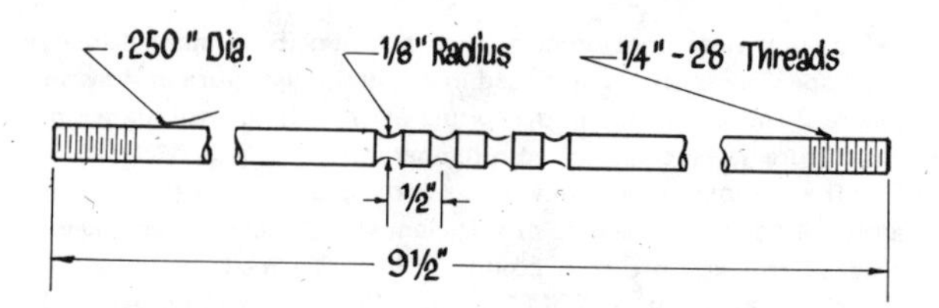

FIGURE 9 — 4 Notch tensile stress corrosion weld test specimen. Test conducted in vertical position with aqueous media covering 1/2 of test specimen.

TABLE 1 — Effect of Cast and Wrought Structures in AISI 310 Stainless Steel on Resistance to Stress Cracking in Boiling 45% $MgCl_2$ Using the 4 Notch Tensile Test (Stressed at 483 MPa; 70,000 psi)

Form	Condition	Composition C	Cr	Ni	Hours to Failure
Bar	Anneal	0.11	27.21	21.51	10.0
Bar	Anneal	0.11	27.21	21.51	8.5
All Weld Metal	As-Welded	0.11	27.33	21.38	17
All Weld Metal	As-Welded	0.11	27.33	21.38	15

concentration factor of 1.3 at the notches. The stress is developed through a system of lever arms and weights. The load remains constant until the cross section is decreased by cracking. This causes an increase in unit stress followed by rapid progression of the crack. This is converse to all of the prior systems discussed—where unit stresses decrease as cracking progresses, causing a decrease in crack progression.

The 4 notch tensile system does not lend itself well to the testing of weldments in their transverse direction because of the difficulty in locating the notches. However, all-weld metal samples have been tested and the results in boiling 45% $MgCl_2$ at 54 C (310 F) indicate AISI 310 stainless steel weld metal is significantly more resistant than the same alloy in wrought form, as shown in Table 1.

A relatively new system of testing specimens for SCC is the application of a **constant strain rate.**[12-14] A specimen used by U.S. Steel in low alloy steel work is shown in Figure 10. Testing consists of selecting a strain rate that is rapid enough to prevent reformation of a protective film on newly exposed surfaces and yet slow enough to preclude ductile failure. Susceptibility to SCC is determined by load at failure, reduction of area, or area under stress-strain curve, and metallographic examination of surface at failure. Time to failure may also be used as a measure of susceptibility to stress corrosion.

Constant strain tests differ from most other tests in that they are conducted under a dynamic load and this inevitably causes failure. The test is severe, having the capability of uncovering

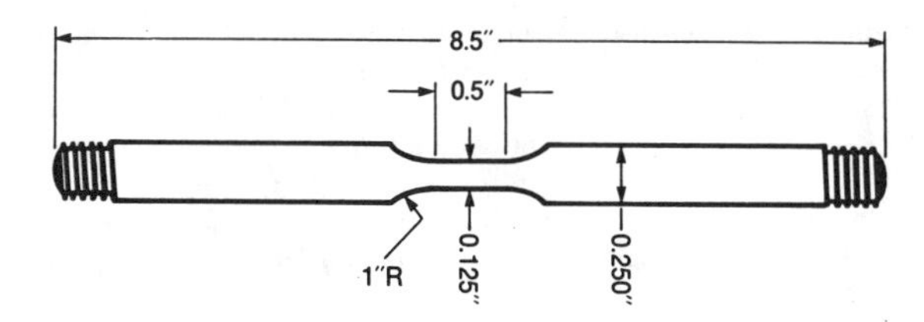

FIGURE 10 — Constant strain rate stress corrosion weld test specimen. Test conducted in vertical position with aqueous media covering entire gage length.

sensitivity to SCC sometimes not revealed by static systems. While constant strain in one sense does not duplicate actual service conditions, it can, in another sense, be looked upon as duplicating a precracked test, one that is getting more attention as fracture mechanics becomes increasingly important in today's study of metal failures.

Because this test method is rapid and the results only relative, it finds use mostly in determining trends in the effects of environmental and/or alloy changes.

The testing of transverse weldments using the U.S. Steel test configuration could be applied providing the weld area was less than the 13 mm (1/2 inch) gage length. For weldments having a weld area greater than 13 mm (1/2 inch), a modification of the test specimen gage length would appear quite feasible. All weld deposits could easily be tested in the same manner as used with the four notch tensile specimen.

A constant strain rate test, referred to as a constant extension rate test (CERT), has been used by Clarke, Cowan, and Danko[15] for nuclear boiling H_2O reactor materials performance. Application of this same test for weldments has been reported by Kass, Henry, Pickett, and Walker.[16]

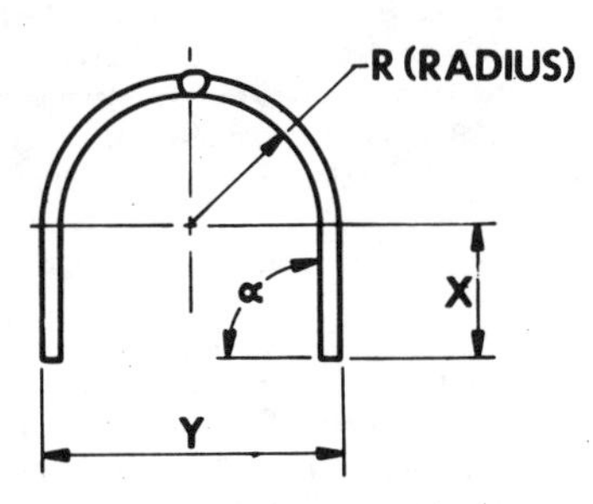

FIGURE 11 — U-bend stress corrosion weld test specimen. Specimen shown has weld perpendicular to bend direction. Weld may also be parallel to bend direction.

Class III—U-Bend. This system consists of applying cold work by bending a weldment after the weldment has been made as shown in Figure 11. After bending, the weldment is stressed to a precalculated level and exposed to a corrosive medium. The weldment may be either parallel or perpendicular to the bend direction.

While the stress applied after bending can be calculated, the actual stress in various parts of the weldment can vary considerably from the desired stress. In applications where severe cold forming is actually carried out after welding, this type of test will provide useful information. Otherwise, U-bend tests of weldments would have to be considered quite severe and could give misleading results. However, they are useful in establishing immunity to SCC under the most severe conditions.

The details of (1) bending and (2) application of stress after bending through proper use of insulated loading devices is described in ASTM designation G30-72. A general discussion on U-bends for single as well as the two stage U-bend suggested here is included in the ASTM designation.

In developing Fe-Cr-Mo alloys with 2% Ni added for good resistance to pitting in SO_4 containing media, Streicher[17] describes

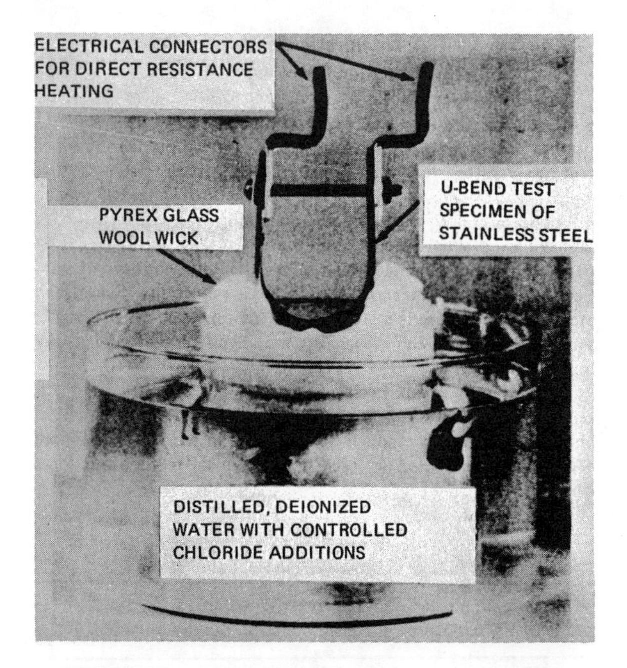

FIGURE 12 — Apparatus used for wick test as described in ASTM C692-77. Standard method for evaluating the influence of absorptive thermal insulations on the chloride stress corrosion cracking tendency of austenitic stainless steel.

the wick test,[18] using a neutral wick with a standard NaCl solution as shown in Figure 12. A crack-free bend with this NaCl solution at 100 C (212 F) resembling industrial exposure, showed the Ni addition to be satisfactory. Similar U-bends in the stronger boiling 45% $MgCl_2$ at 155 C (310 F) cracked in 16 hours.

Stress Corrosion Cracking Test Media

The test media for SCC testing of weldments may vary from actual service conditions requiring long time exposure, to laboratory media that produce highly accelerated corrosive attack in relatively short times. Common accelerated test media for the martensitic stainless steels used by Bloom[19] are H_2S-acetic-NaCl solutions at room temperature (NACE TM-01-77). For the austenitic stainless steels and nickel base alloys, the boiling 45% $MgCl_2$ at 155 C (310 F) solution is frequently used.

The standard recommended practice for "Performing Stress Corrosion Cracking Tests in a Boiling $MgCl_2$ Solution" is given in the ASTM designation G36-73. The specimen may be completely immersed in an aqueous medium or it may be placed in such a manner that part of the specimen is in the aqueous medium and part in the vapor phase. In other tests, the media may be cycled so that the sample is completely immersed for a period of time, followed by a period in the vapor phase.

The wick test,[18] used first by Dana and Delong,[20] was developed to simulate a hot stainless steel pipe covered with insulation and exposed to rain water. The test setup shown in Figure 12 has a glass wool wick which is used for checking insulation to be used with stainless steel. This same setup is also used for metal evaluation in alloy development work.

Still another test designed to simulate aerospace exposure is the dip-dry test reported by Humphries[21] in which the specimen is dipped in an aqueous medium at room temperature, removed, and dried under an infrared light. The times of dip and dry are preset and the cycle is repeated any desired number of times. These tests concentrate salts by evaporation.

Polythionic acid, a medium often encountered in the petrochemical industry, is known to cause grain boundary attack in sensitized austenitic stainless steel when under stress. A standard test often used for determining resistance to cracking in this medium is described in ASTM G35-73.

Interpretation of Stress Corrosion Cracking Test Data

While exposure to actual service conditions gives the most reliable results, the long time required to produce results frequently does not permit such practices in material evaluation. On the other hand, accelerated tests may result in rejection of materials that would perform satisfactorily in less severe corrosive environments.

An ideal situation for any service application is one in which materials are available for selection that are totally resistant to SCC. Often this is not possible and the user is then confronted with selecting a material that will give the best service life. Even when totally resistant materials are available, factors such as economics, reliability for safety, and availability, may significantly influence the choice.

One of the major problems in evaluating a weldment for application in potential SCC media is interpretation of results from short time severe test media in terms of life in potential, but more mild, media. Tabulated in Table 2 is a typical relationship for a textile application. While this tabulation is a good guide for applications at this corrosive level, more information of this type is needed to cover wide ranges of corrosive conditions. A start in this direction could be obtained by relating case histories as reported by the ASTM STP No. 264[22] to basic test data as reported by Denhard.[7]

TABLE 2 – Accelerated Stress Corrosion Cracking Test Data Related to Service Life in Hot Chloride Media

Test Conditions	Type of Stainless Steel and Time to Failure		
	304L	316L	Nitronic 33
Unwelded BM stressed at 50,000 psi (345 MPa) in boiling $MgCl_2$	1/2 hours	2 hours	4 hours
Unwelded BM stressed at 24,000 psi (165 MPa) in boiling $MgCl_2$	2 hours	7 hours	NF[(1)] after 1000 hours
As-welded weldment; bead on bar in boiling $MgCl_2$	4 hours	24 hours	NF after 240 hours
As-welded vessel—service media 3% NaCl cycled daily to 200 F (93.5 C)	3 months	1 year	NF after 5 years

[(1)]NF = No failure.

Summary

The evaluation of weldments for resistance to stress corrosion cracking is best done by using actual weldments rather than by simulating stress and weld heat effect artificially. The type of test selected would of course depend on the application involved. For most applications, a Class I test (circular bead, circular patch and bead-on-bar) is suitable.

For applications where in-service stresses may exceed weld stresses, a Class II test (bent beam, C-ring, split ring, 4 notch tensile, and constant strain rate) may be considered. Class III tests (U-bend) represent the unusual and should be used only where forming of a component is anticipated after welding, or where proof of total immunity is desired.

The test media may vary from actual service exposure to severe, highly accelerated media, *e.g.,* boiling $MgCl_2$. In using severe accelerated media, a problem of interpreting test results in terms of service life exists.

While the various test systems described here provide useful information in selecting materials for weldments to be used in media where stress corrosion cracking potential exists, there is need for added information. More data are needed on actual stress levels in welds of varying geometry, thickness, and materials.

Improved techniques with strain gages and scribed line systems, such as used by Steklov[8] and described by Robelotto[21] should be helpful in providing added information. New tests and/or methods of interpreting test results in terms of service life in various corrosive media are also needed. A survey of case histories[22] and relating these to basic test data[7] could be a first step in this area.

Acknowledgments

The author wishes to thank the Welding Research Council Sub-Committee on Corrosion Resistance of High Alloy Weldments (M. A. Streicher, Chairman) for its many suggestions and comments during preparation of this paper.

References

1. Scharfstein, L. R., Brindley, W. F. Corrosion, Vol. 14, p. 588t (1958).
2. Samans, C. H. Corrosion, Vol. 20, p. 256 (1964) August.
3. Berry, W. E., White, E. L., Boyd, W. K. Corrosion, Vol. 29, p. 451 (1973) December.
4. Espy, R. H. Welding Design and Fabrication, p. 82 (1977) October.
5. Espy, R. H. Residual Stresses and Stress Corrosion Cracking in Stainless Steel Weldments. Proceedings of an International Conference Control of Distortion and Residual Stress in Weldings, November 16-17, 1976, Chicago, Illinois, pp. 68-75.
6. Streicher, M. A., Sweet, A. J. Corrosion, Vol. 25, p. 1 (1969) January.
7. Denhard, E. E. Corrosion, Vol. 16, p. 131 (1960) July.
8. Steklov, O. I. Automatic Welding, Vol. 24, No. 9, p. 44 (1971).
9. Copson, H. R. Physical Metallurgy of Stress Corrosion Fracture, p. 256, Rhodin, T. N., Ed., New York Interscience Publishers, Inc., 1959.
10. Denhard, E. E., Espy, R. H. Austenitic Stainless Steels with Unusual Mechanical and Corrosion Properties. Petroleum Mechanical Engineering with Under Water Technology Conference, September 19-23, 1971, Houston, Texas.
11. Denhard, E. E., Espy, R. H. Metals Engineering Quarterly, Vol. 12, No. 4, p. 18 (1972).
12. Henthorne, M., Parkins, R. N. Stress Corrosion Test Methods, British Corrosion Journal, Vol. 2, p. 186 (1967); and Proceedings of Third International Congress on Metallic Corrosion, Moscow, 1966, p. 309 (available from NACE).
13. Parkins, R. N. British Corrosion Journal, Vol. 7, p. 15 (1972) January.
14. Parkins, R. N. British Corrosion Journal, Vol. 7, p. 153 (1972) July.
15. Clarke, W. L., Cowan, R. L., Danko, J. C. A Dynamic Straining Stress Corrosion Test for Predicting Boiling H_2O Reactor Materials Performance. Proceedings of ASTM Conf., Toronto, Ontario, April, 1977, Symposium on Stress Corrosion Cracking.
16. Kass, J. N., Henry, M. F., Pickett, A. E., Walker, W. L. Corrosion, Vol. 35, p. 229 (1979) June.
17. Streicher, M. A. Corrosion, Vol. 30, p. 77 (1974) March.
18. ASTM C 692-77. Standard Method for Evaluating the Influence of Wicking Type Thermal Insulations on the Stress Corrosion Cracking Tendency of Austenitic Stainless Steels.
19. Bloom, F. K. Corrosion, Vol. 11, p. 351t (1955) August.
20. Dana, A., DeLong, W. B. Corrosion, Vol. 12, p. 309t (1956) July.
21. Humphries, T. S. Process for Externally Loading and Corrosion Testing Stress Corrosion Specimens, NAS TM 5383, ASTM G44, July, 1966.
22. Joint Report ASTM and NACE. Stress Corrosion Cracking of Austenitic Chromium-Nickel Stainless Steels, ASTM STP No. 264, March, 1960.
23. Robelotto, R. Investigation of Magnitude and Distribution of Stresses in Welded Structures, Technical Report AFML TR-67-293, September, 1967.

Note:

Stress Corrosion of Brass: Field Tests in Different Types of Atmosphere

by W. Landegren* and E. Mattsson†

(*Manuscript received* 1 *March*, 1976)

Stress corrosion tests were carried out with deep drawn cup specimens of Cu63Zn37 brass, stress-relief annealed at different temperatures after the deep drawing operation. The stress levels of the variants were determined by a cutting-up method. The specimens were exposed to the atmosphere; indoors in a storehouse with or without heating facilities and outdoors in marine, urban and rural atmospheres with or without shelter from rain. The specimens were inspected regularly during a 10*-year period with respect to stress corrosion cracks, and the following conclusions were reached:*

Cups of α-brass of type CuZn37 with high tensile stresses run very little risk of stress corrosion cracking in dry storage with heating facilities, although the risk of cracking cannot be completely excluded for material with a very high stress level. In a storehouse without heating, where condensation may occur occasionally, the risk of cracking is greater and exists also for material with somewhat lower stress level.

On outdoor exposure the risk of stress corrosion cracking is considerably less in a marine atmosphere than in a rural or urban atmosphere; the stress corrosion risk was found to be somewhat greater under shelter from rain than on open exposure.

Brass cups of the type examined can be made proof against stress corrosion cracking by stress-relief annealing. For outdoor storage the requirement would be, e.g., annealing for 2h at a temperature of at least 300°C. *For indoor storage annealing at* 225°C *for the same time appears to be sufficient.*

Introduction

In 1963 an investigation on stress corrosion of brass was started, its aim being to establish the correlation between field tests in the atmosphere and accelerated tests. This work was carried out within Sub-Committee 9.5 of the Corrosion Committee of the Royal Swedish Academy of Engineering Sciences. Results of the accelerated tests and of the first five years of exposure in different types of atmosphere have earlier been reported by Mattsson, Landegren, Lindgren, Rask and Wennström.[1,2]

After about 10 years' exposure the field tests in the atmosphere have been terminated. The work after 1965 has been carried out under the auspices of Working Group 2 60 05 0 of the Swedish Corrosion Institute. The present authors have examined the specimens remaining on the racks at the end of the exposure period and have collected together in this report all facts and results concerning the field tests.

Experimental

Specimens

The specimens had the shape of deep-drawn cups made of semi-hard brass strip (Fig. 1). All the cups were made from the same strip. The brass was a binary copper–zinc alloy of type CuZn37 containing 62·4% Cu and remainder Zn. The alloy consisted mainly of alpha phase, but contained traces of beta phase. A small hole was drilled in the bottom of all cups for the fastening wire used in the field tests. After manufacture, the cups were vapour-degreased in trichloroethylene. The total number of specimens was divided into groups, each being stress-relief annealed at a certain temperature. After annealing, the specimens were pickled in 10% (by weight) sulphuric acid and finally rinsed in distilled water. The different variants of cups thus obtained contained different proportions of the residual stresses introduced by the deep-drawing process.

The longitudinal and tangential tensile stresses in the cylindrical part of the cups were determined by a 'cutting-up method' described by Andersson & Fahlman[3] for longitudinal stresses and by Hatfield & Thirkell[4] for tangential stresses. The results are given in Fig. 2.

*Gränges Metallverken, S-721 88 Västerås
†Swedish Corrosion Institute, Drottning Kristinas väg 48, S-114 28 Stockholm

The Hatfield–Thirkell method for the determination of the tangential stresses requires a tubular specimen of length 2–3 times the outer diameter of the tube. In this case the length of the cylindrical part of the specimen was smaller than the diameter. Furthermore there is a tendency for this method to give too high a value in cases when the stresses are very high, as in the unannealed specimens. So the results may be regarded as approximate.

For information on how stress-relief annealing affected the mechanical properties, the hardness was determined in cross-sections of the wall in the cylindrical part and in the undeformed bottom of all the variants of specimen cups. The results are shown in Fig. 3.

Field tests

In the field tests the specimens were exposed to different types of atmosphere.

Indoors:

In a storehouse with heating facilities, with the temperature controlled at 20–22°C;

In a storehouse without heating facilities; up to 1940 this building was used as a stable.

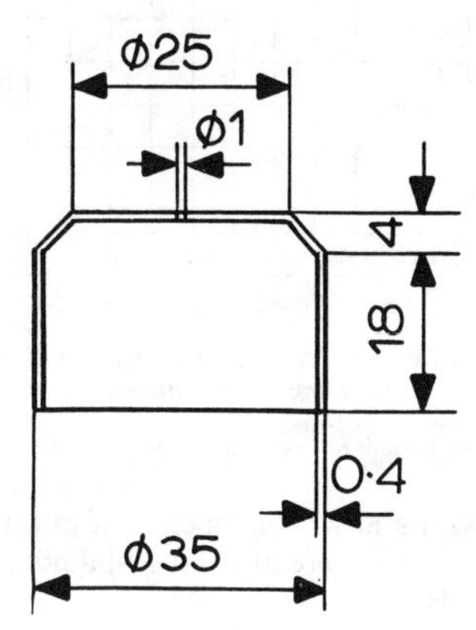

Fig. 1. *Shape of the cup used as specimen in the investigation*
Dimensions shown in mm

Reprinted with permission from *British Corrosion Journal*, Vol. II, No. 21, 1976, 80-85

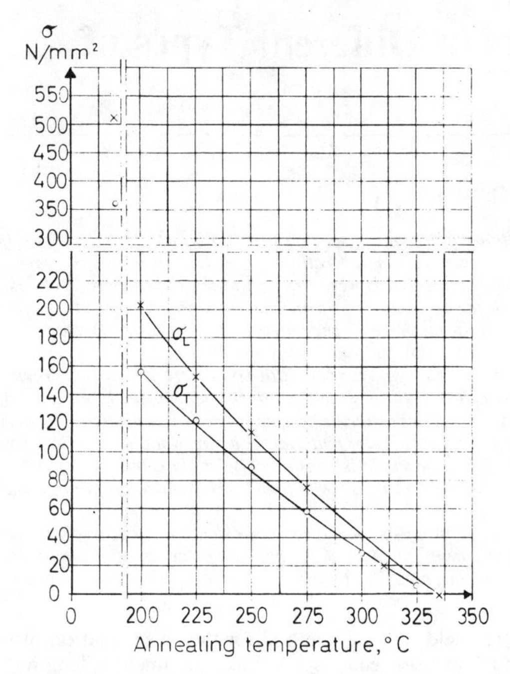

Fig. 2. *Residual tensile stresses vs annealing temperature for the different specimen variants; annealing time 2h*
δ_L Longitudinal stresses
δ_T Tangential stesses

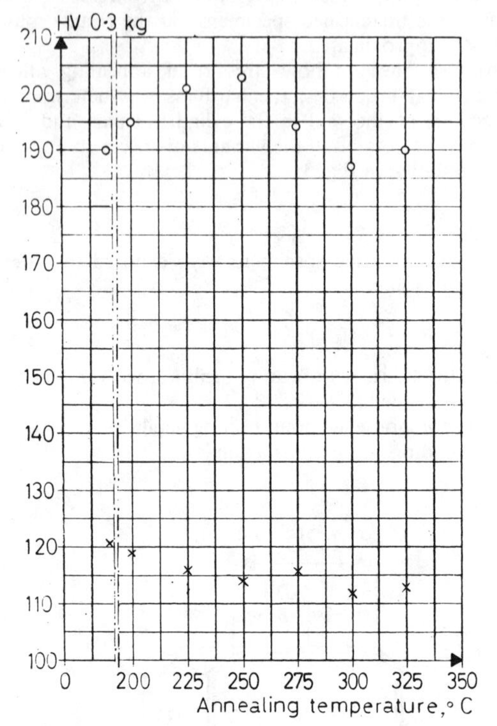

Fig. 3. *Hardness in cross section of specimen wall vs stress-relief annealing temperature*
○ Cylindrical part of specimen
× Undeformed bottom of specimen

Outdoors both with and without shelter from rain:

In a marine atmosphere at Bohus Malmön, on the west coast of Sweden;

In an urban atmosphere at Vanadis in Stockholm;

In a rural atmosphere at Erken in Uppland.

Fig. 4. *Rack with specimens on open exposure (left) and a Stevenson screen for exposure with shelter from rain (right)*

The field tests were made on 20 replicates of each specimen variant on each site. The specimens were fastened in a vertical frame of copper tubing with copper wire (Fig. 4). The tubes in the frame were wrapped in Teflon foil in order to protect the specimens from rainwater contaminated by the frame. Exposure with shelter from rain was done in a Stevenson screen, i.e., a type of cage used by meteorologists for thermometers, etc. (Fig. 4).

Exposure on the racks of specimens of the main series was started in May 1963 and inspection for cracks was carried out after 1·5, 3, 5, 9, and 12 months of exposure, and thereafter generally once a year. When cracking was observed in a specimen, this was removed from the site. Metallographic examination was made on cracked specimens typical of the different variants.

In October 1963 exposure of another series of specimens was started, this series comprising only the variant stress-relief annealed at 250°c. The purpose was to study whether the season in which exposure is started affects the stress corrosion.

At the outdoor test sites climatic parameters were recorded and chemical analysis was made continuously of contaminants in the air and precipitation. Significant results for the sites during the period of exposure are given in Table I. Characteristic features are the high chloride content of the marine atmosphere (Bohus Malmön), and also the rather high sulphur content of the urban atmosphere (Vanadis).

Results

The specimens were examined during exposure on the racks as well as after exposure in the laboratory.

Visual inspection

The accumulated percentage of cracked specimens on each inspection is shown in Table II. The typical appearance of cracked cups is shown in Fig. 5. As the figure indicates, the characteristic crack pattern was dependent on the stress-relief annealing given to the cup.

When the exposure test was terminated after 121 months, the remaining uncracked specimens were taken in for examination in the laboratory. The typical appearance of specimens from different sites can be seen in Fig. 6. It is especially worth noting that at Vanadis a heavy grey surface deposit had developed on the specimens sheltered from the rain. The deposit partly flaked off on occasions. Then a new deposit was formed where flakes had fallen off. A minor deposit of a similar kind developed on specimens sheltered from the rain at Bohus Malmön. The specimens exposed outdoors without shelter from rain had a dark, smooth surface coating.

TABLE I

Climatic conditions and significant amounts of principal atmospheric contaminants and pH values of precipitation on the test sites

Test site	Mean annual temp., °C	Precipitation, mm/year	Contaminants				pH value of precipitation
			In air, μg/m³		In precipitation, mg/m². month		
			S	Cl	S	Cl	
Rural (Erken)	5·8	550	8	2	50	20	4·9
Marine (Bohus Malmön)	6·5	720	10	50	100	800	5·1
Urban (Vanadis)	6·6	550	70	4	180	60	4·7

To permit a more careful examination the uncracked variants with the highest stress level were pickled in 10% (by weight) sulphuric acid for 10 min at room temperature. After pickling, these specimens were again inspected for cracking. The results of this inspection are also included in Table II. As will be seen, additional cracks were generally not revealed by the pickling. Exceptions, however, are the sheltered specimens at Bohus Malmön which, after pickling, showed additional fine, short cracks.

The outdoor exposure of specimens for which inspection results are shown in Table II was started in May 1963. The other set of specimens first exposed on the same sites in October 1963 reached the same state of cracking as the May specimens in less than one year. In fact the degree of cracking became slightly higher for the October specimens. The difference, however, does not seem to be of any practical importance.

Analysis of surface deposits

The deposits collected on the specimen surface were examined by X-ray diffraction analysis. Analysis was in most cases carried out on deposits remaining on the metal surface. In some cases the analysis was carried out on deposit removed from the surface and attached to a Millipore filter.

The results are shown in Table III. The presence of metallic Cu in excess of the amount occurring in the brass indicates dezincification at all the outdoor sites, most at Vanadis. The amount of Cu_2O was greatest on the unsheltered specimens. $Cu_2(OH)_3Cl$ occurred on the Bohus Malmön specimens, carbon on those exposed at Vanadis and Erken. The presence of $2FeCl_3.7H_2O$ on Bohus Malmön sheltered specimens is difficult to explain. The iron compound may originate from a rusty steel fence in the neighbourhood.

Metallographic examination

To check whether microcracks occurred in the specimens which appeared to be free from cracks on visual inspection, a metallographic examination was carried out. This check was restricted on each site to one specimen of the uncracked variant with the highest stress level. The metallographic examination was carried out on a cross-section of the cylindrical part of the cup. At a magnification of 90 × none of the specimens examined showed any microcracks.

In cracked specimens the nature of the cracks was studied metallographically at a magnification of 80–450 ×. Specimens from all the sites were studied, both unannealed variants and variants annealed at 200, 250 and 300°C. In all cases the cracks were thin, unbranched and predominantly intergranular (Fig. 7). They always appeared to have started from the outside of the wall.

The depth of dezincification was also determined by metallographic examination. In these cases a longitudinal cross-section through the wall of the cylindrical part of the cup was examined. The sloping part of the wall and one part of the cup bottom were also included in this cross-section. The dezincification was found to have the character of localised attack (Fig. 8), so the maximum depth of dezincification was determined. From each site an examination was made of one specimen of the variant annealed at 275°C and, in addition, some other variants from the Bohus Malmön site. The specimens examined had with one exception been exposed for 121 months. The results are shown in Table IV.

Discussion

Influence of atmosphere

Comparing the results of the two *indoor exposures* the advantage of the storehouse with heating facilities is evident. There only 10% of the unannealed cups cracked within 121 months. The cracking tendency was much larger in the storehouse without heating facilities. The reason was probably that in the latter case the metal surface was occasionally covered with a film of moisture formed through condensation as a result of temperature changes.

As regards the *outdoor atmosphere*, it is remarkable that open exposure to the marine atmosphere led to cracking virtually only in the unannealed specimens, while open exposure to a rural atmosphere caused considerable attack on the 225°C level and in an urban atmosphere even on the 250°C level. Thus, it appears as though Cl^- or some chlorine compound, possibly $Cu_2(OH)_3Cl$, inhibits stress corrosion cracking in brass, while the pollutants of the urban atmosphere, i.e., sulphur compounds or soot, accelerate it.

The finding that the stress corrosion tendency is considerably smaller in a marine atmosphere than in a rural or urban atmosphere is in agreement with results obtained in the USSR.[5] There it has been found that the stress corrosion tendency of brass is much smaller in the Bering Sea and the Black Sea atmospheres than in industrial and urban atmospheres in the central part of the USSR.

Influence of shelter from rain

In outdoor exposure, shelter from rain has caused cracking at a lower stress level than was the case for open exposure. This was true for all the sites.

The dezincification was also greater in sheltered conditions than on open exposure, at least in urban and rural atmospheres. In the marine atmosphere this conclusion is valid only for the cylindrical surfaces exposed vertically, while in the sloping parts and the horizontal bottom surface the difference was less pronounced.

The greater attack under shelter from rain especially in an urban atmosphere may be due to the great amount of deposits collecting on the surface under these conditions, which help to keep the surface moist.

Table II

Results of field tests; figures give the percentage of cracked specimens out of 20 replicates after the indicated time of exposure

Specimen variant	Inspection without pickling														Inspection after pickling
	Exposure time, months														
	1·5	3	5	9	12	17	25	36	52	61	78	87	101	121	121
STOREHOUSE WITH HEATING FACILITIES (ARBOGA)															
Without annealing	0	0	0	0	0	5	5	5	5	5	5	5	10	10	Pickling not required
Annealing 2 h at temperature 200°c	0	0	0	0	0	0	0	0	0	0	0	0	0	0	
225°c	0	0	0	0	0	0	0	0	0	0	0	0	0	0	
250°c	0	0	0	0	0	0	0	0	0	0	0	0	0	0	
275°c	0	0	0	0	0	0	0	0	0	0	0	0	0	0	
300°c	0	0	0	0	0	0	0	0	0	0	0	0	0	0	
325°c	0	0	0	0	0	0	0	0	0	0	0	0	0	0	
STOREHOUSE WITHOUT HEATING FACILITIES (KARLSTAD)															
Without annealing	0	0	0	5	5	10	10	15	90	95	95	100	—	—	Pickling not required
Annealing 2 h at temperature 200°c	0	0	0	0	0	0	0	0	0	0	0	0	0	50	
225°c	0	0	0	0	0	0	0	0	0	0	0	0	0	5	
250°c	0	0	0	0	0	0	0	0	0	0	0	0	0	0	
275°c	0	0	0	0	0	0	0	0	0	0	0	0	0	0	
300°c	0	0	0	0	0	0	0	0	0	0	0	0	0	0	
325°c	0	0	0	0	0	0	0	0	0	0	0	0	0	0	
RURAL ATMOSPHERE (ERKEN) WITHOUT RAIN SHELTER															
Without annealing	95	100	—	—	—	—	—	—	—	—	—	—	—	—	—
Annealing 2 h at temperature 200°c	0	0	0	100	—	—	—	—	—	—	—	—	—	—	—
225°c	0	0	0	50	80	85	85	90	90	90	90	90	90	90	90
250°c	0	0	0	0	5	5	5	5	5	5	5	5	5	5	5
275°c	0	0	0	0	0	0	0	0	0	0	0	0	0	0	0
300°c	0	0	0	0	0	0	0	0	0	0	0	0	0	0	0
325°c	0	0	0	0	0	0	0	0	0	0	0	0	0	0	0
RURAL ATMOSPHERE (ERKEN) WITH RAIN SHELTER															
Without annealing	35	100	—	—	—	—	—	—	—	—	—	—	—	—	—
Annealing 2 h at temperature 200°c	0	25	100	—	—	—	—	—	—	—	—	—	—	—	—
225°c	0	5	70	100	—	—	—	—	—	—	—	—	—	—	—
250°c	0	0	0	65	75	75	75	75	75	75	75	75	75	75	75
275°c	0	0	0	0	10	10	10	10	10	10	10	10	10	10	10
300°c	0	0	0	0	0	0	0	0	0	0	0	0	0	0	0
325°c	0	0	0	0	0	0	0	0	0	0	0	0	0	0	0
URBAN ATMOSPHERE (VANADIS) WITHOUT RAIN SHELTER															
Without annealing	100	—	—	—	—	—	—	—	—	—	—	—	—	—	—
Annealing 2 h at temperature 200°c	5	70	100	—	—	—	—	—	—	—	—	—	—	—	—
225°c	0	10	100	—	—	—	—	—	—	—	—	—	—	—	—
250°c	0	0	20	100	—	—	—	—	—	—	—	—	—	—	—
275°c	0	0	0	0	0	0	10	10	10	10	10	10	10	10	10
300°c	0	0	0	0	0	0	0	0	0	0	0	0	0	0	0
325°c	0	0	0	0	0	0	0	0	0	0	0	0	0	0	0
URBAN ATMOSPHERE (VANADIS) WITH RAIN SHELTER															
Without annealing	100	—	—	—	—	—	—	—	—	—	—	—	—	—	—
Annealing 2 h at temperature 200°c	100	—	—	—	—	—	—	—	—	—	—	—	—	—	—
225°c	85	90	90	100	—	—	—	—	—	—	—	—	—	—	—
250°c	0	20	45	95	95	95	100	—	—	—	—	—	—	—	—
275°c	0	0	0	10	30	30	75	85	90	95	95	95	95	95	95
300°c	0	0	0	0	0	0	0	0	0	0	0	0	0	0	0
325°c	0	0	0	0	0	0	0	0	0	0	0	0	0	0	0

(continued)

Specimen variant	Inspection without pickling														Inspection after pickling
	Exposure time, months														
	1·5	3	5	9	12	17	25	36	52	61	78	87	101	121	121
MARINE ATMOSPHERE (BOHUS MALMÖN) WITHOUT RAIN SHELTER															
Without annealing	0	0	0	60	60	60	60	70	70	70	75	85	85	95	95
Annealing 2 h at temperature 200°C	0	0	0	0	0	0	0	0	0	0	0	0	0	0	5
225°C	0	0	0	0	0	0	0	0	0	0	0	0	0	0	0
250°C	0	0	0	0	0	0	0	0	0	0	0	0	0	0	0
275°C	0	0	0	0	0	0	0	0	0	0	0	0	0	0	0
300°C	0	0	0	0	0	0	0	0	0	0	0	0	0	0	0
325°C	0	0	0	0	0	0	0	0	0	0	0	0	0	0	0
MARINE ATMOSPHERE (BOHUS MALMÖN) WITH RAIN SHELTER															
Without annealing	5	15	30	30	35	35	55	80	85	90	95	100	—	—	—
Annealing 2 h at temperature 200°C	0	0	0	0	0	0	0	25	25	25	25	25	25	40	75
225°C	0	0	0	0	0	0	0	0	0	0	15	15	15	25	50
250°C	0	0	0	0	0	0	0	0	0	0	0	0	0	0	0
275°C	0	0	0	0	0	0	0	0	0	0	0	0	0	0	0
300°C	0	0	0	0	0	0	0	0	0	0	0	0	0	0	0
325°C	0	0	0	0	0	0	0	0	0	0	0	0	0	0	0

TABLE III

Results of X-ray analysis of deposits on specimens exposed at the outdoor sites; + indicates exposure under rain shelter

Test site	Annealing temperature, °C	Time of exposure, months	Rain shelter	Identified components in deposits
Bohus Malmön	—	36	—	Cu, Cu_2O, $Cu_2(OH)_3Cl$ (trace)
Bohus Malmön	—	36	+	Cu, Cu_2O, $Cu_2(OH)_3Cl$, $2FeCl_3.7H_2O$
Erken	275	25	—	Cu, Cu_2O, C (trace)
Erken	275	36	+	Cu, Cu_2O, C (trace)
Vanadis	225	17	—	Cu, Cu_2O, C (trace)
Vanadis	275	12	+	Cu, Cu_2O, C

TABLE IV

Maximum dezincification depth found on metallographic examination of exposed samples; + indicates exposure under rain shelter

Specimen characteristics				Max. dezincification depth, μm					
Annealing temp., °C	Exposure site	Rain shelter	Time of exposure, months	Cylindrical part		Sloping part		Bottom	
				Outside	Inside	Outside	Inside	Outside	Inside
275	Arboga		121	0	0	0	0	0	0
275	Karlstad		121	0	0	0	0	0	0
275	Bohus Malmön	+	121	170	70	70	40	40	55
275	Bohus Malmön	—	121	55	40	40	40	55	55
275	Erken	+	121	140	40	40	0	40	70
275	Erken	—	121	40	30	0	15	0	15
275	Vanadis	+	121	270	185	100	55	100	130
275	Vanadis	—	121	85	40	30	30	0	55
—	Bohus Malmön	+	85	140	140	40	30	30	40
—	Bohus Malmön	—	121	55	40	70	55	55	55
200	Bohus Malmön	+	121	230	70	55	40	40	70
200	Bohus Malmön	—	121	100	55	70	40	55	40
325	Bohus Malmön	—	121	55	40	40	40	40	40

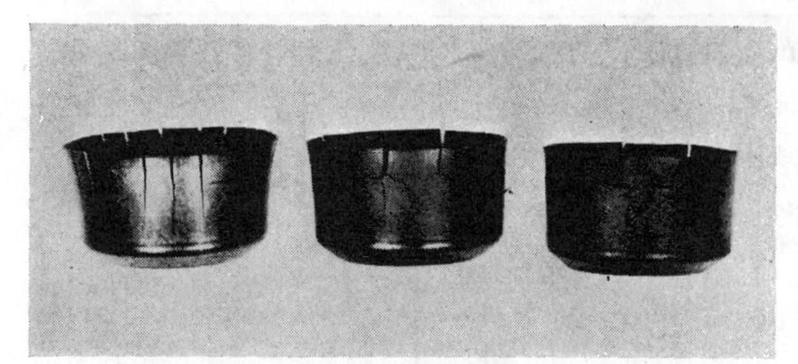

Fig. 5. *Characteristic crack patterns on specimens exposed in an urban atmosphere*
Unannealed specimen; 1·5 months' exposure (left)
Specimen annealed at 225°c for 2h; 5 months' exposure (middle)
Specimen annealed at 250°c for 2h; 4 months' exposure (right)

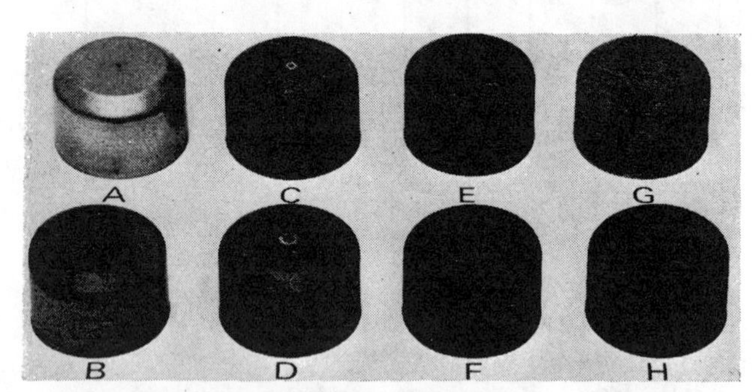

Fig. 6. *Appearance of specimens (annealed at* 300°c *for 2h) after exposure in:*
A Storehouse with heating facilities (Arboga)
B Storehouse without heating facilities (Karlstad)
C Marine atmosphere with shelter from rain (Bohus Malmön)
D Marine atmosphere without shelter from rain (Bohus Malmön)
E Rural atmosphere with shelter from rain (Erken)
F Rural atmosphere without shelter from rain (Erken)
G Urban atmosphere with shelter from rain (Vanadis)
H Urban atmosphere without shelter from rain (Vanadis)

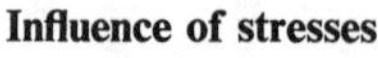

Influence of stresses

As expected, the stress level is a determining factor for the cracking. The stıess corrosion risk for the deep-drawn cup can be eliminated by annealing. If an annealing time of 2h was chosen, an annealing temperature of at least 300°c was required for exposure in an urban or rural atmosphere and at least 225°c for a marine atmosphere (Table II). Such annealing will decrease the tensile stresses in the present type of object to 30–40 N/mm² at the higher temperature, and to 120–150 N/mm² at the lower temperature (Fig. 2).

To eliminate the stress corrosion risk for indoor storage, even in storage where occasionally condensation may occur, annealing for 2h at a temperature of at least 225°c was required, corresponding to a stress level of 120–150 N/mm².

Conclusions

The results of the field tests lead to the following conclusions:

Cups of α-brass of type CuZn37 with high tensile stresses run very little risk of stress corrosion cracking in dry storage with heating facilities, although the risk of cracking cannot be completely excluded for material with a very high stress level.

In a storehouse without heating, where condensation may occur occasionally, the risk of cracking is greater and exists also for material with a somewhat lower stress level.

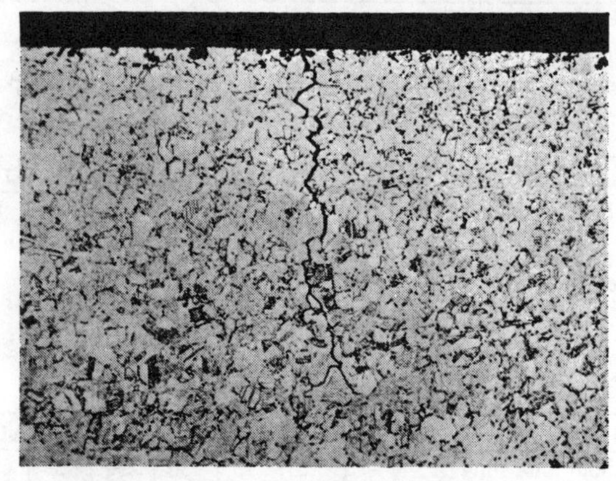

Fig. 7. *Characteristic stress corrosion crack; specimen annealed at* 250°c *for 2h; exposed for 9 months in urban atmosphere; etching agent–ammonium persulphate* × 133

Fig. 8. *Section through specimen wall showing dezincification; specimen annealed at* 275°c *for 2h; exposed for* 121 *months; unetched* × 33

Out-of-doors the risk of stress corrosion cracking is considerably less in a marine atmosphere than in a rural or urban atmosphere.

On outdoor exposure the stress corrosion risk was found to be somewhat greater under shelter from rain than on open exposure.

Brass cups of the type examined can be made proof against stress corrosion cracking by stress-relief annealing. For outdoor storage the requirement would be, e.g., annealing for 2h at a temperature of at least 300°c. For indoor storage annealing at 225°c for the same time appears to be sufficient.

It should be pointed out that these tests have been made with cups of a certain shape having residual tensile stresses and with a copper–zinc alloy not containing any other alloying constituents. Without checking, the results cannot be applied to other shapes, to other stress conditions to brasses with other constituents, or to bronzes.

References

1. Mattsson, E., Lindgren, S., Rask, S., & Wennström, G., *Proc. 4th Scand. Corros. Congr., Helsinki*, 1964, p. 171
2. Mattsson, E., Landegren, W., & Rask, S., *Proc. 5th Scand. Corros. Congr., Copenhagen*, 1968, p. 9-1
3. Andersson, R. J., & Fahlman, E. G., *J. Inst. Metals*, 1924, **32**, 367
4. Hatfield, W. H., & Thirkell, G. L., *ibid.*, 1919, **22**, 67
5. Klark, G. B., 'Corrosion cracking of brass in different climatic regions of the USSR. Intercrystalline corrosion and corrosion of metals under stress'. Ed. I. A. Levin. 1962, p. 311 (New York: Consultants Bureau)

Stress-Corrosion Cracking of Steels In Agricultural Ammonia★

By A. W. LOGINOW* and E. H. PHELPS*

Abstract

Stress-corrosion cracking has been determined as the cause of failure of carbon steel tanks in agricultural ammonia service. Effects of ammonia contaminants, such as air, water, carbon dioxide, oil, etc. on failure rates were studied. In general, air contamination increased stress-corrosion cracking and water in small amounts inhibited the attack.

For prevention of such failures, it was recommended that high residual stresses in ammonia vessels be avoided by the use of stress relieving treatments, that air be eliminated from agricultural ammonia systems and that the ammonia should have a minimum water content of 0.20 percent. 3.5.8

Introduction

ANHYDROUS AMMONIA HAS been used extensively for many years in the heat-treating, chemical and refrigeration industries. Its use in agriculture as a nitrogen fertilizer, however, is comparatively new. Because of excellent results with direct application of anhydrous ammonia to the soil, its use as a fertilizer has expanded rapidly and a large number of steel tanks have been placed in service for agricultural ammonia storage, transport and application. Many failures of these tanks by cracking have occurred, although the chemical and refrigeration industries had not reported any appreciable difficulties with steel equipment handling anhydrous ammonia. A survey indicated that 3 percent of the tanks in agricultural service had failed during an average of three years' service,[1] because of minor leaks from what proved to be stress-corrosion cracking.

The Agricultural Ammonia Institute Research Committee was formed in 1954 to determine the cause of and methods for preventing cracking in agricultural-ammonia vessels. This work has consisted of (1) reviews of all reported failures of ammonia vessels, (2) detailed metallurgical examinations, (3) a sponsored program at Georgia Institute of Technology, and (4) cooperative programs between the Committee and companies concerned with the problem.

From metallurgical examinations of failed vessels, the Committee diagnosed the cause of unexpected failure of properly fabricated vessels as stress-corrosion cracking.[1] An extensive research program was begun to confirm this diagnosis and to establish the agents in ammonia that were responsible for the cracking. In addition, methods were studied for preventing further vessel failures.

★ Submitted for publication January 2, 1962. A paper presented at the 18th Annual Conference, National Association of Corrosion Engineers, March 19-23, 1962, Kansas City, Missouri.

* U. S. Steel Corporation, Applied Research Laboratory Monroeville, Pa.

This article summarizes the results of stress-corrosion research by the Committee.

Scope of the Research Program

Metallurgical examinations of failed ammonia vessels had usually revealed numerous fine cracks on the interior surfaces, principally in the heads near the head-to-shell weld, and a branching pattern of cracking within the metal. These are typical indications of stress-corrosion cracking. The observation that cracking occurred in vessels with cold-formed heads but not in those which had been stress-relieved after fabrication indicated that residual stresses were contributing to the cracking.

Examinations of failed vessels, however, furnished only indirect evidence as to the cause of failure. Therefore, the first aim of the Committee's research program was to devise and test steel specimens to establish whether stress-corrosion cracking was actually occurring. After this had been shown with stressed specimens, the pro-

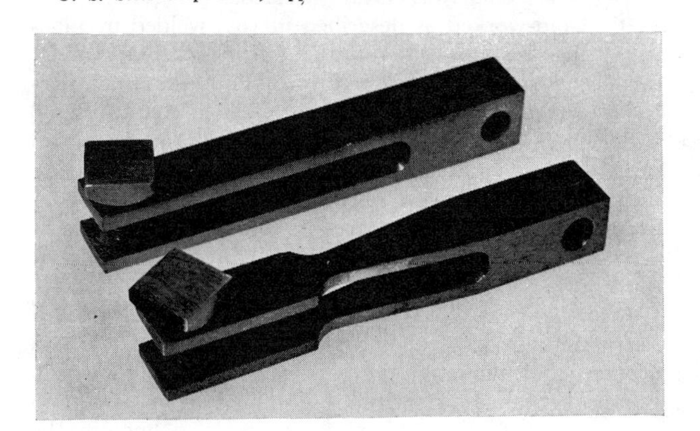

Figure 1—Tuning-fork specimens used in stress-corrosion studies. Approximately ⅔ actual size.

Figure 2—Tuning-fork specimens. Conditions (left to right): as machined; cold worked; stressed; cold-worked, welded and stressed. Approximately ½ actual size.

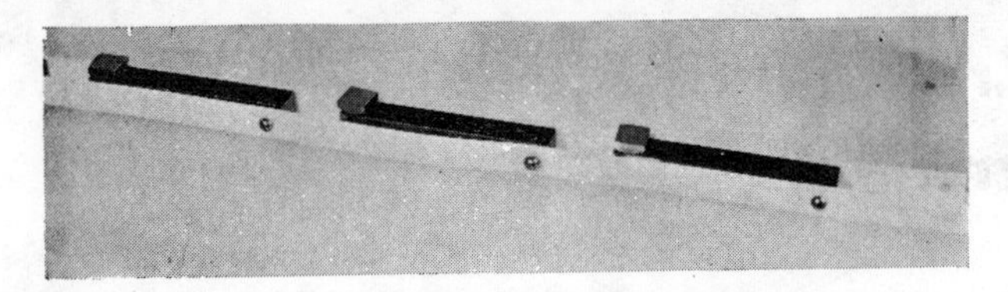

Figure 3—Holders used to suspend specimens in ammonia vessels and cylinders. Approximately ¼ actual size.

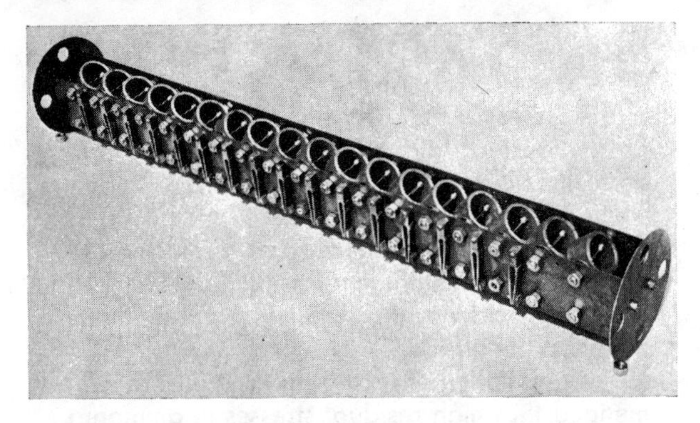

Figure 5—Tensile specimens and tuning-fork specimens assembled on rack for exposure in ammonia.

gram was broadened to determine the following:

1. Effects of storage and handling techniques on stress-corrosion cracking.
2. Specific ammonia contaminants causing the cracking.
3. An inhibitor that could be added to ammonia in small amounts to prevent stress-corrosion cracking.

This program was conducted by exposing stressed test specimens in agricultural ammonia tanks in field tests, and in cylinders containing refrigeration-grade ammonia with various contaminant and/or inhibitor additions. The various stress-corrosion tests were conducted through the cooperation of the Southeastern Liquid Fertilizer Company, the Phillips Petroleum Company, the United States Steel Corporation, and as part of the Research Committee's sponsored program at the Georgia Institute of Technology.

Experimental Procedures

Tuning-Fork Test Specimens

Specimens were required that could be placed in ammonia cylinders and in ammonia vessels, such as nurse and applicator tanks, which did not have large openings. The tuning-fork specimens shown in Figure 1 were suitable. These specimens were machined from ½-inch plate and when finished were 3½ inches long, ½ inch wide and ½ inch deep. The tines were ⅛ inch thick and about 2½ inches long. Some of the tuning forks had tines with parallel edges and others had tines with tapered edges (upper and lower specimens, respectively, in Figure 1). The tapered tines were machined partially along the sides of a triangle with its apex at the center of the bolt and its base coinciding with the base of the tines. For all of the tuning-fork specimens, the machining was done so that the original rolled surface of the plate corresponded to the flat surface of the tines. The bolt was machined from the same steel used in making the specimen, so that any possible galvanic effect would be avoided. These specimens could be

Figure 4—Tensile specimen with ring used for stressing. Approximately ⅔ actual size.

inserted through the ¾-inch openings in ammonia vessels and cylinders. The specimens were stressed by closing the specimen tines and restraining them in the closed position with the bolt at the tine ends. The amount of closure was determined from the empirical relationship:

$$S = 54{,}300 \frac{\Delta}{t},$$

where S is the maximum stress in pounds per square inch (psi) in the outer fiber of either tine, Δ is the total amount of closure in inches at the tine ends, and t is the thickness of the tines in inches. The stress on tuning forks with parallel sides reaches a maximum in a small area at the base of the tines. The above relationship was derived from data obtained with strain gauges placed at this location on calibration specimens with parallel sides. On the tuning forks with tapered tines, the maximum stress extends uniformly along the tapered section. The above relationship was applicable in stressing both types of tuning forks.

To simulate the various metallurgical conditions present in ammonia vessels with cold-formed heads, the tuning forks were subjected to one of the following conditions:

(a) Stressed to a definite stress level in the outer fibers of the tines. This condition is herein referred to as "stressed only."

(b) Cold-worked by bending the tines outward until an appreciable permanent set was achieved, then stressed by closing the specimen tines. Referred to as "cold-worked and stressed."

(c) Cold-worked as described in (b), welded by placing a weld bead on the outside of one tine at its base, then stressed by closing the tines. Referred to as "cold-worked, welded and stressed."

Specimens illustrating the above procedures are shown in Figure 2. The actual stress in the outer fibers of specimens can be estimated only in condition (a). In conditions (b) and (c), the stress is not known because of work-hardening that occurred during cold working of the tines.

After being stressed, the tuning forks were degreased in trichloroethylene vapor, washed with water, dried with acetone and mounted on plastic holders as shown in Figure 3. Stainless steel bolts, insulated from the specimens with plastic tubing, were used to mount the specimens. The assembled holders were then suspended in ammonia vessels or cylinders.

TABLE 1—Compositions of Steels

Steel	Composition, Percent										
	C	Mn	P	S	Si	Cu	Ni	Cr	Mo	V	B
ASTM A-212* Grade B	0.33	0.77	0.017	0.030	0.24	0.04	0.03	0.04	....	....	....
ASTM A-285*	0.17	0.53	0.009	0.031	0.063	0.04	....	....	....	....	....
ASME Case 1056*	0.30	1.13	0.017	0.030	0.03	0.04	0.11	0.07	....	....	....
ASTM A-202 Grade B	0.19	1.46	0.016	0.020	0.77	0.04	0.05	0.45	....	....	....
"T-1" Steel*	0.16	0.88	0.017	0.019	0.22	0.34	0.84	0.53	0.48	0.05	0.002
AISI 4130	0.33	0.54	0.009	0.018	0.28	....	....	0.98	0.19	....	....
Springs	0.52	0.71	0.007	0.040	0.13	0.20	0.09	0.05	0.08	0.002	....

* More than one heat of steel tested; compositions shown are averages.

TABLE 3—Welding Electrodes Used for Preparing Welded Specimens

STEEL	Electrode
ASTM A-212 Grade B	E6015
ASTM A-285	E6010
ASME Case 1056	E7016
ASTM A-202 Grade B	E7016
"T-1" steel	E12015
AISI 4130	E12015

TABLE 2—Mechanical Properties of Steels

Steel	Yield Strength, psi	Tensile Strength, psi	Elongation, percent		Hardness
			In 2 Inch	In 8 Inch	
ASTM A-212* Grade B	51,000	83,000	34	23	RB 84
ASTM A-285*	37,000	59,000	45	31	RB 66
ASME Case 1056*	48,000	79,000	49	25	RB 85
ASTM A-202 Grade B	57,000	93,000	41	18	RB 94
"T-1" Steel*	116,000	125,000	26	12	Rc 25
AISI 4130**	119,000	134,000	19	..	Rc 29
Springs	......	......	..	..	Rc 55

* More than one heat of steel tested; mechanical properties shown are averages.
** Austenitized at 1650 F for 30 minutes, oil quenched, tempered at 1100 F for 2 hr, air-cooled, stress-relieved at 500 F, furnace-cooled.

Tensile Specimens

In the later stages of the program, special cylinders were constructed with a large opening in one end. These cylinders were made of tubes about 8 feet long by 9 inches in diameter. The necessary fittings, safety valve and pressure gauge were mounted on a blind flange comprising one end of the tube. The other end was closed by welding a hemispherical cap to the tube.

The tensile specimens used were of standard 0.252-inch diameter, slightly modified at the threaded ends. They were contained in a ring as shown in Figure 4, and were stressed by tightening the nuts on the specimen ends against the ring. The rings were machined from a 1.5-inch length of carbon steel seamless pipe with an outside diameter of 2.4 inches and a wall thickness of 0.22 inch. The amount of stress was determined by measuring the distance between two reference points on the specimen surface.

The tensile specimens were used in both the welded and unwelded conditions. The welded specimens were machined from a butt-welded ½-inch plate, so that the heat-affected zone on one side of the weld was approximately in the center of the specimen gauge-length. Both welded and unwelded specimens were stressed either slightly above the yield point or appreciably above it. The latter was done by prestraining the specimens in a tensile machine to 5 to 10 percent plastic strain, then stressing them in the rings until that strain was slightly exceeded.

Because of the mode of stressing and the strain-measuring equipment, an accurate knowledge of the direction and magnitude of the applied stresses is not possible with this type of specimen. Tightening the nuts introduced unmeasured torsional stresses in addition to the measured axial stresses. The tensile specimens were stressed above the yield point to obtain a relatively high state of stress; therefore, it is believed that the additional torsion did not materially influence the results obtained.

After being stressed, the tensile specimens were cleaned by the procedure used to clean the tuning forks and were then assembled on racks (Figure 5). Two assembled racks were then exposed to ammonia in each of the special cylinders.

Spring Specimens

As will be discussed later, failures of tuning-fork and tensile specimens of steels used in ammonia vessels did not always yield reproducible or consistent results. Therefore, an effort was made to develop a specimen that would be highly sensitive to stress-corrosion cracking in ammonia. Preliminary studies showed that stressed steel springs were of this type.

The springs used were purchased commercially and were heat-treated. They were austenitized at 1550 F for 30 minutes, oil-quenched, and tempered at 500 F for 30 minutes. After the heat treatment, they were stressed by expanding the helix until a permanent set was achieved, and in the expanded state exposed to ammonia. Some tests with springs were conducted in specially constructed small tubes that were about 3 inches in diameter and 4 feet long.

Steels Evaluated

Several types of steels were included in the investigation. Their chemical composition and the mechanical properties are shown in Tables 1 and 2, respectively. Several heats of the same steel were tested in most instances, as indicated in the tables, and the values listed as averages.

The reasons for selection of the various steels were as follows:

1. ASTM A212 Grade B, ASTM A285 and ASME Case 1056 steels because they represent structural carbon steels commonly used to fabricate agricultural ammonia vessels.
2. ASTM A202 Grade B and USS "T-1" constructional alloy steels because they represent steels commonly used to fabricate ammonia-transport vessels. ASTM A202 Grade B steel is a chromium-manganese-silicon alloy steel intended for boilers and other pressure vessels. "T-1" steel is a high yield-strength, quenched and tempered, constructional alloy steel, approved for pressure-vessel applications under ASME Case 1204.
3. AISI 4130 steel was selected as a control steel because it can be heat-treated, as shown in Table 2, to approximately the same strength as "T-1" steel. AISI 4130 is not used for the fabrication of ammonia vessels.
4. As discussed previously, spring steel was chosen for its high susceptibility to stress-corrosion cracking. Its tensile properties were not determined.

The electrodes used in welding the tuning-fork and tensile-type stress-corrosion specimens made of the various steels are shown in Table 3. The manual metal-arc

Figure 6—Appearance of stress-corrosion cracks on tuning-fork specimens with non-tapered tines. Approximately ⅔ actual size.

process was used to place the weld bead on the tuning-fork specimens and to butt-weld the plates from which the welded tensile specimens were machined.

Exposure Procedures

Exposure tests in the field were conducted in 1000-gallon nurse tanks, used to transport ammonia from storage tanks to applicator tanks, and in applicator tanks of from 100- to 250-gallon capacity.

Standard ammonia cylinders of 100-pound capacity, as well as the special cylinders and small tubes, were used to study the effects of various contaminants on stress-corrosion cracking in ammonia. To achieve air and water contamination, the cylinders or tubes were steam-cleaned, dried with air, and the necessary amount of distilled water added. Then the stress-corrosion specimens were inserted and the valve or flange securely fastened into the cylinder or tube head. Ammonia was then pumped into the cylinders or tubes until they were half full; the air within the cylinders or tubes was not allowed to escape. The "air-free" condition, without water addition, was achieved by first pumping about 5 pounds of ammonia into the cylinders or tubes, then letting it evaporate through the valve, thus expelling air. The cylinders or tubes were subsequently half-filled with ammonia.

Cylinder and tube tests were also conducted to determine the effects of other contaminants. These were added to the cylinders or tubes in definite amounts which were then treated as above to obtain the following conditions: the contaminant without air or water, the contaminant with air but without water, and the contaminant with air and water.

For the tests in standard ammonia cylinders, 6 to 10 specimens were suspended within the cylinders with half of the total number exposed to the liquid phase and half to the vapor phase. The cylinders were tipped once each day (except weekends) to wet the vapor-phase specimens. Half of the total number of specimens in the special cylinders and tubes were also exposed in the liquid phase and half in the vapor phase; however, the special cylinders and tubes were not tipped. The cylinders and tubes were stored outdoors.

TABLE 4—Summary of Stress-Corrosion Tests* at Ammonia Distributing Plant

Specimens		
Exposed	Failed	Failures, %
161	82	50.7

Exposure Time, weeks	Specimen Failures, %
0–50	27.3
51–100	6.9
101–150	8.1
151–200	3.7
201–250	4.7
	Total 50.7

* Tuning-fork specimens of various steels cold-worked, welded and stressed; cold-worked and stressed; and stressed only.

TABLE 5—Effect of Specimen Condition and Stress Level
(Summary of stress-corrosion tests at ammonia distributing plant)

	No. of Specimens			Number of Specimens* Failed in Number of Weeks				
Specimen Condition	Exposed	Failed	Failures, %	0-50	51-100	101-150	151-200	201-250
Cold worked, welded, stressed	61	37	60.6	20	2	9	1	5
Cold worked, stressed	22	12	54.5	10	0	1	1	0
Stressed only	78	33	42.5	14	9	3	4	3
Stress Level, psi**								
113,000	14	8	57.0	2	1	2	2	1
88,000	12	12	100.0	11	0	0	0	1
57,000	17	7	41.2	1	4	1	1	0
47,000	12	0	0.0	0	0	0	0	0
30,000	23	6	26.1	0	4	0	1	1

* Tuning-fork specimens of various steels.
** Data obtained with stressed-only specimens.

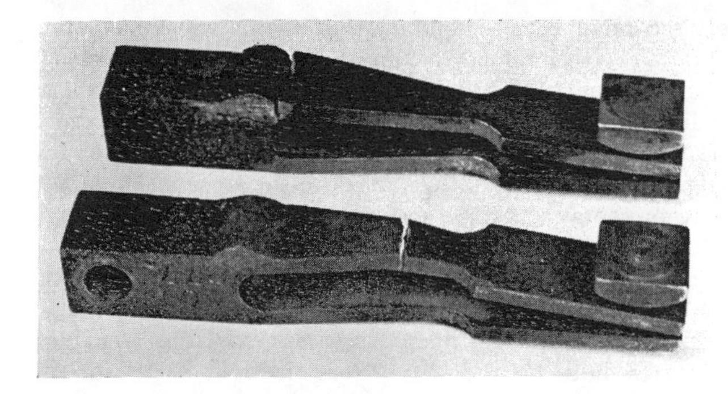

Figure 7—Appearance of stress-corrosion cracks on tuning-fork specimens with tapered tines. Approximately actual size.

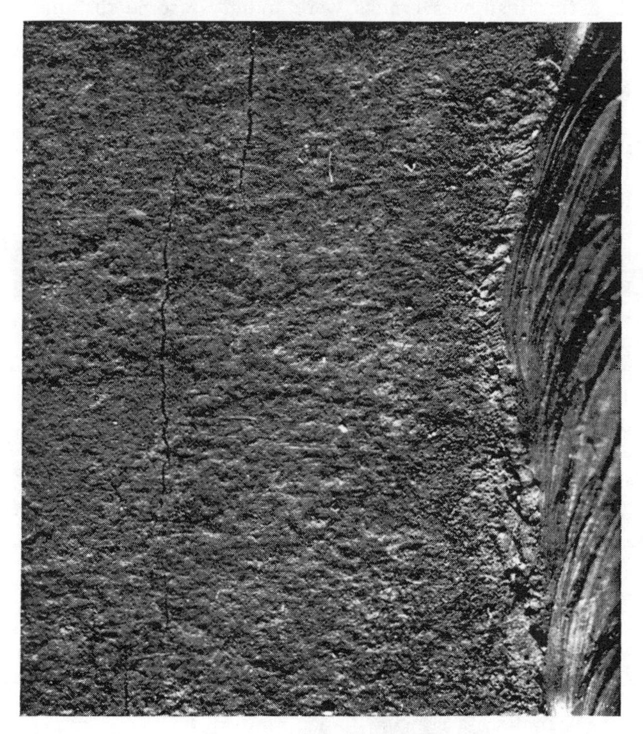

Figure 9—Top view of a cracked tuning-fork specimen with the cracks away from the weld bead. X7.

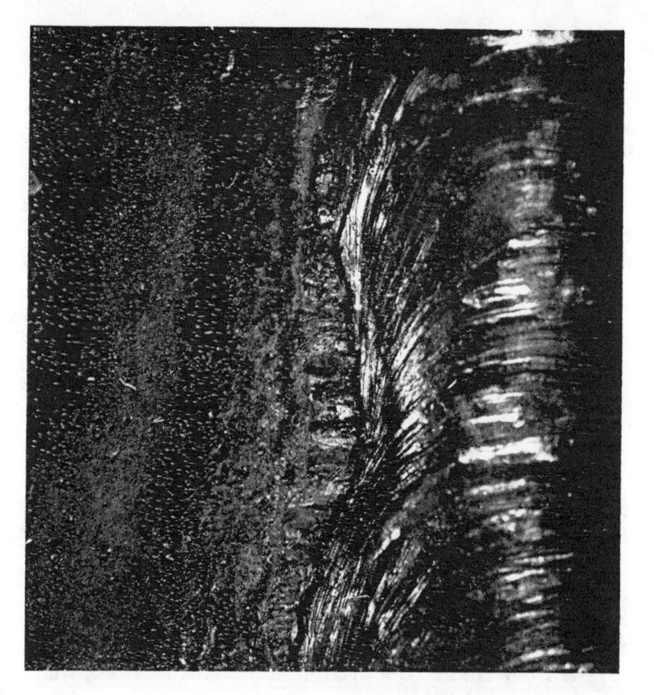

Top View

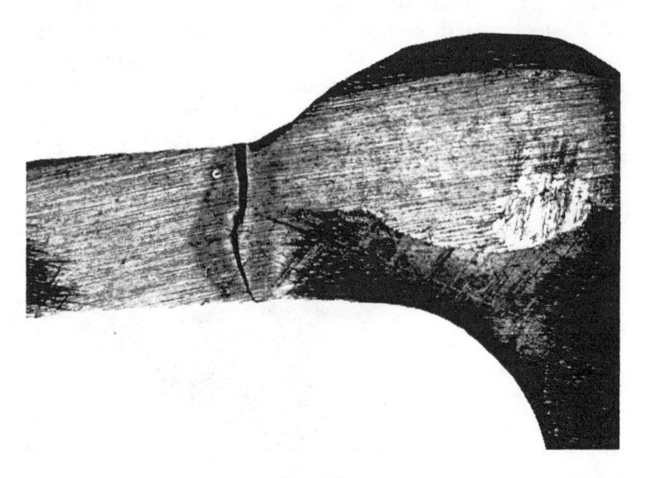

Side View

Figure 8—Tuning-fork specimen with a crack adjacent to the weld bead. Approximately 7 times actual size.

The various specimens exposed in ammonia were examined at intervals that ranged from three weeks to one year. The average interval between inspections, however, was about three months. A specimen was considered to have failed if it fractured or if it contained a crack visible under low magnification (approximately 5X). When specimen failures occurred in any particular test or exposure, the failed specimens were usually replaced with new ones. Therefore, for each test, the specimens were not all exposed at the same time.

Results and Discussion

Confirmation of Stress-Corrosion Cracking

Stressed test specimens were first exposed in agricultural ammonia in 1956 at an ammonia distributing plant.* Totals of 30 tuning-fork specimens each of "T-1" steel and ASTM A212 steel were exposed in nurse and applicator tanks. No failures were detected during an inspection of 18 of these specimens after 7 weeks' exposure, but one failed "T-1" steel specimen (cold-worked, welded and stressed) was found in another 18 of these specimens after 10 weeks' exposure. It is believed that this failure represents the first time that stress-corrosion cracking has been produced on a test specimen exposed in ammonia. The occurrence of this and subsequent failures on test specimens permitted the exposure program to be expanded so that the effects of steel composition, stress level and contamination of ammonia could be studied.

The exposure tests at the ammonia distributing plant have continued since 1956. The results obtained to date in these tests are summarized in Table 4. Eighty-two of a total of 161 stressed tuning-fork specimens exposed at this test site failed during the 5-year exposure period.

Table 5 shows the effects of distributing plant specimen condition and stress level on the time for stress-corrosion cracking. Most severe was the cold-worked, welded and stressed condition, in which the steel is at a very high stress level and a heat-affected zone is present. Next most severe was the cold-worked and stressed condition, followed by the stressed-only condition. The severity of

* Southeastern Liquid Fertilizer Co. (SELFCO), Albany, Ga.

cracking generally decreased with the stress level. The failure of all specimens exposed at a stress level of 88,000 psi, however, is not consistent with this pattern, possibly because unusually active ammonia may have been in contact with these specimens during the first year of exposure, when 11 of the 12 specimens failed. Table 4 also shows that none of the specimens failed at a stress level of 47,000 psi. All these specimens were ASTM A212 steel, which did not exhibit failure in any of the conditions tested.

The steels exposed in the distributing plant tests included "T-1" steel, ASTM A202 and A212 steels, and ASME Case 1056 steel. Failures were observed with specimens of all of these steels except ASTM A212 steel. Failures occured more readily in the "T-1" steel, with a high yield-strength of 116,000 psi, than in the other steels with substantially lower yield-strengths. For all specimen conditions, 74 percent of the "T-1" steel specimens failed (73 of 99 exposed), 30 percent of the ASME Case 1056 steel specimens failed (3 of 10 exposed), and 27 percent of the ASTM A202 steel specimens failed (6 of 22 exposed). The difference in behavior between steels is attributed to the differences in yield strength; the yield strength determines the maximum amount of stress that can be applied in the stress-corrosion specimens. The absence of any failures in stressed specimens of ASTM A212 steel tested is not consistent with the reported stress-corrosion failures of vessels made of this steel. This suggests that some heats of this steel are susceptible to stress-corrosion cracking whereas others are not.

From the standpoint of correlation with service performance, it is noteworthy that no stress-corrosion cracking failures of "T-1" steel ammonia-transport vessels have occurred. "T-1" steel vessels with one to two years' service have been examined thoroughly with magnetic and X-ray techniques and no cracks have been detected. The excellent service record of vessels made of this steel indicates that stress-relieving after fabrication, which is standard practice for "T-1" steel ammonia vessels, is an effective method of preventing stress-corrosion cracking.

Representative failed specimens from the distributing plant tests are shown in Figures 6 and 7. On the specimens with non-tapered tines, cracking generally occurred in the highly stressed areas at the root of the tines, Figure 6. On the specimens with tapered tines, cracking occurred both at and away from the root of the tines, Figure 7. On welded specimens, the cracks usually occurred at the toe of the weld bead, Figure 8, although in a few instances cracks were found at some distance from the weld bead, Figures 7 and 9. On non-welded specimens, one main crack and several secondary cracks were usually evident.

Photomicrographs illustrating the appearance of cracks in tuning-fork specimens are shown in Figures 10 through 14. Figure 10 shows a highly branched crack in a cold-worked, welded and stressed "T-1" steel specimen. This crack started at the toe of the weld bead (visible in the upper left corner of Figure 10A), and progressed through the heat-affected zone into the base metal. Cracks usually progressed intergranularly in the heat-affected zones of "T-1" steel, Figure 11, but both intergranularly and transgranularly in the base metal, Figure 12. Predominantly transgranular cracks with relatively little branching are shown progressing through the heat-affected zone into the base metal of ASME Case 1056 steel in Figure 13. Branched and predominantly intergranular cracking in the heat-affected zone of ASTM A202 steel is shown in Figure 14.

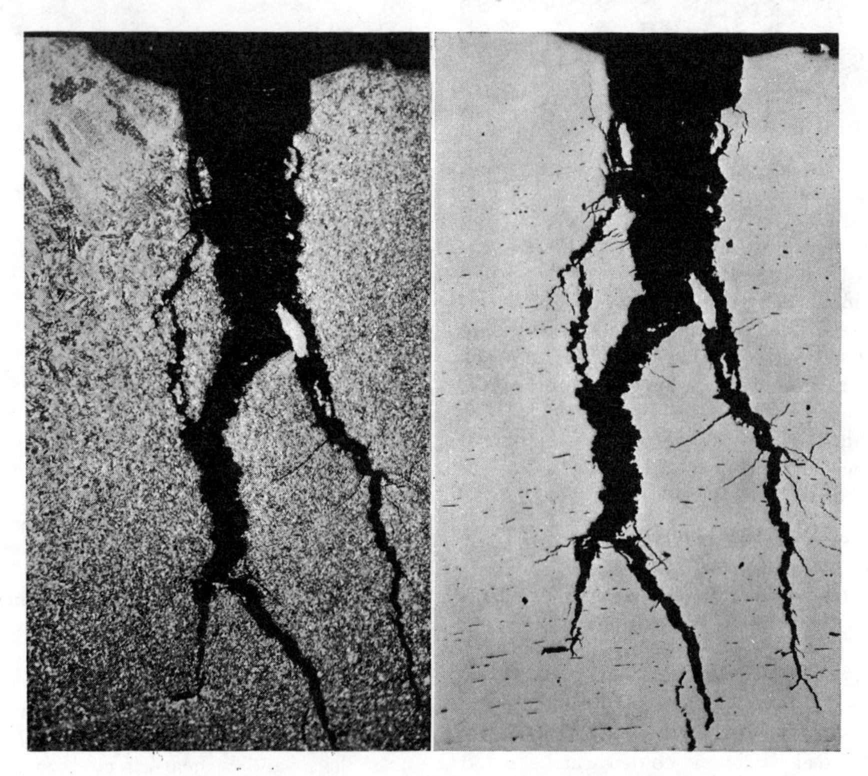

A. Etched in picral B. Unetched

Figure 10—Cross-section showing stress-corrosion cracking in "T-1" steel. X40

Figure 11—Cross-section showing intergranular stress-corrosion cracking in heat-affected zone of "T-1" steel. Etched in nital. X165

Results of the distributing plant tests show that agricultural ammonia as used in the field is capable of causing stress-corrosion cracking of stressed steel specimens.

Effects in Transfer and Storage Systems

At this distributing plant, ammonia is transported in 1000-gallon nurse tanks from storage tanks to the applicator tanks in the field. The nurse tanks are filled by means of compressors so that the ammonia vapor and liquid move in a closed system. The applicator tanks, however, are filled by the bleeding method. A hose connection is established between the liquid phase in the nurse tank and the applicator tank, and the vapor of the applicator tank is vented ("bled") to the atmosphere. The pressure in the nurse tank, determined by the temperature of the liquid ammonia, is appreciably above atmospheric; in the applicator tank, because of the bleeding, the pressure is only slightly above atmospheric. This pressure difference drives the liquid from the nurse tank into the applicator tank.

In the closed-system method of transfer, any non-condensible gases, such as air, remain in the system. In the bleeding method the air would be expected to escape from the applicator tank because the ammonia boils at the lower pressure of the applicator tank, purging the ammonia of air.

During the first few months of the distributing plant tests, all of the specimen failures occurred in nurse tanks. It was thought that the specimens in the applicator tanks were less susceptible to failure because of the transfer method used to fill these tanks. However, as the tests progressed, specimen failures occurred in the applicator tanks; this indicated that transfer of ammonia by bleeding does not prevent cracking.

The ammonia at this distributing plant is stored at ambient temperature and elevated pressure. Some other distributors, however, store ammonia at low temperature and approximately atmospheric pressure. To maintain these conditions, vapor from the storage tank is compressed to liquid and returned to the storage tank. The non-condensible portion of this vapor, mostly air, is discharged from the system. Thus, in the low-temperature storage system, air is continuously purged from the storage tanks.

Figure 12—Cross-section showing intergranular and transgranular stress-corrosion cracking in non-welded "T-1" steel. Etched in nital. X400

For comparison with the results obtained at the fertilizer plant, a set of tuning-fork specimens was exposed in nurse tanks of a distributor in Mississippi who used a low-temperature storage system. Tuning-fork specimens were also exposed in tanks of two distributors other than the fertilizer plant who store ammonia at ambient temperature. These distributors were located in Louisiana and Oklahoma. After one year of exposure, none of the specimens (36 total) exposed in nurse tanks of the distributor in Mississippi using the low-temperature storage system had failed. One specimen failure occurred in the test in Louisiana (30 specimens exposed) and in Oklahoma (18 specimens exposed), where ambient temperature storage was used.

The results obtained suggest that ammonia which has been stored in a low-temperature system does not cause cracking; however, a definite conclusion is not warranted because it was not established that the particular ammonia used by the Mississippi distributor was originally capable of causing stress-corrosion cracking. The results obtained in tanks of the Louisiana and Oklahoma distributors using ambient temperature storage show that ammonia capable of causing stress-corrosion cracking was present in their systems during the tests. The frequency of cracking at these locations, however, did not approach that obtained at the distributing plant.

In none of the tests in nurse or applicator tanks was it known whether the specimen failures were caused by exposure to one, a few or all of the batches of ammonia in the tanks between inspections. The tanks were used repeatedly during the fertilizing seasons and stood

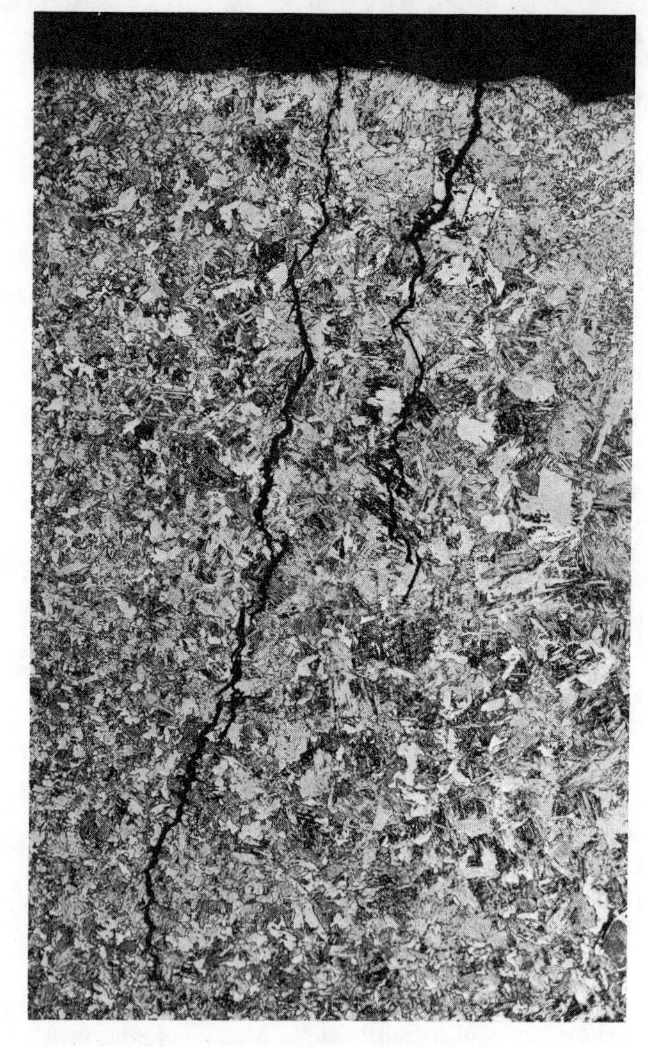

Figure 13—Cross-section showing stress-corrosion cracking in ASME Case 1056 steel. Etched in picral. X75

Figure 14—Cross-section showing stress-corrosion cracking in heat affected zone of ASTM A202 steel. Etched in picral. X400

idle (usually full of ammonia) during the rest of the year. At the distributing plant, there was some indication that the incidence of specimen failures was higher during the fertilizing seasons than during periods of lesser activity; however, no consistent pattern could be established.

Effects of Various Contaminants and Development of an Inhibitor

Exposure tests to determine the effects of various contaminants and to develop an inhibitor for stress-corrosion cracking in ammonia were conducted in cylinders at the Georgia Institute of Technology, at the distributing plant, and at an ammonia plant.*

At the Georgia Institute of Technology, the effects of air, water and other possible contaminants were studied in refrigeration-grade ammonia. In the initial series of tests started in 1956, tuning-fork specimens were exposed in cylinders containing ammonia that had been contaminated with air, and with air and 0.5, 1.0, 2.0, and 4.0 percent water, respectively. After about one year of exposure, specimen failures had occurred in the cylinder contaminated with air only, but not in those with both air and water. These results led to the belief that air in ammonia caused stress-corrosion cracking, and that water was an effective inhibitor.

*Coulee ammonia plant of the Phillips Pacific Chemical Co.

TABLE 6—Effects of Air and Water Additions to Refrigeration-Grade Ammonia
(Cylinder tests at Georgia Institute of Technology)

Additions to Ammonia		Specimens*		
Air	Water, %	No. Exposed	No. of Failures	Frequency, %
No	No	14	3	21.4
Yes	No	44	10	22.8
Yes	0.1	4	0	0
Yes	0.25	4	0	0
Yes	0.5	14	0	0
Yes	1.0	14	0	0
Yes	2.0	10	0	0
Yes	4.0	10	0	0

* Results obtained with "T-1" steel and AISI 4130 alloy steel tuning-fork specimens in the cold-worked, welded, and stressed condition. Test duration, 115 to 160 weeks.

TABLE 7—Effect of Various Contaminants, Air and Water in Refrigeration-Grade Ammonia
(Cylinder tests at Georgia Institute of Technology)

		No. of Specimens*					
		No Air No Water		Air No Water		Air 0.5% Water	
Contaminant	Concentration, wt %	Exposed	Failed	Exposed	Failed	Exposed	Failed
Compressor oil..	1.0	4	1	4	0	4	0
Ethyl mercaptan.	0.005	4	0	4	0	4	0
Sodium cyanide..	0.01	4	1	4	1	4	0
Sodium chloride.	1.0	4	0	4	0	4	0
Copper sulfate...	0.1	4	2	4	1	4	0

* Results obtained with "T-1" steel and AISI 4130 alloy steel tuning-fork specimens in the cold-worked, welded, and stressed condition. Test duration, 115 weeks.

TABLE 8—Effect of Air and Water Additions to Agricultural Ammonia
(Cylinder tests at ammonia distributing plant)

Additions to Ammonia		Specimens*		
Air	Water	No. Exposed	No. of Failures	Frequency, %
No	No	43	7	16.3
Yes	No	43	1	2.3
Yes	0.25%	43	0	0

* Results obtained with "T-1" steel and AISI 4130 alloy steel tuning-fork specimens in the cold-worked, welded and stressed condition. Test duration, 48 weeks.

TABLE 9—Summary of Stress-Corrosion Tests with Tuning-Fork and Tensile-Type Specimens in Contaminated Ammonia
(Cylinder tests of 7 to 33 weeks' duration at ammonia plant)

Contamination	Specimen Failures	Type of Steel
None	None	
5 psi air*	None	
30 psi air	None	
100 psi air	None	
200 psi air	None	
200 psi air + 0.25% water	None	
CO_2 added as ammonium carbonate		
100 ppm CO_2	None	
30 psi air + 20 ppm CO_2	5	"T-1" steel (4),** Case 1056 (1)
100 psi air + 1000 ppm CO_2	9	"T-1" steel (6), Case 1056 (1) A-202 (2)
100 psi air + 1000 ppm CO_2 + 0.25% water	None	
CO_2 added as dry ice		
100 psi air + 1000 ppm CO_2	1	"T-1" steel (1)
100 psi air + 1000 ppm CO_2 + 0.25% water	None	

* The air used in all experiments was scrubbed with caustic to remove CO_2.

** The figures in parentheses indicate the numbers of specimens of each type of steel that failed. A total of 16 specimens (cold-worked, welded and stressed tuning forks, and unwelded and welded tensile specimens) of "T-1" steel, Case 1056 steel, A202 steel, A285 steel and A212 steel was exposed in each environment.

TABLE 10—Results of Short-Term Tests with Heat-Treated Springs
(Tests in small tubes at ammonia plant)

Tube No.	Type of Contamination*	No. of Springs Failed**		
		Test A	Test B	Test C
1	Air + CO_2 + water	0	0	1
2	Air + CO_2	18	18	18

* Tube No. 1. Ammonia with 100 psi air (CO_2-free), 1000 ppm CO_2 and 0.25 percent water.
Tube No. 2. Ammonia with 100 psi air (CO_2-free) and 1000 ppm CO_2.
** 18 specimens exposed in each tube.

A second series of tests was started at the Georgia Institute of Technology in 1957 to determine the effects of air, water and other possible contaminants in ammonia. These other contaminants were compressor oil, ethyl mercaptan, sodium cyanide, sodium chloride and copper sulfate. Tuning-fork specimens were exposed in cylinders containing these contaminants with no air or water, with air but no water, and with air and 0.5 percent water. Specimens were also exposed in refrigeration-grade ammonia without any contaminants, with air but no water, and with air and 0.1, 0.25, 0.5 and 1.0 percent water.

The results of tests to determine the effects of air and water are summarized in Table 6, those for air, water and the other contaminants are shown in Table 7. In these tests, specimen failures occured only with the higher strength steels ("T-1" steel and AISI 4130 alloy steel).

Table 6 shows that failures did not occur with "T-1" steel or AISI 4130 steel specimens when 0.1 percent or more water was added to air-contaminated refrigeration-grade ammonia. Whether air is a necessary causative factor was not established in these tests because the frequency of the observed failures was about the same in the uncontaminated ammonia as it was in the ammonia contaminated with air. However, the refrigeration-grade ammonia used was not analyzed for air content and may have contained some air. The results obtained in these tests are considered inconclusive for establishing the effect of contamination with air.

Table 7 confirms that water effectively inhibits stress-corrosion cracking of steel in contaminated ammonia. Specimen failures were obtained in cylinders containing some of the contaminants studied, with and without intentional air contamination. Here again, however, it is not known whether air was present in the cylinders that were not intentionally contaminated. Because the frequency of cracking was not high in any of these tests, they indicate that the various contaminants studied are not very active causes of stress corrosion.

The effects of air and water additions were also studied in cylinder tests at the distributing plant. The results show that water effectively inhibits stress-corrosion cracking of steel in agricultural ammonia (Table 8). Specimen failures occurred in the ammonia with and without intentional air contamination. The previous tests had already established that intentional contamination was not required to cause cracking at that location.

That fewer specimen failures occurred in ammonia intentionally contaminated by air than in uncontaminated ammonia, however, was unexpected and is inconsistent with other results of air contamination presented herein. The tests in Table 9 were started on a hot, humid day; by calculation, contamination with air under these conditions would introduce 0.01 to 0.02 percent water into the ammonia. This water, along with a small amount initially present in the ammonia, apparently was sufficient to inhibit cracking partially.

Additional tests on the effects of contaminants were conducted at the ammonia plant. This site was chosen because of the high purity of ammonia produced there and because the possibility of test ammonia contamination during transportation to some other site could be eliminated. The water content of ammonia at this plant was approximately 0.007 percent; the oil content was 1.6 parts per million (ppm).

Three approaches were used in the ammonia plant tests to increase the frequency or reproducibility of specimen failures: (1) tensile-type stress-corrosion specimens were exposed to determine whether this type of specimen would be more effective in producing cracking than the tuning-fork specimens; (2) tests were conducted with stressed springs to determine whether these would give reproducible results; (3) in addition, the cylinders were contaminated with air at pressures up to 200 psi to produce greater air concentrations than had been obtained in the previous tests.

Tuning-fork and tensile-type specimens, and stressed springs were exposed in cylinders containing ammonia (a) without air contamination; (b) contaminated with air at 5, 30, 100 and 200 psi; and (c) contaminated with air at 200 psi with 0.25 percent water. These tests did not produce any failures of tuning-fork or tensile-type specimens (Table 9), although failure of spring specimens did occur, as will be described later. The absence of failures in tuning-fork or tensile-type specimens was unexpected because, in previous tests at the Georgia Institute of Technology and the distributing plant, failures did occur in air-contaminated ammonia. However, a significant clue to the cause of cracking developed from these tests when it was realized that by passing the air through caustic scrubbers, carbon dioxide had been removed from the pressurized air used to contaminate the cylinders. Thus, it seemed probable that the carbon dioxide content of air was an important cause of cracking.

Additional tests were conducted in which specimens were exposed in cylinders contaminated with various amounts of air (CO_2-free), carbon dioxide (added either as ammonium carbonate or dry ice) and water. In these tests (Table 9), stress-corrosion cracking of tuning-fork and tensile-type specimens occurred in a 17-week exposure when the cylinders were contaminated with air and carbon dioxide, but did not occur in cylinders contaminated with carbon dioxide only or with air, carbon dioxide and 0.25 percent water. These results indicate that in ammonia the combined presence of air and carbon dioxide is necessary for the occurrence of stress-corrosion cracking. The inhibiting effect of water was again confirmed.

In 7- to 26-week exposures at Coulee, the heat-treated springs were found to be highly sensitive to stress-corrosion cracking. Failures of springs did not occur in uncontaminated ammonia, but were found in ammonia contaminated with 5-, 30- and 200-psi air (CO_2-free), with 200-psi air and 0.25 percent water, and with 30-psi air (CO_2-free) plus 20 ppm carbon dioxide. These results indicate that air (CO_2-free) or air with carbon dioxide as a contaminant in ammonia are capable of causing stress-corrosion cracking of the heat-treated springs, and that 0.25 percent water does not effectively inhibit the cracking. The appearance of springs before and after exposure is shown in Figure 15.

Additional tests of shorter duration were conducted with springs in small tubes. In two- to four-week tests (Table 10), all spring specimens failed in ammonia containing air and carbon dioxide but only one failure occurred in ammonia contaminated with air, carbon dioxide and water. These tests show that water delays the occurrence of cracking in heat-treated springs.

Mechanism of Corrosion and Inhibition in Ammonia

In the mechanism of corrosion in air-contaminated ammonia, it is believed that oxygen acts by an electrochemical reduction or cathodic process, such as

$$O_2 + 2\,NH_4^+ + 4e \rightarrow 2OH^- + 2\,NH_3$$

The anodic reaction can be represented as

$$2\,Fe \rightarrow 2\,Fe^{++} + 4e$$

for an over-all corrosion reaction of

$$O_2 + 2\,NH_4^+ + 2\,Fe \rightarrow 2\,Fe^{++} + 2OH^- + 2\,NH_3$$

The effect of carbon dioxide may result from the formation of ammonium carbamate

$$2\,NH_3 + CO_2 \rightarrow NH_4CO_2NH_2$$

which could increase the ammonium ion concentration:

$$NH_4CO_2NH_2 \rightarrow NH_4^+ + NH_2CO_2^-$$

It is not understood why the corrosion of steel in air-contaminated ammonia localizes to the extent required for stress-corrosion cracking. However, the steel and contaminated ammonia system has characteristics similar to those of other systems in which stress-corrosion cracking occurs: corrosion proceeds at a relatively slow rate; corrosion product films can form, serving to localize the attack; ammonia as a polar solvent permits electrochemical reactions generally associated with stress-corrosion processes.

Corrosion inhibitors function by forming protective

Figure 15—Spring specimens before and after exposure to ammonia. Approximately 1.5 actual size.

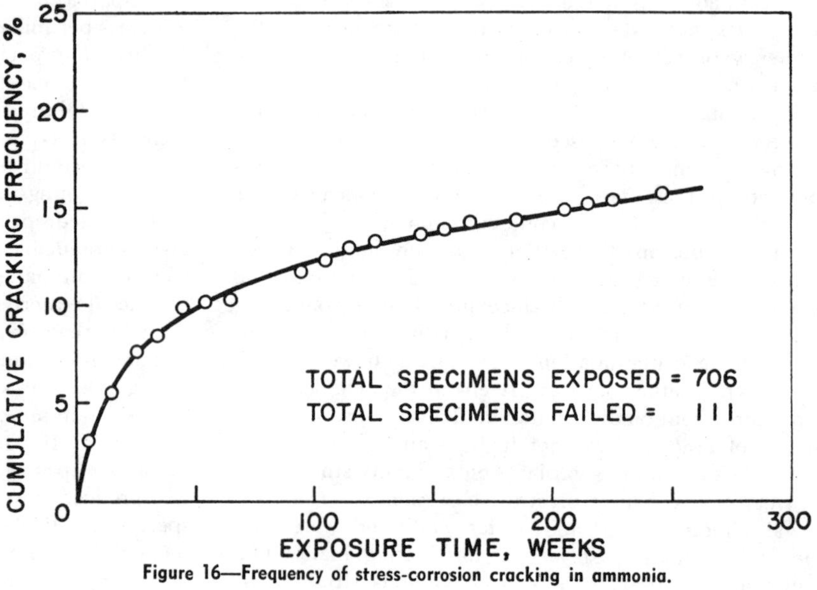

Figure 16—Frequency of stress-corrosion cracking in ammonia.

films or by reacting with corrosive species within an environment. The inhibiting effect of water in air-contaminated ammonia is believed to be the result of protective film formation. It may not be generally recognized that there are many non-aqueous environments in which added water acts as a corrosion inhibitor. For example: magnesium corrodes rapidly in anhydrous methanol, but the addition of 0.1 percent water greatly decreases the rate of attack[2]; anhydrous ethyl alcohol at the boiling point corrodes aluminum, but a small amount of water acts as an inhibitor[3]; titanium is severely corroded by dry chlorine, but is resistant to corrosion in chlorine containing at least 0.013 percent water[4]; corrosion of titanium is also inhibited by the addition of 1 to 2 percent water to fuming nitric acid.[5]

Summary of Results

The results of the tests in nurse and applicator tanks show that agricultural ammonia as used in the field can cause stress-corrosion cracking of steel. The tendency to failure was substantially higher in the higher strength steels, which were capable of withstanding a high stress.

The results obtained in the contamination studies are believed to indicate that the presence of contaminants in ammonia is necessary for the occurrence of stress-corrosion cracking. In arriving at this belief, more credence has been placed on the results obtained in the Coulee tests than on those at Georgia Institute of Technology, where contaminants might have been present in the test ammonia not purposely contaminated. Air of normal carbon dioxide content is believed to be the most important contaminant in causing stress-corrosion cracking of ammonia vessels in service.

Water in amounts greater than 0.1 percent was found to be an effective inhibitor of stress-corrosion cracking of constructional steels used in the fabrication of ammonia vessels. Failures of specimens made of these steels did not occur in any of the tests in which water was added. Water additions delayed but did not prevent cracking of heat-treated springs.

One of the characteristics of stress-corrosion cracking of steel in ammonia as established by these tests is that failures continue to occur after long exposure periods. In Figure 16, the cumulative cracking frequencies of tuning-fork specimens exposed in ammonia without water at the various testing locations are plotted against exposure time. The summation of results in this figure shows that the highest rate of cracking occurs during the first year, but that failures continue even after five years' exposure. After the first year, failures occurred each year in about 1 percent of the total number of specimens initially exposed. The continued occurrence of specimen failures after long exposure parallels reported failures of ammonia vessels after long service.

Prevention of Stress Corrosion Cracking By Agricultural Ammonia

The Research Committee recommends that stress-corrosion cracking of agricultural-ammonia vessels be prevented as follows:

1. Tanks over 36 inches in diameter for agricultural ammonia service should be either fully stress-relieved or fabricated with heads that are hot-formed or stress-relieved. (Service experience has indicated that applicator tanks with diameters less than 36 inches are relatively insensitive to cracking.)
2. Extreme care should be used to eliminate air from agricultural ammonia systems. When they are new or have been open to the atmosphere, ammonia vessels should be thoroughly purged to eliminate air contamination.
3. Agricultural ammonia should contain water to inhibit stress-corrosion cracking of ammonia vessels. The minimum water content should be 0.20 percent and the maximum water content should be consistent with a final product marketable as containing 82 percent nitrogen. Distilled water or steam condensate should be added by the ammonia producer.

Each of these recommendations has definite limitations when considered individually: (1) many non-stress-relieved vessels are now in service and it is probable that all of them could not be stress-relieved; (2) purging is considered an effective method of preventing further contamination, but not for removing air from contaminated ammonia systems; (3) during the fertilizing seasons, it is conceivable that refrigeration-grade ammonia might be diverted to agricultural use and this ammonia would not contain the inhibitor.

Since each recommendation individually has definite limitations, all three should be followed to avoid stress-corrosion cracking of agricultural ammonia vessels. The Research Committee plans to study the effectiveness of these measures by reviewing failure reports from the field. Water addition may be sufficiently effective that stress-relieving someday may not be necessary.

Acknowledgment

The authors express their appreciation to the other members of the Research Committee of the Agricultural Ammonia Institute, without whose cooperation and assistance the research described in this committee report would not have been possible.

References

1. T. J. Dawson. *Welding Journal*, 35, 568-574 (1956).
2. Corrosion Handbook. Edited by H. H. Uhlig, John Wiley and Sons, 1948, p. 242.
3. Metals Handbook. Published by American Society for Metals, 1961, p 931.
4. M. G. Fontana. Corrosion: A Compilation, The Press of Hollenback, 1957, p 121.
5. J. B. Rittenhouse and C. A. Papp. *Corrosion*, **14**, 283t-284t (1958) June.

Shot Peening Prevents Stress Cracking in Aircraft Equipment

SUMMARY

Stress corrosion-cracking occurs in metal if the surface is stressed and subjected to a corrosive environment. Shot peening, which has been used for over 30 years to prevent fatigue failures of metal parts, induces an extremely high residual compressive stress in the surface. The proper controls of shot peening are an important aspect in obtaining the full benefits. The author discusses these controls, but not before presenting recent tests and their results, along with case histories that help to illustrate shot peening as a means of preventing cracking.

WHILE MANY INDUSTRIES are faced with the need to control or eliminate stress corrosion, aircraft manufacturers are the most intimately concerned. Recently, there have been some definitive studies made to show what can be done, within the limits of current technology, to understand and control this kind of equipment failure. This paper discusses some recent tests and their results, along with case histories that help to illustrate the effect of shot peening as a means of preventing stress corrosion cracking.

Case Histories

Aluminum Forging

Figure 1 is an example that did not appear as a problem for approximately 15,000 hours of operation. It is one of the largest aluminum forgings used in aircraft, weighing approximately 700 lbs. Part of the main landing gear fitting, the material is 7075 used in the T6 condition. Original failure appeared as a crack starting in a trunion hole into which a bushing is pressed. The opposing fracture faces are shown in Figure 2. Three areas were chosen for subsequent examination with an electron microscope.

The first two showed clearly the brittle intercrystalline mode of failure with ready evidence of corrosion products. The third area was one of transition from brittle to ductile failure. A survey of the area near the failure point showed many other sources of

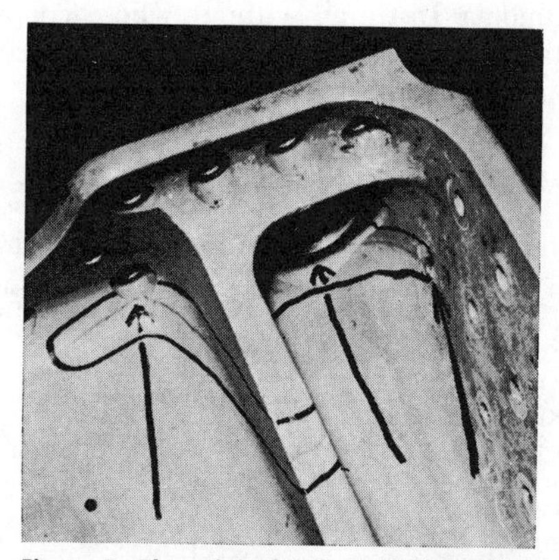

Figure 1—The original failure of the forging in the main landing gear fitting appeared as a crack starting in a trunion hole into which a bushing is pressed. The corrosion did not appear for approximately 15,000 hours.

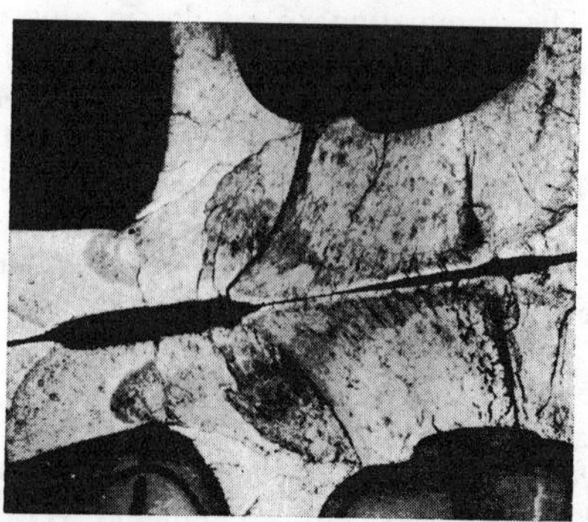

Figure 2—Opposing fracture faces of the corroded aluminum forging.

stress corrosion cracks. The failure could have originated from any of several points.

After careful study of the fracture, the decision was made to require controlled shot peening on all new parts before putting them in service.

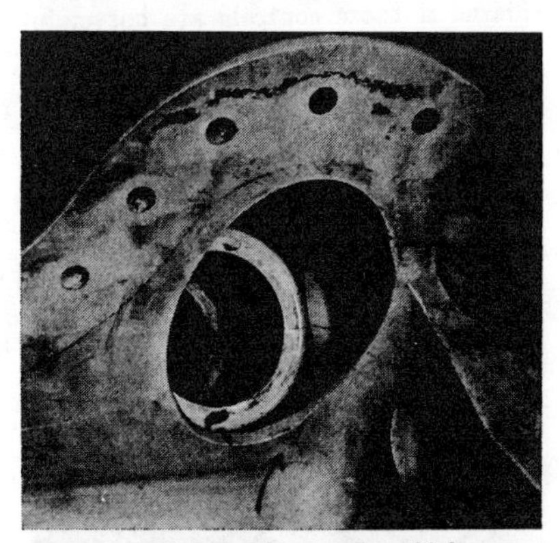

Figure 3—This massive crack, which seems to have two origins, extended completely through the forging.

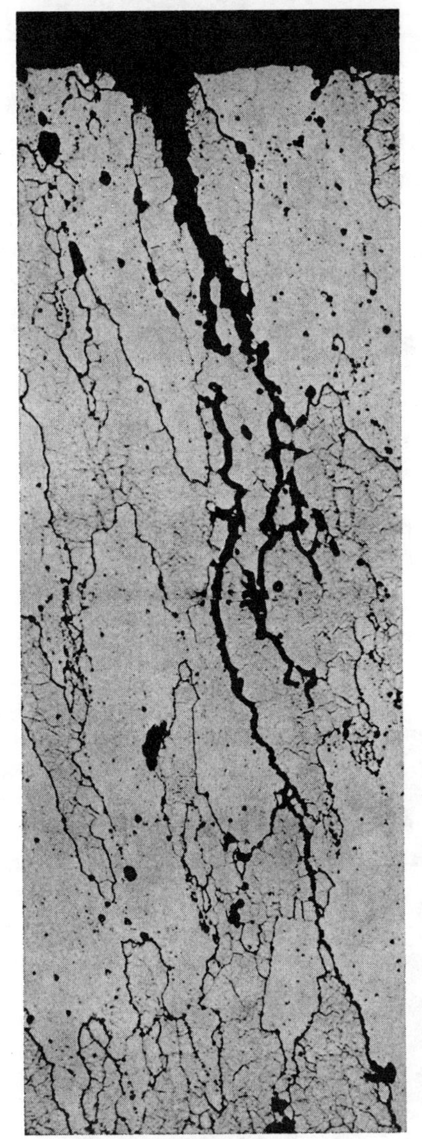

Figure 4—Photomicrograph of a crack found in the 7079-T6 forging. 500X.

While investigating another problem, careful inspection uncovered a crack in a different area (see Figure 3). The massive crack, extending completely through the section, apparently had two points of origin. Following the discovery the decision was made to shot peen all surfaces of finished forgings. In addition, forgings already in service were shot peened in place to avoid a difficult and involved removal procedure.

Wing Spar

The next example, a channel-shaped wing spar, precipitated a massive shot peening program. The component was approximately 8 ft in length with a 5-inch dimension at the widest end, tapering down to approximately 1¼ inches at the narrower end. Section thickness varied from approximately 0.125 to 0.060. Material was 7079-T6 aluminum and originated as a forging.

This wing spar first came to the attention of the metallurgical engineering group as the result of a request to evaluate the effects of pitting corrosion that apparently had occurred during anodizing. During the investigation, numerous cracks were noted in the flange. These cracks were closely checked and the decision was made to check other parts subject to the same processing. An inspection was made of all the parts formed from 7079-T6. Many were found to have similarly cracked areas. Sections taken through the cracks showed them to be primarily intergranular. This is shown in Figure 4.

Figures 5 and 6 are electron fractographs. The first of these shows the brittle, intergranular mode of failure. In the second, grain boundary oxidation is evident. Undoubtedly these were stress corrosion failures. (It is well to emphasize that these components had never seen service).

The failures were caused by lack of adequate protection during processing and the existence of residual tensile stresses. The method was a comprehensive program of surface protection during manufacturing processing and the shot peening. Of course, in this and in all other examples cited, the use of shot peening automatically infers the use of adequate controls to insure maximum benefit.

Table 1—Test Specimens after Exposure to the Alternate Immersion Cycling.

Conditions[1]	Results
T6, press fit	Failure in 120 hrs
T73, press fit	No failure
T6, shrink fit	Failure in 120 hrs
T73, shrink fit	No failure
T6, press fit, anodized, baked fluid, resistant primer	Failure in 216 hrs
T6, press fit, shot peened	No failure

[1] 0.004 interference

Landing Gear Cylinder

This next example of the use of shot peening in a stress corrosion application also comes from the aircraft industry. The part is the main landing gear cylinder of a small fighter-bomber. It is a forging made from 7075-T6 aluminum. Approximately 100 units from the vendor were put into service with no apparent problems. The next group of parts came from another vendor.

The cylinder was removed from service. While on the ground, the cylinder lost its hydraulic capability and the strut flattened. A crack could be seen running the length of the cylinder and into the trunion hole. Another cylinder was removed from service in the course of a regular maintenance check. Crack location is at the forging parting line, which puts the short transverse grain structure in the worst possible orientation.

Figure 7 is a photograph of a section through the cylinder. The surface was etched to bring out the forging lines; failure originated in the area indicated by the arrow. Here again, the method was to shot peen all surfaces. This was approached in steps, with the crack moving each time from a peened area to one that had not been specified as requiring peening. There were serious problems in maintaining accurate dimensions in some of the bored holes. Only through careful control of the peening variables was this possible.

Previous examples were all field failures. Had the proper information (including information on peening) been in the hands of the design and material people, these failures would never have originated. Using shot peening as a design metal process instead of as an aid to get out of trouble would help in minimizing service difficulties.

Some recent work has been done supporting previous efforts to establish the validity of shot peening under controlled conditions.

Aluminum Bores

One test evaluated the susceptibility to and the effectiveness in preventing stress corrosion in aluminum bores having steel sleeves. The ring material was 7075 in both the T6 and the T73 tempers.

Table 1 shows the test specimens after exposure to the alternate immersion cycling. The surface corrosion clearly indicates the severity of the test and the accompanying validity of the stress corrosion results.

Obviously, the bare material in the T6 condition would not be satisfactory in a stress corrosion application. Protective coatings do little better, if for no other reason than their susceptibility to surface damage. Shrink or interference fit made no apparent difference. The only two means that produced immunity to stress corrosion failure were the T73 heat treat or shot peening.

Source: *Materials Protection*, September 1968, 39-42

With the T73, though, temper would involve an accompanying loss in strength.

High Strength Steels

While much of the field-gathered information concentrates on aluminum, stress corrosion is certainly not limited to aluminum alloys. High strength steels are one of the most critical considerations. In fact, one of the influential government laboratories required, as a matter of course, thorough preventative measures in all steel fittings used in the high strength range (260,000 psi and above). They now have comparatively little difficulty with stress corrosion failures.

As new high strength steels are made available, they are tested to determine their susceptibility to stress corrosion. Recently examined was the HP 9-4-45 alloy produced by Republic Steel. Stress corrosion testing was done using alternate immersion in synthetic sea water under sustained load of 80% of the material's yield strength. (Small variations were made in the test specimens for other test conditions.) Included in the test of stress corrosion influences were welding, grinding, drilling, cadmium plating and shot peening. Residual stress measurements on better than 25% of the specimens were made using X-ray diffraction techniques.

The results indicate that only one process produced immunity to stress corrosion. All others experienced some degree of failure. Only shot peening was 100% successful. Both the grind-to-burn and the drill-to-burn samples indicated some improved resistance from the oxide layer produced by the machining. However, these were only limited improvements.

It is interesting to note that merely the presence of tensile stress did not promote failure. A stress diagram for several of the variables notes that as the magnitude of the net surface tensile stresses increase, the time for failure decreases. Under load, even the shot peened pieces experienced exposure to tensile stress. Although high stresses (approaching the yield point) are generally needed for stress corrosion cracking, frequently stresses that are small relative to the yield produce failure. Apparently though, there is a critical limit below which stress corrosion does not occur. The team that did the evaluation set the threshold value between 89 ksi and 127 ksi.

Titanium

A third metal, titanium—that element so vital to the aerospace industry—also has its stress corrosion problems. Small titanium tank shells used for liquid propellant were found ostensibly to fail prematurely due to material pressures. On investigation it was proven that the failures were not due to pressures, but to stress corrosion.

Tests were made for possible solutions to stress corrosion. Three tanks were used; the first served as a control and was used as received; it failed after approximately 115 hours at 105 F (76 C). The second tank was vibratory cleaned and failed after approximately 200 hours at the same temperature. The third tank was glass bead peened and successfully withstood 720 hours of exposure at 105 F.[1]

Process Control

In each of the examples cited, whether field service or laboratory originated, shot peening successfully arrested stress corrosion. It is a well known fact that cold plastic surface deformation converts harmful tensile surface stresses to compressive stresses. It is also evident that for stress corrosion cracking to occur, surface and subsurface tensile stresses must be present. Therefore, if the net harmful tensile stress can be lowered or converted to beneficial compressive stresses, failures can be minimized. The benefits will only be fully realized if the proper *control* of the process is exercised. The control of the shot peening process depends on four important variables: shot materials, shot size and uniformity, shot velocity, and shot coverage.

These four variables are collectively referred to as "intensity." Presently there are no economic non-destructive means for checking a part for its intensity. The only accepted standard is an arbitrary one called the Almen strip.

The Almen strip is a steel strip 3 inches long and ¾ inch wide. Its thickness depends on the intensity of the specified peening. The strip is positioned so that it may simulate as closely as possible the surface to be peened. The strip is then processed with the peening variables held at the same value as for the part. Since the strip is peened only on one surface, it curves. This curvature is measured on an Almen gauge. When the strip is fully saturated, the arc height, measured over a chord of 1¼ inches, is expressed in thousandths and termed the intensity. Evidently controls are essential in order to maintain uniformity.

Just what are the dangers to be considered if these controls are not maintained? Consider the mechanism by which shot peening provides stress corrosion immunity. Simply stated, the benefits accrue from the imposition of a compressive stress. The magnitude of this stress must be sufficiently high to afford protection after the application of the service pattern produced by

Figure 5—Electron fractograph showing the intergranular type brittle failure. 12,000X.

Figure 6—Electron fractograph showing grain boundary oxidation. 12,000X.

Figure 7—The corroded surface was etched to bring out the forging lines. Failure originated in the area indicated by the arrow. The surface was shot peened.

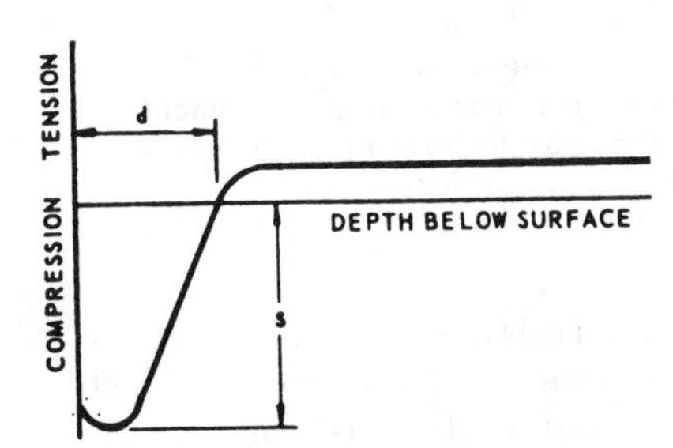

Figure 8—Typical stress pattern produced by peening. There is a high compressive stress at the peened surface, increasing slightly just below the surface.

peening (Figure 8). There is a high compressive stress at the peened surface, increasing slightly just below the surface.

The magnitude of the residual compressive stresses induced by shot peening is a function of the yield strength of the material. The level of compression decreases until the point of zero stress is reached. Then the stress changes to one that is tensile. As has been mentioned, the benefits of shot peening are due to the compressively stressed layer and the depth induced by the process. The depth must be compatible to the thickness of the part. Generally the induced surface compressive layer should not be greater than 20 to 25% of the cross section of the thinnest section, nor should the layer be too thin for it could corrode or wear away, thereby negating the benefits of peening. The photomicrograph shown in Figure 9 shows a single dimple indentation that might result from one piece of shot impacting a metal surface.[2] The metal was heat treated to promote grain growth in the section affected by the cold work. It also provided a representative picture of the typical stress pattern. The large grains portray the compressive zone. Immediately below this zone is a tensile layer. A group of dimples close together provide an even compressive layer. If an area has no dimples (or dimples that are of lower intensity), the surface stress will either be a lower compressive or a tensile stress. This then leaves the material vulnerable to stress corrosion.

Peening is not enough. *Controlled* peening is essential to adequately retard stress corrosion.

Acknowledgment

The author acknowledges Thomas Croucher and Edward Lauchner of Northrop Norair; from McDonnell-Douglas, Long Beach, Robert Gassner, Richard Turley and Charles Avery, who have spent many hours with David Zawolkow, consultant, and without whose help this paper could not have been completed.

References

1. W. B. Lisagor, Charles R. Manning, Jr. and Thomas T. Bales. Stress Corrosion Cracking of Ti6 Al-4V Titanium Alloy in Nitrogen Tetroxide. NASA Langley Research Center.
2. K. B. Valentine. Recrystallization As A Measurement of Relative Shot Peening Intensities. *A. S. M.*, 40, 420 (1948).

Bibliography

S. K. Coburn. Designing to Prevent Corrosion—Protection of Equipment Should Begin at the Drawing Board. *Materials Protection*, 6, 33 (1967) February.

T. R. Croucher. Stress Corrosion Testing—Various Techniques and Environments Compared. *Materials Protection*, 6, 44 (1967) August.

E. D. Verink, Jr. and D. B. Bird. Designing With Aluminum—Alloys for Various Corrosive Environments. *Materials Protection*, 6, 28 (1967) February.

M. E. Holmberg and T. V. Bruno. Metallurgy and the New Corrosion Engineer: What He Should Know About Corrosion Induced Brittle Failures. *Materials Protection*, 5, 8 (1966) May.

NACE Technical Committees Progress Reports. Corrosion of Metals. *Materials Protection*, 5, 3 (1966) May.

A. W. Longinow. Stress Corrosion Testing of Alloys. *Materials Protection*, 5, 33 (1966) May.

J. H. MILO is Assistant Vice President of Technical Service of the Metal Improvement Co. in New Jersey. He has recently taken over his duties in the corporate office, having first joined the company in 1961 as manager of technical service. Other assignments included manager of Metal Improvement's Connecticut and New Jersey divisions. Although a graduate in industrial engineering, most of his professional career has been spent in the sales engineering field. He is a member of ASM.

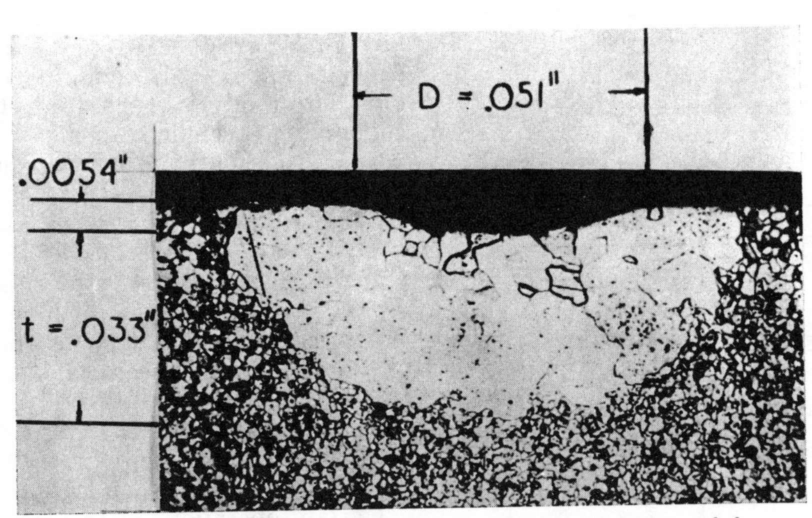

Figure 9—Photomicrograph showing a single dimple indentation that might result from one piece of shot impacting a metal surface.

SUMMARY

Discusses hydrogen embrittlement as a nuisance, danger and expense for anyone working with structural materials. This problem is described and four types of embrittlement discussed: hydride, true hydrogen, hydrogen attack and permanent damage. Also explains how hydrogen pickup occurs in certain operations in fabrication of iron and steel products.

HYDROGEN EMBRITTLEMENT in Engineering Materials*

Harry C. Rogers
General Electric Company
Schenectady, New York

HYDROGEN IN STRUCTURAL metals is like dust in a house—it is extremely difficult to get rid of completely and everything that is done seems to produce a little. Moreover, a little hydrogen is often all that is needed to produce a serious metal failure. In fact, laboratory experiments have shown that failures can occur as a result of amounts of hydrogen undetectable by normal analytical techniques, leaving the door open for blaming many failures of unknown origin on hydrogen embrittlement. The well documented effects of hydrogen on iron and steel products leave little doubt, however, that hydrogen can be a source of considerable annoyance, danger, and expense throughout industry.

The principal structural metals with which hydrogen is a problem fall into two major categories: the so-called "exothermic occluders" and "endothermic occluders." The former are characterized by the large quantities of hydrogen absorbed (sometimes as many as three atoms of hydrogen to one of metal), the tendency to form definite hydrides, and the negative heat of solution which results in a solubility *decrease* with *increasing* temperature at constant pressure. Among these are vanadium, titanium, zirconium, tantalum, thorium, and cerium. Conversely, the endothermic occluders, such as iron, copper, cobalt, and nickel, have a lower solubility for hydrogen, which increases as the temperature is raised, indicating a positive heat of solution.

Hydrogen enters metals in the atomic form and, in general, diffuses interstitially. Any process that produces atomic hydrogen at the metal surface normally will induce considerable hydrogen absorption in the metal, although a large fraction of these atoms recombines to form gaseous molecular hydrogen, which is not absorbed.

Hydrogen Pickup

Certain operations in the fabrication of iron

---
* Reprinted from General Electric Research Laboratory Bulletin, Fall, 1961.

and steel products are especially prone to the pickup of hydrogen. The principal ones are steelmaking, heat treatment and hot working, welding, chemical pickling, and electroplating.

In steelmaking, the principal source of hydrogen contamination is the reaction of iron with the moisture in the charge or in the humid atmosphere to produce hydrogen and iron oxide. It is in this phase of production that some of the most serious problems of hydrogen embrittlement today must be prevented. An example is the damage to large forgings from hydrogen. Attempts to keep water vapor to a minimum during steel manufacture have met with partial success, but often there are still a few parts per million of hydrogen, which is an amount sufficient to produce permanent damage. An apparent solution to this problem has been found recently, however, in the pouring or casting of the steel under vacuum.

In arc welding, moisture is again one of the major causes of hydrogen pickup. However, the primary one is undoubtedly the high-temperature decomposition of the electrode coating and the reaction of the molten metal with these products.

Small parts and structures, particularly of high-strength steels, probably pick up the major amount of hydrogen from chemical pickling or cleaning operations used to remove scale and oxide in preparation for further treatment or finishing and from electroplating in aqueous solutions.

In electroplating, hydrogen ions in the solution act as metal ions and are "plated" onto the cathode along with the desired plate. If there is a high rate of co-deposition, the plate itself may be embrittled, but normally the effect is primarily that of embrittling the base metal. The embrittlement problem is further complicated by the slow rates of diffusion through most of the metal plates. This tends to "trap" the hydrogen and make its elimination that much more difficult.

It is known from thermodynamics that gaseous hydrogen is composed primarily of molecular hydrogen with a very small percentage of atomic hydrogen in equilibrium. This amount increases both with temperature and pressure to the point where there is a significant absorption of hydrogen during heat treatment of iron and steel in an atmosphere containing hydrogen. In many instances hydrogen is picked up not in the fabrication of the structure but during its use. The processing and containment vessels and equipment of the petroleum and chemical industries are prime examples. Hydrogen may be picked up from hot hydrogenous gases, par-

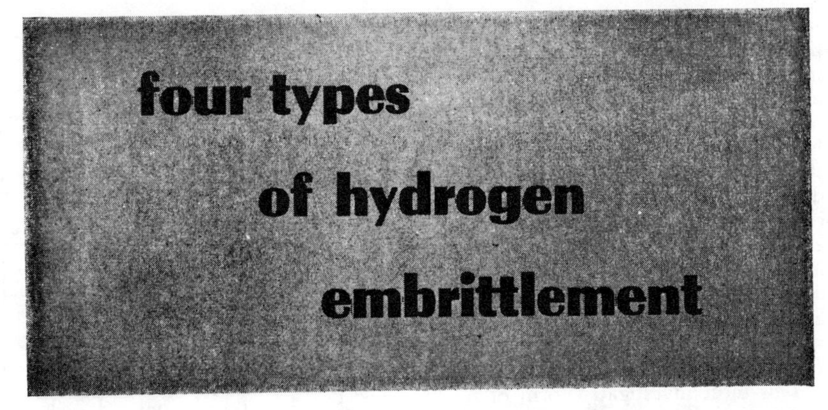

ticularly in pressure vessels, or it may be the result of a corrosion reaction which produces atomic hydrogen at the surface of the metal.

Types of Embrittlement

There appear to be four distinct ways in which hydrogen may embrittle metallic structures, or at least cause some damage to the structure which may increase its susceptibility to other forms of embrittlement. It is of primary interest to the engineer to be able to distinguish these from each other, since the approach to a solution of each problem is different.

Hydrogen Attack

This term has been applied to a permanent type of embrittlement that is the result of a chemical reaction of hydrogen with a second phase or a high local concentration of a minor element. The commercial materials in which this has been recognized as a severe problem are silver, copper, low-carbon iron and steel. In the first three materials, the element with which the hydrogen reacts is oxygen, which has concentrated primarily at the grain boundaries. The formation of water vapor and the resulting oxygen depletion in the metal results in an intergranular embrittlement, which cannot be eliminated by removing the hydrogen from the metal.

The principal source of hydrogen attack in steel has been the high-temperature reaction of hydrogen with the carbides in the steel to form methane. This formation appears to take place primarily at grain boundaries with the buildup of high pressure of the gas and intergranular embrittlement. Because of the high diffusion rate of carbon at these temperatures, there is a breakup of the carbide structure in the pearlite and a subsequent decarburization. The structure then becomes weak and spongy with considerable permanent loss in ductility. Van Ness states that although increased pressure increases the rate of hydrogen attack in steel, it is unlikely that it will take place at any pressure below about 400 F.

Prevention is the only solution to this problem. Either one must use a material which is not susceptible to this form of attack—such as austenitic stainless steels or, in the case of copper, oxygen-free copper—or care must be taken to prevent the exposure to the hydrogen-producing atmosphere. For pressure vessels where economics dictate the use of such materials, steels which are resistant to hydrogen attack can be produced by alloying with strong carbide-forming elements, such as titanium and vanadium. Chromium and molybdenum are also effective when the additions are large.

Hydride Embrittlement

A second form of embrittlement which plagues many exothermic occluders is hydride, or impact, embrittlement. Titanium and its alloys are unquestionably the outstanding examples of this type of embrittlement in commercial materials, but vanadium, tantalum, and columbium also will undoubtedly present a problem in this respect as they become more popular. This form of embrittlement is characterized by an increasing tendency to brittleness with increasing hydrogen content, decreasing temperature and increasing strain rate, similar to the characteristics of the low-temperature brittleness in iron.

It appears in titanium when the solid solubility of hydrogen in titanium is exceeded, a hydride phase precipitating on cooling. Thus, the α titanium alloys are plagued with this type of embrittlement more than the $\alpha + \beta$ and β alloys since the solubility of hydrogen in the beta phase is much greater than that in the α-phase; however, if the solid solubility is exceeded in any of the phases, this form of embrittlement begins to appear. Microexamination indicates that the fracture in such alloys begins at the alpha titanium-hydride interface.

The metallurgical approach to a solution of this problem is either to heat-treat to produce a fine hydride precipitate or small grain size, or to increase the hydrogen tolerance by alloying. This can be accomplished by increasing the solid solubility of hydrogen in the alpha phase or by stabilizing the β phase, in which hydrogen is more soluble. Of the many elements which accomplish this result, aluminum appears to most effectively increase the α-solubility, while molybdenum is the most effective β stabilizer.

True Hydrogen Embrittlement

This most investigated form of hydrogen embrittlement, known as "low strain rate embrittlement" in the titanium industry, occurs in iron and steel, titanium and its alloys, and vanadium; it will also probably be found in other exothermic occluders as they are investigated. Once again the commercial materials which are little susceptible to this form of embrittlement are the austenitic stainless steels. True hydrogen embrittlement is characterized primarily by the necessity for hydrogen to be present during the deformation and by a decreasing ductility with decreasing strain rate, which is contrary to the behavior in most other types of embrittlement.

There is a ductility minimum for iron and steel near room temperature. There seems to be no embrittlement when these metals are tested above about 350 to 400 F. The degree of embrittlement also increases with increasing hydrogen content (charging time), as shown in Figure 1, which also shows that the greater the strength level of the steel of a given composition and approximately the same structure the greater the sensitivity to hydrogen embrittlement.

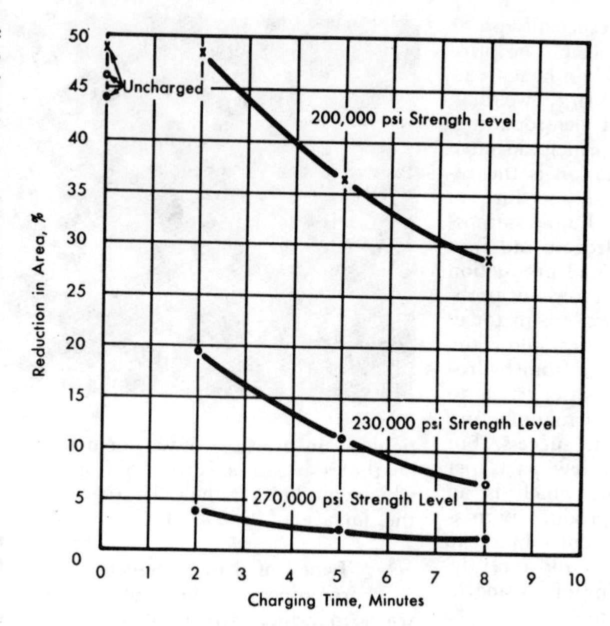

Figure 1—Effect of charging time (hydrogen content) on ductility of AISI 4340 steel of various strengths.

This brings into focus one of the major problems in understanding all the details of this type of embrittlement: the difficulty in distinguishing between an effect of true hydrogen embrittlement, presumably recoverable on removal of the hydrogen, and permanent damage incurred during the absorption of hydrogen. Figure 2 shows the results of tensile tests of AISI 1020 steel at a series of temperatures. The uncharged steel shows a gradual loss in ductility as the temperature is lowered. The aging treatment had no effect on the ductility of uncharged steel at any temperature. The cathodically charged specimens, on the other hand, show the characteristic embrittlement in the neighborhood of room temperature, with the ductility approaching that of the uncharged specimens as the test temperature is lowered and reaching it at about —160 C. However, at liquid nitrogen temperature the ductility is again much lower than that of the uncharged steel.

Aging of such charged steel results in complete recovery at those higher temperatures where the uncharged steel is ductile but continues to show the low ductility of the charged steel at liquid nitrogen temperature. The explanation, substantiated by a return

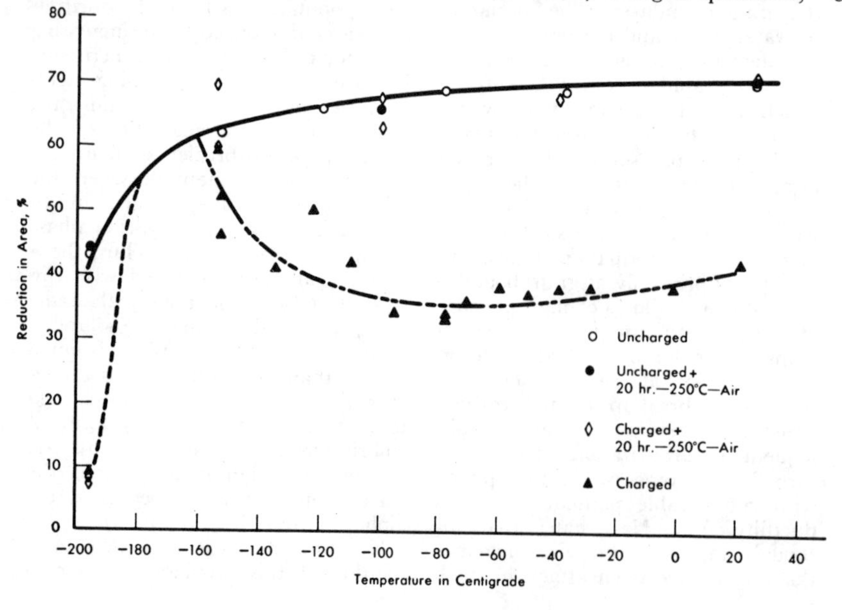

Figure 2—Effect of temperature on ductility of AISI 1020 steel in various conditions of charging and aging.

of complete ductility at this temperature upon removal of the surface layers, is that during cathode charging the surface was permanently blistered and cracked, these defects causing brittle failure in the steel, which is notch sensitive at the low temperature. This occurs whether or not the hydrogen is in the metal at the time of the test. Thus the recovery from true hydrogen embrittlement on aging is obscured by the permanent damage associated with the takeup of hydrogen. This differentiation becomes more difficult in the harder materials which are more sensitive to structural defects.

Cold work markedly increases the embrittlement on hydrogen charging and also decreases the ability of the metal to recover on aging. It is not at present clear whether this is a permanent effect or whether the cold work merely reduces the rate of recovery.

There is a marked effect of structure on this form of embrittlement. For example, in the uncharged condition a chrome-molybdenum steel with a tempered martensite structure has twice the ductility it has when the structure is coarse bainite, the ductility of other structures ranging between these two limits. However, after cathodic charging with hydrogen, the ductilities vary from zero percent reduction of area for untempered martensite to 71.6 percent for the spheroidized structure. In all cases, the ductility was recovered by aging for 1000 hours under mercury, so that the effect can be attributed to true hydrogen embrittlement.

From metallographic observations of cracks and fractures in various embrittled structures, it appears that in martensitic structures the fracture path follows the prior austenite grain boundaries, while in ferritic structures, it follows a cleavage plane. The latter is not particularly well confirmed, however.

1. Delayed Failure. The most spectacular of all the phenomena associated with true hydrogen embrittlement is that of delayed failure, also called static fatigue. The general characteristics of this type failure are the low stresses at which it occurs and the long waiting period after stressing before the relatively brittle fracture takes place. During this time apparently nothing is happening. This type of failure is most often observed in high-strength steel parts and structures, where failure takes place well below the yield stress. However, it has been shown to occur in lower strength steels such as rotor-forging steels at stress levels above the yield stress, and in plain carbon steel and low-strength iron-nickel alloys when continuously charged under stress. Two things stand out as most significant in this type of failure:

Delayed failure can occur when the same material with the same hydrogen content shows no embrittlement in a tensile test.

The stress that results in failure may be either applied or residual, such as a transformation or quenching stress. The failure under the action of a residual stress gives an appearance of spontaneity to the failure.

Troiano and his associates have made the most extensive laboratory investigations of delayed failure, both when the distribution of hydrogen is highly nonuniform, such as would normally result from pickling or electroplating, and when it is made uniform so that the mechanism can be more easily studied. Figure 3 shows the general shape of the time for failure as a function of the stress level and hydrogen content for a uniform distribution of hydrogen.

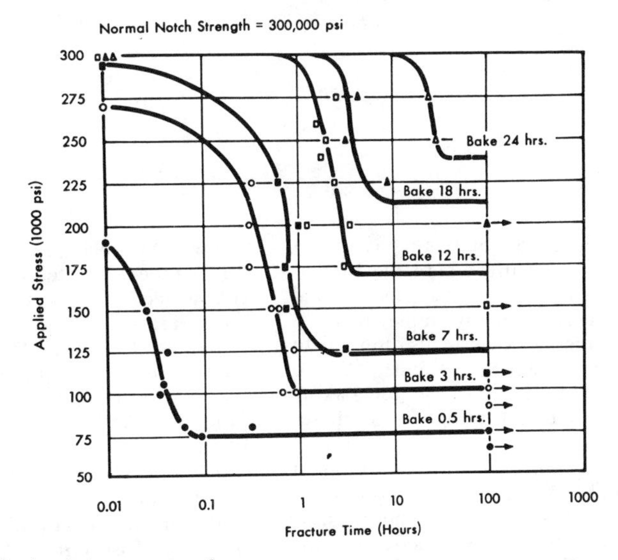

Figure 3—Static fatigue curves for various hydrogen concentrations (corresponding to different baking times). Steel used was notched AISI 4340, heat-treated to 230,000 psi strength.

The shorter the bakeout time listed, the higher the hydrogen content. The most important feature is that for every testing condition there exists a stress called the *static fatigue limit,* below which failure does not occur even after an extensive waiting period. This stress decreases as notch severity increases, and is a relatively lower proportion of the uncharged notch tensile stress as the strength level in a particular steel increases. This phase of hydrogen embrittlement is also structure sensitive, the martensitic structure again being most sensitive to this type of failure.

Troiano and associates have also shown that there exists an incubation period during which damage is recoverable, but beyond which time the amount of recoverability is a function of how sensitive the steel is to an existing crack. When the distribution is highly nonuniform, the situation is somewhat different. The hydrogen-rich case cracks almost instantaneously on stressing and the slow crack growth follows the inward diffusion of the hydrogen front. Hence, once the part has been stressed, there is some non-recoverable damage. In both cases, if no stresses exceeding the static fatigue limit have been applied and if the material contains no residual stresses which have had time to act since the piece contained hydrogen, the embrittlement should be recoverable.

This problem has plagued users and suppliers of high-strength and high-hardness parts, such as springs and lockwashers, for some time, but has now become even more severe with the highly stressed high-strength steel parts used in aircraft. Although a large amount of hydrogen may be picked up in the chemical pickling used to clean off scale in these parts, it is the plating which frequently follows the cleaning which causes the major amount of trouble. Not only is a large amount of hydrogen introduced into the metal during the plating operation, but the metals which are used as plating materials in general have a much lower permeation to hydrogen than does the base metal. Thus, when an attempt is made to

Source: *Materials Protection,* April 1962, 26-33

remove hydrogen by baking at an elevated temperature where the rate of diffusion is much higher, the normal times used for baking unplated metal are ineffective. One is also limited to the temperature (about 350° to 400°) that one can use for baking and still retain the high strength required.

Three approaches to a solution of this problem in addition to using some other method of scale removal, such as grit blasting and increasing the plating efficiency, are as follows:

Electroplate only a flash coat of cadmium on the metal, which, because of its thinness, could be baked out in a short time. The temperature dependence of metallic diffusion is such that, upon dropping back to room temperature, this thin coat of cadmium acts as an effective barrier to further penetration of hydrogen during any subsequent plating operations carried out to build up the required thickness of plate.

Control plating conditions so that the plate is porous. Since the cadmium or zinc plate is a sacrificial plate, this has no effect on its usefulness. During the baking operation, the hydrogen can then escape from the base metal through the pores, and the baking time is considerably reduced.

Vacuum deposit a plate which may or may not be subsequently electroplated.

2. Flakes and Weld Cracking. The two other large areas in which the effects of true hydrogen embrittlement are highly damaging are in the production of large forgings and in welding, particularly electric arc welding. The type of damage encountered in forgings is called hairline cracking or flaking. A section through the forging reveals a multitude of fine cracks. Although when detected the damage from these defects is permanent, it is still a form of true hydrogen embrittlement which occurs during, or as a result of, the processing, rather than during the use of the embrittled structure.

The two types of defects are similar in cause and hence will be discussed together. In both cases, the hydrogen pickup occurs primarily while the metal is molten. When the austenite decomposes as the temperature is lowered, the structure produced has a lower hydrogen solubility and becomes supersaturated. It is this hydrogen which reduces the ductility of the transformed structure when subjected to the severe thermal stresses induced by the rapid quenching of the weld by the plate or to the transformation and quenching stresses produced in the forging on cooling.

The normal procedure for preventing flaking is soaking the billet at high temperature to allow the hydrogen to diffuse out before transforming the steel. However, since the rate of reduction of hydrogen would vary roughly as the square of the smallest linear dimension, it becomes impractical as well as uneconomical to carry out satisfactory thermal treatments on very large forgings used today because of the long times involved. Hence, the necessity for the prevention of hydrogen absorption in the metal by means of vacuum pouring.

The reduction of hydrogen damage in welding is obtained principally by reducing the hydrogen pickup by using dry electrodes of the type shown to produce the lowest hydrogen and still fulfill the requirements for the weld and by reducing the rate of cooling in the critical temperature ranges below 300 C. particularly in the neighborhood of 150 to 120 C. Preheating is very effective in reducing hydrogen damage, while a postweld heat treatment immediately after welding has some beneficial effect. This is not as effective as preheating, since it does not prevent the severe quenching stresses nor allow more time for higher temperature transformations, but it does tend to bake out the hydrogen and afford a minor stress relief.

Permanent Damage

In addition to the clear-cut permanent damage resulting from hydrogen attack or from true hydrogen embrittlement, there exists another extremely prevalent form of permanent damage. This is apparently produced by hydrogen alone, without the necessity of stresses or a second phase, although these do promote it. This type of damage appears primarily in the form of surface blisters and cracks.

As mentioned previously, there is an equilibrium balance between the pressure of molecular hydrogen and atomic hydrogen. There is a similar equilibrium between the hydrogen in solution in the metal and gaseous molecular hydrogen at its surface. For the solubilities normally encountered, the equilibrium pressure may be 200,000 psi or more. Thus, when the hydrogen in solution comes to any surface, it will tend to diffuse out, recombining to form molecular hydrogen. It will continue to do this until the pressure of the gas has reached a value in equilibrium with the remaining lattice hydrogen.

Such a surface may be either an external or an internal surface. Blisters, blowholes and cracks are considered to be merely the result of a buildup of molecular hydrogen under high pressure at an internal surface. If this discontinuity is near the outside metal surface, the pressure can actually expand the metal plastically, finally erupting to the outside surface by tearing the blister wall. When the surface is deep inside, the large mass of metal surrounding the defect prevents significant plastic deformation, and cracking is the result.

Since this damage is permanent it must be prevented or at least inhibited in commercial structures, where corrosion is the primary cause of such damage. This is accomplished by using corrosion-resistant steels and by using chemicals to inhibit corrosion in one way or another, the latter being the more economical way whenever effective. In electrochemical processes, reducing the current density reduces the blistering.

Theories of true hydrogen embrittlement and variations on them are almost as numerous as the investigators who have studied this type of failure. Since none is consistent with all the accumulated experimental evidence gathered both in the laboratory and from industrial experience, none is operationally very useful to the engineer. He should rather apply the general rules derived from this mass of empirical knowledge to the solution of his problems, bearing in mind the several types of hydrogen embrittlement with differing modes of solution. It may also be necessary to implement the available data with further tests designed to evaluate a specific material or particular condition.

Factors Influencing the Hydrogen Cracking Sensitivity of Pipeline Steels*

E. M. MOORE and J. J. WARGA, *Arabian American Oil Company, Dhahran, Saudi Arabia*

Tests are described which permit qualitative assessment of the probable sensitivity of pipeline steels to stress cracking by hydrogen. Investigation of the possible causes of hydrogen stress cracking and the results of test programs indicates that cracking susceptibility is a function of composition and deoxidation practice which produces Mn sulfide inclusions of a geometry leading to cracking. Sulfur content does not appear to be significant, nor do variations in mechanical properties. Authors conclude semi-killed steels are markedly less susceptible, that addition of 0.25 to 3% Cu reduces cracking and blistering in fully killed steels. Use of recommended metallurgical controls should not, however, reduce necessity for inhibitor injection or other operational controls.

IN 1974, three service ruptures occurred in a spirally welded, API 5LX-Grade 42 sour gas transmission pipeline between 4 and 7 weeks after commissioning. A subsequent hydrotesting program revealed extensive blistering and cracking over more than 10 km of this 90 km pipeline. All the failures were due to stepwise cracking associated with elongated manganese sulfide inclusions in the steel. Blistering was often, but not always, evident in the cracked areas. The failures were always located near the spiral weld but were never identified with weld defects of any kind. Blistering and incipient cracking occurred randomly on the bottom and sides of the affected sections of the pipeline. No defects were found in the top quadrant of these pipes.

Hydrogen blistering and hydrogen induced stepwise cracking in sour gas transmission pipelines have been reported previously in the literature.[1,2] A method of testing the hydrogen blistering and cracking sensitivity of pipeline steels has been developed by H. C. Cotton of the British Petroleum Company.[3] Because of its success in duplicating service blister formation and stepwise cracking in specimens of susceptible line pipe, the authors have used this test to evaluate those factors in the manufacture and composition of steel pipe that contribute to the susceptibility to cracking in sour gas service.

*Presented during Corrosion/76, March 22-26, 1976, Houston, Texas.

Reprinted with permission from *Materials Performance*, June 1976, 17-23, © 1976 National Association of Corrosion Engineers

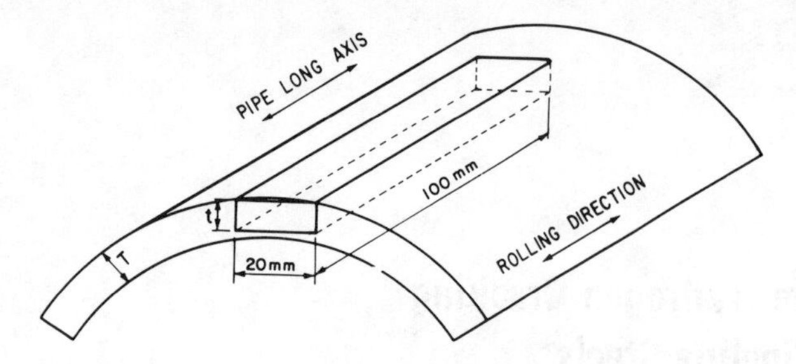

FIGURE 1 — Orientation of test coupons taken from seamless pipe and from the parent material of longitudinally welded pipe.

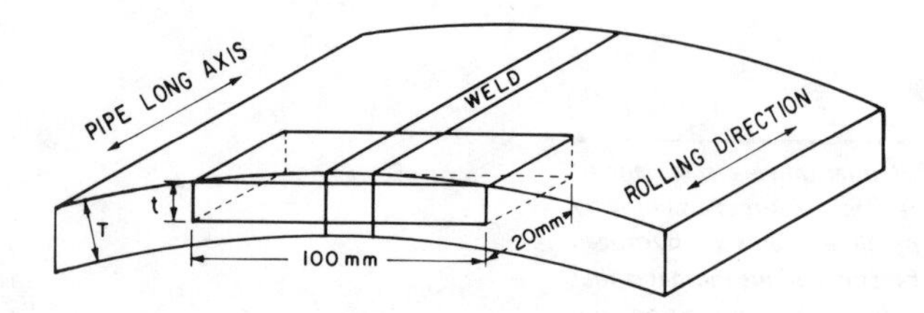

FIGURE 2 — Orientation of test coupons taken from the weld area of longitudinally welded pipe.

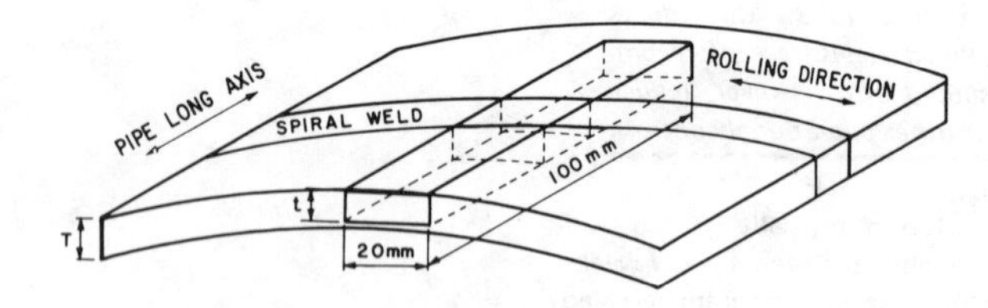

FIGURE 3 — Orientation of test coupons taken from the weld area of spiral welded pipe.

This paper describes the test procedures and test matrix used to evaluate the cracking tendency of selected pipeline steels, the test results that were obtained, and correlation of the cracking with various microstructural and compositional factors. The conclusions drawn are presented to stimulate discussion, and to initiate further studies of those factors that sensitize pipeline steels to hydrogen blistering and cracking.

The Hydrogen Cracking Sensitivity Test

In the Cotton test, carefully prepared specimens are immersed for 96 hours in H_2S saturated synthetic sea water with no applied stress. The resulting surface blistering, internal stepwise cracking, and diffusible hydrogen content are then measured.

Specimen Preparation

For each test, at least two coupons are cut from the material to be tested and one coupon is cut containing a weld. For seamless pipe, three metal coupons are used. The shape of the coupons has been arbitrarily standardized as a rectangle 20 by 100 mm. The top and bottom faces are lightly machined until they are flat. The resulting thickness should not be less than the pipe wall thickness minus 2,5 mm. If the pipe diameter is such that this criterion cannot be met, the coupon should be as thick as possible.

The long axis of the coupon is oriented parallel to the axis of the pipe, as illustrated in Figure 1. The test specimens should be taken from the side opposite the weld. They should be from locations at least 100 mm apart (measured either longitudinally or circumferentially). Whenever possible, the samples should be taken from a length of pipe fabricated from steel that came from the segregation zone (*i.e.,* "pipe") section of an ingot.

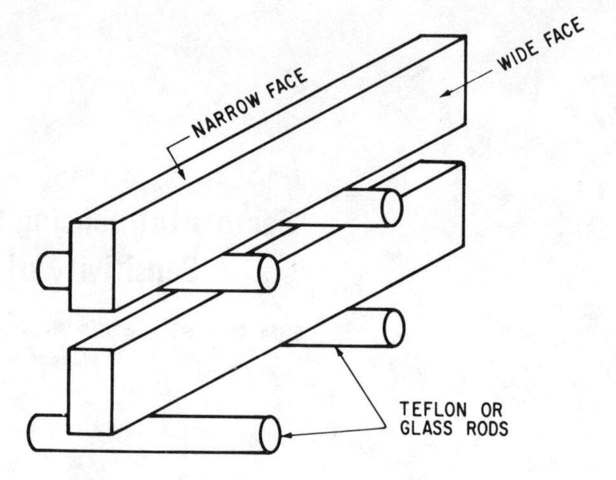

FIGURE 4 — Method of stacking coupons in the test solution.

TABLE 1 — Standard Test Conditions for Hydrogen Cracking Sensitivity Test

Temperature	25 C ± 3 C
H_2S Concentration	2300-3500 ppm (saturated)
pH of H_2S saturated solution	4.8-5.4
Applied Stress	None

For longitudinally welded pipe, a coupon containing the weld is cut transverse to the weld as shown in Figure 2. For spiral welded pipe, a coupon containing the weld is cut with its long axis parallel to the pipe axis and having the weld running diagonally across, as shown in Figure 3.

When coupons are taken from the plate rather than from finished pipe, three coupons are cut from the middle of the plate with a distance of at least 100 mm between each coupon. Whenever possible, the plate chosen for testing should be fabricated from the "pipe" section of the ingot. The long axis of the coupon is oriented parallel to the plate rolling direction.

The cut coupons are ground on a wet endless belt, and finish ground on dry 320 grit silicon carbide paper. The specimens are then degreased in acetone. Thereafter, extreme care must be taken not to contaminate the coupons. They should be handled only with degreased tongs or clean gloves. Finger grease on the coupon surfaces has been reported to affect the results of the test.[4]

Standard Test Solution

Synthetic sea water is prepared in accordance with the requirements of ASTM D1141-52. The ratio of the specific volume of solution to the surface area of the specimen should be in the range of 3 to 6 ml/cm^2. The initial pH of the solution should be 8.2. The solution is deaerated in a closed container by bubbling nitrogen through it at a rate of 100 ml/l/minute for 1 hour.

The specimens are then placed horizontally in the solution with their wide faces vertical and their narrow faces horizontal. The lower face is raised from the cell bottom on bars of teflon or glass. When stacked, the specimens are separated by similar bars (Figure 4). The specimens should be placed in the solution quickly to limit oxygen pickup. The solution is then saturated by bubbling H_2S of at least 99.5 Vol% purity at the rate of 2 to 5 l/minute for 1 hour through an open ended tube with a 5 mm internal diameter. Upon reaching saturation, the H_2S gas flow rate is reduced to 100 ml/minute for a 10 liter cell, or pro rata, and maintained thereafter at this rate. It is advisable to maintain a small positive pressure of H_2S in the test cell by the use of an outlet trap to prevent oxygen contamination from the air. The standard test conditions at the start of the test are shown in Table 1. The pH of the solution at the end

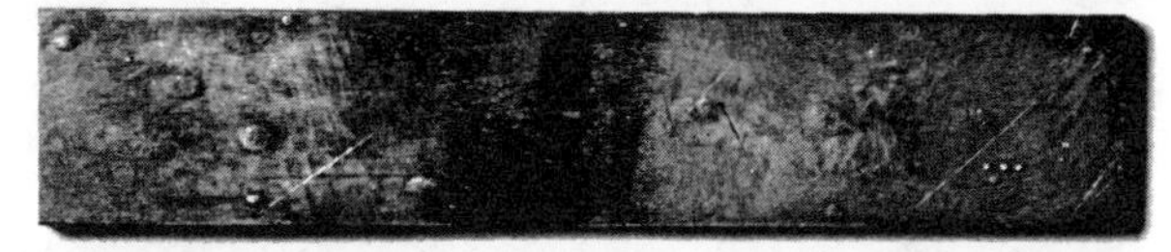

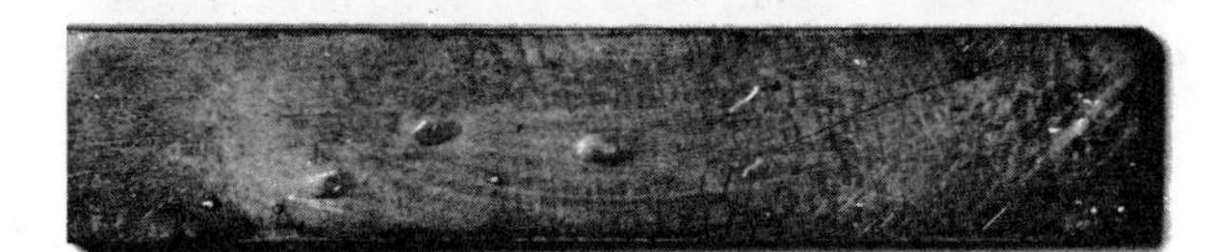

FIGURE 5 — Examples of blistering produced by exposure to the hydrogen cracking sensitivity test and by exposure to service conditions that resulted in a pipeline rupture. Upper: blisters in test coupons, 0.81X, and lower: blisters on ruptured pipeline, 0.2X.

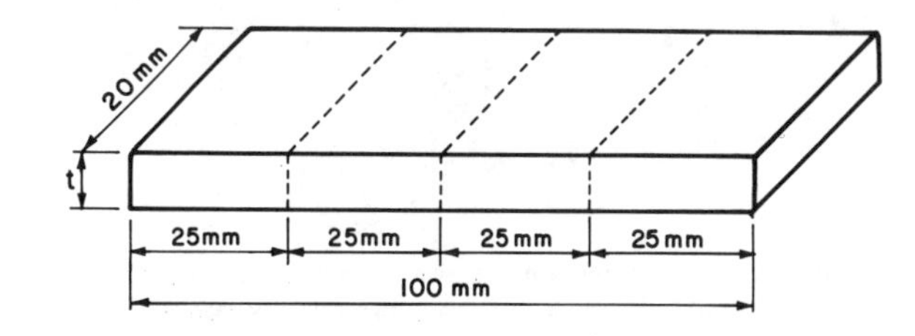

FIGURE 6 — Sectioning procedure for cutting test coupons from seamless pipe and the parent material of welded pipes.

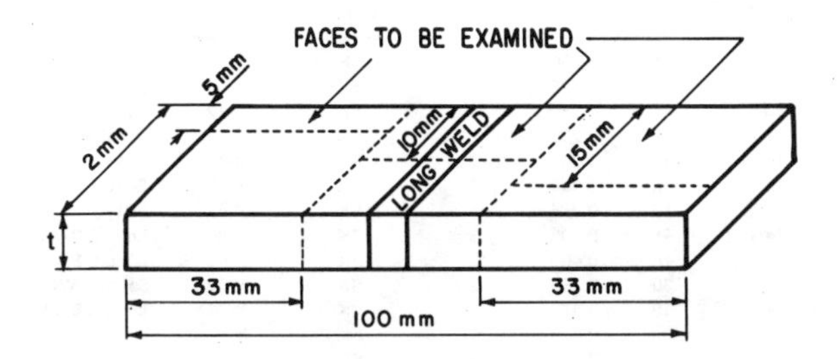

FIGURE 7 — Sectioning procedure for cutting test coupons from the weld area of longitudinally welded pipe.

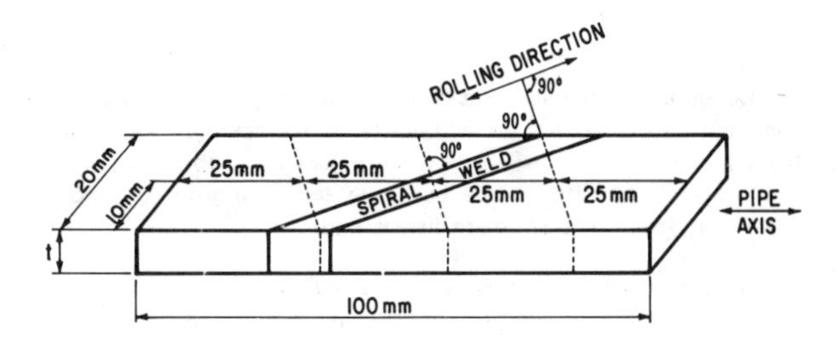

FIGURE 8 — Sectioning procedure for cutting test coupons from the weld area of spiral welded pipe.

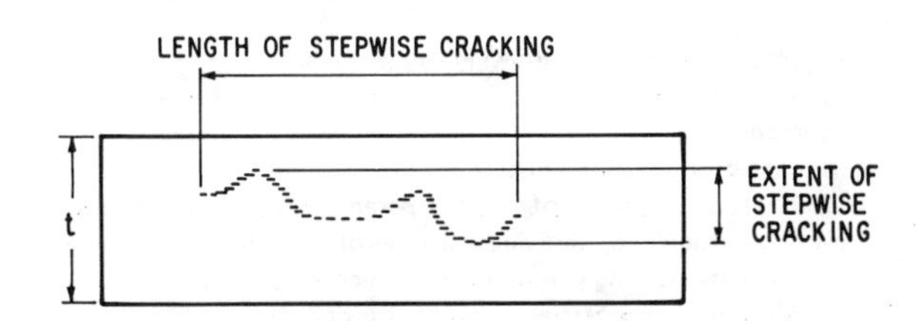

FIGURE 9 — Schematic illustrating how the length and extent of stepwise cracking are defined. "t" indicates the through-thickness direction in the plate or pipe sample.

of the test is measured and recorded. At the same time, the H_2S concentration in the solution is determined by iodometric titration. It is necessary to collect the test sample rapidly and keep it in a stoppered bottle or H_2S losses to the air may be appreciable.

Evaluation and Reporting of Results

After testing, the specimen is removed from the solution, washed in running water, wire brushed to remove loose deposits, washed in acetone and dried in petroleum ether and cold air. It is particularly important that the coupon be thoroughly dried. The coupon is then placed in a glass cell and the evolved hydrogen is collected over glycerine at a temperature of 45 C for 72 hours. The specimen should be placed in the hydrogen determination cell as rapidly as possible, unless it is stored in liquid nitrogen between the water washing and wire brushing stages.

The two wide faces of each coupon are photographed to record any blistering that might have occurred. An example is shown in Figure 5a. Note that material from the ruptured pipeline has similar blisters (Figure 5b).

Coupons are sectioned transversely at three points as shown in Figure 6. Coupons cut containing a weld from longitudinally welded pipe are sectioned as shown in Figure 7. Coupons cut from either the parent material or a weld area in spiral welded pipe are sectioned as shown in Figure 8. The intention of the sectioning procedures is to examine for cracks on a plane transverse to the rolling direction.

The sections are then mounted in epoxy resin (or equal) and polished. Cracking is then estimated by eye, and by microscopic examination at magnifications of 30 and 100X. For each crack observed, the length and extent of stepwise propagation (Figure 9) is reported.

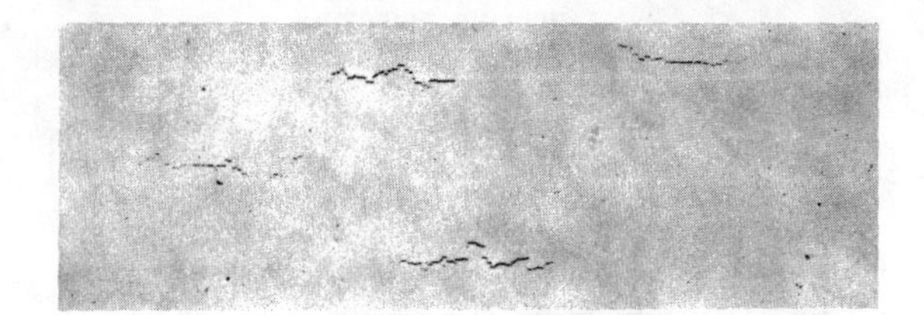

FIGURE 10 – Example of stepwise cracking in a susceptible steel exposed to the hydrogen cracking sensitivity test. 6X

TABLE 2 – Identification of Materials Tested

Sample Number	Pipe Size (In) Diam.	Wall	Supplier	Deoxidation Practice(1)	Nominal Grade	Method of Fabrication
8	24	0.250	1	SK	B	Spiral SAW
9	24	0.375	2	SK	X-52	Spiral SAW
10	24	0.250	3	FK	X-42	Spiral SAW
11	22	0.250	1	SK	X-42	Spiral SAW
12	16	0.250	1	SK	X-42	ERW
13-1	16	0.344	1	SK	X-42	ERW
13-2	16	0.344	1	SK	X-42	ERW
14-1	46	0.400	3	FK	X-60	Longit. SAW
14-2	46	0.400	3	FK	X-60	Longit. SAW
16	30	0.250	2	SK	X-52	Spiral SAW
17	30	0.312	4	SK	X-52	Longit. SAW
19		0.430	5	FK	N/A	Plate
20		0.370	5	FK	N/A	Plate(2)
21		0.512	3	FK	N/A	Plate

(1) See the text. The term "fully killed" (FK) means killed with both silicon and aluminum. The predominant inclusions are Al_2O_3 and Type II manganese sulfides. "Semi-killed" (SK) includes the so-called fully silicon killed steels. The predominant inclusions are silicates and Type I manganese sulfides.
(2) Plate containing SAW weld.

Coupons containing cracks are photographed showing the complete transverse sections with examples of cracking. An example from Sample Number 14-1 is shown in Figure 10. In this evaluation, cracks associated with surface blistering and having no part more than 1 mm from the surface are disregarded.

Summary of Data to be Recorded for Each Test

1. Location and dimensions of coupons, and whether taken from pipe or plate.
2. pH of H_2S saturated brine at the end of the test, and also the H_2S concentration.
3. Diffusible hydrogen content of each coupon in ml/100 g steel.
4. Photos of coupons, showing any blisters.
5. Results of cracking in cross sectional evaluation, with photographs.
6. Composition of material tested.
7. Photomicrographs of typical parent material structures as follows: (a) Unetched, showing the type of inclusions in the steel, and (b) etched, showing parent material microstructure.
8. Mechanical properties of material tested.

Steels Tested

Samples from 14 pipeline steels were evaluated for sensitivity to hydrogen cracking in accordance with the procedures just described. One of the samples, Number 17, was taken from an operating sour gas transmission line that had been in service 16 years. Three other samples (Numbers 10, 11, and 12) were taken from line pipe stocks used in fabricating the ruptured line. Seven of the remaining samples were selected from pipelines under construction. Samples 19, 20, and 21 were taken from material proposed for use in future pipelines.

The pipe samples covered the size range between 16 and 46 inches (41 and 117 cm) diameter and API Grades B through X-60. The steels were produced by 5 different suppliers in Europe and Japan, and were fabricated by either submerged arc (SAW) or electric resistance welding (ERW) procedures. Table 2 shows the relevant information for the samples.

TABLE 3 – Ranking of Materials Tested in Order of Increasing Susceptibility to Hydrogen Cracking

Sample Number	Cracking Sensitivity Ratio (AC/TA)X1000(1)	Number of Sections Cracked(2)	Extent of Blistering(3)	Size of Blisters(4)	Diffusible H_2 ml/100 gm(5) PM	WM	pH, End of Test
13-1	0	0/9	O	N/A	0.6,1.0	2.6	5.4
13-2	0	0/9	O	N/A	0.5,0.9	2.6	5.4
21	0	0/9	O	N/A	0.3	N/A	5.2
11	0	0/9	E	S-M	2.0,2.2	1.1	4.9
12	0	0/9	S	M	7.4,3.0	4.8	5.1
17	0.71	1/9	E	S-M	1.9,5.1	4.0	5.2
8	2.00	1/12	E	S	2.3,1.8	1.9	–
16	2.10	1/9	S	S-M	2.8,2.5	2.1	4.9
9	2.42	3/12	E	S	3.8,3.9	3.3	–
20	2.55	6/9	E	S-M	4.3,5.0	3.6	4.9
14-1	5.38	6/9	E	S-L	1.8,3.0	2.0	5.1
14-2	7.92	5/12	E	S-L	3.5,2.7	2.6	5.1
19	14.60	6/9	E	S-L	4.6	N/A	5.1
10	99.52	7/9	E	S-L	3.5,1.8	1.9	4.9

(1) AC = Total cracked area. The sum of the individual cracked areas which are defined as length of stepwise propagation times extent of stepwise propagation (Figure 9), and TA = Total cross-sectional area of sectioned coupon faces examined for cracking.
(2) X/N = X sectioned faces out of a total of N sectioned faces examined showing cracking.
(3) O = none, S = small, and E = extensive.
(4) S = small, M = moderate, L = large, and N/A = not applicable.
(5) PM = parent metal, and WM = weld metal.

Chemical composition, weld, and parent metal microstructure, inclusion morphology, and mechanical properties were determined for all the samples.

Test Results

Relative Cracking Sensitivity

Table 3 ranks the materials tested in order of increasing susceptibility to hydrogen cracking. The ranking parameter used was ratio of the cracked area to the cross-sectional area examined on the sectioned coupons. With the exception of Number 14-2, a ranking in terms of the number of sections that cracked would be identical. This table also shows that while the blistering susceptibility roughly parallels cracking susceptibility, it is not a very sensitive parameter. The three samples that did not blister did not crack either. However, two samples that did not crack showed blistering. With the exception of Number 16, samples that cracked showed numerous blisters, but the size ranges were variable.

The measured diffusible hydrogen contents are shown in this table for the sake of completeness, but the authors believe these data are not significant. The scatter is large, and there is no consistent relationship between weld metal and parent metal results. The authors believe a more reliable measuring technique is needed.

Finally, Table 3 shows the pH of the test solution at the end of the tests. Other experimenters have noticed a definite relationship between pH and cracking susceptibility.[5,6] For the tests reported here, the solution pH averaged 5.1 with a standard deviation of 0.2. No relationship was found between pH and the test results.

Effect of Steel Composition and Deoxidation Practice

The compositions of the test materials were determined spectrographically, with confirmations of individual elements made from time to time using other procedures. The Wt% of 17 elements were determined; iron forms the remainder. Only those 13 most commonly present in line pipe steels are shown in Table 4. Boron, lead, tin, and cobalt were determined in all analyses to be at or below 0.001, 0.01, 0.01, and 0.01 Wt%, respectively. The carbon equivalent of each test material was calculated from the composition according to the following relationship:

$$Ceq = C + \frac{Mn}{6} + \frac{Cr + Mo + V}{5} + \frac{Ni + Cu}{15}$$

Table 5, which is a combination of Tables 2, 3, and 4, shows that the sensitivity of these pipeline steels to hydrogen induced cracking is markedly dependent on the deoxidation practice used during steelmaking. With a single exception (Number 21), all the fully killed steels were more susceptible to cracking than the semi-killed steels. In this paper, the term "fully killed" means deoxidized with both silicon and aluminum. The term "semi-killed"

TABLE 4 — Composition of Materials Tested

Sample Number	C	S	P	Si	Al	Mn	Ni	Cr	Mo	V	Cu	Nb	Ti	Carbon Equivalent
13-1	0.21	0.012	0.01	0.01	0.012	0.76	0.01	0.01	0.01	0.01	0.01	0.005	0.01	0.34
13-2	0.28	0.010	0.013	0.01	0.006	0.83	0.01	0.01	0.01	0.01	0.01	0.005	0.01	0.42
21	0.10	0.005	0.01	0.23	0.031	0.92	0.13	0.30	0.01	0.01	0.27	0.011	0.01	0.34
11	0.17	0.011	0.012	0.01	0.005	0.92	0.01	0.01	0.01	0.01	0.03	0.005	0.01	0.33
12	0.26	0.013	0.008	0.01	0.015	0.60	0.02	0.01	0.01	0.01	0.01	0.005	0.01	0.36
17	0.19	0.026	0.070	0.38	0.006	1.06	0.06	0.06	0.01	0.01	0.22	0.005	0.01	0.36
8	0.15	0.012	0.014	0.04	0.005	0.75	0.01	0.01	0.01	0.01	0.02	0.005	0.01	0.28
16	0.17	0.011	0.010	0.01	0.005	1.28	0.01	0.01	0.01	0.01	0.03	0.021	0.01	0.39
9	0.15	0.009	0.011	0.01	0.005	1.21	0.03	0.01	0.01	0.01	0.003	0.02	0.01	0.36
20	0.12	0.020	0.017	0.17	0.018	0.72	0.01	0.01	0.01	0.01	0.02	0.005	0.01	0.24
14-1	0.12	0.005	0.014	0.35	0.039	1.31	0.02	0.15	0.03	0.01	0.01	0.014	0.01	0.38
14-2	0.13	0.007	0.014	0.34	0.021	1.34	0.03	0.21	0.03	0.01	0.02	0.015	0.01	0.41
19	0.13	0.020	0.016	0.17	0.022	0.72	0.02	0.01	0.01	0.01	0.04	0.005	0.01	0.25
10	0.15	0.007	0.12	0.22	0.036	1.01	0.01	0.01	0.01	0.01	0.01	0.005	0.01	0.32

TABLE 5 — Sensitivity to Hydrogen Cracking as a Function of Inclusion Types and Steelmaking Practice

Sample Number	Cracking Sensitivity Ratio, X1000	Description of Inclusions(1)	Steelmaking Practice(2)
13-1	0	I MnS, I duplex, silicates, occasional II MnS	SK
13-2	0	I MnS, I duplex, silicates	SK
21	0	II MnS, Al_2O_3	FK(3)
11	0	I MnS, I duplex, silicates	SK
12	0	I MnS, I duplex, silicates, occasional II MnS	SK
17	0.71	I MnS, I duplex, silicates	SK
8	2.00	I MnS, I duplex, silicates	SK
16	2.10	I MnS, I duplex, silicates	SK
9	2.42	I MnS, I duplex, silicates	SK
20	2.55	II MnS, Al_2O_3	FK
14-1	5.38	II MnS, Al_2O_3	FK
14-2	7.92	II MnS, Al_2O_3	FK
19	14.60	II MnS, Al_2O_3	FK
10	99.52	II MnS, Al_2O_3	FK

(1)"I MnS" means Type I manganese sulfides, "I duplex" means Type I duplex sulfide-silicates, "II MnS means Type II manganese sulfides.
(2)"SK" means semi-killed. "FK" means fully killed with both silicon and aluminum.
(3)Special copper bearing steel.

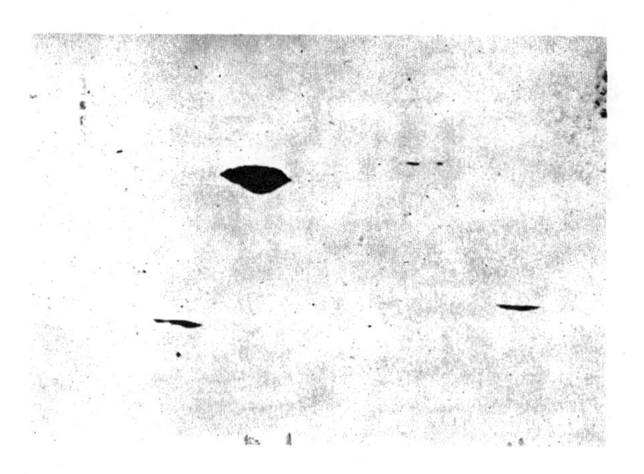

FIGURE 11 — Typical ellipsoidal shape of a Type I inclusion. 687X

includes the so-called fully silicon killed steels. Number 21 was a specially made copper bearing steel. The effect of copper additions is considered in detail in the Discussion section.

For the materials tested, a Pearson's Product Moment analysis[7] was made to obtain the correlation between various constituents in the steel and the susceptibility of the steels to hydrogen cracking. Each constituent was correlated with two measures of hydrogen cracking susceptibility. In every case, the correlation coefficient obtained with percentage of specimens exhibiting cracking was more positive than the correlation coefficient obtained with the cracking sensitivity ratio. Correlation coefficients less than 0.60 are statistically insignificant at a 95% confidence level for the number of samples tested. Data from the special copper bearing steel (Number 21) were not included in the analysis. The correlation coefficients were as follows, with cracking sensitivity listed first: sulfur: -0.26, -0.06; manganese: 0.08, 0.20; silicon: 0.24, 0.61; and aluminum: 0.60, 0.83. Hence, the only consistently significant correlation was with aluminum. Since it was shown in Table 5 that all the fully killed steels (except Number 21) were more crack sensitive than the semi-killed steels, this result for aluminum is reasonable.

Effect of Microstructure

Specimens for the test materials were sectioned and polished in the direction parallel to rolling for examination of the types of inclusions present. Additional specimens were selected and polished to examine the microstructure of the parent material (PM) and the heat affected zone (HAZ) and weld metal (WM) of the submerged arc or electric resistance welded seams. Inclusions were classified by composition, type, and method of formation in accordance with Sims' notation.[8] The inclusions that were observed were either Type I or II sulfides, Type I duplex sulfide-silicates, silicates, or aluminum oxide, the occurrence being dependent on the deoxidation practice, *i.e.,* whether semi-killed or fully killed. The predominant inclusions are Al_2O_3 and Type II manganese sulfides. Silicates and Type I sulfides are absent. The predominant inclusions are silicates and Type I sulfides and duplex sulfide-silicates. Al_2O_3 and Type II sulfides are generally absent, although occasional Type II sulfides may occur when semi-killing is achieved using aluminum additions, as for instance in Numbers 12 and 13-1. The predominant inclusion morphology is still Type I, however. The types of inclusions found in the various steels are shown in Table 5. The only steel containing predominantly Type II MnS inclusions that was not sensitive to hydrogen induced cracking was Number 21.

Examples of the ellipsoidal Type I sulfide and duplex sulfide-silicate inclusions are shown in Figure 11. Elongated, Type II sulfides are shown in Figure 12.

Parent metal microstructures, as well as weld area microstructures, where applicable, were examined for all the samples tested. There appeared to be a slight correspondence between cracking sensitivity and pronounced banding in the parent material. There did not appear to be any correlation between weld or heat affected zone structures and cracking sensitivity. Indeed, no hydrogen cracking was observed in these areas. A typical parent metal microstructure illustrating pearlite and ferrite banding is shown in Figure 13.

Effect of Mechanical Properties

Tensile properties were determined transverse to the rolling direction of the plate material using the specimen sizes and

FIGURE 12 — Typical elongated shape of Type II MnS inclusions. 544X

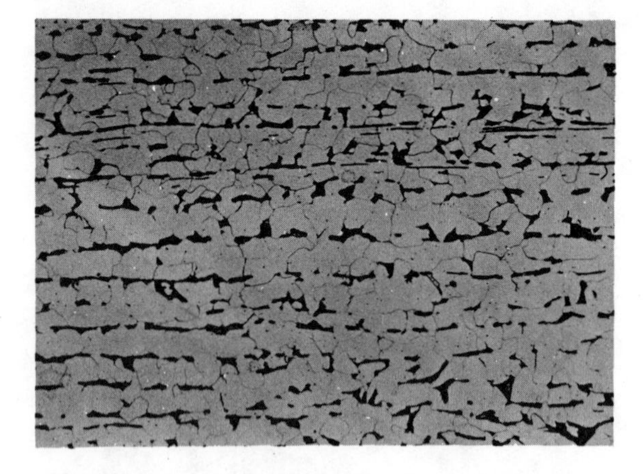

FIGURE 13 — Typical parent metal microstructure of line pipe steels showing pearlite and ferrite banding. 179X

procedures recommended in ASTM Standard A370. Yield strength, ultimate tensile strength, % elongation, % reduction in area, and the hardness of parent metal, weld metal, and heat affected zone are shown in Table 6.

Hardness traverses were conducted across the weld and heat affected zone areas using a Vickers diamond pyramid indentor and 2.5 kg load. In general, the hardness differences across the weld were relatively minor, the greatest difference being 20 points. No hardness reading greater than DPH 219 was measured (this hardness corresponds approximately to Rockwell B94).

There was no correlation between the hydrogen cracking sensitivity and the mechanical properties of the steels tested.

TABLE 6 — Mechanical Properties of the Test Materials

Sample Number	YS(1)	UTS(2)	% Elong.	% R/A(3)	Average Hardness Vickers 2.5 kg(4) PM	HAZ	WM
13-1	50.1	67.1	33	62	183	182	183
13-2	54.6	71.4	30	48	192	191	193
21	57.3	72.6	35	55	N/A	N/A	N/A
11	49.6	68.0	31	53	162	165	158
12	49.9	68.3	32	58	183	175	165
17	49.6	76.4	29	44	176	177	170
8	51.4	73.0	36	72	164	162	178
16	62.5	82.6	24	44	203	199	193
9	64.5	79.6	31	67	188	188	196
20	46.0	66.1	29	56	149	155	168
14-1	66.6	81.0	29	66	215	197	217
14-2	69.2	84.2	24	57	216	206	219
19	42.1	63.8	39	57	N/A	N/A	N/A
10	54.3	73.2	34	59	165	175	175

(1)Yield strength, ksi.
(2)Ultimate tensile strength, ksi.
(3)Percent reduction in area.
(4)PM = parent metal, HAZ = heat affected zone, and WM = weld metal.

Discussion

The results of the hydrogen cracking sensitivity tests indicate that cracking sensitivity is strongly influenced by the steelmaking practice used; specifically, by the shape of the manganese sulfide (MnS) inclusions formed by that practice. The incidence of cracking was either nil or low in the semi-killed steels where the sulfides were of the Type I (*i.e.,* ellipsoidal) kind. Cracking was prevalent in the fully killed steels, in which the sulfides were of the Type II (*i.e.,* elongated stringer) kind.

In recent years, two concurrent changes in the steelmaking practice used to produce high strength line pipe have influenced the sulfide shape in the final product. The first is the use of aluminum additions to refine the ferrite grain size in order to lower the ductile-to-brittle transition temperature and increase the yield strength. The second is the use of controlled rolling at temperatures near the critical temperature to further refine the grain size. Unfortunately, both these processes are conducive to the formation of Type II MnS inclusions.

Sims[8] has explained how deoxidation increases the solubility of sulfur in the melt, and how increasing deoxidation results in a conversion from Type I MnS to Type II MnS. In semi-killed steels, the sulfides solidify before the rest of the melt and consequently assume a globular shape. In fully killed steels, however, the sulfides are among the last products to solidify. They freeze as thin films in austenite grain boundaries and between dendrites.

Controlled rolling is conducive to the formation of elongated MnS stringers because the deformability of MnS increases with decreasing temperature.[4,9] Also, Type II MnS is softer than Type I MnS.[8,10] Thus, controlled rolling of fully killed steels results in exceptionally long, thin MnS stringers in the final product. On the other hand, Type I MnS inclusions, which are harder than Type II MnS and furthermore are initially spherical, are rolled out into lozenge shapes in the final product rather than elongated stringers.

Recent work in Japan[11] shows a positive correlation between hydrogen cracking sensitivity and the total length of inclusions in a unit cross-sectional area. There appears to be no doubt that Type II MnS stringers strongly sensitize a steel to hydrogen cracking.

Steels intended for use in sour gas service are often specified to have a low sulfur content. For the steels reported here, sulfur content had no effect on cracking sensitivity over the range from 0.005 to 0.026%. The Japanese work just cited[11] suggests that reducing sulfur contents below 0.010 to 0.015% is not effective in further reducing cracking susceptibility. This is reportedly due to the fact that significant segregation of sulfur (along with phosphorus, silicon, and manganese) occurs in the "pipe" section of ingots even for very low sulfur steels. Although the remainder of the ingot is relatively less sensitive to hydrogen cracking, the sensitivity of the "pipe" section remains high. The larger the ingot, the more crack sensitive material there will be. In this regard, continuously cast steels, which have a very narrow segregation zone, appear attractive for use in sour gas service.

Although fully killed steels are more prone to hydrogen cracking than semi-killed steels, it is often desirable to make use of their higher impact and yield strengths, and at least two commercial steelmaking practices are available that considerably reduce their cracking sensitivity. The first is sulfide shape control by the use of rare earth metals, especially cerium, to achieve spherical sulfides. The mechanisms by which rare earth metal additions alter sulfide morphology have been discussed in the literature.[12,13] Some of the steel manufacturers with whom the authors are familiar are reluctant to use rare earth metal additions because control of amount of alloying element added to the melt is critical. Furthermore, many of them are reluctant to use rare earth metal additions for steels of strength levels below about API 5LX-60. Currently at least, sulfide shape control by rare earth metal additions is usually restricted to relatively high strength steels.

A second method of decreasing the susceptibility of fully killed steels to hydrogen cracking involves alloying with copper, and can be employed by most steel manufacturers. Other investigators[5,14] have reported that the addition of copper in amounts on the order of 0.25 to 0.30% eliminates cracking in otherwise highly susceptible steels in the standard hydrogen cracking sensitivity test. Copper additions less than 0.20% apparently have no effect. A recent hydrogen cracking sensitivity investigation of 24 line pipe samples from 5 heats, fully killed, and containing 0.26 to 0.30% copper showed no cracking in any of the samples.[15] In the testing program reported here, Sample No. 21 contained 0.27% copper, and although it was fully killed, it did not crack or blister. All the other fully killed steels tested exhibited widespread blistering and cracking.

The mechanisms by which copper additions reduce the hydrogen cracking susceptibility of steels are not fully known. They appear to be some combination of improved corrosion resistance and reduced ability of transient hydrogen to adsorb onto the steel surface.[11,14]

Copper additions in the order of magnitude needed to reduce hydrogen cracking susceptibility have no detrimental effects on the weldability of the finished product. In addition, tests made by the authors using plate from Sample No. 21 showed no measurable (*i.e.,* less than 0.001 mV) potential difference between the plate material and superimposed weld beads made from E-6010 and E-7018 electrodes with the samples immersed in brackish water. No fabrication, field welding, or subsequent galvanic corrosion problems between pipe and weld beads are expected if copper bearing steel is used for wet or sour gas service. In this regard, it should be recalled that Sample No. 17 contained 0.22% copper, and the sour gas pipeline from which it was taken has operated for 16 years without any of these problems.

Conclusions

1. The hydrogen cracking sensitivity test described in this paper reproduces the blistering and cracking observed to occur in service in wet, sour gas pipelines. Furthermore, it does so in a conveniently short time (96 hours), and with relatively simple equipment and sample preparation. It provides a quick method for separating very crack sensitive steels from very insensitive steels. Since it is a fairly qualitative test, however, different laboratories may obtain different results for moderately crack sensitive steels. Comparisons of results obtained by different laboratories should be made with caution.

2. The most important factor affecting the hydrogen cracking sensitivity of steels is the MnS inclusion morphology. Elongated (*i.e.,* Type II) MnS stringers render the steel very crack sensitive. Lozenge shapes (*i.e.,* Type I) MnS inclusions are much less harmful. Since all fully killed steels except those treated with rare earth materials necessarily contain Type II MnS inclusions, they are inherently more susceptible to hydrogen induced cracking than semi-killed steels.

3. Alloying with 0.25 to 0.30% copper eliminates or significantly reduces hydrogen cracking and blistering in fully killed steels without impairing their other properties.

4. Reducing the sulfur content below about 0.015% does not further reduce the susceptibility of line pipe steels to hydrogen cracking.

5. Although steels can be made that are relatively insensitive to hydrogen cracking, a completely immune steel is probably not feasible. Therefore, additional operational controls such as inhibitor injection should continue to be used in sour gas service. The use of crack insensitive steels should be viewed as insurance during times of operational upsets, and not as an alternative to good operational practice.

Acknowledgments

Many individuals and organizations have contributed to this paper. The authors are particularly grateful to the Arabian American Oil Company for permission to present data used in the paper; to The Welding Institute for performing the hydrogen cracking sensitivity tests; to Mr. H. C. Cotton of the British Petroleum Company for making the details of the test available to us; and to the following steel companies for extensive technical discussions and exchanges of information: Sumitomo Metal Industries Ltd., Nippon Kokan K.K., Nippon Steel Corporation, and Kawasaki Steel Corporation. J. G. Kerr and D. J. Truax of the Standard Oil Company of California have provided technical assistance and are presently running a further testing program at the Richmond Laboratories as part of a joint Aramco/Socal project. D. J. Moore made the statistical correlations.

References

1. Paredes, F., Mize, W. W. Oil and Gas Journal, Vol. 52 (1954) December.
2. Naumann, F. K., Spies, F. Praktische Metallographic, Vol. 10, p. 475 (1973).
3. Private Communication with H. C. Cotton (British Petroleum Company) (1975).
4. Moore, E. M., Wiesman, J. C. Discussions with Japanese Steel Manufacturers, October 7-11, 1974, Tokyo, Japan.
5. Private Communication with Sumitomo Metals Industries, Ltd. (1974).
6. Private Communication with J. G. Kerr (Standard Oil Company of California) (1974).
7. Minium, E. W. Statistical Reasoning in Psychology and Education, Wiley, p. 138 (1970).
8. Sims, C. E. Transactions AIME, Vol. 215, p. 367 (1959).
9. Private Communication with D. J. Truax (Standard Oil Company of California) (1974).
10. Private Communication with J. L. Robinson (The Welding Institute) (1974).
11. Myoshi, E., Tanaka, T., Terasaki, F., Ikeda, A. Hydrogen Induced Cracking of Steels Under Wet Hydrogen Sulfide Environment, Preprint of Paper Presented to 30th Annual Meeting, ASME (Pet. Div.), Tulsa, Oklahoma, September, 1975.
12. Kepka, M., Skala, J. Hutnicke Listy, Vol. 27, No. 7, p. 484 (1972). (Translated by V. E. Riecansky, Translation No. VR/98/72, Technical Translations, 8 High St., Linton, Cambridge, England.)
13. Wilson, William G. Optimum Rare Earth Additions for Improved Mechanical Properties, Paper 5/1, Proceedings of the Conference "Inclusions and Their Effect on Steel Properties," Leeds, England, September, 1974.
14. Private Communication with Nippon Kokan K.K. (1974).
15. Private Communication with Sumitomo Metals Industries, Ltd. (1975).

SECTION V

Special Topics

COPPER FAILURES

STEEL PILING

CONCRETE

MISCELLANEOUS

Early Corrosion Failures in Copper Heat Exchanger Tubing*

J. O. EDWARDS, R. I. HAMILTON, and J. B. GILMOUR
Physical Metallurgy Research Laboratories, CANMET,
Department of Energy, Mines, and Resources, Ottawa, Ontario, Canada

Early failure of 1.9 cm (0.75 inch) deoxidized copper, Al finned heat exchanger tubes in large commercial building heating and air conditioning systems was investigated. Pin-hole leaks in tubing of 3 of 7 units after 2 years exposure to about 4 psi steam were attributed to a lack of proper slope of the tubes, lack of response of a vacuum release valve, and cavitation attack as a result of cycling pressure-vacuum in the tubes. Pimples on exterior tube surfaces coinciding with locations of inside pits supported the cavitation premise, because internal pressures were too low to account for them. Another investigation involved early failure of chilled water tubes in a large building attributed to breakdown or contamination of antifreeze residues left after pressure testing. The tubes were stored outside with end stoppers for 12 months before installation. Attack by oxygen or carbon dioxide from condensation was not ruled out.

TWO UNRELATED FAILURES in recently installed air conditioning systems in large buildings are considered worthy of record because of the short service life and unusual features of the failures.

Failure in a Modulated Steam System

General

After about two years service, leaks were experienced in three of seven similar units in a large multistory building. The incidence of leaking was a function of the time the individual units had been in service, and it was considered that eventually all units might experience the same problem.

The "pin-hole" leaks occurred in 19 mm (3/4 inch) phosphorus deoxidized copper tubes (CDA 122 alloy) with aluminum fins, which form the reheater unit of a system comprising, in line: a steam preheater [7 C (45 F) winter], filter, cooling unit [16 C (60 F) summer], and a reheater unit [16 C (60 F) winter]. In fact, all units are "on line" both summer and winter with appropriate thermostatic valves modulating the supply of steam or chilled water. Low pressure steam at 69 kPa (10 psi) is provided from a 700 kPa (100 psi) line, but the pressure in the reheaters rarely exceeds 28 kPa (4 psi) because of the action of the control valves. A condensate drain is provided downstream, and there is a vacuum release valve designed to admit air in the event of condensation causing a partial vacuum in the tubes. Air passing over the reheater tubes is circulated to areas in the building where final temperature adjustment is achieved by individual room coils.

Examination

All sections of leaking tubes examined showed the following to a greater or lesser extent:

1. The insides of the tubes were covered with a black, adherent deposit, of variable roughness which, in some areas, showed "water line" markings, indicating condensate had lain on the bottom of the tubes.

2. A sharply defined "line of attack" could be seen, which was most pronounced at the 6 o'clock position but sometimes showed at the 12 o'clock, or at both.

*Voluntary manuscript submitted for publication December, 1976.

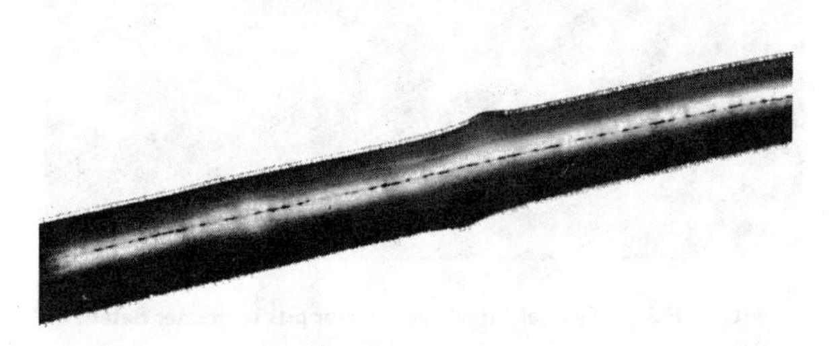

FIGURE 1 — Inside of leaking reheater tube showing well defined line of pits on bottom (6 o'clock) of tube.

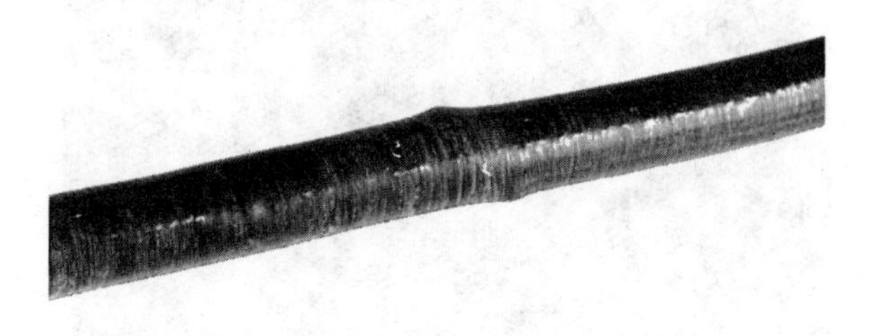

FIGURE 2 — Outside of tube section shown in Figure 1. Note exterior "pimples" corresponding to deepest interior pits.

3. In some areas, characteristic deep pits occurred in the "line of attack" at 6 o'clock, and these were often accompanied by corresponding "pimples" on the outside of the tube which could be readily seen when the aluminum fins were stripped off.

Figures 1 and 2 show the linearity of the attack, the pits on the inside, and the corresponding pimples on the outside. Figure 3 shows the pits in greater detail, illustrating their small diameter and their grouping to form "slots". Figure 4 is a cross section of a typical pit (sample nickel plated to preserve edge detail). Shallower pits occasionally observed at the 12 o'clock position had the same type of structure as Figure 4. There was no evidence of manufacturing deficiencies in the tubes in the form of laps, seams, inclusions, etc.

Discussion

From the deformed metal structure at the base of the pits shown in Figure 4 (also confirmed by microhardness testing), it is apparent that the damage was caused either by impingement or cavitation. In many cases, the small pits were quite discrete. There was little evidence of surface roughening, the 6 and 12 o'clock orientation effects do not directly relate to flow, and the medium is low pressure (and hence, low velocity) steam. Impingement, therefore, seems unlikely, so we are left with cavitation damage

FIGURE 3 – Typical "line" of interior pits in greater detail. 9X

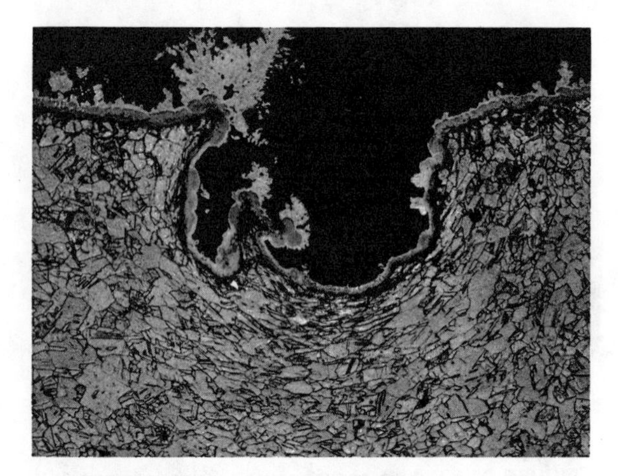

FIGURE 4 – Cross section of typical pits in reheater tube showing marked deformation of grain structure in the immediate area of the pit. (Sample nickel plated to preserve edge detail.) Etched in alcoholic ferric chloride. 165X

which was caused by the generation of bubbles of gas or vapor within a liquid in areas of low pressure and their subsequent collapse as they moved away to higher pressure areas. The collapse of these bubbles can cause a microjet of high velocity [460 m/s (1500 feet/s)] and high dynamic impact [1035 MPa (150,000 psi)[1]]. It is considered that such cavitation conditions could apply in the reheater units because of the absence of slope on the horizontal runs allowing condensate to accumulate, and the known vacuum generating effect of such condensation processes. Thus, depending on the functioning and dynamic response time of the thermostatic steam valve and vacuum breaker valves, etc., it is considered that alternating cycles of vacuum and pressurization could exist within the tubes causing bubble formation and collapse within the condensate layer. Because of the geometry of this accumulated layer, bubbles would form at the 6 o'clock position, and once pits had begun to form at this position, the pits would become preferential bubble sites, thus perpetuating the situation and leading to selective pitting and perforation.

The presence of the "pimples" on the outside of the tube corresponding to the location of the deeper pits on the inside is considered strong support for the cavitation hypothesis. How else could such pimpling be obtained with an internal steam pressure of only 28 kPa (4 psi), and pits essentially free of corrosion product? The hypothesis does not readily explain the similar but much less extensive damage observed on the 12 o'clock position of some tubes unless they were filled with condensate, and vapor bubbles had time to float to the top of the tubes where they would then collapse on the pressure cycle.

FIGURE 5 – Interior of cleaned tube from chilled water coil in each area of leaks showing extensive pitting and "rain drop" staining. 10X

The examination indicates that the mere presence of vacuum breaker valves in the system is no guarantee that vacuum conditions cannot exist at some point in the system for short periods. The presence of such transient conditions could be readily monitored by transducers and would be expected to vary with the season, *i.e.,* with the steam load on the tubes. Even so, it is probable that the extensive damage observed could have been avoided if the tubes had been installed with a positive slope to drain the condensate, as is commonly recommended.

Corrosion Pitting of Chilled Water Tubes

General

Samples of 19 mm (3/4 inch) copper tube from newly installed chilled water units of area air conditioning services from a large public building were examined. The units had been manufactured, pressure tested with antifreeze, drained, after which they were plugged to prevent ingress of foreign matter. They were wrapped in plastic and shipped for storage on-site for about 12 months. On installation, they were pressure tested and found to show a high incidence of random "pin-hole" leaks. Again, the copper tubes had exterior aluminum fins to form the heat exchange surface with air.

Examination

The interior tube surfaces were blue-black to "rusty", and showed no obvious defects corresponding to the leaks. After cleaning the tubes with dilute sulfuric acid, an overall mottled "rain drop" effect was apparent on the interior together with extensive pitting in some areas, as shown in Figure 5. Heaviest pitting coincided with the occurrence of leaks, and this was confirmed by

FIGURE 6 — Complete penetration of tube by pit showing presence of corrosion product, probably cuprous oxide. Bottom = tube interior. 55X

FIGURE 7 — Branching pit which probably also completely penetrated the chilled water tube. 55X

microexamination. Deep pits containing corrosion deposit (probably cuprous oxide) are shown in Figures 6 and 7. Samples submitted to SEM examination showed Cu and Fe in the corrosion deposits together with smaller quantities of P, S, and Ca. Corrosion deposit picked off a pitted area showed statistically significant amounts of Cl_2, probably from chlorides.

Discussion

The fact that the units had leaked immediately following installation showed that corrosion must have occurred in the storage period subsequent to manufacture. Units were stored outdoors in plastic wrapping under tarps with the ends of the tubes sealed with plastic plugs. As pitting is usually caused by conditions which are only mildly corrosive, it was thought initially that condensation within the units and a high O_2-CO_2 concentration in this condensate might have been the corrosive media. However, the number and depth of pits and the relatively short exposure period all indicated more severe conditions. It was subsequently found that immediately after manufacture, the units had been tested with an antifreeze solution and that "residual amounts remain inside the coil", according to the manufacturer's label.

As all the units had by that time been installed, filled with water, and pressure tested, there was no opportunity to examine a unit as it came from storage.

It is suggested, however, that the pitting arose from breakdown of the antifreeze inhibitors and/or the possibility of high O_2 and CO_2 concentrations in the liquid on the "wet" surface of the closed tubes. The Cl_2 detected by SEM also indicates that the antifreeze solution may have become contaminated with chlorides although this could also arise from chlorinated Ottawa water.

Solutions to the problem would appear to be thorough rinsing and drying of the tubes before plugging, coating the tubes with a more adequate inhibitor, or possibly shipping the tubes full of inhibited antifreeze to prevent the ingress of air.

Summary

Short service life corrosion failures of copper tubing in two heat exchange units, one heating and one cooling, were examined.

The heating unit showed unusual, highly distinctive, linear pitting on the inside confined mainly to the 6 o'clock position. The damage showed all the characteristics of cavitation failure and was attributed to some lack of response of the vacuum breaker valve allowing vacuum pressure cycles to be established, and the fact that there was no slope on the tubes to drain condensate.

The failure in the chilled water units, which was observed immediately after installation, was attributed to pitting corrosion during the prior 12 month storage when the tubes contained residual antifreeze from the manufacturers' tests.

Acknowledgments

The authors wish to acknowledge assistance freely given by various members of the staff in discussion, sample preparation, etc., and particularly K. M. Pickwick who performed the SEM analysis and examination.

References

1. ASM Metals Handbook, Vol. 10, Failure Analysis and Prevention (1975), p. 160-167, Liquid Erosion Failures.

Sea Water Corrosion of 90-10 and 70-30 Cu-Ni: 14 Year Exposures*

K. D. EFIRD, *The International Nickel Co., Inc., Wrightsville Beach, North Carolina*
and
D. B. ANDERSON, *The International Nickel Co., Inc., New York, New York*

Fourteen year data for 90-10 and 70-30 cupronickel alloys exposed in sea water at the F. L. LaQue Corrosion Laboratory, Wrightsville Beach, North Carolina are reported. Corrosion rates for both alloys in quiet and flowing as well as in the tidal zone tended to become linear after the first 4 years' exposure. Initially, corrosion rates for 90-10 tended to be much higher in flowing than in either quiet or tidal zone exposures, but at 14 years, rates in all environments were about the same, 0.05 mils per year. Similarly, 70-30 had high initial rates in flowing water, but at 14 years, rates were about the same for all three exposures, 0.03 to 0.08 mils per year.

THE Cu-Ni ALLOYS are becoming increasingly more important in marine engineering systems.[1,2] This is principally due to their good corrosion resistance, ductility, weldability, useful resistance to fouling, and general immunity from stress corrosion cracking in marine environments.

The complete solid solubility of copper and nickel results in a metallurgically *simple* system with an essentially homogeneous structure, and corresponding freedom from corrosion problems associated with heterogeneous alloys, *e.g.,* pitting, stress corrosion cracking, etc.

The alloy compositions in use today were developed over a number of years, with research efforts concentrating on the effects of iron additions.[3-6] These studies confirmed the beneficial effects of dissolved iron in the copper-nickel alloys.

Previous long-term corrosion data for 90-10 and 70-30 Cu-Ni were presented by May and Weldon,[7] Lennox, *et al,*[8] and Reinhart,[9] with all of these being less than 3 years' total exposure. Some 7 year corrosion data were presented by Anderson and Efird,[10] but treatment was not extensive.

Experimental Procedure

Tests were conducted at the Francis L. LaQue Corrosion Laboratory, Wrightsville Beach, North Carolina. The characteristics of sea water at this site are given in Table 1.

*Submitted for publication April, 1974.

TABLE 1 — Characteristics of Sea Water at the Francis L. LaQue Corrosion Laboratory, Wrightsville Beach, North Carolina

Major Characterization

	Max.	Min.	Avg.
pH	8.1	7.8	8.0
T, C	29	6	18
Cl^-, g/l	19.8	18.1	19.0
O_2, mg/l	9.3	5.0	6.4

Average Analysis, mg/l

Cations		Anions	
Na	10,006	Cl	19,000
Ca	398	SO_4	2510
Mg	1204	HCO_3	133
K	369	NO_3	1.2
Cu	0.015	PO_4	0.01
Fe	0.001	F	1.5
Zn	0.012	Br	61
		I	0.16

Hardness ($CaCO_3$) 5970 mg/l
Dissolved Solids 38,255 mg/l

Multiple specimens of 90-10 Cu-Ni and 70-30 Cu-Ni [4 x 6 x 1/4 inch (10 x 15 x 0.6 cm)] were used in this investigation with compositions and mechanical properties given in Table 2. Specimens were cut from a single plate of each alloy, and identified by both stencil and notch code to assure positive identification after exposure. The materials were cleaned before exposure by scrubbing with a pumice slurry, rinsing in distilled water, air-drying, and then weighed.

The specimens were exposed in three locations: (1) quiet sea water, (2) flowing sea water, and (3) the tidal zone. The specimens exposed in quiet sea water and the tidal zone were mounted vertically on Monel alloy 400 racks using porcelain insulators, and suspended beneath the laboratory wharf at appropriate levels. The

Reprinted with permission from *Materials Performance,* November 1975, 37-40, © 1975 National Association of Corrosion Engineers

TABLE 2 – Properties of the Alloys Tested

Alloy	Composition, %					
	Cu	Ni	Fe	Mn	Zn	Pb
90-10 Cu-Ni	Bal	10.21	1.74	0.29	<0.10	<0.02
70-30 Cu-Ni	Bal	30.68	0.61	0.45	0.05	0.01

Alloy	Mechanical Properties				
	Yield Strength		Tensile Strength		Elongation
	ksi	MN/m^2	ksi	MN/m^2	in 2 Inches
90-10 Cu-Ni	29.6	290	45.9	316	42%
70-30 Cu-Ni	32.4	318	58.7	404	43%

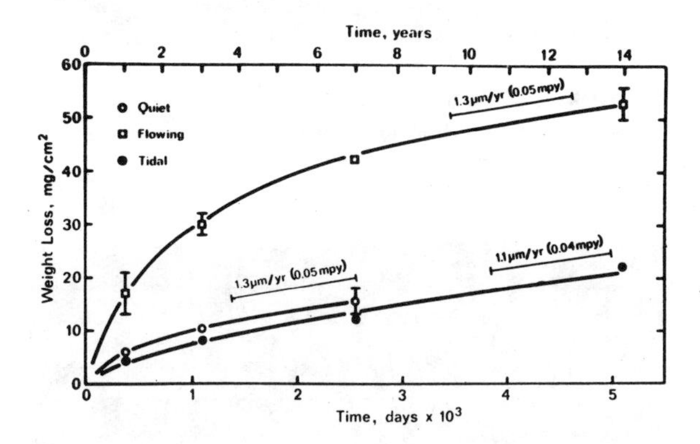

FIGURE 1 – Chronogravimetric curves for 90-10 Cu-Ni in quiet, flowing, and tidal zone sea water.

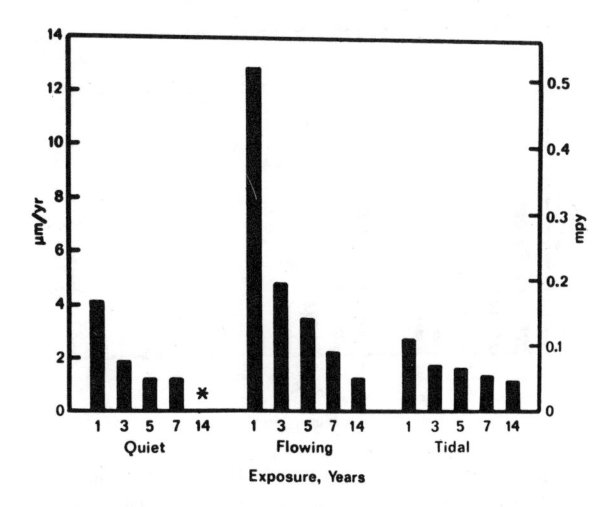

FIGURE 2 – The change in corrosion rate with time for 90-10 Cu-Ni in quiet, flowing, and tidal zone sea water. Rates calculated from the slope of the chronogravimetric curve.

depth of immersion for the quiet sea water rack ranged from 2 to 6 feet depending upon the tide, while the tidal zone rack was alternately immersed with each changing tide. Specimens for exposure in flowing sea water were mounted horizontally on racks constructed of 1/4 inch (6 mm) wide Micarta[(1)] strips. This exposure was in a covered trough designed to give a constant sea water velocity of 2 fps (0.6 m/s) past the specimens. Removals were made after 1, 3, 7, and 14 year exposures in all 3 locations.

After removal, all specimens were cleaned in 10% sulfuric acid, rinsed in distilled water, and air dried. Measurements of weight loss and depth of localized attack were made on all samples and corrosion rates calculated.

Results and Discussion

Corrosion Resistance

Corrosion data for 90-10 Cu-Ni in quiet sea water, flowing sea water, and the tidal zone are given in Figure 1 as chronogravimetric curves (weight loss vs exposure time). The initial weight loss in flowing sea water is much higher than the other two exposures, reflecting a lower rate of protective film formation under these highly aerated and somewhat turbulent flow conditions. However, it should be noted that after long exposure times, the corrosion rates for all 3 zones are similar. Unfortunately, the 14 year panels of 90-10 Cu-Ni in quiet sea water were lost; however, 7 year data for this exposure are available. The 14 year corrosion rates for 90-10 Cu-Ni, taken as the slope of the chronogravimetric curves, are 1.3 µm/yr (0.05 mpy) in flowing sea water and 1.1 µm/yr (0.04 mpy) in the tidal zone. Of specific interest is the time required for the corrosion rate to become linear. For 90-10 Cu-Ni, this occurs after approximately 4 years' exposure in quiet sea water and 8 years' exposure in the tidal zone. It is still decreasing and has not stabilized after 14 years' exposure in flowing sea water. This change in corrosion rate with time, taken from the curves, is shown graphically in Figure 2. The appearance of the long-term 90-10 copper-nickel exposure panels, after cleaning, is shown in Figure 3. Of particular note is the lack of localized attack on the samples after these long exposure periods.

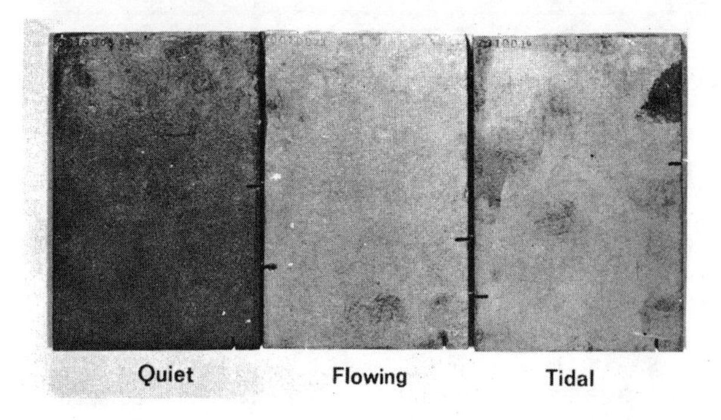

FIGURE 3 – 90-10 Cu-Ni panels after 7 years' exposure in quiet sea water and 14 years' exposure in flowing and tidal zone sea water.

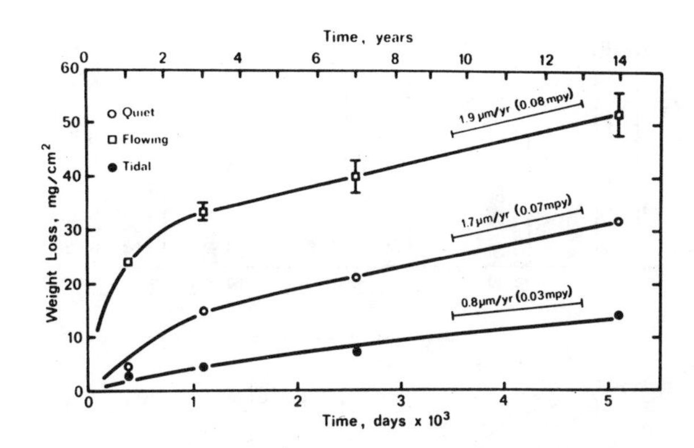

FIGURE 4 – Chronogravimetric curves for 70-30 Cu-Ni in quiet, flowing, and tidal zone sea water.

Corrosion data for 70-30 copper-nickel in quiet, flowing, and tidal sea water are shown in Figure 4. After 14 years' exposure, the

(1) Registered tradename, Conroy and Knowlton, Inc., Los Angeles, California.

Source: *Materials Performance*, November 1975, 37-40

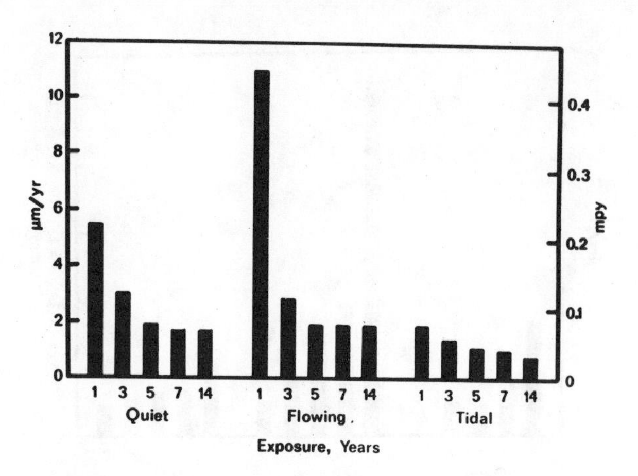

FIGURE 5 – The change in corrosion rate with time for 70-30 Cu-Ni in quiet, flowing, and tidal zone sea water. Rates calculated from the slope of the chronogravimetric curve.

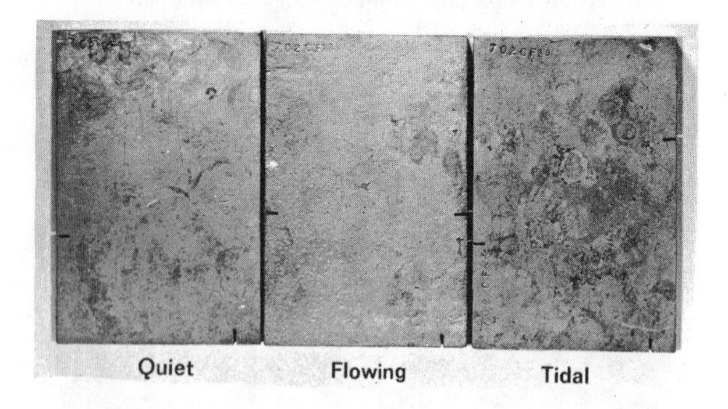

FIGURE 6 – 70-30 Cu-Ni panels after 14 years' exposure in quiet, flowing, and tidal zone sea water.

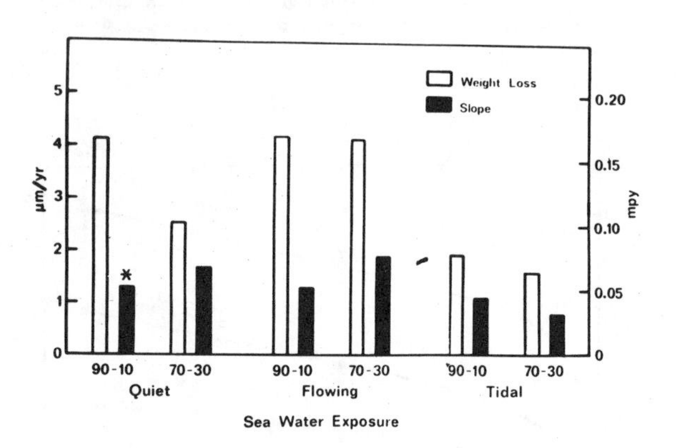

FIGURE 7 – Fourteen years' corrosion rate data.

corrosion rate of 70-30 Cu-Ni, taken from the slope of the curve, is 1.7 µm/yr (0.07 mpy) for quiet sea water, 1.9 µm/yr (0.08 mpy) for flowing sea water, and 0.8 µm/yr (0.03 mpy) for tidal zone exposures. Stabilization of the corrosion rate occurred after approximately 14 years' exposure in quiet sea water and flowing sea water, but is still decreasing slightly after 14 years' tidal zone exposure. This is shown diagrammatically in Figure 5. The appearance of 70-30 Cu-Ni samples after 14 years' exposure in all 3 zones is shown in Figure 6. There was no significant localized attack.

The corrosion rates after 14 years for 90-10 and 70-30 Cu-Ni in quiet, flowing, and tidal zone sea water are compared in Figure 7.

FIGURE 8 – 70-30 Cu-Ni after 14 years' exposure in quiet sea water before cleaning. Light fouling has occurred consisting of anomia, serpula, crisia bryozoa, and barnacles.

FIGURE 9 – 90-10 Cu-Ni (left 2 panels) and 70-30 Cu-Ni (right 2 panels) after 14 years' exposure in flowing sea water before cleaning. Very light fouling consisting of serupla, anomia, and small barnacles has occurred.

Comparison is also made between the corrosion rate determined from the weight loss measurements (average corrosion rate over the total exposure period) and as determined from the slope of the chronogravimetric curve (the corrosion rate at a specific point in time). The greatest difference in these two methods of corrosion rate determination occurs when the initial weight loss is very high, as was observed in these flowing sea water exposures. Thus, while the steady state corrosion rates are equal, and in some cases less in flowing sea water than in quiet sea water or tidal zone exposures, the corrosion rate taken from total weight loss measurements is considerably higher. The steady state values calculated from the chronogravimetric curves are more applicable to accurately predict metal loss during continuing exposures.

Fouling Resistance

Of special interest, particularly due to the low corrosion rates observed after long time exposure, was the condition of the specimens on removal with respect to fouling. Uncleaned samples of 70-30 Cu-Ni exposed 14 years in quiet sea water are shown in Figure 8. The 90-10 Cu-Ni samples were unfortunately lose. Even on the 70-30 Cu-Ni samples, however, fouling is minimal after 14 years' quiet sea water exposure, consisting of serpula, anomia, small barnacles, and crisia bryozoa.

The as-removed condition of both 90-10 and 70-30 Cu-Ni samples after exposure in flowing sea water for 14 years is shown in Figure 9. Here again fouling attachment is minimal with barnacles,

FIGURE 10 — 90-10 Cu-Ni (left) and 70-30 Cu-Ni (right) after 14 years' exposure in the tidal zone, before removal from the exposure rack, showing heavy oyster growth.

FIGURE 11 — Panels in Figure 10 after removal from the rack, before cleaning. The 90-10 Cu-Ni (left) has one barnacle attached, with no evidence of oyster attachment. The 70-30 Cu-Ni (right) has several small barnacles and mussels attached, and evidence of some oyster attachment.

anomia, and some serpula present, although none developed to significant size.

The fouling condition of 90-10 and 70-30 Cu-Ni after 14 years' tidal zone exposure is shown in Figures 10 and 11. In Figure 10, the samples, before being removed from the exposure racks, show a great deal of oyster fouling. However, after removal from the racks (Figure 11), it can be seen that most of the oysters are not attached but had bridged over from the Monel alloy 400 support racks. Other fouling consists of small serpulids and barnacles. The differences in the fouling communities which developed in the 3 exposure conditions reflect abilities of specific organisms to adapt to natural ecological variables, particularly intermittent versus continuous immersion and the presence or absence of sunlight.

Conclusions

1. The measured corrosion rates of 90-10 Cu-Ni are 1.3 μm/yr (0.05 mpy) after 7 years' exposure in quiet sea water, 1.3 μm/yr (0.05 mpy) after 14 years' exposure in flowing sea water, and 1.1 μm/yr (0.04 mpy) after 14 years' exposure in the tidal zone.
2. Stabilization of the corrosion rate for 90-10 Cu-Ni occurs after approximately 4 years' exposure in quiet sea water, and 8 years' exposure in the tidal zone, and is still decreasing after 14 years' exposure in flowing sea water.
3. The measured corrosion rates for 70-30 Cu-Ni after 14 year exposures are 1.7 μm/yr (0.07 mpy) in quiet sea water, 1.9 μm/yr (0.08 mpy) in flowing sea water, and 0.8 μm/yr (0.03 mpy) in the tidal zone.
4. Stabilization of the corrosion rate for 70-30 Cu-Ni occurs after approximately 4 years' exposure in quiet sea water and flowing sea water, and is still decreasing after 14 years' tidal zone exposure.
5. The initial corrosion rate in flowing sea water for both 90-10 and 70-30 Cu-Ni is higher than for exposures in quiet sea water and the tidal zone, but stabilizes at equally low values after long exposure times.
6. The use of the slope of the chronogravimetric curve provides more useful corrosion rate data than the total weight loss method, particularly where the initial weight loss is high followed by significant decrease in the rate of weight loss.
7. 90-10 and 70-30 Cu-Ni continue to provide useful fouling resistance after 14 years' exposure in quiet, flowing, and tidal zone sea water.

References

1. Burlow, C. L. Naval Engineers Journal, Vol. 77, p. 470 (1965).
2. Hunt, J. R., Bellware, M. D. Transactions of the Third Annual MTS Conference, San Diego, California (1967).
3. Tracey, A. W., Hungerford, R. L. Proceedings of ASTM, Vol. 45, p. 591 (1945).
4. Bailey, G. L. J. Inst. Metals, Vol. 79, p. 243.
5. Stewart, W. C., LaQue, F. L. Corrosion, Vol. 8, p. 259 (1952).
6. Krafak, Karla, Franke, Erich. Werkstoffe u. Korrosion, Vol. 4, p. 310 (1953).
7. May, T. P., Weldon, B. A. Proceedings: Congress International de la Corrosion Marine et des Salissures, Cannes, France, p. 141-156 (1965).
8. Lennox, T. J., Jr., Peterson, M. H., Grover, R. E. Materials Protection and Performance, Vol. 10, p. 31 (1971) July.
9. Reinhart, F. M. Naval Civil Engineering Lab., Technical Note No. N-961, AD-835-104, 58 p. (1968).
10. Anderson, D. B., Efird, K. D. Proceedings: Third International Congress on Marine Corrosion and Fouling, p. 264-276 (1972).

Source: *Materials Performance*, November 1975, 37-40

Figure 1. H-PILINGS removed after 18-year exposure underground are laboratory tested for extent of corrosion. The 139 foot beams were tested in three sections to demonstrate corrosivity in various strata. At left, the section exposed at water table zone in fill material; center, the section in clay soil at —30 feet and right, section exposed to coarse sand and gravel stratum at —110 to —126 feet.

Long Exposure Underground Has Little Effect On Steel Pilings

Summary

The strength and useful life of steel piling driven into the ground are not significantly affected by corrosion, according to a recent survey by the National Bureau of Standards. Steel sheet pilings used in construction of offshore drilling rigs and dock curtain walls exhibit the same resistance to corrosion over long periods of exposure. These results contrast sharply with earlier findings which showed that metal structures such as pipelines, buried in backfilled trenches and excavations, exhibit corrosion in varying degrees. These disparate results are attributed to the differences in oxygen content in the "undisturbed" piling environment and in the "disturbed" pipeline environment.

EXTENT OF CORROSION in driven steel piling was recently investigated by the National Bureau of Standards. Data obtained from different geographical areas show that the strength and useful life of the driven piling are not materially affected by corrosion.[1] These findings are in sharp contrast to those of earlier corrosion studies in which metal specimens such as pipelines that are buried under "disturbed" soil exhibit varying amounts of corrosion.[2]

The present survey was conducted by Melvin Romanoff of the Bureau's metal reactions laboratory and was sponsored by the American Iron and Steel Institute. The U. S. Army Corps of Engineers, which maintains a number of floodwall and dam installations, cooperated in the study.

Previous studies of the corrosivity of soils toward metals have been restricted to the behavior of specimens in trenches or excavations which were dug and backfilled after installation. These studies revealed that the corrosion of ferrous metals varied from negligible to severe in different soil environments.[2] The major cause of corrosion was a nonuniform distribution of oxygen and moisture along the surface of the buried metal. This resulted in the formation of oxygen concentration cells that initiated an electrochemical corrosion process.

Soil Properties Significant

Earlier investigations revealed at least a rough correlation between corrosion and soil properties such as pH, resistivity and chemical composition. The present survey was made to determine whether these properties also affected the behavior of steel piling used to resist lateral pressures from earth and water or to transmit loads to lower levels.

For examination of entire lengths of piling, some of which had been exposed from 32 to 40 years, H-piling used as load bearing foundations and sheet piling used as structural members of dams, several floodwalls and bulkheads

Reprinted with permission from *Corrosion*, December 1962, 45-47, © 1962 National Association of Corrosion Engineers

were extracted from eight locations (See Table 1). At floodwall and dam installations whose existing structures could not be disturbed, adjacent test holes were excavated for inspection of piling in service in various soil environments. Pile sections and soil samples were tested subsequently in the laboratory (see Figure 1).

The inspections provided information on the behavior of steel piling over a wide range of conditions. Backfill material varied in content from riprap, cinders and slag to combinations of sand, silt and clay. Soils around driven piling varied from well drained sands to impervious clay. Soil resistivities ranged from 300 ohm-cm (indicating a high concentration of soluble salts) to over 50,000 ohm-cm (indicating absence of soluble salts). Soil pH varied from 2.3 to 8.6.

Localized Pitting Found

Limited corrosion in the form of highly localized pitting occurred in some cases below the water table zone. Measurements of a number of the specimens revealed only small or negligible reduction in wall thicknesses. The pitting type corrosion is of major importance in pipelines or other metal structures designed to carry fluids, but in piling structures is not as serious as a uniform reduction in thickness of a large area of structural surface.

Sections of piling exposed to fill soil above or in the water table zone appeared to be most vulnerable to corrosion. However, only localized pitting was found and such sections were accessible for protective measures.

The survey showed that soil environments which normally are corrosive to specimens buried in "disturbed" soils are not corrosive to steel piling driven into "undisturbed" soils. Apparently these soils are deficient in oxygen at levels above the water table zone or ground line and steel piling is not corroded in any of the different soil conditions.

Sheet steel piling which had been driven since 1933 and pulled for replacement had so little corrosion evident on the surface that it was re-used in construction of a dock curtain wall as shown in Figure 2. The piling, consisting of 25,550 feet of steel sheet, was unprotected during the 28 years of service.

It is concluded that the type, drainage, resistivity, pH or chemical composition of soils is of no practical value in determining the soils' corrosiveness toward steel piling driven into the ground and such data should not be used to estimate the length of service of piling installed in this manner.

References

1. Melvin Romanoff. Corrosion of Steel Piling in Soils. *J. Research NBS*, **66c** (Eng. & Instr.) No. 3, 223-244 (1962) July-Sept.
2. Melvin Romanoff. Underground Corrosion. NBS Circular **579**, Superintendent of Documents, U. S. Government Printing Office, Washington 25, D. C.; also Underground Corrosion, a summary of 45 years of research. *NBS Tech. News Bull.*, **42**, 181 (1958).

TABLE 1—Steel Pilings Extracted From Location

Structure	Location	Age (Years)	Type	Length Exposed Below Ground (ft.)
Bonnet Carre Spillway.....	New Orleans, La.	17	H	122
Test Piling...............	Sparrows Point, Md.	18	H	136
Corps of Engineers Lock and Dam No. 8.........	Ouachita River, Ark.	40	Sheet	15
Grenada Dam Spillway....	Grenada, Miss.	12	Sheet	14
Sardis Dam Outlet........	Sardis, Miss.	20	Sheet	3.5
Simpson-Long Bridge Retaining Wall.........	New Orleans, La.	32	Sheet	33
Wilmington Marine Terminal, Pile Jetty........	Wilmington, Del.	23	Sheet	60 & 100
Lumber River Bridge, Cofferdam..............	Boardman, N.C.	37	Sheet	17.5

Figure 2. AFTER 28 YEARS' EXPOSURE to brackish water without a corrosion resistant coating, this piling was re-used to build a dock curtain wall at Wilmington, Delaware. Examination proved the piling had lost little or none of its strength from corrosion. It was scraped and redriven without protective coating.

New Data Show That

Steel Has Low Corrosion Rate During Long Sea Water Exposure

C. P. Larrabee
retired
Applied Research Laboratory
United States Steel Corporation
Monroeville, Pennsylvania

Introduction

THE CORROSION RATE most commonly used for carbon steel in sea water is 5 mils per year (0.005 inch) and may be considered "linear with time."[1] Since publication of these data and statements in 1948, carefully controlled tests[2] over an eight-year period showed a decrease of corrosion rate with time: 6 mils per year after a one-year exposure and 3.2 mils per year after eight years. Results are given below of an investigation of steel piles that had been immersed in sea water for 23.6 years.

A steel pier near Santa Barbara, California (Figure 1), used for offshore oil wells, was being dismantled because new methods permitted wells to be drilled from the shore. Those sections of the 10-inch H piles that had been exposed below low tide and five feet above the ocean bottom were reported by engineers to be in such good condition that they were being sold for reuse. The investigation described below was made to determine quantitatively the average corrosion rate of these piles in that zone.

Field and Laboratory Work

Thickness measurements were made with a 0.25-inch-diameter, flat-end-shaft micrometer in the field of the flanges of each of 20 piles at a

. . . average 2 mils per year for first 20 years, then about 1 mil per year as subsequent rate

Reprinted with permission from *Materials Protection*, December 1962, 95-96,

Figure 1—Pier at Santa Barbara, Cal., during dismantling operations. Steel pilings had been exposed to unpolluted sea water for over 23 years but were in such good condition that they were being sold for reuse.

random distance from one end. Each flange was measured at a distance of 3 inches from each of the four edges. A piece, 14 to 18 inches long, was cut from one end of each of nine piles, returned to the Applied Research Laboratory, cleaned in a sodium hydride bath and weighed. Then thickness measurements with the flat-end micrometer were made on both the flanges and webs.

A 16-inch long piece from the same purchase of piling was found encased in concrete. The concrete and intact rolling scale also were removed by sodium hydride. The original thickness and weight per linear foot of the piling were thus determined. The losses of the nine pieces were calculated after machining their ends perpendicular to their edges to obtain an accurate weight-per-linear-foot measurement.

Results and Discussion

By statistical methods, it was determined that the thickness measurements, as made in the field, were sufficient in number and reproducibility to establish the average thickness of 95 percent of the H piles within ±20 mils.

The average corrosion rates calculated from the losses of thickness and the rate as determined by loss of weight are given in Table 1.

Conclusion

Many investigations have shown that steel immersed in sea water loses several mils during the first few years of exposure. From these previous observations and from the present data obtained as described here, it seems conservative to assume that steel structures immersed in unpolluted sea water will lose about 40 mils from each surface so exposed (average two mils per year) during the first 20 years and that the subsequent corrosion rate will be about one mil per year.

References

1. H. H. Uhlig. Corrosion Handbook, pp. 383-388. John Wiley & Sons, Inc., New York City, 1948.
2. Corrosion of Metals in Tropical Environments Part 3. Naval Research Laboratory, Report 5153.

TABLE 1—Average Corrosion Rate (mpy) of Steel Piling Immersed for 23.6 Years in Sea Water

	Measured Loss of Thickness	Loss of Weight
In field, flanges of 20 piles..........	1.53 ± 0.23	....
In Laboratory, flanges of 9 piles.....	1.34	....
In Laboratory, webs of 9 piles.......	1.72	....
In Laboratory, weighed average*....	1.46 ± 0.25	1.73 ± 0.21**

* Area relationships of flanges and web taken into account.
** Nine piers, each over 14 inches long.

Offshore Platform Shows

Corrosion Rate of Carbon Steel In Gulf of Mexico Exposure*

SALVAGE OF an offshore platform in the Gulf of Mexico provided an opportunity for close inspection of corrosion damage to carbon steel which had been exposed in 20-foot water for approximately nine years.

A vertical section 1 by 12 feet from a 24-inch OD steel conductor pipe was used to check the corrosion rate. This section included surfaces exposed from —6.0 inches to +11.5 feet mean Gulf level. A scissors caliper with a dial indicator was used to measure the average or effective thickness across the removed section.

The average corrosion rates taken at 1 to 12-foot elevations (Figure 1) are considered to be average rates for uncoated carbon steel not subjected to mechanical damage such as barges, boats, etc., and submerged in the Gulf of Mexico.

Maximum average corrosion rate (0.021 ipy) was from 9.5 to 10 feet mean Gulf level. On submerged dolphin piling templates which were not cathodically protected, maximum corrosion rate was 0.018 ipy.

★ Revision of a paper titled "Corrosion Rate of Carbon Steel in the Gulf of Mexico" by Dean Patterson, Phillips Petroleum Co., Bartlesville, Okla., submitted April 6, 1959, for publication.

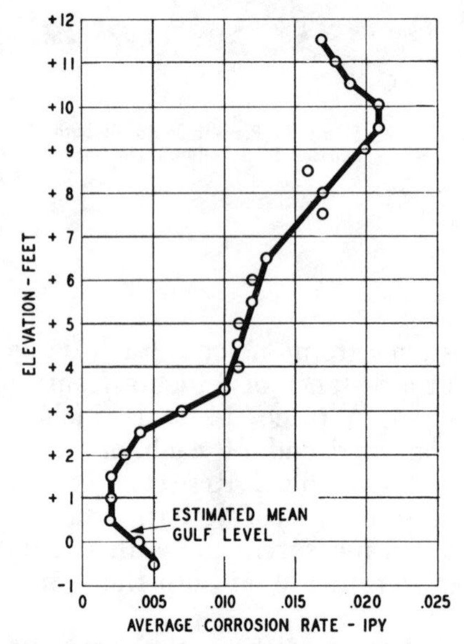

Figure 1—Corrosion graph showing relation of corrosion rate of 24-inch carbon steel conductor pipe exposed on an offshore platform in the Gulf of Mexico.

Splash Zone Sheathing Provides Long Term Protection

With increasing demands on the resources of the nation and the world, protection of offshore structures from the ravages of the marine environment is becoming more and more important. The splash zone region is receiving special attention of corrosion engineers at the design stage of many structures.

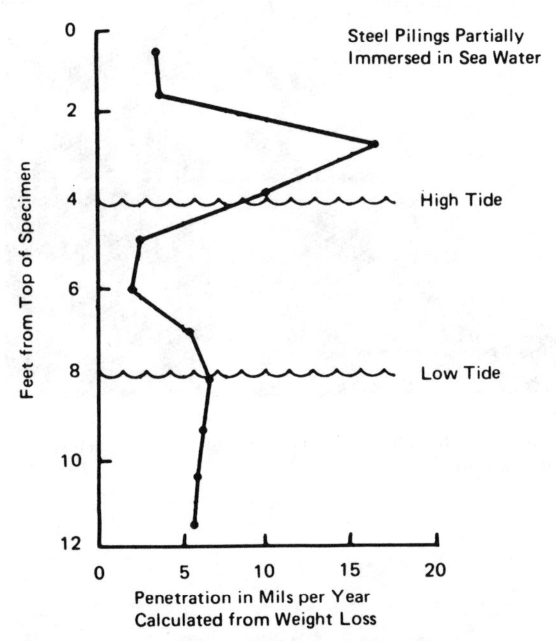

Figure 1 – Corrosion in the splash zone contrasted to the remaining portions of structure in marine environment.

Corrosion rates of steel in the splash zone region may be as high as 55 mils per year (mpy) in the Gulf of Mexico and twice that rate at Alaska's Cook Inlet. Figure 1 shows severe corrosion at the splash zone compared with other portions of a steel structure.

For more than 20 years, Monel* alloy 400 sheathing has provided protection to steel structures in the highly corrosive splash zone. One of the first installations incorporating the nickel-copper sheathing was a drilling platform built in 1949 offshore Louisiana. This structure used 0.062-inch thick sheet material. Numerous units were built during the 1950's and 1960's using both 0.050-inch and 0.062-inch thick Alloy 400 welded directly to the steel leg structures. Figure 2 shows the condition of the alloy sheathing on one of these drilling rigs located which was sheathed in 1951. The rig is located in the Gulf of Mexico and is still receiving adequate protection from the nickel-copper alloy sheathing.

After 15 years of service protecting pier structures at INCO's Francis L. LaQue Corrosion Laboratory at Wrightsville Beach, N. C., (front cover photo), the alloy sheathing exhibited a corrosion rate of 0.1 mpy. Figure 3 shows sheathed and unsheathed areas of the same type of steel piling after 15 years' exposure.

Today, 0.018-inch thick nickel-copper alloy sheathing can be attached to structures by mechanical fasteners. Experiments at the LaQue Laboratory with sheathing applied by both welding and fasteners indicate that it is unnecessary to completely seal the steel piling from the marine environment in the splash zone region. The alloy

*A tradename of The International Nickel Co., Inc., New York, N. Y.

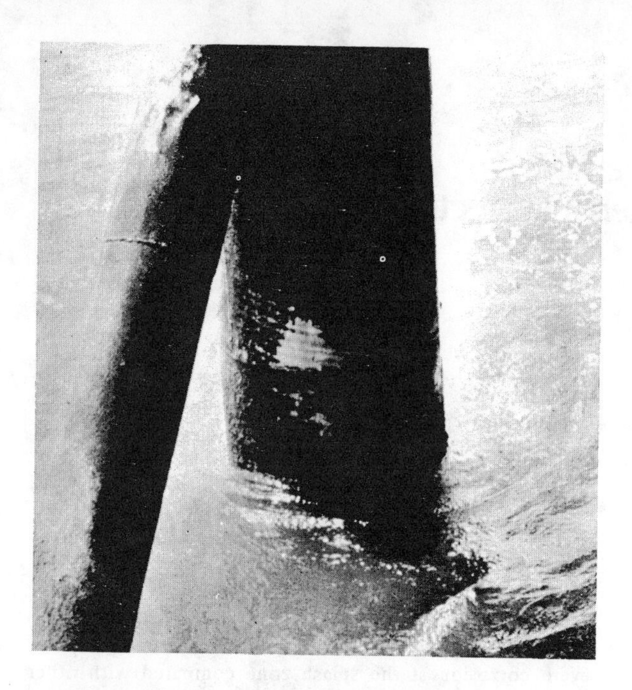

Figure 2 – Condition of nickel-copper alloy sheathing on drilling rig in the Gulf of Mexico. Sheathing applied in 1951.

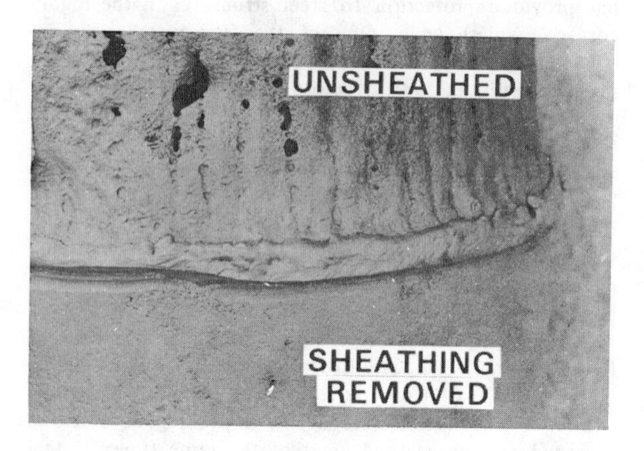

Figure 3 – Sheathed and unsheathed areas of steel piling after 15 years of exposure.

sheathing prevents the removal of surface oxides by mechanical and abrasive action of the waves. The oxide prevents further corrosion of the steel underneath the sheathing and provides a natural seal between the piling and sheathing.

The marine contractor has the option of using mechanical fasteners to attach the metallic sheathing without welding. A typical method is illustrated in Figure 4 where Alloy 400 sheathing is being attached to a steel structure with explosively driven rivets.

According to INCO's Huntington Alloy Products Division, for new or unprotected existing structures, a cold, bitumastic coating should be applied to steel piling before sheathing to offer additional protection to the steel, allow for a tighter fit, and overcome any pulsing action from the ocean behind the structure.

Figure 4 – Sheathing being attached to a steel structure with explosively driven rivets.

Investigation of Corrosion in the Steel H-Piles Supporting the Richmond-San Rafael Bridge

BEN BALALA, *California Division of Bay Toll Crossings, San Francisco, Calif.*

Engineering literature is full of daring and novel designs. A large share of the literature describes in detail the execution of innovations of note. It is a great deal harder to ascertain the degree of success or failure of such innovations than it is to find a description of the execution. Such innovations may involve calculated risks probably in some proportion to the amount of time and money that they save, and some of the risks may be inadvertent. Finally, of course, such innovations are recognized as the path of progress. The final measure of success or failure of any design, however, is the length of the service life of the construction product. Short of failure, predictions of projected life are appraised by periodic inspection during use. This is the story of a method of appraisal of the Richmond-San Rafael Bridge substructure.

THE SUBSTRUCTURE of the Richmond-San Rafael Bridge was constructed in 1953 to 1956 somewhat by the same general method as two major bridges on the Eastern seaboard of the US in the late '40s. This method was very economical when compared to the caisson type of construction for foundations bearing at this depth. The piers of the Richmond-San Rafael Bridge are supported on piles driven to bed rock or deep gravel beds to tip elevations as much as 250 ft below mean sea level.

The construction was performed in the following sequence:

1. An excavation was made in the bay bottom, usually mud, to a depth of 10 to 12 ft.
2. Temporary piles (untreated wood) were driven to hold and support an underwater template (base grid) of precast concrete.
3. Steel H-piles were driven through and grouted into the base grid which was 1 ft thick and reinforced.
4. The first part of the shell which formed and contained subsequent tremie concrete fill was placed and partly filled with approximately 5 ft of tremie concrete to develop enough bond on the steel H-piles to carry the subsequent construction loads.
5. The balance of the forms reaching above the water surface were then placed and followed by tremie fill to above the surface of the bay. Construction then proceeded in accordance with normal overwater procedures. The underwater forms were structural steel, which was considered expendable, or precast concrete which contained the reinforcement required by the pier design.

The substructure contract also provided for an item of pier backfill material to fill the remainder of the excavation in the bay bottom, but actual backfill was deferred until the action of the tidal currents on the new construction could be studied. When scour developed to the stage where 3 to 5 ft of water was reported under various base grids by the diver, the backfill operation was begun, placing graded rock with a maximum size of 3 inches to a minimum height of 3 ft above the base grid and at its natural slope of repose to the sides of the excavation.

Early Corrosion Study

During the construction of the piers, when it was determined that a pile corrosion problem was present, a simple system of measuring electrical potentials was installed on the advice of consulting corrosion engineers at six piers. These devices were installed before the broken stone backfill was placed. In a typical installation, one electrical conductor was fastened to a foundation pile, another conductor to a steel pipe buried in the bay mud, and a third to a steel pipe suspended in the bay water. In addition, a section of steel bar cut from a steel pile was suspended beneath the precast concrete in which the steel piles were embedded.

In the opinion of the consulting corrosion engineers in 1956, from the fact that the electrical potentials observed among the three conductors decreased appreciably following the backfill operations, it was logical to assume that any progressive corrosion progress had been arrested.

Reprinted with permission from *Materials Protection and Performance*, September 1972, 30-32, © 1972 National Association of Corrosion Engineers

The electrical potential readings were not continued over the intervening years by the maintenance forces, and maintenance efforts were confined to maintaining the broken stone backfill to the necessary surface configuration. Relatively small quantities of broken stone backfill were added twice during the 15 years following completion of construction.

Present Study

During the 15-year interval, the substructure for the high level spans of the San Mateo-Hayward Bridge was constructed in a similar manner. However, when preliminary designs were considered for the Southern Crossing of San Francisco Bay in 1969 and 1970, some of the steel bars cut from steel pile segments suspended under the Richmond-San Rafael Bridge piers were recovered, washed, calipered and weighed disclosing loss of section as high as 15%. As a consequence, a committee was formed representing the Construction, Design, Maintenance and Planning Branches of the Division of Bay Toll Crossings with the writer as chairman to study the problem of corrosion of the steel H-piles under the Richmond-San Rafael Bridge.

A cursory examination of the plans showing cylindrical pier bases and the knowledge that the broken rock backfill was simply dumped into place would show that no rock was expected to be placed under the pier base except that which rolled in from the edge. The space between the concrete pier base and the excavation before backfill varied but was on the order of 5 ft in the most extreme cases.

A preliminary attempt was made to appraise the problem by sending down a diver in June 1970 to inspect the backfill with the thought that some piers might have the backfill in such a condition that the diver could enter under the pier and caliper the piles. No such areas were found. However, with a minor amount of excavation at an existing low spot in the backfill, it was possible to expose one pile where calipers and tactual inspection disclosed a pit near the edge of one flange.

This pit coupled with the loss from the recovered steel bars was sufficient to result in a committee consensus that a more thorough inspection was necessary.

Inspection Contract

The immediate problem of inspection was approached as the inspection of two piers. One was Pier 57 on which backfill had periodically required replenishment over the 15-year period. The other was Pier 27 on which no further backfill had been placed following the original contract in 1956.

The inspection under a contract awarded April 1971 to the James Marsh Co. of San Carlos, Calif. was performed by digging a hole tnrough the backfill to an elevation below the originally excavated bottom alongside the pier and jetting material from around the piles to be inspected into the pit thus provided. Since the turbidity of the water precluded visual examination, a thickness survey was made on the piles by ultrasonic methods which were checked manually by calipers used in conjunction with a step gauge.

There were six piles investigated, three at each of the two piers selected, Pier 27 and Pier 57.

The length of piling selected for investigation was from that point unaffected and embedded in the concrete of the pier to a point several feet below the elevation of the original broken rock backfill. This was a length of approximately 9 ft.

Inspection Findings

At Pier 27, the bottom was mud, and the space underneath the center of the cylindrical pier base was virtually filled with mud. This indicates that either the silt had settled out of the water under the protection of the broken rock backfill or that the broken rock backfill had squeezed the mud under the pier during the placement operation. It was determined from the original construction records that this pier base had been backfilled virtually as soon as construction had been completed.

At Pier 27, very little, if any, loss of metal was found on the piles. The average thicknesses are shown in Table 1. The piles were covered with a tight, hard coating of oxidized metal or rust.

An interesting sidelight developed during the 1970 diver's inspection. The diver reported that he found warm water trapped underneath the pier in the recess in the slot through which the pile had been driven. During the 1971 inspection, the diver carried a recording maximum-minimum thermometer to record the temperature of the normal bay water and the temperature in the pile slot. A difference of four degrees Fahrenheit was recorded. A plausible explanation of the temperature difference is that the pile is embedded in a mass of concrete which is still hydrating. The heat is conducted out of the mass by the steel pile which heats the water. The warm water in turn is trapped in the pile slot.

At Pier 57, the bottom was mostly broken stone, and it was only with great difficulty that the diver was able to expose the piles to a depth of 7 ft below the pier concrete. The backfill appeared to be within a foot or two of the concrete at the middle of the circular pier base. Loss of section of 7 1/2% was recorded as almost uniform from the concrete to the bottom of the length exposed. This pier had stood unbackfilled for about a year after completion.

Further Findings

In studying the data relating to metal loss, it is necessary to realize that the rolled pile sections nominally .616 inches thick, are subject to a ±2.5% mill tolerance and that the ultrasonic recordings accumulated are not considered reliable in spite of a control steel tab 1/2 inch thick used for calibration. This is ascribed partly to equipment difficulties and partly to the difficulty of underwater

TABLE 1 – Average Thickness in Inches
Column Headings Indicate Dates of Readings Taken in 1971

		Pier 57					Pier 27	
		Ultrasonic		Caliper			Ultrasonic	Caliper
Pipe	Station	5-27	6-7	5-27	6-7	6-22	6-3	6-22
1	1	.469				.555	.586	.617
	2	.464					.581	
	3	.471				.532	.582	.607
	4	.468					.578	
	5	.472				.535	.579	.615
2	1	.457				.582	.577	.598
	2	.458					.574	
	3	.461				.585	.567	.616
	4	.457					.574	
	5	.459				.598	.569	.612
3	18" ir conc.	.463	.600				.566	
	A	.465	.600		.560		.571	
	B	.454	.598		.560		.562	
	1	.471	.564	.554	.540	.600	.567	.602
	2	.476	.501	.525	.545		.567	
	3	.478	.469	.520	.545	.562	.567	.615
	4	.460	.461	.540			.568	
	5	.480	.456	.530		.560	.561	.618

work. The two sets of caliper readings were made by two different divers, and the earlier readings are considered more reliable. Accordingly, the estimated loss of 7 1/2% is considered to be an order of magnitude figure. If the currently continuing readings indicate that this is an ongoing rate and not something that occurred before the pier was backfilled, some remedial measures must be considered, and other piers will also require investigation.

When the actual pile measurements at Pier 57 were studied with the corrosion engineer and showed no return to full section as expected at the bottom of the excavation, a reason was sought in the resident engineer's report for the substructure contract completed in 1955. The corrosion engineer stated that the only explanation for this phenomenon could be that the pier site had been excavated and backfilled, and the corrosion would be present in the backfill and not in the undisturbed foundation material. This was completely confirmed by the resident engineer's 1956 report which showed an upward adjustment of the footing elevation at three piers due to rock which was higher than had been expected. Part of this adjustment was made by backfilling the over-excavation with quarry waste before performing most of the pile driving. Three piers were reportedly so adjusted due to rock which was higher than expected.

Instrumentation

From the findings which briefly showed no loss at Pier 27 and considerable loss at Pier 57, from the difference in underlying material, and the difference in exposure before backfilling, a positive effort to determine the present rate of metal loss, if any, at Pier 57 appeared necessary.

Since the electrical potential readings made at the time of construction are at present considered inconclusive so far as quantitative results are concerned, a new effort was made to select and install a monitoring system that would allow observation of metal loss after replacement of the broken rock backfill following the inspection and instrumentation.

Three types of devices were installed: a permanent ultrasonic probe, a strain gauge instrumented bolt, and a vibrating wire strain gauge. Each was installed within the limitations of underwater work by the diver according to the recommendations and under the supervision of the vendor.

In addition to the measuring devices, two empty ducts were installed to reach above the broken stone backfill both under and outside of the pier for possible later installation of remedial material. The inspection pits at both piers were then backfilled with broken stone with a maximum dimension of 3 inches.

The principal findings from the installation could be of great importance in further studies of foundation pile corrosion problems. It is hoped that the three systems (the ultrasonic by direct measurement of thickness and the two strain gauges by indirectly measuring the change in cross sectional area) will check each other at least in degree of magnitude or that at least two of these systems will confirm that observations are reliable. (Unfortunately, the stressed bolt devices have produced no usable data.) Additionally, it is hoped that it will be possible to ascertain whether metal loss is occurring at the present time and if it is occurring, to numerically determine the rate. Further, if remedial backfill materials are deemed necessary and are possible to install through the ducts provided, their effect on the corrosion rate may also be appraised before making decisions on measures to be adopted at other locations. Some materials tentatively proposed to backfill the water filled void that may occur under the concrete pier base are portland cement-bentonite slurry and polyurethane foam. Others will be suggested and considered if and when remedial measures prove necessary.

Conclusions

On the subject of steel foundation piles, the literature does not accentuate the effect of concrete and sea water. On the other hand, the extensive information on steel bulkhead piles and piles that protrude from sea water to the superstructure proper is definitive enough to estimate corrosion rates; further, such corrosion is subject to control by cathodic protection. The subject of steel foundation piles capped by portland cement concrete is best covered in the literature by M. Romanoff's Monograph 58, Corrosion of Steel Pilings in Soils, published by the National Bureau of Standards in 1962. The science of corrosion engineering is a fast-growing specialty. It is best known in bridge work from the present attention to and the increasing amount of literature on the present problem of reinforcing bars corroding in concrete decks under the influence of deicing salts.

The application of cathodic protection to the steel foundation piles of the Richmond-San Rafael Bridge is extremely complex and possibly not financially feasible. However, it is the writer's belief that the information obtained from this test program and from related test programs and studies will serve to develop construction procedures that will minimize corrosion losses at moderate or low cost. The proven economies of this method of pier construction are such that more protective or positive backfill procedures or materials may be employed without pricing this construction method out of use.

Acknowledgment

The State of California owns the Richmond-San Rafael Bridge which it operates under the direction of E. R. Foley, Chief Engineer. The corrosion engineer, Richard F. Stratfull, was assigned to this investigation by special arrangement with John L. Beaton, Materials & Research Engineer for the California Division of Highways.

***BEN BALALA** is the principal bridge engineer for the State of California Division of Bay Toll Crossings, San Francisco, Calif. His 40 years of experience in design and construction supervision of bridges, buildings, and highways provide a valuable resource for knowledge. A registered civil engineer and structural engineer in the State of California, Balala is a Fellow of the American Society of Civil Engineers.*

Source: *Materials Protection and Performance*, September 1972, 30-32

Corrosion of Steel in Prestressed Concrete★

I. Cornet
and
B. Bresler
University of California
Berkeley, California

SUMMARY

The ratio of surface area to volume of metal is much greater for prestressing wires than for reinforcing bars. As compared to reinforced concrete, prestressed concrete is more sensitive to the influence of chemical reactions—which occur during setting and early curing stages—on concrete-metal bond. This article discusses preliminary tests with concrete beams prestressed with black steel and with galvanized steel wire. The tests indicated that performance under cyclic loading and loading to destruction was similar for the two types of wire.

SOME reports have stated that when galvanized steel is embedded in concrete, hydrogen may be liberated at the zinc-concrete interface under certain conditions, and the bond between metal and concrete may be impaired.[1,2] Recent work has shown, however, that under other conditions the bond of galvanized steel to concrete may be equal to the bond of black steel to concrete, and the galvanized steel may be significantly superior to black steel in resisting corrosion in concrete.[3] It is important to understand the conditions which affect bond and corrosion performance of metal in concrete.

As part of a continuing study* of the performance of galvanized steel in concrete, a small prestressed concrete beam using wires was designed and prepared for corrosion studies. The beam selected had no end plates and no gripping devices on the ends of the wires. Compression of the concrete depends only on transfer of stress from the wires to the concrete due to bond between them. The ratio of surface area to volume of metal is much greater for small diameter prestressing wires than for heavy reinforcing bars. As compared to reinforced concrete, prestressed concrete can furnish a more sensitive indication of the effect on bond of concrete to metal due to chemical reactions occurring during setting and early stages of curing of concrete. There will also be sensitivity if deterioration of bond occurs or if the wire section is reduced due to corrosion.

Materials, Apparatus, Procedure

The beam specimen selected was six feet three inches long, three inches by five inches in section, with two wires 0.148 inch in diameter, 1½ inches from the bottom of the section, as shown in Figure 1.

Materials

Difficulty was encountered in obtaining steel and galvanized steel wires of similar properties. The steel wire selected was a high tensile strength oil-tempered wire conforming to ASTM-A-229 specifications. A batch of this wire was hot-dip shop galvanized. Galvanizing increased the wire diameter from 0.148 to 0.160 inch. Tension tests were run on three specimens each of the black wire and of the galvanized wire. The maximum load for the black wire was 3550 pounds, for the galvanized wire 3480 pounds, corresponding to maximum stresses of 210 and 173 ksi respectively. Elongations in a 10-inch length were 0.60 for black and 0.49 inch for galvanized wire. Proportional limits were 110 ksi (black) and 140 ksi (galvanized). The modulus of elasticity was 28,100 ksi (black) and 24,000 ksi (galvanized). Concrete was designed to have a 28-day compressive strength of 6000 psi. A water/cement ratio of 0.45 was used. Proportions used were, by weight: 1 part Santa Cruz Type I cement (Table 1); 0.45 parts water; 2.27 parts Fair Oaks ¼ inch to ½ inch gravel; and 2.45 parts Elliott sand.

Apparatus

Forms for casting also served as the prestressing bed. Two specimens were

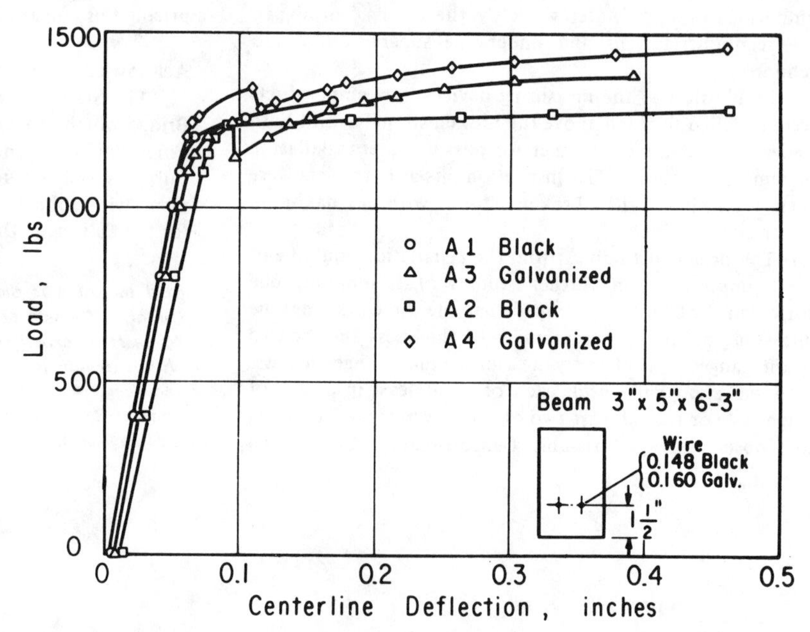

Figure 1—Final cycle load-deflection curve for both black and galvanized wire prestressed beams. Notice that the curves are similar.

★ Revision of a paper entitled "A Note on Prestressed Concrete Beam Specimens With Steel and With Galvanized Steel Wires" which was presented at the Western Region Conference, National Association of Corrosion Engineers, October 27-30, 1964, Phoenix, Ariz.

* This study was supported in part by a research grant of the International Lead and Zinc Research Organization.

cast in each set of forms. Each wire was prestressed separately by means of a jacking system at each end of the form. Calibrated load cells permitted accurate tensioning. The bottom of the form was plastic-coated ¾-inch plywood; the sides of the form were the backs of 7-inch steel channels. Forms and prestressing wires were cleaned with carbon tetrachloride, and all form joints were sealed with masking tape. Concrete for 4 beams and 14 control cylinders was mixed in a single batch, transported to the casting area in a single buggy, and cast. Consolidation of the concrete was handled effectively by clamping a form vibrator to the center of the metal forms. Beams were cured seven days in the forms with moist burlap covering the exposed top surface. After seven days the prestress force was transferred to the concrete by releasing the jacking plates. The beam specimens were then removed from the forms and cured an additional seven days in moist burlap. Then the burlap was removed and the specimens were air dried for 28 days. Control cylinders were cured in the same manner as the beam specimens.

Procedure

Before casting the beam, prestress wires were overstressed for two minutes to 150 ksi; then the prestress level was reduced to 125 ksi (based on black wire). Initial overstress was to reduce relaxation losses in the wire. After casting, load cell strain indicator readings were taken with time. In the seven days of initial curing under wet burlap, the force in the load cell fell from about 2160 pounds at casting time to about 2080 pounds, both for black and galvanized wires.

After seven days the beams were removed from the forms, and the force on the wires was transferred to the concrete and could no longer be recorded with the load cells. From 7 to 28 days, the change in the concrete strain was measured with a Whittemore gage. From 7 to 14 days, under moist cure conditions, the prestress force dropped about 30 pounds in both black and galvanized wires.

When the beams were 14 days of age, the wet burlap was removed and air dry cure commenced. Shrinkage and creep of concrete then resulted in about 100 pounds decrease in prestress force for the black wire and about 85 pounds decrease for the galvanized wire, within about 24 hours.

Subsequent air dry curing, for 28 days, was accompanied by further shrinkage and creep of concrete and the prestress force declined about 220 pounds, both the amount and the rate of decline being substantially the same for black and galvanized wire. Force on the prestressed wire was independent of type of wire involved, but stresses on the galvanized wire were based on the original black wire cross section in applying prestresses.

When the beam specimens were 28 days old, they were taken to a 60 K Baldwin Hydraulic Testing Machine for test. Beams were simply supported five inches from each end, and were loaded at mid-span, bearing on a spherical block assembly. The loading procedure used was first to apply 100-pound increments until the first flexural cracks appeared. The load then was removed in 200-pound increments. Subsequent cycle load increments were 0, 400, 800, 1000, cracking, 800, 400, 0 pounds. This loading cycle was repeated three times. On the fifth cycle, the beam was loaded as before to the cracking load; after the cracking load, smaller increments of load were applied until the concrete crushed in the region of maximum moment. After each increment of load was applied to the beam, mid-span deflection, concrete strain at third points, and cracking pattern were noted. On the fifth loading cycle only deflection and crack progression were recorded after the cracking load was reached.

TABLE 1—Analysis of Cement Used—Type 1*

SiO_2	21.72%
Fe_2O_3	2.56
Al_2O_3	5.69
CaO	63.73
MgO	1.94
SO_3	2.40
Ignition Loss	1.00
Insoluble	0.20
C_3S	46
C_2S	27
C3A	10.6
C_4AF	7.7
$CaSO_4$	4.0
K_2O	0.72
Na_2O	0.55
Total Alkalies as Na_2O	1.02
CrO_3	0.002
Specific, Wagner	1792
Surface, Blaine	3245
Autoclave Expansion	0.16
Tensile Strength Lb Per Sq In	
3 days	345
7 days	435
28 days	---
Compressive Strength Lb Per Sq In	
3 days	2308
7 days	3291
28 days	---

*Cement meets ASTM Specification C-150-60, Type 1

ISRAEL CORNET is a professor of mechanical engineering at the University of California at Berkeley. He teaches corrosion courses and has published articles on corrosion fatigue, corrosion of metals in concrete, and corrosion as a mass transfer phenomenon. An NACE member since 1953, he is active in NACE Unit Committee T-3K and in the educational efforts of the NACE Western Region. In 1957, he held a Guggenheim Fellowship and studied at the University of Cambridge in England. In 1960, he was research professor, Miller Institute for Basic Research in Science at the University of California. He is a registered professional engineer in California.

BORIS BRESLER is professor of civil engineering at the University of California and acting director of the structural engineering materials laboratory. He teaches courses on concrete and steel structures and has published articles on concrete and reinforced concrete. In 1959, he was the recipient of the Wason Medal for Research from the American Concrete Institute. In Fall 1961, he held a National Science Foundation Postdoctoral Fellowship, and in Spring 1962, a Guggenheim Fellowship, while conducting research at Imperial College, England.

Results, Discussion

For the two black wire prestressed beams, first flexural cracking occurred at loads of 1175 and 1250 pounds with mid-span deflections of 0.056 and 0.058 inches. For the two galvanized wire prestressed beams, first flexural cracks occurred at loads of 1125 and 1200 pounds with mid-span deflections of 0.056 and 0.054 inch. Load deflections curves were quite similar for both black and galvanized wire prestressed beams for all of the cycles to final failure (Figure 1). Curves for load versus third-point concrete strain were also similar. Average depth of neutral axis at failure was about ½ inch. Each of the specimens failed by yielding of the steel and eventual crushing of concrete (tensile failure) under the point of load application (maximum moment). Loads at failure were 1280 and 1300 pounds for black wire and 1380 and 1460 pounds for galvanized wire prestressed specimens.

Considering the accuracy of the experimental data, there were no significant differences in the behavior of beams prestressed with black or with galvanized steel wires. Beams pretensioned with the same initial force experienced similar losses. Load-deflection and load-strain plots were not noticeably different for black and galvanized wire specimens.

This preliminary investigation of prestressed concrete beams was to provide a specimen suitable for future corrosion studies. However, there has been some question whether or not deleterious corrosion might occur during the initial set of concrete about galvanized wire. In part, this stems from misunderstanding of some publications of C. E. Bird.[1]

Bird pointed out that galvanized steel annealed at 400 C (752 F) for two hours may have zinc-iron alloy extending to the outer surface and that such improperly heat treated material is unsuitable for embeddment in concrete. Also, it probably is unacceptable as wire or bar because it will lack ductility. Normally galvanized steel has an outer layer of zinc, and with modern mill-produced wire, the alloy layer is thin. A later paper by Bird elaborates on the first paper and includes laboratory studies in sodium hydroxide solutions as well as cement pastes and concrete.[2] Again improperly annealed galvanized steel, 350 C (662 F) for three hours, is observed to be reactive. Bird also observed that pastes made with cements containing less than 65 ppm of CrO_3 could liberate hydrogen from galvanized steel.

Analysis, using the method described by Bird, showed that the cement used in this investigation contained 20 ppm (0.002%) CrO_3. However, CrO_3 in sand, gravel, mixing water, or on the surface of the galvanized steel could also inhibit hydrogen liberation. The liberation of hydrogen and formation of spongy concrete in practice has not been reported in the literature and may be related to an exceptional quality of some South African cement. The present investigation is consistent with work of D. A. Lewis and J. A. P. Laurie, who observed that black and galvanized steel give similar bond strengths.[4]

Acknowledgments

The authors gratefully acknowledge the assistance of F. E. Peterson and M. S. Lin, graduate students.

References

1. C. E. Bird. Bond of Galvanized Steel Reinforcement in Concrete, *Nature*, 194, No. 4830, p.798 (1962) May.
2. C. E. Bird. The Influence of Minor Constituents in Portland Cement on the Behavior of Galvanized Steel in Concrete. *Corrosion Prevention and Control*. No. 7, pp 17-21 (1964) July.
3. I. Cornet and B. Bresler. Corrosion of Steel and Galvanized Steel in Concrete, a paper presented at the NACE Western Region Conference, Phoenix, Ariz., October 28, 1964.
4. Private communication with J. A. P. Laurie, National Building Research Institute, South African Council for Scientific and Industrial Research. Pretoria (1963).

D. A. Hausmann
American Pipe and Construction Co.
South Gate, Calif.

SUMMARY

Though cathodic protection of steel in concrete is normally not required, its application to reinforced concrete structures is sometimes justified to prevent or arrest corrosion in high chloride environments. In such cases, the passivating effect of portland cement should be recognized, and, as illustrated by the experiments reported in this paper, the criteria for cathodic protection of reinforced concrete structures are entirely different from those for bare or organically coated steel structures.

Criteria for Cathodic Protection of Steel in Concrete Structures

STEEL CAST in concrete quickly develops a passivating iron oxide film that prevents further corrosion and makes cathodic protection unnecessary for most reinforced concrete structures that are not defective or subject to severe electrolysis. Passivation of steel in concrete may be destroyed by high concentrations of chloride ions;[1] consequently, cathodic protection of sound concrete structures is sometimes justified to insure extended protection against corrosion in environments high in chlorides.[(1)]

There are no widely accepted criteria for cathodic protection of steel in concrete. In the absence of such criteria, reinforced concrete structures are sometimes cathodically protected as if they were bare or organically coated steel structures. Thus some, applying a widely used criterion for buried steel pipelines,[2] assume that steel in concrete is cathodically protected only if sufficient current is applied to maintain a minimum steel potential of —0.85 v referred to a close copper sulfate electrode. As will be shown, this criterion is not applicable to steel cast in concrete, and its use generally results in overprotection.

Previous Experiments

The electrochemical behavior of steel reinforcement in moist or saturated portland cement concrete can be studied with bare steel rods in a saturated lime solution.[3] This is possible because calcium hydroxide is the principle soluble component of hydrated portland cement, and the pH of portland cement concrete is at least as high as that of a saturated lime solution[4] (about 12.5).

Experiments[1] with lime solutions containing chlorides have shown that there exists a threshold concentration of chloride ions above which corrosion of bare steel rods is initiated. The threshold concentration of sodium chloride causing pitting corrosion of mild steel with free oxygen present is about 0.02M (moles per liter), or 700 ppm chloride ion. At the threshold of corrosion, Cl^- activity varies approximately in direct proportion to OH^- activity, as shown by Curve 1 in Figure 1.

Leckie and Uhlig[5] have defined a boundary between pitting and inhibition of steel in alkaline solutions containing chlorides. This boundary, shown by Curve 2 in Figure 1, defines a much higher Cl^- activity at pitting than does Curve 1, but the experiments were made with 18-8 stainless steel in deaerated solutions.

The existence of boundaries similar to those shown in Figure 1 explains why steel in concrete rarely corrodes unless exposed to high concentrations of chlorides. For example, if the OH^- and Cl^- activities in a given reinforced concrete structure are as defined by Point A in Figure 1, pitting corrosion will not occur, and cathodic protection will not be necessary. On the other

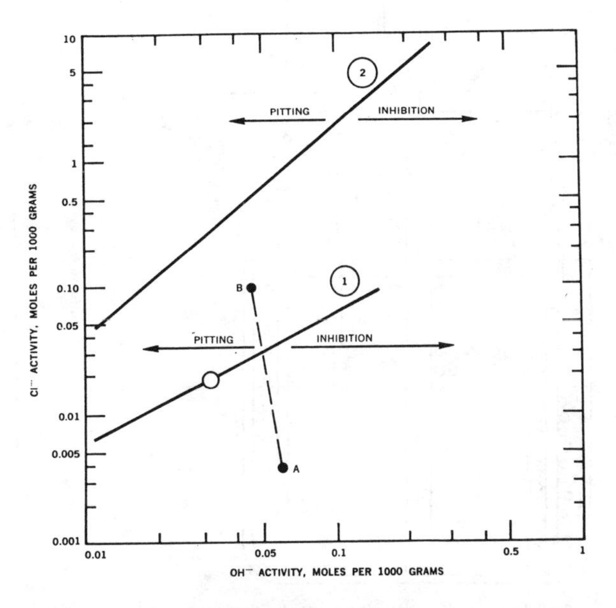

Figure 1—Boundaries between pitting and inhibition in alkaline aqueous solutions containing chlorides. Curve 1 is for steel with air bubbles trapped against the surface; the encircled point represents a saturated lime solution. Curve 2 is for 18-8 stainless steel in deaerated solution.

(1) Bromide, iodide, and sulfide ions can also destroy passivity of steel in concrete, but in concrete structures, these anions are rarely encountered in concentrations sufficient to cause corrosion.

hand, if the concrete structure is located in a high chloride environment, the Cl^- activity at steel surfaces may eventually reach a high level, as shown by Point B in Figure 1, where pitting corrosion is probable. In this case, cathodic protection may be required to prevent corrosion.

Cathodic Protection Experiments

The experiments described in this paper were designed (1) to provide an estimate of the minimum level of cathodic protection required to prevent steel corrosion in concrete exposed to a high-chloride environment and (2) to investigate the effectiveness of cathodic protection in arresting steel corrosion in chloride-contaminated concrete. The latter experiment was conducted at cathodic protection potentials between −0.71 and −0.81 v to a copper sulfate electrode, a range suggested by Scott[3] from the potential of iron in equilibrium with saturated ferrous chloride.

Experimental Procedures and Results

All experiments were made with a cathodic protection cell (Figure 2) which simulates a defective reinforced concrete structure. Twelve bare steel rods representing steel reinforcement in concrete are immersed in a saturated lime solution which provides a chemical environment comparable to that of moist or saturated concrete, and limestone gravel packed around the rods serve to trap large air bubbles against steel surfaces when the electrolyte is added.

A carbon anode was inserted in the center of the cell. The bare steel rods, used as cathodes, were arranged symmetrically around the anode and connected to the negative terminal of a rectifier. Sodium chloride was added to the saturated lime electrolyte in concentrations of 0.08, 0.16, 0.32, or 0.64M (4 to 32 times the minimum concentration causing steel corrosion in saturated lime solutions with free oxygen present). Instrumentation was provided to measure applied voltage, individual and total cathode current, and electrode potential measured to a saturated calomel electrode. Cathode potentials were measured frequently with current applied, and again with current temporarily interrupted.

The rectifier for each cell was initially set to produce an interrupted cathodic potential (hereafter referred to as polarization potential) of about −1,000 mv to the saturated calomel electrode (SCE). This potential was maintained for one week and later reduced to less negative values in steps of about 25 mv. Each new level was held for two to four days, or until the polarization potential had changed less than 10 mv in 24 hrs. Depolarization was continued until one of the 12 rods in the test cell corroded. Corrosion was identified by reversal of current to the corroding rod, and the anodic rod was removed from the cell and inspected for visible corrosion. Single small pitting type anodes were found at random orientation on all rods removed.

When the corroding rod was removed, the polarization potential of the remaining rods became from 20 to 100 mv more negative because the current, previously applied to 12 rods, was now being applied to only 11 rods. The rods were again depolarized slowly, this time in decrements of 10 mv or less, until another rod corroded. This procedure was repeated until several rods had corroded in each of the four test electrolytes. The most negative polarization potential at which corrosion occurred in each electrolyte is hereafter referred to as the critical polarization potential.

Critical Polarization Potential as a Function of Chloride Concentration

The critical polarization potential is plotted as a function of chloride ion concentration in Figure 3. The curve is discontinuous at a Cl^- concentration of 0.02M, the threshold concentration previously determined for passivation of mild steel in saturated lime solution. At a Cl^- concentration of 0.64M (4.1% sodium chloride), the critical polarization potential is −435 mv (SCE), or approximately −0.50 v to a copper sulfate electrode.

Current-Potential Relationships

During depolarization, the cathode potential decreased as a logarithmic function of current density. This relationship, shown characteristically in Figure 4, is consistent with the observations of Tomashov;[6] it held until the cathode was depolarized to its critical polarization potential and corrosion occurred. At initiation of corrosion the current reversed direction and subsequently increased in magnitude. The

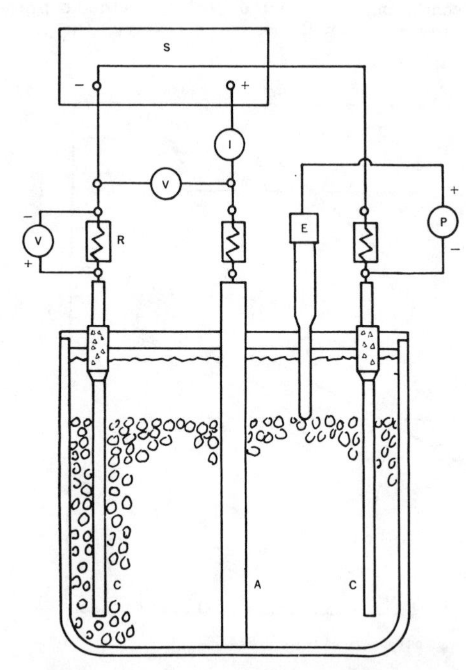

Figure 2—Cell used for cathodic protection experiments. A—carbon anode; C—bare steel cathodes; E—saturated calomel electrode; L—ammeter; R—resistor; S—d-c power supply; P—potentiometer; and V—voltmeter.

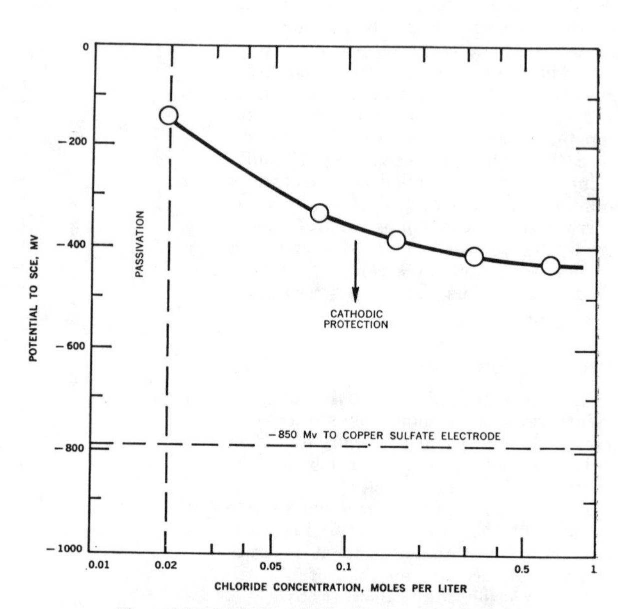

Figure 3—Critical polarization potential for cathodic protection of steel in saturated lime solutions containing sodium chloride.

anodic potential also appears to be a logarithmic function of current.

Extensions of the cathodic and anodic current-potential curves intersect at the unpolarized (base) potential of bare steel in saturated lime solution.[1] The current at the projected intersection is shown as zero in Figure 4, with the log-current scale discontinued at about 0.1 ma/sq ft. At lower current densities, the cathodic depolarization curve may be considered linear, as suggested by Tomashov.[6]

Cathodic Protection to Arrest Corrosion

Cathodic protection was applied, and the electrolyte added simultaneously in order to prevent initial corrosion of the rods. In a supplementary experiment, cathodic protection was delayed 24 hrs after the electrolyte was added to determine whether corrosion could be arrested by cathodic protection in the potential range from —650 to —750 mv (SCE). The saturated lime electrolyte used in this cell and in an identical control cell contained 0.64M Cl^-. Potential measurements indicated that anodes developed on all rods in both cells within 15 minutes after the electrolyte was added; average potentials were —195 mv (SCE) at start of test, —425 mv after 15 minutes, and —515 mv after 24 hrs.

Six rods were withdrawn from the control cell (no cathodic protection) after one day; the remaining six rods were withdrawn after 30 days. Six rods were withdrawn from the cathodically protected cell after 30 days. Anodic areas on all 18 rods were cleaned and pit depths measured.

The pit depth data, indicated in Table 1, show that cathodic protection applied after one day was completely effective in stopping pitting corrosion during the following 29 days. The cathodic protection potential for this test (Figure 5) was generally maintained less negative than —700 mv (SCE), and occasionally dropped as low as —640 mv. Operating and polarization potentials were identical after the first few hours of cathodic protection.

Current density, also shown in Figure 5, decreased logarithmically with time from 5 to 3 ma/sq ft during the first 10 days. Between the tenth and twentieth days, the current fell sharply to about 0.25 ma/sq ft. During this transitional period, the polarization potential became erratic, shifting to more negative values and requiring frequent adjustment of the rectifier output. The rapid change in current with time after ten days may have resulted from oxygen depletion in the cell, which may occur either by direct reduction at the cathode or by reaction with hydrogen.

Discussion and Conclusions

Cathodic protection is one of the most useful corrosion control techniques available to the corrosion engineer, but its indiscriminate use can be costly and damaging. Misapplication of cathodic protection is especially likely with reinforced concrete structures if the passivating effect of portland cement is not recognized.

Consider, for example, the cathodic protection of a reinforced concrete structure to prevent damage by stray current electrolysis. If the structure is properly designed and constructed so that local cell corrosion is unlikely, free current discharge will be prevented by making the entire structure a cathode. Cathodic protection of embedded steel is assured in this case if polarization potential is shifted to a minimum value of —0.40 v to a copper sulfate electrode[7] — about 0.45 v less negative than the minimum potential generally maintained for cathodic protection of bare or organically coated steel.

Different criteria for cathodic protection should be applied if a reinforced concrete structure with (1) the steel reinforcement uncorroded, or (2) the steel reinforcement corroded is exposed to a high chloride environment. The two cases studied in the cathodic protection experiments reported in this paper indicate (1) that corrosion of steel can be prevented in concrete exposed to a high chloride environment if sufficent current is applied to shift the steel polarization potential to a minimum value of —0.50 v to a copper sulfate electrode, and (2) that corrosion of steel can be arrested in chloride-contaminated concrete if sufficient current is applied to shift the polarization potential to a minimum value of —0.71 v to a copper sulfate electrode.

References

1. D. A. Hausmann. Steel Corrosion in Concrete. *Materials Protection,* 6, No. 11, 19-23 (1967) November.
2. NACE Technical Unit Committee T-2C. Criteria for Adequate Cathodic Protection of Coated, Buried or Submerged Steel Pipe Lines and Similar Steel Structures. *Corrosion,* 14, No. 12, 561t (1958) December.
3. G. N. Scott. Corrosion Protection Properties of Portland Cement. *J Am. Water Works Assoc.,* 57, No. 8, 1038-1052 (1965) August.
4. T. C. Powers. The Nature of Concrete. Am. Soc. Testing Materials, STP No. 169-A, 61-72 (1966).
5. H. P. Leckie and H. H. Uhlig. Environmental factors Affecting the Critical Potential for Pitting in 18-8 Stainless Steel. *J. Am. Electrochemical Soc.,* 113, No. 12, 1262-1267 (1966) December.
6. N. D. Tomashov. *Theory of Corrosion and Protection of Metals.* (1966) 172-176. The Macmillan Co., New York.
7. D. A. Hausmann. Electrochemical Behavior of Steel in Concrete. *J. Am. Concrete Inst., Proceedings,* 61, No. 2, 171-188 (1964) February.

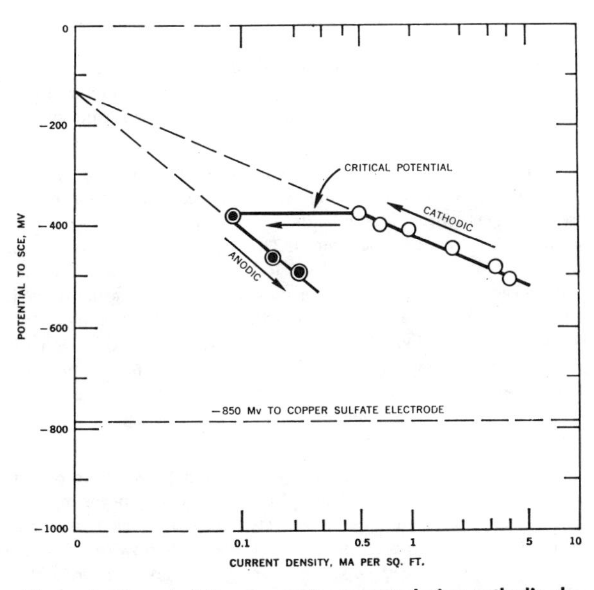

Figure 4—Characteristic change in current during cathodic depolarization of steel in a saturated lime solution containing 0.64M sodium chloride. Corrosion occurred, and the current quickly reversed direction when the critical polarization potential was reached.

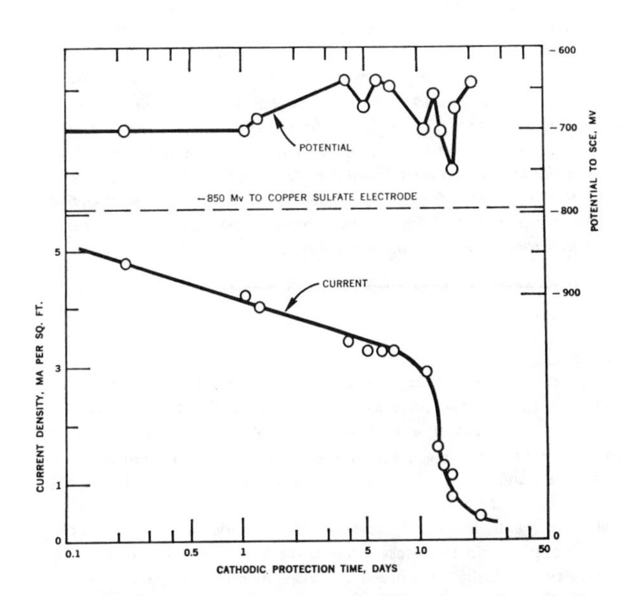

Figure 5—Current and potential as functions of time for cathodically protected bare steel rods in a saturated lime solution containing 0.64M sodium chloride. Rods were corroded 1 day before cathodic protection was applied.

Effect of Use of Galvanized Steel on the Durability of Reinforced Concrete*

HAJIME OKAMURA and YOSHIHIRO HISAMATSU
University of Tokyo, Tokyo, Japan

Results of atmospheric exposure with cyclic application of NaCl solution on prestressed black and galvanized steel reinforcing bars are reported. Fatigue tests were conducted after 8 months. High strength reinforcing steel bars require limited crack widths for proper durability, so galvanized bars have an added safety factor in aggressive exposures. After vibration testing to failure, black bars were rusted at cracks while galvanized bars showed little attack. Concrete adhered tightly to galvanized bars near cracks. Tests showed galvanized bars lost less fatigue strength than black bars, and that they could tolerate greater crack widths.

REINFORCING STEEL BARS in concrete usually have little chance to become corroded as long as widths of cracks in the concrete are within a certain limit.[1] When reinforced concrete develops an excessive crack width while being subjected to stress, bars in concrete may become corroded and affect its durability. Corrosion of reinforcing bars depends not only on crack widths, but also on crack directions, qualities of the concrete, the reinforcing bar material, and thickness of concrete over the bars. The exposure conditions of the reinforced concrete member, in particular, is a very influential factor affecting corrosion. Thus, allowable limits of crack width in reinforced concrete, which are intended to ensure durability, are determined in degree according to the exposure conditions for the member. For example, the European Concrete Committee[1] specifies allowable crack widths of 0.3, 0.2, and 0.1 mm for reinforced concrete members under 3 different conditions: (1) indoor, (2) outdoor, and (3) in a highly corrosive atmosphere, respectively.

The widths of cracks increase proportionally to the applied stress on the reinforcing bars involved. This means that the effective use of high tensile strength deformed bars is limited by the allowable limit of the crack width. The use of galvanized steel bars may be cited as a practical solution to this problem. In other words, it has been pointed out that the corrosion rate of the reinforcing steel bars in concrete can be reduced by galvanizing treatment.[2,3] However, the relationship between crack width and corrosion of bars in concrete reinforced with galvanized steel bars is not yet definite, and it remains to be clarified how much the allowable limit of the crack width can be increased using galvanized steel bars, rather than black steel bars.

Accordingly, research on the relationship between corrosion and crack width was carried out. Reinforced concrete beams in a cracked condition, with galvanized steel bars or black bars, were exposed to a 3% solution of sodium chloride sprayed twice a day for 1 year. The reinforced concrete beams that had undergone the exposure test were then subjected to a fatigue test. After this, they were crushed to investigate the relationship between corrosion of the galvanized steel bars and crack width compared with black steel bars.

*Voluntary manuscript submitted for publication October, 1975.

Reprinted with permission from *Materials Performance*, July 1976, 43-47, © 1976 National Association of Corrosion Engineers

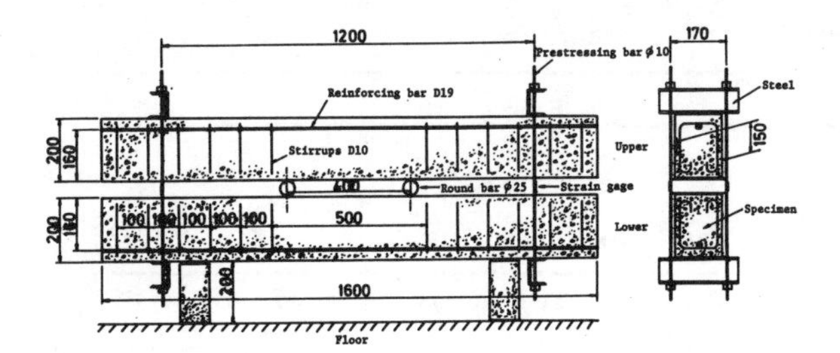

FIGURE 1 — Test specimens (all dimensions are in mm).

TABLE 1 — Crack Widths of Beams at Loading

Specimen No.	Stress of Rebar MPa[2]	Number of Cracks	Width of Each Crack[1] 1/100 mm								Average of 3 Main Cracks
62	308	7	32	28	22	21	19	19			27.3
			33	30	27	27	26	19			30.0
38	308	8	35	26	23	23	20	19	19		28.0
			43	32	30	30	29	15	20		35.0
60	265	8	27	25	19	19	10	8			23.7
			32	27	18	16	14	14	13	13	25.7
61	265	7	26	25	22	18	17	14	11		24.3
			29	27	25	24	21	18	13		27.0
59	316	7	32	29	28	23	17				29.7
			34	29	27	23	21	18			30.0
37	316	7	31	28	28	27	22	21	16		29.0
			41	39	35	33	30	27	17		38.3
53	267	8	27	25	25	18	14	12	6		25.7
			36	32	30	24	18	11	11		32.7
57	261	8	30	29	28	24	22	14			29.0
			40	39	29	27	26	15			36.0
54	267	8	26	25	22	18	15	14	5		24.3
			32	29	28	24	22	17	10		29.7
58	261	7	35	26	26	22	16	15			29.0
			39	30	30	28	20	14			33.0
51	312	7	26	25	21	15	15	12	12		24.0
			31	30	24	23	21	17	12		28.3
55	298	7	34	29	21	21	17	16			28.0
			29	28	26	26	26	17			27.7
52	312	7	24	23	22	20	20	12	11		23.0
			31	28	27	23	22	15			28.7
56	298	6	29	28	26	26	18	6			27.7
			38	35	35	31	21	13			36.0
43	257	8	25	24	18	16	14	13	7		22.3
			32	31	29	25	20	17	14		30.7
47	269	7	34	29	23	23	16	9			28.7
			37	33	27	24	19	17			32.3
44	257	7	27	26	21	20	15	14			24.7
			32	28	28	28	23	18			29.3
48	269	6	32	26	22	21	10				26.7
			39	34	29	27	16				34.0
39	302	8									
45	303	9	26	22	22	22	17	11	10		23.3
			34	29	28	21	20	20	13		30.3
49	298	7	26	26	24	23	22	18	12		25.3
			34	31	31	30	25	22	20		31.3
40	302	7	43	34	32	23	20	16			39.7
			–	41	41	–	24	–			
46	303	7	36	31	23	21	16	12			30.0
			40	36	28	26	16	15			34.7
50	298	6	33	32	30	28	27	19			31.7
			40	37	36	35	26				37.7

(1) Widths at the side of the specimen and the bottom (Figure 2).
(2) 1 MPa = 0.145 ksi.

Materials

High strength deformed bars with a yield strength of 600 MPa (87 ksi) were used. The bar diameter was 19 mm, the angle of inclination formed by lugs and bar axes was 90°, and arcs at the bases of lugs with radii of lug-height were provided. These two parameters have been recognized to affect the fatigue properties of deformed bars. The zinc coating weight on the galvanized bars was 0.6 kg/m^2 (2 oz/ft^2).

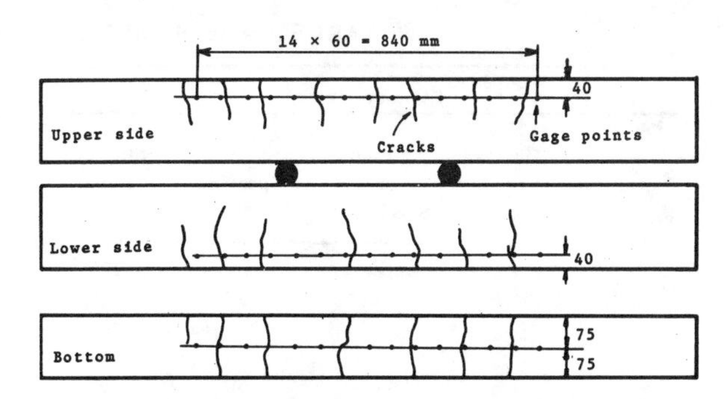

FIGURE 2 — Schematic representation of cracks developed in concrete beams and sites of gage points to measure the crack width.

FIGURE 3 — Exposure test.

The concrete was made with high early strength portland cement, whose compressive strength was about 40 MPa (5.8 ksi) after 28 days. However, the cement content was relatively low, or only 276 kg/m^3 (17 lbs/cu ft). Yet, the concrete itself had a relatively high durability against freezing and thawing or other atmospheric action because it had a low unit water content with the addition of a surface active agent,[4] with an air content of about 3%. Concrete with 25 mm maximum sized coarse aggregate had a water-cement ratio of 0.55, and a slump of approximately 40 mm. Specimens were stripped from the molds at 2 days, and concrete surfaces were covered with wet cloth which was covered again with vinyl sheet for moist curing until testing.

Specimens and Test Procedures

Each specimen consisted of a set of two rectangular beams, each measuring 150 mm in width, 200 mm in height, and 1600 mm in length, arranged in such a manner that one beam was put on the other, with two 25 mm diameter round bars in between, and bending moment was imparted to them stretching 4 prestressing steel bars (Figure 1). The prestressing bars were checked for strain

TABLE 2 – Results of Fatigue Tests After the Exposure Tests

Exposure Tests					Fatigue Tests					
Period of Exposure (Months)	Galv. or Black	Position of Specimen	Specimen No.	Stress of Rebar MPa(2)	Stress Range MPa(2)	Cycles N x 10^3	N_{20} x 10^3	ΣN_{20} x 10^3	Log ΣN_{20}	fs(3) MPa(2)
		Upper	62	308	196	771	771	771	5.887	172
3	Black	Lower	38	308	196	1960	1960			
					235	2230	10370	12330	7.091	242
		Upper	60	265	196	2310	2310			
					235	340	1580	3890	6.590	213
6	Black	Lower	61	265	196	2310	2310			
					235	57	265	2575	6.411	203
		Upper	59	316	196	1950	1950	1950	6.290	195
8	Galv.	Lower	37	316	196	1950	1950			
					235	536	2490	4440	6.480	207
		Upper	53	267	157	1970	425	425	5.628	157
			57	261	157	1920	414	414	5.617	156
			54	267	157	1970	425			
		Lower			196	220	220	645	5.810	168
			58	261	157	1760	379	379	5.579	154
	Black		51	312	157	2360	508			
					196	810	810	1318	6.120	185
		Upper	55	298	157	2630	565			
					196	704	704	1269	6.104	184
			52	312	157	2360	508			
		Lower			196	593	593	1101	6.046	181
			56	298	157	1664	358	358	5.554	152
			43	257	157	2300	495			
					196	2030	2030			
		Upper			235	1840	3960	6485	6.812	226
			47	269	157	2250	484			
					196	987	987	1471	6.168	188
12			44	257	157	2300	495			
					196	2030	2030			
		Lower			216	1215	2620	5145	6.712	221
			48	269	157	2250	484			
					196	90	90	574	5.759	164
			39	302	157	2340	505			
	Galv.				196	630	630	1135	6.055	181
			45	303	157	2260	486			
		Upper			196	1460	1460	1946	6.289	195
			49	298	157	2230	480			
					196	731	731	1211	6.083	183
			40	302	157	2340	505			
					196	483	483	988	5.995	178
			46	303	157	2260	486			
		Lower			196	2410	2410			
					216	1830	3940	6836	6.835	227
			50	298	157	2230	480			
					196	89	89	569	5.755	164

(1) See Figure 2.
(2) fs = fatigue stress.
(3) 1 MPa = 0.145 ksi.

by wire strain gages on the bars to determine two cases where tensile stress acting on reinforcing bars in beams would become about 275 MPa (40 ksi) and about 314 MPa (45 ksi), and the tensile force against prestressing steel bars was adjusted accordingly. After imparting tensile force to the prestressing bars, they were fixed with nuts and subjected to the exposure tests. The tensile force was a little reduced when the bars were fixed, and calculated values of the tensile stress acting on the reinforcing bars were as shown in Table 1.

Due to the bending moment each beam developed 6 to 9 transverse cracks as shown in Figure 2. Crack width is one of the most important factors affecting the corrosion of reinforcing bars. Therefore, crack widths on the surface of the concrete beams were measured with contact type strain gages at two levels: at the center of the bottom, and on the side of the specimen 40 mm from the bottom where the bar was located (Figure 2). Elongations between gage points which had been put on the surface of concrete were measured, and these were assumed to be the widths of cracks. Because, the elongations of uncracked sections were very small, the elongations were considered to be concentrated in the cracks. Table 1 shows widths of the main cracks developed in the beams. The width of the widest crack on the side and bottom of each beam

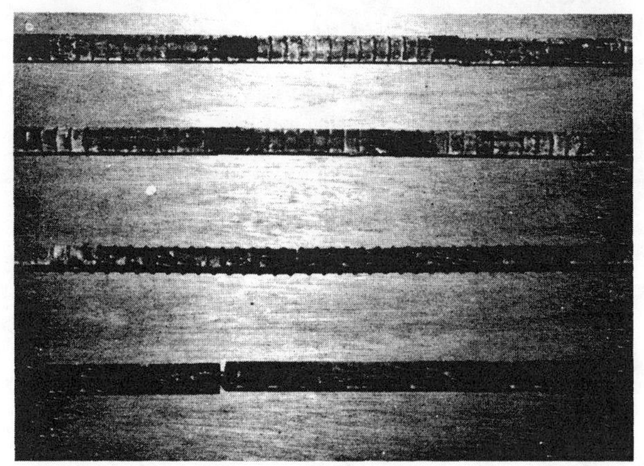

FIGURE 4 — Reinforcing bars corroded near cracks. Top: galvanized steel, and bottom: black steel.

ranged from 0.24 to 0.43 mm, and 0.27 to 0.41 mm, respectively, which was considerably larger than the allowable width limits for ordinary reinforced concrete.

The exposure test was conducted at the University of Tokyo as shown in Figure 3. The beams were exposed directly to sunshine and rain. Each pair of specimens was sprayed with 1.0 liter of 3% sodium chloride solution every morning and afternoon, with a flower watering pot. The spraying was daily, regardless of weather, except Sundays and holidays. It was assumed that as the solution penetrated through the cracks to deprive the concrete of alkalinity and supply oxygen, corrosion of the reinforcing bars would be considerably accelerated. Corrosion would be far more rapid if drying and wetting was repeated inside the cracks, than in a condition where the cracks were always filled with solution, and accordingly, the spraying method was employed for the test.

The exposure test began in June, 1972, and it was scheduled that, of the total of 12 pairs of beams, 3 pairs would be subjected to a fatigue test at 3, 6, and 8 months during the exposure test. From the results obtained, it was decided to terminate the test after 12 months. The pairs of beams were not inverted during the experiment. After the fatigue tests, the specimens were crushed to observe the degree of corrosion on the galvanized and black steel bars. This was compared to investigate the effect of crack width on corrosion.

The bending fatigue test apparatus consisted of a jack of 10 ton capacity and cycle of 5 Hz. The beams were supported by rollers on ball bearings. Tests were conducted by two point loading at span of 1200 mm, with 400 mm between loading points. The tests were generally continued to failure, but the maximum loads were increased at 2 million cycles if failure was not reached by that level (Table 2). The load used in the fatigue test was determined based on reinforcing bar stress calculated from elastic theory. The minimum load was selected to obtain a reinforcing bar stress of 39 MPa (5.6 ksi), while maximum loads were varied in several stages (Table 2).

Results and Discussion

The results of the fatigue tests conducted after exposure to the sodium chloride solution are shown in Table 2. All beams failed at last due to fatigue rupture of the reinforcing steel. Therefore, from the test results, the relation between the calculated stress ranges (σ_{max}-σ_{min}) of the reinforcing bars and cycles of load to rupture was investigated. A 2 million cycle fatigue stress range of each specimen was estimated by applying Miner's rule.[5] For calculation of the fatigue stress range, the following hypotheses were used, considering the previous test results.[6]

$$\mathrm{Log}N_{f_2} = \mathrm{Log}N_{f_1} + (f_1 - f_2)/59$$

$$f_{sr} = f_2 + 59\,[\mathrm{Log}\,\Sigma N_{f_2} - \mathrm{Log}\,(2 \times 10^6)]$$

where f_1 = a certain stress range, MPa; f_2 = a certain stress range, MPa; N_{f_1} = cycles under stress range of f_1; N_{f_2} = equivalent cycles under stress range of f_2 corresponding to N_{f_1} under f_1; ΣN_{f_2} = total equivalent cycles under stress range of f_2; and f_{sr} = 2 million cycle fatigue stress range, MPa.

The results of the calculations are shown in Table 2. Before the exposure tests, the averages of the 2 million cycle fatigue stress for galvanized steel was 250 MPa, while that of black steel was 267 MPa. These results may indicate that the effect of galvanizing treatment on fatigue strength of bars will be negligible compared with the remarkable effect of the difference in the deformations on bar. Therefore, it is important to select a bar with good deformations for fatigue resistance, and there will be practically no need to consider the effect of galvanizing treatment on the fatigue strength of reinforced concrete beams.

The first thing to be pointed out from Table 2 is that the longer the duration of exposure, the lower the fatigue strength of a bar due to corrosion of the bars near the crack openings. In the case of black steel, the 2 million cycle fatigue stress before the exposure test was 267 MPa (39 ksi), but reduced to about 200 MPa (29 ksi) at 6 months' exposure, and to 167 MPa (24 ksi) after exposure of 1 year. In the case of galvanized steel, the reduction of fatigue stress due to corrosion was also seen and the 2 million cycle fatigue stress was 200 MPa (29 ksi) for 8 months exposure and 195 MPa (28 ksi) for 1 year. However, the degree of the reduction in fatigue strength was smaller for galvanized steel. It seems that galvanizing will alleviate the corrosion of reinforcing steel in cracked concrete.

Figure 4 shows an example of reinforcing bars which were taken from the beam after the fatigue tests. Red rust was found on the black steel bars near the position of each crack in the concrete. On the other hand, little such rust was found on the galvanized bars. However, galvanized steel seemed to react with the concrete near the cracks, as concrete near the cracks adhered tightly to the steel bars.

Figure 5 shows the relationship between width of the maximum crack and the fatigue stress. The crack width at the bottom of a specimen seems to have a remarkable effect on the fatigue strength of bars subjected to exposure tests compared to the crack width at the side of a specimen. Although the rupture of some bars occurred near wide cracks, this did not happen in every case.

All the beams had a constant depth of concrete cover. There is evidence now that surface crack widths alone are not a sufficient indicator of corrosion resistance.[7]

Summary

Fatigue tests were conducted on 60 stressed concrete beams reinforced with galvanized bars or black steel bars. Twenty-four of these were tested after exposure to sodium chloride solution for a duration of 6 to 12 months in cracked condition. Within the scope of the experiment, the following can be said:

1. The longer the duration of exposure to sodium chloride solution, the lower the fatigue strength of a bar in concrete due to corrosion of the bar near the crack openings. However, the degree of

reduction of fatigue strength was less for galvanized steel bars. The durability of reinforced concrete against sodium chloride solution or sea water will be improved by using galvanized reinforcing bars. By using galvanized steel, the durability of concrete members with cracks of about 0.3 mm width will have the same durability as ordinary reinforced concrete with crack width of about 0.2 mm.

2. The use of galvanized steel does not significantly affect the fatigue strength of reinforced concrete beams.

Acknowledgment

Financial assistance for this investigation was provided by the International Lead Zinc Research Organization and is gratefully acknowledged.

References

1. European Concrete Committee, Recommendations for an International Code of Practice for Reinforced Concrete (1964).
2. Bresler, B., Cornet, I. Materials Protection, Vol. 5, No. 4, p. 69 (1966).
3. Cornet, I., Ishikawa, T., Bresler, B. Materials Protection, Vol. 7, No. 3, p. 44 (1968).
4. Mielenz, R. C. Proc. Fifth International Symposium on Concrete, Tokyo (1968).
5. Sandor, B. I. Fundamentals of Cyclic Stress and Strain, p. 69, The University of Wisconsin Press (1972).
6. Okamura, H. Fundamental Study on Use of High Strength Reinforcing Bars, Concrete Journal, Vol. 4, No. 6 (1966) (In Japanese).

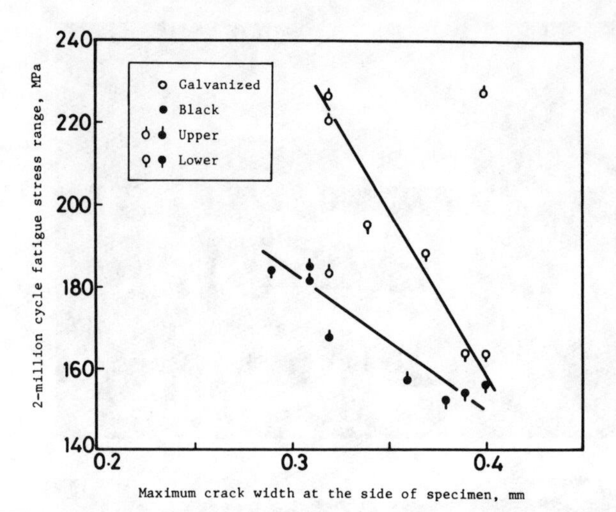

FIGURE 5 – Effects of maximum crack width in beam on the fatigue strength of reinforcing bars in concrete exposed to a solution of sodium chloride for a duration of 1 year.

7. Atimtay, E., Ferguson, P. M. Materials Performance, Vol. 13, No. 12, p. 18 (1974).

Performance of alloys against erosion-corrosion attack

G. SCHIEFELBEIN, Stainless Foundry & Engineering, Inc., Milwaukee, Wis.

The high frequency of pump repair and replacement in the pulp and paper industry prompted a research program to develop an alloy that would resist the attack of abrasive slurries in corrosive media. The comparative performance of alloys commonly used in paper mills is described.

INCREASED APPLICATIONS of stainless specialty alloys in the pulp and paper industry have been attributed to better knowledge of the behavior of new alloys and economic factors within the industry which justify their use. Many recommendations for the application of stainless specialty alloys are based on static laboratory corrosion tests, field corrosion data, and extrapolated experience obtained from the chemical industry.

A new dimension in laboratory testing was developed by Beck[1] at Ohio State University for an Alloy Casting Institute project. The equipment was designed to measure erosion-corrosion, the mechanism which causes problems for the pulp and paper industry. A variable speed motor and test cell determine the effect of velocity or rotational speed on the corrosion of the specimens immersed in the test media. A constant speed apparatus provides data to measure long term specimen weight loss. Data obtained is converted to corrosion rate in mpy.

Most applications are not abrasive enough to produce accelerated test data. The naturally occurring abrasive must be fortified with a controlled addition of abrasive metal oxides. Results obtained from the accelerated test are biased. Therefore it is important that the same bias be applied to all alloys being exposed, allowing the measured corrosion rate to be a comparison of the relative performance of the tested materials. The test duration is generally a continuous 168 hours.

Erosion-Corrosion in Sulfite Processing

The abrasives occurring naturally in wood vary according to geographical area and mineral content of the soil in which the trees grew. An analysis of a typical hardwood spent sulfite liquor, containing about 12% by weight solids, is shown in Table 1.

Erosion-corrosion measurements of three alloys, Illium[(1)] PD, Illium 98, and a modified CN-7M (Alloy 20) widely used in the paper industry (Table 2), were obtained in this liquor which was fortified with a controlled amount of aluminum oxide to accelerate the test. Table 3 shows the relative erosion-corrosion rate factors for each of the alloys studied. This data indicate that the combined abrasion and chemical resistant properties of Illium 98 and Illium PD are superior to the modified CN-7M (Alloy 20) tested in this application.

(1) A tradename of Stainless Foundry and Engineering, Inc., Milwaukee, Wis.

Figure 1—Modified Alloy 20 pump which failed after 12 months in a paper mill.

Reprinted with permission from *Materials Protection*, June 1970, 11-13, © 1970 National Association of Corrosion Engineers

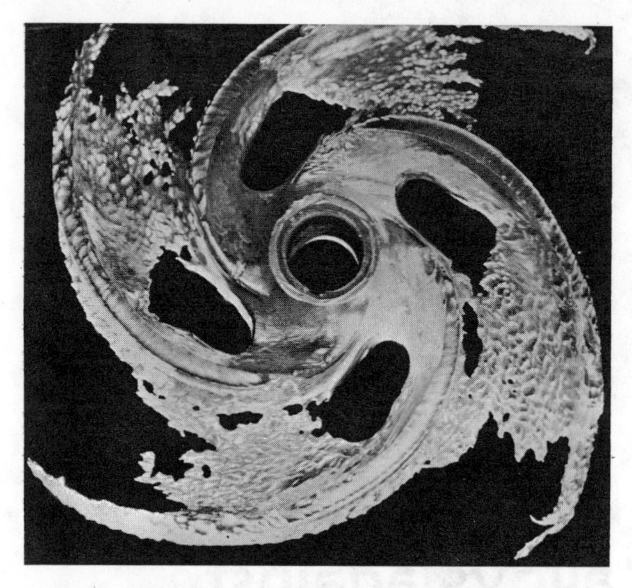

Figure 2—Modified Alloy 20 impeller which failed after 12 months of service pumping 11% NaOH at 185 F.

Figure 3—Illium PD pump impeller as it appeared after use in this application, 4368 hrs. Illium PD has a projected life of two years.

The potential presence of sulfuric acid in the sulfite process also poses a threat to alloys in this system. For this reason, alloys with a proven record of resistance to sulfuric acid are important in providing optimum service life from pumps, valves, and other high frequency replacement components in sulfite process equipment. Illium 98 and Illium PD are capable of withstanding the corrosive attack of sulfuric acid at temperatures to the boiling point in concentrations in excess of those normally encountered in this application.

Erosion-Corrosion in the Soda and Kraft Processes

Sodium hydroxide liquor is common to the soda and Kraft processes. This highly caustic solution (pH 12 to 13) dissolves the lignins in wood chips and ruins pumps. A specialty alloy pump impeller removed after twelve months of service in this environment (Figures 1, 2) was cast in modified CN-7M (Alloy 20). The pump was interposed between the causticizing tank and the clarifier. The solution pumped was approximately 11% by weight sodium hydroxide and contains abrasive solids (calcium carbonate) at a temperature of 185 F (85 C).

TABLE 1—Hardwood Spent Sulfite Liquor from Digester Typical Analysis

Specific Gravity (25 C)	1.048
pH	2.30 to 3.0
CaO	4.52% on Solids
MgO	0.18% on Solids
Total Sulfur (%S)	5.95% on Solids
Acetic Acid	6.37% on Solids
Formic Acid	1.26% on Solids
OCH	8.74% on Solids
Reducing Sugars	25.00% on Solids
Sulfonic Sulfur (%S)	4.60% on Solids

The causticizing process is an important economic step in reclaiming liquor for re-use in the digestion process. Soda liquor separated from the pulp in the blow tank is concentrated to about 40° Bé. This concentrate is converted to a black ash in a kiln. The black ash containing sodium carbonate is dissolved in water and separated from the insolubles of the ash. Makeup sodium carbonate is blended into the stock solution to bring the concentration to 20% by weight. This feed material is mixed with calcium hydroxide at 185 F to form sodium hydroxide and calcium carbonate.

Results of laboratory erosion-corrosion studies of the five alloys (Table 2) in a controlled alumina enriched sample of this caustic liquor are shown in Table 4. The performance of Illium PD appears to be significantly better than the other alloys studied. From these results, it was reasoned that the projected life expectancy for this alloy should be nearly twice that of the modified CN-7M alloy which was in service only twelve months when it failed.

Field Test Procedure

Laboratory test results indicated that Illium PD outperformed the more expensive modified CN-7M alloy by

TABLE 3—Accelerated Erosion-Corrosion in Hardwood Spent Sulfite Liquors

Alloy	Specimen Weight (gms) Before	After	Relative Erosion-Corrosion Rate (mpy)
Illium 98	32.2792	32.2789	0.51
Illium PD	29.8990	29.8944	0.82
Modified CN-7M (Alloy 20)	32.1817	32.1722	1.73

TABLE 2—Mechanical Properties of Specialty Alloys Subjected to Erosion-Corrosion Studies

Designation	Description	Tensile ($lbs/inch^2$)	Yield ($lbs/inch^2$)	Elongation (%)	Hardness BHN
CF-8M	Cast 316 Alloy	80,000	42,000	50	165
Modified CN-7M	(Alloy 20)	75,000	30,000	35	140
Illium P	Fe Base Cr-Ni-Cu Alloy	115,000	78,500	20.5	255
Illium PD	Fe Base Cr-Ni-Co Alloy	100,000	70,000	37	210
Illium 98	Ni Base Cr-Mo-Cu Alloy	54,000	41,000	18	160

TABLE 4—Accelerated Erosion-Corrosion in Causticized Liquor

Alloy	Specimen Weight (gms) Before	After	Relative Erosion-Corrosion Rate (mpy)
CF-8M (316)	30.3667	30.3646	0.65
Modified CN-7M	32.1459	32.1437	0.57
Illium P	29.7435	29.7420	0.41
CN-7M (S-20)	31.4510	31.4491	0.37
Illium PD	29.8805	29.8793	0.31
Illium 98	32.5120	32.5109	0.27

Figure 4—Modified CN-7M alloy pump impeller which failed after 2916 hrs.

TABLE 5--Comparative Chemistry of Tested Impellers

	Illium PD (%)	Modified CN-7M (%)
Nickel + Cobalt	13	23.5
Chromium	27	19.5
Silicon	–	3.25
Molybdenum	2.5	2.75
Copper	–	1.75
Iron	Balance	Balance

two to one in both the Kraft and sulfite processes. In order to test the validity of the data, it was necessary to determine the comparative performance of both alloys operating in a pulp and paper mill. A Kraft mill in northern Wisconsin cooperated by allowing a test pump to be installed in place of an existing pump. Differences in pump design can affect the life of a pump. In order to compare only alloy performance, wear components were made of the two test alloys (Table 5). The pump casing and one impeller were cast in Illium PD. A duplicate impeller was cast in the modified CN-7M alloy. The intent was to operate the pump with one impeller for six months, change impellers, and operate the second impeller for the same period of time.

The pump replaced was adjacent to the causticizing tank and had pumped a sodium hydroxide solution containing abrasive calcium carbonate solids. This slurry, maintained at a temperature of 185 F, was pumped from the causticizing tank to the clarifier, where the solids were removed.

The pump with the Illium PD impeller was installed August 5, 1968. An elapsed time-hour meter was installed in the pump motor circuit to determine precise operating time. Six months later, the pump was shut down for inspection and impeller change. The hour-meter reading at the time of the shutdown was 4368 hours. The Illium PD impeller was removed and replaced with the modified CN-7M impeller. The pump was returned to operation the same day. Approximately five months later, the pump was shut down after only 2916 hours of operating time.

Results of Field Test

The wear pattern of the modified CN-7M alloy impeller was identical to that exhibited in Figure 2. A comparison of Illium PD, and the modified CN-7M alloy pump impellers as they appeared after use in this application is shown in Figures 3 and 4.

Detailed dimensional analysis and weight loss measurements of the impeller demonstrated that Illium PD outperformed the modified CN-7M alloy by a ratio of nearly 5 to 1. Erosion-corrosion rates based on dimensional changes were 0.052 inch per year for Illium PD, while the rate for the modified CN-7M (Alloy 20) measured 0.270 inch per year (Table 6).

Conclusions

According to field test results, Illium PD extended pump life to nearly two years, compared with five months obtained from the modified CN-7M alloy.

Reference

1. F. H. Beck. Alloy Casting Institute Project 50, "Erosion-Corrosion Resistance of Cast Stainless Alloys."

GLENN SCHIEFELBEIN is technical director with Stainless Foundry & Engineering, Inc., Milwaukee, Wis., where his responsibilities involve research and engineering. An NACE member, Schiefelbein holds a BS in Physics from Marquette University.

TABLE 6—Results of In-Plant Evaluation

	Weight Loss (lb)	Dimensional Change (inch)	Operating Time (hrs)	Measured Erosion-Corrosion Rate (ipy)
Illium PD	0.300	0.026	4368	0.052
Modified CN-7M	0.850	0.085	2916 (failed)	0.270

Prevention of Condenser Inlet Tube Erosion-Corrosion

JORAM LICHTENSTEIN, *Southern California Edison Co., Rosemead, Calif.*

ONE of the major problems encountered with surface condensers, whether utilizing sea water, fresh water, or brackish water, is the problem of inlet tube corrosion and erosion or impingement attack. This problem is caused by the relatively high water velocity and the turbulence of the water as it enters the tube. In many cases, the rapid localized attack is also attributed to the ingestion of air trapped in the water box. The air entrainment and turbulence can disrupt the normally present protective film or scale on the tube.

Erosion-corrosion impingement can be detected by the localized attack on the inlet ends of copper based tubes, usually within the first six inches of the inlet end. The metal is thinned with gouges or ripples, and frequently a "horseshoe" shaped cavity or pit with the toe pointing in the upstream direction appears.

Of course, the severity of the attack depends on the water velocity in the tube, the cooling media (*i.e.*, salt water, brackish water, etc.) and on the tube material or alloy. In terms of relative resistance to impingement attack, the alloys in order of increasing resistance are

Arsenical Copper	(CDA 142)
Admiralty Brass	(CDA 443)
Aluminum Brass	(CDA 687)
90-10 Copper-Nickel	(CDA 706)
70-30 Copper-Nickel	(CDA 715)

This is a general list; there are numerous other more costly materials that perform better, such as titanium, etc. In addition, in some cases, all of the above copper base alloys have been reported to suffer attack in various degrees under severe circumstances.

With a given set of conditions, when erosion-corrosion problems arise and upgrading the alloy does not correct the problem, a relatively inexpensive method of preventing the attack and prolonging the life of the tube would be to coat the inlet end in the attacked region (normally 6 to 10 inches inside the tube) and coating the tube sheet. The coating material is usually an epoxy material or a high quality organic material.

Experience with Coatings

Sprayed-in epoxy has been used successfully by a major West Coast oil company for more than 10 years with virtually complete success. The success is attributed to the resistance of the coating in sea water to velocities above 25 fps. Since the coating is applied on a corrosion resistant alloy, the presence of corrosion products which might cause an undercutting problem for the coating is minimal, and the life of the coating appears to be almost indefinite. Users report the coating to be as good as new after four or more years of service.

The writer sees only one possible problem with a sprayed-in epoxy coating. If the coating material after application and during set up time runs or flows from the top of the tube to the bottom of the tube, an accumulation of material could cause accelerated cavitation in the tube immediately beyond the coating. The same problem might occur if the spraying procedure is incorrect.

Coating Application

1. Dry the condenser or heat exchanger with compressed or hot air. Normal draining

Reprinted with permission from *Materials Performance*, March 1974, 17-18, © 1974 National Association of Corrosion Engineers

CRAFTSMAN spraying tube ends.

and drying may suffice; however, the condenser must be absolutely dry before coating.

2. Sandblast each tube with short and quick bursts using ordinary sandblast equipment. A dwell period of 1 to 3 seconds is normally sufficient. Individual experience will soon tell a maintenance crew the required time to produce a clean surface.

Some specialized sandblasting equipment operates on a ram-jet principle, sending a short burst of sand through the tube. Five to 10 bursts usually cleans the tube end entirely.

Prior to cleaning individual tubes, the tube sheet should be brushed off and blast cleaned to remove loosely adhering contaminants.

3. Each tube is then coated internally by inserting a special spray nozzle to the depth desired, pulling the spray gun trigger and rapidly withdrawing the gun.

The major problem that the writer can see in this procedure, however, assuming that a nonsagging coating is used, is that the workman spraying the tubes might pull the spray gun trigger before he has started withdrawing the gun. If this happens, the excessively thick ring of coating will cause a new erosion problem. This, incidentally, is the problem with most tube inserts that are used to overcome inlet end erosion problems. If the insert does not conform completely at the inner end, secondary erosion will occur just beyond the end. In the experience of the writer, full conformation of the inserts is seldom achieved, due to the fact that the tube itself is not perfect.

4. A better application procedure is to use the extrusion method developed by one coating applicator on the West Coast where it has been used successfully for several years. This procedure involves the application of the coating to the tube by withdrawing a plunger which has been loaded with epoxy material. This is a difficult, slow and costly operation; however, the coating is applied uniformly throughout the tube, and this procedure also uniformly fills all the pitted areas in the tube. The excess material that flows out of the tube after extraction of the plunger is used in coating the tube sheet.

5. Any high quality catalyzed epoxy or modified epoxy will work satisfactorily; however, it is absolutely necessary that the coating material be viscous enough to avoid sagging or flow after application.

This method of protecting the tube ends has withstood the test of time. It is felt that it is a positive economical method of overcoming tube end damage, and is much less expensive than using plastic or metallic inserts. At an average speed, it takes about 3 seconds to coat a tube, including moving, inspection, etc.

JORAM LICHTENSTEIN is the corrosion engineer for the Southern California Edison Co., Rosemead, Calif. He received BS and MS degrees from California State University, Northridge, where he also taught for two years in the Mechanical Engineering Department. Lichtenstein is accredited as an NACE Corrosion Specialist and is currently serving as chairman of the NACE Los Angeles Section. He is chairman of the Southern California Cathodic Protection Committee and past chairman of the Western States Corrosion Seminar.

Source: *Materials Performance*, March 1974, 17-18.

Robert F. Joy
Bethlehem Steel Corp.
Bethlehem, Pa.

A BIN of corroded bolts can represent a huge investment loss.

Fastener Corrosion—An Expensive Problem

SUMMARY

From two tests conducted under widely differing severe industrial atmospheric conditions, aluminum coatings for fasteners indicated superiority over zinc and other coatings. For less severe conditions, a zinc coating can give adequate protection and possibly is most economical. Though initial cost is important in the selection of a coating, replacement costs, ease of removal by ordinary wrenches, safety of installation, and reliability of production equipment and processes also must be considered.

THE purchase price of fasteners is only a fraction of the ultimate cost. Expenses such as labor cost for the time of one or more men to remove and replace a corroded $0.25 bolt, transportation to the job site, and downtime on production equipment may be involved. Bolting is generally given little consideration during initial design and erection or as a maintenance item. Bolts sometimes corrode to the extent (Figure 1) that removing them for critical maintenance becomes impossible. After a single plant turnaround, scrap bins may be filled with discarded corroded bolts.

Industry has long recognized the need for improved corrosion resistance and longer life of fasteners. The application of zinc coatings, either hot dipped or electroplated, was followed by electroplated cadmium for specialized applications. About ten years ago, after many years of laboratory experimentation, commercial aluminum coated fasteners and poleline hardware items were made available as standard products.

As early as the 1930s, ASTM Subcommittee XVI of A-5 reported the superiority of aluminum coatings over zinc coatings, especially under severe atmospheric conditions. With this as a background, production zinc and aluminum coated fasteners and poleline hardware items were exposed at various locations under widely varying degrees of atmospheric contamination. A special attempt was made to test coatings of commercial quality with comparable thicknesses so that direct comparisons would be meaningful. Results of a 10-year exposure are recorded under Test Series 1.

Several years ago, Gatewood Norman, Corrosion Engineer, Firestone Tire & Rubber Co., Orange, Texas, became interested in fastener coatings and exposed specimens for a 21½-month period in the Firestone chemical complex. The specimens consisted of B-7 studs, both uncoated and with various metallic coatings, assembled in pipe flanges. The

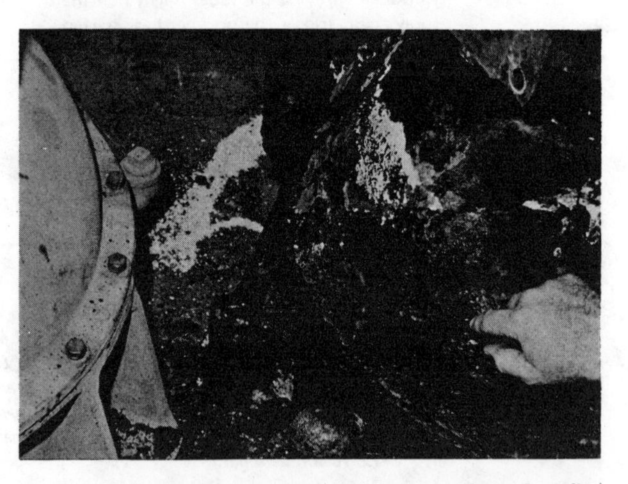

Figure 1—Corroded bolts in the head of a gear box of cooling tower equipment.

Reprinted with permission from *Materials Protection*, December 1969, 63-67, © 1969 National Association of Corrosion Engineers

flanges also were given various coatings and used in different combinations to evaluate effects, if any, of galvanic corrosion. Results after 21½ months of weathering will be evaluated under Series 2.

Series 1 Tests

In 1956, test racks consisting of duplicate samples of zinc and aluminum coated items including guy strands and carefully weighed washers were exposed at several test locations. At periodic intervals specimens were removed for laboratory examination and determination of coating loss. Test locations:

1. Philadelphia, Pa. — This location was in the middle of an oil refinery producing many petroleum byproducts. Because of the diversity of aggressive atmospheric contaminants, it was considered a severe industrial location.
2. Bethlehem, Pa. — The test rack was placed in the middle of a fully integrated steel plant within 300 yds of the blast furnaces and on a level with the top of the stacks. The site was considered moderate industrial.
3. San Francisco, Calif. — A small steel mill located on San Francisco Bay and subjected to ocean fogs. The location was considered mild industrial-marine.

Philadelphia Oil Refinery Exposure

In the Philadelphia oil refinery, all items originally zinc coated were badly corroded, whereas those aluminum coated, although discolored, had not rusted. The structural steel angle crossarm originally zinc coated was 100% rusted. The head of the originally galvanized lag bolt fastening the angle to the wood crossarm was badly corroded. Not only had the zinc completely disappeared, but the head had corroded to about ½ the original thickness. The original aluminum coated bolt was in prime condition.

Typical comparative clamp assemblies from the refinery location are shown in Figure 2. Obviously, the rusted strand, originally Class A zinc coated, had practically no strength remaining; a laboratory pull test of that aluminum coated proved the strand retained 100% breaking strength.

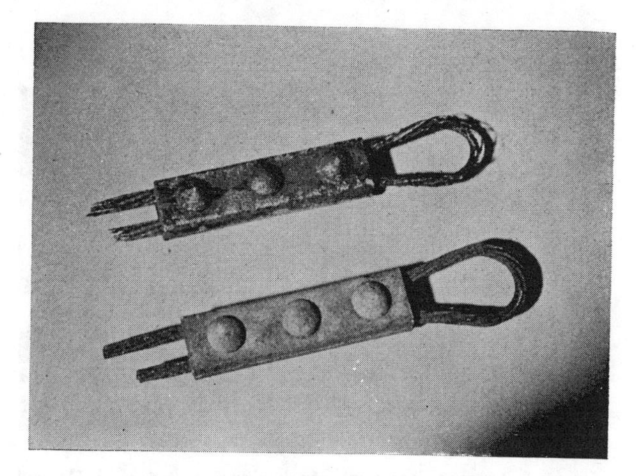

Figure 2—Comparative clamp assemblies.

Zinc coated strands assembled in aluminum coated guy clamps were comparable with those in zinc coated guy clamps, indicating little if any galvanic action between these metals under these atmospheric conditions. Actually, the zinc and aluminum couple had been used for years in the electrical utility field with no serious consequences. Most aluminum conductors currently in use are fastened with galvanized hardware.

Coating Characteristics

Coating Characteristics

Electroplated Zinc

While it is possible to produce heavy coatings by electroplating, this is not done commercially because of excessive cost. The heaviest (thickest) electro-zinc coating recognized by ASTM Specification A164 is less than ½ that of the minimum of a hot dip zinc coating (ASTM A153); hence, life expectancy is less than half, since corrosion of zinc is directly proportional to the amount present. Where corrosion is not severe, or long life is not required, the main advantages of electrodeposited coatings are more positive control of coating thickness, improved appearance, and improved thread fit.

Hot Dip Zinc & Aluminum

Since the weight of aluminum is only about 37% that of zinc, it is preferable to use coating thickness rather than ounces per sq ft when comparing the two coatings. ASTM A153 specifies a minimum zinc coating of 1.25 ounces per sq ft or 2.1 mils, which is comparable to the aluminum which has a specified 2.0 mil minimum, 2.3 mil average coating. The threads of aluminum coated fasteners are somewhat cleaner, allowing for improved thread fit and less oversize tapping of nuts. Aluminum coated nuts are generally tapped 0.010-inch oversize vs 0.020-inch for zinc coated. Both types of coated nuts will break the bolt before stripping occurs.

Zinc Silicate

This is a zinc silicate coating and results from the mixture of a reactive liquid with finely divided zinc powder. While not a paint, it can be applied to properly sandblasted steel surfaces by conventional spray equipment. Following coating and prior to service, the coating must be cured either by a surface chemical treatment or by baking at 350 F (177 C) for one hour. The resulting gray, matte coating has good weather resistance and affords a degree of galvanic protection to the base steel at damaged spots. The coating becomes costly for small parts such as fasteners because of individual treatment.

Inorganic Zinc Coating

This is a zinc rich coating utilizing an inorganic polymer of silicon and oxygen. It can be applied by brush or spray gun over a commercial or white sandblasted surface. There is no special curing treatment required, and the coating becomes water insoluble 20 minutes after application. The coating is sacrificial and will protect steel galvanically.

Moly Coating

The primary purpose of this coating is for lubrication, and it acts to prevent galling. A molybdenum disulfide solution is sprayed over a phosphate coating base and then baked. Molybdenum disulfide has a very low coefficient of friction.

Phenolic-Moly Coating

A molybdenum disulfide coating is applied over an epoxy-phenolic coating. The phenolic coating imparts corrosion resistance, while the moly adds lubricity and resists galling.

Source: *Materials Protection*, December 1969, 63-67

Figure 3—Exposed specimens of aluminum coated steel (A) and zinc coated steel (B).

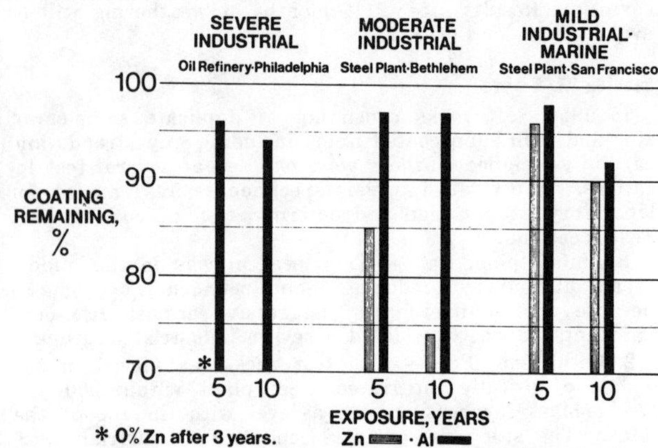

Figure 4—Corrosion comparison of zinc and aluminum coatings. Five and 10-year atmospheric exposure.

Bethlehem Steel Plant Exposure

The specimens from the Bethlehem location followed patterns similar to those of the refinery but to a lesser extent. Rusting of some of the zinc coated items began after ten years' weathering.

San Francisco Steel Mill Exposure

All items exposed at the San Francisco location were in good condition after ten years. There was some discoloration of the aluminum coated parts which is characteristic of the early stages of weathering of this coating.

Possibly due to the small percentage of iron (less than 2%) in the coating or to small pits developing through the coating, a yellow or red cast may develop on the surface relatively early. After additional weathering in industrial atmospheres, aluminum coatings usually turn dark, and the color cast disappears. The parts shown in Figure 3 were exposed by a public utility at San Francisco, adjacent to the ocean where sea spray and fog had free access—a more aggressive location than the steel mill test site. The aluminum coated steel discolored within six months. After three years, the zinc coated specimens started rusting, and this is the appearance after five years. Although discolored, the aluminum coating protected the integrity of the steel, whereas zinc did not. Corrosion severity can change radically in a distance of a few miles, as indicated by these two San Francisco test sites.

Included in the test specimens were carefully weighed steel washers with zinc and aluminum coatings. Sets of these washers were removed periodically, carefully cleaned, and reweighed to determine loss of coating. In the refinery atmosphere, all the zinc coating (originally averaged 5.5 mils) was depleted in 2 to 3 years, hence 0% at 5 years. These data are presented in Figure 4.

Zinc and aluminum coatings do not react the same during exposure. Zinc corrodes relatively uniformly, the rate being dependent upon the severity of the atmosphere. Since the corrosion products formed are soluble and washed away by rains, they do not become increasingly protective with time as occurs with some other metals. The more zinc, the longer the life, in direct proportion.

Because of this uniform corrosion, it is possible to predict the approximate life of zinc coatings at any location, provided coating thickness and corrosion rate are known. From the washer specimens, it has been possible to determine corrosion rates and make life predictions for zinc coatings for the three test locations. A 1.25 oz per sq ft (2.1 mils) zinc coating (ASTM A394 for bolts) would last only about two years in the oil refinery, about 12½ years in Bethlehem, and over 40 years in the mild industrial-marine atmosphere of the San Francisco Bay area. The 2.3 oz per sq ft coating (ASTM A123) specified for structural material would have a life to 100% rust of about 3, 23, and 70 years, respectively.

These results clearly indicate that while a coating may not be suitable for one location or atmosphere, it may be excellent for another. Indications are that zinc coating fasteners has limited value in refinery atmospheres, possibly acceptable life in moderate industrial atmospheres, and is excellent in mild industrial-marine locations. Type of coating selected is entirely a matter of economics and life expectancy.

Aluminum coating corrosion differs from zinc in that it does not corrode uniformly but rather by pitting. These pits may be as much as 3 mils in depth but usually are less and in many instances stop at the iron-aluminum alloy bonding layer. Aluminum corrosion products are not water soluble nor are the complex iron-aluminum corrosion products formed at scratches or other damaged spots in the coating. The tightly adherent corrosion products, apparently impervious to moisture, seal off further attack. There does not appear to be any undercutting or flaking of the coating.

Series 2 Tests

March 15, 1966, ASTM 181 Grade 1 slip-on pipe flange specimens utilizing 8⅝ x 3½-inch B-7 studs with hexagon nuts were exposed on a water cooling tower in the center of the Firestone Synthetic Rubber & Latex Co. plant, Orange, Texas. Not only were the specimens exposed to the spray and steam of the cooling tower, but also to prevailing salt laden air from the Gulf and other atmospheric contaminants typical of a chemical plant complex with adjacent refineries and chemical complexes—a severe industrial area complicated by sea salt. Each of the six specimens consisted of two pipe flanges separated by asbestos gaskets bolted together with studs having different protective coatings. Nuts of the same coating as the studs were screwed hand tight. The stud and nut coatings exposed in pipe flanges were identified: (1) uncoated, (2) electrogalvanized, (3) moly coat (molybdenum disulfide), (4) inorganic zinc, (5) aluminum, (6) hot dip galvanized, (7) epoxy-phenolic moly coating, and (8) zinc silicate.

Specimen 1 consists of two zinc silicate coated flanges; Specimen 2, one galvanized and one zinc silicate; Specimen 3, two uncoated; Specimen 4, one uncoated and one galvanized; Specimen 5, two galvanized; Specimen 6, one uncoated and one zinc silicate. In this manner, it was hoped to resolve galvanic coupling effects. Figure 5 shows typical specimens after three months' exposure. The base steel flanges pictured indicate the severity of the corrosive atmosphere. The specimens were removed on January 30, 1968, or after approximately 21½ months of exposure. Results of careful examination of the bolts and nuts are shown in Table 1. There is a definite galvanic reaction between zinc coated and uncoated items. There is loss of zinc coating from fasteners 4, 6, and 8 where rusting is starting from the bottom of nuts which are in direct contact with the uncoated pipe flange. Where zinc coated flanges are in contact with uncoated or nonmetallic coated fasteners, the flange surface is corroded.

Assuming the original coatings were of equal thickness, zinc applied by hot dipping dissipates more quickly either by weathering or galvanically than zinc in a silicate binder. This is most noticeable in comparing the pipe flanges. There is more severe rusting of the hot dip galvanized coating around the corroded nuts than on a zinc silicate flange. Also, this is evidenced by the hot dipped galvanized nuts on the uncoated flanges starting to rust from the bottom, whereas those coated with a zinc silicate show only minor rust.

A rating of the fasteners indicates the uncoated poorest, with the moly coat practically as bad, and the epoxy-phenolic a very close third. Electrogalvanized, although rusted, is substantially better than the uncoated and moly coated. The hot dip galvanized fasteners all showed white rust with some minor red spots indicating approaching

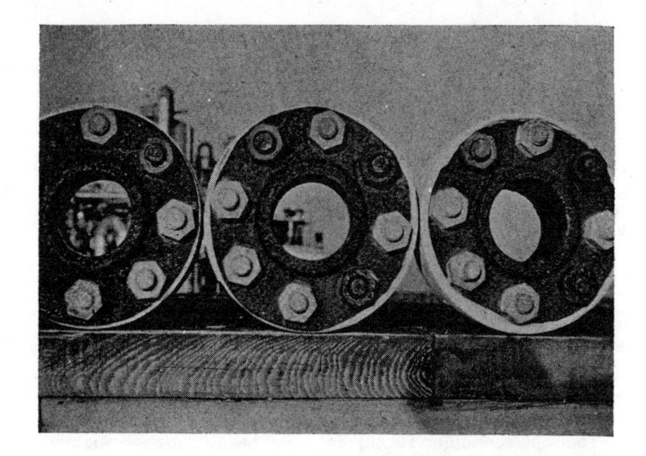

Figure 5—Specimens after three months of exposure.

Coating Procedures

A. Electroplated Zinc
1. Degrease—Alkali bath to remove manufacturing lubricants, grease, and soil.
2. Water Rinse.
3. Pickle—H_2SO_4 for removal of mill scale and rust.
4. Water Rinse.
5. Electroplate—Acid bath with soluble zinc anodes.
6. Water Rinse—Cold (twice).
7. Water Rinse—Hot.
8. Supplemental Chromate coating where specified.

B. Hot Dip Zinc and Hot Dip Aluminum
1. Degrease—Alkali bath.
2. Water Rinse.
3. Pickle—H_2SO_4.
4. Water Rinse.
5. Flux—Zinc ammonium chloride solution at 200 F (93 C) in the case of zinc. Fused salt at 1250/1300 F (677/704 C) aluminum. The chloride-fluoride salt also preheats parts and prevents oxidation until immersed in the molten aluminum.
6. Hot dip zinc is held at approximately 850 F (454 C), aluminum is maintained between 1250/1275 F (677/690 C). Length of time in the bath affects coating thickness.
7. Centrifuge.
8. Water Quench.

C. Zinc Silicate[1]
1. Parts are dry sandblasted to provide a surface free of all mill scale, rust, rust scale, grease, paint or other foreign matter. A 40-80 grade flint or silica sand is used with a minimum of 200 CFM airblast at 100 psi.
2. Blasted surface must be an even, gray-white color and all dust must be removed by brushing or by vacuum.
3. Zinc silicate must be applied as soon as possible after blasting. Do not allow the metal to remain uncoated overnight.
4. Mix zinc silicate and apply by spray to optimum thickness of 2-5 mils, depending on surface roughness of base metal. Pot life of mixed zinc silicate is 4 hours.
5. When parts are dry (usually ½ to 1 hour), coating is cured by
   (a) baking at 350 F (177 C) for 1 hour or
   (b) chemical cure by application of curing solution. Coating must be free of moisture until curing solution is applied.
6. After curing solution has remained on surface for at least 24 hours, wash residue from the surface. Coating is now ready for service.

D. Inorganic Zinc[2]
1. Remove oil and grease deposits.
2. Commercial blasting satisfactory for atmospheric exposure but white blasting required for emersion service.
3. Coating can be brushed or sprayed and can be applied over slightly damp surfaces.
4. Air curing is required prior to service. Curing time is temperature dependent; varying from 6 hours at 85 F (29 C) to 24 hours at 0 to 40 F. Pot life is 12 hours at 75 F (24 C).

E. Moly Coating[3]
1. Degrease.
2. Sand blast to white metal.
3. Phosphate coat—Iron-manganese phosphate.
4. Spray phosphoric acid base primer.
5. Dry.
6. Spray molybdenum disulfide lubricant.
7. Bake at 325 F (163 C) for 2 to 3 hours after part has reached temperature.

F. Epoxy-Phenolic Moly Coating[3]
Steps 1 through 5 as above.
6. Two spray coats of an epoxy-phenolic resin. Dry between coats. Aimed total coating thickness equals two mils.
7. Dry for about 20 minutes at 325 F.
8. Spray moly-disulfide lubricant.
9. Bake at 325 F for 2 to 3 hours after part has reached temperature.

[1] From Dimetcote 3 instruction booklet issued by Amercoat Corporation, South Gate, Calif.
[2] From instruction booklet published by Carboline Corp., St. Louis, Mo.
[3] These procedures from Coatings, Inc., Houston, Texas.

TABLE 1—Results of Examination of Bolts and Nuts

Stud Coating	Uncoated	Flange Coating Galvanized	Zinc Silicate
1 Uncoated	100% R & BF	100% R & BF	100% R & BF
2 Electrogalvanized	100% R & F	100% R & F	100% R
3 Moly	100% R & BF	100% R & BF	100% R & BF
4 Inorganic zinc	RB	RS & WR	WR
5 Aluminum	OK	OK	OK
6 Galvanized	Minor RB & WR	WR	WR
7 Epoxy-phenolic	100% R & F	100% R & F	100% R & F
8 Zinc silicate	Minor RB	OK	OK

(1) R = rust, BF = bad flaking, F = some flaking, RS = rust spots, WR = white rust, and RB = rust starting at bottom of nut.

end of the coating's life. After 21½ months, the aluminum coating and the zinc silicate were in excellent condition.

From an economic standpoint, the best coating appears to be the aluminum coating. This coating is readily available commercially and is stocked at several Gulf Coast locations. It does not appear economical to produce zinc silicate coatings on such small items as fasteners where each piece requires individual handling from initial sandblasting to baking or chemical treatment.

Acknowledgment

The author expresses appreciation to Mr. Gatewood Norman and Firestone Tire & Rubber Co., Orange, Texas for permission to use data related to Test Series 2.

ROBERT F. JOY is a development engineer, New Products Group, Sales Engineering Division, Sales Department, Bethlehem Steel Corp. A native of Philadelphia, Pa., he graduated from the University of Pennsylvania with a BS in civil engineering. He holds a degree in metallurgy from Pennsylvania State University and a second civil engineering degree from the University of Pennsylvania.

He joined Bethlehem Steel in the research department in 1936 and has held various positions in research and development, engineering, market research and sales. A member of the ASM and the Wire Association, Joy is the author of numerous technical papers on metal coatings, wire and wire products, and effects of corrosion on steel.

Corrosion by Microbiological Organisms in Natural Waters*

GREGORY KOBRIN, *E. I. du Pont de Nemours & Co., Inc., Beaumont, Texas*

Characteristics of microscopic organisms known to cause metal attack are reviewed and several classes identified. Particular attention is given to sulfate reducing bacteria. Several case histories are given describing attacks on metals resulting from microbial activity, including water side attack on Ni heat exchanger tubing resulting from differential aeration cells under mud and silt deposits. Concentration of manganese and iron concentrating microbes near weld zones in austenitic stainless steel piping and tanks resulting from inadvertent exposures to contaminated well water caused leaks and/or deep pitting. Pits were cleaned and filled with appropriate weld metal. As a consequence of this experience, the practice at this plant is to use demineralized or high purity steam condensate or natural fresh water—in descending order of desirability—for hydrostatic testing. In all cases, water is removed after testing, and piping and tanks are blown or wiped dry. Bimetallic tubing is used for heat exchangers when the water side is exposed to natural waters likely to be contaminated with bacteria.

MICROBIOLOGICAL ORGANISMS, OR MICROBES as they are more commonly known, may be classified in 4 general groups: (1) bacteria, (2) fungi, (3) algae, and (4) yeasts. These groups broadly represent all the flora and fauna of the microbiological world. Certain species from all groups are known to cause corrosion of metals. In virtually all cases, this corrosion is highly localized and, thus, difficult to monitor and detect before serious damage occurs.

The literature contains several notable examples of catastrophic corrosion attributed to microbes. In the 1950's and 1960's, microbial deposits in aluminum alloy fuel tanks on jet aircraft plugged fuel lines and pitted and perforated the tanks and structural members. Investigation showed bacteria, fungi, and yeasts which had contaminated the water phase of kerosene type fuels to be responsible.[1,2]

In the metal working industry, there have been numerous cases of microbe contamination of emulsions, lubricants, and coolants used in machining, wire drawing, rolling, and deep drawing operations. The results were serious corrosion of drawing dies and wire and sheet products, and staining or alteration of surface finishes.[3]

Thus, it is apparent that microbial corrosion is a serious and expensive problem of concern to the petrochemical industry and many other industries.

This paper will describe in general terms the classes of microbes which cause corrosion of metals, along with some of the postulated corrosion mechanisms. Case histories from the author's experience of catastrophic corrosion of nickel, high nickel alloys, and austenitic stainless steels, by microbes in natural waters serve as examples. Finally, guidelines used by the author's company for avoiding corrosion by waters contaminated with microbes are presented.

*Presented during Corrosion/76, March 22-26, 1976, Houston, Texas.

Classes of Corrosive Microbes

The role of microbes in corrosion processes is not always clear. The mechanisms are not completely understood, and even today, much controversy and conflicting data exist. Frequently, the possibility of microbial influence in metallic corrosion is either completely overlooked or goes unrecognized.

However, at least this much is known about microbes. They can:

1. Produce acids—inorganic, such as sulfuric, as well as organic, such as formic and acetic.
2. Destroy protective coatings.
3. Create corrosion cells—differential aeration (oxygen) and ion concentration cells are notable examples.
4. Produce hydrogen sulfide.
5. Concentrate anions and cations.
6. Oxidize metal ions.
7. Depolarize cathodic sites by consumption of hydrogen.
8. Foul equipment—cooling towers, water lines, heat exchangers, etc.

Corrosive microbes are classified in 6 general groups:

1. **Acid producers.** Some microbes can oxidize sulfur compounds to sulfuric acid; a pH as low as 2 has been recorded where sulfur oxidizing microbes are active. Others can produce organic acids from organic compounds.
2. **Mold growers.** These are primarily fungi.
3. **Slime formers.** Certain algae, yeasts, bacteria, and fungi fit this class. The deposits they form create concentration cells and foul equipment.
4. **Sulfate reducers.** Perhaps the most publicized class of corrosive microbes, they reduce sulfates to sulfides and depolarize cathodic sites on metal surfaces by consuming hydrogen.
5. **Hydrocarbon feeders.** Virtually all microbes feed on hydrocarbons, but those of specific concern to corrosion engineers disbond or destroy organic coatings and linings.
6. **Metal ion concentrators/oxidizers.** Iron and manganese bacteria are important examples of this class. They generally form thick, bulky deposits which create concentration cells or harbor other corrosive microbes.

Aerobic microbes thrive in air and actually need air to survice. Anaerobic microbes, however, can exist and multiply in the absence of air.

There is considerable overlap within these classes of microbes. For example, certain mold growers also produce acids and form slimes.

Microbes are present in virtually all natural soils and fresh and salt waters. Some may also be air borne from one location to another.

The remainder of this paper will be limited to a discussion, including case histories, of corrosion by sulfate reducing and metal ion concentrating/oxidizing microbes in natural waters.

Sulfate Reducing Microbes

Review

Sulfate reducers are found all over the world in soils and waters. They are among the most ancient of living organisms on earth.

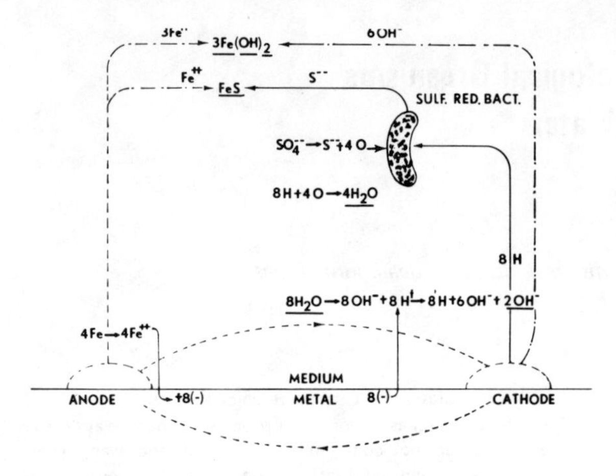

FIGURE 1 – Graphical presentation of the hydrogenase theory of corrosion by sulfate reducing microbes.[6]

Some of the earliest investigations of corrosion by sulfate reducers were made by von Wolzogen Kuhr and his associates, of the Holland Water Works, in the 1920's and 1930's. His publications are considered classics in the field of corrosion science.

Von Wolzogen Kuhr[4] expressed his theory of the mechanism by which sulfate reducing microbes corrode metals with these equations:

$$8H_2O \rightarrow 8OH^- + 8H^+ \quad (1)$$
(electrolytic dissociation of water)

$$4Fe \rightarrow 4Fe^{++} + 8e^- \text{ (anode reaction)} \quad (2)$$

$$8H^+ + 8e^- \rightarrow 8H \text{ (cathode reaction)} \quad (3)$$

$$SO_4^= + 8H \rightarrow S^= + 4H_2O \quad (4)$$
(cathodic depolarization by sulfate reducing microbes)

$$Fe^{++} + S^= \rightarrow FeS \quad (5)$$
(corrosion product at anode sites)

$$3Fe^{++} + 6(OH)^- \rightarrow 3Fe(OH)_2 \quad (6)$$
(corrosion product at anode sites)

$$4Fe + SO_4^= + 4H_2O \rightarrow FeS + 3Fe(OH)_2 + 2(OH)^- \quad (7)$$
(summarized equation)

Equation (4) is the controversial one in the theory. Some of the more recent investigations both support and deny the role of microbes as cathodic depolarizers. For example, investigators have found that FeS will depolarize cathodic sites in the absence of microbes.[5] However, there is little or no disagreement with the fact that iron corrodes in the presence of sulfates and certain anaerobic microbes.

Another theory involves the activity of the enzyme hydrogenase.[6] This enzyme is present in all known sulfate reducers plus many other microbial species. Investigators believe hydrogenase acts as a biological catalyst in the reduction of sulfates by hydrogen at cathodic sites on metal surfaces. This frees oxygen from the sulfate radical which acts as a cathodic depolarizer and increases the corrosion rate of the metal. This mechanism is shown graphically in Figure 1.[6] Note that the equations are similar to those of von Wolzogen Kuhr, except that oxygen is shown as a cathodic depolarizer.

Anaerobic sulfate reducers can exist under conditions normally thought to be strongly aerobic. Sulfate reducers have been found under slime deposits and iron bacteria colonies in aerated water systems such as those served by cooling towers. The deposits shield the sulfate reducers from air, thus allowing them to thrive.

Corrosion attributed to sulfate reducers is virtually always highly localized pitting. Some characteristics are:

1. The pits are generally filled with black corrosion product which, when treated with hydrochloric acid, liberates hydrogen sulfide gas.[5]

2. Metal surfaces beneath the corrosion products are often bright and active.[5]

3. Pits are round at the outer surface, and conical in cross section, with concentric rings inside the pits.[7]

Case History

Heat exchangers at one plant site were cooled by once-through river water. Prior to 1970, the only treatment was silt settlement in a shallow holdup pond. Typical heat exchanger materials, which were required for process corrosion, were nickel and nickel-base alloys such as nickel-copper Alloy 400 and nickel-molybdenum Alloy "B". The process flowed through the tubes, and cooling water was circulated through the exchanger shells.

A few nickel tubes started to leak in 1970 after 18 to 24 months in service, and were plugged. The cause of failure was unknown, but process side corrosion was suspected. Finally, some leaking tubes were removed from an exchanger for inspection. Numerous pits on the external or water-side surfaces were found under heavy mud and slime deposits. Corrosion products in pits were tested for sulfides by treatment with HCl and the results were inconclusive. However, most of the pits were round, conical in cross section, and contained concentric rings.

Microbiological analysis of the cooling water showed the presence of intense microbial activity, including sulfate reducers. Coincidental with the first tube failures was the dredging of the inlet water canal which disrupted the mud and silt deposits, and increased the solids and microbial contents of the water.

Essentially all water-cooled nickel-tubed exchangers were pitted to varying degrees. One tube bundle failed after only 8 weeks in service. Alloy 400 and Alloy "B" tubes were also pitted, but not as severely as nickel.

At this time, an extensive corrosion testing and water treatment program was initiated. The cooling water was treated with chlorine and a solids suspensoid, and miniature test heat exchangers were installed to monitor corrosion of existing tube materials and evaluate alternates.

The results showed that the overall water treatment did not significantly reduce pitting. However, 90-10 copper nickel, nickel-chromium-molybdenum Alloy "C" and AISI Type 316 stainless steel were found to be resistant to pitting under all water treatment conditions.

The following recommendations were made and accepted by plant management:

1. Retube heat exchangers as they fail with bimetallic tubes: 90-10 copper nickel outer (water side) component and Alloy 400 inner (process side) component (determined to be the optimum material for process corrosion resistance).

2. Continue solids suspensoid treatment.

3. Maintain chlorination on a "shot" basis to control slime deposits and clam growth.

The bimetallic tubing has been in service for up to 4 years to date with no failures. This includes the unit where nickel tubes failed in only 8 weeks.

Metal Ion Concentrating/Oxidizing Microbes

Review

This class of microbes is also very abundant worldwide, particularly in natural waters. Essentially all species are aerobes.

The most common are iron and manganese microbes. Bergey's Manual of Determinative Bacteriology[8] presents the following on this type:

1. Grow in iron/manganese-bearing waters.

2. Generally are encapsulated by iron/manganese compounds.

3. Form voluminous, reddish-brown, slimy deposits.

One proposed mechanism for corrosion by iron/manganese microbes[6] is shown graphically in Figure 2. The bulky microbial

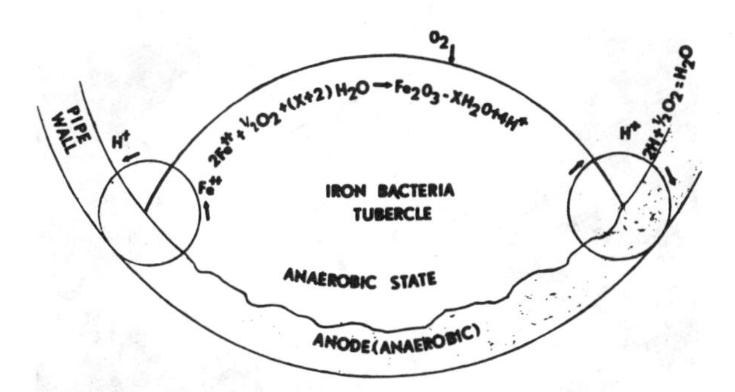

FIGURE 2 – Graphical presentation of corrosion by iron/manganese microbes.[6]

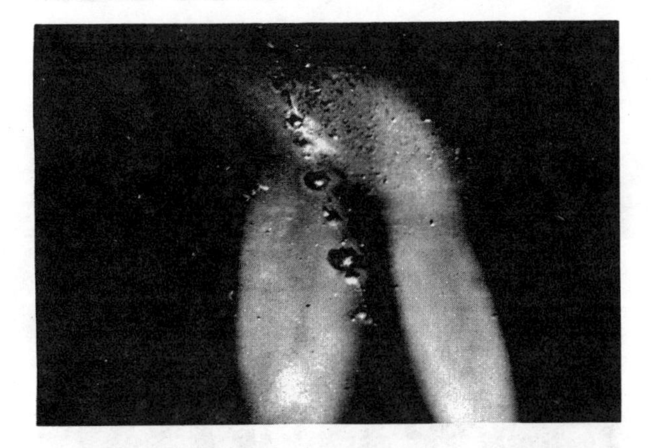

FIGURE 3 – Mound-like deposits along weld seam in bottom of Type 304L stainless steel tank after several months exposure to well water at ambient temperature.

deposits are believed to create differential aeration (oxygen concentration) cells on the metal surface.

Other investigators believe the deposits create anaerobic conditions which harbor sulfate-reducing microbes which actually cause the corrosion. This is one example of anaerobic microbial activity under apparently aerobic conditions.

A third mechanism may involve microbial concentration and oxidation of iron/manganese compounds which contain harmful anions such as halides, on metal surfaces. Ferric and manganic chlorides, for example, are known to cause severe and highly localized attack of stainless steels by penetration of the protective oxide films.

Case History

At one plant site in the 1960's, several projects required a large investment in austenitic stainless steels, primarily Types AISI 304L and 316L, for resistance to nitric, formic, and acetic acids, and to avoid product contamination.

In order to minimize crevice corrosion and gasket leaks, flanges were eliminated wherever possible. Thus, nearly all of the piping was fabricated by butt welding, and hydrostatically tested in the field after erection.

Additionally, about 15 stainless steel storage tanks were too large for shop fabrication, so they were erected and hydrostatically tested in the field.

Early in the construction stages of the various projects, the best quality water available in sufficient quantities for hydrotesting was drawn from two wells on plant property. Although the water was fairly high in chlorides (about 200 ppm), it was potable and, after a sodium softening treatment, was used for powerhouse boiler feed. There were no reports of any unusual incidents of corrosion of piping and equipment by this water. (In the later construction

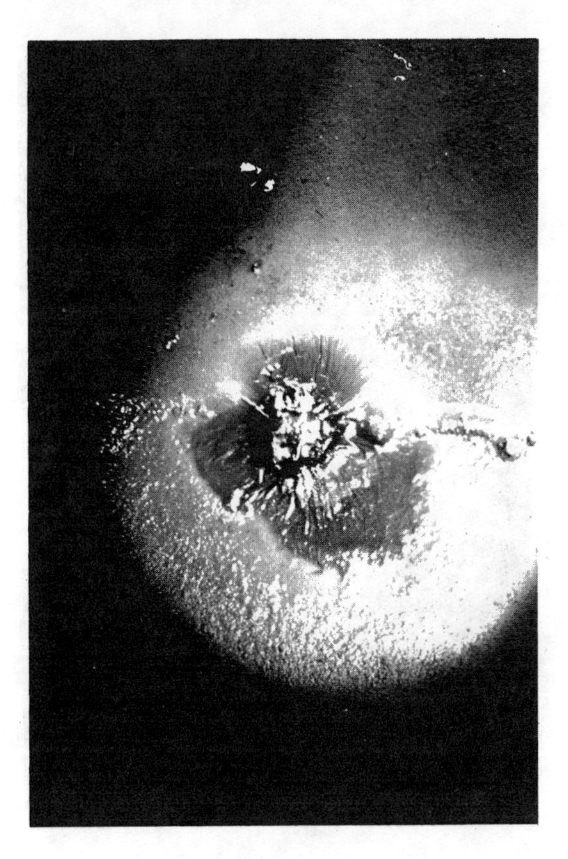

FIGURE 4 – Closeup of a wet deposit.

stages, large quantities of demineralized water became available for testing.)

After the pipelines were field tested, there was little or no attempt made to drain and dry the lines. The tanks were drained soon after testing. However, well water was added to a depth of 2 to 3 feet as ballast because of a hurricane threat. None of the tanks were drained after the threat passed.

The first indication of a problem was water dripping from butt welds in several long horizontal runs of 304L piping about 1 month after hydrostatic test with well water. Inspection showed the nominal 3 mm (1/8 inch) thick pipe walls were perforated by pits adjacent to butt welds. Large, reddish-brown deposits covered the pits.

The second indication was a similar observation on a 316L stainless steel pipeline, except perforation of the nominal 3 mm (1/8 inch) wall occurred about 4 months after the hydrotest.

At this time, the tanks which contained ballast well water were opened for inspection. In the first tank (304L stainless steel), the water had evaporated nearly to dryness, leaving a silt-like residue on the bottom. As shown in Figure 3, reddish-brown deposits were found strung out along all of the bottom butt weld seams.

A closeup of a typical wet mound-like deposit is shown in Figure 4. Note the gelatinous, slimy appearance. The overall deposit is 7 to 10 cm (3 to 4 inches) in diameter.

Figure 5 is a closeup of a dried mound deposit.

The deposit in Figure 5 was wiped clean. The underlying surface was examined visually with the aid of a 10X hand lens. A dark ring-like stain outlined the area under the deposit. The surface inside the ring was bright and lightly etched. However, there was no evidence of severe corrosion or pitting, even after light sanding with emery cloth (Figure 6). Finally, careful probing with an icepick uncovered a large deep pit (Figure 7, arrow) at the edge of the weld.

A radiograph of this weld seam is shown in Figure 8. Note the large pit which consumes almost the entire weld bead width, plus several smaller pits at the edge of the weld.

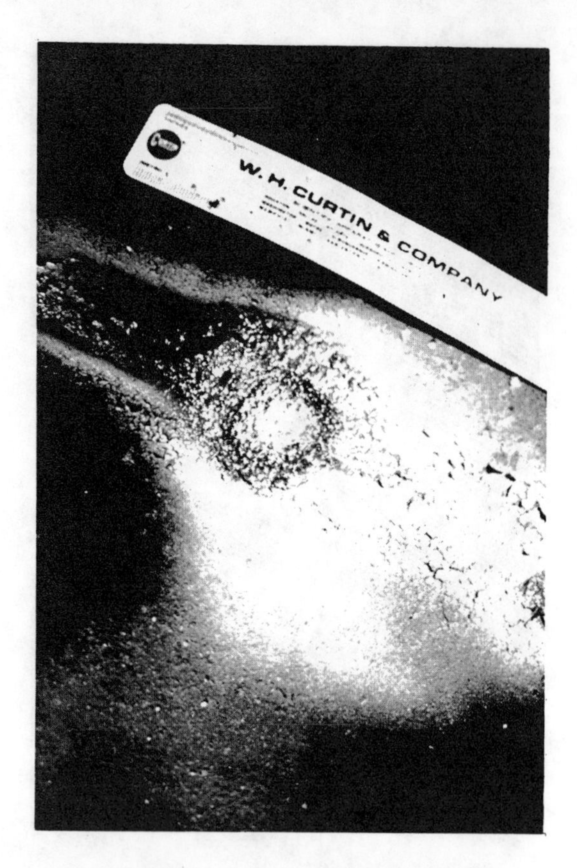

FIGURE 5 – Closeup of a dry deposit.

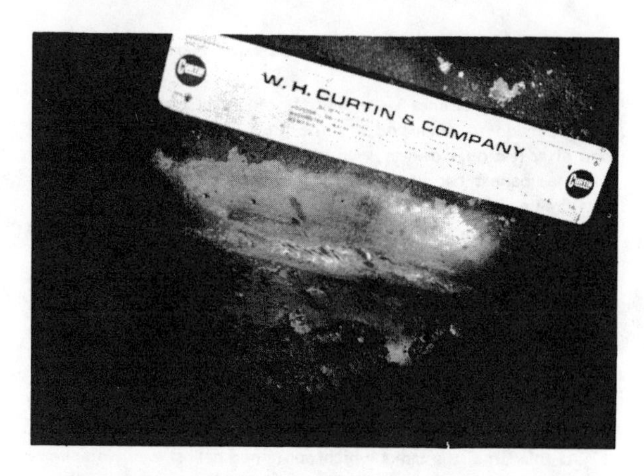

FIGURE 6 – Same as Figure 5, after removal of deposit and light sanding of underlying stainless steel surface.

Samples along pitted welds were taken for metallographic examination. A cross section through a severe pit is shown in Figure 9 [the plate thickness was a nominal 9 mm (3/8 inch)]. The pit was characterized by a minute mouth (arrow) at the surface. A thin shell of metal covered a huge bottle-shaped pit which had consumed both weld and base metal. This shows why the pits were hard to find once the deposits were wiped away.

In this particular case, there was no evidence of intergranular or interdendritic attack of base or weld metal. However, metallographic examination of pitted weld seams in a 316L stainless steel tank showed evidence of preferential corrosion of delta ferrite stringers in weld metal (Figure 10).

This particular 316L tank was drained about a month after hydrotest with well water. The bottom was severely pitted along the weld seams. However, rust-colored streaks were found normal to the horizontal weld seams (Figure 11) in the sidewall. Close examination showed one or more pits at the edge of the weld for every rust-colored streak (Figure 12).

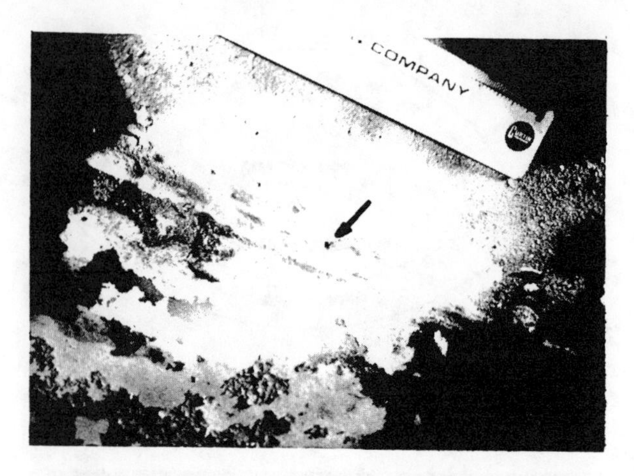

FIGURE 7 – Large pit at edge of weld in Figure 6 disclosed by probing with icepick.

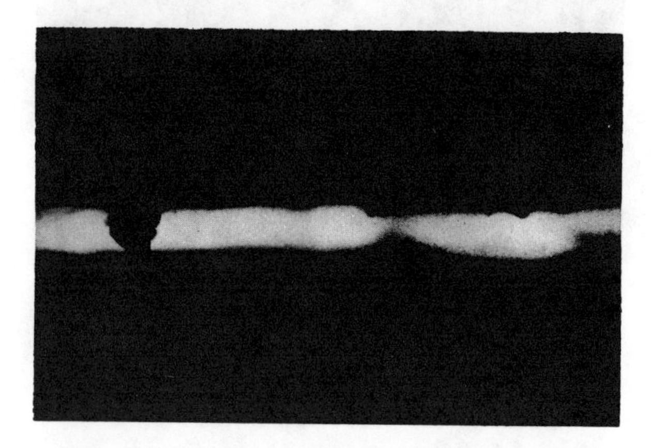

FIGURE 8 – Radiograph of a pitted weld seam in a 304L tank.

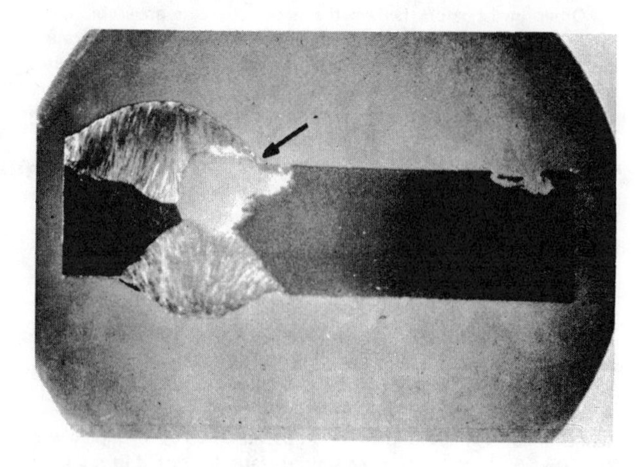

FIGURE 9 – Metallographic mount of a cross section through a pitted weld seam.

Microbiological analysis of the well water showed high iron and manganese bacterial counts. Active colonies of these bacteria were also found in the mound-like deposits. No other corrosive microbes

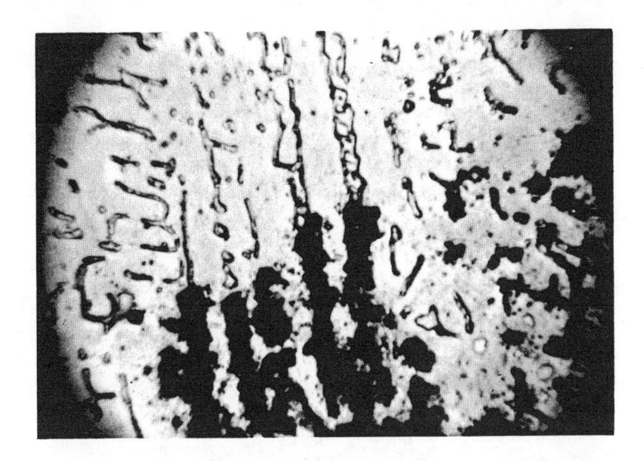

FIGURE 10 – Photomicrograph showing preferential attack of delta ferrite stringers in 316L weld metal. 250X

FIGURE 11 – Rust-colored streaks normal to horizontal weld seams in sidewall of 316L tank.

FIGURE 12 – Closeup of streaks shown in Figure 11.

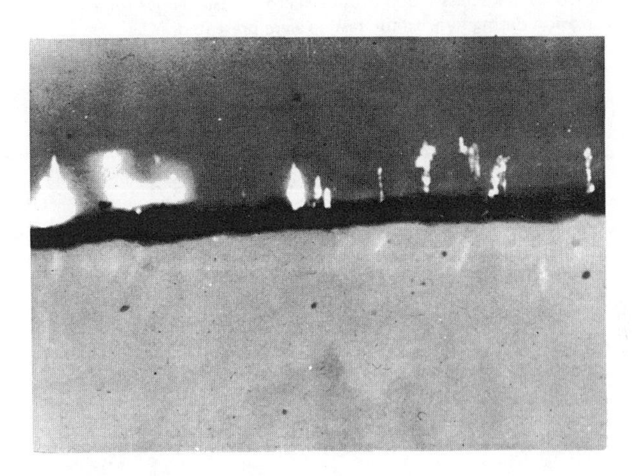

FIGURE 13 – Radiograph of weld seam in bottom of 316 stainless steel solvent tank showing unusual subsurface tunnel pits.

were found. Chemical analysis of typical deposits showed high concentrations (several thousand ppm) of iron, manganese, and chlorides.

We believe the iron and manganese bacteria in the stagnant well water at ambient temperature concentrated iron and manganese compounds containing chlorides on the stainless steel surfaces. The bacteria oxidized iron and manganese to ferric and manganic ions, thus creating ferric and manganic chlorides which are well known as severe pitters of stainless steels.

As indicated, nearly all pits were found at the edges of, or very close to, weld seams. Apparently, the minute surface imperfections typically associated with welds (oxide or slag inclusions, porosity, ripples, etc.) attracted microbial activity.

All of the affected piping was replaced in kind before the facilities were placed in service. The tanks, however, were repaired by first sandblasting to uncover all pits, grinding out each pit to sound metal and welding with the appropriate stainless steel filler metal. The piping and tanks have been in corrosive service for about 8 years to date with very few leaks, an indication that the inspection and repair program which followed initial discovery of this problem was effective.

More recently, the same plant experienced unusual pitting in a repaired 316 stainless steel tank used for storing a recycle solvent stream. Pitting had perforated the tank bottom at several locations after about 6 years service. The pits had the characteristic minute "mouths" at the surface, connected to huge subsurface tunnel-shaped cavities. Figure 13 is a radiograph which shows pits strung out in base metal on one side only of a weld seam. None of the pits were located at prior weld repairs. When the tank was being prepared for inspection, a reddish-brown sludge layer was noted on the bottom. Unfortunately, it was washed out before a sample could be taken for analysis. However, we believe the water phase of the solvent stream had become contaminated with iron/manganese microbes which caused pitting similar to the aforementioned problem associated with well water.

Hydrostatic Testing Guidelines

As a result of these and other experiences with corrosion by microbes in natural waters, particularly at Gulf Coast plant sites, our current practice for hydrostatic testing stainless steel piping and equipment is as follows:

1. First choice—demineralized water. • Drain and dry at the earliest opportunity after testing.

2. Second choice—high purity steam condensate. • Drain and dry at the earliest opportunity after testing.

3. Last choice—natural fresh water, such as river, pond, canal, well, etc. • Drain immediately after test, and • flush with demineralized water or steam condensate, and/or • blow or mop dry within 3 to 5 days of the hydrotest.

We have experienced no corrosion of austenitic stainless steel equipment and piping attributed to hydrostatic test water since the above practice was adopted.

Summary

Microbiological organisms which are known to cause corrosion of metals have been classified, and corrosion mechanisms described briefly. Case histories from the author's experience of pitting of nickel and high nickel alloy heat exchanger tubing and austenitic stainless steel tanks and piping, which was attributed to microbial activity in natural waters, serve to illustrate catastrophic damage caused by two classes of corrosive microbes. Guidelines for avoiding corrosion during hydrostatic testing were presented.

Acknowledgments

The invaluable assistance and advice of R. B. Hitchcock, F. T. Wyman, and H. A. Norman are gratefully acknowledged.

References

1. Miller, R. N. Materials Protection, Vol. 3, No. 9, p. 60 (1964).
2. Elphick, J. J. Microbial Corrosion in Aircraft Fuel Tanks, Chapter 6, Microbial Aspects of Metallurgy, American Elsevier Publishing Co. (1970).
3. Hill, E. C. Microbial Infections in Relation to Corrosion During Metal Machining and Deformation, Chapter 5, Microbial Aspects of Metallurgy, American Elsevier Publishing Co. (1970).
4. von Wolzogen Kuhr, C. A. H. The Unity of the Anaerobic and Aerobic Iron Corrosion Process in the Soil, presented at the Fourth National Bureau of Standards Soil Corrosion Conference, Washington, D. C., 1937.
5. Miller, J. D. A., Tiller, A. K. Microbial Corrosion of Buried and Immersed Metal, Chapter 3, Microbial Aspects of Metallurgy, American Elsevier Publishing Co. (1970).
6. Sharpley, J. M. Corrosion, Vol. 17, p. 386t (1961).
7. Troscinski, E. S., Watson, R. G. Controlling Deposits in Cooling Water Systems, Chemical Engineering, Vol. 77, p. 125 (1970) March.
8. Breed, R. S., Murray, E. G. D., Smith, N. R. Gergey's Manual of Determinative Bacteriology, Seventh Edition, Williams and Wilkins Co., pp. 214-219, 227 (1957).

Valve Corrosion

SUMMARY

Presents 14 case histories of valve corrosion problems, explaining what caused the failure in each case and how the problem was solved. Several types of valves are included: gate, safety relief, plug, and ball. Valve services involve steam, dry chlorine gas, acetic acid, sulfuric acid, hydrogen fluoride, chlorides, and river water. Materials discussed include stainless steels (18-8, 316, and 304), carbon steel, cast iron, chrome plated carbon steel, titanium, and a nickel-molybdenum.

Figure 1—External corrosion on this gate valve's stem started the corrosion attack that resulted in considerable reduction of the stem's diameter. Being on the suction side of a dry chlorine compressor, the valve was subject to condensation and ice formation which caused some external corrosion, especially on the stem. When this roughened stem area was inserted and withdrawn through the packing gland during opening and closing of the valve, dry chlorine leaked onto the stem. When this chlorine combined with the atmospheric moisture, a severely corrosive medium was formed.

Reprinted with permission from *Materials Protection*, June 1965, 28-35, © 1965 National Association of Corrosion Engineers

ONE helper grabbed a valve from the stock room, rushed it to the line where a corroded valve had already been removed. Although inexperienced, he was helping install the valve. All available engineers, operators, technicians, and helpers were rushing to make the necessary maintenance repairs. The chemical plant was down for repairs. The men were trying to cut down-time to a minimum.

When the plant went back on stream, some maintenance men were checking the valve for leaks. They watched with amazement as the hot process acid ate completely through the valve body in about five minutes.

After the chaos and confusion of replacing the valve, the plant corrosion engineer determined what had happened—why the valve lasted only five minutes.

The valve was made of titanium—an exceptional metal with high corrosion resistances but not recommended for the particular hot acid service in this plant. The catastropic failure resulted from a chain of events during the rush of maintenance repair. Valves lined up for replacement during the shut-down were not labelled as to material. How was the inexperienced helper to distinguish between a titanium valve and one of stainless that was supposed to be installed? He didn't even know that a titanium valve would weigh considerably less than a stainless steel valve. And how should he know that the hot acid would devour titanium in minutes?

This catastrophic failure can be explained and possibly excused when one realizes that all technicians, engineers, and experienced men in the plant were busy during the down-time, trying to get all the repair work done in a minimum of time.

This particular valve corrosion problem is probably unique, but the millions of dollars spent each year by American industry to replace corroded valves indicate that other valve problems are not so unique.

Many Problems Involved

Valves, like pumps and other pieces of process equipment, must be designed and specified for corrosion resistance not only to the material being handled but also the environment. This is particularly true of a process stream which becomes highly corrosive when mixed with air or water, as happens on valve stems when the packing fails or when gaskets give way.

Such an example is shown in Figure 1, a gate valve that internally was giving good service but suffered rather severe corrosion externally. The valve was on the suction line of a dry chlorine compressor. Condensate and ice formed on the exterior, causing corrosion on the stem, this resulting in a roughened surface. During valve closings and openings, this roughened area on the stem caused minor leakage as it passed in and out of the valve packing. Any chlorine leakage would then make the corrosion more severe, as evidenced by the reduced diameter of the stem (Figure 1). Type 316 stainless stems are being used to solve this problem.

Rather severe film undercutting also has occurred on the external coating of the valve body and wheel, as shown in Figure 1.

Many Valve Types Available

The diversity of problems that corrosion engineers and operating personnel confront can be assumed to be in some mathematical proportion to the types of valves which range from gate, globe, angle, needle, check, and ball to such specialized ones as diaphragm valves. And each of these types have variations in their trim (replaceable parts of the valve).

For example, gate valves are available with four basic disc designs: double wedge, ball and socket wedge, cored wedge, and solid wedge. And there are four principal designs for globe valve discs: modified plug, ball plug, renewable disc, and v-port plug.

When valves are being specified for particular applications, not only must many valve types and designs be considered but also many types of corrosion resistant materials now available. Other choices involve rubber and plastic linings for valve parts or all-plastic valves and components.

When proper consideration is given to (1) selection of material, (2) effect of various corrosion processes, and (3) design for specific service conditions, the corrosion engineer can expect his specification for a particular valve to give him a valve installation that is as maintenance-free as possible with a long service life.[1]

Case Histories Presented

This article presents some case histories of valve corrosion problems where improper specifications were issued, where proper specifications were made but not produced, and where the corrosion engineer's best laid plans went awry, sometimes for reasons beyond his control. These histories are a meager sampling from only a few plants and installations. Thousands of examples could be found—if all engineers and plant managers would openly discuss their valve experiences.

Some Valve Stem Problems

A mistaken assumption that a Hastelloy C stem for a motor valve would resist the erosion caused by 600 pounds of steam resulted in a failure shown in Figure 2. An 18-8 stainless stem had failed in a similar manner. The problem was solved by using a hardened Type 410 stainless stem and a Stellite plug.

In less than one year, severe corrosion caused replacement of a bronze stem in a gate valve (Figure 3). The attack was localized at the packing area where hot acetic acid and air mixed. Good condition of the interior threads on the stem shows that process corrosion by low oxygen-containing acetic acid was no problem. Hastelloy C stems were used for a temporary solution. Final solution, however, was to change from a gate valve to a ball valve. The limited travel and motion of the ball valve stem reduces the likelihood of acid leaks and has eliminated this problem.

Another example of stem corrosion is similar to that shown in Figure 3. In this case, however, the stem is Type 316 stainless and the acid 98% sulfuric acid. The attack shown in Figure 4, although not too severe, was sufficient to create a problem. The acid leaked out to the air and absorbed enough atmospheric moisture to be diluted to a concentration that would attack the stainless steel. A braided Teflon packing improved the sealing of the stems and reduced this problem.

The operator in one oil refinery probably wondered where his valve stem went (see Figure 5). The stem corroded off completely, necessitating insertion of a threaded steel rod before the valve could be operated.

The valve and its stem in Figure 5 were carbon steel, used in a refinery's strong sulfuric acid service. Carbon steel valves in this service had lasted an average of six months. Where high velocities were encountered, however, the valves were scrapped after only six to eight weeks service.[2]

These failures occurred because strong acid remained on the stem when the valve was opened. This strong acid picked up atmospheric moisture and attacked the stem, resulting in a roughened surface. Once the smoothness of the stem had been attacked, acid leakage could not be stopped.[2]

Another example of corrosion (Figure 6) on the outside of a valve reminds one of the philosophical saying "For want of a nail . . . a kingdom was lost." The $200 Type 316 stainless valve in Figure 6 became inoperable because the $1 hand wheel almost corroded away. Often, expensive valves will fail because of corrosion of the small insignificant parts such as packing gland bolts, stem packing or bonnet gaskets. Experience has shown that a valve to be used in the corrosive atmosphere of a chemical plant should be made of materials compatible with the environment involved.

Experience has taught one chemical plant that bolts on valves should be installed head down so that any severe corrosion on the threaded end extending beyond the nut can be observed. Several bolts had been so severely corroded that the nuts had dropped off, but this condition could not be seen unless maintenance men stooped over and checked underneath.

Valve Plugs Turn Into Gravel

Strange things can happen with valves. An example is steel plug valves in a hydrogen fluoride alkylation process. Some of the plugs were in pieces; others were almost like gravel.

Because hydrogen fluoride was considered a hazardous chemical, a minimum number of flanges and connections were used in the plant. Where

Figure 2—Stem made of Hastelloy C for a motor valve was eroded by steam at 600 pounds. The Hastelloy stem was specified after an 18-8 stainless stem had failed in a similar manner. A hardened 410 stainless stem with a Stellite plug solved this problem.

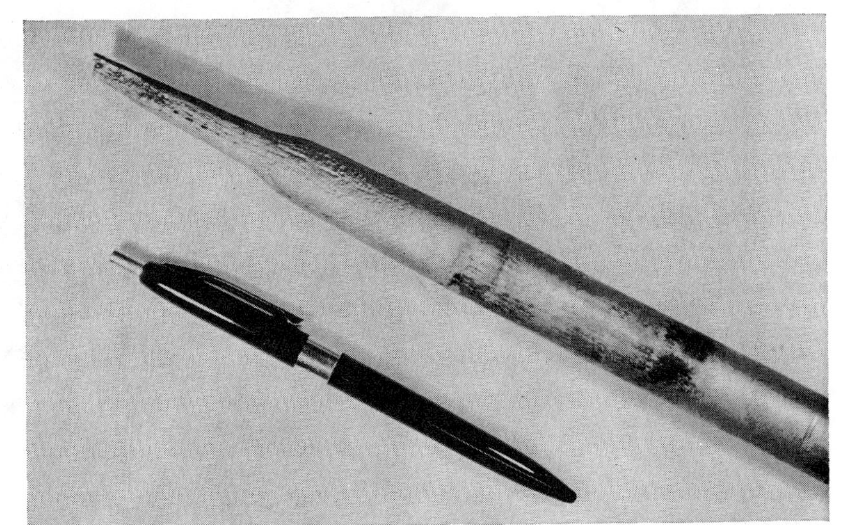

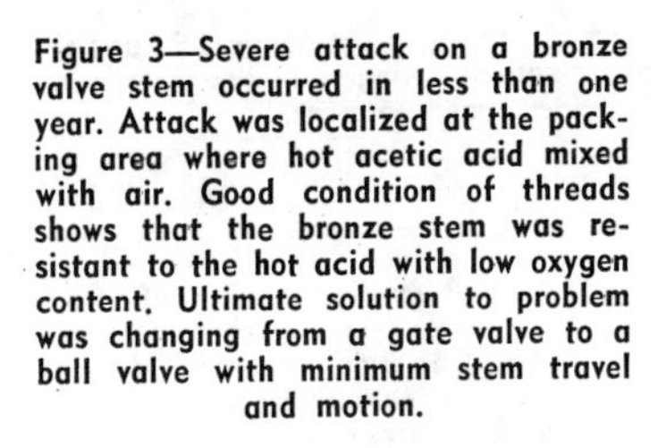

Figure 3—Severe attack on a bronze valve stem occurred in less than one year. Attack was localized at the packing area where hot acetic acid mixed with air. Good condition of threads shows that the bronze stem was resistant to the hot acid with low oxygen content. Ultimate solution to problem was changing from a gate valve to a ball valve with minimum stem travel and motion.

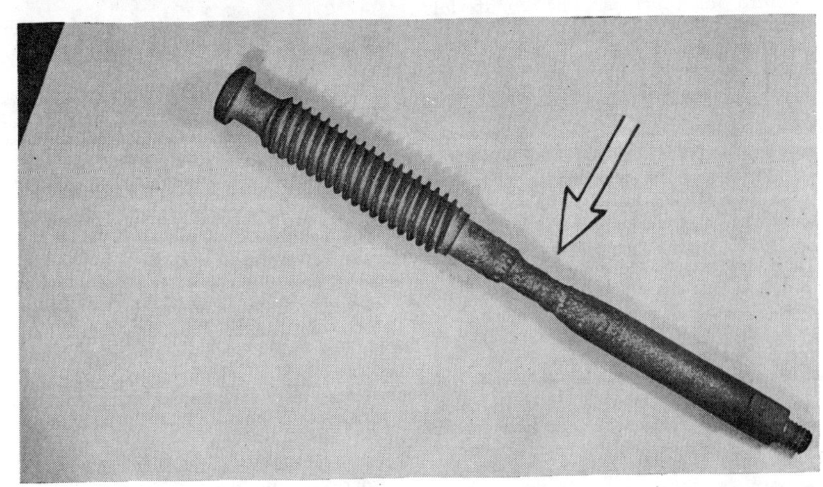

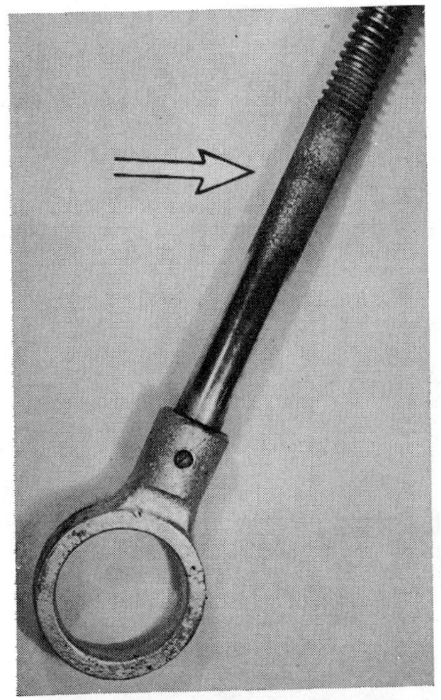

Figure 4—Attack of 98% sulfuric acid on a Type 316 stainless steel stem. Use of braided Teflon packing has improved the sealing of stems on this type valve so that the acid does not leak and dilute with moist air to form a highly corrosive medium.

Figure 5—Stem corroded away on this carbon steel valve when sulfuric acid leaked through the stem packing. A threaded rod had to be inserted to open the valve.

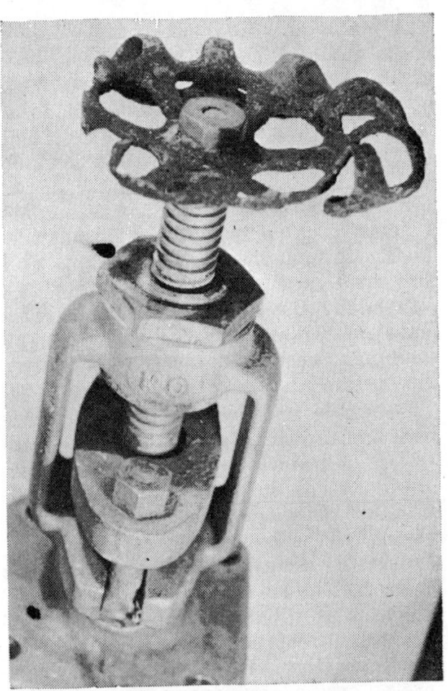

Figure 6—A $200 stainless valve became inoperable when corrosion almost destroyed the hand wheel. This example suggests that all parts of a valve should be of corrosion resistant materials if the valve is to be exposed to a plant's corrosive atmosphere.

Source: *Materials Protection*, June 1965, 28-35

valves were required, grease-sealed cocks were used.[3]

These cocks quickly developed stress corrosion cracking. Although a plug in hydrocarbon acid service might be in excellent working order when the valve was closed following plant start-up, the cock might be completely inoperable a month later when the plug had to be operated to isolate a piece of equipment.

The plugs, made of steel with about 0.45% carbon, had been hardened to 500 Brinnel or more by heating and quenching. Quenching the large plug in a tub of water caused shrinkage of the outside skin while the inside was still hot. After the plugs cooled to a uniform temperature, the outside was under relatively high compression. Any shrinkage cracks which formed usually were invisible.[3]

This problem was solved by using Monel lined, non-hardened plugs with Stellite lined bodies. These materials eliminated the plug cracking and still gave the highly desirable difference in hardness between plug and valve body for tight shut-off.[3]

Figure 7—Male and female discs taken from several gate valves in acetic acid and formic acid service. Honeycomb corrosion was caused by carbide stringers in the metal's microstructure, resulting from improper heat treatment.

Another Metallurgical Problem

Improper heat treatment of an expensive stainless steel valve can cut its service life from several years to eight months, as shown by the next case history. All that was needed to correct the problem was a high temperature anneal heat treatment.

Results of this fabrication oversight were honeycomb attacks as observed on discs taken from several different size valves in acetic acid and formic acid service (see Figure 7). The honeycomb corrosion was caused by carbide stringers resulting from improper heat treatment.

A detailed study of a 2-inch gate valve made of Type 316 stainless steel established the cause of this honeycomb attack. This valve was exposed to a mixture of acetic acid (15 to 30%) and formic acid (1 to 2%, remainder water) at 260 to 270 F (127 to 132 C).

An exploded view of the valve components (Figure 8) shows crevice attack at the threaded end of the stem. The disc arm was almost completely destroyed by the honeycomb attack. The male and female discs also show honeycombing. These same discs are shown at the bottom of Figure 7.

The male and female discs, the disc arm, and the stem were fabricated from Type 316 stainless steel rod stock. The carbide stringers seen in the microstructure (Figure 9) indicate that the original rod or bar stock was hot drawn. During the drawing operation, the carbides precipitated out, parallel to the direction of drawing. This resulted in chromium depletion of the areas adjacent to the carbide stringers. The chromium depleted areas consequently became anodic to the surrounding matrix and were subject to selective attack in acid media. The typical honeycomb attack of the discs and other components is the result of selective corrosion of the chromium depleted areas.

The dramatic change in the microstructure of the stainless material by heat treating to 1850 F (1010 C) and water quenching is shown in Figure 10.

Corrosion of the bonnet of a 2-inch stainless valve (see Figure 11) is the result of interdendritic segregation and porosity from shrinkage during casting operations. Proper casting techinque could minimize the possibility of this type of attack in acid services.

Corrosion engineers at the plant where this failure occurred recommended that the honeycomb attack could be prevented in austenitic stainless steels by proper heat treatment of the bar stock after hot drawing or by heat treatment of the parts after fabrication. The chromium carbides should be redissolved by heating to 2050 F (1121 C) and then water quenched to prevent their reprecipitation.

Exceptional Metals No Panacea

Often, when a valve corrosion problem is encountered, the natural reaction is to specify one of the exceptional metals that usually have high resistance to corrosion. Unless such a specification is first verified by corrosion testing or checked against previous experience records, the more expensive materials may not be the answer at all.[4]

Although these high alloy materials can solve many difficult corrosion problems, selection of a more expensive alloy or metal does not assure that a good application will result. These materials must be specified for valve service with good engineering discretion and with the realization that they are subject to the same corrosion effects as the

Figure 8—Exploded view of corroded parts in a 2-inch gate valve made of Type 316 stainless steel. Honeycomb attack on the plugs and general corrosion occurred in only eight months service in a mixture of acetic acid and formic acid with water at 260 to 270 F (127 to 132 C).

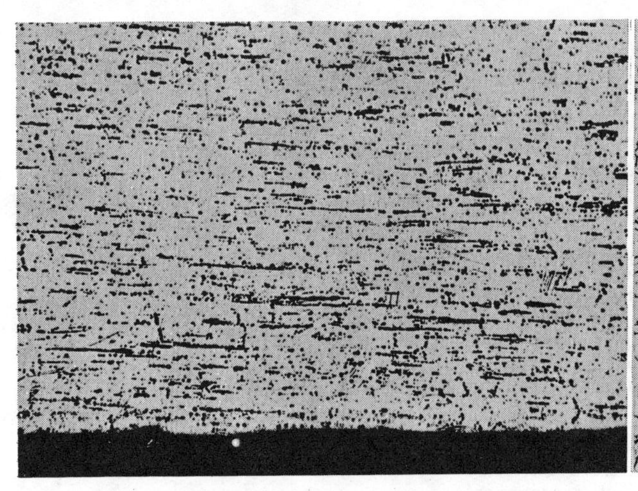

Figure 9—Microstructure of Type 304 stainless disc from a gate valve, showing chromium carbide stringers oriented in the direction of rolling of the stainless bar stock (to the right). 120X, oxalic acid etch.

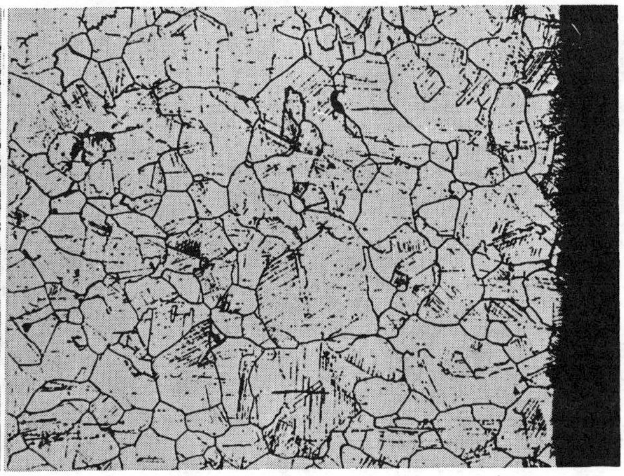

Figure 10—Dramatic change in microstructure of Type 304 stainless steel as compared with Figure 9. This specimen was heat treated to 1850 F (1010 C) and water quenched, 120X, oxalic acid etch.

Figure 11—Corrosion in bonnet of 2-inch gate valve which was cast in Type 316 stainless steel. Attack probably was caused by interdendritic segregation and porosity from shrinkage during casting operations.

more common and less expensive materials of construction.[4]

An example of what can happen is shown in Figure 12, showing the drastic failure of two nickel-molybdenum valves after only five or six weeks service.

Selection of the alloy to handle a process stream consisting of 50% sulfuric acid, various hydrocarbons, and some sulfur dioxide at 275 F (135 C) seemed justified. Sludge contamination, especially with oxidizing agents, was one possible explanation of the failure but did not seem to account for the drastic attack. Erosion-corrosion was ruled out: no high velocities. The possibility of galvanic attack was indicated because the valve had been installed in contact with carbon pipe liners.

Tests showed that contamination (largely copper) accounted for about a 10-fold increase in the corrosion rate of the sludge as compared to chemically pure sulfuric acid. Further tests, however, showed that coupling of the alloy with graphite through chemically pure acid gave a 100-fold increase in corrosion rate over the uncoupled rate. Coupling the nickel-molybdenum with amorphous carbon increased the rate 500 times.[4]

The catastrophic attack shown in Figure 12 should be a warning against coupling nickel-molybdenum alloys with carbon or graphite under like conditions.

Another example of valve corrosion when an exceptional metal was used is shown in Figure 13. This titanium motor valve plug lasted only one month in an organic chloride environment containing free sodium chloride salt. The titanium plug was installed after a Type 316 stainless steel and a Hastelloy C plug were found unsatisfactory due to corrosion. The titanium was resistant to corrosion, as expected, but the plug failed because of erosion from the salt particles.

Figure 12—Drastic failure of these nickel-molybdenum valve parts occurred in only five or six weeks. The exceptional metal alloy was used to resist 275 F (135 C) process stream containing 50% sulfuric acid, various hydrocarbons, and some sulfur dioxide. Tests showed failure was caused by galvanic coupling of the alloy with carbon pipe liners.

River Water and a Ball Valve

Only three or four months were required to severely corrode a ball valve handling river water. Two views of the chrome plated, carbon steel ball are shown in Figure 14.

The river water varied in corrosivity due to upstream contamination; river water pH varied from 5.8 to 6.5, depending on the periodic contamination (acid sulfates). Water temperature was about 60 to 65 F (16 to 18 C); velocity varied from 12 to 30 fps.

Chrome plate on the surface of the ball was almost eliminated in the 4-month service period. The exposed carbon steel was also disappearing due to erosion-corrosion. The valve body, also made of carbon steel, was badly corroded.

Solution to this problem was use of an all stainless (Type 316) ball valve which is giving good service. However, the installation is being closely watched. A final replacement will be a rubber lined diaphragm valve which will be designed to eliminate the corrosion-erosion problem caused by the contaminated river water.

Cracking of a Safety Valve

Not one would deny that all valves are important to a system and that corrosion is dangerous wherever it occurs in the process stream. But corrosion attack on a safety relief valve will make any corrosion engineer become wide eyed. The safety engineer, of course, would be even wider eyed.

Stress corrosion cracking developed over a period of years in a Type 304 stainless safety valve in 300-pound steam service when chloride-bearing salts collected in the stagnant hollow interior of the plug. The cracking developed to the point that the plug broke away from the stem, as shown in Figure 15. This could have been prevented by a design change to eliminate hollows or crevices where the salts could collect.

Failure at Gasket

Even when the valve design is perfect and the proper materials have been selected for the best corrosion resistance to the process stream, gasket failure or even minor gasket leaks can cause severe problems.

An example is the cast iron valve body shown in Figure 16. The gasket developed a leak and permitted sulfuric acid of intermediate concentration to contact the exposed cast iron flanged edge of the valve. The plastic lined valve was handling 30% sulfuric acid and additional chemical reagents. The gasket loosened between the rubber lined pipe flange and the raised face of the plastic lined valve flange. As a result, the sulfuric acid worked its way past the loosened gasket and down over the flange and flange bolting.

Final result was that the cast iron quickly corroded away in the areas shown in Figure 16. The plastic lined area was unaffected.

Solutions to this gasket problem was to replace the valve body with another cast iron body but to take better care in gasketing the pipe joint. More frequent maintenance inspections were made to insure corrected gasket pressure on the flange joint.

Valve Coatings—Another Problem

If a valve is to be protected externally from the corrosive environment in a given plant, the customer normally applies a coating when the whole piping system is coated. Many manufacturers apply a quick drying enamel coating to make their valves look pretty and smooth, realizing that

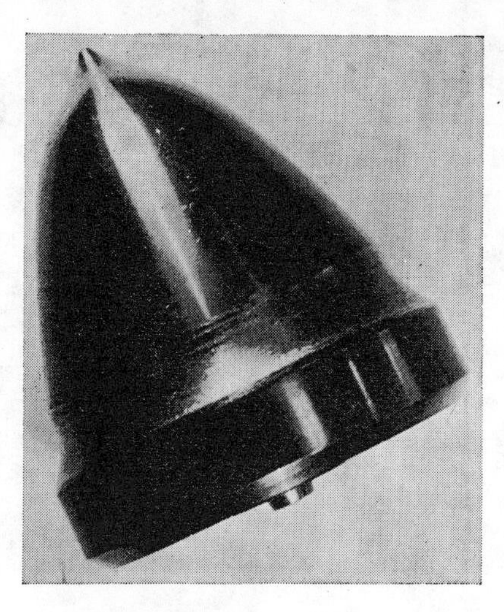

Figure 13—Titanium plug for a motor valve lasted less than one month in an organic chloride environment. Titanium resisted the corrosion but failed because of erosion from the salt particles.

Figure 15—Type 304 stainless safety relief valve failed because of stress corrosion cracking. Chloride bearing salts in the steam collected in the stagnant hollow interior of the plug and triggered the corrosion cracking. The problem could have been prevented by designing the plug without hollow spaces or crevices to collect the salts.

Figure 14—Two views of a chrome plated, carbon steel ball which failed in throttled river water. The chrome plating has been almost eliminated and the exposed carbon steel has been attacked by erosion-corrosion. Replacement was with a ball and valve body made of Type 316 stainless steel.

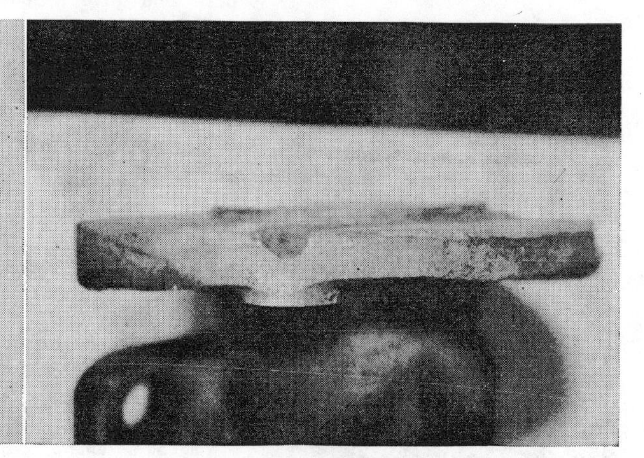

Figure 16—Two views of the corrosion attack which occurred on a cast iron valve body when the gasket permitted 30% sulfuric acid to leak onto the flange area. A new valve body and more stringent maintenance inspections for proper gasket pressure have solved this problem.

this coating is not adequate for corrosive situations. But while one valve customer who prefers a better coating must remove the manufacturer's enamel before applying a high performance film, another may complain if the manufacturer eliminates his quick drying enamel.

But the manufacturer's coating problem is relatively minor and only temporary in comparison to the constant, major problems which face the corrosion engineer and his colleagues in trying to prevent valve corrosion problems and to repair failures that get ahead of the plant's prevention program.

Acknowledgments

The MP editorial staff thanks the following persons and their management for contributing some of the case history data presented in this article and for their permission to publish them:

C. C. Weikal and J. A. Pierce, Chemcel Plant, Celanese Chemical Company, Bishop, Texas.

R. McFarland, Hill-McCanna Company, Carpentersville, Ill.

W. G. Ashbaugh, Chemicals Division, Union Carbide Corporation, Texas City, Texas.

References

1. E. G. Holmberg. Valves for Corrosive Fluids. *Corrosion*, 11, 406t (1955) September.
2. V. J. Groth and R. J. Hafsten. Corrosion of Refinery Equipment by Sulfuric Acid and Sulfuric Acid Sludges. *Corrosion*, 10, 368 (1954) November.
3. D. J. Bergman and G. W. G. McDonald. Solving Corrosion Problems in HF Alkylation Units. *Corrosion*, 17, 9 (1961) April.
4. F. D. Weisert. High Alloys to Combat Corrosion. *Corrosion*, 13, 659t (1957) October.

Other Articles of Interest

Corrosive Hydrogen Cyanide Handled by Stainless Valves. *Materials Protection*, 1, 69 (1962) May.

F. W. Jessen and Ricardo J. Molina. Laboratory and Field Tests on Titanium for Oil Field Pump and Valve Parts. *Corrosion*, 17, 16 (1961) November.

J. C. Spurr. Pump and Valve Corrosion in a High Temperature Hot Water Heating System. *Corrosion*, 16, 9 (1960) May.

J. E. Prior. Corrosion of Valve Seat Ring Results in Flexible Metal. *Corrosion*, 15, 1t (1959) January.

C. H. Allen and M. J. Tauschek. Corrosion of Gasoline Engine Exhaust Valve Steels. *Corrosion*, 12, 39t (1956) January.

Mortimer Schussler. Minimizing Stress Corrosion Cracking of Cylinder Valves. *Corrosion*, 11, 105t (1955) March.

L. M. Rasmussen. Corrosion by Valve Packing. *Corrosion*, 11, 155t (1955) April. Discussion: 559t (1955) December.

Source: *Materials Protection*, June 1965, 28-35

CORROSION FATIGUE CRACKING OF OIL

SUMMARY

The six laws governing fatigue cracking of a metal are presented. A literature survey concerning the cracking process provides verification of electrochemical action involved in fatigue cracking. A procedure used to measure stress potential transient in an oil well sucker rod is also described. The stress potential transient is useful for screening fatigue inhibitors that are successful in preventing rod breaks.

A METAL UNDER CYCLIC stress may suffer certain metallurgical changes, particularly in surface layers, and cracks may be formed transverse to the applied force.[1-14] An oil well sucker rod, almost entirely subjected to tensile cycling stress will develop narrow crevices that are half-moon shaped in the transverse section and horn shaped in the longitudinal section.[6, 15-17] When a crack has penetrated through the surface, the underlying metal eventually ruptures during a tension period.[17, 18] Usually, this type of cracking is known as "fatigue" if the environment is air, and "corrosion fatigue" if the environment is a corrosive aqueous phase.[6, 19]

This discussion deals with attack in an aqueous phase, however, and the term "fatigue" is used for both types of failure because dry and wet fatigue are usually the same mechano-electrochemical phenomena which occur in either an air formed oxide or aqueous solution.[20]

The discussion describes a procedure for testing aqueous systems to determine their tendencies to accelerate fatigue. The procedure illustrates the phenomenon used to slow the electrochemical step of fatigue crack growth. The results are particularly valuable for comparing the effectiveness of different corrosion inhibitors in suppressing fatigue of oil well sucker rods.

Characteristics of Fatigue

Considerable information is available on fatigue of alloys, and numerous curves showing stress vs logarithm-of-number-of-cycles-to-failure have been developed showing practical fatigue or endurance limits.[3, 6, 17, 21] Optical microscopic views of cracks are also common in the literature, and hysteresis loops in metal stress-strain and SN curves have been studied to identify Bauschinger and strain-aging effects.[6, 14, 22-27]

A discussion of the subject in the published literature is warranted because a knowledge of the cracking mechanism is mandantory to an understanding of how to prevent it.

Location of Crack Initiation

One simple experiment reported by Gough in 1932 demonstrates what happens to a metal structure during fatigue.[6] The first step involves bending the metal sharply. Succeeding steps include bending the specimen back and forth repeatedly, in this same region, until the metal breaks. A fatigued wire or light bar will always exhibit a "hook" on one side of the failure, indicating that the highest degree of cold work is adjacent to the break.[6]

This demonstration illustrates the first law of fatigue:

A fatigue crack forms and grows most rapidly, not where the cold work is greatest, but close by where differential effects are at a maximum.[3, 28-29]

Stages of Penetration

Observations of sucker rod crack penetration as a function of reversal accumulation are possible using a bending apparatus and a magnetic fluorescent powder technique. Penetration vs reversal curves resemble the one shown in Figure 1 when the stress is well above the endurance limit.[14, 30-31]

During bending, no penetration is apparent in the first 40-60% of the specimen's fatigue life, even though intrusions and extrusions may form earlier.[32, 33] A crack eventually appears and progresses through the specimen. When the penetration reaches a certain percentage of the cross section, the cracking accelerates until catastrophic failure occurs.[6, 28, 30]

These distinct divisions in the development of a failure have established

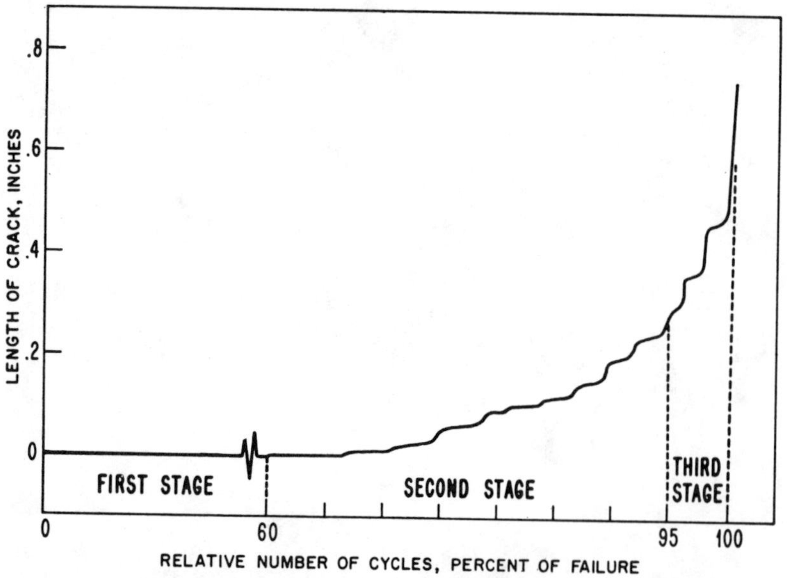

Figure 1—Typical progress of a crack at high stress plotted against number of cycles showing stages in the fatigue process.

WELL SUCKER RODS

Joseph F. Chittum,
Chevron Research Co., La Habra, Calif.

a second law of fatigue: *Fatigue penetrates in three stages: crack initiation, crack growth, and crack acceleration.* The problems of crack development in the three stages are so different that this law by no means eliminates the idea that the mechanism is identical for each. The slow step in the mechanism may be different in the three stages. An initial weakness created in the original metal surface, by differential cold work, may occur very slowly and may be almost completely independent of the environment.[34, 35]

Sequence of Penetration

Careful examination of fatigue failed surfaces shows corrugations. Study of these ripples has disclosed that they represent "crack arrest" lines.[22, 36, 37] Crack arrest lines support the third law of fatigue: *A crack penetrates a specimen discontinuously during the crack growth stage.*

This third law has far reaching implications as to what happens to a metal when a crack penetrates from one crack arrest line to the next. The first suggestion is that each unit of penetration is in a regular sequence of initiation and growth steps. Another is that the final step in a unit penetration strengthens the specimen so that cycling must restart the sequence.

Modes of Development

A fatigue crack develops by three different modes.[22, 38] Two of these are identified from the fundamental ideas about the effect of stress on dislocation motion in a crystalline solid. Edge dislocations converge on the high stress line at a crack tip. This pile-up of dislocations, in itself, amounts to crack development by the slip plane motion. As soon as triaxial stresses dominate the motion, a cleavage develops in a transverse plane which redirects the crack transversely across the specimen. The third mode sets in (for a ductile solid) as soon as the triaxial forces operate. The free surface energy draws metal along the tip of the advancing crack, giving the internal effect of cold working the crystals in front and the external effect of depressing the specimen along the sides. Thus, the fourth law of fatigue indicates *that the three modes of crack propagation are slip, cleavage, and ductile strain.*

An illustration of all three modes is shown in Figure 2, which is a picture of a new unit breaking out near a fatigue crack tip. The new crack bypassed the cold worked metal crystals at the tip and followed a slip plane for a distance while dominated by the slip. Cleavage then straightened it out in the transverse direction. Ductile strain caused the metal around the advancing crack to be depressed until the surface was almost out of

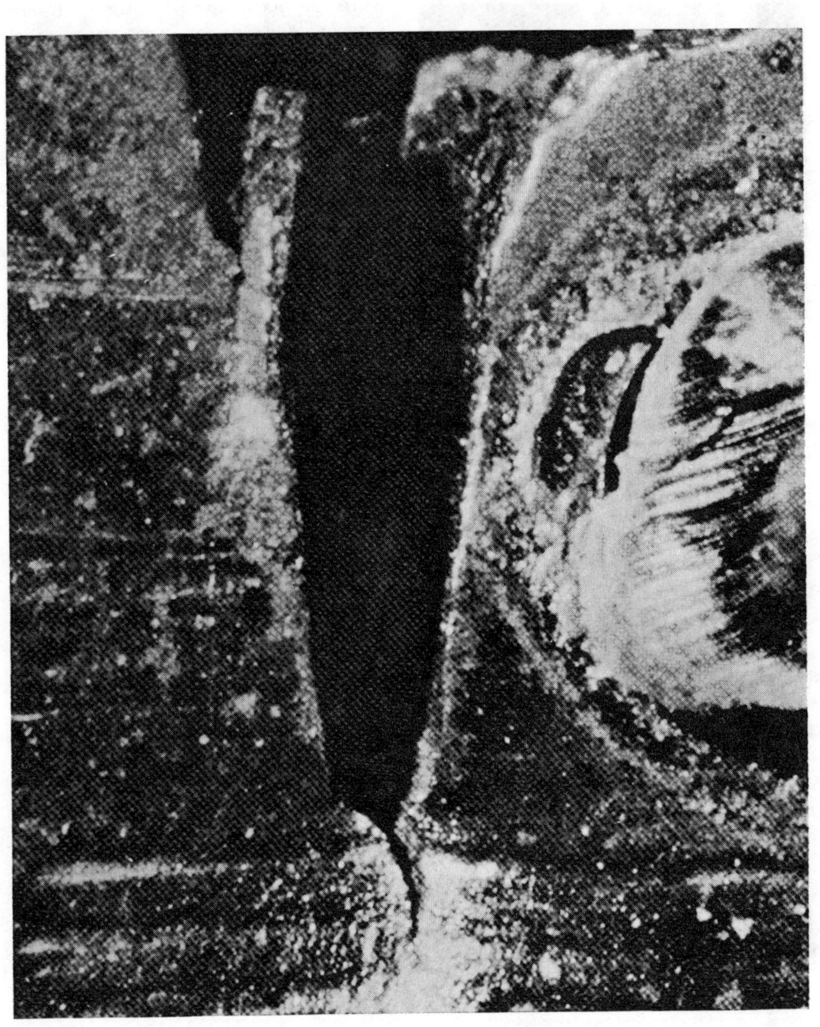

Figure 2—Fatigue crack opened up slightly to show the three modes of crack propagation.

Source: *Materials Protection*, April 1968, 30-34

focus, as can be seen in the photograph.

Mechanical Action

Mechanical action alone produces all of the effects of a fatigue failure.[2-6] However, to obtain "pure" fatigue, it is necessary to cycle the specimen in an inert atmosphere, such as helium gas, at high stress levels.

An experiment, in which an extremely thin section of sucker rod prepared for transmission electron microscopy was subjected to cyclic stress in helium, produced a crack that penetrated along subgrain boundaries.[38] This fact, *that cyclic mechanical action alone at high enough stress levels is sufficient to cause cracks,* is the fifth law of fatigue.

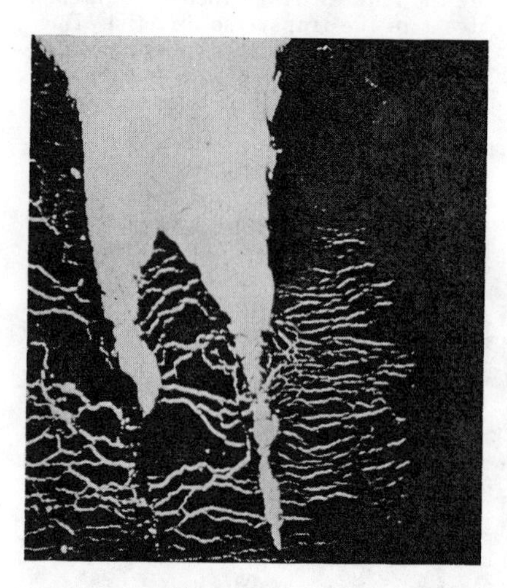

Figure 3—Negative of double fatigue crack using stress induced electrochemical action to sensitize film.

Electrochemical Action

When an electrolyte and corrodents are present on a metal surface, as on a sucker rod, electrochemical action hastens fatigue failure.[3, 6, 13, 19] A relatively simple experiment demonstrates the presence of electrochemical action.[39, 40] It requires a longitudinal section of a sucker rod cut through a lateral fatigue crack (not a fresh metal surface) and autoradiograph photographic film with a means for floating it on water.

The floating process applies the film to the sucker rod section in the absence of oxygen and activating radiation. Also, in a helium atmosphere and in the dark, a desication process dries the film on the rod section. The rod section is then stretched and released after which the film is removed and developed. The results of one test are shown in Figure 3. White lines show where the film wrinkled and lost contact with the metal, thus demonstrating electrochemical rather than chemical activation of the latent image. This demonstration, plus the proven fact that anodic action is concentrated near the tip of an advancing crack, illustrate the sixth law of fatigue: *Corrosion hastens the progress of a crack failure by electrochemical action.*

Fatigue Theory

Fatigue consists of repetitive sequences, each having four steps:

1. Cyclic action cold works the metal differentially in the region where stress is concentrated initially at the original metal surface and later near the tip of a crack. This process creates subgrain boundaries that divide minute oriented fragments from slightly disturbed metal grains.

2. Coalescence of dislocations in the subgrain region by the Stroh process, in which the new boundary supports the defect compression action, almost forms a brittle crack.[41] Electrochemical action where dislocations are coalescing causes an additional loss of metal atoms which completes the brittle crack formation.

3. The brittle crack serves as a sharp weakness that causes high stress to spread across the remaining transverse section.

4. Stress above the yield point results in strain in the transverse section which occurs near the crack tip by the three modes of propagation. During the process, free surface energy forces cause the strain to strengthen the transverse section by increasing crack radius and filling in any irregular extensions into the metal. Therefore, the process must repeat the steps.

During the first stage of the fatigue process, Step No. 1 must be rate limiting because Step No. 2 cannot commence until sufficient differential cold work occurs. The original metal surface usually has sufficient uniformity of grain structure to preclude the build-up of dislocations to form the first brittle crack.

After the first brittle crack is initiated, No. 2 is the slow step in the process and electrochemical action is the slowest part of this step. Thus, the effect of corrosion can be illustrated with curves of stress vs logarithm-of-number-of-reversals-to-failure for sucker rod steel (Figure 4). Corrosion accelerates cracks propagation, so the SN fatigue curve drops from AB to CD, as shown in the figure. Deceleration of the slow stage with a corrosion inhibitor will raise the SN fatigue curve from CD to EF.

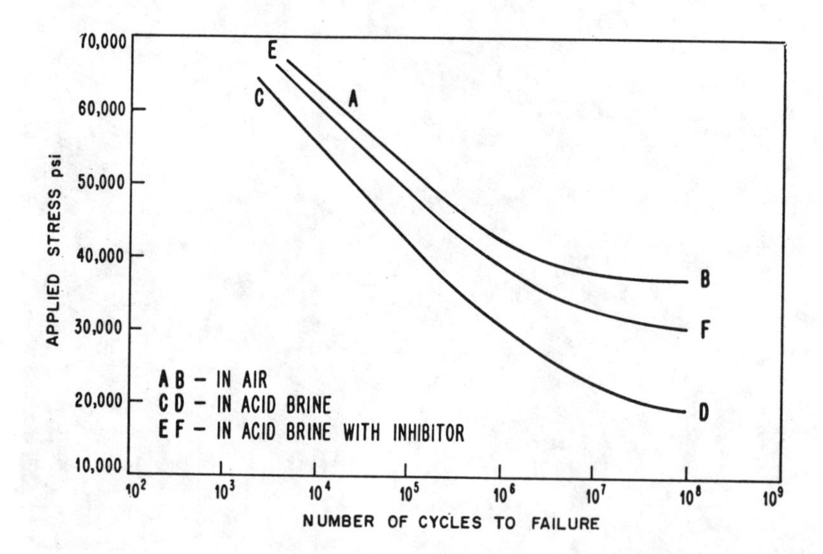

Figure 4—Effect of corrosion and corrosion inhibitors on the SN curves for steel.

Two factors in electrochemical attack need explanation. These are contribution of the metal and contribution of the aqueous phase. The partial coalescing of dislocations largely isolates metal atoms so they have an increased tendency to go into solution. Therefore, the assembling of dislocations on the metallic side of the interface where stress is greatest concentrates anodic action near the tip of a fatigue crack.

A solution in contact with the metal contributes substantially to the electrochemical action in two ways: (1) by supplying ions or molecules that adsorb on any exposed metal surface where either dislocation coalescence isolates metal atoms or a solution-formed solid layer has cracked under stress, and (2) by initially forming the surface layer that tends either to crack or otherwise to open up, exposing metal when any strain occurs in the underlying base. Therefore, chemical treatment of the solution to suppress electrochemical action is a reasonable method for retarding fatigue. The principal problem involved in developing a suitable treatment has been the lack of a method to identify chemicals that are effective fatigue inhibitors.

A specific test method, however, is available for screening these inhibitors. Application of substantial stress to a sucker rod produces in the metal surface a transient mixed potential decrease (change in the sodium direction) that results from the two electrochemical factors. Any inhibitor that suppresses this transient potential should retard the fatigue process.

Evaluation of Inhibitors

Laboratory Testing

The electrochemical action in fatigue must result from a local stress potential transient.[42] Therefore, an initial study developed a procedure for producing and measuring a transient stress potential, comparable to that which causes final initiation of a fatigue crack in a sucker rod.[42] This procedure has been applied by Chevron Research Co. as a means for identifying effective fatigue inhibitors.

The apparatus consists of (1) a device for applying sudden stress to a standard 4-ft pony rod, as shown in Figure 5 (with the rod in place); (2) an electrochemical cell made from a plastic cylinder, which contains the corrosive aqueous phase through which the stressed pony rod passes and into which an unstressed rod is immersed as reference; and (3) an oscilliscope with a camera attachment for observing and photographing curves of the transient potentials measured between stressed and unstressed rods.

For uniformity of structure and composition, each inhibitor test involved use of a new ¾-inch diameter carbon manganese pony rod from one manufacturer. Surface preparation involved degreasing and sand blasting the rod with 325 mesh alundum powder to an over-all velvet appearance.

The test procedure included: (1) inserting the rod through the plastic cylinder; (2) mounting the rod in the apparatus; (3) assembling the cell; (4) filling the cell with corrosive aqueous phase, usually 3% NaCl solution in distilled water, with and without inhibitor; (5) bubbling CO_2 through the solution until oxygen was eliminated and pH was decreased to a value (usually 3.9) constant for a series of tests; (6) stretching the rod at stresses near its yield point; and (7) recording the transient potential with the oscilliscope camera.

Results of these tests are shown in Figure 6. The curve which has the smallest potential-time loop is for the inhibitor which has the greatest ability to retard fatigue.

Field Testing

Corrosion inhibitors were selected that were most effective in suppressing stress potential transients of a standard pony rod (Figure 6D). They were tested in oil fields where rod breaks were frequent.

In two of the fields, each corrosive well had received inhibitor treatment for the previous five years, and the resulting over-all savings amounted to 70-75% of previous maintenance costs. In spite of this record, 19 wells in each field were still considered rod problem wells. Each of these wells, under the original inhibition program, had experienced an average of more than two rod pulling jobs per year and more than eight rod replacements per year for three years.

At the start of one test, the 19 wells in the first field received batch treatment twice a week totalling 25 ppm of an inhibitor that suppressed the stress potential transient satisfactorily. However, this inhibitor was difficult to disperse in the produced brine. The first year of treatment showed a 50% decrease in rod pulling jobs and a 30% decrease in rod replacement. During the next year the same 19 wells received batch treatments twice a week totalling 25 ppm of another inhibitor that dispersed better in the brine, with the result that rod jobs and replacements decreased another 30%.

The results of these field trials were so satisfying that the stress potential transient screening test has been used by Chevron for several years as a means for selecting fatigue inhibitors. The inhibitors selected on the basis of their performance in the stress potential transient test have proved successful in controlling rod breaks when their application presented no dispersion problem.

Stress potential transient suppression, however, should not be the final consideration for rod problem well treatment. Adequate dispersion of the

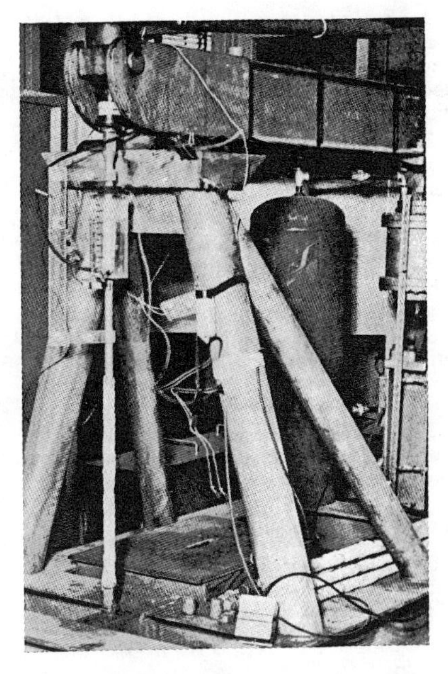

Figure 5—Pony rod stretcher with stress transient cell in place.

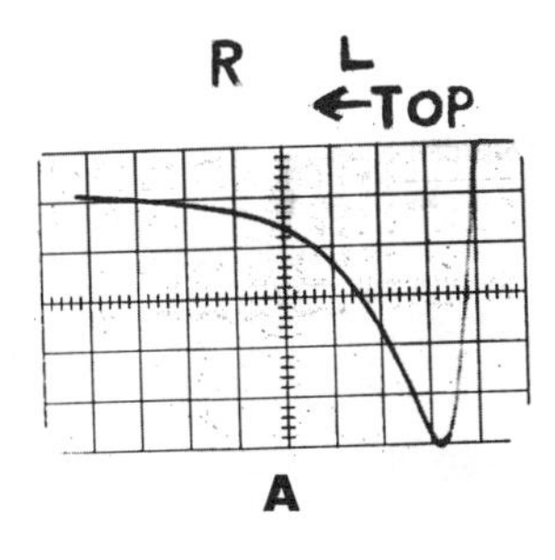

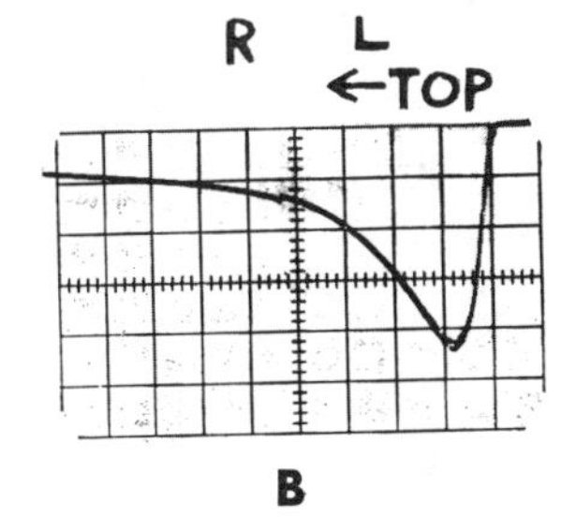

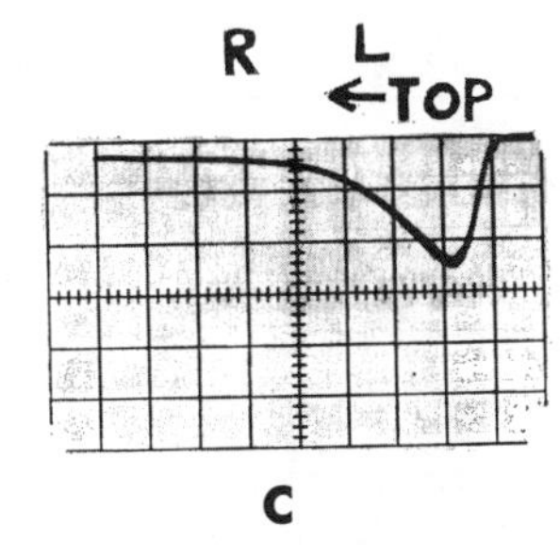

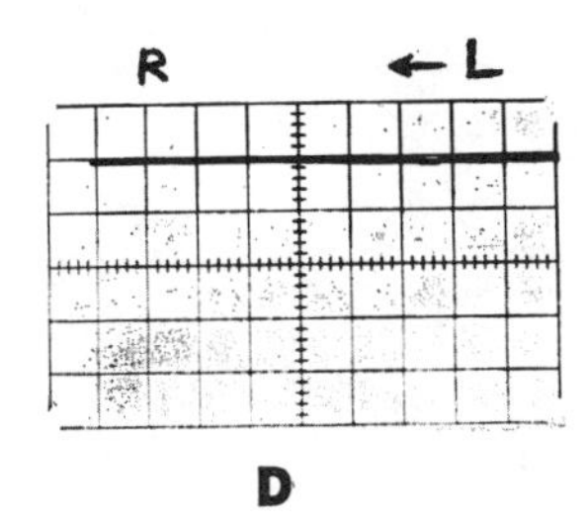

Figure 6—Potential transient curves for sand blasted steel rods in 3% NaCl solution, pH—3.8. A: Blank. B, C, and D: Treated with three different inhibitors.

chemical in the brine is an important accomplishment for two parts of the treating process: (1) The inhibitor must mix well with the flushing water to carry the chemical to the bottom of a well, and (2) the mixture must stay dispersed when it contacts the oil well brine or it will separate and become ineffective.

Conclusions

1. Fatigue of oil well sucker rods occurs in a series of unit penetrations, and electrochemical corrosion plays a decisive role in each slow step of a unit crack penetration.

2. Therefore, electrochemical measurement of a stress potential transient permits relative fatigue susceptibility to be estimated for sucker rods in oil well brines.

3. Highly dispersible corrosion inhibitors can be rated for their effectiveness in preventing fatigue of sucker rods on the basis of their ability to suppress the stress potential transient.

Acknowledgment

The author acknowledges the assistance of H. E. Englander in conducting the physical tests on specimens and H. Kieron for taking the electron and optical micrographs.

References

1. J. A. Ewing and J. C. W. Humphrey. *Phil. Trans. Royal Soc.*, (London) **A200**, 241 (1903).
2. B. P. Haigh. *Jour. Scotland* and *Iron and Steel Inst.*, **23**, 17 (1915).
3. D. J. McAdam, Jr. *Proc Amer. Soc. Test Mat*, **26**, 11, 224 (1926).
4. H. J. Gough. *The Fatigue of Metals*. Benn, London (1926).
5. C. G. Horst and B. B. Westcott. *Petroleum Eng.*, Nov. (1931).
6. H. J. Gough. *Jour. Institute of Metals*, **49**, 17 (1932).
7. B. B. Westcott and C. N. Bowers, *Proceedings API Section IV*, Bul. 212, 29 (1933).
8. H. J. Gough and W. A. Wood. *Proc. Royal Soc.*, London, A **154**, 510 (1936).
9. E. Orwan. *Proc. Royal Soc.*, A **171**, 79 (1939).
10. J. A. Bennett. *Proc. Amer. Soc. Test. Mat.*, **46**, 693 (1946).
11. P. J. E. Forsyth. *Jour. Institute of Metals*, **80**, 181 (1952).
12. W. Weibull. *Fatigue and Fracture of Metals*. Wiley and Sons, New York, (1952).
13. R. Cazaud, *Fatigue of Metals*. Translated by A. J. Fenner, Chapman and Hall, Ltd., London (1953).
14. N. Thompson and N. J. Wadsworth. *Advances in Physics*, **7**, 72 (1958).
15. A. H. Cottrell. *The Mechanical Properties of Matter*. John Wiley and Sons, Inc., London, p. 368 (1964).
16. N. Thompson. *Proceedings of International Conference on Fatigue of Metals*, London-New York, 527, p. 155 (1956).
17. W. H. Harris. *Metallic Fatigue*. Pergamon Press, London, p. 155 (1961).
18. *Corrosion of Oil and Gas Well Equipment*. American Petroleum Institute, p. 12 (1958).
19. U. R. Evans. *The Corrosion and Oxidation of Metals*. Edward Arnold, Ltd., London, p. 701 (1960).
20. J. F. Chittum. *Memoires Scientifiques Revue Metallurg*, LXII, 15, Mai (1965).
21. J. A. Pope. *Metal Fatigue*. Chapman and Hall, Ltd., London (1959).
22. P. J. E. Forsyth. *Proceedings of the Crack Symposium*, Cranfield 1, 76 (1961).
23. N. E. Frost, J. Holden, and C. E. Phillips. *Proceedings of the Crack Symposium*, Cranfield, 1, 166 (1961).
24. J. Bauschinger. *Mitt Mech-tech Lab Munch* 13, k (1886).
25. J. Masing, *Wiss Veroff*, Siemens **3**, 231 (1923).
26. S. N. Buckley and R. M. Entwistle. *Act.' Met.* **4**, 352 (1956).
27. T. J. Dolan. *Internal Stresses and Fatigue of Metals*. edited by G. M. Rossweiler and W. L. Grube, Elsevier Publishing Co., p. 284 (1959).
28. H. de Leiris. *Proc. International Conference on Fatigue of Metals*, London-New York, p. 118 (1956).
29. M. Hempel, *Proc. International Conference of Fatigue of Metals*, London-New York, p. 543 (1956).
30. S. R. Valluri. Preprint Joint Meeting of IAS and ARS (1961).
31. T. Yokobori. *The Strength, Fracture, and Fatigue of Materials*. Translated by S. Matsuo and M. Inoae, P. Noordhoff, Groningen, The Netherlands, p. 209 (1965).
32. N. J. Wadsworth, and J. Hutchings. *Phil. Mag* **3**, 1154 (1958).
33. E. R. Parker and D. M. Fegredo. *Internal Stresses and Fatigue of Metals*. Edited by G. M. Rossweiler and W. L. Grube, Elsevier Publishing Co., p. 263 (1959).
34. H. J. Gough and D. G. Sopwith. *J. Inst. Metals*, **56**, 55 (1935).
35. N. Thompson, N. J. Wadsworth and N. Louat. *Phil Mag* **1**, 113 (1956).
36. R. H. Christensen. *Proceedings of the Crack Symposium*, Cranfield, **2**, 326 (1961).
37. C. D. Beachem. Defense Documentation Center Bulletin AD 623567, September 28 (1965).
38. J. C. Grosskreutz. *Proceedings of the 10th Sagamore Army Materials Research Conference*. Syracuse University Press, p. 27 (1964).
39. A. H. Meleka and W. Barr. *Nature* **187**, 232 (1960).
40. J. C. Grosskreutz. *J. Applied Physics* **33**, 2653 (1962).
41. A. N. Stroh. *Proceedings of International Conference on Fracture*, Swampscott, Edited by B. L. Averbach, et al. Chapman and Hall, Ltd. London, p. 117 (1959).
42. A. J. Gould and U. R. Evans. *J. Iron and Steel Inst.* Second Report Alloy Steel Research Committee (1939).

JOSEPH F. CHITTUM is a retired senior research associate with Chevron Research Co. His responsibilities included completing unfinished fundamental research which was interrupted by his retirement. He received a BS from Iowa Wesleyan College in 1924 and a PhD from the University of Chicago in 1928. An NACE member for about six years, he also belongs to ACS, and the American Association for the Advancement of Science.

Corrosion Problems in Solar Energy Systems*

PAUL D. THOMPSON and MICHAEL B. HAYDEN
Beltsville Agricultrual Research Center,
U.S. Department of Agriculture, Beltsville, Maryland

Corrosion problems associated with Al liquid filled solar energy systems for small scale ambient temperature operation are listed and suggested preventives and control itemized. Comments pertain mainly to systems such as that developed at the Beltsville Agricultural Research Center which consists of Al piping and components. Suggestions are made concerning water controls, inhibition, exclusion and scavenging of oxygen, pH control and design, and operation modes tending to control access of oxygen to the system. Also considered are such factors as flow velocity, elimination of galvanic couples, deleterious ion solution in water, controls for displaced vapor, avoidance of freezing, and possibility of cathodic protection. A nontoxic inhibitor, such as a silicate is recommended. Economic considerations and some magnitudes are discussed.

SOLAR HEATING is an attractive technology for applications that require the use of heat at moderate temperatures as a means of supplying loads characterized by fairly uniform demands over their annual cycles. By "moderate temperatures" is meant below 65 C (150 F), and, by "uniform demands over their annual cycles", is meant the mean demand in any 1 week period is not more than twice the annual mean demand.

The reasons for these limitations are not technological, but economic. For example, it would be technologically feasible to construct a solar air conditioning system that would operate 3 months out of each year and in which solar collectors would supply 93 C (200 F) water to an absorption-cycle refrigeration system. Suppose that such a system were sized such to provide heat to the refrigeration plant at a rate of 3500 watts (1 ton of refrigeration if the coefficient of performance were 1). If the system runs 8 hours a day for 90 days a year, the total energy produced would be 9122 MJ (2534 kWh) per year. In comparison, suppose the same capital investment were used for the construction of a solar water heating system that would produce 54 C (130 F) water 365 days a year. The collectors would be about twice as efficient at the lower operating temperature, and, of course, the additional operating days would not affect the capital investment. This water heating system would collect 74,000 MJ (20,500 kWh) of energy per year, which would produce about eight times the annual output of the air conditioning system at the same investment cost.

What does the above comparison have to do with corrosion? We have tried to show that the practicality of solar heat is largely determined by the annual energy production for a given capital investment. The important characteristic of the capital investment is

*Presented during Corrosion/77, March, 1977, San Francisco, California.

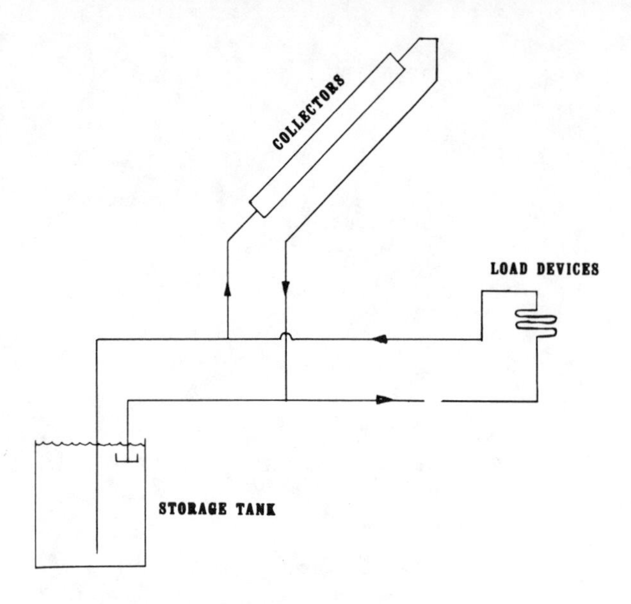

FIGURE 1 – The main elements of any solar heating system are the solar collectors, the heat storage tank, and the load devices. All are interconnected by appropriate piping.

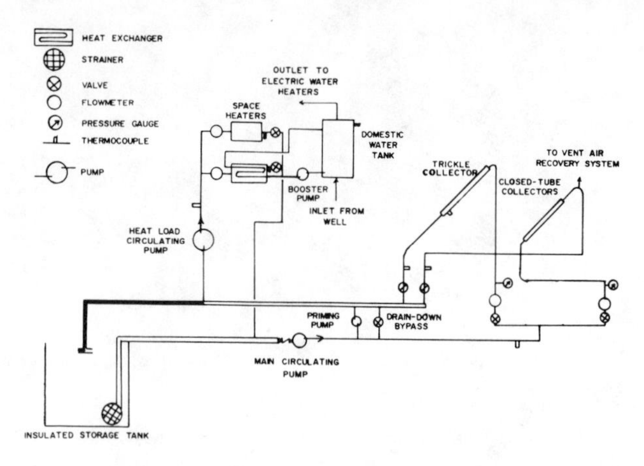

FIGURE 2 – System diagram of the solar heating system installed at the Beltsville Agricultural Research Center.

not the initial value of the investment, but the annual cost for amortization of the loan or for establishment of a reserve for equipment replacement. These annual costs are largely determined by the expected lifetime of the equipment. For example, if a reserve were to be established, at 10% interest the capital recovery factor for a 5 year expected lifetime would be 0.26; for a 10 year lifetime, it would be 0.16. Thus, if the expected life could be increased from 5 to 10 years, the effective annual cost of the energy produced would decline by 39%. Since the primary annual cost of solar heating systems is the recovery of investment, their expected lifetimes are much more important than in conventional systems in which the primary annual cost is purchase of fuel. Further, the high-value components of solar heating systems are passive (collectors, tanks, and piping), so the primary factor that determines life expectation is corrosion rather than wear.

Characteristics of Solar Heating Systems

Solar heating systems generally include four main elements: solar energy collectors, a thermal energy storage unit, loading devices such as space heaters, and the interconnecting piping, pumps, and control devices. The working fluid can be liquid or gas. Air is universally used in gas filled systems, which will not be discussed in this paper. Liquid filled systems are filled with water or organic fluids, or mixtures of both. Many of the elements have analogies in conventional hot water heating systems. Figure 1 illustrates the basic arrangement. For fail safe drainage of collectors in freezing weather, they should be located higher than the storage tank. Best efficiency is obtained if the collectors are supplied from the cold (lower) stratum of the storage tank and the loads are supplied from the warm (upper) stratum.

The characteristic component of solar heating systems is, of course, the solar collector. Essentially, this collector is an absorber that is painted black or otherwise coated so that it absorbs light energy from the sun. It then converts the absorbed light energy to thermal energy in the form of a temperature increase of the absorber. Loss of this heat is prevented by one or more layers of glass or plastic glazing in front of the absorber and by insulation behind it. The heat is transferred for use by circulation of the working liquid through tubes in the absorber or by flow of the working liquid over the front of the absorber.

The storage element of a solar heating system carries the system through periods of darkness or cloudy weather by storing heat. The system stores heat by warming a large mass of liquid or stone or by melting an organic material or low melting point salt. The load devices are usually equivalent to those of conventional systems. For space heating, liquid to air heat exchangers are used. These might be radiant heating tube bundles, baseboard convectors, or fan-coil heaters. The criteria for the design of circulating pumps, piping, and controls are the same as for conventional systems.

The system piping layout for the solar heating system at the Beltsville Agricultural Research Center is shown in Figure 2. This heating system includes solar collectors, hot water storage, load devices (space heaters and water heating heat exchanger), and the associated piping and controls. Comparisons of several collector designs and corrosion reducing strategies are permitted by the system.

Solar and conventional heating systems differ in several ways. The first difference is that the volume of liquid in a solar system is much greater than in a conventional system if the heat transfer liquid is used as the heat storage medium. Such a design is the most efficient since heat exchangers are not used between the storage unit and the circulating liquid. A typical volume of liquid for a residential system is 76 kl (2000 gallons). Because of the large volume, the circulating fluid (including any corrosion inhibitor or antifreeze) must be inexpensive.

A second difference is that solar heating systems must be protected against freezing since the collectors will be at or below outside air temperature during darkness. Since protection against freezing by lowering the freezing point of the circulating liquid is expensive and thermally inefficient, any portion of the system that is susceptible to freezing should be drainable. When solar heating systems have been drained and refilled, they are more likely to have dissolved oxygen in the circulating liquid than would conventional heating systems, which are not drained.

A third difference between solar and conventional heating systems is that domestic water heating is much more likely to be included as part of a solar heating system. A solar heating system is most economical if water heating is included as a means of leveling the annual demand cycle; whereas, conventional heating systems are most efficient if separate units supply seasonal demands. A fourth difference is that a solar system has a wider variation in operating temperature than does a conventional system. In solar heating systems, loading devices must be designed so that they will operate when the circulating liquids are at the lowest possible temperature. This is because both the heat capacity of the storage device and the efficiency of collectors are improved by a loading device capability for low temperature operation. On the other hand, most collectors are capable of flashing water into steam under zero flow conditions, so piping components must tolerate 100 C (212 F) temperatures.

A final difference between solar and conventional heating systems is the use of aluminum water passages, particularly in the absorbers of solar heating collectors. Although copper absorbers are available, as are iron or plastic ones, aluminum absorbers are less expensive. Particularly popular are absorbers made of Roll-Bond[1] aluminum, in which the liquid passages are cold formed in the absorber sheet. Aluminum presents a much greater corrosion problem than do conventional piping materials.

Corrosion Problems Characteristic of Solar Heating Systems

The most pressing problem in solar heating system design is control of corrosion in aluminum absorbers. Absorbers cannot be opened for inspection, coating, or descaling, so corrosion treatment must be foolproof. Any extremes of pH, dissolved heavy metal or chloride ions in the water, "hard" water, and dissolved oxygen increase the risk of corrosion in aluminum absorbers.

A second problem is the difficulty of building single metal systems. Although it is possible, for example, for one to build an all-iron conventional hot water heating system, the plumbing components for a solar heating system generally include some copper or brass (particularly in valve bodies, piping components, and heating coils), some iron (tanks and pump housings), and some aluminum (collectors). Plastic components such as fiberglass or PVC pipe can be used to some extent, but their procurement in a full range of sizes and configurations is difficult. Low temperature plastics such as polyethylene or PVC are unsuitable for solar heating systems.

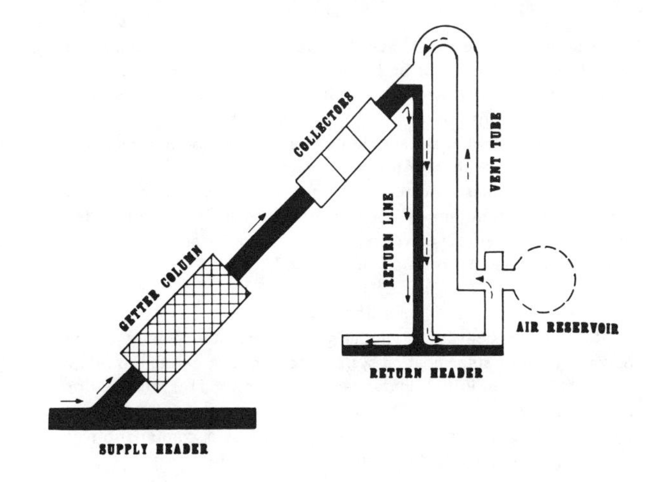

FIGURE 3 — A practical way to arrange the piping to a single group of collectors.

General Approaches to Corrosion Protection for Solar Heating Systems

Water treatment is potentially the best protective measure. The water initially used in system filling should be low in chloride, dissolved oxygen, and hardness. Untreated well or surface water has low chloride in most locations, and commercial water treatment systems can be used for reduction of hardness. One can deoxygenate initial fill water by heating it and passing it over iron wool, but, practically speaking, probably the best that can be done is to avoid aeration during filling. Chlorine and oxygen interact, and, thereby, increase corrosion rates since chloride ions penetrate the oxide films of aluminum or steel. Hardness, itself, is not a corrosion factor, but the resulting scale accelerates corrosion by permitting formation of localized concentration cells where the scale coating is incomplete or broken. Scale formation also degrades collector performance by reducing the transfer of heat into the circulating fluid.

Once the system is filled, treatment of the water with chemicals should include pH control and maintenance of a suitable dose of inhibitor such as sodium silicate. Chromate inhibitors are effective, but the possibility of potable water contamination or of eventual discharge of the system's treated water should be carefully considered before chromates or other toxic compounds are selected. In the solar heating system at Beltsville, the pH is maintained between 8 and 9.5, because aluminum is amphoteric and will be corrosively attacked under strongly acidic or basic conditions. Further, a corrosion inhibitor level of 200 ppm proprietary inhibitor (Wrico 84—a blend of a sodium silicate with a modified carbohydrate), is maintained in the water. Weekly monitoring of pH, inhibitor concentration, and dissolved oxygen is desirable. Most solar heating systems have closed cycles, and no makeup water is added to the primary loop. Therefore, continuing treatment after the initial treatment is required only if one needs to compensate for chemical reactions of the circulating fluid or for the addition of untreated fluid after leaks and spills.

Control of dissolved oxygen is achieved by proper system design. Each vented tank, surface flow collector, or drain down air vent is a potential source of dissolved oxygen. One can cover the surface of water in storage tanks with plastic film or, perhaps, float an appropriate oil on the surface to avoid oxygen pickup. Collectors should be sealed so that no exchange takes place between the outside air and the gas that is in contact with the circulating water. It will generally be necessary for collector flow loops that will be drained for protection from freezing to have a vent at a high point so they will not develop an air lock when they are drained. This same vent may also serve as a means of breaking the vacuum created by the column of water that flows down the return pipe as the collector operates. The vent system should be closed so that additional oxygen is not introduced into the water through the vent system during normal operation or during drain down.

A "getter column" can be installed in the feed to a bank of collectors. This column is essentially a chamber that contains replaceable pads of an electrochemically active sacrificial material. The active material might be, for example, pads of zinc wool followed by pads of aluminum wool. The condition of the last aluminum wool pad before the discharge into the collectors should indicate the rate of corrosive attack on the collector. In addition to serving as a sacrificial material, the column also serves as a fine filter for the removal of bits of such materials as solder and metal filings. The local galvanic couple created by a scrap of copper or solder in an aluminum collector can cause failure of the aluminum in a few weeks.

Figure 3 shows a practical way to arrange the piping to one group of collectors. A getter column passivates water supplied to the collectors. The vent tube acts as a vacuum breaker for the return line during operation, vent gas being recirculated through the partially filled return header back into the vent line. The air reservoir acts in combination with the vent line to take up the volume changes associated with filling and emptying the collectors at the beginning and end of each operating cycle.

Flow velocity through aluminum water tubes has a definite effect on corrosion rate. As velocity increases, erosion of the relatively soft aluminum increases. If flow is too slow [below 90 cm/s (3.5 fps) linear velocity] corrosion increases. Correct flow velocity is achieved by correct design of the water passages in the absorber plate. The designer of a solar heating system has only limited flexibility in this regard since the total volumetric flow through a solar collector must be controlled such that it will give the designed temperature rise on one pass of the circulating liquid through the collector. The designer of collectors should size the water passages for correct flow velocities at the expected operating volumetric flow.

The usefulness of cathodic protection, in which either conventional zinc plugs or electrical current from an external source is used, does not appear to have been studied. The long, small diameter tubes of most solar heat collectors would be unsuitable for these techniques, but consideration might be given to their use in new collector designs. It is, of course, essential that one isolate dissimilar metals from each other to avoid the creation of simple galvanic couples. One can use short lengths of high temperature hose to connect collectors to other piping components, and dielectric couplings between dissimilar metals should be used throughout the system in accordance with normal plumbing practice.

---

(1)Mention of a trade name, proprietary product, or specific equipment does not constitute a guarantee or warranty by the U.S. Department of Agriculture and does not imply its approval to the exclusion of other products that may be suitable.

The authors' personal feeling is that toxic corrosion inhibitors should be avoided in solar heating systems since they must be discharged when the system is emptied. If they are used, potable water supplies must be completely protected from contamination by use of such devices as back flow preventers, siphon breakers, and double wall heat exchanger tubes.

Several commercial water treatments are available. The essential requirement for obtaining good water treatment is to obtain the service of a qualified corrosion engineer and to ensure that the problems involved are mutually understood.

Summary

Solar heat collection will gain wide acceptance only if the operating life of solar heating systems can be long enough to repay the cost of the initial investment. The key factor that limits operating life is corrosion, particularly the corrosion of aluminum absorbers. The prevention of corrosion requires a combination of correct system design and appropriate water treatment.

Bibliography

The applicable literature seems to be mainly in the field of cooling towers. The following books were useful:

Bregman, J. I. "Corrosion Inhibitors", The MacMillan Co., Inc., New York, New York (1963).

Gleason, M. N. et al. "Clinical Toxicology of Commercial Products—Acute Poisoning".

Third Edition, Williams & Wilkins, Baltimore (1969).

Nathan, C. C. "Corrosion Inhibitors", p. 290-346 in Vol. 6 of Mark, H. T., McKetta, J. J., and Othmer, D. F. "Encyclopedia of Chemical Technology", 2nd Edition, Interscience Publishers (1965).

Powers, R. A. and Roebuck, A. H. "Corrosion", p. 289-316 in Vol. 6 of Mark, H. F., McKetta, J. J., and Othermer, D. F. "Encyclopedia of Chemical Technology", 2nd Edition, Interscience Publishers (1965).

Editor's Note: In the production of Roll-Bond aluminum heat exchange panels, the parting agent is frequently graphite which causes pitting of aluminum in water unless adequately inhibited. A chromate parting agent can be used which does not have this disadvantage, but I do not know if it is on the market.

HPG

Practical Fundamentals for Field Use of Instruments and Equipment

JAMES R. COWLES, *Agra Engineering Company*

This paper outlines the basic characteristics of cathodic protection instruments primarily as applied to underground structures. Emphasis is placed on the reasons that certain varieties of meters are used, and the limitations that are imposed by what meters can and cannot do.

AS a basis for the discussion which follows the following analogies, definitions and symbols are given. The symbols may seem awkward, but these are the ones conventionally used, so they must be mastered.

E = potential difference stated as volts. This is equivalent to pressure when dealing with fluids. While potential and potential difference is the correct terminology in practical work, the word "voltage" is used.

I = current stated as amperes. This is the equivalent of gallons per minute. This analogy is a very close one, as current represents the number of electrons per second passing a given point.

R = resistance stated in ohms. Here electricity differs from fluid flow as the pressure drop in fluid flow varies with flow, whereas in simple electrical circuits it does not. The unit called the ohm is indispensable from a practical standpoint, but actually it is simply the volts required to force one ampere through a particular resistance. (This is, of course, a simplification, as a current flow of one ampere is not always practical.)

Quite often the basic quantities, such as the volt or ampere, are too large or too small to be convenient, so prefixes as follows are used: (1) micro = millionths, (2) milli = thousandths, (3) kilo = thousands, and (4) mega = millions. These prefixes can be added to amperes, ohms, or volts, though for inexplicable reasons the term kilohms is seldom seen. The existence of it is acknowledged by the use of "K" to stand for thousands. Thus, 3000 ohms is often marked on instruments or written as 3K. No other such marking is common except M. It is virtually worthless, except in the factory where it is stamped on the resistor, as it has been used for both thousands and millions.

Note: Reprinted from *Proceedings of the 16th Annual Applachian Underground Corrosion Short Course;* West Virginia University (1971).

The Nature of Indicating Meters

The DC meter used in the majority of meters is a special form of a DC motor stalled by a spring. It differs from an ordinary motor in that only the copper coil moves. All of the iron parts remain stationary. The more electricity present the harder the armature pushes against the spring, hence the farther the hand attached to the armature moves. Figure 1 shows cutaway pictures of two common types of meter movements. Note that the iron core inside the coil is stationary and does not rotate as it would in a motor.

The first generalization about cathodic protection instruments is that they are generally operated by very small amounts of power, so that their design must be pushed to practical limits. In a power plant a meter can use 10 watts and the loss will never be noticed, whereas in cathodic protection work the process being studied may only involve minute wattages, so almost every task requires a meter designed specifically for it. In other words most of the meters used in cathodic protection have to have "high sensitivity."

There are two possible extremes when winding the moving coil for a meter. It can be wound with the finest wire made, in which cases meters result which operate on from 5 to 20 microamperes, but the resistance of the coil wire will be so high that 30 millivolts upward will be

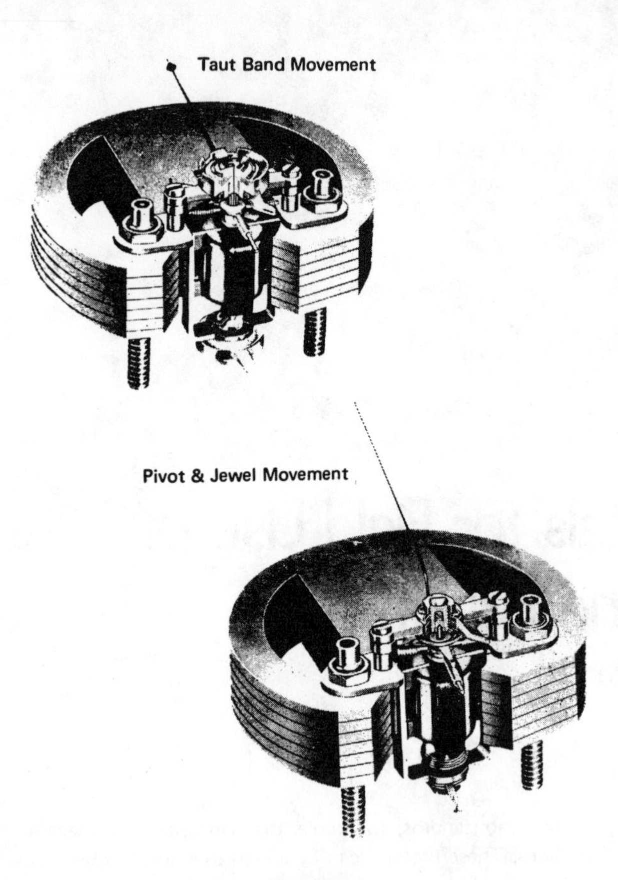

Figure 1*

required to force the required current through the coil. Such a meter cannot possibly have a full scale reading of one or two millivolts. Such meters are known as high resistance meters, and such a meter can have from 50,000 to 200,000 ohms per volt sensitivity. On the other hand the coil in a meter can be wound with the coarsest wire, which is compatible with the cross-sectional area of the springs that must carry the current to the bobbin. In conventional high sensivity meters this yields a 10 to 5 ohm, 2 to 1 millivolt meter as a practical limit. The only way to avoid these limitations is to employ vacuum tube or transistor arrangements, which is not generally necessary.

An important factor in what sensitivity a meter can have is the area which the moving coil encompasses. This is of importance in cathodic protection instruments because it makes it impossible to have highest sensitivity in small meters. A two and one half inch meter cannot have as wide and tall a coil as a six inch meter, in which advantage is taken of the larger size to use a larger moving coil, and so cannot be as sensitive. A three and one half inch meter is the smallest suitable meter.

Another limitation on meter sensitivity is the friction in pivots and jewels. There is simply a minimum practical torque below which the movement of the coil is unreliable. The newer taut band construction, which will be discussed later, presents no friction problems. It is now possible to have 200,000 ohms per volt meters in a 3 1/2-inch meter, which formerly was not generally offered in sensitivities higher than 50,000 ohms per volt. Since small meters cost less than large meters, until miniaturization problems occur, this factor has lowered the cost of instruments for certain purposes.

Ohm's Law is the basic rule governing all DC meters. Such meters have the characteristics of all using the same basic style of movement. In fact, if a person is not close enough to a meter to read the wording on the dial, he cannot distinguish between an ammeter and voltmeter. By the same token, a meter may be made to perform as several meters, in which case the unit is called a multi-meter.

Since Ohm's Law is the basic rule governing meters it is necessary that it be defined.

The three forms of Ohm's Law are:

$$I = \frac{E}{R} \quad , \quad R = \frac{E}{I} \qquad E = IR$$

When using prefixed quantities and Ohm's Law care must be taken to remember that unprefixed quantities are required, though certain short cuts can be taken, such as dividing millivolts by milliamperes, since the thousandths involved divide out.

Galvanometers

The simplest type of meter is the galvanometer. When this term is used in the laboratory or shop, it implies a very delicate unit of one of several possible constructions. In the meters used in corrosion control it is merely a zero center meter, or some modification thereof, with the desired characteristics. Thus, the ammeters that used to be used to show whether the battery on automobiles was charging or discharging was, in essence, a calibrated galvanometer. Generally though the scale divisions on a galvanometer do not represent any particular current or voltage. The purpose of a galvanometer is to permit the making of adjustments so that two points are at the same voltage, in which case another meter or dial setting gives the answer desired. In several cathodic protection instruments on the market the function of a galvanometer is performed by a meter which is also used for other purposes. In some such cases advantage is taken of the fact that the meter hand will move to the left of zero as well as up scale. When the meter is serving as a galvanometer the instrument circuit is adjusted until the hand on this particular meter is brought back to zero.

The average corrosion engineer does not purchase galvanometers, except in an instrument. The fine points of selection of the proper type are, therefore, a manufacturer's design problem. In passing, it perhaps should be pointed out that they can have the characteristics of being high resistance low current, high current low resistance types of unit, or of course, of intermediate characteristics. These characteristics must be chosen on the basis of the circuit in which the galvanometer is used, as must be done with any meter.

Basic Meter Movements

For the purpose of discussing voltmeters and ammeters it will be well to select a specific basic meter movement. By

*Courtesy Simpson Electric Co.

basic movement the author means a meter with no resistors in it and no numbers on the dial. Such a meter is by no means a fictitious thing. The trend or style is for manufacturers to furnish just such meters to authorized repair and sales firms. These modification points then tailor them for purposes for which a stock meter will not serve, thus relieving the manufacturer of many special orders. Also, the local firm can often discuss the meter with the ultimate user and simplify the transaction. Time, too, is a factor. Current meter manufacturer deliveries are a month, or more. Modification requires very little time. This procedure can be valuable to corrosion engineers replacing rectifier meters.

For a basic movement suitable for a pipe to soil voltmeter and ammeter a 20 microampere, 30 millivolt unit is a good choice, remembering that there are no numbers or words on the dial. This means that when enough current is passed through the movement to make the hand rest on the last scale mark, or the right side of the dial, it will be found to be 20 microamperes, and the voltage between the terminals will be found to be 30 millivolts. Obviously, appropriate numbers, and a caption of either microamperes or millivolts, could be printed on the dial. By Ohm's Law the meter has a resistance of:

$$R = \frac{E}{I} = \frac{\frac{30}{1,000}}{\frac{20}{1,000,000}} = 1,500 \text{ ohms.}$$

The basic meter movement, therefore, is either a voltmeter or an ammeter with a definite basic resistance. If so desired, it can be used as a galvanometer as is, however, if this is to be its basic use the hand will be moved to the center of the scale, and this point marked zero.

Voltmeters

If a one volt meter is desired, a resistor, called a multiplier, is added in series with the basic meter. Correct readings will be obtained when the resistor has a value such that one volt will force 20 microamperes through it and the meter. Then there will be 1 volt–0.03 volt (30 millivolts) lost across the multiplier, or a 48,500 ohm resistor will be used. The fact that the meter will have a resistance of 50,000 ohms for one volt, and incidentally 100,000 ohms for 2 volts, gives rise to the term "ohms per volt," and is, through the use of Ohm's Law, a measure of the current rating of the basic movement. Voltmeters draw some current, and this can disturb the circuit. Generally, the less current the less disturbance, so that low current drain is desired. Thus, the higher the ohms per volt the better the meter is in principle for, say, pipe to soil voltage. This does not necessarily mean that a very high resistance meter is always the best available. For example, a master meter for checking other meters would not be high resistance, because more accuracy is possible with lower resistance. Meters of high ohm per volt characteristics are of the high sensitivity type, so in the case of voltmeters a high ohm per volt rating is synonymous with high sensitivity.

Going back to the basic movement, temperature has more effect on meters than might be imagined. The springs and magnets are very slightly weakened or strengthened by a change in temperature in such a manner that a certain amount of compensation exists. Between the small change which occurs and the self-correcting nature of the pair, this effect is only a curiosity outside the laboratory. The copper coil is a very different matter. In a meter movement, without any resistors, it contributes all of the resistance, except that in the springs. (These, too, will have a high coefficient). Its coefficient of resistance is approximately 0.22% per °F. Thus, a measurement, made with the basic movement alone, on a 20°F day can be approximately 17.6% higher than one made on a 100°F day. This is certainly no minor matter. With one or two millivolt meters (used for measuring line currents), where there is no possibility of a multiplier, this fact should be remembered.

In regard to the basic movement selected for discussion, i.e., a 30 millivolt meter, the problem will exist on a 30 millivolt meter, but if a higher range meter is created by using a multiplier, it may not be serious. Multipliers in good meters are made of materials whose resistance changes very little with temperature. If the multiplier to meter resistance ratio is high, then temperature has essentially the effect on the meter that it does on the multiplier, which is neglibible. This is temperature compensation by "swamping." This problem can be eliminated another way, if the first meter multiplier can be made of a material whose resistance goes down with temperature to the degree that the meter movement's resistance goes up. Relatives of transistors, known as "thermistors," do this. With a suitable "thermistor" the basic movement, being used for illustrative purposes, can be compensated if a few millivolts can be spared. Thus, by not trying to make less than a 40 millivolt meter out of it, it can be a temperature compensated meter.

The foregoing covers the characteristics of meters using the voltmeter as a basis for discussion. Since a simple meter is not always the best solution to a measurement problem, instruments, combining meters with other components, need to be covered before discussing field voltage measurements.

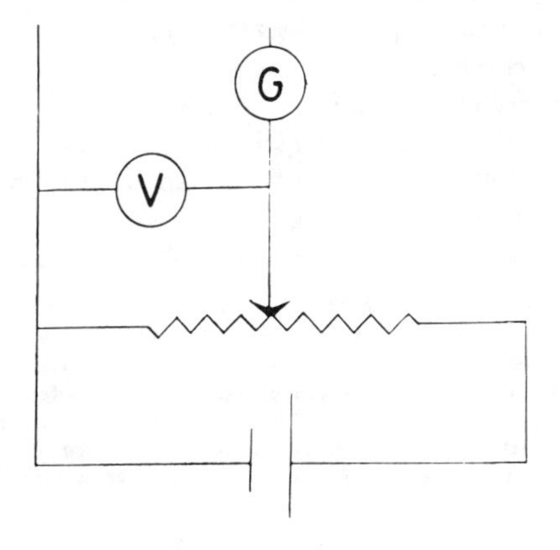

Figure 2

Other Voltage Measuring Instruments

The potentiometric circuit is quite common in corrosion work. The automotive generator ammeter and battery circuit, without a voltage controller, make a good analogy. If the engine is speeded up or slowed down, a speed can be selected, such that the generator is neither charging nor discharging the battery. Then the voltage of each is the same, and further, no current is flowing between them, as evidenced by the fact that the ammeter (galvanometer) so reads.** Therein is the value of potentiometric measurement. If no current is flowing in a circuit, then resistance has no effect (E = IR = OR = Zero).** This is particularly valuable in pipe to soil voltage measurement, where the copper sulfate electrode to soil resistance is unknown. Its main disadvantage is its inability to follow varying voltages. One advantage is that it has a memory, i.e. the reading stays in the meter until a knob is turned. Another disadvantage is that a balance must be made for each reading, otherwise it is possible to record the last reading indefinitely if a lead breaks.

The potentiometer-voltmeter reads the "generator" voltage off a voltmeter. It is quite popular, particularly since the voltmeter can also be made to serve as an ammeter with suitable characteristics. Figure 2 shows the scheme used for the potentiometer-voltmeter.

The true potentiometer has a "slide wire" in which a current flows. This current is adjusted until the graduations on the dial represent the voltages marked theron, usually by equating the voltage across a resistor in series with the slide wire to that of a standard cell*** via the galvanometer. (The fixed voltage reference may also be a Zenner diode, particularly in the future.) The slide wire then takes the place of the generator in the automotive system analogy. In practice the potentiometer is not the most popular instrument. It is essentially a laboratory instrument, and can be very accurate (much more so than needed). It is rather bulky for the use it will serve, and the standard cell is subject to freezing. It is also an example of asking too much from a meter movement or galvanometer. It is usually made to read from a millivolt or 2 to 1.6 volts, but the galvanometer does not respond too well to such a wide range of voltages.

The vacuum tube voltmeter is also a suitable voltage measuring instrument for pipe to soil voltages because most designs operate on virtually no current, making IR in the electrode circuit practically zero. Meter input resistance should be checked, however, as some units are no better than ordinary indicating meters. The low voltage ranges are a major problem. Few such instruments have ranges below about one volt. Some have one side grounded to the case, which severely limits their use. Many vacuum tube voltmeters also have to be reset occasionally. With solid state devices replacing the tubes, there can well be developments in this field. One advantage of the instrument is that a faster moving meter can be used than is possible when a high resistance meter is used. This has value in stray current work.

The Copper Sulfate Electrode

As important as any meter is the copper sulfate electrode. It can have many shapes, but a popular one is a plastic tube about 8 inches long with a wood or ceramic plug in one end and a plastic screw top on the other. Inside the tube is a saturated copper sulfate solution (which must contain some excess crystals of copper sulfate) and a copper rod. The rod usually extends through the plastic top, and the exposed end is the terminal. The end with the wood plug is inserted into the ground. The resistance between the terminal and the earth in general varies from a few hundred to perhaps 10,000 ohms. Some of this resistance is in the porous plug, but the soil around the tip also contributes. Pipe to soil voltage is the voltage between the terminal and the pipe.

One of the main assets of the copper sulfate electrode is its durability. It seldom needs care. Cleaning once or twice a year is in order, but does not usually change the voltage reading, unless a few millivolts is important as in two electrode surveys. In most areas drinking water is sufficiently pure for the solution. The only practical check on accuracy is comparison with several other electrodes, preferably electrodes that are used only for this purpose. Two electrodes, with their plugs touching, should not exhibit more than about 10 millivolts difference.

Field Voltage Measurements

Pipe to soil voltages set the instrument or meter specifications for most voltage measurements. Straight voltmeters can serve in many areas. If soil resistivities are exceptionally high, they would not be desirable. If a high resistance meter errs, it must read low, never high, so that there is a safety factor. Erring on the low side can call for more cathodic protection current, however. The more or less established sensitivity for such meters is 200,000 ohms per volt. High resistance voltmeters are not suitable for two electrode surveys. Here a range at least as low as 100 Mv must be used. A 200,000 ohms per volt meter has a resistance of 20,000 ohms on this range, and this no longer swamps electrode resistance, which is twice normal when two electrodes are used.

The advantages and disadvantages of potentiometers, potentiometer voltmeters and vacuum tube voltmeters for pipe to soil measurements have been discussed. The most popular method for an engineer, who must be equipped for any adverse condition, is the potentiometer voltmeter.

On occasion the current in pipes must be measured by taking the voltage over, for example, a 100 ft span. This yields voltages from a fraction of a millivolt up. A sensitive meter, with a one or two millivolt scale, is necessary. There is no temperature compensation on this range, and small thermoelectric voltages can occur. The heaviest practical leads are needed, as the meter will draw enough current that the voltage drop in the lead wires can be serious. These readings

**This type of measurement is known as a "null measurement." While it is not strictly true, this term generally implies a freedom from resistance problems in the circuit being measured.

**The penalty paid for excessive resistance in a potentiometric measurement is that the galvanometer becomes insensitive. As long as the galvanometer responds well enough to set the instrument voltage at a definite voltage, the instrument is accurate.

***This is a special cell used as a basis for establishing voltage accurately. Properly cared for its voltage does not change appreciably for years.

can be converted to amperes, but in trouble shooting, such as looking for a contact, the readings are recorded only in millivolts.

For other voltage measurement use is made of existing meters. Two precautions are in order. Highly sensitive meters must usually be in the horizontal position. This can be important around rectifiers where this is inconvenient. When testing batteries, have the equipment that the battery feeds operating. A voltage, obtained with a high resistance meter on an unloaded battery, can be very misleading about battery condition.

Ammeters

To make an ammeter out of the basic meter a proportioning device, called a shunt, is used. This resistor parallels the meter, and passes all of the current, except that part needed to operate the meter. For a one ampere meter and the basic movement chosen, 0.99998 amperes go through the shunt, so that the basis can be shifted to that of choosing a resistor that will develop 30 millivolts across it when one ampere flows through it. The current flowing through the meter can be neglected. This shunt is rated one Amp, 30 Mv. Shunts are usually so rated, and their actual resistance is seldom stated, except in the corrosion field. In general in electric work meter current is considered, and when the shunt is outside the meter the meter resistance is measured from the ends of the lead wires, which attach to the shunt, but the high sensitivity meters that the corrosion engineer uses seldom require such consideration, except where very long lead wires are used, as on line currents, where the pipe is used as a shunt. Even then it is desirable that the meter be calibrated at its terminals, since the same lead wire will not always be used.

If a corrosion engineer has a meter with a 50 Mv range, matching his meter to commercial shunts is no problem; he merely needs to buy a shunt which is compatible with a scale on his meter, i.e. if has a 2.5 volt range, it will match a 0.25, 2.5, 25.0 A, etc. shunt.

If a corrosion engineer has a meter which does not have a 50 Mv range, he must calculate current flow from Ohm's Law. Thus, 0.001, 0.01, etc. ohm shunts are popular, as they minimize arithmetic. Since the ohmage of the shunt says nothing about the current that it can safely carry, this must be stated separately when a shunt is rated in ohms. Thus 0.001 ohm shunts of 25 Amp capacity are available, whereas in conventional ratings the nearest equivalent would be a 50 Amp 50 Mv shunt, which would necessarily be larger.

Voltage multipliers for meters are simple devices comparatively speaking. If they are made of wire, it will be quite a few feet long, and cutting it to the proper length is difficult. In fact, the effect of adding or subtracting an inch may be impossible to see on the meter. Not so with shunts. Generally, they are wires or strips not over three inches long. Even the amount of solder used at the ends of the wire or strip can make a major change in its resistance. Shunts should never be repaired in the field.

When ammeters are being purchased expect shunts up to 35 or 50 amperes, depending on the manufacturer, to be contained in the meter. Shunts for larger currents are mounted outside the meter. Generally speaking milliammeters use 100 millivolt shunts† and ammeters 50, though 75 and 100 Mv external shunts are readily available. Electronic multimeters often have 250 millivolt or more shunts.†† The choice of 50 millivolts as an industry standard in ammeters leaves room for temperature compensation, which should be insisted on, particularly in rectifiers. It also reduces the likelihood of thermoelectric voltages, possibly a fraction of a millivolt, from being too serious. A scheme, which many corrosion engineers miss, is that higher rating shunts can be used on lower ranges to advantage. Take the case where the corrosion engineer, using a portable or external shunt from his tool box, is making a temporary test, say with a welding machine. He will never miss 50 millivolts. Thus, if he purchases a 100 ampere, 100 millivolt shunt, he also has a good 50 ampere, 50 millivolt shunt. This scheme can considerably reduce his shunt requirements.

The foregoing arrangement brings up a point which many corrosion engineers miss when confronted with the shunts available for permanent installation in junction boxes. Many of these shunts develop less than 10 millivolts when applied in the customary manner, and many corrosion men simply do not own a suitable meter for these low millivolt measurements. If they do, battery clip contact resistances and thermoelectric voltages between different metals in their lead wire circuits, as well as lack of temperature compensation in their instrument, may cause large errors. Nor is there much economy. At 3 cents per KWH and 50% rectifier efficiency, a 50 millivolt, 5 ampere shunt consumes 13 cents worth of electricity per year. The economy realized by going to low millivolt shunts is not much, and over $100 may have to be spent for a suitable meter for each man expected to test at this point. Shunts, which will develop enough voltage to be read on a 50 millivolt meter, are suggested.

Resistance Measurements

Since ohms may also be called volts per ampere, resistance measurements are ratio measurements of voltage to current. This makes schemes other than measuring current and voltage desirable. The simplest in common corrosion prevention use is the Wheatstone Bridge. Figure 3 shows this circuit. Some reasonable values of a bridge, such as is used, are shown, as there are many ways to set such a bridge up. Note that a battery is used for power. A buzzer, for example, can be used to generate A C. As shown the slide wire is at mid point, so half of the battery voltage exists at this point. If X is 500 ohms, then half of the battery voltage will exist on the left side of the galvanometer too. Then the galvanometer will read zero with the slide wire pointer at 500. This is a null measurement, but unlike the potentiometer there is current flow through the instrument terminals at all times. With an inappropriate bridge arrangement this current can cause damage. For example, attempting to measure a meter's resistance with a bridge will almost always destroy the meter.

†This is a point to be watched if a stock meter is purchased. Sacrificial anodes yield about 750 millivolts to do their job. If 100 of these are consumed in a meter during the tesing of anode current, serious error can result.

††One manufacturer now makes a 50 Mv meter, the basis being that a low meter loss is desirable for transistor circuits.

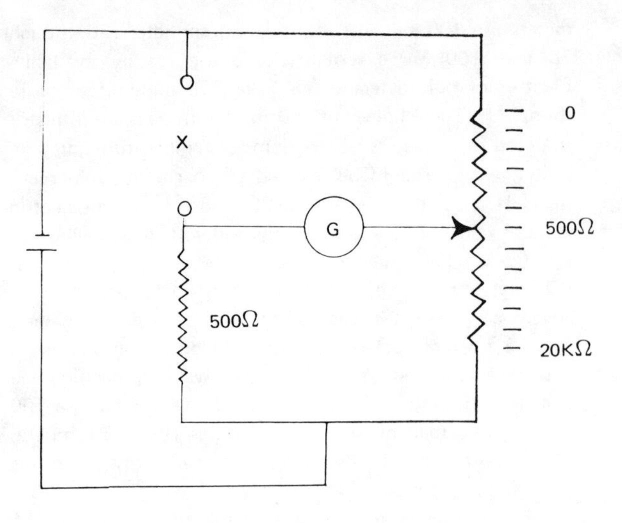

Figure 3

AC vs DC Resistance

Except in circuits where there is no inductance or capacity, differences in AC and DC resistances are to be expected. AC instruments are the most practical for in-place soil resistivity measurement, particularly if the depth is modest, and for soil boxes; but such instruments should be considered questionable for uses other than those for which they were designed.

Taut Band Meters

These meters use taut metal ribbons or bands to hold the moving coil in place instead of pivots and jewels. (See Figure 1). Since they are used mostly on high sensitivity meters they are important in corrosion instruments. As pointed out earlier this construction permits more sensitivity and is better suited for galvanometer construction. Some peculiarities are showing up. Vibration can cause the pointer to wander. They may be slightly easier to burn out. It is becoming very clear that they do not require the repair that conventional meters do, and for this reason are cheaper in the long run, even though they cost more.

Checking Meters

There is considerable misunderstanding about the proper way to compare meters. It is wrong to substitute one meter for another, and use this as a basis for comparison. Each meter can disturb the circuit to a different degree, and thus affect its reading accordingly. Voltmeters should be kept in parallel during comparison, and ammeters should be kept in series. Another precaution to be observed is meter position. Many high sensitivity meters are position sensitive. As a rule they are designed to be used in a horizontal position. This is often inconvenient around rectifiers, but very important.

When checking uncompensated rectifier ammeters, temperature should be taken into account. For example, if the ammeter is found to be one division high on a cold day, when compared with a compensated instrument, it would be well to leave it that way. It is probably the most accurate reading that the meter is capable of, and readjusting it will throw it further off when the weather becomes warm than it would be if left as found.

The taut band meter poses a new problem, roughly like metallic thermometers versus glass. When a glass thermometer is damaged it is obvious, but this may not be true in the case of the metallic. Taut bands are so stout that parts within the meter may move before the band breaks. Also, taut band meters do not have to be repaired as often as pivoted and jeweled meters, so they are seldom subject to repair station inspection. Checking, therefore, becomes more important.

It is important that meters be checked from time to time. Meters do develop troubles that may not be obvious, and which can prove disastrous.

Care of Instruments

Handling is the first consideration. Naturally, no one intends to drop an instrument, but this invariably happens sometime or other. The first precaution is simply to not take unnecessary instruments out into the field where they may be dropped. Both accidental damage and the continued carrying of an instrument in a car probably represent the hardest usage that an instrument receives. The foregoing does not mean that a suitable assortment of instruments for any work that might be expected should not be carried. It is aimed at carrying such meters as long line current millivoltmeters, which clearly may not be used on a particular day or on a particular job. Two more precautions in carrying instruments involve the matter of position. When it can be avoided an instrument, with pivots and jewels, should not be carried with the pivots horizontal. This seems to subject them to more damage than when they are carried in a vertical position. Also, the shocks encountered during travel can sometimes be placed on the top jewel and pivot in a meter by carrying the meter upside down. (This precaution is based on the fact that meter operation is much less dependent on the top jewel than on the bottom jewel.) A meter that would be carried this way would be a particularly sensitive meter. Felt padding is, of course, nice where pickup is used, and the meters cannot be carried on the seat. Sponge rubber should be avoided, as sulfur fumes may corrode the silver switch points which exist in most instruments.

One common fault in using meters is tapping the glass to overcome any possible friction. An examination of the manner in which meters are constructed will show that the glass is not in a particularly advantageous position to mechanically jar friction out of the movement. Tapping along side the glass is equally satisfactory and will not tend to knock out the glass. Tapping will not help with taut band meters. It may cause the hand to jump.

There are several common meter difficulties which can be recognized very easily. The first, because it sometimes occurs during shipment from the manufacturer to the customer, is one or more tangled springs. Severe jarring can cause a hairspring to tangle, and the parcel post clerks succeed in accomplishing this fairly regularly by the manner in which they throw parcels. If it is granted that the meter was zeroed, or nearly so when the manufacturer packed it, the first thing to be expected is the meter will be from an eighth to a quarter inch off zero. The meter may swing perculiarly and upon testing will read low.

Another effect of jarring is to affect balance, but this generally requires dropping. The pointer of a good meter of ordinary sensitivity should stay on zero when it is not in use, regardless of position. Ultra sensitive meters should stay very nearly on zero when tilted 30° from horizontal.

The last precaution the author suggests concerns meters being shipped to a repair shop, particularly if the instrument is only slightly damaged and does not need a complete overhaul. Any broken glasses should be Scotch taped, or the opening taped closed with cardboard. If excelsior, or any dirty packing is used, the instrument should be wrapped and taped. Finally, there should be two inches of packing on all sides of the instrument. This is a parcel post regulation for insuring delicate items, and experience shows that if it is followed it is almost impossible to damage the contents, except possibly tangling the springs, which presents no great problem to a good repairman. If the instrument must be crated, as for water shipment, use screws, not nails, to close the crate.

Meters and Equipment Required for Non-Stray Current Work

Meters

- Pipe to soil voltmeter with ranges 50 Mv to 10 V
- Millivoltmeter, 0-2 Mv for line currents. Should have short, open and 2 volt (for checking against P/S meter) range switch positions. Ammeter and shunts to cover ranges from 100 Ma to perhaps 100 A.
- Vibroground
- AC bridge and soil resistivity rod

Pipe Locators

- Audio frequency for locating contacts and insulating joints
- Radio frequency for fast locating in non-congested areas

Miscellaneous

- Copper sulfate electrodes, two each
- Flange tester, compass type (also serves as current source for some tests)
- Probe bars
- Coiled toaster element for use as rheostat
- Soil box
- Wire reels
- Interrupter

Conclusions

The cathodic protection worker is totally dependent upon his instruments, so they deserve considerable attention. With good care they will serve him well. Meters are expensive, and once purchased must be used for years. A man just entering cathodic protection work should wait to purchase instruments until he has worked with an experienced engineer, as he will usually want to use the brand of instruments on which he trains.

OPS INTERPRETATIONS

• *Weld Acceptability*

Question: How would a company comply with the provisions of Paragraph 192.229(c) when its welders are qualified under Section 3 of API Standard 1104 by passing the multiple qualification test? Would a production weld have to be cut out of a butt weld joint and a branch weld joint, or would one or the other suffice?

OPS Interpretation: For welders qualified to make both butt and fillet welds by the multiple qualification test of API Standard 1104, either type of weld can be destructively tested to comply with Section 192.229(c) since only one weld is required to be tested. However, a butt weld does not have to be destructively tested to comply with Section 192.229(c) if it is nondestructively tested and found acceptable under Section 6 of API Standard 1104.

• *Transmission Line Definitions*

Question: In Subparagraph (1) under the definition of "transmission line" as found in Paragraph 192.3, would the term "storage facility" be defined to mean only pipe-type and bottle-type holders, or would this include storage in a natural underground cavity as well as low pressure system holders?

OPS Interpretation: The term "storage facility" in the definition of "transmission line" includes storage in a natural underground cavity. A low pressure holder would also be considered as a "storage facility."

• *Component Certification*

Question: For components manufactured to industry standards, how may the certification be made that a component was tested to at least the pressure required for the pipeline to which it is being added when the required pipeline pressure is not known at the time of manufacture?

OPS Interpretation: A certification meeting the requirements of Section 192.505(d) could be made in one of several ways. Where a component has been individually tested or manufactured under the required quality control system, the component itself could be marked with the manufacturer's name, specification to which manufactured, pressure rating, and test pressure to which the component (or prototype) was subjected. Alternatively, the certification requirement could be met, for a qualified component, by a statement in the manufacturer's catalogue that the component, suitably identified, meets the requirements of an accepted industry standard.

As an example, API Standard 6D requires Class 300 valves (pressure rating 720 psig) to withstand a hydrostatic shell test of 1100 psig. An operator who needed to install a valve in an existing pipeline (Class 3 location) which has a design pressure of 720 psig and which has been hydrostatically tested to 1080 psig (720 x test factor of 1.5 for a Class 3 location) could install a valve purchased off the shelf, certified to API 6D, even though the valve was not manufactured with that particular installation in mind. The operator is responsible for selecting and using a component that has been certified to a pressure at least as high as the test pressure for the pipeline in which it is to be installed.

With regard to an appropriate quality control system, the standard to be met is set forth in Section 192.505(d)(2). This is stated as a performance type requirement since it would be impracticable to set forth a detailed quality control system for each type of component being manufactured. However, any acceptable quality control system would consist of two basic elements: continuous inspection by qualified personnel utilizing appropriate inspection equipment, and periodic testing (both mechanical and chemical) of materials going into the components.

NOTE: The above OPS Interpretations were reprinted from OPS Advisory Bulletins 72-10 and 72-11.

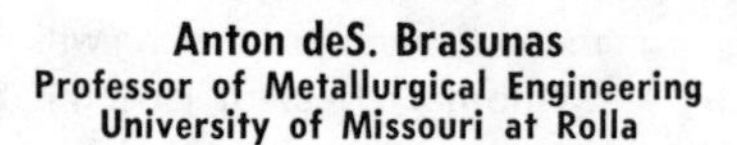

Anton deS. Brasunas
Professor of Metallurgical Engineering
University of Missouri at Rolla

A Critical Review of

CORROSION

WHEN CORROSION OCCURS, it can be expected to recur in exactly the same manner only if all conditions are exactly duplicated. This duplication, in the field or laboratory, is difficult to achieve. Corrosion seldom proceeds uniformly over the same specimen, and hence is even less likely to proceed uniformly on sets of supposedly identical specimens. This, essentially, is the major problem of reproducibility, spread of data, and appropriate reporting of corrosion testing.

Tests vary in environment, nature of attack, time of testing, and other conditions, such as velocity and metal composition. The key to proper corrosion testing methods lies in the definition of such terms as "duplicate" and "identical."

If all the factors that can affect corrosion were compiled, the researcher might be amazed that the good agreement, such as is frequently observed, is ever achieved. In many cases, however, some of the possible controlling factors may remain dormant, exercise a negligible effect, or two effects may almost nullify each other.

Classification of Tests

Corrosion occurs in gaseous, liquid, or solid environments. Gaseous environment tests are typified by atmospheric exposure tests often made using special racks on rooftops, along highways, or along the oceanfront. Variables include humidity or contamination, flue gases, chemical fumes, temperature and pressure. Many atmospheric tests are conducted in test chambers.

Liquid environment tests are perhaps most common and are conducted with specimens submerged or partly submerged in various waters, in industrial process lines or in laboratory situations such as boiling acid tests. Nonaqueous liquid environments include organic solvents, molten metals, molten salts and other homogeneous and even heterogeneous mixtures (slurries).

The solid environments are few and may be typified by underground corrosion of pipelines or the attack of metals embedded in concrete. These are not truly solid environments, since the presence of moisture and anions largely determine the possibility and extent of corrosion. An illustration of a true solid environment is a metal in contact with another metal or solid body producing solid state diffusion.

Variables

The four major variables which affect corrosion rates are listed below.

Composition of the Environment

This includes nature of medium, concentration, impurities and pH. Even small differences in these variables are significant, such phenomena as inhibition and stress corrosion cracking may be strongly influenced by very small chemical environment changes.[1]

Chemical Composition of the Metal

The tremendous variations in corrosion rates of different metals is well known, but variations in corrosion of metals where only slight differences in composition exist is not well understood. The beneficial effects of small amounts of copper in steel on atmospheric corrosion resistance[1] is striking, but the similar effects or extremely small amounts of phosphorus[2] in the vicinity of 0.05% is truly remarkable. Similar small variations in steel composition, however, have little effect in underground corrosion.[3]

Temperature

In most cases increasing temperature increases the severity of corrosion. An exception is in the vicinity of 200 F (93 C) where oxygen is removed from ordinary water in aqueous tests. The increase in corrosion rates with increasing temperature[4] is most frequently illustrated with an Arrhenius type equation which may be expressed:

$$\text{Rate of Corrosion} = \text{A e} - \frac{\text{B}}{\text{RT}}$$

where all symbols except T are constants; T is the absolute temperature, B is the activation energy, R is the gas constant and e is the base of natural logarithms.

In some instances anode-cathode potentials may reverse at elevated temperature. For instance, iron, which is cathodic to zinc at room temperature, becomes anodic to zinc[5] at temperatures above 140 F (60 C).

Time

Since corrosion rates are time dependent in almost all instances, it is a significant variable.

A fifth factor which can be of extreme importance in certain specific environments, but only for certain alloys and temperature ranges is:

State of Stress

Tensile stresses, whether internally or externally generated, may result in unexpected failures with minimal weight loss (stress corrosion cracking). This is restricted to environments in which corrosion in the absence of stress is extremely low. This important subject has been discussed often[6] and is now receiving much attention because of its relationship to the deep diving submarines, rockets and other pressurized systems.

Ten other important variables may be added to the list.

SUMMARY

Methods of corrosion testing are described and analyzed. The reliability and interpretation of significant test variables are examined with respect to accuracy and reproducibility. New test techniques are suggested to remedy situations where results ares typically dubious.

Reproducibility and Interpretation of

TEST PROCEDURES

Presence of Films or Foreign Matter on the Surface

This can result in a very significant increase in corrosion (as effects of fingerprints or inorganic contaminants[7]), or decrease in corrosion (films on passive stainless steel[8]).

Velocity Effects

Increased velocity within limits can have either a beneficial effect if it eliminates non-homogeneities or harmful effect by eliminating polarization.

Presence of Electrical Potentials

In general, cathodic potentials are considered desirable and anodic potentials undesirable except in unusual cases where anodic passivation develops.[9] The Pourbaix diagrams[10] show such areas.

Particles or Bubbles in the Environment

These factors coupled with high velocity can result in damaging erosion-corrosion.[11]

Chemical Homogeneity of the Alloys and Presence of Other Phases in the Metal Structure

Lack of homogeneity is known to result in local potential differences on the surface and increase local action cells; i.e. sensitized stainless steel.[8]

Homogeneity of the Environment

The detrimental effects of differences in oxygen concentration as well as certain anions is well known as a source of damaging potentials leading to corrosion currents.

Fluctuations in Temperature or Temperature Gradients

Fluctuations at elevated temperatures can cause spalling of protective oxide layers with consequent increase in corrosive attack. Temperature gradients are known to cause potential differences and solubility differences.

Surface Roughness

The true effect of surface roughness is not clearly established. A uniform and reproducible surface, however, is generally desirable.

Pressure

In atmospheric corrosion the partial pressure of a particular component of the gas atmosphere is analogous to concentration of substances in aqueous solution.

Grain Size

If corrosion is not intergranular in nature, grain size is not generally considered to be an important factor; although a recent paper[12] shows approximately 65% more life in stress corrosion tests with fine grain size stainless steel.[11] Orientation has long been known to affect the initial rate of corrosion in some environments.[13, 14]

The Specimen

The size, shape and weight of the specimen to be used in a corrosion test is often largely determined by the purpose of the test. Many tests are made on finished shapes, such as sheet, wire and tubing, and many of the dimensions of the specimen may already be fixed. However, a large surface to volume ratio increases sensitivity. If an alloy is to be evaluated for use in a specific environment, choice of size and shape may be unlimited.

Corner and edge effects are generally unavoidable except with spherical specimens. Therefore, cylindrical specimens may have certain advantages over plate or sheet specimens.

Where maximum precision is desired, the weighing device should be extremely sensitive. This generally means a relatively light specimen under 200 gms. Further, a practical maximum surface area to fixed weight ratio suggests a thin sheet specimen.

Surface Condition

A relatively smooth stress-free surface may be prepared by electropolishing a specimen that has been brought to almost-finished size by mechanical means. Electrolytically polished surfaces have areas nearer those that are calculated from physical dimensions although they are still not at the ideal ratio of one. If a specimen surface cannot be brought to the elusive ideal condition, a reproducible surface should be developed such as that obtained by liquid honing or abrading all exposed metal surfaces. Surfaces should be free of all foreign matter.

One aspect, sometimes overlooked, is the localized surface disturbance caused by machining and abrading, which annealing or pickling may either remove or minimize. However, the surface formed after annealing or pickling may become even less easily defined.

Surface Area

The extent of corrosion is generally evaluated by determining a weight change per unit surface area, but two important aspects are generally overlooked:

(1) The area calculated from the initial geometric dimensions yields an "apparent area" which differs from the "true area." Surface irregularities (hills and valleys on a micro scale) of an abraded surface have been reported to increase the actual surface area by 3 to 50 times.[15] True area is determined by the adsorption of gas molecules on a

Source: *Materials Protection*, December 1967, 20-24

surface or by electrolytic polarization. Both the disturbed condition of the metal surface layer and high surface area are important factors in causing the high initial corrosion rates frequently noted.

(2) The area of the metal undergoing corrosion diminishes with time as it shrinks in size due to corrosion. In certain cases, however, as in intergranular attack, this area may actually increase.

The instantaneous corrosion rate should be measured in cognizance of these factors. Their proper use may sometimes yield much higher corrosion rates than those otherwise obtained. This was the case in affirming that a corrosion rate was actually increasing when a conventional plot of corrosion vs. time inferred that the rate was decreasing.[16]

Testing Procedures

Test procedure is important if it affects results. Therefore, committees of individuals familiar with all aspects of certain types of tests have carefully set down procedures intended to yield reasonably reliable and reproducible results. Allowances are made for variations to permit evaluation of certain variables or test shapes. The *ASTM Book of Standards*[17] is typical of such documentation.

These procedures are good though not entirely satisfactory and should be revised as new materials, equipment, refinements and techniques become available. Further, no standard or tentative methods exist for many types of corrosion tests, such as those at elevated temperatures. In certain instances, standardization may not be desirable for many years.

In various laboratories interesting techniques are employed. For example, some high temperature corrosion and mechanical property tests (such as tensile or creep testing) are conducted using an electric current to achieve high temperature. Although this technique may eliminate furnace problems, it introduces entirely new ones.

As an illustration, sudden variations in line voltage may cause an unexpected increase or decrease in temperature. Just prior to failure current density may become excessively high, and the temperature may rise to the melting temperature as shown in Figure 1. In such cases the actual failure is due to melting rather than corrosion or lack of adequate solid state strength at the "test temperature."

Repeat Tests

Attempts to make significant tests with the materials and procedures available often require multiple tests and statistical analysis to obtain meaningful results.

If 1001 specimens were simultaneously exposed to identical tests in three different laboratories, a different data spread would probably develop although the average values might be identical. This can be typified by a regular distribution about the average value as shown by curves A, B, and C in Figure 2. With a truly normal or symmetrical distribution the average value and median value should be identical. The central value in 1001 values would be the 501st value or median, if listed in order of magnitude.

The variables appear to be much better controlled in Lab. A, where results varied from a' to a., and least controlled in Lab. B, where results gave values between c' and c. Thus, if a single data point were available, one would question its reliability from Lab. C more seriously since it may vary or deviate from the average value, v, by as much as (c-v).

The term "standard deviation" indicates the degree of spread which includes slightly over 68.26% of all the data on either side of the average figure. Mathematically the standard deviation, σ, can be calculated by:

$$\sigma = \sqrt{\frac{\Sigma\,(d^2)}{n}}$$

where σ is the standard deviation (quadratic mean)
d is the actual deviation of a given experimental point from average (arithmetic mean)
n is the total number of experimental points.

In the above illustration the standard deviation σ_a for Lab. A was small because d values were small and for Lab. C, σ_c was greatest. Thus, by carefully derived relationships between σ and the scatter curve (often referred to as a probability curve), 68.26% of all points should fall within $v \pm \sigma$, where v is the average value and σ is the standard deviation; 90% of the points fall within $v \pm 1.65\,\sigma$, and 99.93% of the data fall within $v \pm 3\,\sigma$. This is illustrated in Figure 3.

Numerous texts on the subject, one of which is available as a paperback,[18] sections in comprehensive corrosion texts,[9,19] even ASTM procedures[17] and papers over 35 years ago[19] are concerned with this subject.

Selection of Units

The extent of corrosion may be expressed by several methods. Frequently convenience dictates the method used rather than accuracy. Actually, there is no single correct way, and probably a combination of several methods should be used.

The classification of methods of reporting corrosion rates may be summarized as follows:

1. Gravimetric
   (a)—weight loss per unit area
   (b)—weight gain per unit area
   (c)—% weight change
   (d)—amount of corrosion product in environment

Figure 1—Upper segment of internally heated high temperature corrosion test specimen indicating that actual failure was caused by melting rather than excessive corrosion.

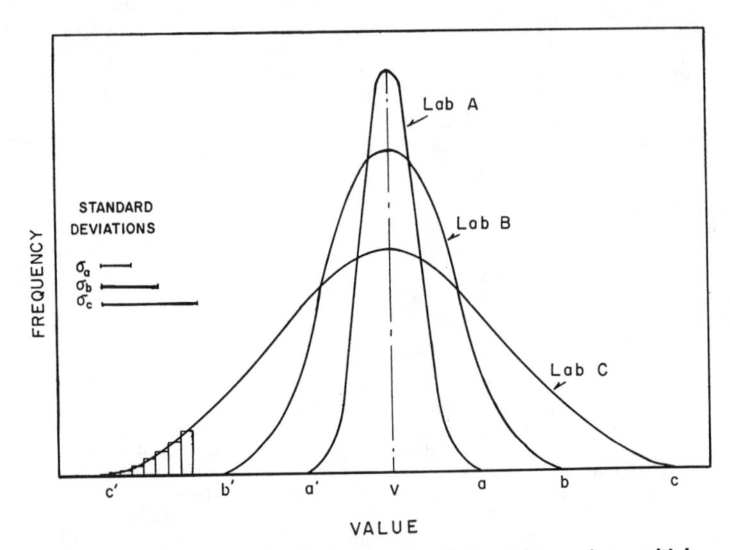

Figure 2—Symmetrical distribution of hypothetical data using multiple specimens from 3 laboratories. An identical average value is assumed for each set of data.

2. Changes in electrical conductivity
3. Changes in residual mechanical properties
4. Radiographic determinations
5. Environmental changes
   (a)—increase or decrease in certain substances in solution
   (b)—decrease in pressure of a gas (as in constant volume, closed-chamber atmospheric and gas corrosion tests)
6. Volumetric changes—constant pressure[16]
7. Direct surface recession measurements
   (a)—thickness changes
   (b)—pit depth determination
   (c)—surface recession from reference point[21]
8. Metallographic studies—microsection selected for study must be carefully chosen and polished to the proper level to reveal typical or most severe area and should be recognized as such.
9. Optical changes on surface—corrosion product film thicknesses have been determined from interference color changes[14] (index of refraction must be known) and from the ellipticity of polarized light.[22]
10. Electron microscopy is a relatively new tool to study surfaces and very thin metal sections.

Gravimetric

Gravimetric changes usually represent the most common mode of determining the extent of corrosion. Two classes of units are commonly used, namely:

A. weight change per unit of original surface area per unit time
   (1) milligrams per square decimeter per day—mdd*
   (2) milligrams per square decimeter per test duration*
   (3) grams per square inch per day
   (4) grams per square centimeter per hour*
   (5) milligrams per square millimeter per hour
   (6) milligrams per square centimeter per hour*

B. surface recession, calculation—(based on uniform metal removal)
   (1) inches penetration per year—ipy*
   (2) inches per month—ipm*
   (3) mils per year—mpy
   (4) furlongs per fortnight—fpf
   (5) centimeters per year*
   (6) microns per year

Those with the asterisk are found in various sections of the Corrosion Handbook.[3, 19] Authors are advised to conform to milligrams per square decimeter per day, abbreviated mdd; and mils per year, mpy, so that data can be compared more easily.

The conversion from one type of recommended unit to the other must involve density. Therefore, the change from mdd to mpy is made as follows:

$$\text{mpy} = \text{mdd} \times \frac{1.437}{\text{density}}$$

where density is expressed as g/cc or specific gravity. For Type 304 stainless steel, for instance, the density is reported[23] to be 7.92; therefore, the above expression becomes:

$$\text{mpy} = 0.181 \text{ mdd} \quad \text{or} \quad \text{mdd} = 5.52 \text{ mpy}$$

However, such calculations assume uniform metal removal from all exposed surfaces. If this is true, such conversion is perfectly suitable. However, if metal loss were uneven, such as in pitting, intergranular corrosion, or other form of highly localized attack, reported data based on a uniform metal removal base would be misleading.

Other units found in the Corrosion Handbook are "% wt. loss" or simply "milligrams" of weight loss without reference to surface area. These are poor choices since specimen shape can largely determine values obtained.

An illustration involves a low corrosion rate of 0.005 ipy. Two specimens of identical composition were tested for a period of one year; both weighed 10 gms. One was a sphere and the other was a foil 0.010" thick. On the basis of % wt. loss, the foil would be 100% lost, and the sphere, less than 15% lost, a difference of about seven to one.

Weight gain is usually associated with corrosion reactions which build up corrosion product layers, such as for hot gas corrosion. A continuous record of weight gain vs. time can be made with a single specimen suspended in a furnace from an arm of a gravimetric balance.

However, errors in such weight gain measurements would develop if:

a) any of the scale layer flaked off, or

b) evaporated as a result of volatility (sulfides, chlorides, or oxides such as CO_2 and MoO_3 are quite volatile).

Weight losses, on the other hand, require well cleaned corroded specimens completely free of corrosive product. Volatile or soluble surface products present no problem, but surface layers made up of entrapped corrosion product make accurate weight loss measurements impossible. Here even the meaning of a weight gain measurement would be difficult to interpret. The correlation between weight gain and weight loss has been analyzed by the author in an earlier publication.[16] However, neither type could adequately express the extent of corrosion without an accompanying photomicrograph.

Other Methods

Measurement of the extent of corrosion has also been determined by measuring changes in electrical conductivity. As the effective cross section of the metal is reduced by corrosion, and especially if it were intergranular in nature, the electrical resistance changes sufficiently so that correlations can be made to the average depth of attack. Care must be exercised in applying such data since other factors may enter into measured resistance changes. In the ASTM wire test,[24] a 10% increase in electrical resistance is regarded as the limit for useful life.

Similarly, dimensional changes caused by corrosion can influence such mechanical properties as ductility or tensile strength. These changes can also permit in-place nondestructive determination of residual metal thickness or pits by radiographic techniques.[25]

Systems could be monitored to detect specified levels of corrosion product, like Fe^{+++} ions, and the *Corrosion Handbook* plots (p 211) extent of corrosion by indicating as a measure of corrosion, not weight loss, but mg Pb per ml of solution. The use of radioactive tracers are now being used to increase sensitivity to smaller concentrations.

Corrosion tests in gaseous oxygen have been made by determining the

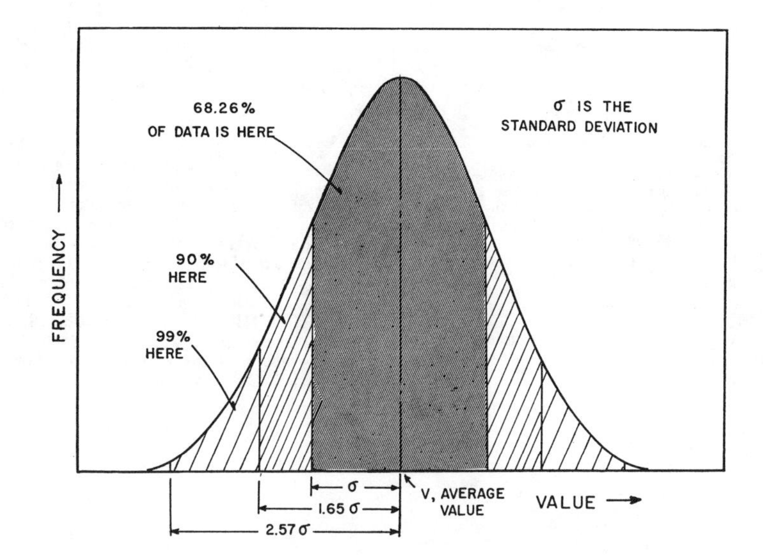

Figure 3—Normal distribution curve showing proportion of data points embraced by "plus or minus" sigma, 1.65 and 2.57. The confidence level is high when σ is low.

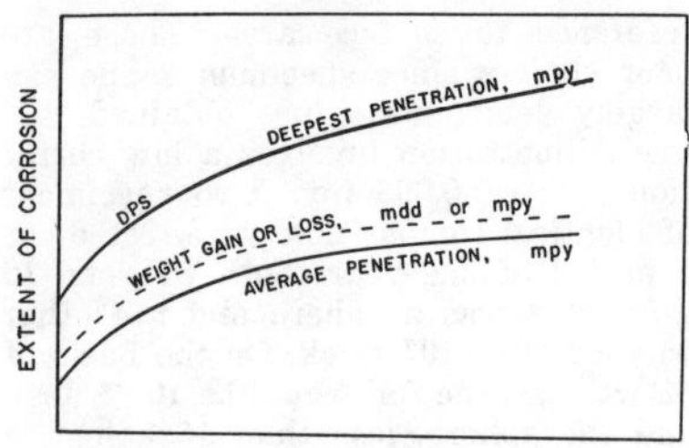

Figure 4—Plot showing weight loss, expressed as mdd or mpy, and actual penetration of pits or intergranular attack as average and deepest penetration.

volume of oxygen absorbed as a function of time. This permits plotting the progress of attack without disturbing the specimen.[16]

Without doubt, the most satisfactory way to determine the extent of corrosion is to determine residual metal thickness after corrosion has occurred and be assured that no damaging reactions have occurred to the underlying base metal. Such assurance can only be obtained by some form of mechanical test coupled with thorough microscopic examination.

Recommendations

Popular practice of reporting merely the average weight gain or weight loss is not entirely improper, but should be supplemented with necessary information added to portray an accurate analysis.

If pitting or intergranular attack occurs, for example, in addition to gravimetric data, a report of the average depth of pitting or intergranular attack as well as deepest penetration should be reported. If other localized attack occurs, the deepest penetration should be reported, as well as the frequency of attack per unit length or area. This requires close visual inspection and depth determination with a suitable probe, radiographic technique or metallographic technique focussed on suitable areas.

Thus, a report on corrosion versus temperature or some other variable may be illustrated as in Figure 4.

Other parameters of corrosion damage may be desirable. For instance, if strength is important, the residual tensile strength should be related, if possible, to the gravimetric data. Measurement of deepest penetration from the original surface (dps) is important in determining residual strength or life of pressurized tanks.

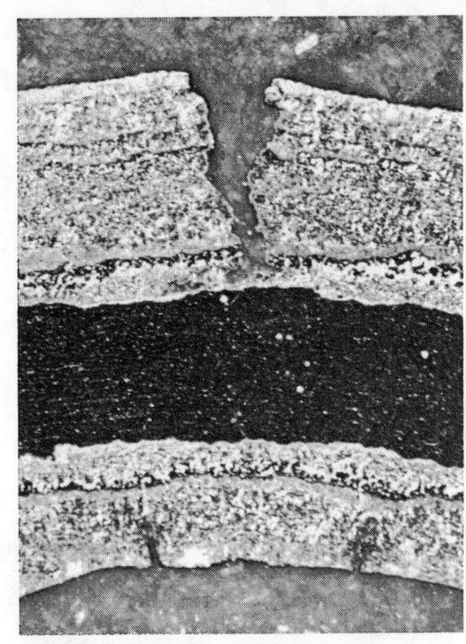

Figure 5—Dezincified brass revealing unaffected ductile brass in center and dezincified layer at top and bottom. Note evidence of brittleness. Magnification 6 X.

When specimens are inspected for corrosion damage or are mounted for metallographic study, bending the specimen may provide additional useful information. In Figure 5 a brass specimen which suffered parting attack (dezincification in this case) left the surface area embrittled. Corrosion reactions at the surface sometimes have no apparent effect, when in reality some embrittlement may have occurred. Even bending, unless it is very sharp and severe, may not disclose this damage. In such cases microhardness indentations can be very helpful especially if the polished and etched surface area reveals no visible changes.

Since corrosive attack occurs in a wide variety of ways, leaving the corroded metal damaged to a relatively insignificant or serious extent, a proper evaluation should always consider the various ways that the metal could have been affected.

References

1. F. L. LaQue. *Corrosion Testing, Proc. ASTM,* 51, 495, 1951
2. C. H. Lorig and D. E. Krause. Phosphorus as an Alloying Element in Low Carbon Low Alloy Steels III, *Metals and Alloys,* 7, 69, 1936
3. H. K. Logan. Corrosion by Soils, *Corrosion Handbook,* 446, 1948, John Wiley & Sons
4. H. H. Uhlig and A. deS. Brasunas. Effect of Magnetic Transformation at the Curie Temperature on Oxidation Rates of Chromium Iron Alloys, *Journal Electrochemical Society,* 97, 448, 1960
5. D. M. Geehan. Theoretical Consideration of Zinc Electrode Behavior in Pressurized Aqueous Systems, PhD Thesis, Case Inst. Technology, 1962
6. W. D. Robertson. Stress Corrosion Cracking and Embrittlement, 1956, John Wiley and Sons
7. G. W. Cunningham and A. deS. Brasunas. The Effects of Contamination by Vanadium and Sodium Compounds on the Air Corrosion of Stainless Steel, *Corrosion,* 12, 389t (1956) August
8. N. A. Nielsen. Passivation of Stainless Steel, Symposium on Corrosion Fundamentals, 148, 1956, University of Tennessee Press.
9. U. R. Evans. The Corrosion and Oxidation of Metals, 900, 1959, Arnold Press, London
10. M. Pourbaix and G. Covaerts. The Work of Cebelcor's Commission des Études Fondamentales et Applications (CEFA), First International Congress on Metallic Corrosion, 96, 1962 Butterworths, London
11. M. G. Fontana. Erosion Corrosion, *Industrial and Engineering Chemistry,* 39, 87, June, 1949
12. V. L. Barnwell, J. R. Myers, and R. K. Saxer. Effect of Grain Size on Stress Corrosion of Type 302 Austenitic Stainless Steel, *Corrosion,* 22, 261, (1966) Sept.
13. F. W. Young, J. V. Cathcart, and A. T. Gwathmey. The Rates of Oxidation of Several Faces of a Single Crystal of Copper, *Acta Metallurgica,* 4, 145, March, 1956
14. G. Tammann. The Determination of Crystallite Orientation, *Journal Institute of Metals,* 44, 29, 1930
15. S. Dushman. Scientific Foundation of Vacuum Techniques, 2nd Ed., 417, 1962, John Wiley and Sons, Inc.
16. A. deS. Brasunas and N. J. Grant. Accelerated Oxidation of Metals at High Temperatures, Trans. *ASM,* 44, 1117, 1952.
17. ASTM Standards, ASTM Designations; E 178-61T, Dealing with Outlying Observations, E 177-61T, Use of Terms Precision and Accuracy Part 30, 1965
18. H. Arkin and R. Colton. Statistical Methods, 1966, Barnes and Noble, Inc.
19. G. G. Eldredge. Statistical Methods, *Corrosion Handbook, 1083,* 1948, John Wiley and Sons, Inc.
20. C. H. Humes, R. F. Passano, and A. Hayes. A Study of the Error of Averages and its Application to Corrosion Tests. *Proc. ASTM,* 30, 11, 448, 1930
21. A. Brasunas. Catastrophic Corrosion Resulting from Vanadium Pentoxide in Fuel Oil Ash, *Corrosion,* 11, 17, (1955) Jan.
22. A. B. Winterbottom. Optical Methods of Studying Films on Reflecting Bases Depending on Polarization and Interference Phenomena, Trans. *Faraday Soc.* 42, 487, 1946
23. C. J. Smithells. *Metals Reference Book,* 11, 3rd Edition, 708, 1962, Butterworth & Co. London
24. A deS. Brasunas and H. H. Uhlig. Oxidation of Alloys by the Wire Test Method, *ASTM Bulletin,* 182, p 71, May 1952
25. R. S. Herman. Radiographic-Photographic Method for Measuring Depth and Distribution of Pitting, *Corrosion,* 20, 361t, (1964) Nov.

ANTON deS. BRASUNAS is associate Dean of Engineering and director of the St. Louis Graduate Engineering Center, University of Missouri at Rolla, Mo. From 1955 until September, 1964, he was director of the Metals Engineering Institute. Prior to that he was associate professor of metallurgical engineering at the University of Tennessee. He also has been associated with the Oak Ridge National Laboratory and Battelle Memorial Institute. A member of NACE, he received his DSc from MIT, his MSc from Ohio State University, and his BS from Antioch College. He also is a member of ASM, AIME, NSPE, and ASEE. . .

The Concept and Development of Corrosion Monitoring*

P.J. MORELAND and J. G. HINES
ICI Ltd, Mond Division, Runcorn, Cheshire, England

The development of corrosion monitoring from its origin in laboratory corrosion tests and plant inspection techniques with the objective of reducing maintenance and shutdown costs is described as well as possible future developments. Techniques are summarized with regard to measurement type, degree of use, and application area, together with criteria for selection. Economic advantages have been estimated to be as much as 25 percent of maintenance costs saved. Specific savings from $250,000 to $1.3 million have been documented.

THE CONCEPTS of corrosion monitoring techniques covered in this paper are those developed in the *Handbook of Industrial Corrosion Monitoring* [1] prepared in part by the above authors on behalf of the United Kingdom Department of Industry. Parts of this paper are also based upon the handbook to which the authors are contributors.

Philosophy

The concept of corrosion monitoring has developed from two distinct areas, plant inspection techniques, and laboratory corrosion testing techniques, with the original aim of assessing or predicting the corrosion behavior of plant and equipment between shutdowns. There are, similarly, two sets of objectives which are essentially distinct, although related and to an extent overlapping. The first is to obtain information on the state of operational equipment—to permit the better scheduling of maintenance work, to ease the inspection load during shutdowns and, of course, to avoid unplanned shutdown occurring because of unforeseen deterioration of plant. The second objective is to obtain information on the interrelation between corrosion processes and operating variables — to help diagnosis of the problem and to allow improved control of corrosion and more efficient operation of the plant.

Not only are the objectives related, but some of the techniques available may be applied to either. Few of the techniques can answer all possible questions but all will provide valuable information in the right situation. However, each technique has its limitations, and the spectrum of methods should be considered to be complementary rather than as alternatives.

Undoubtedly the incentive to use corrosion monitoring is substantially financial. With the increasing trend towards integrated plants, where unexpected failure of a relatively minor item may incur large cost penalties due to loss of production, the financial agruments cannot be expected to diminish. However, at present there is an increasing concern for the safety of personnel in industry, and a growing public awareness of environmental considerations. The increased confidence which can be provided in both these areas by carefully implemented corrosion monitoring is a benefit which though not easily quantified is nevertheless considerable.

Development of Monitoring Methods

Until perhaps 25 years ago laboratory corrosion studies were carried out almost entirely by measuring change in weight over a given period, or some equivalent measurement. It was sometimes possible, but rarely convenient, to make the time period short and follow the change in rate with time, but this approach is experimentally tedious and may not allow straightforward interpretation. Consequently, electrochemical techniques were developed to measure instantaneous corrosion rates, and a wide range of properties can now be studied. Such techniques are much more convenient, and have the additional advantage of clarifying some of the situations in which weight loss tests give widely erratic results. As a result of this type of work, most corrosion processes are now relatively well understood, at least in general terms.

Corrosion measurements in plant traditionally followed the classical pattern of laboratory experiments. The only methods available were to introduce and remove specimens at shutdown, or to carry out detailed examination of items of plant while shut down. Two features make this approach more difficult than laboratory work. The first is that the test period (the interval between shutdowns) is set by production or maintenance requirements and may be far from suitable for a corrosion experiment. The second is that the conditions within a plant may vary considerably during this period. This is particularly true of factors which may be vital to the corrosion test but are not important from a process point of view, and hence are not recorded. In consequence, the result merely integrates the total corrosion over a period during which considerable changes in rate may have occurred. It is not surprising that results from this type of plant investigation are often difficult to interpret, or that behavior in a second apparently identical plant may be completely different.

The earliest variations in method were to make control of the test period independent of plant operation. This may be achieved by introducing a small test pot or perhaps a trial heat exchanger in a by-pass, so that the specimens are exposed to the process stream but can be removed at will by isolating the by-pass. A second and more representative method is to arrange to introduce or remove the specimens while the plant is on-line. In a typical case, specimens are attached to a rod which passes through a gland, and which can be pushed into or retracted from the process stream through a wide opening valve. This whole system can even be installed with the plant on-stream by using one of the commercially available "hot tapping" systems.

Another well established approach has been to make measurements on the plant while it is operating. For example, by using ultrasonics it is possible to take thickness readings on line at regular intervals, even at temperatures of the order of 500 C. An alternative and slightly older method involves electrical resistance measurements in which the change in thickness of the test element due to corrosion can be estimated from its change in resistance. Both methods allow more frequent determinations to be made than do the first two. However, the methods are only useful when corrosion is reasonably uniform, and it is difficult to get a sufficiently high sensitivity to follow changes in corrosion rates unless these are relatively slow.

A later but well established approach is based on the polarization resistance technique, which allows almost instantaneous measurement of corrosion rate in the form of an electrical signal which can be used in conjunction with conventional instrumentation to provide warning or even to operate control equipment. This provides a double advantage. It is possible to follow quite rapid changes in corrosion rate which in turn can be correlated with the

*Presented during Corrosion/78, March, 1978, Houston, Texas.

plant conditions, and corrosion information can be used for plant control in just the same way as temperature and pressure measurement have been used for many years.

Other electrochemical techniques used in the laboratory do not at first seem very suitable for plants, but in practice are now proving useful. For example, when potential measurements are made in a laboratory, it is normal to use an accurate, and usually delicate, reference electrode and measure the potential on a high resistance electrometer to better than 1 mV. The reference electrode will pass only very low currents, and the high impedance of the system makes it necessary to use screened cables to avoid spurious signals. Such careful precautions are necessary for fundamental electrochemical studies, but they are not necessary for most corrosion experiments, and the techniques and apparatus are used simply because they are already available. In fact, it is often sufficient to make quite crude measurements so that experiments can be done with a less precise reference electrode, such as a platinum button, and it is then fairly easy to design electrode assemblies which are similar to a thermocouple and equally robust.

A wide range of corrosion monitoring techniques is now available (Table 1) allowing determination of total corrosion, corrosion rate, corrosion state, analytical determination of corrosion product or active species, detection of defects or changes in physical parameters. Associated costs can be small where simple instrumentation and a few measurements are appropriate but in some cases may be extremely costly and require expert skills.

Much of the progress which has been made in the past few years has been due to advances in electronics which have allowed multiprobe measurement and recording at a tolerable cost. Instantaneous feedback of corrosion information can be obtained from various parts of the plant, which can be fed to the plant control room and/or plant computer to permit control of the necessary process variable to provide corrosion control.

The Case for Corrosion Monitoring

Estimates have been made of the potential benefits of more extensive use of corrosion monitoring techniques, including possible developments in this area, and of more effective use of the information. Large potential savings have been postulated, such as 25% of the costs associated with maintenance (including production losses) in major petrochemical, metallurgical or power industry complexes. Such savings would be several millions of dollars annually in an individual complex.

Estimates of this type are subject to error, but the evidence that substantial savings are possible, in principle, is solid. There is no doubt that the need to inspect major items at shutdown, and delays arising through rectification work not anticipated, cause a substantial part of the maintenance costs on a large plant. Moreover examples are known where applications of corrosion monitoring have led to savings of $200,000 to $2,000,000 or even more, for expenditures which are trivial in comparison. While these last two points refer to large plants, and hence to a situation which is not entirely typical, it is clear that major returns are possible from developments both in the large plant situation and in repetitive situations where the individual return is less spectacular.

Corrosion monitoring may be used for a variety of reasons, and it is often difficult to obtain the information needed to make a meaningful assessment of the benefits on a strict economic basis. One difficulty in the development of a general economic case is that confidentiality limits the extent to which information about many of the best examples can be disclosed. Occasionally this is because the facts could be embarrassing, but typically the reason is simply that a successful application contributes valuable "know how". There are cases where proprietary process depend in part on monitoring for their success, and others where the knowledge gained by monitoring provides an important competitive advantage. A more common problem, however, is that the necessary economic information is simply not available. Existing accounting systems are rarely suitable for isolating the appropriate costs, which can be expensive, to devise a special system for the purpose. In any event, corrosion monitoring is typically only one of a number of actions taken in a given situation, and even if the total benefit were known with certainty, the proportion of this benefit which should be balanced against the costs of corrosion monitoring would be at best a matter of opinion. It is therefore relatively unusual to be able to compute the costs and savings due to corrosion monitoring with any degree of certainty. In consequence, while those who know the background of a particular case well are able to appreciate the significance of a semiquantitative argument, it is most difficult to construct a strictly economic assessment which is completely convincing to others. A number of examples have been described in the literature, and in particular in two symposia,[2,3] and the *Handbook of Industrial Corrosion Monitoring.*[1]

It is helpful to analyze some of the reasons for adopting monitoring techniques, and to expand the earlier comments on objectives. The reasons are:

As a Diagonstic Tool

The most common use of on-line monitoring is to provide information for the solution of a corrosion problem. Knowledge of the pattern of behavior (*e.g.*, the uniformity or time-dependence of corrosion rate) can be very valuable. Frequently several variables could be the significant ones, and the ability to correlate corrosion with the variable that is significant under specific circumstances may be vital. This may provide the basis of a solution, but in any case, a protective system is best designed to counter the real conditions.

Corrosion monitoring is one of several tools available to the investigator, but is particularly valuable in that it may provide information which is not easily obtained in other ways, and thus lead to a better or speedier solution. Since the work will normally be under the direct control of a specialist, whose interpretative skill may be vital, relatively simple equipment is often adequate.

To Monitor the Effectiveness of a Solution

A logical extension of the diagnostic application is to use corrosion monitoring techniques to establish whether a solution has been effective. This can be done simply by continuing the original investigation, but more permanent installations are being used to an increasing extent to provide long term assurance. Such equipment is likely to be more sophisticated, since the information is recorded with other operational data and interpreted, in the first instance at least, by staff with a more limited corrosion knowledge.

To Provide Operational or Management Information

Corrosion can often be controlled by maintaining a single operational variable (*e.g.*, temperature, pH, humidity) within limits determined by prior monitoring or other investigations. If the significant variable is measured for other reasons, this measurement can be used directly for corrosion control. If the variable is not otherwise measured, or in more complex cases where several variables interact, corrosion monitoring information can be used by plant operators to control plant operation so as to control corrosion.

A similar application is in process optimization, or to examine the effect of variation of raw material. Any process change may have significant effects on corrosion, and corrosion monitoring techniques allow full scale trials to proceed with a minimum of risk to plant.

In all these cases, the information is used by nonspecialists, who may indeed be nontechnical staff. A high standard of equipment reliability is essential, and the display must be appropriate to the particular circumstances. The equipment is therefore likely to be more sophisticated, and costly, than in an installation controlled by a specialist, and a detailed understanding of the corrosion behavior of the system is essential before such systems are designed. Further, clear and unambiguous instructions must be provided for action by the plant operators.

As Part of a Control System

An extension of the use of monitoring techniques to provide operating information is to use the monitoring information directly to control plant, or parts of plant. The use of potential measurements to control anodic or cathodic protection systems is an example, as is the more recent development of using monitoring signals to control inhibitor or other additions to cooling water systems. Practical applications of this type have been confined to

TABLE 1 — Instrumentation for Corrosion Monitoring

Method	Measures or Detects	Notes	Use
Linear polarization (polarization resistance)	Corrosion rate is measured by the electrochemical polarization resistance method with two or three electrode probes.	Suitable for most engineering alloys providing process fluid is of suitable conductivity. Portable instruments at modest cost to more expensive automatic units available from Magna Instruments[1] and Petrolite.[2]	Frequent
Electrical resistance	Integrated metal loss is measured by the resistance change of a corroding metal element. Corrosion rates can be calculated.	Suitable for measurements in liquid or vapor phase on most engineering metals and alloys. Probes as well as portable and more expensive multichannel units available from Rohrback.[3]	Frequent
Potential monitoring	Potential change of monitored metal or alloy (preferably plant) with respect to a reference electrode.	Measures directly state of corrosion of plant, e.g., active, passive, pitting, stress corrosion cracking via use of a voltmeter and reference electrode.	Moderate
Corrosion coupon testing	Average corrosion rate over a known exposure period by weight loss or weight gain.	Most suitable when corrosion is a steady rate. Indicates corrosion type. Moderately cheap method with corrosion coupons and spools readily made.	Frequent
Analytical	Concentration of the corroded metal ions or concentration of inhibitor.	Can identify specific corroding equipment. Wide range of analytical tools available. Specific ion electrodes readily used.	Moderate
Analytical	pH of process stream	Commonly used in effluents. Standard equipment available through robust pH responsive electrodes such as antimony, platinum, tungsten can be preferable to glass electrodes. Solid Ag/AgCl is useful reference electrode.	Frequent
Analytical	Oxygen concentration in process stream.	Useful where oxygen control against corrosion using oxygen scavengers such as bisulfite or dithionite is necessary. Electrochemical measurement.	Moderate
Radiography	Flaws and cracks by penetration of radiation and detection on film.	Very useful for detecting flaws in welds. Requires specialized knowledge and careful handling.	Frequent
Ultrasonics	Thickness of metal and presence of cracks, pits, etc. by changes in response to ultrasonic waves.	Widely used for metal thickness and crack detection. Instrumentation is moderately expensive but simple jobs contracted out at fairly low cost.	Frequent
Eddy current testing	Uses a magnetic probe to scan surface.	Detects surface defects such as pits and cracks with basic instrumentation of only moderate cost.	Frequent
Infrared imaging (thermography)	Spot surface temperatures or surface temperature pattern as indicator of physical state of object.	Used most effectively on refractory and insulation furnace tube inspection. Requires specialized skills and instrumentation is costly.	Infrequent
Acoustic emission	(a) Leaks, collapse of cavitation, bubbles vibration level in equipment. (b) Cracks: by detection of the sound emitted during their propagation.	A new technique capable of detecting leaks, cavitation, corrosion fatigue pitting and stress corrosion cracking in vessels and lines.	Infrequent
Zero resistance ammeter	Galvanic current between dissimilar metal electrodes in suitable electrolyte.	Indicate polarity and direction of bimetallic corrosion. Useful as dewpoint detector of atmospheric corrosion or leak detection behind linings.	Infrequent
Hydrogen sensing	Hydrogen probe used to measure hydrogen gas liberated by corrosion.	Used in mild steel corrosion involving sulfide, cyanide and other poisons likely to cause hydrogen embrittlement.	Frequent in petrochemical industry
Sentinel holes	Indicates when corrosion allowance has been consumed.	Useful in preventing catastrophic failure due to erosion at pipe bends, etc. Leaking hole indicates corrosion allowance has been consumed.	Infrequent

[1] Magna Instruments (now Rohrback Corp.), Santa Fe Springs, California.
[2] Petrolite Corporation, Houston, Texas.
[3] Rohrback Corporation (formerly Magna Instruments), Santa Fe Springs, California.

systems whose behavior is relatively simple. Some progress has been made in handling more complex cases however, and the increasing use of computers for process control is likely to accelerate such developments. Clearly, equipment used in this way requires a high standard of reliability, and some form of "self-testing" feature, and a detailed understanding of the corrosion behavior of the plant is essential to the successful design of the control system.

As Part of a Management System

The data from corrosion monitoring can be useful to management in several ways. It can supplement other inspection techniques, and improve the management of maintenance and its coordination with production schedules. This approach is used to varying degrees of sophistication in a number of industries. The same information can also be used with benefits in optimization and other investigations, though this is less widely appreciated. Developments in computers and data handling have been an important stimulus in these areas, since a substantial mass of data has to be processed to a manageable form.

Selecting a Technique

Many techniques have been used for corrosion monitoring, and it is clearly possible to develop others. Consequently when a possible new application is being considered, a problem arises in choosing the most appropriate technique. Each has its strong points and its limitations, and none is the best for all situations.

Any monitoring technique can provide only a limited amount of information, and the techniques should be regarded as complementary rather than competitive. Where more than one technique will give the information required, the information is obtained in different ways; a cross-check can be valuable and differences in detail can add meaning.

A corrosion monitoring technique rarely gives wrong information—unless of course the equipment used is faulty. "Nonsense" results arise because the information is correct, but irrelevant in the corrosion sense. The polarization resistance method, for example, measures the combined rate of any electrochemical reactions at the surface of the test sample. If the main reactions are the corrosion ones, the rate measured is the corrosion rate. If however, other reactions are possible at rates that are comparable or greater, the measured rate includes the other reactions. Useful deductions can still be made provided it is recognized that the corrosion rate has not been measured.

It might appear that the choice of a monitoring technique is a complex problem requiring expert knowledge. This is true to a point, but in practice the choice for a given situation is usually relatively clear cut once the requirements are defined. The first essential is to establish what type of information is needed. This necessarily involves an input from the management of the plant in question. It may well be that this decision will, in effect, decide which monitoring technique should be used. The information below will give general guidance.

Where the Primary Objective is Diagnosis in a New Situation

Typically the nature of the corrosion processes involved and the controlling parameters are uncertain. It may be difficult to decide on the most appropriate technique, but it is in any case often advantageous to use more than one. The factors that actually prove to be significant are not always those which would have been expected.

One approach is to undertake a laboratory study to determine which parameters are likely to be important, the information being used both to decide which techniques should be used on the plant and to aid interpretation of the results obtained on the plant. Alternatively, monitoring can be undertaken directly. The choice between these approaches depends on the availability of suitable laboratory facilities and staff with the necessary experience, and on the extent to which the problem is understood. In either case, it is sensible to check the information obtained by monitoring; by inspection before and after, or other means. Expert help is often necessary in interpretation of the results and may be desirable in planning the work and selection of techniques. However, successful interpretation requires *knowledge of the plant and process in question* as well as expertise in monitoring techniques and knowledge of corrosion.

Where the Primary Objective is to Monitor the Behavior of a Known System.

Applications of this type often follow one of the diagnostic type; alternatively the problem resembles other cases where monitoring has been used successfully. In either case, the choice of technique is based on past experience. Expert assistance may well be unnecessary even in interpretation of the results, unless unusual features appear.

In addition to choosing the technique, it is necessary to decide the degree of complexity that is appropriate. The basic monitoring equipment for most techniques is relatively simple, comprising a probe (the sensing element) and a measuring instrument. The equipment cost is relatively modest, as is the labor cost if only a few readings are required. The amount of information that can be obtained by this approach is limited, but may be sufficient. If not, additional probes can be installed and or more complex instrumentation introduced to enable automatic scanning, automatic recording or regular readings from one or more probes and control panel displays. These more complex installations are more expensive—possibly considerably so—but much more information is obtained more conveniently and it is more likely to be practicable to use the information for control purposes.

Eight criteria on which the choice of a technique depends are summarized in Table 2 for the various corrosion monitoring methods and described as follows:

1. Time for Individual Measurement—Some techniques provide information that is effectively instantaneous, while others are necessarily slower in this respect.

2. Type of Information Obtained—Some techniques provide a measurement of corrosion rate, others measure total corrosion, or the remaining thickness, which is not exactly equivalent; yet others provide information on the distribution of corrosion on the corrosion regime.

3. Speed of Response to Change—Techniques which do not provide an individual measurement quickly are obviously unsuitable for situations where a fast response is required. Not all techniques that provide effectively instantaneous information are however capable of a fast response. Where the measurement is of rate, of corrosion regime, a fast response can be obtained, but if the measurement is of total corrosion, remaining thickness or distribution of corrosion, the speed of response is limited by the ability of the technique to discriminate between successive readings.

4. Relation to Plant Behavior—Many of the more effective techniques provide information on the behavior of a probe inserted into the plant, which does not necessarily reflect the behavior of the plant itself. The information obtained is in fact a measure of the corrosivity of the environment, from which plant behavior can be inferred. Other techniques provide an indication of the total corrosion in the system, with little or no indication of its distribution. Others give an accurate picture of a local corrosion pattern of the plant itself, but no information on what is happening elsewhere.

5. Applicability to Environments—A fast response is most readily obtained from electrochemical measurements which require that the environment is an electrolyte; a high electrolytic conductivity is not always necessary however. Nonelectrochemical measurements can be used in gaseous environments, or nonconducting fluids, as well as in electrolytes.

6. Type of Corrosion—Most corrosion monitoring techniques are best suited to situations where corrosion is general, but some provide at least some information on localized corrosion.

7. Difficulty of Interpretation—Interpretation of the results is often relatively straightforward if the technique is used within its limitations. The interpretation of the results obtained by some techniques is however, more difficult, and this is true of all techniques if they are used near the limits of their applicability.

8. Technological Culture—Some techniques are inherently tech-

TABLE 2 – Characteristics of Corrosion Monitoring Techniques

Technique	Time for Individual Measurement	Type of Information	Speed of Response to Change	Relation to Plant	Possible Environments	Type of Corrosion	Ease of Interpretation	Technological Culture Needed
Electrical resistance	Instantaneous	Integrated corrosion	Moderate	Probe	Any	General	Normally easy	Relatively simple
Polarization resistance	Instantaneous	Rate	Fast	Probe	Electrolyte	General	Normally easy	Relatively simple
Potential measurement	Instantaneous	Corrosion state and indirect indication of rate	Fast	Probe or plant in general	Electrolyte	General or localized	Normally relatively easy but needs knowledge of corrosion. May need expert	Relatively simple
Galvanic measurements (zero resistance ammeter)	Instantaneous	Corrosion state and indication of galvanic	Fast	Probe or occasionally plant in general	Electrolyte	General or unfavorable conditions localized	Normally relatively easy but needs knowledge of corrosion	Relatively simple
Analytical methods	Normally fairly fast	Corrosion state, total corrosion in system item corroding	Normally fairly fast	Plant in general	Any	General	Relatively easy but needs knowledge of plant	Moderate to demanding
Acoustic emission	Instantaneous	Crack propagation and leak detection	Fast	Plant in general	Any cavitation	Cracking, cavitation and leak detection, pitting	Normally	Crack propagation specialized, otherwise relatively simple
Thermography	Relatively fast	Distribution of attack	Poor	Localized on plant	Any. Must be warm or sub-ambient	Localized	Easy	Specialized and difficult
Optical aids (closed circuit TV, light tubes, etc.)	Fast when access available, otherwise slow	Distribution of attack	Poor	Localized on plant	Any	Localized	Easy	Relatively simple
Visual, with aid of gauges	Slow. Requires entry on shutdown	Distribution of attack indication of rate	Poor	Accessible surfaces	Any	General or localized	Easy	Relatively simple but experience needed
Corrosion coupons	Long duration of exposure	Average corrosion rate and form	Poor	Probe	Any	General or localized	Easy	Simple
Ultrasonics	Fairly fast	Remaining thickness or presence of cracks and pits.	Fairly poor	Localized on plant	Any	General or localized	Easy	Simple
Hydrogen probe	Fast or instantaneous	Total corrosion	Fairly poor	Localized on plant or probe	Nonoxidizing electrolyte or hot gases	General	Easy	Simple
Sentinel holes	Slow	Go/no go remaining thickness	Poor	Localized on plant	Any, gas or vapor preferred	General	Easy	Relatively simple
Radiography	Relatively slow	Distribution of corrosion	Poor	Localized on plant	Any	Pitting, possibly, cracking	Easy	Simple but specialized radiation hazard.

nically sophisticated; this tends to limit their use to organizations with a strong technological culture. Most others are much less demanding in this respect.

In principle, the available techniques could be ranked in an order or merit for each of these eight criteria. In practice, the relative merits change with circumstances so that a formal treatment of this type is potentially misleading. The most useful general approach is therefore, to consider the strengths and weaknesses of the techniques individually, and Table 2 provides a reasonable starting point.

Possible Future Developments

The most obvious line of development is a continuation of the past trend, whereby additional techniques are devised for particular purposes. A second trend which is likely to continue is refinement of the measuring element itself to achieve greater reliability or to simulate plant conditions more precisely. Many corrosion monitoring techniques, for example, utilize probes which are geometrically very different to the plant surfaces and may behave differently for this reason, and for some applications, it would be desirable for example to use a heated probe to simulate heat transfer surfaces. Another may be multipurpose probes exploiting both polarization resistance and electrical resistance.[3]

Another obvious line of development is improvements to measuring instruments arising from developments in electronic engineering. More sophisticated and more reliable instruments can now be devised than were possible until quite recently, often at lower cost. It also seems likely that some techniques will become available for plant use that have previously been confined to the laboratory because the available instruments were too delicate or could only be used by operators skilled in their use.

The developments in electronic engineering may be even more significant by removing some of the practical limitations common to many existing techniques. There are two related possibilities, the inclusion of minicomputer facilities and modified forms of display.

Existing techniques often give only limited information and are practicable only under a restricted range of circumstances. Many also have the weakness that "nonsense" results are difficult to distinguish from valid ones and can cause the wrong conclusions to be reached; in other words, a measurement is obtained that is valid in itself but does not correlate with the quantity of interest. These difficulties could be overcome by using the technology of minicomputers, which should permit relatively inexpensive equipment incorporating logic and calculation facilities. The instrument might make several measurements, by the same or quite different techniques, and perform calculations based on algorithms that could be complex, displaying a derived quantity. Some, if not all, nonvalid measurements could be rejected and the fact indicated. The UMIST polarization impedance technique is an example in which measurements at a range of frequencies are analyzed to separate the component representing corrosion rate from others that may interfere with conventional polarization resistance measurements.[3] The analysis is possible only if certain conditions are fulfilled so that a test for validity can be incorporated. Developments of this type can be envisaged with many other techniqes.

Obviously, corrosion monitoring is not a panacea, but since its adoption from the late 1950's, certain industries have recognized it as likely to give significant returns. The state of the art is such that a wide variety of corrosion monitoring tools can be used in the almoury, and although the limitations of existing techniques need to be recognized, they are within the scope of future developments which should make the applications more practical for a wider range of users.

References

1. Handbook of Industrial Corrosion Monitoring, HMSO, London (1978).
2. Corrosion Testing and Monitoring, Inst. Corr. Sc. Tech Conference, Westfield College, London, 30 March-1 April 77, Chameleon Press Ltd. (1977).
3. On Line Surveillance and Monitoring of Process Plant. Soc. Chem. Ind. Symposium City University, London (1977) September.

Effect of Tin Content on the Corrosion Resistance of Terne-Coated Steel*

D. M. SMITH
Armco Steel Corporation, Middletown, Ohio

The electrochemical behavior of various Terne alloys having tin levels of 0-17% was studied. Anodic and cathodic polarization curves were developed for each alloy in a chloride solution. The polarization behavior of these alloys as it relates to the corrosion behavior of Terne coated steel is discussed. Linear polarization curves were used to determine the corrosion rates of the Terne alloys. A marked increase in corrosion rate was observed when the tin level dropped below 3-4%. The behavior of Terne alloy-steel couples was also investigated. The behavior of the various couples as predicted by the polarization data was confirmed by galvanic current and couple potential measurements.

TERNE PLATE consists of steel sheet coated with a lead-tin alloy, combining the strength of steel with the corrosion resistance of the alloy coating. The addition of tin enhances the corrosion resistance of lead in many environments and allows the coating to adhere to the steel.[1] Depending upon the end use, the tin content can vary from a few percent up to about 20%.

The major uses of Terne plate lie in the automotive industry. By far the largest use is in fuel tanks for cars, trucks, and tractors. Terne plate has several advantages for this particular application. First, the lead-tin coating is resistant to corrosion, especially by gasoline from inside and by salt from outside, the two main enemies of fuel tanks.[2] Corrosion products inside the tank are minimized and are fine enough not to block the fuel line. Terne plate, though relatively soft, is tough and can withstand impact from stones, grit or collisions without shattering and releasing its inflammable contents. Finally, the material is relatively easy to form and can be fabricated either by soldering or by welding.

The level of tin in Terne plate for fuel tanks was 18% prior to World War II. However, during the war a strong effort was made to reduce the tin content, primarily by substituting other alloying elements. Post World War II production facilities utilizing exit rolls encountered de-wetting problems if the tin level was not above 9 to 10%. However, the tin content remained at a relatively high 14% until air knives or nozzles on production equipment eliminated the exit roll problem, allowing the tin composition to drop safely to 12%. The escalating price of tin ($1.60 to $4.60 per pound in 1971) made it necessary to drop the level to 8%, particularly since the technology was available to continuously employ lower tin coatings. However, this change did not come until 1973, primarily because automotive specifications required tin levels greater than 8% in Terne plate for fuel tanks.

*Presented during Corrosion/78 (Paper 14), March, 1978, Houston, Texas.

With the price of tin still above $5 per pound, and lead at approximately $.30 per pound, there was a strong incentive to study the behavior of lower tin alloys as it relates to the performance of fuel tanks. One must consider two specific aspects of fuel tank performance: (1) pinholing, and (2) corrosion of the coating in saline environments.[3] This investigation describes the work relating to the corrosion in saline environments and does not address the pinholing.

Procedure

Materials

The Terne alloys used in this study had tin levels of 0, 1, 2, 3, 4.75, 6.75, 8, 12, and 17%. The alloys were made by melting commercially pure lead and tin, and casting into 1.27 cm diameter rods. The cast samples were machined to 0.635 cm diameter rods, 1.27 cm in length for polarization studies.

The electrolyte for this study was a 5% NaCl solution made from reagent grade sodium chloride and deionized water. This solution is representative of water contaminated with highway deicing salts. The chloride solution was deaerated or aerated, depending upon the electrochemical test employed. Deaeration was achieved by continuously purging nitrogen through the solution. All tests were conducted at 25 C.

Electrochemical Tests

Potentiodynamic anodic polarization curves were developed for the lead-tin alloys using standard techniques.[4] All samples were tested in the deaerated sodium chloride solution. Samples were prepolarized for 1 hour at the corrosion potential. After prepolarization, an anodic scan was made at a rate of 60 mV/minute.

Potentiodynamic cathodic polarization curves were developed in a similar manner, except, an aerated chloride solution was used. Corrosion potentials were measured after 1/2 hour and 1 hour prepolarization and recorded. A cathodic scan was then made at a rate of 60 mV/minute.

Potentiodynamic linear polarization curves were determined to measure the corrosion rates of the various lead-tin alloys in the deaerated chloride solution. After 1 hour prepolarization, the potential was displaced 30 mV anodic to the corrosion potential and a cathodic scan was made at 12 mV/minute to a potential 30 mV cathodic to the corrosion potential.

A specimen of each of the lead-tin alloys was coupled to a carbon steel specimen of equal area to study the behavior of such a couple in the aerated chloride solution. The samples were placed 5 cm apart in an electrochemical cell. The galvanic current of each couple was measured using a potentiostat as a zero resistance ammeter. The potential of the couple was measured with an electrometer. The outputs from both of these instruments were continuously monitored with a two-pen strip chart recorder.

Results

The corrosion potentials (ϕ_{corr}) of the various Terne alloys in the aerated chloride solution are summarized in Table 1. It is clear that the corrosion potential is independent of the tin level in this solution. Also, a stable potential is attained in a relatively short period.

The anodic polarization curves are depicted in Figures 1-3. Active-passive behavior is indicated by the curves, but considerable corrosion occurs before the critical current density for passivity (i_c) is attained. There is no significant difference in the critical current density (i_c) as a function of tin content. Passive current densities seem to decrease with decreasing tin levels, but the magnitude of these passive currents is relatively high suggesting corrosion rates that are not realistic of Terne alloys in chloride solutions. Therefore, corrosion rates at ϕ_{corr} would be more representative of the behavior of Terne alloys in chloride solutions.[5]

The corrosion rates determined from the linear polarization curves are presented in Table 2. Figure 4 depicts the linear polarization curve for the 2% tin alloy. A technique developed by Mansfeld and Oldham[6] permits the calculation of corrosion rates utilizing the shape of the curve without separate experimental determinations of Tafel slopes. A plot of corrosion rate versus tin level is shown in Figure 5. Dissolution rates increased steadily as the tin level dropped below 3 to 4%.

The cathodic polarization curves for the lead-tin alloys in aerated chloride solution are shown in Figures 6 to 8. The shape of

TABLE 1 – Corrosion Potentials of Terne Alloys (5% NaCl, aerated, 25 C)

Tin Level (%)	ϕ, V SCE (1/2 Hour)	ϕ, V SCE (1 Hour)
0	-0.520	-0.518
1	-0.517	-0.518
2	-0.517	-0.517
3	-0.518	-0.518
4.75	-0.518	-0.518
6.75	-0.518	-0.518
8	-0.518	-0.517
12	-0.517	-0.516
17	-0.518	-0.517

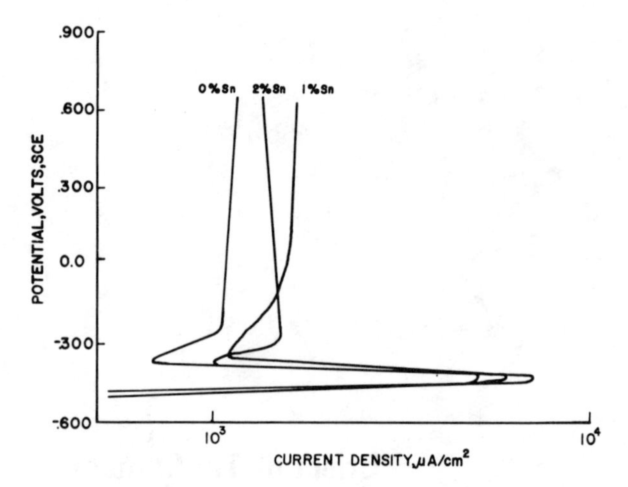

FIGURE 1 – Anodic polarization curves of lead-tin alloys in 5% NaCl, deaerated, 25 C.

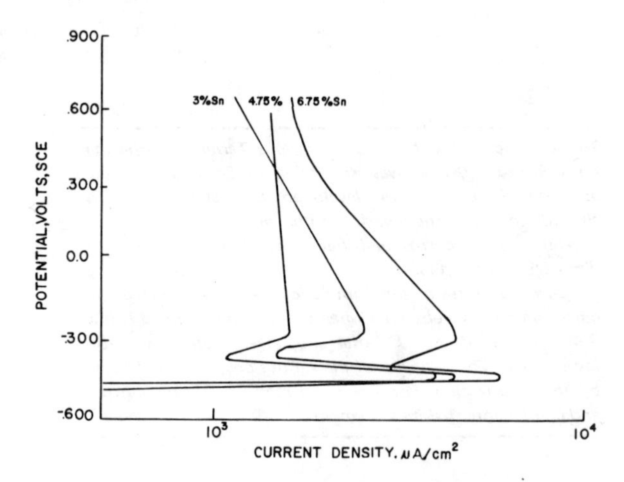

FIGURE 2 – Anodic polarization curves of lead-tin alloys in 5% NaCl, deaerated, 25 C.

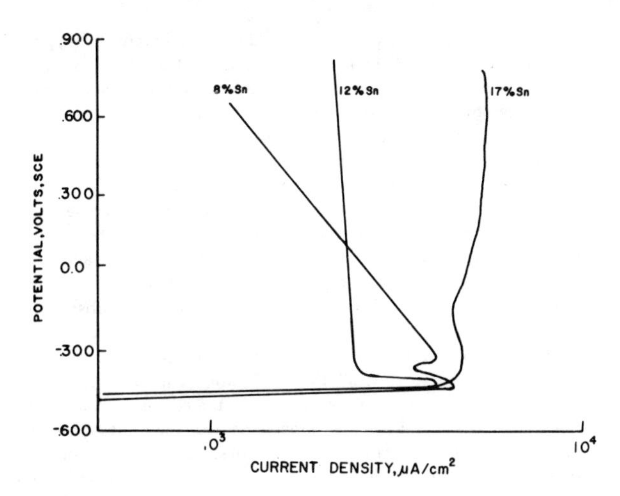

FIGURE 3 — Anodic polarization curves of lead-tin alloys in 5% NaCl, deaerated, 25 C.

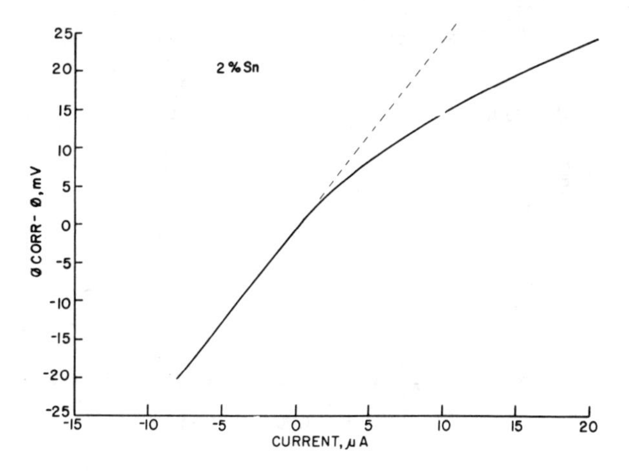

FIGURE 4 — Linear polarization curve of 2% Terne alloy in 5% NaCl, deaerated, 25 C.

TABLE 2 — Corrosion Rates of Terne Alloys (5% NaCl, deaerated, 25 C)

Tin Level (%)	Corrosion Rate μm/y	mpy
0	154.9	6.1
1	73.7	2.9
2	53.3	2.1
3	45.7	1.8
4.75	19.6	0.77
6.75	5.6	0.22
8	7.1	0.28
12	14.0	0.55
17	6.4	0.25

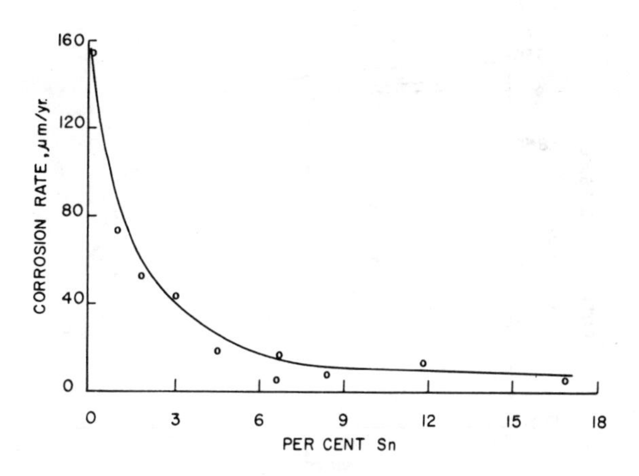

FIGURE 5 — Corrosion rates of various Terne alloys in 5% NaCl, deaerated, 25 C.

these curves suggests the cathodic reaction is diffusion controlled oxygen reduction which has a limiting diffusion current density (i_L) of 100 $\mu A/cm^2$.[7] This current density can be determined from:

$$i_L = \frac{DnFC_B}{X} \quad (1)$$

where i_L is the limiting diffusion current density, D is the diffusion coefficient of the reacting species, C_B is the concentration of the reacting species in the bulk solution, and X is the thickness of the diffusion layer. The data clearly shows an i_L of 100 $\mu A/cm^2$ and is independent of the tin content in the alloy.

The potentials and current densities of the Terne alloy-steel couples are reported in Table 3. In all instances, the Terne alloy was active to the steel. The potential of the Terne alloys was shifted approximately 100 to 150 mV in the active direction when coupled to steel. However, the couple potential and current are independent of the tin composition.

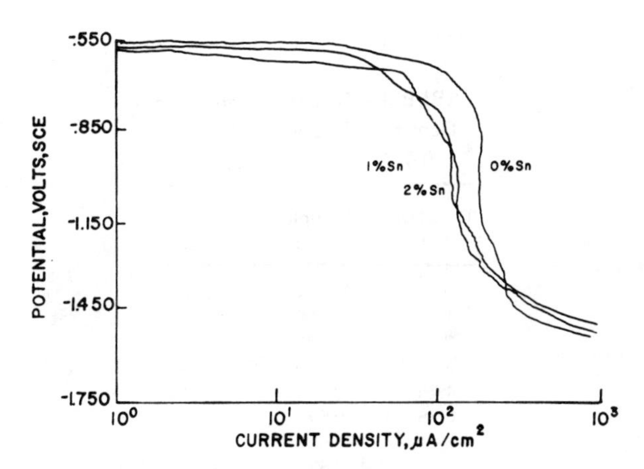

FIGURE 6 — Cathodic polarization curves of lead-tin alloys in 5% NaCl, aerated, 25 C.

Discussion

The polarization behavior of the Pb-Sn alloys has been presented, but the relationship between the polarization phenomena and the corrosion of Terne coated steel is still needed. Terne plate can be considered to be a galvanic couple between a Terne alloy and steel with the area ratio of Terne coating to base steel being quite large because of inherent pinholes in the product.

A galvanic couple is strictly more than electrical contact between two dissimilar metals in which one is the anode and the other is the cathode. A galvanic couple consists of two metals, where both oxidation and reduction reactions are occurring on both metals. The cathode of the couple has a net reduction current (cathodic), while the anode has a net oxidation current (anodic).

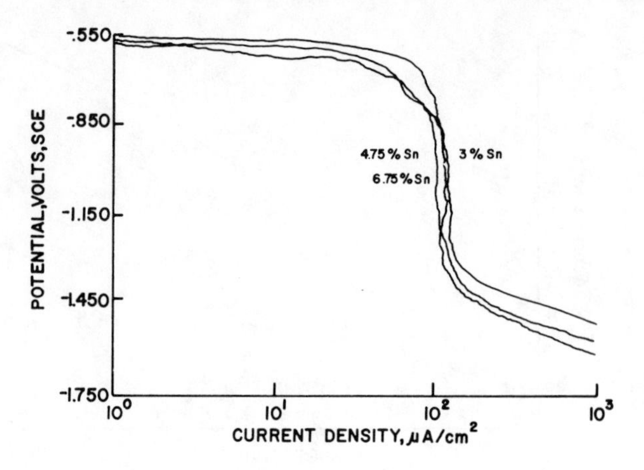

FIGURE 7 – Cathodic polarization curves of lead-tin alloys in 5% NaCl, aerated, 25 C.

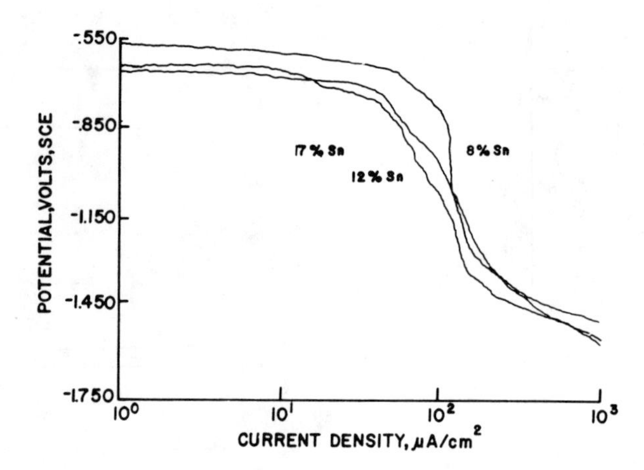

FIGURE 8 – Cathodic polarization curves of lead-tin alloys in 5% NaCl, aerated, 25 C.

TABLE 3 – Potentials and Current Densities of Galvanic Couples (5% NaCl, aerated, 25 C)

Tin Level (%)	ϕ couple, V SCE	I (μA/cm^2)
0	-0.659	61.8
1	-0.700	56.1
2	-0.670	52.6
3	-0.650	59.6
4.75	-0.660	54.5
6.75	-0.670	52.3
8	-0.650	49.4
12	-0.670	56.1
17	-0.620	49.1

The corrosion behavior of a galvanic couple is determined by the reversible potential of the actual reactions involved, their exchange current densities, their polarization behavior, and the relative areas of the two metals. Galvanic corrosion behavior cannot be predicted accurately on the basis of emf potentials.[7]

The galvanic couple data revealed the Terne alloys were active to the steel, meaning there was a net cathodic current on this metal. The polarization behavior, both anodic and cathodic curves, was virtually independent of tin level in the Terne alloys, indicating the behavior of the galvanic couple between Terne metal and steel would be identical regardless of the tin content of the alloy. The potentials and current densities of the various couples substantiate this conclusion. In the case of Terne coated steel, the large relative area of the coating compared to the steel must be considered. This area effect tends to shift all the polarization curves of the Terne alloys to the right causing an increase in the corrosion rate of the steel. Once again, this effect is independent of the tin level because of the identical polarization characteristics. Furthermore, once the steel initially corrodes and corrosion products form, the corrosion current is then controlled by the IR drop through the saline environment and corrosion products in the pore.

There will be areas on a Terne plate fuel tank that will not take part in the galvanic couple because of the lack of throwing power of the cell and the remoteness from a pinhole. In these areas, the corrosion rate of the Terne alloy becomes important. The linear polarization data shows the effect of tin level on the corrosion rate of the alloys. Tin levels below 3 to 4% exhibit marked increase in dissolution rate.

Thus the behavior of Terne coated steel can be explained utilizing polarization phenomena of various Terne alloys. Furthermore, the role of tin level in the coating as it relates to the corrosion resistance of Terne coated steel has been established. The effect of such a variable as it relates to pinholing, however, must be explored in order that the service ability of Terne coated steel is completely known.

Summary

1. The polarization characteristics and galvanic couple data clearly demonstrated that the corrosion of base steel in Terne plate in sodium chloride solution is independent of the tin level of the coating.

2. Linear polarization data indicated there was a marked increase in dissolution rates of the Terne alloys when the tin level dropped below 3 to 4%.

3. The anodic polarization curves of the various Terne alloys were virtually identical except for the passive current densities.

4. The cathodic polarization curves of the various Terne alloys were identical.

Acknowledgments

The author expresses his gratitude to D. J. Robbins for conducting the experimental work and would like to acknowledge F. C. Dunbar and J. Kolts for their contributions through discussions.

References

1. MacKay, C. A., and Evans, C. J. Tin and Its Uses, Vol. 105, p. 104 (1975).
2. Long, J. B. Tin and Its Uses, Vol. 79, p. 5 (1968).
3. Smith, R. Sheet Metal Industries, p. 761 (1972) December.
4. Greene, N. D. Experimental Electrode Kinetics, RPI (1965).
5. von Frauhofer, J. A. Anti-Corrosion Methods Materials, Vol. 16, No. 5, p. 21 (1969).
6. Oldham, K. B., and Mansfeld, F. Corrosion Science, Vol. 13, p. 813 (1973).
7. Fontana, M. G. and Greene, N. D. Corrosion Engineering, McGraw-Hill, Inc., p. 330 (1967).

"Exotic" Metals

Frederick H. Vorhis
The Pfaudler Co.
Div. of Ritter Pfaudler Corporation
Rochester, N. Y.

their use as construction materials

Use of the so-called "exotic" or reactive metals in chemical processing equipment is discussed. Gives physical properties and corrosion resistant characteristics of titanium, tantalum, and zirconium. Also discusses corrosive environments that attack these reactive metals.

THE CORROSION engineer today has at his disposal an ever growing list of construction materials to solve his corrosion problems. As a result, the technology of corrosion resistance has become a never ending study to evaluate the economics and effectiveness of these materials for use in many hostile environments. The manufacturer of chemical process equipment must maintain a progressive program to develop economical designs and methods of manufacture that will make it possible for the corrosion engineer to utilize the most suitable material.

Exotic Metals

In recent years, several of the so-called "exotic" metals have become very important materials of construction for handling difficult corrosion environments. In this category fall the reactive metals—tantalum, columbium, titanium, and zirconium. The purpose of this article is to discuss some of the properties and applications of the reactive metals.

The term "reactive" metals includes all of the less common metals of construction which react with moisture, carbon, oxygen, and nitrogen upon heating to temperatures normally encountered in fabrication. In its generally accepted sense, it is taken to include titanium, zirconium, tantalum, molybdenum, columbium, vanadium, tungsten, and a number of others. Of these, tantalum, titanium, and zirconium merit attention because of their usefulness to the chemical industry.

Titanium

Numerous titanium alloys are available containing combinations of aluminum, manganese, vanadium, chromium, tin, iron, molybdenum, or columbium, all having varying properties. Because of the desirable characteristics offered, the Navy is currently evaluating several of the titanium alloys for submarine and other deep sea applications. The commercially pure grades of titanium, however, possess greater corrosion resistant properties and are the most desired by chemical industries.

Zirconium

The major application for zirconium has been in atomic power applications because of its low thermal neutron absorption cross section and corrosion resistance. It is used as cladding material for fuel elements as well as structural material for components of nuclear reactors.

Zirconium and its sister metal, hafnium, have similar chemical and mechanical properties. Hafnium, however, has a high thermal neutron absorption cross section and therefore is not used for nuclear applications. Hafnium-free zirconium is referred to as "reactor grade zirconium." For greater strength, reactor grade zirconium is alloyed with 1.5% tin, 0.12% iron, 10% chromium, and 0.05% nickel. This alloy is referred to as Zircalloy II.

The three basic grades of zirconium, commercial grade with 2.5% hafnium, reactor grade (hafnium-free), and Zircalloy II all have excellent corrosion resistance to most acids. To date, however, the commercial grade has had the greatest usage for chemical applications.

Tantalum

Although tantalum can be and is often referred to as a "reactive" metal, it more properly should be classified in the family of refractory metals along with molybdenum, tungsten, and columbium because of its high melting temperature (5400 F or 2871 C). The first ductile tantalum was produced in the early 1900's. Its major application has been and still is in the electronics industry.

In recent years, because of their ability to withstand high temperatures, tantalum and tantalum alloys have played an important role as construction materials for jet and aerospace applications.

Tantalum's universal corrosion resistance has long been recognized; however, its use as a construction material for chemical process equipment has been limited because of its high cost. Since 1960, as the result of improved methods of production and competition by metal producers, the price of average tantalum mill products has been reduced 40 to 50%. The combination of lower cost, improved quality of mill product, and development of improved fabricating and welding techniques has increased its importance as a material of construction for chemical process equipment in severely corrosive applications.

Fabrication of Reactive Metals

The reactive metals can be readily fabricated or formed into the many complicated shapes required for chemical process equipment.

Since titanium and zirconium have yield strengths which are 80 to 90% of their ultimate tensile strength, they possess a high degree of spring-back characteristics. Fifty to sixty percent spring-back is not at all uncommon in many forming operations. Fortunately, because their yield strength drops rapidly as temperature increases, heating to 600 to 800 F (316 to 427 C) will permit forming by most conventional methods—rolling, press braking, spinning, bending, deep drawing, swaging, etc.

In contrast, tantalum is a very ductile material which can be formed into complicated shapes at room temperature. Because it readily reacts with oxygen and hydrogen, heating can not be tolerated. Stress relieving usually is not required; however, if it is, it must be done in high vacuum or strict atmosphere control.

The high cost of tantalum dictates the use of thin cross sections, 0.013-inch to 0.030-inch, usually in the form of linings. Handling this soft, thin material presents the greatest problem in fabrication. It buckles, dents and scratches very easily and, therefore, must be handled with extreme care;

Reprinted with permission from *Materials Protection*, August 1966, 21-24, © 1966 National Association of Corrosion Engineers

constant cleanliness is a must. A dent or a scratch in 0.013-inch material can be a major defect.

Corrosion Resistant Properties

The corrosion resistance of reactive metals is the result of a very passive oxide surface film. As long as conditions are such that this film is maintained, the metals remain passive. Tantalum has the most passive film, tantalum pentoxide, which, in some cases, is as inert as gold or platinum in aggressive corrosive environments. The oxide films on titanium and zirconium are somewhat less passive; however. Each has unique corrosion resistant properties in certain environments. The following can be used as a general guide to applications for which the reactive metals may be considered.

Tantalum

Tantalum is inert to nitric acid including red fuming, hydrochloric acid at all concentrations (although at 375 F or 190 C it becomes embrittled in 30% HCl), aqua regia, perchloric acid, chromic acid, nitrogen oxides, chlorine oxides, hypochlorous and wet chlorine, hydrobromic acid and wet bromine, most organic acids, hydrogen peroxide, and inorganic and organic chlorine compounds.

The halogens, chlorine, bromine, and iodine in liquid do not react with the metal up to 300 F (149 C), and as gases up to 480 F (249 C). Tantalum resists attack by all salts except those which hydrolyze to strong alkalies. Almost all gases with the exception of fluorine and nascent hydrogen have no effect on tantalum under 300 F.

Among the few environments which attack tantalum at room temperature are fluorine, hydrofluoric acid, sulfur trioxide, and strong alkalies above 5% concentration. Tantalum begins to be attacked slowly and uniformly by 98% sulfuric acid at about 338 F (170 C) and by 85% phosphoric acid at about 374 F (189 C), the rate of attack increasing as the temperature rises. Likewise, as the concentration of these acids is decreased, the corrosion rate at a given temperature becomes less. Fuming sulfuric acid containing sulfur trioxide attacks tantalum at room temperature.

Titanium

The only acid to which titanium is generally resistant is nitric, except in the red fuming condition. It is readily attacked by hydrochloric and sulfuric acid at moderate concentrations and temperatures. Phosphoric and formic acids attack the metal at elevated temperatures and concentrations.

The resistance of titanium to acidic environments can be improved considerably by the presence of oxidizing agents or certain metal ions. Its excellent resistance to aqua regia is a good example of attack inhibited by an oxidizing agent. Cupric, nickel, ferric, and many other metal ions in low concentration have been found to inhibit attack by acids. Titanium alloyed with 0.2% palladium greatly improves its performances in most strong acid enviroments.

One of the outstanding features of titanium is its resistance to environments containing inorganic chlorides. With the exception of hot aluminum chloride, it is virtually unaffected by salt solutions, sea water, brines and marine atmospheres, and it has good resistance to many of the metallic chlorides, such as iron and copper. In such environments, it is not subject to pitting attack or intergranular attack of the heat affected weld zones as is the case with stainless steel.

The most successful applications for titanium result from its corrosion resistance to wet chlorine gas, bleach solutions containing sodium and calcium hypochlorite, and chlorine dioxide. Chlorine producers and industries such as paper, plastics, and detergent, which use wet chlorine, furnish a rapidly growing market for titanium equipment.

Zirconium

Zirconium is corrosion resistant to a wide variety of corrosive media, both acid and alkaline. Its resistance to hydrochloric acid is outstanding—less than 1 mil per year in all concentrations and temperatures through the boiling point.

In sulfuric, zirconium is resistant to all concentrations up to 75% at room temperature. As temperatures are increased, concentration should be reduced.

It has good to excellent corrosion resistance to nitric, chromic and phosphoric acids, depending upon concentration and temperature. It shows excellent resistance to most of the organic acids. Its behavior in organic chloride environments is excellent with the exception of ferric and cupric chlorides. Ferric and cupric chloride solutions of about 1% concentration corrode zirconium.

Selecting Reactive Metals

Because of the high initial cost of reactive metal chemical processing equipment, the decision to use a given metal should be made only after a carefully conducted corrosion test program which should include:

1. A review of published corrosion data for the selection of the metal or metals to be tested.
2. A laboratory corrosion test on welded samples. This should be considered as a screening test to determine if the metal or metals perform as indicated in the published data.
3. An in-process corrosion test under actual operating conditions. This is usually the most difficult to obtain; however, its importance can not be overemphasized. There are many case histories of disappointments because this test was not run. The many process variables that can not be duplicated in the laboratory can greatly influence corrosion test results. The test piece can be a welded test tab, a tube in an existing heat exchanger, a component of a vessel such as a thermo well, baffle, agitator, or, in the case of a major installation, a small pilot vessel.

If, for a given corrosive environment, one of the reactive metals shows promise of being suitable, it then requires an economic evaluation to determine how, where, and if the metal can be used. There are two basic questions that should be considered:

1. Does the environment have varying corrosive conditions in vapor phase, in liquid phase, or at varying temperatures in the operation? Can a combination of materials be used that will reduce initial costs? To date the use of combinations has been one of the most successful ways of using this group of metals. For example:

(a) Reactive metal accessories such as agitators, baffles, coils, etc., in a plastic lined vessel.

(b) Glass, plastic, or rubber lined bonnets on a reactive metal heat exchanger.

(c) Reactive metal sections of a column used in areas of severe corrosion whereas less expensive materials of construction are used in areas of less severe service.

2. If the environment is such that only the reactive metals can be used, what type of construction should be used—loose liner, solid, or integrally clad? In order to fully evaluate this question, the engineer must consider the following:

(a) What are the actual operating conditions of pressure and temperature? Because of past practices using less expensive construction materials, there is a tendency to establish design specifications which are two to three times greater than actual operating conditions. For such materials as carbon steel or stainless steel, such inflated ratings may not represent a great expenditure of money; however, with reactive metals, the cost of materials may be considerable. The design engineer should work as closely to actual working conditions as is consistent with good safety practices.

(b) Is vacuum involved? If so, can the process be altered to eliminate the need for low pressures?

(c) Is heat transfer important, and if so, can internal coils be used?

Selecting the type of construction should not be on the basis of initial cost alone. Rather, it should be based on the method that will give maximum service life for minimum cost.

There is no magic formula that can be used, since many factors must be considered. The following comments on construction methods cover a few of these important factors.

Loose-Lined Construction

The loose-liner construction has been the most economical approach for reactive metal chemical process equipment; however, as the price of mill products drop, as in the case of titanium, the picture changes for many applications. This method of construction combines the strength and low cost of carbon steel with the excellent corrosion resistance of the reactive metals. Liner thicknesses vary from 0.015 to 0.030-inch for tantalum to 0.040 to 0.125-inch for titanium and zirconium.

Generally, carbon steel loose-lined construction should be limited to pressures under 100 psi and to temperatures not exceeding 350 F (177 C).

There are certain disadvantages with loose-lined construction that should be recognized:

1. A loose-liner is not suitable for applications involving vacuum operating conditions.

2. Loose-lined construction is not suitable for applications involving heat transfer. The small air space between the liner and carbon steel external shell acts as an insulator reducing heat transfer efficiency.

3. Failures in the liner material or welds are difficult and costly to repair. Should the liner material become contaminated by corrosion products, attempts to repair it without first cleaning it properly result in further damage. Usually, it is necessary to return the equipment to the manufacturer for repair.

Solid Construction

Because of a significant drop in the price of mill products in the past five years, there are many services involving pressure and temperature where vessels of solid construction become more economical. This is particularly true today for titanium and zirconium. The following are suggested applications for solid construction:

1. Vessels under 100 gallons capacity.
2. Vessels with operating pressures under 50 psi and temperatures under 300 F.
3. Columns of all sizes operating from vacuum to 50 psi internal pressure at temperatures under 300 F.
4. Accessories such as agitators, baffles, thermometer wells, dip pipes, coils, etc.
5. Heat exchanger bonnets.

The initial cost of the solid construction in many cases will be somewhat higher; however, looking at the overall picture, it might well be the most economical approach. For example:

1. The limitations of loose-lined construction are eliminated, thus less maintenance and greater versatility.

2. In the event of failure, repairs can generally be made on the job site by personnel trained in reactive metal fabrication. Likewise, there is not the problem of attempting to clean inaccessible surfaces as in the case of loose liners.

3. Rearrangement of or additional nozzles can be accomplished on the job site.

4. There is greater flexibility in design, particularly so with columns when internal supports are required.

To summarize, a few extra dollars spent at the outset for solid construction may be returned several-fold in trouble-free operation.

Integrally Clad

Reactive metals can be integrally clad with less expensive backing materials such as carbon steel and stainless steel by three basic methods:

1. Brazing of the two materials together in a vacuum and under pressures using a material such as silver or an alloy which is compatible with both. This method produces a good product for some applications; however, the bond is limited by the strength and the melting point of the bonding materials.

2. Sandwich rolling of the two materials at elevated temperatures in controlled atmospheric conditions. This method is essentially the same as that used for producing stainless clad plate.

3. By metallurgical bonding of the two materials using a carefully controlled explosion process. No heat or intermediate material is used and, therefore, the characteristics and properties of the two bonding materials remain unchanged. The resultant bond has excellent integrity with respect to shear strength and temperature. This process has unlimited possibilities as to the materials that can be joined together. Explosively-bonded titanium and zirconium-carbon steel plate are commercially available and are used today for the manufacture of chemical process equipment and offer an exciting future for chemical applications.

Methods of fabricating and welding have been developed for titanium and zirconium clad steel plate which are similar to those used for the respective reactive metal involved. Chemical process equipment fabricated from these materials has a minimum cladding thickness of 0.083-inch and is a commercial reality. The following are applications where clad materials should be considered:

Jacketed vessels with the advantage that jacket sealers or attachments can be welded to the carbon steel backing material provided the reactive metal cladding is protected during the welding operation.

Vessels and columns in applications where high pressure and temperature or vacuum are involved. These become a matter of economics and the cost should be compared to that of solid construction.

Heat exchanger tube sheets where high pressures and temperatures are involved.

Because of the extremely high melting temperature of tantalum as compared to that of the carbon steel backing material and because of tantalum's negative tolerance for iron contamination, methods of welding tantalum-clad steel must be developed. Experimental work is being done and investigators are confident that the problem will be resolved.

In conclusion, reactive metals have become very important materials of construction to the corrosion engineer. Process equipment fabricated from these materials is expensive at the outset; however, if the following steps are taken, many dollars can be saved in reduced maintenance and improved efficiency:

1. Careful selection of the right material by good test techniques under actual operating conditions.

2. A proven design for the intended operating conditions which is also compatible with good fabricating practice.

3. Careful selection of the manufacturer based on his fabricating experience with the reactive metals.

4. A review of the actual process conditions to insure they are the same as those under which the metal was tested.

FREDERICK H. VORHIS is product manager, alloy equipment, for the Pfaudler Co., division of Ritter Pfaudler Corp., Rochester, N. Y. He is responsible for reactive metal and alloy vessel sales to the chemical industry. A graduate from Cornell University with a BS in administrative engineering with mechanical engineering option, he is a member of NACE and the American Welding Society.

Source: *Materials Protection*, August 1966, 21-24

EDITOR'S NOTE

During the 23rd Annual Conference of the National Association of Corrosion Engineers last March in Los Angeles, eight authorities in the field of corrosion engineering conducted a panel discussion on the catastrophic corrosion behavior of titanium in chemical process service. This discussion was recorded, transcribed, and is now presented in edited form. The panel moderator was L. W. Gleekman (Wyandotte Chemicals Corp., Wyandotte, Mich.). Members were P. J. Gegner (Chemicals Division, PPG Industries, Barberton, Ohio), E. G. Bohlmann (Oak Ridge National Laboratory, Oak Ridge, Tenn), F. H. Vorhis (Pfaudler Co., Elyria, Ohio), A. O. Fisher (Monsanto Co., St. Louis, Mo.), H. B. Bomberger (Reactive Metals, Niles, Ohio), W. K. Boyd (Battelle Memorial Institute, Columbus, Ohio), and E. V. Kunkel (Celanese Corp. of America, Corpus Christi, Texas). Photographs of the panel members appear at right.

Gleekman

Gegner

THE CATASTROPHIC CORROSION BEHAVIOR

A COMMON FEELING among persons concerned with the use and protection of titanium in the chemical industry is that much of the printed information on the corrosion mechanism of titanium is out of date.

Developments in the properties and performance of titanium are constantly occurring. In fact, technological change in this field is so rapid that data developed today can be out of date within six months. As a result, there is little up-to-date information on which the corrosion engineer can base his decisions concerning the use and protection of titanium chemical process equipment.

The engineer has always sought a panacea for his corrosion problems, and when titanium was first commercially introduced, he was told that his problems would be solved. But, as he soon discovered, the metal needed extensive testing and improvement. Over the years, titanium has been improved and it is now considered "the" material for certain applications. It is, however, far from being a panacea for the corrosion engineer, and it still needs extensive testing and improvement.

A "Reactive" Metal

Titanium is inherently a "reactive" metal, and like stainless steel, aluminum, magnesium, and some other metals, it forms a protective oxide film on its surface. Certain chemicals improve the stability of this protective film; others tend to degrade it. The metal, for example, is unstable in low pH reducing environments and in solutions which promote soluble corrosion products, including those which contain fluoride or oxylate ions.

The surface film, however, is generally strengthened by oxidizing agents and heavy metal ions. Experiments by H. H. Uhlig and others indicate that heavy metal ions are adsorbed on the surface of titanium. Furthermore, oxidizing agents and noble metal ions serve as cathode depolarizers which promote anode polarization over the surface—hence, corrosion resistance.

Use of an anodic voltage is another method for strengthening the film. By the constant application of a small voltage, an excellent condition of passivity can be maintained. In addition, passivity of the metal in chlorine gas and red fuming nitric acid can usually be maintained by the addition of a small quantity of water in the product stream. The minimum amount of water required depends on temperature and the chemical environment.

Performance in Various Environments

Because experience and test data were not always available on this new material, some unwise applications have been made. Table 1 shows corrosion data on commercially pure titanium. A few of these environments have been troublesome. Aluminum chloride, calcium chloride, and magnesium chloride are acidic at high concentrations, and if these solutions are not aerated with air (oxygen), or some other inhibitor is not present, corrosion is likely to result at high temperatures. Even in sodium chloride, titanium may corrode at high temperatures (250 F or 121 C) if deep restricted crevices exist which receive poor aeration.

The corrosion rate of titanium in nitric acid solutions can also vary. This variation is apparently caused by chemical make-up of the nitric acid. Traces of fluoride ion are detrimental. The volume of the solution, and the temperature and time in the test are also important because silicon from glass containers tends to inhibit corrosion. Furthermore, the tetravalent titanium ion, which may slowly accumulate, is also an inhibitor.

Practical steps can be taken, however, to improve the corrosion resistance of titanium in reducing environments. Data are given in Table 2 which show that unalloyed titanium is not resistant to 25% aluminum chloride but the 0.2% palladium-titanium alloy performs quite well in this solution. Alloys containing molybdenum or tantalum also have excellent corrosion resistance, and the same is true for these alloys in hydrochloric acid, sulfuric acid, and other strong reducing solutions. However, the resistance of the palladium alloy is not as pronounced as that of those alloyed with molybdenum or tantalum. Small additions of nickel appear to provide excellent resistance to crevice corrosion in chloride salt solutions.

The addition of heavy metal ions, such as ferric ion in the form of a chloride salt, to the acid solution will inhibit the unalloyed titanium. This addition, however, does not seem to have a marked effect on the palladium alloy

TABLE 1—Nominal Corrosion Resistance of Commercially Pure Titanium in Chloride Solutions

SOLUTION (1)	WT %	F TEMPERATURE	CORROSION RATE MPY
Aluminum Chloride	10	212	<1
	20	212	1 to 100 (2)
	25	212	260 to 2000 (2)
Calcium Chloride	5 to 20	212	<1
	40	230	<1
	62	310	<1 to 60 (2)
	70	310	<1 to >2000 (2)
Cupric Chloride	1 to 20	212	<1
	55	240	<1
Ferric Chloride	1 to 30	212	<1
	50	300	<1
Magnesium Chloride	5 to 30	212	<1
	50	390	<1 to >2000 (2)
Potassium Chloride	36	230	<1
Sodium Chloride	1 to 30	212	<1
Zinc Chloride	20	220	<1
	50	300	<1
	75		
	90	390	<1 to 3000 (2)

(1) Solutions not aerated

(2) Borderline passivity

Bohlmann

Vorhis

Fisher

Bomberger

Boyd

Kunkel

OF TITANIUM IN CHEMICAL PROCESSES

and it tends to accelerate corrosion of the molybdenum alloy.

Anode Polarization

Figure 1 shows anode polarization curves for unalloyed titanium, the palladium alloy, and the molybdenum alloy. Under anodic voltage, these metals do not readily corrode until a breakdown voltage is reached. This breakdown or pitting voltage varies with different alloys and decreases with increasing temperature. For example, the 8% manganese-titanium alloy breaks down in chloride solutions at about four volts at room temperature. This alloy has, unfortunately, been used with poor results, in anodizing applications. Unalloyed titanium should be recommended for such applications because of its higher breakdown voltage of approximately 10 volts.

Cathode polarization curves are interesting in this case because these high anode voltages are not normally encountered where corrosion is evident.

Cathode Polarization

Figure 2 gives cathode polarization data on the same materials. In 5% hydrochloric acid aerated with oxygen, the two alloys and the unalloyed titanium were fairly stable; however, when the solution was aerated with argon and the potential changed approximately 0.2-volt cathodically, the unalloyed titanium corroded quite rapidly. These data show that in reducing solutions the protective oxide film of the unalloyed metal is destroyed.

Titanium, however, performs satisfactorily in most neutral and oxidizing environments. It can also be used in reducing environments if: (1) natural inhibitors are present, (2) an inhibitor is added, (3) a small anodic emf is applied, or (4) the titanium is alloyed for resistance to the environment.

A Major Pitfall Crevice Corrosion

Crevice corrosion has long been recognized as a major pitfall of titanium applications in wet chlorine gas service. Considerable effort has been exerted to solve the problem as it relates to design of equipment in which water saturated chlorine gas is allowed to become stagnant, creating conditions that cause moisture to drop to the level below that required for effective inhibition. Attempts have been made to define this phenomenon and numerous papers have been published describing the work performed. Still, this area requires more laboratory and field testing.

A crevice cannot simply be defined as an improperly sealed joint or an area in which product is allowed to penetrate. It may be an infinitesimal groove at a metal-to-metal interface. A unique type of crevice corrosion, for example, is that which occurs at tube-to-tube sheet joints in heat exchangers used in processing both wet chlorine and brine. To illustrate, consider one case in which an exchanger was used to process brine at temperatures varying from 120 to 220 F (49 to 104 C). The brine consisted of 29% solids (primarily calcium chloride and sodium chloride), with a small amount of hydrogen sulfide. It also contained approximately 0.1% chlorine in the form of hypochlorite. The pH on the tube side was 6.3 and 5.6 on the shell side. This heat exchanger, representing an investment of $25,000, failed after only 14 days of service.

On the shell side (in back of the tube sheet in the jacket area), a ring of corrosion appeared approximately 1/16-inch wide and 0.015 to 0.014-inch deep, which appeared as though it had been carefully machine grooved. Corrosion also occurred on the shell side of the tube sheet face and extended the full length of the tube sheet joint.

In order to determine the cause of the corrosion, a one-tube test heat exchanger was constructed. The jacket was formed of carbon steel, and the one-tube bundle was designed so it could be removed readily for inspection and replaced easily and quickly. The test exchanger was placed next to the

TABLE 2—Effect of Alloying on the Corrosion Behavior of Titanium

SOLUTION, WT %	TEMPERATURE	CORROSION RATES IN MILS PER YEAR[1]		
		UNALLOYED TITANIUM	Ti-.2Pd	Ti-30Mo
25% NaCl	Boiling	Nil	Nil	Nil
25% $AlCl_3$	Boiling	2,000	1	Nil
30% $FeCl_3$	Boiling	1	Nil	5
5% HCl	Boiling	1,100	7	Nil
20% HCl	R.T.	25	4	Nil
20% HCl	Boiling	5,000	770	5-10
20% HCl + 1% $FeCl_3$	Boiling	110	115	1,500
5% H_2SO_4	Boiling	1,900	20	Nil
40% H_2SO_4	R.T.	65	9	Nil
40% H_2SO_4	Boiling	13,000	----	2-10
40% H_2SO_4+1% $Fe_2(SO_4)_3$	Boiling	110	90	750
65% HNO_3	R.T.	Nil	Nil	Nil
65% HNO_3	Boiling	2-20	25	50
10% Oxalic	212 F	3,500	4,800	1.5

(1)48-hour tests in 500 ml of solution, not aerated.

line and both processed and raw brine were throttled through at the same proportional flow rate and at the same temperature as previously mentioned. Corrosion was duplicated in exactly 14 days after installation.

At the hot end of the test heat exchanger, on the back of the tube sheet, a small crevice had formed and at the face of the heat exchanger, the tube had begun to corrode. Corrosion also occurred in the rolled joint area of both the tube and the tube sheet.

Because the problem was first suspected to have been caused by stray current, a second test section was installed and grounded. The tube sheet also was grounded and the test run for an additional 14 days. Corrosion again was encountered which indicated that stray current was not the culprit.

The next step was to determine whether or not this corrosion was due to a leaking joint. Pressure tests indicated the joints had not leaked. Additional tests ultimately led to the conclusion that this particular type crevice corrosion had been caused by an ion concentration cell. Due to the relatively high temperature of the tube sheet, the low pH of the raw brine solution, and the presence of tramp elements (such as fluoride ions), a breakdown of the oxide film occurred in the area between the tube and the tube sheet face. Minute amount of brine then contacted the unprotected titanium, thus initiating crevice corrosion. As the corrosion continued, metal ions were swept away by the brine solution on the outside of the crevice; whereas, on the inside the crevice (which slowly increased), there was a metal ion build-up, and an electrolytic cell was formed.

A characteristic of titanium crevice corrosion in saline waters at elevated temperatures which deserves emphasis is that the attack, once inititated, often continues well beyond the confines of the original crevice. Apparently, the aggressive environment produced in the crevice is preserved under the rutile corrosion product which is formed, and rapid corrosion continues and spreads. This phenomenon is illustrated by the failure of a 1½-inch Schedule 40 pipe section of a titanium corrosion loop.

A diagram of the loop is shown in Figure 3, and the failure occurred at the point labeled C in the diagram.

The large shallow pit which resulted is shown in Figure 4. Penetration of the 0.2-inch pipe wall occurred after a total of 8468 hr of operation principally with 1 and 2 *M* NaCl solutions with pH in the range of 5.5 to 7.2. The temperature history was 2808 hr at 100 C, 5374 hr at 150 C, and 286 hr at 200 C.

A careful examination of the loop revealed additional smaller areas of attack at the locations indicated by "C" in Figure 3. A representative example is shown in Figure 5.

The large pit may have initiated in a crevice between a sample holder and the pipe wall, but the smaller pits initiated in a region of high turbulence at a bulk flow velocity of approximately 17 fps.

In all cases, however, the attack spread widely from the point of initiation. The consistent occurrence of the attack in association with flange joints suggests an explanation for the initiation of attack in these areas after thousands of hours of operation without difficulty.

The repeated assembly of the flanged joints in connection with specimen insertion and removal results in mechanical deformation of the pipe wall, as can be seen in Figure 4. Such deformation may cause cracks in the heavy oxide scale which builds up in the loops with resultant formation of microcrevices in which the attack initiates. This is consistent with many observations of initiation of attack at microcrevices at laps, fissures, inclusions, etc., present as a result of metal fabrication.

An extreme example of the self-sustaining characteristics of the attack is shown in Figure 6. This is a pump impeller from the loop shown in Figure 3. As a result of the method of fabrication, tight crevices were formed between the front shroud and the vanes of this impeller. The attack shown apparently initiated in one of these crevices and continued well beyond the confines of the crevice in spite of the fact that the lineal velocity at this point was about 120 fps. Exposure conditions were: near neutral 1 *M* NaCl, about 1000 hr at 100 C and about 1000 hr at 150 C.

The manner in which crevice corrosion initiates in titanium appears fairly standard. Due to the formation of a differential aeration cell, the oxygen in the crevice is consumed, and the protective oxide is destroyed. Hydrolysis of the resultant titanium ions produces hydrogen ions, and an active-passive cell is set up. Corrosion then proceeds rapidly; rates have ranged from 200 to 2000 mpy at 100 to 200 C.

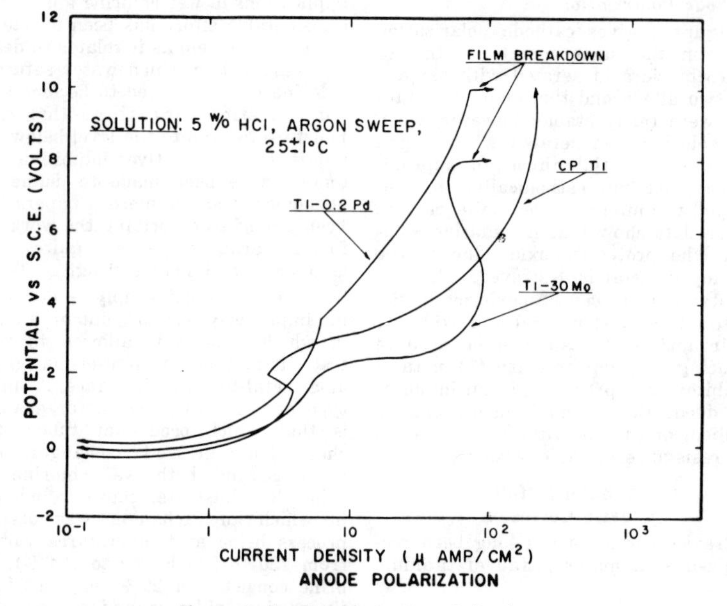

Figure 1—Anodic polarization of titanium.

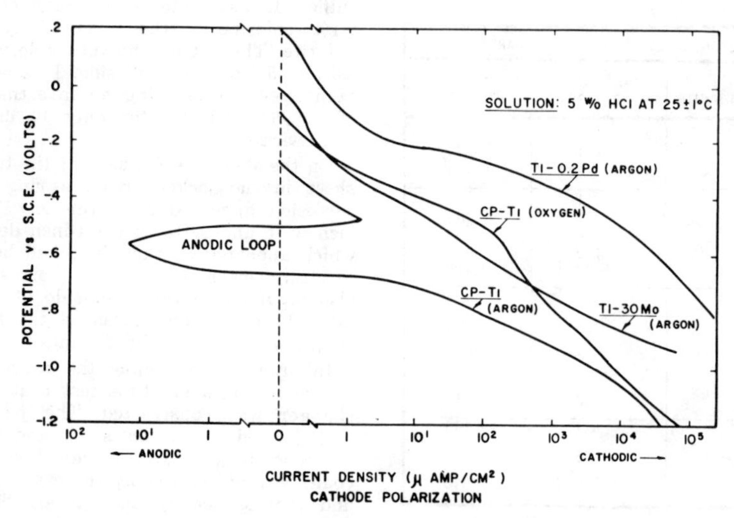

Figure 2—Cathodic polarization of titanium.

Such rates are consistent with corrosion rates observed on anodic polarization of active titanium in 0.1 to 1.0 *M* HCl + NaCl at 100 C.

A recent observation concerning the behavior of titanium alloys in saline water at elevated temperature also deserves mention. It is generally known that most titanium alloys have very high (10 volts) pitting potentials in such environments at ambient temperatures and so are not susceptible to pitting attack. It has been observed, however, that the pitting potentials of many titanium alloys are inversely dependent on temperature, as shown in Figure 7. As a consequence, pitting attack appears to be a more likely problem at temperatures above 200 C.

Case Histories
Successful Applications

The subject of this discussion concerns the catastrophic corrosion behavior of titanium, but the successful performance is as important here as are its shortcomings. The following case studies will not only help to put the success and failure of titanium into proper perspective, but will also help to illustrate the erratic performance of titanium in various environments.

Ammonia Recovery Still

The largest use of titanium in the manufacture of soda-ash is in ammonia recovery stills. In this particular process, the Chemicals Division of PPG Industries has 8000 titanium tubes in service, ranging from 1¾-inch to 4½-inches OD, and from 9½ to 10½ ft in length with 0.030-inch wall thicknesses. The oldest have been in service for 4½ years and are still in perfect condition. Some have been slit open, miked, inspected and no corrosion has been noted. The chemical environment to which the tubes are exposed is free ammonia and ammonium chloride at approximately 175 F (79 C) on the inside of the tubes and saturated CO_2 and ammonia at approximately 190 F (88 C) on the outside.

Two other soda ash manufacturers have not been so lucky and have had repeated erratic catastrophic failure of similar tubes in identical service.

Chlorine Cooler

Another titanium application at PPG Industries is a chlorine cooler (a fairly common use for the metal). The cooler contains over 1000 titanium tubes, each 1-inch in diameter, 8½-ft long, with a a wall thickness of 0.020-inch. When installed, the tubes were packed in a 2-inch thick phenolic-filled asbestos tube sheet, using a double O-ring seal. Because crevice corrosion was anticipated to occur under or near the O-rings, the tubes were pulled after four years of service. No evidence of crevice corrosion was observed. The tubes now have been in service for over five years.

Other users of titanium chlorine coolers have been plagued with failure at the tube sheet joint, until a welded joint is used.

Heating Coil

A specific application of titanium-palladium alloy is in a 163-ft heating coil, which for four years was used in 62% calcium chloride service at 310 F (154 C) with no corrosion problems. Over a year ago, the calcium chloride concentration was increased to 75% and the temperature upped to 350 F (177 C). To date, the coil has experienced no significant corrosion.

Other manufacturers report rapid failure of the alloy in this service.

Titanium is Often Abused

Because titanium has a reputation of

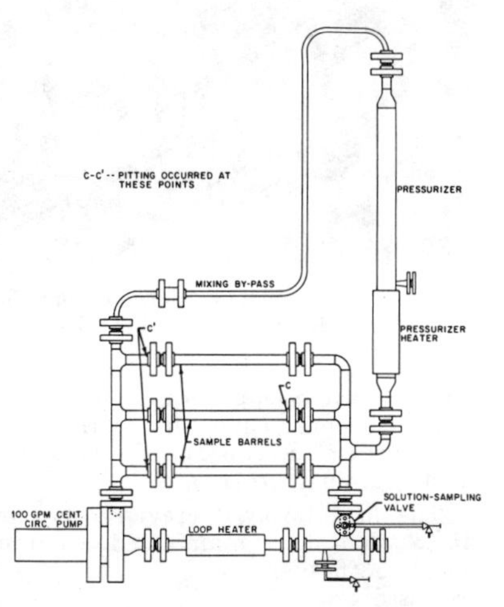

Figure 3—Titanium corrosion test loop. Failure occurred at the point labeled "C."

SYMBOL | ALLOY
110-AT
0.15% Pd
0.15% Pd
150 A
150 A
55 A
Commercial Ti

ELECTRODE POTENTIAL (volts vs. S.C.E.)
TEMPERATURE (°F)
TEMPERATURE (°C)

Figure 7—Effect of temperature on pitting potentials of titanium alloys in 1 M NaCl.

Figure 4—Example of pitting which occurred in the titanium corrosion test loop shown in Figure 3. Penetration indicated by "P."

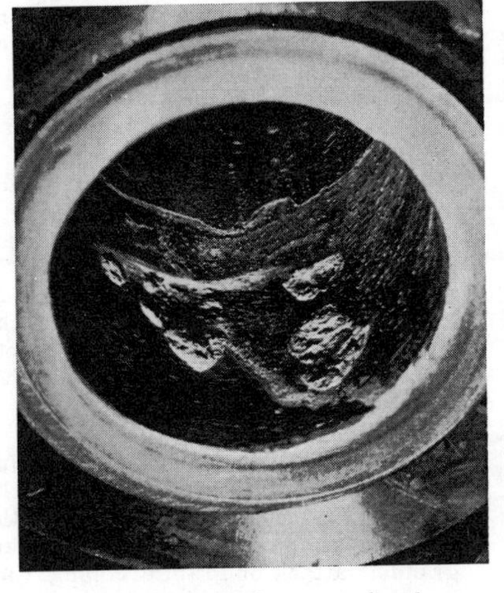

Figure 5—Representative example of corrosion in a titanium corrosion test loop.

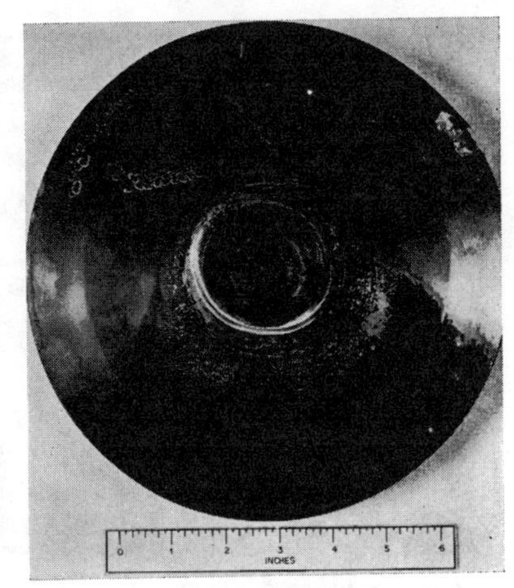

Figure 6—Corrosion of a titanium pump impeller.

Source: *Materials Protection*, October 1967, 22-27

being an extremely durable metal, it is often used improperly. It is frequently used in applications that stretch its capabilities, and as a result, problems develop. Such instances reflect on the economic use of titanium, since it is a relatively expensive metal. Properly used, however, the metal can be economically applied, as indicated in the case studies above.

Specific Failures

Loose Liners

Much of the titanium used by Celanese Corp. is installed as loose liners in carbon steel shells. At first, the company encountered problems in welding the titanium liners. Even though jigs were used to maintain a close fit between the liner and the shell, the welding operation caused slight contraction of the titanium; consequently, the liner separated from the shell. During pressure testing, the liner would crack and, in some cases, pull apart at the seams.

In one critical application, this problem was eliminated in an unusual manner. A resinous, flexible epoxy was developed and injected into the annulus between the titanium liner and the steel shell. The epoxy was cured in place to a semi-hard, durable consistency. This filler acted as a supporting cushion and allowed the liner to hold up during pressure testing. This method was used to reclaim one defective vessel.

Rigid specifications have been written for titanium loose liner fabrication and the company now follows the specifications closely during construction of titanium lined steel vessels. These rigid procedures preclude the use of cushioning material in the annulus.

The following minimum thicknesses are used in liner construction. On vessels less than 24 inches in diameter, a 1/16-inch minimum thickness is used if a good fit-up is obtainable; otherwise, a 1/8-inch liner thickness is used. Vessels from 24 to 96 inches require 1/8-inch liner plates, and vessels greater than 96 inches require 3/16-inch plates.

Chlorine Contaminated Equipment

Monsanto Co's experiences with titanium have been generally good. However, in connection with the theme of this discussion, two specific failures with titanium are related.

Figure 8—Cross section showing corrosion on the inside of a titanium tube (2½X).

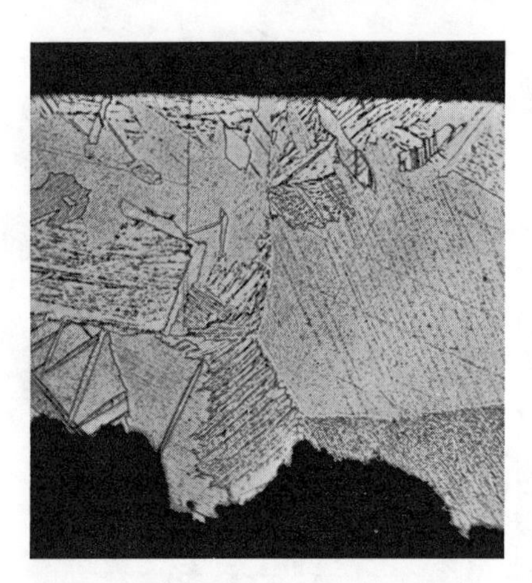

Figure 9—Photomicrograph showing enlarged grains in the weld area of the tube shown in Figure 8 (100X).

The first involved crevice corrosion in a chloride contaminated environment. A header and tube sheet with recessed grooves was involved. The environment was 96% aromatic hydrocarbon, with the remainder being 2% water and slight traces of organic chloride. The total chloride content was 320 ppm, and the pH varied from 6.7 to 8.4. Failure of the tube sheet due to corrosion in this case was presumably caused by both the high steam temperature (340 F or 171 C) and the fact that chlorides were concentrated while water was driven out. This caused crevice corrosion in the recessed grooves of the tube sheet.

Titanium Preheater

A second example was of a different nature. Failure of a titanium preheater handling 56% nitric acid solution at 220 F (104 C) occurred after one year of service. Visual examination revealed that corrosion was confined to the weld on the inside of the tubes which were in contact with the nitric acid. Seal welds on the tube sheet of the unit's 4-pass heat exchanger were only slightly etched and that etching was observed only on the outlet of the hottest pass. The situation indicated that titanium can be used successfully in processing nitric acid solutions, but that the type of weld may be a factor.

Figure 8 shows a cross section of a failed tube which reveals that attack was primarily on the inside. Figure 9 is a photomicrograph showing enlarged grains in the weld area and a profile of the corroded inside surface. Figure 10 at the same magnification shows the microstructure of the tube metal. Note the large difference in grain size between weld zone in Figure 9 and parent metal in Figure 10.

All attacked areas showed the same corrosion pattern—etching along the weld zone. Visual examination of the corroded surfaces revealed no intergranular attack or grain dropping. Caliper measurements of wall thickness indicated that corrosion rate in the severely attacked area was over 100 mpy.

The failures were suspected to have been related to metallurgical conditions. A check with the tube supplier revealed that the tubes were 16 British Wire Gauge welded tubes, supplied in accordance with ASTM B-338-61-T, and stress relieved after welding. The supplier also stated that tubes were cold-formed, butted and TIG-welded, and rotor straightened after annealing. Although metallographic examination of a failed tube section revealed an abnormally large grain structure in the weld zone, no hydrogen cracking, porosity, or hydride formations could be detected.

For record purposes, hydrogen and oxygen content of the tubing was as follows:

	Parent Metal	Weld
Oxygen ppm	990	1007
Hydrogen ppm	82	57

The failures appeared to originate from two possible sources: (1) the presence of abnormally large grains which were anodic to the smaller grains and the parent metal, causing galvanic attack, and (2) improper gas shielding during welding, causing the formation of a thin film of contaminated metal.

A question arising from this experience is whether or not all welds in titanium welded tubing contain large grains. Experience has indicated that heliarc welding of titanium does produce large grains because of the slow cooling rates involved. Another question is whether susceptibility to corrosion is merely the result of improper gas shielding, which permits contamination of the surface. Still another question is whether nitric acid is unique in revealing metallurgical abnormalities in titanium. Some tube suppliers relate that several unexplained failures of titanium in nitric acid have occurred and that some of these do not involve welds.

Although sufficient evidence does not exist at this time to fully answer these questions, the short-term solution is to use seamless titanium tubing for nitric

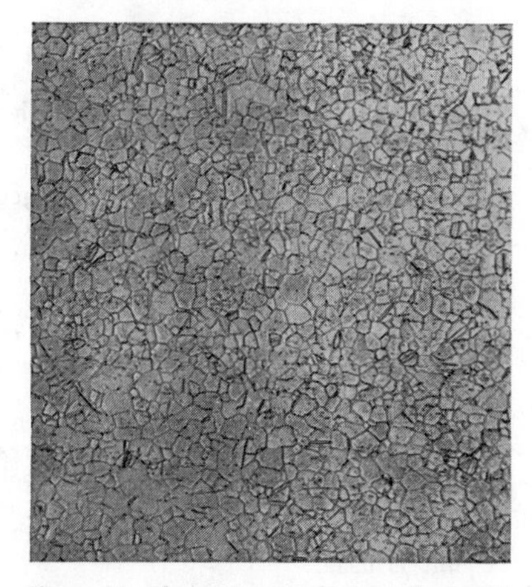

Figure 10—Photomicrograph show the metal structure of the tube in Figure 8 (100X).

acid service at temperatures above 220 F (104 C).

PPG Industries has not experienced catastrophic failures *per se* in titanium equipment. Some, however, have occurred and the fact that these failures have not been catastrophic or major does not detract from their importance, and a study of them will lead to a better understanding of titanium corrosion in chemical processing.

Brine Feed Orifices

A series of failures occurred in small brine feed orifices constructed of titanium. These orifices controlled feed to chlorine cells. No chlorine was contained in the brine passing through these orifices, and the pH was generally over 7. Yet, the orifices failed completely in from one week to four months. Stray current was suspected to have caused the failures.

This experience has been duplicated by others in this field.

Contact Cooler Sparger

A close-up photograph in Figure 11 shows corrosion and complete perforation in a contact cooler sparger pipe. This area was inside a ceramic liner of the contact cooler, which was brick lined. The failure occurred in less than one year. The ceramic sleeve fitted closely around the sparger pipe. A crevice formed and soon corrosion was quite wide-spread. The chemical environment was hot chlorinated water and hot chlorine gas. This type failure has become fairly common in a wet chlorine environment.

Figure 12 shows a typical crevice failure in chlorinated brine service. A titanium pump shaft sleeve over which was applied a rubber sleeve was used in the process. In service, crevices formed and caused the situation shown within 30 days.

Another failure of titanium in chlorine service occurred 15 months after installation of an interior liner for a concrete chlorine cell cover. Some of the failure was probably caused by faulty welds, but other evidence indicated that at least part of the failure could be attributed to stray current.

Conclusion

Even though some of these case studies indicate to the contrary, titanium is an excellent material of construction. It will not, however, solve all corrosion problems any more than will stainless steel, the aluminum alloys, zirconium, or other strong and durable materials. These metals were carefully tested and improved before they became widely used. Their shortcomings were evaluated as carefully as were their advantages. Enough testing and evaluation of titanium, however, has not been done, and it has come into sudden wide use without adequate preparation and testing.

Titanium fabricators and users should realize that because it is a reactive metal, it requires special handling and special fabrication procedures. Fabricators must determine what is to be done to the metal to make it as desirable and useful as possible.

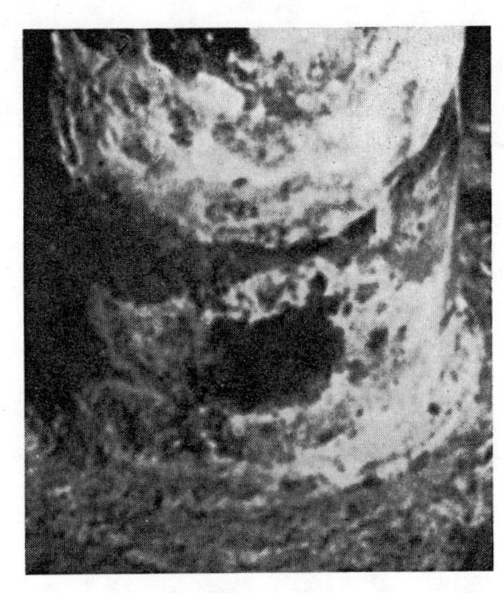

Figure 11—Corrosion in a contact cooler sparger pipe.

The major area in need of specific research is in the alloying of titanium for aggressive environments that unalloyed titanium is incapable of handling. Test data indicate this is possible. More research is required, however.

In this discussion of the catastrophic corrosion behavior of titanium in chemical process service, several points of view have been considered—the metallurgist's, the electrochemist's, the thermodynamicst's, and the materials engineer's. Specific conclusions as to fabrication, application, and standardization of titanium as a construction material are not attempted. Rather, the purpose was to present a general discussion of titanium equipment in actual service and to emphasize that an organized, definite research program is desperately needed.

Figure 12—Crevice corrosion of a titanium pump shaft in chlorinated brine service.

DISCUSSION

The following questions and replies evolved at the end of the panel's presentations. Many of the questions were not audible enough to be recorded. These questions, however, offer a sampling of the current interest in titanium.

Question:

Has the blue oxide film on the surface of titanium been investigated?

Reply:

J. P. Cotton has conducted investigations into anodic films on titanium, classifying them according to color, voltage required to produce them, and thickness. The results are expected to be published soon. Also, the USSR has done considerable work in relation to oxides of titanium, which has resulted in the identification of an entire series. The blue color is probably an interference color that is achieved with a very low thin oxide film. The color is a function of the applied voltage.

Question:

Is titanium hydride formation within the metal a normal condition that occurs in the corrosion of titanium, or is it an exceptional condition?

Reply:

The hydride formation within the metal is quite normal if corrosion occurs in a reducing environment. Hydrides are present in some form in the corrosion of titanium.

Question:

Are organics a cause of crevice corrosion in titanium?

Reply:

In one test, fluorocarbon material washers were considered a cause of crevice corrosion of titanium. In another case, both fluorocarbon tape and a silicone rubber material initiated crevice corrosion. In still another case, asbestos caused severe crevice attack. One school of thought is that it does not matter what type of organic materials is used; if a crevice exists, the possibility of crevice attack also exists.

Question:

Is the titanium-palladium alloy entirely resistant to crevice corrosion? If not, under what conditions will it be subject to attack?

Reply:

J. P. Cotton has done work which shows that as the temperature is increased above 212 F into the range of 250 to 275 F, in certain salt solutions, the titanium-palladium alloy has a tendency to crevice corrode.

Question:

How does titanium perform in alcohol solutions?

Reply:

Alloys have been known to crack in methanol systems. The exact mechanisms are not completely understood. However, cracking does not occur unless the stress levels are relatively high, and unless the protective film is ruptured. The working mechanism appears to be electrochemical, at least in part, because the application of a cathodic current will prevent cracking.

Source: *Materials Protection*, October 1967, 22-27

INDEX

D

E

F

G

H

I

T

U

V

W